PROBABILISTIC MODELS IN ENGINEERING SCIENCES

VOLUME II

PROBABILISTIC MODELS IN ENGINEERING SCIENCES

VOLUME II

Random noise, signals, and dynamic systems

Harold J. Larson
Bruno O. Shubert
Naval Postgraduate School

John Wiley & Sons
New York Chichester Brisbane Toronto

Library of Congress Cataloging in Publication Data:

Larson, Harold J 1934–
Probabilistic models in engineering sciences.

Includes index.
CONTENTS: v. 1. Random variables and stochastic processes.
—v. 2. Random noise, signals, and dynamic systems.
1. Engineering—Statistical methods. 2. Probabilities.
3. Stochastic processes. I. Shubert, Bruno O., joint author.
II. Title.
TA340.L37 620′.001′5192 79–755
ISBN 0-471-01751-5 (v. 1)
ISBN 0-471-05179-9 (v. 2)

Printed in the United States of America

10 9 8 7 6 5 4 3 2 1

PREFACE

This is the second of a two-volume introduction to applied probability theory. It deals with further topics in the theory and applications of stochastic processes. The unifying theme for this volume (apart from Chapter Three) is the Doob-Meyer decomposition of a process into its "signal" and "noise" components. We consider both continuous path processes, which lead naturally to white Gaussian noise, and point processes, which lead to the Poisson noise process. This, together with the frequent references made to the canonical state-space representation of a disturbed dynamic system, explains our choice of a subtitle for this volume.

Since this is the second of a two-volume set, ideally students who use this book should have first completed Volume I. This is, however, not necessary; Chapter One is written especially for those who have taken a postcalculus probability course using some other textbook. In this chapter we introduce our notation and review the important topics from Volume I. They include several topics that may not have been covered in a first course from a different text (e.g., the concepts of a martingale, stationarity, correlation function, the Markov property, etc.), especially a text that is prerequisite to a course in mathematical statistics. Study of Chapter One and its exercises are sufficient to allow easy access to Chapters Two through Seven for anyone who has successfully completed a postcalculus probability course. Those students who have completed Volume I can skip Chapter One entirely, or perhaps can use it to sharpen their recollection of the basics.

Our philosophy regarding the presentation of material in Volume II is the same as that for Volume I: we illuminate and illustrate important concepts, and rely on intuition without totally slighting rigor. The great majority of probabilistic models used in modern engineering sciences involve stochastic processes rather than time-isolated random variables. It is unfortunate that a completely rigorous development of the theory of stochastic processes requires an extensive background in measure theory and functional analysis. Also, a basic understanding of many complex physical systems is not possible without some familiarity with rather advanced results from the theory of stochastic processes. The resulting dilemma of accessibility versus rigor has frequently been resolved by allocating the study of many useful stochastic processes to doctoral-level programs, thus severely and artificially limiting the accessibility of this material. Although many results in the theory of stochastic processes are difficult to prove with complete rigor and in their complete generality, they are not difficult to understand; indeed, many of them are fairly easy to derive on a semiheuristic level. A prime example is afforded by the concept of a martingale which, although it has been around for decades and is not difficult to understand and use, is still relatively inaccessible to those who lack a strong measure-theoretic background. We have thus chosen a semiheuristic method of approach, together with paraphrased real-world examples, in this book. Occasionally this has led to a certain wordiness and length of presentation. On the other hand, we are able to derive and discuss many important results that otherwise would be inaccessible to students at this level, albeit with something less than full generality.

The main topics we discuss in this volume include:

Chapter Two Linear transformations of discrete time processes, as well as their decomposition and innovation representation.

Chapter Three Discrete-state Markov chains, in both discrete and continuous time.

Chapter Four Mean square calculus and linear transformations of continuous-time processes.

Chapter Five Spectral representation of stationary processes.

Chapter Six Markovian diffusion processes.

Chapter Seven Point processes and their specification through either counting or waiting time processes.

Many of these chapters are independent of one another, so that instructors can exercise choices in terms of the number (or the order) of topics to be covered. Diagram I, which follows the Preface, shows the interrelationships of the 28 sections in Chapters Two through Seven. Note that Chapters Two and Three are totally independent of the others and that Chapter Seven depends only on the first three sections of Chapter Four, and can even be made completely independent of them if some derivations are omitted. All of the material in Volume II can be covered in a two-quarter, four-hour course. As is indicated in the interrelationship diagram, several options are available for the selection of topics for a considerably shorter course.

The prerequisites necessary to use of this volume (in addition to Volume I or some other postcalculus course plus Chapter One) are a good working knowledge of calculus, together with the rudiments of complex variables, Fourier transforms and differential equations. The level of knowledge required is essentially the same as that required for circuit theory, mechanics, electromagnetic theory, and many other engineering courses. Throughout both volumes we have adopted the notation most commonly encountered in the engineering literature, including capital letters for random variables, f for density functions, and boldface type for vectors and matrices. The single exception to standard usage is our choice of the lowercase Greek letter iota, ι, to represent the imaginary unit. Because this letter is graphically similar to both i and j, the two most frequently used symbols for

$\sqrt{-1}$, and because its usage then allows free usage of i and j for indices, we believe this exception is justifiable.

Each of the 28 sections in Chapters Two through Seven is concluded with a set of exercises. We have deliberately made these exercises realistic, to further illustrate applications and to help avoid the impression that unnecessary generality was used in the development of the theory. As a result, some of these exercises are perhaps more difficult than those encountered in other textbooks at this level. They are not, in the main, difficult conceptually, but, rather, are somewhat demanding in terms of the length of the mathematical manipulation required to arrive at the solution. But, then, real engineering problems are typically burdened by lengthy expressions and seldom result in neat formulas that involve only simple functions.

For convenience, many basic mathematical facts are summarized in Appendix Three. Appendix One, the Dirac δ-function, is the same as Appendix One to Volume I and is included for those who have not studied Volume I. Appendix Two gives a listing of facts regarding many standard, frequently used probability distributions. We have not included tables of any probability distributions because, like powers, roots, and logarithms, their evaluation is readily available by using hand-held calculators.

We have not included references to other sources in the text because most of the results we discuss are typically treated with a considerably higher degree of rigor. In fact, even the statements of these results, done rigorously in references available, require knowledge of concepts we have not taken to be prerequisites. A sample of books currently available that treat the material we discuss on an advanced mathematical level include the following:

Loève, Michel. *Probability Theory I, II*. Fourth Edition. Springer-Verlag, New York, 1977.
Breiman, L. *Probability*. Addison-Wesley, Reading, Mass., 1968.
Cramér, H. and M. R. Leadbetter. *Stationary and Related Stochastic Processes*. John Wiley & Sons, New York, 1967.
Karlin, S. and H. M. Taylor. *A First Course in Stochastic Processes*. Second Edition. Academic Press, New York, 1975.
Wong, E. *Stochastic Processes in Information and Dynamical Systems*. McGraw-Hill, New York, 1971.
Lamperti, J. *Stochastic Processes*. Springer-Verlag, New York, 1977.
Arnold, L. *Stochastic Differential Equations*. John Wiley & Sons, New York, 1974.
Snyder, D. L. *Random Point Processes*. John Wiley & Sons, New York, 1975.

We thank Rosemary Lande and Rosemarie Stampfel for their excellent typing of the manuscript.

Bruno O. Shubert
Harold J. Larson

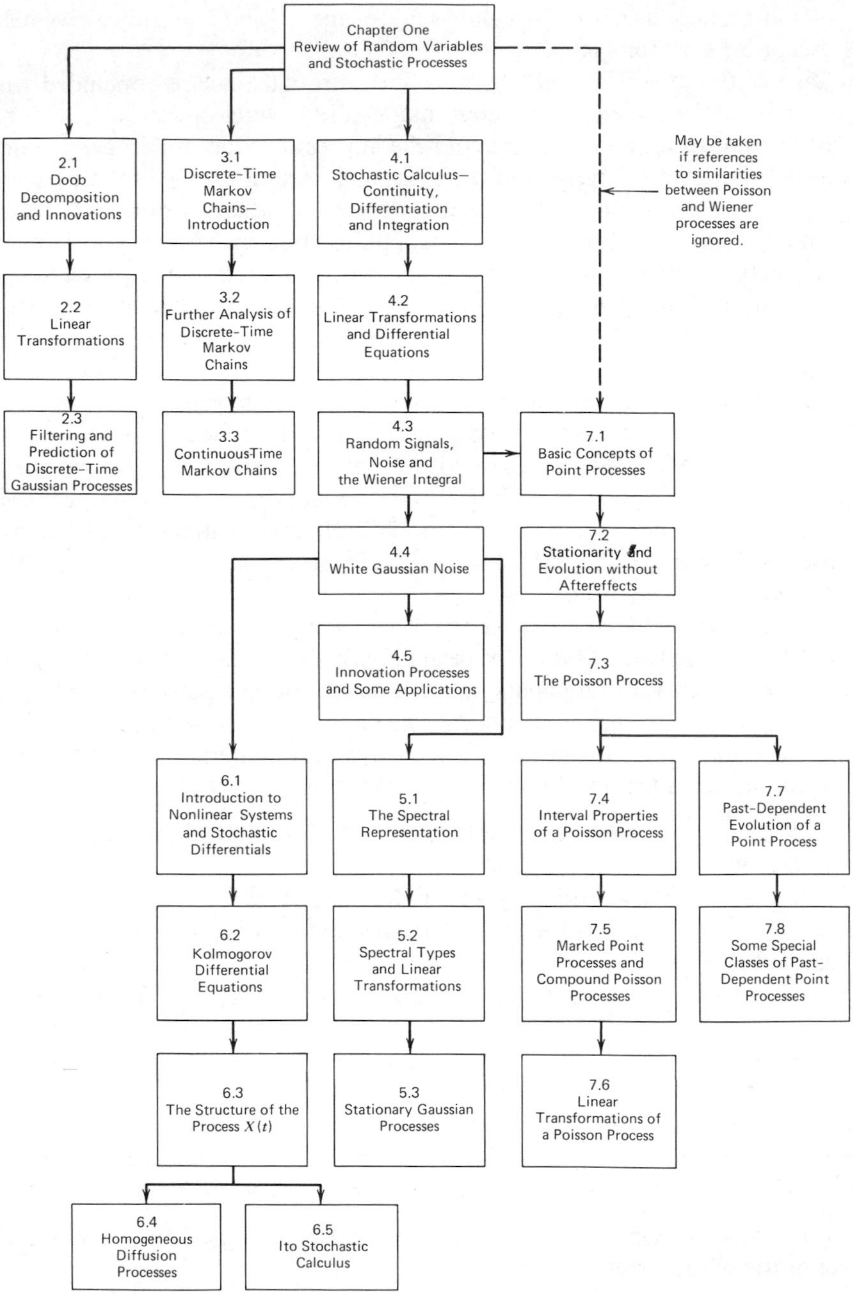
Chapter One
Review of Random Variables and Stochastic Processes
May be taken if references to similarities between Poisson and Wiener processes are ignored.
2.1 Doob Decomposition and Innovations
2.2 Linear Transformations
2.3 Filtering and Prediction of Discrete-Time Gaussian Processes
3.1 Discrete-Time Markov Chains—Introduction
3.2 Further Analysis of Discrete-Time Markov Chains
3.3 Continuous-Time Markov Chains
4.1 Stochastic Calculus—Continuity, Differentiation and Integration
4.2 Linear Transformations and Differential Equations
4.3 Random Signals, Noise and the Wiener Integral
4.4 White Gaussian Noise
4.5 Innovation Processes and Some Applications
7.1 Basic Concepts of Point Processes
7.2 Stationarity and Evolution without Aftereffects
7.3 The Poisson Process
6.1 Introduction to Nonlinear Systems and Stochastic Differentials
6.2 Kolmogorov Differential Equations
6.3 The Structure of the Process X(t)
6.4 Homogeneous Diffusion Processes
6.5 Ito Stochastic Calculus
5.1 The Spectral Representation
5.2 Spectral Types and Linear Transformations
5.3 Stationary Gaussian Processes
7.4 Interval Properties of a Poisson Process
7.5 Marked Point Processes and Compound Poisson Processes
7.6 Linear Transformations of a Poisson Process
7.7 Past-Dependent Evolution of a Point Process
7.8 Some Special Classes of Past-Dependent Point Processes

CONTENTS

PROBABILISTIC MODELS IN ENGINEERING SCIENCES

VOLUME II

CHAPTER ONE

REVIEW OF RANDOM VARIABLES AND STOCHASTIC PROCESSES

A general mathematical description of a random phenomenon or dynamic system is provided by the *probabilistic model* of the phenomenon under investigation. Each such phenomenon, also referred to as an experiment, is characterized by a numerical quantity or quantities whose values can be observed when the experiment is performed. This quantity can be a single number, a collection of numbers (a vector), or a function of a parameter t usually interpreted as time and describing the temporal evolution of the phenomenon. Unlike deterministic phenomena or systems, the inherent randomness causes the observed quantity to take on generally different values each time the phenomenon or system is observed, even if all external controllable factors are kept constant.

Thus, in order to describe the phenomenon mathematically, it is necessary to specify the collection of all possible values the observed quantity can take on. This collection, or set, is called an *ensemble* and is denoted by $\mathcal{E}_X$, where the subscript X (or Y, Z, U, etc.) is the symbol for the observed quantity.

If the observed quantity X is a single number, it is called a *random variable*, and its ensemble $\mathcal{E}_X$ is generally a subset of the set of all real (or complex) numbers.

If $\mathbf{X}$ is a vector† $\mathbf{X} = (X_1, \ldots, X_m)'$ with components $X_1, \ldots, X_m$, then it is called a *random vector*, and the ensemble $\mathcal{E}_{\mathbf{X}}$ is a subset of m-dimensional Euclidean space.

If X is a function of the time parameter t, to be denoted by $X(t)$, then we call it a *stochastic process*. The ensemble $\mathcal{E}_{X(t)}$ of a stochastic process $X(t)$ is then a collection of time functions $x(t)$ called *sample paths* of the process. The time parameter t is assumed to range through some *time domain* $\mathcal{T}$, which is usually one of the sets

$$\mathcal{T} = [0, \infty), \qquad \mathcal{T} = (-\infty, \infty),$$

or

$$\mathcal{T} = \{0, 1, \ldots\}, \qquad \mathcal{T} = \{\ldots, -1, 0, 1, \ldots\}.$$

† Vectors will henceforth be denoted by boldface capital letters and considered column vectors; $\mathbf{X}'$ is the transpose of $\mathbf{X}$.

In the former two cases the time t is thought of as evolving continuously, while in the latter the process is assumed to evolve or is being observed only at discrete time instances. Accordingly, we call it a *continuous-time process* or a *discrete-time process*.

Just as a random vector $\mathbf{X} = (X_1, \ldots, X_m)'$ can be regarded as an m-tuple of random variables (with components $X_1, \ldots, X_m$), so can a stochastic process $X(t)$ be regarded as an infinite collection of random variables $X(t)$, $t \in \mathcal{T}$, indexed by the time parameter t. The discrete-time stochastic process with time domain $\mathcal{T} = \{0, 1, \ldots\}$ (or $\mathcal{T} = \{1, 2, \ldots\}$) can also be referred to as a *stochastic sequence*, that is, an infinite sequence of random variables $X(0), X(1), \ldots$. We shall often label the discrete-time parameter by n instead of t and write the stochastic sequence as $X(n)$, $n = 0, 1, \ldots$.

Frequently, a stochastic process is used to model the temporal evolution of a dynamic system, in which case $X(t)$ for a fixed time instant t represents the random *state* of the system at that time instant. Typically, such a state is restricted to be one of some particular set of states $\mathcal{S}$, for instance $\mathcal{S} = \{0, 1\}$ for a binary system. The set of states $\mathcal{S}$ is then called the *state space* of the process $X(t)$, $t \in \mathcal{T}$. Note carefully the distinction between the state space $\mathcal{S}$ and the ensemble $\mathcal{E}_{X(t)}$ of the stochastic process $X(t)$, $t \in \mathcal{T}$.

A stochastic process can also be *vector valued*, $\mathbf{X}(t)$, $t \in \mathcal{T}$, in which case its ensemble $\mathcal{E}_{\mathbf{X}(t)}$ is a collection of vector-valued time functions $\mathbf{x}(t)$, and its state space is then a subset of a Euclidean space of appropriate dimension. A random motion of a particle in space would be an example of such a three-dimensional vector-valued process.

The ensemble of a random variable, random vector, or a stochastic process constitutes the first part of the probabilistic model for the random phenomenon in question. The second part is the *probability law*,† which, roughly speaking, describes the chances of observing some particular value or values of a random variable or vector or a sample path of a stochastic process with some specific properties.

An occurrence or nonoccurrence of some specific values of a random variable or of some sample path with a specific property is called a (*random*) *event*. Specification of an event may be quite simple or quite complicated. The only criterion necessary is that the occurrence or nonoccurrence of a specific event must be well defined once the value of the random variable or the sample path of the stochastic process has been observed. There is a one-to-one correspondence between events and subsets of the ensemble—namely to each event A, there corresponds a subset $\mathcal{A}$ of those elements of the ensemble that result in occurrence of the event A.

The probability law is now defined as an assignment of a number $P(A)$ to every conceivable event A concerning the random variable or stochastic process in

† Equivalently referred to as the *probability distribution*.

question. The number $P(A)$ is called the *probability* of the event A, provided that the assignment is made to conform with the following three axioms:

(*i*) $P(A) \geq 0$ *for each event* A.
(*ii*) $P(\Omega) = 1$ *for the sure event* Ω.
(*iii*) $P(A_1 \cup A_2 \cup \cdots) = P(A_1) + P(A_2) + \cdots$
whenever $A_1, A_2, \ldots$ *is a finite or infinite sequence of mutually exclusive events.*

Thus the *probability law is an assignment of probabilities to all conceivable events defined in terms of a random variable, random vector or a stochastic process.*

Except in a few trivial cases the number of possible events in a particular probabilistic model is infinite, and thus one is not able to specify the probability law by actually listing the probabilities for each event. Fortunately, the axioms governing assignments of probability values allow the specification of the probability law by means of a simple device—the *probability density function.*

For a single random variable X the *probability density* (or simply the *density*) $f_X(x)$ is a real-valued function defined for all real x, $-\infty < x < \infty$, such that

(i) $f_X(x) \geq 0$ for all x,

(ii) $\int_{-\infty}^{\infty} f_X(x)\, dx = 1.$

The density allows one to calculate the probability $P(A)$ for any event $A = [X \in \mathcal{A}]$, $\mathcal{A} \in \mathcal{E}_X$ by the general formula

$$P(A) = \int_{\mathcal{A}} f_X(x)\, dx.$$

By taking $\mathcal{A}$ to be an infinitesimal interval $\mathcal{A} = [x, x + \Delta x]$, we get the approximate relation

$$f_X(x) \doteq \frac{P(x \leq X \leq x + \Delta x)}{\Delta x},$$

which shows the origin of the term "density" for the function $f_X(x)$.

If the random variable X is of the *discrete type*, that is, if its ensemble $\mathcal{E}_X$ consists of isolated points on the real line, then the above ratio is either infinity or zero, depending on whether or not the interval $[x, x + \Delta x]$ contains a point of the ensemble. Nevertheless, the density can still be defined by employing the Dirac delta function,† in particular

$$f_X(x) = \sum_{k \in \mathcal{E}_X} P(X = k)\, \delta(x - k), \qquad -\infty < x < \infty,$$

† See Appendix One.

with the k's being the elements of the ensemble. The nonnegative sequence

$$P(X = k), \qquad k \in \mathcal{E}_X,$$

appearing in the above expression is called the *probability mass distribution* for the discrete random variable X and satisfies the condition

$$\sum_{k \in \mathcal{E}_X} P(X = k) = 1.$$

Thus the density of a discrete random variable is a weighted linear combination of delta functions, the weights being furnished by the values of the probability mass distribution. It follows that the probability law of a discrete random variable can be equivalently and more conveniently defined by means of the probability mass distribution.

If the density $f_X(x)$ contains no delta functions, the random variable X or, more precisely, its probability law is called *continuous*. In this case, it is necessary that $P(X = x) = 0$ for all real x. A random variable X is said to have a *mixed-type probability law* if its density $f_X(x)$ can be written as

$$f_X(x) = g(x) + \sum_k P(X = k)\, \delta(x - k), \qquad -\infty < x < \infty$$

where $\sum_k P(X = k) = q < 1$ and $g(x)$ is an ordinary (free of delta functions) nonnegative function such that $\int_{-\infty}^{\infty} g(x)\, dx = 1 - q$.

Some frequently encountered probability laws for a single random variable arising in connection with some basic probabilistic models are summarized in Appendix Two.

The probability law of a random vector $\mathbf{X} = (X_1, \ldots, X_m)'$ can again be most conveniently specified by the probability density function. Since the ensemble $\mathcal{E}_\mathbf{X}$ is now a subset of an m-dimensional space the density $f_\mathbf{X}(\mathbf{x})$ is a function of m real variables $\mathbf{x} = (x_1, \ldots, x_m)'$, each ranging from $-\infty$ to $+\infty$. Again, the density $f_\mathbf{X}(\mathbf{x})$ must satisfy the conditions

(i) $f_\mathbf{X}(\mathbf{x}) \geq 0$ for all $-\infty < x_j < \infty, \qquad j = 1, \ldots, m$

(ii) $\displaystyle\int_{-\infty}^{\infty} \cdots \int_{-\infty}^{\infty} f_\mathbf{X}(\mathbf{x})\, dx_1 \cdots dx_m = 1,$

and the probability law is obtained by integration

$$P(A) = \int_{\mathcal{A}} \cdots \int f_\mathbf{X}(\mathbf{x})\, d\mathbf{x}$$

for any event $A = [\mathbf{X} \in \mathcal{A}]$, $\mathcal{A} \subset \mathcal{E}_\mathbf{X}$. When the random vector $\mathbf{X} = (X_1, \ldots, X_m)'$ is regarded as an m-tuple of random variables, with components $X_1, \ldots, X_m$, the density $f_\mathbf{X}(\mathbf{x})$ is referred to as the *joint density* of $X_1, \ldots, X_m$.

The density of any single component (or any smaller group of components) of the random vector $\mathbf{X}$ is then referred to as a *marginal density* and is obtained from the joint density by integrating it over the remaining variables from $-\infty$ to $+\infty$, for instance

$$f_{X_1}(x_1) = \int_{-\infty}^{\infty} \cdots \int_{-\infty}^{\infty} f_{\mathbf{X}}(\mathbf{x})\, dx_2 \cdots dx_m .$$

The probability law of a random variable X can equivalently be specified by the *distribution function* $F_X(x)$, $-\infty < x < \infty$, defined by

$$F_X(x) = P(X \leq x).$$

It is easily seen that the relations between the distribution function and the density of X are

$$F_X(x) = \int_{-\infty}^{x+} f_X(x')\, dx' \qquad \text{and} \qquad f_X(x) = \frac{d}{dx} F_X(x).$$

Note that each jump discontinuity in the distribution function contributes a delta function to the expression for the density. (See Appendix One.) Similarly for a random vector $\mathbf{X} = (X_1, \ldots, X_m)'$ the (joint) distribution function is defined by

$$F_{\mathbf{X}}(\mathbf{x}) = P(X_1 \leq x_1, \ldots, X_m \leq x_m),$$

$$\mathbf{x}' = (x_1, \ldots, x_m), \qquad -\infty < x_1 < \infty, \ldots, -\infty < x_m < \infty.$$

Its relation to the joint density is analogously

$$F_{\mathbf{X}}(\mathbf{x}) = \int_{-\infty}^{x_1+} \cdots \int_{-\infty}^{x_m+} f_{\mathbf{X}}(\mathbf{x}')\, d\mathbf{x}', \qquad f_{\mathbf{X}}(\mathbf{x}) = \frac{\partial^m}{\partial x_1 \cdots \partial x_m} F_{\mathbf{X}}(\mathbf{x}).$$

The specification of the probability law for a stochastic process $X(t)$, $t \in \mathcal{T}$, is a little bit more involved. Since the stochastic process consists of an infinite collection of random variables indexed by $t \in \mathcal{T}$, no single density function would suffice. However, if we define the joint densities for each k-tuple of random variables

$$X(t_1), X(t_2), \ldots, X(t_k)$$

obtained by fixing the time instants

$$t_1 < t_2 < \cdots < t_k$$

from the time domain $\mathcal{T}$, then it can be shown that the probability law of the entire process is indeed specified. Of course, it is necessary to do so for *all* choices of the times $t_1 < \cdots < t_k$ and *all* $k = 1, 2, \ldots$, so we now have an infinite family of joint densities to deal with. For each such $k = 1, 2, \ldots$ the joint densities of

$X(t_1), \ldots, X(t_k)$ are called the kth *order densities* of the process, and to simplify notation we will use the symbol

$$f_{t_1, \ldots, t_k}(x_1, \ldots, x_k),$$

rather than

$$f_{X(t_1), \ldots, X(t_k)}(x(t_1), \ldots, x(t_k)).$$

Although this infinite family of densities must be *consistent* in the sense that, for example,

$$f_{t_1}(x_1) = \int_{-\infty}^{\infty} f_{t_1, t_2}(x_1, x_2)\, dx_2 \qquad \text{for all } -\infty < x_1 < \infty$$

(since $f_{t_1}(x_1)$ is the marginal density of $f_{t_1, t_2}(x_1, x_2)$), the complete specification of the probability law of a stochastic process still seems a formidable task. Fortunately, the majority of stochastic processes used in practice are usually assumed to satisfy some further conditions that simplify the specification of their probability law.

In describing the probability law of a stochastic process $X(t)$, $t \in \mathcal{T}$, representing the temporal evolution of a dynamic system, it is often more natural to prescribe the density of some future value $X(t_k)$ under the assumptions that some past and present values $X(t_1) = x_1, \ldots, X(t_{k-1}) = x_{k-1}$, $t_1 < \cdots < t_{k-1} < t_k$ have already been observed. For instance, it may be quite difficult to say directly what the joint density of $X(t_1)$ and $X(t_2)$, $t_1 < t_2$, ought to be inasmuch as the probability law of the "future state" $X(t_2)$ may depend on the value $x(t_1)$ of the "present state" $X(t_1)$. At the same time the nature of this dependence on $x(t_1)$ may be relatively easy to postulate from the physical evolution of the system.

This leads to the concept of the *conditional density* of one random variable Y *given* the value x of another random variable X. It is defined as the ratio

$$f_{Y|X}(y \mid x) = \frac{f_{X,Y}(x, y)}{f_X(x)}, \qquad -\infty < y < \infty$$

for all values of x for which $f_X(x) > 0$. The definition applies equally well to random vectors; the conditional density of $\mathbf{Y}$ given $\mathbf{X} = \mathbf{x}$ is

$$f_{\mathbf{Y}|\mathbf{X}}(\mathbf{y} \mid \mathbf{x}) = \frac{f_{\mathbf{X},\mathbf{Y}}(\mathbf{x}, \mathbf{y})}{f_{\mathbf{X}}(\mathbf{x})}.$$

Conditional densities provide an alternate and often more convenient specification of the probability law of a stochastic process $X(t)$, $t \in \mathcal{T}$. The infinite family of conditional densities†

† We again simplify the notation by writing, e.g., $f_{t_2|t_1}(x_2 \mid x_1)$ instead of $f_{X(t_2)|X(t_1)}(x(t_2) \mid x(t_1))$.

$$
\begin{aligned}
&f_{t_1}(x_1)\\
&f_{t_2|t_1}(x_2\,|\,x_1),\\
&f_{t_3|t_1,t_2}(x_2\,|\,x_1, x_2),\\
&\quad\cdots\\
&f_{t_k|t_1,\ldots,t_{k-1}}(x_k\,|\,x_1, \ldots, x_{k-1}),
\end{aligned}
$$

where $t_1 < t_2 < \cdots < t_k < \cdots$ are arbitrary time instants from the time domain $\mathcal{T}$ again yields a complete specification of the probability law of the stochastic process $X(t)$, $t \in \mathcal{T}$. For instance, the kth order density $f_{t_1,\ldots,t_k}(x_1, \ldots, x_k)$ of the process can be obtained by forming the product

$$
f_{t_k|t_1,\ldots,t_{k-1}}(x_k\,|\,x_1, \ldots, x_{k-1})f_{t_{k-1}|t_1,\ldots,t_{k-2}}(x_{k-1}\,|\,x_1, \ldots, x_{k-2})
\cdots f_{t_3|t_1,t_2}(x_3\,|\,x_1, x_2)f_{t_2|t_1}(x_2\,|\,x_1)f_{t_1}(x_1)
$$

as is readily verified from the definition of a conditional density. This approach to the specification of the probability law of a stochastic process leads immediately to the definition of a very important special class of stochastic processes.

Definition 1.1
A stochastic process $X(t)$, $t \in \mathcal{T}$, is called a *Markov process* if for any sequence of times $t_1 < t_2 < \cdots < t_k < t_{k+1}$, $k = 1, 2, \ldots$ from the time domain the conditional densities satisfy the *Markov property*:

$$
f_{t_{k+1}|t_1,\ldots,t_k}(x_{k+1}\,|\,x_1, \ldots, x_k) = f_{t_{k+1}|t_k}(x_{k+1}\,|\,x_k)
$$

for all $x_1, \ldots, x_{k+1}$ from the state space $\mathcal{S}$.

In words, the Markov property says that the conditional density of $X(t_{k+1})$ given "all the past and present values" $X(t_1) = x(t_1), \ldots, X(t_k) = x(t_k)$ depends only on the present (most recent) value $X(t_k) = x(t_k)$. Interpreting $X(t)$ as the state of a dynamic system, this is exactly what is needed for the canonical state-space representation—the future evolution of the system depends only on its current state.

The probability law of a Markov process is completely characterized by the conditional probabilities

$$
f_{t|t'}(x\,|\,x'), \qquad t' < t, \qquad x, x' \in \mathcal{S},
$$

called *transition densities*, and by the first order density $f_{t_0}(x)$, $x \in \mathcal{S}$, for some fixed time instant $t_0 \in \mathcal{T}$. If $\mathcal{T} = [0, \infty)$ we usually take $t_0 = 0$ and refer to the first order density $f_0(x)$ as the *initial density*. The transition densities of a Markov process, however, cannot be completely arbitrary, since they must satisfy the *Chapman-Kolmogorov equation*:

$$f_{t_3|t_1}(x_3|x_1) = \int_{-\infty}^{\infty} f_{t_3|t_2}(x_3|x_2) f_{t_2|t_1}(x_2|x_1)\, dx_2$$

for all $t_1 < t_2 < t_3$ and all $x_1 \in \mathcal{S}$, $x_3 \in \mathcal{S}$.

Remark: This equation is easily derived by noting that by the definition of conditional densities

$$f_{t_3, t_2|t_1}(x_3, x_2|x_1) = f_{t_3|t_2, t_1}(x_3|x_1, x_2) f_{t_2|t_1}(x_2|x_1)$$

and using the Markov property

$$f_{t_3|t_2, t_1}(x_3|x_1, x_2) = f_{t_3|t_2}(x_3|x_2),$$

and the fact that

$$f_{t_3|t_1}(x_3|x_1) = \int_{-\infty}^{\infty} f_{t_3, t_2|t_1}(x_3, x_2|x_1)\, dx_2.$$ Я

Since by the definition of the conditional density

$$f_{t|t'}(x|x') = \frac{f_{t, t'}(x, x')}{f_{t'}(x')}, \qquad t' < t,$$

it follows that the *probability law of a Markov process is completely specified by the first and second order densities only*. However, these densities must be such that the Chapman-Kolomogorov equations are always satisfied (they must be consistent).

Let us just mention briefly that if the state space of a Markov process is discrete, the transition probabilities can be replaced by the transition probability mass distributions $P(X(t) = x \mid X(t') = x')$, $t < t'$, and the integral in the Chapman-Kolmogorov equation then becomes a sum

$$P(X(t_3) = x_3 \mid X(t_1) = x_1) = \sum_{x_2 \in \mathcal{S}} P(X(t_3) = x_3 \mid X(t_2) = x_2) P(X(t_2) = x_2 \mid X(t_1) = x_1).$$

The conditional probability density is useful in expressing various kinds of dependence between random variables. Frequently, however, we have a collection of random variables representing some physical quantities that we would like to consider independent of one another. That is, we wish to express our belief that the value or values of any particular random variable of the collection has no influence whatsoever on the values of the rest of these variables.

Formally, we say that the random variables $X_1, \ldots, X_m$ are, by definition, *mutually independent* if their joint density is the product of their marginal densities; in symbols,

$$f_{X_1, \ldots, X_m}(x_1, \ldots, x_m) = f_{X_1}(x_1) f_{X_2}(x_2) \cdots f_{X_m}(x_m)$$

for all $-\infty < x_1 < \infty, \ldots, -\infty < x_m < \infty$.

Independence of random variables is a *family property* in the sense that independence of $X_1, \ldots, X_m$ implies independence of any subcollections of these random variables but not conversely. It is also a *hereditary property*, since if $X_1, \ldots, X_m$ are mutually independent random variables and if

$$Y_1 = g_1(X_1, \ldots, X_{j_1}), \qquad Y_2 = g_2(X_{j_1+1}, \ldots, X_{j_2}), \ldots, Y_k = g_k(X_{j_k}, \ldots, X_m)$$

are new random variables, each a function of nonoverlapping groups of the X_i's, then the random variables $Y_1, \ldots, Y_k$ are again mutually independent. They inherit their independence from $X_1, \ldots, X_m$. Note that by the definition of a conditional density $f_{X_1|X_2}(x_1 \mid x_2) = f_{X_1}(x_1)$ and $f_{X_2|X_1}(x_2 \mid x_1) = f_{X_2}(x_2)$ if and only if X_1 and X_2 are independent random variables.

A particularly simple and yet quite important discrete-time stochastic process $X(n)$, $n = 1, 2, \ldots$, is one where for any positive integer m, the random variables $X(1), X(2), \ldots, X(m)$ are mutually independent. As a result, the probability law of such a process is completely specified by just the first order densities $f_n(x_n)$, $n = 1, 2, \ldots$. We refer to such a process as a *sequence of independent random variables*. Furthermore, if all these first order densities are identical, we call the process $X(n)$, $n = 1, 2, \ldots$, a sequence of *independent, identically distributed random variables*, abbreviated i.i.d.

One such particular process occurs so often that it deserves a special designation.

Definition 1.2
A sequence $X(n)$, $n = 1, 2, \ldots$ (or $n = 0, 1, \ldots$) of i.i.d. Bernoulli† random variables with parameter p, $0 < p < 1$, is called a *Bernoulli process*.

Often we are given the probability law of one random variable, vector, or stochastic process X and wish to derive the probability law of a new random variable, vector, or process Y obtained from the former by some deterministic transformation. The basic idea here is quite simple; any such transformation represents a mapping g from the ensemble $\mathcal{E}_X$ of the former random variable to the ensemble $\mathcal{E}_Y$ of the latter, and so to obtain the probability $P(B)$ of some event $B = [Y \in \mathcal{B}]$, $\mathcal{B} \subset \mathcal{E}_Y$, we merely have to find the corresponding subset $\mathcal{A}$ of the ensemble $\mathcal{E}_X$ for which $X \in \mathcal{A}$ whenever $Y = g(X) \in \mathcal{B}$. Then we just set $P(B) = P(A)$ with $A = [X \in \mathcal{A}]$. Doing this for every event B, the probability law of Y is

† See Appendix Two.

established. Of course, in practice the difficult part is to derive the exact relationships between the subsets $\mathcal{A}$ and $\mathcal{B}$, especially for complicated mappings g and "large" ensembles. Therefore, only some special cases will be summarized here.

If X is a random variable with density $f_X(x)$ and $y = g(x)$ is a function with differentiable inverse $x = h(y)$, then the density $f_Y(y)$ of the random variable $Y = g(X)$ is given by the formula

$$f_Y(y) = f_X(h(y))\left|\frac{dh(y)}{dy}\right|.$$

In particular if Y is a linear function of X,

$$Y = aX + b, \qquad a \neq 0, \qquad b \text{ constants,}$$

then

$$f_Y(y) = \frac{1}{|a|} f_X\left(\frac{y-b}{a}\right).$$

Remark: If X and hence also Y is discrete, then the fact that $\delta(u/a) = |a|\delta(u)$ (see Appendix One) implies that the probability mass distribution of Y is

$$P(Y = y) = P\left(X = \frac{y-b}{a}\right). \qquad \text{Я}$$

The conditional densities transform in exactly the same way; if $y = g(x)$ is a function with differentiable inverse $x = h(y)$, then the conditional density of $Y = g(X)$, given $Z = z$, is related to the conditional density of X, given $Z = z$, by the formula

$$f_{Y|Z}(y\,|\,z) = f_{X|Z}(h(y)\,|\,z)\left|\frac{dh(y)}{dy}\right|.$$

If $\mathbf{Y}$ and $\mathbf{Z}$ are random vectors and if $\mathbf{z} = \mathbf{g}(\mathbf{y})$ is a one-to-one transformation between the ensembles $\mathcal{E}_\mathbf{Y}$ and $\mathcal{E}_\mathbf{Z}$, then we also have the identity

$$f_{X|\mathbf{Z}}(x\,|\,\mathbf{z}) = f_{X|\mathbf{Y}}(x\,|\,\mathbf{h}(\mathbf{z}))$$

with $\mathbf{y} = \mathbf{h}(\mathbf{z})$ the inverse transformation of $\mathbf{g}$. Loosely speaking, if $\mathbf{Y}$ and $\mathbf{Z}$ can be uniquely determined from each other, then conditioning by $\mathbf{Y}$ is the same as conditioning by the corresponding value of $\mathbf{Z}$. This shows, for instance, that if g is a *strictly monotonic function and* $X(t)$, $t \in \mathcal{T}$, *is a Markov process, then* $Y(t) = g(X(t))$, $t \in \mathcal{T}$, *is again a Markov process.*

A very useful property of conditional densities in connection with functions of random variables is that the value of the conditioning variable (or variables) can be treated as if it were a constant. That is, if $Z = g(\mathbf{X}, \mathbf{Y})$ for some arbitrary function g, then the conditional density of Z given $\mathbf{X} = \mathbf{x}$ is, for each fixed $\mathbf{x}$, identical with the conditional density of $g(\mathbf{x}, \mathbf{Y})$, given $\mathbf{X} = \mathbf{x}$.

Thus, for example, the conditional density of the sum $S = X + Y$, given $X = x$, is identical with the conditional density of $x + Y$, given $X = x$, and hence

$$f_{S|X}(s\,|\,x) = f_{Y|X}(s - x\,|\,x)$$

by the formula given earlier with $a = 1$ and $b = x$. Upon writing

$$f_S(s) = \int_{-\infty}^{\infty} f_{S,X}(s, x)\,dx = \int_{-\infty}^{\infty} f_{S|X}(s\,|\,x) f_X(x)\,dx$$

we get the formula for the density of the sum of two random variables $S = X + Y$

$$f_S(s) = \int_{-\infty}^{\infty} f_{Y|X}(s - x\,|\,x) f_X(x)\,dx.$$

In the special but very important case, when X and Y are independent random variables, this formula becomes

$$f_S(s) = \int_{-\infty}^{\infty} f_Y(s - x) f_X(x)\,dx \qquad -\infty < s < \infty.$$

The operation represented by this formula, giving the density f_S of the sum $S = X + Y$ in terms of the densities f_X and f_Y of the independent summands, is called a *convolution* and is often abbreviated as

$$f_S = f_X \circledast f_Y.$$

The hereditary property of independence allows one to repeatedly use the convolution formula to obtain the density of a sum of any number of independent random variables. This construction gives several important types of discrete-time stochastic processes, of which we mention just two:

Definition 1.3
The *binomial* (or Bernoulli counting) *process* $S(n)$, $n = 0, 1, \ldots$, is defined by

$$S(n) = X(0) + X(1) + \cdots + X(n),$$

where $X(0) = 0$ and $X(n)$, $n = 1, 2, \ldots$, are i.i.d. Bernoulli random variables with parameter p.

Definition 1.4
The process $V(n)$, $n = 0, 1, \ldots$, defined by $V(n) = X(0) + X(1) + \cdots + X(n)$, where $X(0) = 0$ and $X(n)$, $n = 1, 2, \ldots$, are independent Bernoulli symmetric† random variables, is called the *Bernoulli symmetric random walk*.

† See Appendix Two.

Note that if $S(n)$ is a binomial process with $p = 1/2$, then $V(n) = 2S(n) - n$, $n = 0, 1, \ldots$.

Discrete-time processes constructed as consecutive sums of independent random variables furnish yet another example of processes whose probability law is completely specified by the first order densities.

Suppose next that we have two random vectors $\mathbf{X} = (X_1, \ldots, X_m)'$ and $\mathbf{Y} = (Y_1, \ldots, Y_m)'$ such that there is a one-to-one correspondence $\mathbf{g}$ between the ensembles $\mathcal{E}_\mathbf{X}$ and $\mathcal{E}_\mathbf{Y}$. If $f_\mathbf{X}(\mathbf{x})$ is the joint density of the vector $\mathbf{X}$, then the joint density $f_\mathbf{Y}(\mathbf{y})$ is given by the formula

$$f_\mathbf{Y}(\mathbf{y}) = f_\mathbf{X}(\mathbf{h}(\mathbf{y}))\left|\frac{\partial \mathbf{x}}{\partial \mathbf{y}}\right|,$$

where $\mathbf{x} = \mathbf{h}(\mathbf{y})$ is the inverse transformation of $\mathbf{y} = \mathbf{g}(\mathbf{x})$ and $\partial\mathbf{x}/\partial\mathbf{y}$ is an $m \times m$ determinant, called the Jacobian,

$$\frac{\partial \mathbf{x}}{\partial \mathbf{y}} = \begin{vmatrix} \dfrac{\partial x_1}{\partial y_1}, & \ldots, & \dfrac{\partial x_1}{\partial y_m} \\ \cdots & & \\ \dfrac{\partial x_m}{\partial y_1}, & \ldots, & \dfrac{\partial x_m}{\partial y_m} \end{vmatrix}.$$

Thus if, for example, a stochastic process $Y(t)$, $t \in \mathcal{T}$, is defined as a deterministic transformation of another process $X(t)$, $t \in \mathcal{T}$, we can, at least in principle, calculate all the kth order densities of the process $Y(t)$ from those of the process $X(t)$ and thus obtain the probability law of the former process.

An important special case occurs if g is a *linear* transformation, that is, if

$$\begin{aligned} Y_1 &= a_{11}X_1 + a_{12}X_2 + \cdots + a_{1m}X_m + b_1 \\ Y_2 &= a_{21}X_1 + a_{22}X_2 + \cdots + a_{2m}X_m + b_2 \\ &\cdots \\ Y_m &= a_{m1}X_1 + a_{m2}X_2 + \cdots + a_{mm}X_m + b_m, \end{aligned}$$

where the a's and b's are constants. Calling $\mathbf{a}$ and $\mathbf{b}$ the $m \times m$ matrix and m-component column vector of the coefficients a_{ij} and b_j respectively, that is, writing the above linear transformation as

$$\mathbf{Y} = \mathbf{aX} + \mathbf{b}$$

the formula for the density $f_\mathbf{Y}(\mathbf{y})$ becomes

$$f_\mathbf{Y}(\mathbf{y}) = f_\mathbf{X}(\mathbf{a}^{-1}(\mathbf{y} - \mathbf{b}))\,|\det \mathbf{a}|^{-1},$$

where $\det \mathbf{a}$ is the determinant of the matrix $\mathbf{a}$. (As long as the linear transformation is one-to-one, the determinant is not zero.)

This formula for linear transformations allows us to introduce another important class of stochastic processes whose probability laws are relatively easy to specify, namely processes with independent increments.

Definition 1.5
A stochastic process $X(t)$, $t \in \mathcal{T}$, is said to have *independent increments* if for all time instances $t_1 < t_2 < \cdots < t_k < t_{k+1}$ from the time domain $\mathcal{T}$ and every $k = 1, 2, \ldots$ the random variables

$$X(t_2) - X(t_1), X(t_3) - X(t_2), \ldots, X(t_{k+1}) - X(t_k)$$

are mutually independent.

Since there is a simple linear relationship between the random variables $X(t_1), \ldots, X(t_{k+1})$ and the increments $X(t_2) - X(t_1), \ldots, X(t_{k+1}) - X(t_k)$, and since the joint density of these increments is just the product of their marginal densities, the probability law of a process with independent increments is completely determined by the density of an increment $X(t) - X(t')$ for all $t' < t$ and by the first order density $f_{t_0}(x)$ at some single time instant $t_0 \in \mathcal{T}$.

In fact, the two processes with independent increments defined below are the most fundamental in probability theory.

Definition 1.6
A stochastic process $N_0(t)$, $t \geq 0$, with independent increments such that $N_0(0) = 0$ and such that for each $0 \leq t' < t$ the increment $N_0(t) - N_0(t')$ has the Poisson probability mass distribution

$$P(N_0(t) - N_0(t') = k) = e^{-(t-t')} \frac{(t - t')^k}{k!}, \qquad k = 0, 1, \ldots,$$

is called the *standard Poisson process.*

Definition 1.7
A stochastic process $W_0(t)$, $t \geq 0$, with independent increments such that $W_0(0) = 0$ and such that for each $0 \leq t' < t$ the increment $W_0(t) - W_0(t')$ has the Gaussian density

$$f_{W_0(t) - W_0(t')}(u) = \frac{1}{\sqrt{2\pi(t - t')}} e^{-u^2/2(t-t')}, \qquad -\infty < u < \infty$$

is called the *standard Wiener process.*

Both of these processes can be derived by the passage to a limit of a discrete-time process. The standard Poisson process is obtained from the binomial process by setting the parameter $p = \Delta t$, the time interval between successive discrete-time instances, and letting $\Delta t \to 0$. The standard Wiener process is obtained from the symmetric Bernoulli random walk by first reducing its scale by $\sqrt{\Delta t}$ and then again letting the time intervals Δt between successive time instances shrink to zero.

In spite of this similarity, the two processes are markedly different. The state space of the Poisson process is the set of all nonnegative integers, and a typical sample path is a step function with unit upward jumps occurring at random time instances. On the other hand, the state space of the Wiener process is the interval $(-\infty, \infty)$ and its typical sample paths are continuous, albeit extremely irregular functions. Furthermore, they exhibit *Lévy's oscillation property*:

$$dw(t) \doteq \sqrt{dt},$$

which means roughly that the infinitesimal rate of change $dw(t)/dt$ of a typical sample path is infinite at every time instant.

A (*nonstandard*) *Poisson process* $N(t)$, $t \geq 0$, is obtained from the standard Poisson process by an elementary time transformation

$$N(t) = N_0(\lambda t), \qquad \lambda > 0 \text{ a constant},$$

and is called a Poisson process with *intensity* λ.

Similarly, a *nonstandard Wiener process* $W(t)$, $t \geq 0$, is defined by

$$W(t) = \mu t + \sigma W_0(t), \qquad t \geq 0,$$

where the constant μ is called the *drift parameter*, and the constant $\sigma > 0$ is called the *intensity* of the process.

Neither of these generalizations destroys the independence of increments of these processes, so their probability laws are still easy to specify.

A very important single numerical quantity defined in terms of the probability law of a random variable X is its *expected value* (equivalently *expectation*) or *mean* denoted by

$$\mathrm{E}[X] \qquad \text{or} \qquad \mu_X,$$

and defined by

$$\mathrm{E}[X] = \int_{-\infty}^{\infty} x f_X(x)\, dx,$$

provided the integral converges absolutely, that is, provided that

$$\mathrm{E}[|X|] = \int_{-\infty}^{\infty} |x|\, f_X(x)\, dx < \infty.$$

More generally, if X is a function of a random vector $\mathbf{Y} = (Y_1, \ldots, Y_m)'$,

$$X = g(\mathbf{Y}),$$

then also

$$E[X] = \int_{-\infty}^{\infty} \cdots \int_{-\infty}^{\infty} g(y_1, \ldots, y_m) f_{\mathbf{Y}}(y_1, \ldots, y_m)\, dy_1 \cdots dy_m.$$

These two definitions are consistent in the sense that both yield the same value for the expectation $E[X]$, although the rigorous proof of this statement in full generality is rather delicate.

Being defined as a definite integral the basic properties of expectation reflect those of an integral, in particular:

(i) Linearity:

$$E\left[\sum_{j=1}^{m} a_j X_j\right] = \sum_{j=1}^{m} a_j E[X_j]$$

(ii) Monotonicity:

If $X \geq 0$, then $E[X] \geq 0$

(iii) Convexity:

$$|E[X]| \leq E[|X|]$$

(iv) Schwarz inequality:

$$|E[XY]|^2 \leq E[X^2]E[Y^2]$$

with equality if and only if $aX + bY = 0$ for some constants a and b not both zero.

(v) Triangle inequality:

$$\sqrt{E[(X+Y)^2]} \leq \sqrt{E[X^2]} + \sqrt{E[Y^2]}$$

(vi) Factorization property:

If $X_1, \ldots, X_m$ are independent random variables, then

$$E[X_1 X_2 \cdots X_m] = E[X_1]E[X_2] \cdots E[X_m].$$

The *Chebyshev inequality* asserts that for any $\varepsilon > 0$

$$P(|X| \geq \varepsilon) \leq \frac{1}{\varepsilon^2} E[X^2].$$

A somewhat tighter estimate is provided by the *Chernoff bound*:

$$P(X \geq \varepsilon) \leq \min_{0 \leq u < \infty} e^{-u\varepsilon} E[e^{uX}]$$

again valid for all $\varepsilon > 0$.

Another property of the expectation is the *minimum mean square error property*:

$$\mathrm{E}[(X - \mu_X)^2] \leq \mathrm{E}[(X - a)^2] \qquad \text{for all } -\infty < a < \infty.$$

This last property shows that the expectation $\mu_X = \mathrm{E}[X]$ is the number whose expected squared difference from all values of the random variable X is minimum. The quantity

$$\mathrm{Var}[X] = \mathrm{E}[(X - \mu_X)^2] = \mathrm{E}[X^2] - (\mu_X)^2$$

is then a measure of residual spread or dispersion and is called the *variance* of the random variable X. An alternative notation for the variance is σ_X^2, while the positive square root of this quantity $\sigma_X = \sqrt{\mathrm{Var}[X]}$ is called the *standard deviation.*

The variance of a random variable X, if it exists, is always a nonnegative number and equals zero if and only if $P(X = \mu_X) = 1$, that is, if the random variable X is a constant.

The *covariance* of two random variables X and Y with expectations μ_X and μ_Y is defined by

$$\mathrm{Cov}[X, Y] = \mathrm{E}[(X - \mu_X)(Y - \mu_Y)] = \mathrm{E}[XY] - \mu_X \mu_Y.$$

Remark: The dimensionless ratio $\rho_{X,Y} = \mathrm{Cov}[X, Y]/\sigma_X \sigma_Y$ is known as the *correlation coefficient* and by the Schwarz inequality $|\rho_{X,Y}| \leq 1$. Я

The variance of a linear combination of m random variables then equals

$$\mathrm{Var}\left[\sum_{i=1}^{m} a_i X_i\right] = \sum_{i=1}^{m} a_i^2 \, \mathrm{Var}[X_i] + \sum_{i=1}^{m} \sum_{\substack{j=1 \\ i \neq j}}^{m} a_i a_j \, \mathrm{Cov}[X_i, X_j].$$

If X and Y are independent random variables, then $\mathrm{Cov}[X, Y] = 0$, and hence the *variance of a sum of mutually independent random variables is the sum of their variances*, a result known as the *Bienaymé equality*.

If $\mathbf{X} = (X_1, \ldots, X_m)'$ is a random vector, then the column vector of expected values

$$\boldsymbol{\mu}_{\mathbf{X}} = (\mathrm{E}[X_1], \ldots, \mathrm{E}[X_m])'$$

is called the *mean vector* of $\mathbf{X}$ while the $m \times m$ matrix

$$\boldsymbol{\sigma}_{\mathbf{X}}^2 = \begin{bmatrix} \sigma_{11}^2 & \sigma_{12}^2 & \cdots & \sigma_{1m}^2 \\ \sigma_{21}^2 & \sigma_{22}^2 & \cdots & \sigma_{2m}^2 \\ \vdots & \vdots & & \vdots \\ \sigma_{m1}^2 & \sigma_{m2}^2 & \cdots & \sigma_{mm}^2 \end{bmatrix}$$

with entries $\sigma_{ij}^2 = \text{Cov}[X_i, X_j]$ is called the *covariance matrix* of the random vector $\mathbf{X}$. Since $\text{Cov}[X_i, X_j] = \text{Cov}[X_j, X_i]$, the covariance matrix is *symmetric* and its diagonal entries $\sigma_{ii}^2 = \text{Cov}[X_i, X_i] = \text{Var}[X_i]$ are the variances of the components $X_1, \ldots, X_m$. Furthermore, a covariance matrix is always *nonnegative definite*,

$$\sum_{i=1}^{m} \sum_{j=1}^{m} \sigma_{ij}^2 a_i a_j \geq 0$$

for any choice of constants $a_1, \ldots, a_m$, since the double sum above actually is the variance of the linear combination $a_1 X_1 + \cdots + a_m X_m$ and hence is always nonnegative.

If $X(t)$, $t \in \mathcal{T}$, is a stochastic process, then the time function

$$\mu_X(t) = \text{E}[X(t)], \qquad t \in \mathcal{T},$$

is called the *mean function* of the process $X(t)$.

The time function

$$\sigma_X^2(t) = \text{Var}[X(t)], \qquad t \in \mathcal{T},$$

is called the *variance function*, while the functions of two time variables $t_1 \in \mathcal{T}$, $t_2 \in \mathcal{T}$ defined by

$$R_X(t_1, t_2) = \text{E}[X(t_1)X(t_2)],$$

and

$$\gamma_X(t_1, t_2) = \text{Cov}[X(t_1), X(t_2)]$$

are called the *correlation function* and the *covariance function*, respectively. Note that

$$\gamma_X(t_1, t_2) = R_X(t_1, t_2) - \mu_X(t_1)\mu_X(t_2)$$

and so we shall deal mainly with the correlation function R_X as is customary in the engineering literature. By definition, both the correlation and the covariance functions are symmetric

$$R_X(t_1, t_2) = R_X(t_2, t_1), \qquad \gamma_X(t_1, t_2) = \gamma_X(t_2, t_1)$$

and by the Schwarz inequality

$$|R_X(t_1, t_2)| \leq \sqrt{R_X(t_1, t_1)R_X(t_2, t_2)} = \sqrt{\text{E}[X^2(t_1)]\text{E}[X^2(t_2)]},$$
$$|\gamma_X(t_1, t_2)| \leq \sqrt{\gamma_X(t_1, t_1)\gamma_X(t_2, t_2)} = \sigma_X(t_1)\sigma_X(t_2).$$

Also, both must be nonnegative definite

$$\sum_{i=1}^{m} \sum_{j=1}^{m} R_X(t_i, t_j) a_i a_j \geq 0$$

for any $t_1, \ldots, t_m, a_1, \ldots, a_m$, and $m = 1, 2, \ldots$, and the same is true for the covariance function.

Of course, being defined as expectations, these functions need not always exist. If, however, the stochastic process $X(t)$, $t \in \mathcal{T}$, is such that

$$\mathrm{E}[|X(t)|^2] < \infty \qquad \text{for all } t \in \mathcal{T}$$

then the existence of both the mean function and the correlation function is guaranteed. We shall call a stochastic process satisfying this condition a *finite-variance process.*

Remark: Occasionally, we will deal with stochastic processes that are vector valued or complex valued. For a vector-valued process $\mathbf{X}(t) = (X_1(t), \ldots, X_m(t))'$, $t \in \mathcal{T}$, the definition of the mean and correlation function is quite simple. The mean function

$$\boldsymbol{\mu}_{\mathbf{X}}(t) = (\mu_1(t), \ldots, \mu_m(t))'$$

is vector valued with components

$$\mu_i(t) = \mathrm{E}[X_i(t)], \qquad i = 1, \ldots, m.$$

The variance function $\boldsymbol{\sigma}_{\mathbf{X}}^2(t)$ is matrix valued, being for each $t \in \mathcal{T}$ simply the $m \times m$ covariance matrix of the random vector $\mathbf{X}(t)$. The correlation function is likewise matrix valued, being defined by

$$\mathbf{R}_{\mathbf{X}}(t_1, t_2) = \mathrm{E}[\mathbf{X}(t_1)\mathbf{X}'(t_2)], \qquad t_1 \in \mathcal{T}, \qquad t_2 \in \mathcal{T}$$

so that its (i, j)th entry is the expectation $\mathrm{E}[X_i(t_1)X_j(t_2)]$ and is not, in general, a symmetric matrix. If $X(t)$ and $Y(t)$ are two stochastic processes (scalar or vector valued) with the same time domain $\mathcal{T}$, the *cross-correlation function* between these two processes is a matrix-valued function defined by

$$\mathbf{R}_{\mathbf{X},\mathbf{Y}}(t_1, t_2) = \mathrm{E}[\mathbf{X}(t_1)\mathbf{Y}'(t_2)].$$

Unlike the scalar correlation function, the scalar cross-correlation function is generally not a symmetric function of the two arguments t_1 and t_2.

A complex-valued process $Z(t) = X(t) + \imath Y(t)$, $t \in \mathcal{T}$, where $\imath = \sqrt{-1}$ is the imaginary unit, can be regarded either as one whose state space $\mathcal{S}$ is a subset of the complex plane or equivalently as a two-component vector-valued process, the components being the real part $X(t)$ and the imaginary part $Y(t)$ of the complex-valued random variable $Z(t)$. Its mean function is then complex valued

$$\mu_Z(t) = \mu_X(t) + \imath\mu_Y(t), \qquad t \in \mathcal{T}$$

but the variance function is real

$$\sigma_Z^2(t) = \mathrm{E}[|Z(t) - \mu_Z(t)|^2] = \sigma_X^2(t) + \sigma_Y^2(t), \qquad t \in \mathcal{T},$$

where $|Z(t)| = Z(t)Z^*(t)$ is the modulus of $Z(t)$, and the asterisk denotes the complete conjugate $Z^*(t) = X(t) - \imath Y(t)$. The correlation function is defined by

$$R_Z(t_1, t_2) = \mathrm{E}[Z(t_1)Z^*(t_2)], \qquad t_1 \in \mathcal{T}, \qquad t_2 \in \mathcal{T},$$

and is a complex-valued function such that

$$|R_Z(t_1, t_2)| \leq \sqrt{E[|Z(t)|^2]E[|Z(t_2)|^2]},$$

$$R_Z(t_1, t_2) = R_Z^*(t_2, t_1) \qquad \text{(Hermitian symmetry)},$$

and

$$\sum_{i=1}^{m} \sum_{j=1}^{m} R_Z(t_1, t_2) a_i a_j^* \geq 0 \qquad \text{for all } t_1, \ldots, t_m,$$

all complex $a_1, \ldots, a_m$, and all $m = 1, 2, \ldots$ (nonnegative definiteness). Я

The mean function and the correlation function of a stochastic process generally do not provide enough information to completely specify the probability law of the process inasmuch as they are determined by the second order densities only. There is, however, an important exception to this statement.

Definition 1.8
A stochastic process $X(t)$, $t \in \mathcal{T}$, is called a *Gaussian process* if all its kth order densities $f_{t_1, \ldots, t_k}(x_1, \ldots, x_k)$, $k = 1, 2, \ldots$, are k-variate Gaussian densities.†

Since the kth order density $f_{t_1, \ldots, t_k}(x_1, \ldots, x_k)$ is the density of the random vector $(X(t_1), \ldots, X(t_k))'$, the Gaussian probability law of this vector is completely determined by the means $\mu_X(t_i) = E[X(t_i)]$, $i = 1, \ldots, k$, and by the covariances

$$\gamma_X(t_i, t_j) = \text{Cov}[X(t_i), X(t_j)], \qquad i = 1, \ldots, k, \quad j = 1, \ldots, k.$$

It follows that the knowledge of the mean function $\mu_X(t)$, $t \in \mathcal{T}$, and the correlation function $R_X(t_1, t_2)$, $t_1 \in \mathcal{T}$, $t_2 \in \mathcal{T}$, is sufficient to completely specify all the kth order densities of the Gaussian process and hence also its probability law. Note that every Gaussian process is automatically a finite-variance process. In particular, the *standard Wiener process* $W_0(t)$, $t \geq 0$, can also be defined as a Gaussian process with mean function $\mu_{W_0}(t) = 0$, $t \geq 0$, and correlation function $R_{W_0}(t_1, t_2) = \min(t_1, t_2)$, $t_1 \geq 0$, $t_2 \geq 0$.

Gaussian processes have many nice properties, mainly because linear transformations of Gaussian random vectors again yield Gaussian random vectors. Thus, in general, *any linear operation performed on a Gaussian process results in another Gaussian process.* This fundamental property will be used repeatedly in later chapters.

If a Gaussian process $X(t)$, $t \in \mathcal{T}$, is at the same time a Markov process—

† See Appendix Two.

called in this case a *Gauss-Markov process*—then its covariance function $\gamma_X(t_1, t_2)$ must satisfy the equation

$$\gamma_X(t_3, t_1) = \frac{\gamma_X(t_3, t_2)\gamma_X(t_2, t_1)}{\sigma_X^2(t_2)}$$

for all $t_1 \leq t_2 \leq t_3$ from the time domain $\mathcal{T}$. Note that this equation is somewhat reminiscent of the Chapman-Kolmogorov equation. Conversely, any Gaussian process whose covariance function satisfies this equation is actually a Gauss-Markov process.

Remark: For a vector-valued Gauss-Markov process, this equation becomes

$$\gamma_{\mathbf{X}}(t_3, t_1) = \gamma_{\mathbf{X}}(t_3, t_2)(\sigma_{\mathbf{X}}^2(t_2))^{-1}\gamma_{\mathbf{X}}(t_2, t_1).$$ Я

Among Gauss-Markov processes, the following one plays a prominent role.

Definition 1.9
A Gauss-Markov process $X(t)$, $t \geq 0$, with zero mean function $\mu_X(t) = 0$, $t \geq 0$, and with correlation function

$$R_X(t_1, t_2) = \sigma^2(e^{-\alpha|t_1 - t_2|} - e^{-\alpha(t_1 + t_2)}),$$

$t_1 \geq 0$, $t_2 \geq 0$, is called the *Ornstein-Uhlenbeck process* with parameters $\alpha > 0$ and $\sigma > 0$.

The Ornstein-Uhlenbeck process $X(t)$ can be obtained from the standard Wiener process $W_0(t)$ by means of the transformation

$$X(t) = \sigma e^{-\alpha t} W_0(e^{2\alpha t} - 1), \qquad t \geq 0.$$

It follows that the sample paths of an Ornstein-Uhlenbeck process have a character similar to those of the standard Wiener process—they are continuous functions of time, although extremely jagged.

Many probabilistic models are concerned with the description of a random phenomenon or a dynamic system that has already reached a steady state. The state of such a system, represented by a stochastic process $X(t)$, $t \in \mathcal{T}$, has the property that the probability law of this process is invariant with respect to a shift of time or, in other words, is independent of the choice of the origin of time.

Definition 1.10
A stochastic process $X(t)$, $t \in \mathcal{T}$, is called *strictly stationary* if for any $t_1 < t_2 < \cdots < t_k$ from the time domain $\mathcal{T}$ and any $k = 1, 2, \ldots$ its kth order densities satisfy the condition

$$f_{t_1+\tau, \ldots, t_k+\tau}(x_1, \ldots, x_k) = f_{t_1, \ldots, t_k}(x_1, \ldots, x_k)$$

for all τ for which $t_1 + \tau, \ldots, t_k + \tau$ are in the time domain $\mathcal{T}$.

If the above conditions hold for all kth order densities $k = 1, \ldots, K$ but not for the kth order densities with $k > K$, we call the process Kth order stationary. Clearly, Kth order stationarity implies stationarity of lesser order, while strict stationarity requires stationarity of all orders.

If $X(t)$, $t \in \mathcal{T}$ is a finite-variance second order stationary process, then its mean function $\mu_X(t)$ must be a constant $\mu_X(t) = \mu_X$ for all $t \in \mathcal{T}$, while its correlation function has the property

$$R_X(t_1 + \tau, t_2 + \tau) = R_X(t_1, t_2)$$

for all t_1, t_2, and τ. That is, the correlation function depends on t_1 and t_2 only via the difference $t_2 - t_1$. However, a finite-variance process with constant mean function and with correlation function depending only on the time difference $t_2 - t_1$ need not be even first order stationary.

Definition 1.11

A finite-variance process $X(t)$, $t \in \mathcal{T}$, with constant mean function and with a correlation function such that

$$R_X(t_1 + \tau, t_2 + \tau) = R_X(t_1, t_2)$$

for all t_1, t_2, and τ is called *wide-sense stationary.*

For a wide-sense stationary process $X(t)$, $t \in \mathcal{T}$, it is convenient to define the correlation function as a function of the single time variable $\tau = t_2 - t_1$,

$$R_X(\tau) = \mathrm{E}[X(t)X(t + \tau)]$$

since the right-hand side of this expression does not depend on t. The properties of the correlation function $R_X(\tau)$ follow immediately from the general case; we have

$$|R_X(\tau)| \leq R_X(0),$$

$$R_X(\tau) = R_X(-\tau),$$

and

$$\sum_{i=1}^{m} \sum_{j=1}^{m} R_X(t_i - t_j)a_i a_j \geq 0 \qquad \text{for all } t_1, \ldots, t_m, a_1, \ldots, a_m, m = 1, 2, \ldots.$$

Although wide-sense stationarity does not generally imply stationarity of any order, Gaussian processes are again an exception. *A Gaussian process is strictly stationary if and only if it is wide-sense stationary.*

An example of a stationary Gaussian process is the *stationary Ornstein-Uhlenbeck process* $X(t)$, $-\infty < t < \infty$, which is a Gauss-Markov process with zero mean function and with correlation function

$$R_X(\tau) = \sigma^2 e^{-\alpha|\tau|}, \qquad -\infty < \tau < \infty,$$

with $\sigma > 0$ and $\alpha > 0$ constants. In fact, this is the only nontrivial† stationary process that is Gauss-Markov.

An indispensable concept in analyzing probabilistic models, especially those involving stochastic processes, is the conditional expectation. The definition commonly found in most intermediate-level textbooks uses the conditional density and introduces the conditional expectation of a random variable X given a random vector $\mathbf{Y} = (Y_1, \ldots, Y_m)'$ as an integral

$$\mathrm{E}[X \mid \mathbf{Y} = \mathbf{y}] = \int_{-\infty}^{\infty} x f_{X|\mathbf{Y}}(x \mid \mathbf{y})\, dx$$

However, we shall often need to define the conditional expectation given a portion of a stochastic process, say $(Y(t'), t_1 \le t' \le t_2)$, and since this would involve conditioning on the values of an infinite number of random variables, the above definition cannot be used.

In what follows, let the symbol $\mathcal{Y}$ denote either a single random variable Y, a random vector $\mathbf{Y} = (Y_1, \ldots, Y_m)'$, or a portion of a stochastic process $(Y(t'), t' \le t)$, where t is a fixed time instant. Also let y denote either a particular value $Y = y$ or $\mathbf{Y} = \mathbf{y}$ of the random variable Y or random vector $\mathbf{Y}$ or in the case of the stochastic process a particular sample path $y(t'), t' \le t$.

Definition 1.12
The conditional expectation $\mathrm{E}[X \mid \mathcal{Y} = y]$ of a random variable X given $\mathcal{Y} = y$ is a functional‡ $c(y)$ defined on the ensemble $\mathcal{E}_y$ such that for any other functional $g(y)$ defined on $\mathcal{E}_y$ the expectation

$$\mathrm{E}[(X - c(\mathcal{Y}))^2] \le \mathrm{E}[(X - g(\mathcal{Y}))^2].$$

According to this definition, the conditional expectation $\mathrm{E}[X \mid \mathcal{Y} = y]$ is a functional $c(y)$ defined on the ensemble $\mathcal{E}_y$. But $\mathcal{Y}$ is itself a random quantity (random variable, vector, or process) and so the functional $c(y)$ can be used to *define a new random variable*, say

$$C = c(\mathcal{Y}), \qquad \text{where } c(y) = \mathrm{E}[X \mid \mathcal{Y} = y].$$

Thus, we can actually *regard the conditional expectation itself as a random variable* and hence inquire about its probability law, expectation, and so forth. To distin-

† The process $X(t) = X$ for all $-\infty < t < \infty$ with X a Gaussian random variable is clearly a stationary Gauss-Markov process, but rather a trivial one.

‡ If $\mathcal{Y}$ is a random variable or random vector, read "function" instead of "functional." If $\mathcal{Y}$ is a portion of a stochastic process, a functional defined on $\mathcal{E}_y$ is a rule that assigns a real number to every sample path $y \in \mathcal{E}_y$.

guish between these two viewpoints, we write

$$\mathrm{E}[X \mid \mathcal{Y} = y], \qquad y \in \mathcal{E}_{\mathcal{Y}}$$

if we regard the conditional expectation as a *deterministic functional* and use the notation $\mathrm{E}[X \mid \mathcal{Y}]$ to denote the *random variable* $C = c(\mathcal{Y})$ mentioned above.

An immediate consequence of Definition 1.12 is the theorem known as the *orthogonality principle*:

The functional $c(y) = E[X \mid \mathcal{Y} = y]$, $y \in \mathcal{E}_{\mathcal{Y}}$, *is the conditional expectation of* X *given* $\mathcal{Y} = y$ *if and only if*

$$\mathrm{E}[h(y)(X - c(y))] = 0$$

for any other functional $h(y)$ *on* $\mathcal{E}_{\mathcal{Y}}$ *provided the expectation exists.*

If $\mathcal{Y}$ is a random vector, say $\mathbf{Y} = (Y_1, \ldots, Y_m)'$, then the conditional expectation $c(\mathbf{y})$ as defined in Definition 1.12 can indeed be expressed as an integral

$$c(\mathbf{y}) = \int_{-\infty}^{\infty} x f_{X|\mathbf{Y}}(x \mid \mathbf{y})\, d\mathbf{y}.$$

Even more generally, if a random variable Z is a function of a random vector $\mathbf{X}$, $Z = g(\mathbf{X})$, then the conditional expectation of Z given $\mathbf{Y} = \mathbf{y}$ can be obtained as an integral

$$\int_{-\infty}^{\infty} \cdots \int_{-\infty}^{\infty} g(\mathbf{x}) f_{\mathbf{X}|\mathbf{Y}}(\mathbf{x} \mid \mathbf{y})\, d\mathbf{x}.$$

Thus, as already mentioned, Definition 1.12 does not introduce a different kind of conditional expectation but merely makes the concept more universally applicable.

As far as the properties of conditional expectation are concerned, some are similar to those of ordinary expectation, but others are peculiar to the conditional expectation.

(i) Existence:

$$\mathrm{E}[X \mid \mathcal{Y}] \text{ exists whenever } \mathrm{E}[\,|X|\,] < \infty.$$

(ii) Constant:

If $X = a$, a constant, then $\mathrm{E}[X \mid \mathcal{Y}] = a$.

(iii) Linearity:

$$\mathrm{E}\left[\sum_{i=1}^{k} a_i X_i \mid \mathcal{Y}\right] = \sum_{i=1}^{k} a_i \mathrm{E}[X_i \mid \mathcal{Y}]$$

(iv) Monotonicity:

If $X \geq 0$, then $\mathrm{E}[X \mid \mathcal{Y}] \geq 0$.

(v) Convexity:

$$|\mathrm{E}[X \mid \mathcal{Y}]| \leq \mathrm{E}[|X| \mid \mathcal{Y}]$$

(vi) Expectation:

$$\mathrm{E}[\mathrm{E}[X \mid \mathcal{Y}]] = \mathrm{E}[X]$$

(vii) For any functional $g(\boldsymbol{y})$ defined on $\mathcal{E}_{\mathcal{Y}}$:

$$\mathrm{E}[g(\mathcal{Y})X \mid \mathcal{Y}] = g(\mathcal{Y})\mathrm{E}[X \mid \mathcal{Y}]$$

This is called the *factorization property*, since it says that if a random variable $Z = g(\mathcal{Y})$ is a functional of $\mathcal{Y}$, its values are determined by the conditioning and so it can be factored out of the conditional expectation.

(viii) If $\mathfrak{Z}$ is a random variable, a random vector, or a portion of a stochastic process uniquely determined by the random quantity $\mathcal{Y}$, that is, if $\mathfrak{Z} = g(\mathcal{Y})$, then

$$\mathrm{E}[\mathrm{E}[X \mid \mathcal{Y}] \mid \mathfrak{Z}] = \mathrm{E}[X \mid \mathfrak{Z}]$$

and also

$$\mathrm{E}[\mathrm{E}[X \mid \mathfrak{Z}] \mid \mathcal{Y}] = \mathrm{E}[X \mid \mathfrak{Z}].$$

The first equation expresses the *smoothing property* of conditional expectation: since $\mathcal{Y}$ determines $\mathfrak{Z}$, but not necessarily vice versa, $\mathfrak{Z}$ contains less information than $\mathcal{Y}$ and so conditioning on $\mathfrak{Z}$ destroys or smoothes the conditioning on $\mathcal{Y}$.

(ix) If $\mathfrak{Z}$ and $\mathcal{Y}$ can be uniquely determined from each other then

$$\mathrm{E}[X \mid \mathcal{Y}] = \mathrm{E}[X \mid \mathfrak{Z}]$$

in the sense that the two random variables have identical probability laws.

(x) If the random variable X is independent of $\mathcal{Y}$, then

$$\mathrm{E}[X \mid \mathcal{Y}] = \mathrm{E}[X]$$

so that the conditional expectation is a constant.

Suppose now that $\mathcal{Y} = (Y(t'), t' \leq t_1)$, is a portion of a stochastic process $Y(t)$, $t \in \mathcal{T}$, and the time t_1 represents the present time. By Definition 1.12 the conditional expectation $c(\boldsymbol{y}) = \mathrm{E}[Y(t_2) \mid \mathcal{Y} = \boldsymbol{y}]$, where $t_2 > t_1$ is some future time instant, can then be interpreted as the predictor of the value of $Y(t_2)$ based on the observed past $y(t')$, $t' \leq t_1$. Furthermore, such a predictor is the best in the sense that it minimizes the mean square prediction error $E[(Y(t_2) - c(\mathcal{Y}))^2]$ among all possible predictors based on the same observations.

If $Y(t_2)$ is independent of $\mathcal{Y} = (Y(t'), t' \leq t_1)$, then $E[Y(t_2) | \mathcal{Y} = y] = E[Y(t_2)]$, and so the past and present observation is useless for future prediction. However, it is possible that $Y(t_2)$ is not independent of $\mathcal{Y} = (Y(t'), t' \leq t_1)$ and yet

$$E[Y(t_2) | \mathcal{Y} = y] = E[Y(t_2)], \qquad \text{a constant}, \qquad \text{for all } y \in \mathcal{E}_{\mathcal{Y}}.$$

Definition 1.13

Let $\mathcal{Y}$ be a random variable, random vector, or a portion of a stochastic process. If the conditional expectation

$$E[X | \mathcal{Y}] \qquad \text{is a constant}$$

(whose value then is $E[X]$) the random variable X is called *mean square unpredictable* (abbreviated MS unpredictable) from $\mathcal{Y}$.

MS unpredictability is a weaker property than independence, and in fact, we have the following hierarchy of implications:

X is independent of $\mathcal{Y}$

$\Downarrow$

X is MS unpredictable from $\mathcal{Y}$

$\Downarrow$

X and $\mathcal{Y}$ are uncorrelated†

Reverse implications are generally false, with one important exception: *If the probability law of X and $\mathcal{Y}$ is jointly Gaussian, then concepts of independence, MS unpredictability, and uncorrelatedness are equivalent.*

The concept of MS unpredictability leads to an important class of stochastic processes that generalize processes with independent increments.

Definition 1.14

A stochastic process $X(t)$, $t \in \mathcal{T}$, such that

$$E|X(t)| < \infty \qquad \text{for all } t \in \mathcal{T}$$

and such that

$$E[X(t_2) | X(t'), t' \leq t_1] = X(t_1)$$

for all $t_1 \leq t_2$ from the time domain $\mathcal{T}$ is called a *martingale*.

† Meaning that the covariance between X and any component of the random vector $\mathbf{Y}$ or any random variable $Y(t')$, $t' \leq t$ is zero.

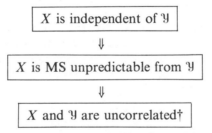

Note that the second condition in this definition can be written as

$$\mathrm{E}[X(t_2) - X(t_1) | X(t'), t' \leq t_1] = 0$$

and so a martingale is, in fact, a process with *MS unpredictable increments.*

For instance, the Bernoulli symmetric random walk or the standard Wiener process, being zero mean processes with independent increments, are automatically martingales, but a martingale need not have independent increments.

Martingales have many remarkable properties; for instance, the mean function of a martingale is always a constant, while the variance function, if it exists, is always nondecreasing. Another useful result is the so-called optional stopping theorem, but before we state this theorem, it is necessary to introduce the concept of a stopping time in connection with a stochastic process.

Definition 1.15

Given a stochastic process $X(t)$, $t \in \mathcal{T}$, a *stopping time* T is a random variable with ensemble $\mathcal{E}_T = \mathcal{T}$ such that for every $t \in \mathcal{T}$, the occurrence or nonoccurrence of the event $[T > t]$ is determined by the portion of the sample path $x(t')$, $t' \leq t$.

Loosely speaking the value $T = t$ of a stopping time T is the time we decide to stop observing the process $X(t)$, provided that this decision is not based on the evolution of the process after we have stopped observing it. A typical example of a stopping time is the random time when the sample path of the process $X(t)$ first reaches some prescribed level.

OPTIONAL STOPPING THEOREM. *Let* $\mathrm{X(t)}$, $\mathrm{t} \geq 0$, *be a martingale, and let* T *be a stopping time for* $\mathrm{X(t)}$, $\mathrm{t} \geq 0$. *If*

$$P(T < \infty) = 1$$

and if there is a constant C *such that for all* $\mathrm{t} \geq 0$

$$\mathrm{E}[X^2(t \wedge T)] \leq C,$$

where

$$X(t \wedge T) = \begin{cases} X(t) & \text{if} \quad T > t, \\ X(T) & \text{if} \quad T \leq t, \end{cases}$$

then $E[\mathrm{X(T)}] = E[\mathrm{X(0)}]$.

This theorem is quite useful in deriving various results for crossing times of stochastic processes as suggested in the exercises at the end of this chapter.

In performing various limiting operations on stochastic processes, topics related to the convergence of sequences of random variables are of prime importance. A stochastic sequence $X(n)$, $n = 1, 2, \ldots$, is, by definition, a whole ensemble of deterministic sequences $x(n)$, $n = 1, 2, \ldots$, (sample paths) together with a probability law, and so the concept of stochastic convergence may be concerned with either the convergence of individual sample paths or the convergence of probabilities of some sequence of events defined by the sequence or, quite frequently, both.

Conceptually the simplest type of stochastic convergence is almost sure convergence.

Definition 1.16

A stochastic sequence $X(n)$, $n = 1, 2, \ldots$, is said to converge *almost surely* if there exists a random variable X (possibly a constant) such that

$$P(X(n) \to X \text{ as } n \to \infty) = 1.$$

Remark: This type of stochastic convergence is encountered, for instance, in connection with martingales; in fact, the *martingale convergence theorem* asserts that a discrete-time martingale $X(n)$, $n = 1, 2, \ldots$, such that for all $n = 1, 2, \ldots$, $\mathrm{E}[|X(n)|] \leq C$, a finite constant, necessarily converges almost surely as $n \to \infty$. A consequence of this is the celebrated *strong law of large numbers*: If $X(n)$, $n = 1, 2, \ldots$, is a sequence of i.i.d. random variables with finite variance then the sequence of consecutive averages

$\bar{X}(n) = \frac{1}{n} \sum_{k=1}^{n} X(k)$, $n = 1, 2, \ldots$, converges almost surely to the expectation μ_X of these random variables. Я

A somewhat weaker type of stochastic convergence is called convergence in mean square, abbreviated *MS convergence.*

Definition 1.17

A finite-variance stochastic sequence $X(n)$, $n = 1, 2, \ldots$, is said to converge in *mean square* if there is a random variable X (possibly a constant) such that

$$\mathrm{E}[(X(n) - X)^2] \to 0 \text{ as } n \to \infty.$$

The random variable X is called the *MS limit* of the sequence $X(n)$, $n = 1, 2, \ldots$, and the above condition is written as

$$X = \underset{n \to \infty}{\text{l.i.m.}}\ X(n).$$

If the stochastic sequence $X(n)$ converges in MS to a random variable X, then so do the sequences of means and variances

$$\lim_{n \to \infty} \mathrm{E}[X(n)] = \mathrm{E}[X],$$

and

$$\lim_{n\to\infty} \mathrm{Var}[X(n)] = \mathrm{Var}[X].$$

The following two criteria are useful in establishing MS convergence of a given stochastic sequence.

Cauchy Criterion:
A stochastic sequence $X(n)$, $n = 1, 2, \ldots$, converges in mean square if and only if

$$\mathrm{E}[(X(n_1) - X(n_2))^2] \to 0 \qquad \text{as both } n_1 \to \infty \quad \text{and} \quad n_2 \to \infty.$$

Loève Criterion:
A stochastic sequence $X(n)$, $n = 1, 2, \ldots$, converges in mean square if and only if the correlation function $R_X(n_1, n_2)$ converges to a constant, $R_X(n_1, n_2) \to c$, as both $n_1 \to \infty$ and $n_2 \to \infty$.

For example, if $X(n)$, $n = 1, 2, \ldots$, is a finite-variance martingale such that $\mathrm{E}[X^2(n)] \le c$, a constant, for all $n = 1, 2, \ldots$, then the Cauchy criterion implies that the martingale converges in mean square to a random variable X.

Remark: Almost sure convergence of a stochastic sequence $X(n)$, $n = 1, 2, \ldots$, to a random variable X does not necessarily imply MS convergence of this sequence. However, if any of the conditions listed below is satisfied, then indeed $\underset{n\to\infty}{\mathrm{l.i.m.}}\, X(n) = X$.

(a) $\mathrm{E}[|X(n)|^2] \to \mathrm{E}[|X|^2]$ as $n \to \infty$.

(b) There exists a random variable Y with finite variance and such that for all $n = 1, 2, \ldots$,

$$|X(n)| \le Y.$$

(c) There are constants c and $\alpha > 0$ such that for all $n = 1, 2, \ldots$,

$$\mathrm{E}[|X(n)|^{2+\alpha}] \le c. \qquad Я$$

When working with stationary stochastic processes, one often wants to measure or estimate the mean and the correlation function of such a process from an actual record of a sample path. Since both the mean and the correlation function are defined as expectations that, in turn, can be interpreted as probabilistic averages over the ensemble (ensemble averages), the idea is to estimate them as the corresponding time averages of the sample path. Specifically, if $X(t)$,

$-\infty < t < \infty$, is a (wide-sense) stationary process with mean μ_X and correlation function $R_X(\tau)$, then we would like to estimate μ_X by the *sample mean*

$$\bar{x} = \frac{1}{t}\int_0^t x(t')\,dt',$$

and $R_X(\tau)$ by the *sample correlation function*

$$\bar{R}(\tau) = \frac{1}{t}\int_0^t x(t')x(t'+\tau)\,dt'$$

for each fixed $\tau \geq 0$, provided the averaging time t is sufficiently large.

If the wide-sense stationary process $X(t)$, $-\infty < t < \infty$, is such that indeed

$$\underset{t\to\infty}{\text{l.i.m.}}\ \bar{X} = \mu_X$$

and for each fixed $\tau \geq 0$

$$\underset{t\to\infty}{\text{l.i.m.}}\ \bar{R}(\tau) = R_X(\tau)$$

then the process $X(t)$ is said to be *weakly ergodic.* For a stationary Gaussian process $X(t)$, $-\infty < t < \infty$, a sufficient condition for its ergodicity is that its covariance function is absolutely integrable

$$\int_{-\infty}^{\infty} |\gamma_X(\tau)|\ d\tau < \infty.$$

For a wide-sense stationary process that is not necessarily Gaussian, this condition guarantees only the ergodicity of the mean

$$\underset{t\to\infty}{\text{l.i.m.}}\ \bar{X} = \mu_X.$$

The third type of stochastic convergence concerns the convergence of probability laws of the individual random variables in the stochastic sequence.

Definition 1.18

A stochastic sequence $X(n)$, $n = 1, 2, \ldots$, is said to *converge in distribution* to a random variable X if, as $n \to \infty$, the sequence of distribution functions $F_{X(n)}(x)$ of the random variables $X(n)$ converges to the distribution function $F_X(x)$ of the random variable X for all $-\infty < x < \infty$ at which the function $F_X(x)$ is continuous.

If a stochastic sequence converges almost surely or in the mean square to a random variable X, then it necessarily converges in distribution to this random variable. The converse is not true; convergence in distribution implies none of the other types, and in this sense it is the weakest type of stochastic convergence. This by no means diminishes its importance, as is demonstrated by the following famous theorem.

CENTRAL LIMIT THEOREM.
If $X(n)$, $n = 1, 2, \ldots,$ *is a sequence of independent random variables with the same variance* $Var[X(n)] = \sigma_X^2 < \infty$, *then the stochastic sequence of standardized sums*

$$\frac{S(n) - E[S(n)]}{\sqrt{Var[S(n)]}} \qquad n = 1, 2, \ldots,$$

where $S(n) = \sum_{k=1}^{n} X(k)$, *converges in distribution to the standard Gaussian random variable.*

Remark: The central limit theorem remains true even if the independent random variables $X(n)$, $n = 1, 2, \ldots,$ do not have the same variance. However, roughly speaking the variances must not be getting too small or too large, which can be guaranteed, for instance, by requiring that for all $n = 1, 2, \ldots,$ the variances are such that $0 < \varepsilon \leq Var[X(n)] \leq C$ for some constants ε and C, or if $|X(n)| \leq C$, $n = 1, 2, \ldots$ by requiring that $Var[S(n)] \to \infty$. Я

One of the frequent applications of the central limit theorem is the *Gaussian approximation formula*, which says that if $S(n) = X(1) + \cdots + X(n)$ is the sum of a "sufficiently large" number of i.i.d. random variables with common mean μ_X and standard deviation σ_X, then for any $-\infty \leq a < b \leq +\infty$

$$P(a < S(n) \leq b) \doteq \Phi\left(\frac{b - n\mu_X}{\sigma_X\sqrt{n}}\right) - \Phi\left(\frac{a - n\mu_X}{\sigma_X\sqrt{n}}\right),$$

where

$$\Phi(z) = \frac{1}{\sqrt{2\pi}}\int_{-\infty}^{z} e^{-t^2/2}\, dt.$$

The central limit theorem also provides justification for the assumption that a random quantity obtained as a sum of a large number of independent contributions (the error in a physical measurement for instance) should obey the Gaussian law.

A useful analytical tool for investigating convergence in distribution is the characteristic function.

Definition 1.19
The characteristic function $\psi_X(\zeta)$ of a random variable X is a complex-valued function of a real variable ζ, $-\infty < \zeta < \infty$, defined by

$$\psi_X(\zeta) = \mathrm{E}[e^{\imath\zeta X}] = \int_{-\infty}^{\infty} e^{\imath\zeta x} f_X(x)\, dx,$$

where $\imath = \sqrt{-1}$ is the imaginary unit.

The characteristic function is actually the inverse Fourier transform of the density $f_X(x)$, and so the density can be recovered from the characteristic function as the Fourier transform

$$f_X(x) = \frac{1}{2\pi} \int_{-\infty}^{\infty} e^{-\imath\zeta x} \psi_X(\zeta)\, d\zeta.$$

The characteristic function of a random vector $\mathbf{X} = (X_1, \ldots, X_m)'$ is defined similarly as a complex-valued function of m real variables $\boldsymbol{\zeta} = (\zeta_1, \ldots, \zeta_m)'$ by

$$\begin{aligned} \psi_{\mathbf{X}}(\boldsymbol{\zeta}) &= \mathrm{E}\left[e^{\imath \sum_{k=1}^{m} \zeta_k X_k}\right] \\ &= \int_{-\infty}^{\infty} \cdots \int_{-\infty}^{\infty} e^{\imath \sum_{k=1}^{m} \zeta_k x_k} f_{\mathbf{X}}(x_1, \ldots, x_m)\, dx_1 \cdots dx_m. \end{aligned}$$

The characteristic function exists and is unique for every random variable or vector, and thus the probability law of a random variable, random vector, or a stochastic process and their properties can also be defined by means of characteristic functions. For instance, a random vector $\mathbf{X} = (X_1, \ldots, X_m)'$ has mutually independent components if and only if

$$\psi_{\mathbf{X}}(\boldsymbol{\zeta}) = \psi_{X_1}(\zeta_1) \cdots \psi_{X_m}(\zeta_m).$$

The characteristic function of a sum $S(n) = X(1) + \cdots + X(n)$ of independent random variables is the product of characteristic functions

$$\psi_{S(n)}(\zeta) = \psi_{X(1)}(\zeta) \cdots \psi_{X(n)}(\zeta)$$

and this allows one to determine the probability law of the sum $S(n)$ in a much simpler way than via the repeated convolution of the densities.

Also the mth moment $\mathrm{E}[X^m]$ of a random variable X, if it exists, can be obtained from the characteristic function $\psi_X(\zeta)$ as the mth derivative

$$\mathrm{E}[X^m] = (-\imath)^m \left.\frac{d^m \psi_X(\zeta)}{d\zeta^m}\right|_{\zeta=0} \qquad m = 1, 2, \ldots.$$

More generally, one can obtain the correlation function of a stochastic process by the formula

$$R_X(t_1, t_2) = -\frac{\partial^2}{\partial\zeta_1\,\partial\zeta_2}\psi_{X(t_1),\,X(t_2)}(\zeta_1, \zeta_2)\Big|_{\substack{\zeta_1=0\\ \zeta_2=0}}.$$

However, the most useful property of the characteristic function in establishing convergence in distribution is the statement of *Levy's continuity theorem*:

Let X(n), n = 1, 2, ..., *be a sequence of random variables with characteristic functions* $\psi_{\mathbf{X(n)}}(\zeta)$. *If for all* $-\infty < \zeta < \infty$

$$\lim_{n\to\infty} \psi_{X(n)}(\zeta)$$

exists and is continuous at $\zeta = 0$, *then the stochastic sequence* X(n), n = 1, 2, ..., *converges in distribution to a random variable* X *with characteristic function*

$$\psi_X(\zeta) = \lim_{n\to\infty} \psi_{X(n)}(\zeta).$$

EXERCISES

1. A weather vane measures the wind direction. Measure the direction it points from the north, so north is 0°, east 90°, south 180° and west 270°. The angle Θ varies continuously.
   (a) For the direction at any specific instant of time, what is the ensemble $\mathcal{E}_\Theta$?
   (b) Assume a continuous record is kept of $\Theta(t)$ for a 24-hour period. Define the time domain $\mathcal{T}$, state space $\mathcal{S}_{\Theta(t)}$, and ensemble $\mathcal{E}_{\Theta(t)}$.
   (c) Assume that in the 24-hour period, the wind was initially from the west, it then shifted clockwise to the northeast, returned to the west, and continued counterclockwise to the southeast. Draw a typical sample path $\theta(t)$. Is it continuous?
2. Assume a single binary digit (0 or 1) is to be transmitted through a transmission channel in the direction from *a* to *b*, schematically pictured in Figure 1.E.1. The waveform received at *b* will be decoded as a 0 or a 1. Because of noise in the channel

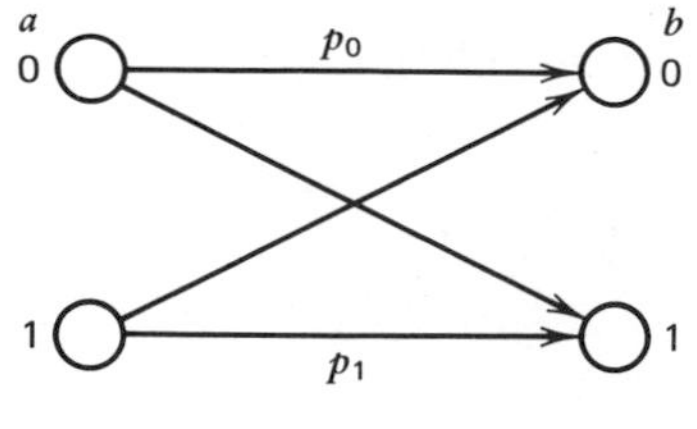

Figure 1.E.1

used, the waveform received may not be decoded correctly. Let X_1 be the digit transmitted, and let X_2 be the digit decoded and assume

$$P(X_1 = 0) = q,$$

$$P(X_1 = 1) = 1 - q = p.$$

These describe the probability the transmitted digit was 0 or 1. We also assume

$$P(X_2 = 0 \mid X_1 = 0) = p_0,$$

$$P(X_2 = 1 \mid X_1 = 0) = q_0 = 1 - p_0,$$

$$P(X_2 = 0 \mid X_1 = 1) = q_1 = 1 - p_1,$$

$$P(X_2 = 1 \mid X_1 = 1) = p_1,$$

which define the probabilities of decoding the received waveform as 0 or 1 in the two cases.

(a) Evaluate the probability that the digit decoded is 1.
(b) Compute the probability an error was made in decoding, given i was transmitted, $i = 0, 1$.

3. Show that if $X(n)$, $n = 1, 2, \ldots$, is a sequence of independent Bernoulli random variables then the event $[X(n) = 1]$ occurs infinitely often if and only if the infinite series $\sum_{n=1}^{\infty} P(X(n) = 1)$ diverges.
(This is known as the Borel-Cantelli lemma.)

4. A filtered Bernoulli process, with ones randomly erased, is defined as follows: Let $X(n)$, $n = 1, 2, 3, \ldots$, be a Bernoulli process with parameter p, and if $X(n) = 0$, let $Y(n) = X(n)$, whereas if $X(n) = 1$, then $Y(n) = X(n)$ with probability α and $Y(n) = 0$ with probability $1 - \alpha$. Specify the probability law for $Y(n)$.

5. Find the mth order probability mass distribution for the binomial process $S(n)$, $n = 0, 1, 2, 3, \ldots$, with parameter p.

6. Assume a system has two states, say operational and failed, represented by 1 and 0, respectively. The time domain is

$$\mathcal{T} = \{0, 1, 2, \ldots\}$$

representing, for example, days, with the initial time (day) labeled 0. Then let $X(n)$, $n \in \mathcal{T}$, be the state of the system and assume $X(n)$ is a Markov process. The initial mass distribution is

$$P(X(0) = x) = \begin{cases} 1 & \text{if} \quad x = 1, \\ 0 & \text{if} \quad x \neq 1. \end{cases}$$

The probability the system moves to the other state, given that it was in either state on the preceding day, is p, so the one-step transition mass distributions are

$$P(X(n+1) = 1 \mid X(n) = 0) = P(X(n+1) = 0 \mid X(n) = 1) = p$$

$$P(X(n+1) = 1 \mid X(n) = 1) = P(X(n+1) = 0 \mid X(n) = 0) = q = 1 - p.$$

The system described is pictured schematically in Figure 1.E.2.

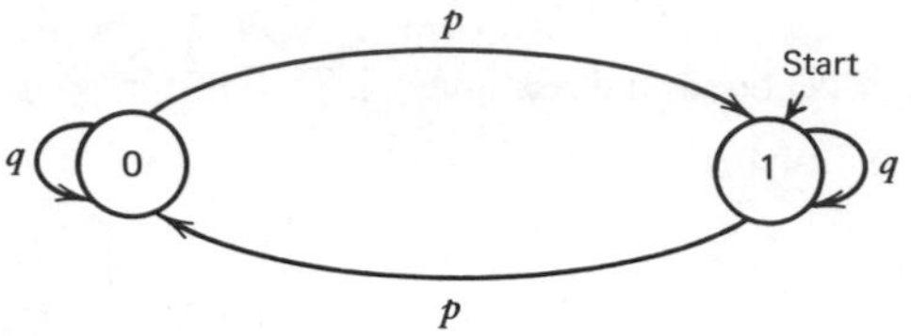

Figure 1.E.2

(a) Show that the two-step transition mass distributions are

$$P(X(n+2) = i \mid X(n) = i) = q^2 + p^2, \qquad i = 0, 1$$

$$P(X(n+2) = |i-1| \mid X(n) = i) = 2pq, \qquad i = 0, 1.$$

(b) Use induction to show that for all n and initial state $j = 0$ or 1

$$P(X(n) = i \mid X(0) = j)$$
$$= (q-p)^{n-1}P(X(1) = i \mid X(0) = j) + \tfrac{1}{2}(1 - (q-p)^{n-1}), \qquad i = 0 \quad \text{or} \quad 1.$$

(c) Use the Chapman-Kolmogorov equation to show that if

$$P(X(0) = 0) = P(X(0) = 1) = \tfrac{1}{2}$$

then, in fact,

$$P(X(n) = 0) = P(X(n) = 1) = \tfrac{1}{2} \qquad \text{for all } n.$$

(d) Generalize to the case

$$P(X(n+1) = 1 \mid X(n) = 0) = p_1,$$
$$P(X(n+1) = 0 \mid X(n) = 1) = p_2$$

with $p_1 \neq p_2$, $0 < p_1 < 1$, $0 < p_2 < 1$. What happens if $p_1 + p_2 = 1$?

7. Show that if $\mathcal{E}_X = \{0, 1, 2, \ldots\}$, then

(a) $\mathrm{E}[X] = \sum_{k=0}^{\infty} kP(X = k) = \sum_{k=0}^{\infty} P(X > k) = \sum_{k=1}^{\infty} P(X \geq k)$,

(b) $\mathrm{E}[X^2] = \sum_{k=0}^{\infty} k^2P(X = k) = \sum_{k=0}^{\infty} (2k+1)P(X > k) = \sum_{k=1}^{\infty} (2k-1)P(X \geq k)$

8. Many fundamental models in statistical mechanics are concerned with the behavior of a large number of particles, the distribution of their positions and momenta, their energies, and similar problems. Imagine a box divided into a large number m of small cells, and consider a random placement of n atoms into these cells. Denoting by X_j the number of atoms placed in the jth cell, the distribution of atoms into cells is then described by a random vector

$$\mathbf{X}' = (X_1, \ldots, X_m).$$

(Note that $\mathbf{X}$ is a discrete random vector and that $X_1 + \cdots + X_m = n$.)

The *Maxwell-Boltzman statistic* is based on the assumption that the atoms are distinguishable (say numbered 1 to n), and any number of them can be placed in any one cell. Show that the probability law for $\mathbf{X}$ is

$$P(\mathbf{X} = \mathbf{k}) = \binom{n}{k_1, k_2, \ldots, k_m} m^{-n}, \qquad \mathbf{k} \in \mathcal{E}_{\mathbf{X}}.$$

This is an example of the *multinomial probability law* with parameters n, $p_i = 1/m$, $i = 1, 2, \ldots, m$ (see Appendix Two).

9. (a) Assume a device consists of m components connected in series; thus the device will fail when any one of the components fails. Suppose further that the time to failure of component i is T_i, $i = 1, 2, \ldots, m$, and that $T_1, T_2, \ldots, T_m$ are i.i.d. with density $f_T(t)$ and distribution function $F_T(t)$. Then the time to failure, Y, of the device is the smallest or minimum of $T_1, T_2, \ldots, T_m$:

$$Y = \min(T_1, T_2, \ldots, T_m).$$

Show that the distribution function for Y is

$$F_Y(t) = 1 - P(Y > t) = 1 - (1 - F_T(t))^m,$$

and thus if each component failure time is exponential with parameter λ, Y is exponential with parameter $m\lambda$ (see Appendix Two).

(b) Now suppose the device consists of m components connected in parallel; the device keeps working until *all* m components fail. If again $T_1, T_2, \ldots, T_m$ are the times to failure of the m components and Y is the time to failure for the device, now Y is equal to the largest or maximum of the T_i's:

$$Y = \max(T_1, T_2, \ldots, T_m).$$

Again let $T_1, T_2, \ldots, T_m$ be independent identically distributed with distribution function $F_T(t)$ and show that

$$F_Y(t) = (F_T(t))^m.$$

10. Assume $X_1, X_2, \ldots, X_n$ are independent, identically distributed, each with distribution function $F_X(x)$, and let $S_n = X_1 + X_2 + \cdots + X_n$.

(a) Given $F_X(x) = 0$ for $x \le 0$, show that

$$P(S_n \le x) \le P(X_1 \le x)P(X_2 \le x) \cdots P(X_n \le x) = (F_X(x))^n$$

for any x, $n = 1, 2, 3, \ldots$.

(b) Show that

$$P(S_1 \le x) \ge P(S_2 \le x) \ge P(S_3 \le x) \ge \cdots$$

for any x.

11. Find the density of the difference of two independent random variables, each Erlang (see Appendix Two) with density

$$f_X(x) = \alpha^2 x e^{-\alpha x} \qquad x \ge 0.$$

12. Two independent Bernoulli processes $X_1(n)$, $X_2(n)$, are running side by side; the probabilities of success for the two are p_1, p_2, respectively. What is the probability that one appears in process 1 earlier than in process 2?
13. Let $S_1(n)$, $S_2(n)$ be two independent binomial processes, both with parameter p. Let $N > 0$ be the first value of the index n such that $S_1(n) = S_2(n)$ and find the probability mass distribution for N.
14. Let $S(n)$ be the binomial process with parameter p. What is the second order probability mass distribution for $S(n_1)$ and $S(n_2)$, where $n_2 > n_1$?
15. X_1 and X_2 are called conditionally independent, given $Y = y$ if

$$f_{X_1, X_2|Y}(x_1, x_2 | y) = f_{X_1|Y}(x_1 | y) f_{X_2|Y}(x_2 | y).$$

Assume X_1 and X_2 are conditionally independent with

$$f_{X_j|Y}(x_j | y) = ye^{-x_j y}, \quad x_j > 0, \quad y > 0, \quad j = 1, 2$$

and

$$f_Y(y) = \alpha e^{-\alpha y}, \quad y > 0.$$

Find

(a) $f_{X_1, X_2}(x_1, x_2)$
(b) $f_{Y|X_1, X_2}(y | x_1, x_2)$

Are X_1 and X_2 unconditionally independent?
16. Let $X_1, X_2, \ldots, X_n$ be uniform, independent on (0, 1). Their ordered values $X_{(1)}$, $X_{(2)}, \ldots, X_{(n)}$ are called the *order statistics* of $X_1, X_2, \ldots, X_n$.
    (a) Find the joint density of the random vector $(X_{(1)}, \ldots, X_{(n)})$.
    (b) Find the joint density for $(X_{(1)}, \ldots, X_{(n)})$ if $X_1, X_2, \ldots, X_n$ are independent, continuous random variables each with density function $f_X(x)$.
17. A stochastic process $X(n)$, $n = 1, 2, 3, \ldots$, has state space $\mathcal{S} = \{x: x \geq 0\}$. The density for $X(1)$ is

$$f_1(x) = e^{-x}, \quad x \geq 0$$

and the joint density of $X(1), X(2), \ldots, X(m)$ is

$$f_{1, 2, \ldots, m}(x_1, x_2, \ldots, x_m)$$
$$= x_1 x_2 \cdots x_{m-1} \exp(-(x_m x_{m-1} + x_{m-1} x_{m-2} + \cdots + x_2 x_1 + x_1)),\dagger$$
$$x_1 \geq 0, x_2 \geq 0, \ldots, x_m \geq 0.$$

Show this is a consistent specification and find the density for $X(2)$. Is $X(n)$ a Markov process?
18. Use the Markov property to show that if $X(n)$, $n = 1, 2, 3, \ldots$, is Markov, then

$$E[X(n+1) | X(1), X(2), \ldots, X(n)] = E[X(n+1) | X(n)].$$

19. Derive explicit formulas for the kth order densities $f_{t_1, \ldots, t_k}(x_1, \ldots, x_k)$ of
    (a) the standard Wiener process.
    (b) the Wiener process with drift μ and intensity σ.
    (c) the Ornstein-Uhlenbeck process.

† We occasionally use the notation exp(a) to represent e^a.

20. Evaluate the mean function, the variance function, the correlation function, and the covariance function of

$$X(t) = W_0^2(t) \qquad t \geq 0$$

where $W_0(t)$ is the standard Wiener process.

21. A random PPM (pulse position modulated) signal $X(t)$ has basic frequency $\nu_0 = 1/t_0$. A unit pulse of constant width b is located "at random" within each cycle (by this we mean that the start of the pulse is uniformly distributed over an interval of length $t_0 - b$, starting at each frequency marker, see Fig. 1.E.3). The random variables $\ldots, \Delta_{-1}, \Delta_0, \Delta_1, \ldots$ are independent and uniform.

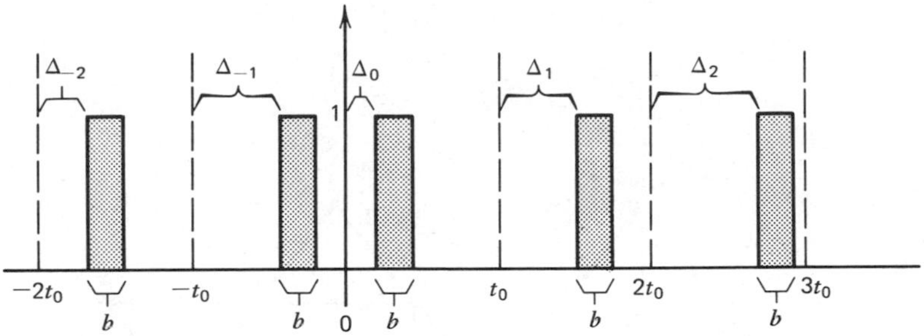

Figure 1.E.3

(a) Show that the mean function $\mu_X(t)$ is as pictured in Figure 1.E.4.
(b) Evaluate the correlation function

$$R_X(t_1, t_2) = \mathrm{E}[X(t_1)X(t_2)].$$

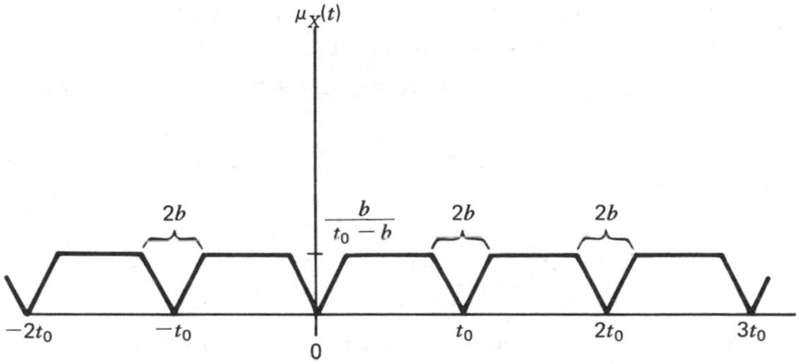

Figure 1.E.4

22. Assume $X(t)$ is a random square wave with fixed frequency t_0. The amplitude A is random, $\mu_A = 0$, $\sigma_A^2 > 0$, and the wave is offset from $t = 0$ by Θ, which is uniform on $(0, t_0)$ and uncorrelated with A (see Fig. 1.E.5). Show that $R(\tau) = \mathrm{E}[X(0)X(\tau)]$ is as pictured in Figure 1.E.6.

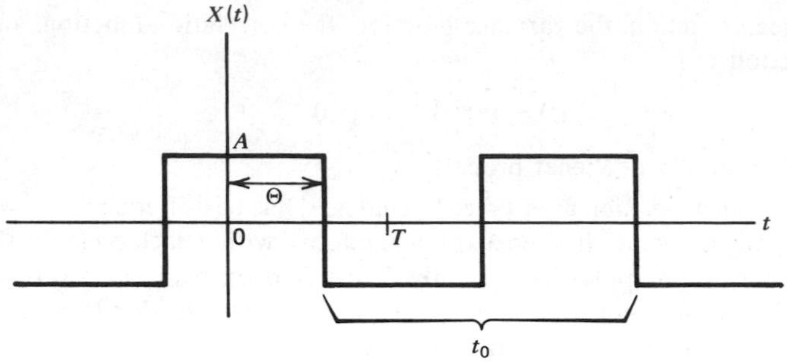

Figure 1.E.5

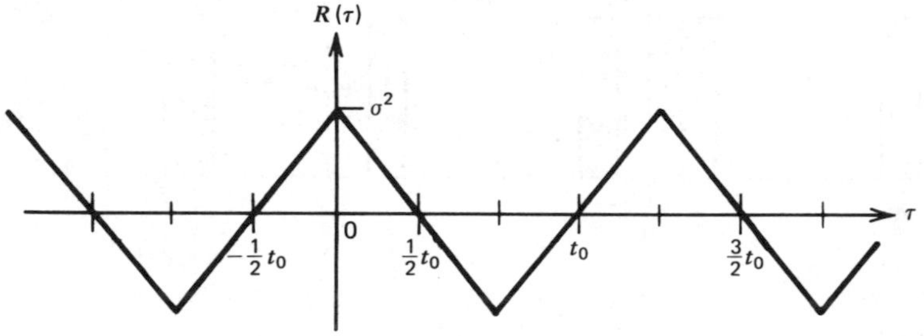

Figure 1.E.6

23. Find the conditional density of $W_0(t)$, given $W_0(1) = 0$, where $0 < t < 1$. This is called the *Brownian bridge*, since it describes densities for paths tied down at both ends.
24. Find the conditional density for $W_0(t_1)$, $W_0(t_2)$, $0 < t_1 < t_2 < 1$, given $W_0(1) = 0$.
25. Let $W_0(t)$, $t \geq 0$, be the standard Wiener process. Show that the first order densities of

$$X(t) = W_0^2(t) - t, \qquad t \geq 0$$

are given by

$$f_t(x) = \frac{1}{\sqrt{2\pi t(x+t)}} e^{-(x+t)/2t}, \qquad x > -t.$$

26. Let $N(t)$, $t \geq 0$ be the Poisson process with parameter λ and define

$$X(t) = \begin{cases} 1 & \text{if} \quad N(t) \text{ is odd} \\ -1 & \text{if} \quad N(t) \text{ is even.} \end{cases}$$

Show that $X(t)$ has the Markov property, and find the first and second order probability mass distributions. $X(t)$ is called a *random telegraph signal.*

27. Let $\mathbf{W}'(t) = (W_{01}(t), W_{02}(t))$, $t \geq 0$ be a vector-valued Wiener process where $W_{01}(t)$ and $W_{02}(t)$ are independent for all t. Find the first order density of the *Rayleigh process*

$$R(t) = \sigma\sqrt{W_{01}^2(t) + W_{02}^2(t)} \qquad t \geq 0$$

where $\sigma > 0$.

28. If $X(t)$ is Gaussian with $\mu_X(t)$ and $R_X(t_1, t_2)$, find the conditional expectation

$$E[X(t_2) \,|\, X(t_1) = x(t_1)], \qquad t_1 < t_2.$$

29. Verify that the process

$$X(t) = \sigma e^{-\alpha t} W_0(e^{2\alpha t} - 1)$$

is indeed an Ornstein-Uhlenbeck process according to Definition 1.8.

30. Show that if $\mathbf{X}_1, \ldots, \mathbf{X}_k$ are mutually independent random vectors with covariance matrices $\boldsymbol{\sigma}_{\mathbf{X}_1}^2, \ldots, \boldsymbol{\sigma}_{\mathbf{X}_k}^2$, then the covariance matrix of $\mathbf{X}_1 + \cdots + \mathbf{X}_k$ is their sum $\boldsymbol{\sigma}_{\mathbf{X}_1}^2 + \boldsymbol{\sigma}_{\mathbf{X}_2}^2 + \cdots + \boldsymbol{\sigma}_{\mathbf{X}_k}^2$. (This is the Generalized Bienaymé equality for vectors.)

31. If $\mathbf{X}$ is a random vector with $\boldsymbol{\mu}_{\mathbf{X}}$ and $\boldsymbol{\sigma}_{\mathbf{X}}^2$ and $\mathbf{Y} = \mathbf{aX} + \mathbf{b}$, show that $\boldsymbol{\mu}_{\mathbf{Y}} = \mathbf{a}\boldsymbol{\mu}_{\mathbf{X}} + \mathbf{b}$ and $\boldsymbol{\sigma}_{\mathbf{Y}}^2 = \mathbf{a}\boldsymbol{\sigma}_{\mathbf{X}}^2\mathbf{a}'$.

32. Show that the correlation function of the standard Wiener process is

$$R_{W_0}(t_1, t_2) = \min(t_1, t_2).$$

33. Let $S = \sum_{j=1}^{N} X_j$, where $X_1, X_2, \ldots, X_N$ are independent, identically distributed random variables and independent of the random variable N; show that

$$\mathrm{E}[S] = \mathrm{E}[N]\mathrm{E}[X],$$

$$\mathrm{Var}[S] = \mathrm{E}[N]\mathrm{Var}[X] + (\mathrm{E}[X])^2\mathrm{Var}[N].$$

34. Let $W_{01}(t)$, $W_{02}(t)$, $W_{03}(t)$, $t > 0$, be independent standard Wiener processes, and, if

$$X(t) = \sqrt{W_{01}^2(t) + W_{02}^2(t) + W_{03}^2(t)}, \qquad t \geq 0$$

show that the conditional density for $X(t + \tau)$, given $X(t) = x(t)$, is

$$\frac{1}{\sqrt{2\pi t}}\frac{u}{v}\left(e^{-(u-v)^2/2\tau} - e^{-(u+v)^2/2\tau}\right)$$

where $x(t + \tau) = u$, $x(t) = v$. From this, then, obtain the first and second order densities for $X(t)$. (*Hint:* Transform the vector $\mathbf{W}_0(t)$ to spherical coordinates.)

35. Find the transition probability density $f_{t_2|t_1}(x_2 | x_1)$, $t_1 < t_2$, for the Ornstein-Uhlenbeck process.

36. Show that if $X(t)$ and $Y(t)$ are independent processes with correlation functions R_X, R_Y, respectively, then $Z(t) = X(t)Y(t)$ has correlation function $R_X = R_X R_Y$. Thus, a product of correlation functions is again a correlation function.

37. To illustrate the smoothing property of conditional expectation, consider the simple random walk pictured in Figure 1.E.7. Here we have a particle whose position at step n is $X(n)$, an integer, and the probability that it makes one step to the left or to the right depends on the direction it came from at the previous step. If it arrived at $X(n)$ *from the left*, the probabilities of moving one step to the right or left are $\frac{3}{4}$ and $\frac{1}{4}$, respectively, while if it arrived at $X(n)$ *from the right*, these same probabilities are $\frac{1}{4}$ and $\frac{3}{4}$. Define the vector $\mathbf{Y}' = (X(n), X(n-1))$, and let $Z = g(\mathbf{Y}) = X(n)$, so Z then is determined by $\mathbf{Y}$, and let $X = X(n+1)$.

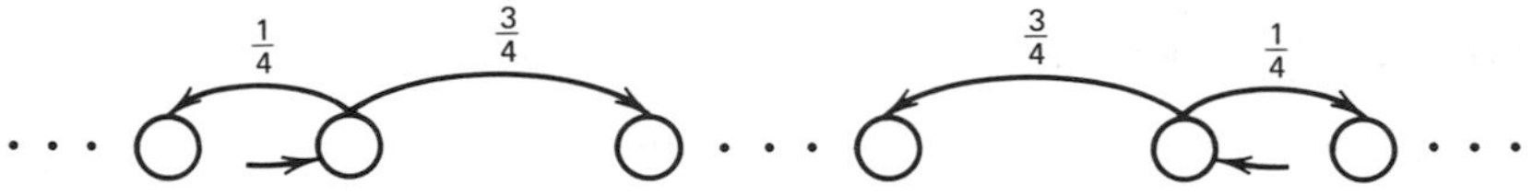

Figure 1.E.7

Show that

$$\mathrm{E}[X \mid \mathbf{y}] = x(n) + \frac{1}{2}$$

if $x(n) - x(n-1) = 1$ (it came from the left) and

$$\mathrm{E}[X \mid \mathbf{y}] = x(n) - \frac{1}{2}$$

if $x(n) - x(n-1) = -1$ (it came from the right), so

$$\mathrm{E}[X \mid \mathbf{Y}] = \frac{3}{2} X(n) - \frac{1}{2} X(n-1).$$

Show that $\mathrm{E}[\mathrm{E}[X \mid \mathbf{Y}] \mid Z] = \mathrm{E}[X \mid Z]$.

38. Use the orthogonality principle to show that the MMSE (minimum mean square error)

$$\varepsilon^2 = \mathrm{E}[(X - \mathrm{E}[X \mid \mathbf{Y}])^2]$$

can also be expressed as

$$\varepsilon^2 = \mathrm{E}[X(X - \mathrm{E}[X \mid \mathbf{Y}])]$$

or as

$$\varepsilon^2 = \mathrm{E}[X^2] - \mathrm{E}[(\mathrm{E}[X \mid \mathbf{Y}])^2].$$

Generalize to the case where $\mathbf{X}$ is a vector and conditioning is on a portion $\mathcal{Y}$ of a stochastic process; that is, show that the MMSE matrix is

$$\begin{aligned}\boldsymbol{\varepsilon}^2 &= \mathrm{E}[(\mathbf{X} - \mathrm{E}[\mathbf{X} \mid \mathcal{Y}])(\mathbf{X} - \mathrm{E}[\mathbf{X} \mid \mathcal{Y}])'] \\ &= \mathrm{E}[\mathbf{X}(\mathbf{X} - \mathrm{E}[\mathbf{X} \mid \mathcal{Y}])'] \\ &= \mathrm{E}[\mathbf{X}\mathbf{X}'] - \mathrm{E}[\mathrm{E}[\mathbf{X} \mid \mathcal{Y}]\mathrm{E}[\mathbf{X} \mid \mathcal{Y}]'].\end{aligned}$$

39. (a) Use conditional expectation to show that if X and Y are independent and $Z = XY$, then

$$\psi_Z(\zeta) = \int_{-\infty}^{\infty} \psi_X(\zeta y) f_Y(y)\, dy$$

$$= \int_{-\infty}^{\infty} \psi_Y(\zeta x) f_X(x)\, dx.$$

(b) Use (a) to find the characteristic function of the product of two independent standard Gaussian random variables. Compare this result with the characteristic function of $X_1 - X_2$, where X_1 and X_2 are independent gamma random variables, each with $\nu = \frac{1}{2}$, $\lambda = 1$.

40. Assume $\mathbf{X}' = (X_1, X_2)$ has the bivariate Gaussian density with $\boldsymbol{\mu} = 0$, $\boldsymbol{\sigma}^2$ arbitrary, and let $c = \rho\sigma_1\sigma_2$ be the covariance between X_1 and X_2. According to *Price's theorem*, if $g(X_1, X_2)$ is an arbitrary function of X_1 and X_2, then

$$\frac{\partial^n \mathrm{E}[g(X_1, X_2)]}{\partial c^n} = \mathrm{E}\left[\frac{\partial^{2n} g(X_1, X_2)}{\partial X_1^n\, \partial X_2^n}\right]$$

Show this result to be true. (*Hint:* $\dfrac{\partial^n}{\partial c^n} f_{\mathbf{X}}(\mathbf{x}) = \dfrac{\partial^{2n}}{\partial x_1^n\, \partial x_2^n} f_{\mathbf{X}}(\mathbf{x})$

and repeatedly integrate $\displaystyle\int_{-\infty}^{\infty}\int_{-\infty}^{\infty} g(x, y) \frac{\partial^{2n}}{\partial x_1^n\, \partial x_2^n} f_{\mathbf{X}}(\mathbf{x})\, d\mathbf{x}$ by parts.)

41. Let $\mathbf{X}$ be 4-variate Gaussian with $\boldsymbol{\mu} = \mathbf{0}$. Show that

$$\mathrm{E}[X_1X_2X_3X_4] = \mathrm{E}[X_1X_2]\mathrm{E}[X_3X_4] + \mathrm{E}[X_1X_3]\mathrm{E}[X_2X_4] + \mathrm{E}[X_1X_4]\mathrm{E}[X_2X_3].$$

(*Hint:* Look at $\dfrac{\partial^4}{\partial\zeta_1\,\partial\zeta_2\,\partial\zeta_3\,\partial\zeta_4}\psi_{\mathbf{X}}(\boldsymbol{\zeta})$.)

42. Let $X(n)$, $n = 1, 2, \ldots$, be a sequence of i.i.d. nonnegative random variables with finite expectation μ. Show that the stochastic sequence $Y(n)$, $n = 1, 2, \ldots$, of products

$$Y(n) = \frac{1}{\mu^n} X(1)X(2)\cdots X(n)$$

is a martingale.

43. Let $X(t) = \sin(\Omega_1 t) \pm \cos(\Omega_2 t)$, $-\infty < t < \infty$, where Ω_1 and Ω_2 are independent random variables uniformly distributed on $(-\pi, \pi)$; the $+$ and $-$ signs are equally likely, independent of Ω_1 and Ω_2. Show that this process is wide-sense stationary but not even first order stationary.

44. Use Lévy's continuity theorem to show that if $X_1, X_2, \ldots$ are independent bilateral exponential random variables with the same parameter (see Appendix Two), then the distribution of

$$Y_n = \frac{1}{\sqrt{n}} \sum_{j=1}^{n} X_j$$

approaches the Gaussian distribution with $\mu = 0$.

45. Find the characteristic functions $\psi_{V(n)}(\zeta)$, where $V(n)$ is the Bernoulli symmetric random walk.
46. Let $\mathbf{X}' = (X_1, X_2)$ be bivariate Gaussian with parameters $\boldsymbol{\mu}$ and $\boldsymbol{\sigma}^2$. Find the equation of the ellipse, centered at $\boldsymbol{\mu}' = (\mu_1, \mu_2)$, that contains $100\gamma\%$ of the volume under the density function. Calculate the coordinates of the center and lengths of the major and minor axes of the ellipse.
47. Let $X(n)$, $n = 1, 2, \ldots$, be a finite variance stochastic sequence with correlation function

$$R_X(n_1, n_2), \qquad n_1 = 1, 2, \ldots, n_2 = 1, 2, \ldots,$$

and let a_n, $n = 1, 2, \ldots$, be a sequence of numbers. Define a new stochastic sequence $Y(n)$, $n = 1, 2, \ldots$, by

$$Y(n) = \sum_{k=1}^{n} a_k X(k).$$

(a) Show that

$$R_Y(n_1, n_2) = \sum_{k=1}^{n_1} \sum_{l=1}^{n_2} a_k a_l R_X(k, l),$$

$n_1 = 1, 2, \ldots, n_2 = 1, 2, \ldots.$

(b) Show that the condition

$$\left| \sum_{k=1}^{\infty} \sum_{l=1}^{\infty} a_k a_l R_X(k, l) \right| < \infty$$

is sufficient for $\operatorname*{l.i.m.}_{n\to\infty} Y(n) = Y$, where

$$\mathrm{E}[Y] = \sum_{n=1}^{\infty} a_n \mathrm{E}[X(n)],$$

$$\mathrm{E}[Y^2] = \sum_{n=1}^{\infty} \sum_{m=1}^{\infty} a_n a_m \mathrm{E}[X(n)X(m)].$$

48. *Branching Process:* Consider a population of objects such as living cells or particles of some kind. Suppose that at each discrete time instant $n = 1, 2, \ldots$, each object present at the time splits into a random number X of new objects independently of everything else. Thus, if $N(n-1)$ denotes the size of the population just before time n and if $X_{j,n}$ is the number of new objects into which the jth object splits at time n (its offspring), then

$$N(n) = \sum_{j=1}^{N(n-1)} X_{j,n}$$

provided only that $N(n-1) \geq 1$. (If $N(n-1) = 0$, then clearly $N(n) = N(n+1) = \cdots = 0$.) Assuming that in the above expression the $X_{j,n}$'s are nonnegative i.i.d.

discrete random variables with common expectation μ_X, show that the stochastic process

$$Y(n) = \frac{1}{\mu_X^n} N(n), \qquad n = 0, 1, \ldots,$$

is a martingale.

49. Show that the random time at which the standard Wiener process, $W_0(t)$, first reaches level a is a stopping time.

50. Consider a sequence $X(n)$, $n = 1, 2, \ldots$, of i.i.d. Bernoulli random variables with $P(X(n) = +1) = P(X(n) = -1) = \frac{1}{2}$.

(a) Show that the symmetric Bernoulli random walk

$$S(n) = \sum_{k=1}^{n} X(k), \qquad k = 1, 2, \ldots, \qquad S(0) = 0,$$

is a martingale sequence.

(b) Let $-\alpha < 0 < \beta$ be two integers, and define T to be the number of steps needed for the random walk to reach either the level $-\alpha$ or the level β, whichever happens first. Formally, T is the first $n = 1, 2, \ldots$ such that either $S(n) = -\alpha$ or $S(n) = \beta$; in symbols

$$T = \min\{n = 1, 2, \ldots : S(n) = -\alpha \text{ or } S(n) = \beta\}$$

Use the optional stopping theorem to show that

$$\mathrm{E}[S(T)] = \mathrm{E}[S(1)] = 0.$$

(c) Define $p_\alpha = P(S(T) = -\alpha)$, $p_\beta = P(S(T) = \beta)$ and use (b) to show that

$$p_\alpha = \frac{\beta}{\alpha + \beta}.$$

(d) Show that

$$Z(n) = S^2(n) - \sum_{k=1}^{n} \mathrm{Var}[X(k)]$$
$$= S^2(n) - n, \qquad n = 1, 2, 3, \ldots$$

is also a martingale and

$$\mathrm{E}[Z(T)] = 0.$$

which, together with

$$\mathrm{E}[S^2(T)] = (-\alpha)^2 p_\alpha + \beta^2 p_\beta$$

and (c) gives $\mathrm{E}[T] = \alpha\beta$.

51. The previous exercise is easily generalized to a nonsymmetric Bernoulli random walk:

$$S(n) = \sum_{k=1}^{n} X(k), \qquad n = 1, 2, \ldots, \qquad S(0) = 0,$$

where now $P(X(k) = +1) = p$ and $P(X(k) = -1) = q$ are no longer equal.

(a) Show that

$$Y(n) = \sum_{k=1}^{n} (X(k) - \mu(k)) = S(n) - n(p - q)$$

is a martingale sequence, and with the stopping time T as in Exercise 50, conclude that

$$-\alpha p_\alpha + \beta p_\beta = (p - q)\mathrm{E}[T].$$

(b) Define the stochastic sequence

$$Z(n) = \left(\frac{q}{p}\right)^{S(n)}, \qquad n - 1, 2, \ldots.$$

Show that this sequence is again a martingale. Argue that optional stopping theorem still applies and that

$$\mathrm{E}[Z(T)] = \left(\frac{q}{p}\right)^{-\alpha} p_\alpha + \left(\frac{q}{p}\right)^{\beta} p_\beta = 1.$$

Conclude that

$$\mathrm{E}[T] = \frac{\beta}{p - q} - \frac{\alpha + \beta}{p - q} \frac{1 - \left(\frac{p}{q}\right)^{\beta}}{1 - \left(\frac{p}{q}\right)^{\alpha+\beta}}.$$

52. Consider a standard Wiener process $W_0(t)$, $t \geq 0$, and let $-\alpha < 0 < \beta$ be two real numbers. We define a stopping time T as the first time the process $W_0(t)$, $t \geq 0$, reaches either the level $-\alpha$ or the level β. Then $P(T < \infty) = 1$. Obviously, $-\alpha \leq W_0(T \wedge t) \leq \beta$, so that the optional stopping theorem applies.

(a) Since the standard Wiener process is a martingale, we have immediately

$$\mathrm{E}[W_0(T)] = \mathrm{E}[W_0(0)] = 0.$$

Show that

$$\mathrm{E}[W_0(T)] = -\alpha p_\alpha + \beta p_\beta,$$

using the fact that a Wiener process has continuous sample paths, which will be proved later. Conclude that

$$p_\alpha = \frac{\beta}{\alpha + \beta} \qquad \text{and} \qquad p_\beta = \frac{\alpha}{\alpha + \beta}.$$

(b) Next show that

$$X(t) = W_0^2(t) - t, \qquad t \geq 0$$

is a martingale and that

$$\mathrm{E}[X(T)] = \mathrm{E}[W_0^2(T)] - \mathrm{E}[T].$$

(c) Conclude that

$$\mathrm{E}[T] = \alpha\beta,$$

the same as for the symmetric random walk.

53. Consider a stochastic sequence $X(n)$, $n = 1, 2, \ldots$, such that $X(n)$ is uniformly distributed on the interval $(-n, n)$. Does the sequence converge in distribution? Does it converge in any other sense?

54. If $X(n)$, $n = 1, 2, \ldots$, is a stochastic sequence converging in distribution to a random variable X and if $y = g(x)$ is a monotone continuous function, show that the stochastic sequence $Y(n) = g(X(n))$, $n = 1, 2, \ldots$, converges in distribution to the random variable $Y = g(X)$.

55. A Gaussian process $X(t)$, $t \geq 0$, has

$$\mu(t) = 0, \; R_X(t_1, t_2) = (t_1 + 1)(t_2 + 1)^2, \; t_1 < t_2.$$

Is $X(t)$ also Markov?

56. Let $X(t)$, $-\infty < t < \infty$, be a stationary Gaussian process with zero mean function and with correlation function

$$R_X(\tau) = \sigma^2 e^{-\alpha\tau} \cos \omega\tau, \qquad -\infty < \tau < \infty,$$

where σ, α, and ω are positive parameters. Show that $X(t)$ is ergodic. Suggest a method for estimating the parameters σ, α, and ω.

57. Let $X(n)$, $n = 0, 1, \ldots$, represent the position of a particle performing a symmetric Bernoulli random walk starting from the origin, and let T be the first time $X(n) = 1$. Then T is a stopping time for the martingale $X(n)$, $n = 0, 1, \ldots$, and since $X(T) = 1$, we must also have $\mathrm{E}[X(T)] = 1$. However, $\mathrm{E}[X(n)] = 0$ for all $n = 0, 1, \ldots$, so that the optional stopping theorem does not apply. Explain what is wrong.

58. Let $X(n)$, $n = \ldots, -1, 0, 1, \ldots$, be a process of independent, identically distributed random variables with common density function $f_X(x)$. A process $Y(n)$, $n = \ldots, -1, 0, 1, \ldots$, is defined by

$$Y(n) = \frac{1}{4} X(n) + \frac{1}{2} X(n-1) + \frac{1}{4} X(n-2)$$

for all n. Verify by calculating the Kth order densities $f_{n, n+1, \ldots, n+K}(y_n, y_{n+1}, \ldots, y_{n+K})$ that the process $Y(n)$ is Kth order stationary for any $K = 1, 2, \ldots$, and hence strictly stationary.

59. Suppose that $X(n)$, $n = 1, 2, \ldots$, is a sequence of independent random variables with $\mathrm{E}[X(n)] = \mu_n$ and $\mathrm{Var}[X(n)] = \sigma_n^2$. Use the Cauchy criterion to show that the sequence of partial sums $S(n) = X(1) + \cdots + X(n)$, $n = 1, 2, \ldots$, converges in mean square if and only if the infinite series $\sum_{n=1}^{\infty} \mu_n$ and $\sum_{n=1}^{\infty} \sigma_n^2$ converge. Can the independence assumption be weakened?

60. Consider a symmetric Bernoulli random walk starting from the origin. Show that the random time T_r of the rth return to the origin (r is a fixed positive integer) is a stopping time.

61. Show that if X_1 and X_2 are identically distributed, each with characteristic function $\psi_X(\zeta)$, then the characteristic function for $Y = X_1 - X_2$ is

$$\psi_Y(\zeta) = \psi_X(\zeta)\psi_X^*(\zeta) = |\psi_X(\zeta)|^2$$

where ψ_X^* is the complex conjugate of ψ_X.

62. Use Lévy's continuity theorem to show that if $X(n)$, $n = 1, 2, \ldots$, is a stochastic sequence such that $X(n)$ is binomial with parameters n and $p_n = \lambda/n$, $\lambda > 0$, then $X(n)$ converges in distribution. Do the same if $X(n)$ is modified negative binomial with parameters $r = n$ and $p_n = 1 - (\lambda/r)$. What is the limiting distribution in either case?

63. Write the characteristic function $\psi_{t_1, t_2, \ldots, t_m}(\zeta_1, \zeta_2, \ldots, \zeta_m)$ for the mth order distributions of

(a) the standard Wiener process $W_0(t)$, $t \geq 0$.
(b) the Wiener process with a drift, $W(t) = \mu t + W_0(t)$, $t \geq 0$.
(c) the Ornstein-Uhlenbeck process.

64. Let $X(t)$, $Y(t)$ be zero mean Gaussian processes and define the complex-valued Gaussian process

$$Z(t) = X(t) + \imath Y(t), \qquad t \in \mathcal{T}.$$

(a) Show that if $X(t)$ and $Y(t)$ are independent and identically distributed then

$$E[Z(t_1)Z(t_2)] = 0 \qquad \text{for all } t_1, t_2 \in \mathcal{T}.$$

(b) Show that

$$E[Z(t_1)Z(t_2)] = 0$$

$$R_Z(t_1, t_2) \text{ is real}$$

together imply $X(t)$ and $Y(t)$ must be independent and identically distributed.

CHAPTER TWO

DISCRETE TIME PROCESSES

As we have already seen, a discrete-time stochastic process

$$X(t) \qquad t \in \mathcal{T}$$

is a process whose time domain $\mathcal{T}$ is an infinite set of discrete time instances $\ldots, t_n, t_{n+1}, \ldots$ such that $\cdots < t_n < t_{n+1} < \cdots$. By replacing the time instances by their labels $\ldots, n, n+1, \ldots$, the process becomes formally identical with a *stochastic sequence*

$$X(n),$$

where $n = 1, 2, \ldots$, or $n = 0, 1, \ldots$, or $n = \cdots -1, 0, 1, \ldots$, whichever is appropriate in the particular case.

We have reviewed some properties of stochastic sequeces in Chapter One. In this chapter we shall concentrate mainly on the transformation of stochastic sequences by dynamic systems. In fact, as a first step we shall show that a stochastic sequence can, under quite general conditions, be regarded as a transformation of a "random noise sequence" or "innovation process" by a dynamic system. Although we begin with quite general discrete-time processes, we shall soon restrict our attention almost entirely to the case where the process in question is Gaussian and the dynamic system operating on it is linear. Linear transformations of discrete-time Gaussian processes are discussed in Section 2.2. This section is independent of Section 2.1, and the reader who is interested only in this topic may go directly to Section 2.2.

A discussion of another important class of discrete-time processes, discrete-time Markov chains, is contained in Chapter Three.

2.1. DOOB DECOMPOSITION AND INNOVATIONS

Let $X(n)$, $n = 0, 1, \ldots$ be a discrete-time process such that

$$\mathrm{E}[|X(n)|] < \infty \text{ for all } n, \tag{1}$$

but otherwise arbitrary.

Consider the one-step forward increment

$$\Delta X(n) = X(n+1) - X(n), \qquad n = 0, 1, \ldots. \tag{2}$$

If we wished to predict the value of the increment $\Delta X(n)$ after observing the past history of the process up to time n, we know that the best mean square prediction (see Chapter One) would be the conditional expectation

$$\Delta Y(n) = \mathrm{E}[\Delta X(n) \mid X(0), \ldots, X(n)], \qquad n = 0, 1, \ldots \tag{3}$$

The difference

$$\Delta U(n) = \Delta X(n) - \Delta Y(n), \tag{4}$$

could then be interpreted as the unpredictable portion of the increment $\Delta X(n)$. Wrting this as

$$\Delta X(k) = \Delta Y(k) + \Delta U(k)$$

and summing over $k = 0, \ldots, n - 1$, we obtain

$$X(n) - X(0) = \sum_{k=0}^{n-1} \Delta Y(k) + \sum_{k=0}^{n-1} \Delta U(k).$$

If we now define two new processes $Y(n)$ and $U(n)$, $n = 0, 1, \ldots,$ by

$$Y(0) = 0, \qquad Y(n) = \sum_{k=0}^{n-1} \Delta Y(k), \qquad n = 1, 2, \ldots, \tag{5}$$

$$U(0) = X(0), \qquad U(n) = \sum_{k=0}^{n-1} \Delta U(k), \qquad n = 1, 2, \ldots \tag{6}$$

we obtain the decomposition of the process $X(n)$, which we can write as

$$X(n) = Y(n) + U(n), \qquad n = 0, 1, \ldots \tag{7}$$

Since by the definition of conditional expectation, $\Delta Y(k)$ is a deterministic function of $X(0), \ldots, X(k)$, the random variable $Y(n)$ is, for each $n > 0$, a deterministic function of the "past history" $X(0), \ldots, X(n - 1)$. In other words, $Y(n)$ can be predicted with certainty from the past values of the process $X(n)$, and therefore the process $Y(n)$ is called *predictable* (*or nonanticipating*) *with respect to the process* $X(n)$.

On the other hand, from (4)

$$\mathrm{E}[\Delta U(n) \mid X(0), \ldots, X(n)] = 0 \tag{8}$$

and

$$\mathrm{E}[\Delta U(n) \mid X(0), \ldots, X(n)] = \mathrm{E}[\Delta U(n) \mid U(0), \ldots, U(n)] \tag{9}$$

since by (3), (4), and (6), each $U(k)$ is a deterministic function of $X(0), \ldots, X(k)$. However, by (6)

$$\Delta U(n) = U(n + 1) - U(n) \tag{10}$$

so that (8)–(10) implies

$$E[U(n+1)\,|\,U(0), \ldots, U(n)] = U(n), \qquad n = 0, 1, \ldots$$

In other words, the process $U(n)$ is a *martingale*. The decomposition (7) of a discrete-time process $X(n)$ into a predictable process $Y(n)$ and a martingale $U(n)$ is called the *Doob decomposition* of $X(n)$.

Example 2.1.1

Suppose that $X(n)$, $n = 0, 1, \ldots$, is a Gaussian discrete-time process with mean function

$$\mu_X(n) = 0, \qquad n = 0, 1, \ldots,$$

and correlation function

$$R_X(n_1, n_2) = \sigma^2 \rho^{n_2 - n_1} \frac{1 - \rho^{2(n_1+1)}}{1 - \rho^2}, \qquad 0 \le n_1 \le n_2,$$

where $\sigma > 0$ and $|\rho| < 1$ are constants.

It is easily verified that for any $0 \le n_1 \le n_2 \le n_3$

$$R_X(n_1, n_3) = \frac{R_X(n_1, n_2) R_X(n_2, n_3)}{R_X(n_2, n_2)},$$

and so the process is also Markov. Consequently

$$\begin{aligned} E[X(n+1)\,|\,X(0), \ldots, X(n)] &= E[X(n+1)\,|\,X(n)] \\ &= \frac{R_X(n, n+1)}{R_X(n, n)} X(n) = \rho X(n), \qquad n = 0, 1, \ldots \end{aligned}$$

and hence in the notation for one-step forward increments

$$\begin{aligned} \Delta Y(n) &= E[\Delta X(n)\,|\,X(0), \ldots, X(n)] \\ &= E[X(n+1)\,|\,X(0), \ldots, X(n)] - X(n) \\ &= (\rho - 1) X(n), \\ \Delta U(n) &= \Delta X(n) - \Delta Y(n) = X(n+1) - \rho X(n). \end{aligned}$$

It follows by (5) and (6) that the predictable process $Y(n)$ and the martingale $U(n)$ in the Doob decomposition

$$X(n) = Y(n) + U(n)$$

are given by

$$Y(n) = (\rho - 1) \sum_{k=0}^{n-1} X(k), \qquad n = 1, 2, \ldots, \qquad Y(0) = 0,$$

and

$$U(n) = (1 - \rho)\sum_{k=0}^{n-1} X(k) + X(n), \qquad n = 1, 2, \ldots, \qquad U(0) = X(0).$$

Note that both are Gaussian processes. We can see that the present value $Y(n)$ of the predictable process can indeed be predicted with certainty from the past history $X(0) = x(0), \ldots, X(n-1) = x(n-1)$, namely

$$Y(n) = (\rho - 1)[x(0) + x(1) + \cdots + x(n-1)]. \qquad \blacktriangle$$

There is another important aspect of the foregoing discussion. Let us introduce a process $V(n)$, $n = 0, 1, \ldots$, by

$$V(0) = X(0), \qquad V(n) = \Delta U(n-1), \qquad n = 1, 2, \ldots$$

From (2), (3), and (4), we have, using the properties of conditional expectation,

$$V(n) = X(n) - \mathrm{E}[X(n) \mid X(0), \ldots, X(n-1)] \tag{11}$$

or by calling $g_n(x_0, \ldots, x_{n-1})$ the deterministic function

$$g_n(x_0, \ldots, x_{n-1}) = \mathrm{E}[X(n) \mid X(0) = x_0, \ldots, X(n-1) = x_{n-1}], \tag{12}$$

$$X(n) = g_n(X(0), \ldots, X(n-1)) + V(n), \qquad n = 1, 2, \ldots \tag{13}$$

This says that the random value $X(n)$ of the process $X(n)$ at time n is obtained by adding a *random innovation* $V(n)$ to a deterministic function g_n of the past history of the process. Furthermore, since by (11)

$$\mathrm{E}[V(n) \mid X(0), \ldots, X(n-1)] = 0, \qquad n = 1, 2, \ldots,$$

the random innovation is *unpredictable* (in the mean square sense) from the past history. In this sense, it is a true innovation of the process.

Let us next explicitly write (13) for $n = 0, 1, 2, 3$. This gives

$$\begin{aligned}
V(0) &= X(0),\\
V(1) &= -g_1(X(0)) + X(1),\\
V(2) &= -g_2(X(0), X(1)) + X(2),\\
V(3) &= -g_3(X(0), X(1), X(2)) + X(3).
\end{aligned} \tag{14}$$

We can imagine a deterministic dynamic system $\mathcal{G}$, which, when fed the sequence $X(0), X(1), \ldots$ at its input, produces the sequence $V(0), V(1), \ldots$ at its output, as pictured in Figure 2.1.1. The important thing to notice is that the system $\mathcal{G}$ is *causal* or physically realizable, since in order to produce an output $V(n)$ at time n, the system does not require inputs which may arrive later.

But the system $\mathcal{G}$ is not only causal but also *causally invertible*. This means

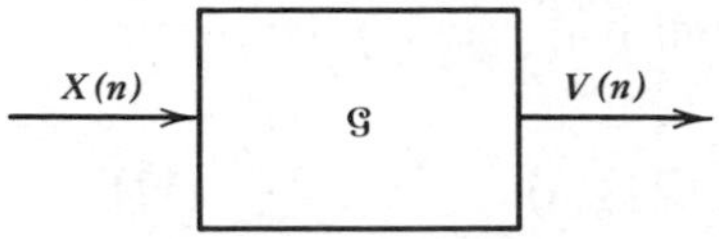

Figure 2.1.1

that it is also possible to have a *causal* system $\mathcal{H} = \mathcal{G}^{-1}$, which produces the sequence $X(0), X(1), \ldots$ at its output when fed with the sequence $V(0)\ V(1), \ldots$ at the input, as shown in Figure 2.1.2. To see this, note that (14) can also be written as

$$\begin{aligned} X(0) &= V(0), \\ X(1) &= h_1(V(0)) + V(1), \\ X(2) &= h_2(V(0), V(1)) + V(2), \\ X(3) &= h_3(V(0), V(1), V(2)) + V(3), \end{aligned} \tag{15}$$

where

$$\begin{aligned} h_1(v_0) &= g_1(v_0), \\ h_2(v_0, v_1) &= g_2(v_0, h_1(v_0) + v_1), \\ h_3(v_0, v_1, v_2) &= g_3(v_0, h_1(v_0) + v_1, h_2(v_0, v_1) + v_2). \end{aligned}$$

Thus the process $X(n)$ is entirely built up from its innovations in a causal manner.

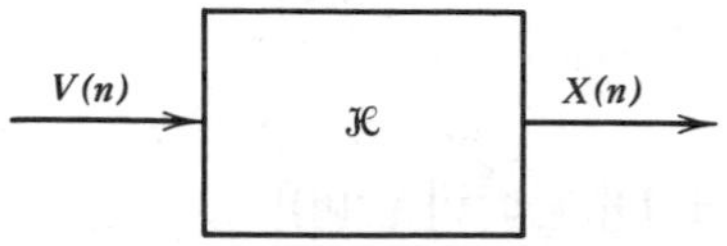

Figure 2.1.2

To summarize we have:

THEOREM 2.1.1 *Let* X(n), n = 0, 1, ..., *be a process such that* $E[|X(n)|] < \infty$ *for all* n = 0, 1, ... *Then the process*

$$V_X(n), \qquad n = 0, 1, \ldots$$

defined by

$$\begin{aligned} V_X(0) &= X(0), \\ V_X(n+1) &= \Delta X(n) - E[\Delta X(n) \,|\, X(0), \ldots, X(n)] \\ &= X(n+1) - E[X(n+1) \,|\, X(0), \ldots, X(n)], \end{aligned}$$

is called the innovation process *of the process* X(n). *It is a mean square unpredictable process, that is,*

$$E[V_X(n) \,|\, V_X(0), \ldots, V_X(n-1)] = 0, \qquad n = 1, 2, \ldots, \tag{16}$$

and is related to the process X(n) *by a causal and causally invertible transformation, in symbols*

$$V_X(n) = \mathcal{G}\{X(n)\}, \qquad X(n) = \mathcal{H}\{V_X(n)\}. \tag{17}$$

Note that also $\mathrm{E}[V_X(n] = 0$ for all $n = 1, 2, \ldots$ (take the expectation of (16)), so that the innovation process has a zero mean function except for the *initial innovation* $V_X(0) = X(0)$. Note also that

$$\sum_{k=0}^{n} V_X(k) = U(n), \qquad n = 0, 1, \ldots$$

where $U(n)$ is the martingale in Doob decomposition (7).

Example 2.1.2 (Continuation of 2.1.1)
The innovation process $V_X(n)$, $n = 1, 2, \ldots$, is by definition $V_X(0) = X(0)$ and $V_X(n+1) = \Delta X(n) - \mathrm{E}[\Delta X(n) \mid X(0), \ldots, X(n)]$, and hence in this case,

$$V_X(n+1) = X(n+1) - \rho X(n), \qquad n = 0, 1, \ldots. \tag{18}$$

It is therefore also a Gaussian process, and since it is always a mean square unpredictable process the random variables $V(n)$, $n = 0, 1, \ldots$, must, in fact, be independent.

Clearly,

$$\mathrm{E}[V_X(n)] = 0, \qquad n = 0, 1, \ldots$$

and from (18)

$$\begin{aligned}
\mathrm{E}[V_X^2(n+1)] &= \mathrm{E}[X^2(n+1)] - 2\rho\mathrm{E}[X(n)X(n+1)] + \rho^2\mathrm{E}[X^2(n)] \\
&= R_X(n+1, n+1) - 2\rho R_X(n, n+1) + \rho^2 R_X(n, n) \\
&= \frac{\sigma^2}{1-\rho^2}[1 - \rho^{2(n+2)} - 2\rho\rho(1 - \rho^{2(n+1)}) + \rho^2(1 - \rho^{2(n+1)})] \\
&= \sigma^2.
\end{aligned}$$

Since we also have $\mathrm{E}[V_X^2(0)] = \mathrm{E}[X^2(0)] = R_X(0, 0) = \sigma^2$, it follows that the innovation process is a sequence of i.i.d. Gaussian random variables with mean zero and variance σ^2.

From (18) the transformation $V_X(n) = \mathcal{G}\{X(n)\}$ has a very simple form:

$$V_X(0) = X(0), \qquad V_X(n) = -\rho X(n-1) + X(n), \qquad n = 1, 2, \ldots.$$

The inverse transformation $X(n) = \mathcal{H}\{V_X(n)\}$ is only slightly more complicated:

$$X(n) = \rho^n V_X(0) + \rho^{n-1} V_X(1) + \cdots + \rho V_X(n-1) + V_X(n),$$

$n = 0, 1, \ldots.$ It is immediately apparent that both transformations are causal. ▲

Example 2.1.3

Suppose that $X(n)$, $n = 0, 1, \ldots$, is a discrete-time Markov process such that $X(0)$ has a bilateral exponential distribution with density

$$f_0(x) = \frac{1}{2} e^{-|x|}, \qquad -\infty < x < \infty, \tag{19}$$

while the conditional density of $X(n)$, given $X(n-1) = x(n-1)$, is

$$f_{n|n-1}(x \mid x(n-1)) = \frac{1}{2} e^{-|x - \text{sign } x(n-1)|}, \qquad -\infty < x < \infty,$$

$n = 1, 2, \ldots$, where the function

$$\text{sign } x = \begin{cases} +1 & \text{if} \quad x > 0, \\ 0 & \text{if} \quad x = 0, \\ -1 & \text{if} \quad x < 0. \end{cases}$$

One way to think about such a process is as the output of a nonlinear discrete-time feedback circuit, as pictured in Figure 2.1.3.

Here, a noise voltage with density (19) is added to the output of a flip-flop with two voltage levels $+1$ and -1. The resulting output voltage is fed into a trigger generator, which reverses the flip-flop each time its input voltage crosses the zero level. Thus the voltage at the output of the flip-flop at time n is $+1$ or -1, depending on whether $X(n-1)$ was positive or negative.

It is easily seen that for $n \geq 1$

$$\text{E}[X(n) \mid X(0), \ldots, X(n-1)] = \text{sign } X(n-1)$$

and so the innovation process $V_X(n)$ is defined by

$$V_X(0) = X(0),$$
$$V_X(n) = X(n) - \text{sign } X(n-1), \qquad n = 1, 2, \ldots,$$

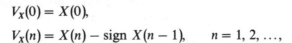

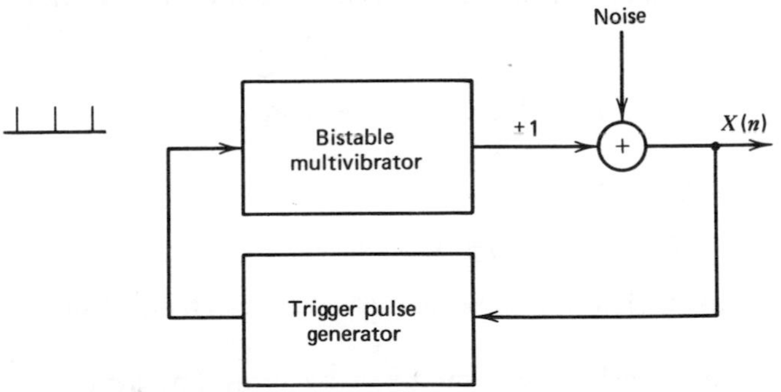

Figure 2.1.3

which defines the causal transformation $V_X(n) = \mathcal{G}\{X(n)\}$. The inverse transformation $X(n) = \mathcal{H}\{V_X(n)\}$ is defined by

$$\begin{aligned}
X(0) &= V_X(0),\\
X(1) &= \text{sign } V_X(0) + V_X(1),\\
X(2) &= \text{sign}(\text{sign } V_X(0) + V_X(1)) + V_X(2),\\
X(3) &= \text{sign}(\text{sign}(\text{sign } V_X(0) + V_X(1)) + V_X(2)) + V_X(3),\\
&\ldots.
\end{aligned}$$

and is also easily seen to be causal. Both these transformations appear nonlinear, and the latter are rather complicated, but this is deceptive; they are actually quite simple.

The reason for the simplicity is the innovation process $V_X(n)$, $n = 0, 1, \ldots,$ is in fact a sequence of independent random variables, each with the bilateral exponential density (19). To see that this is true, notice that the conditional distribution of $V_X(n)$, given $X(0), \ldots, X(n-1)$, $n \geq 1$, is by definition the same as that of $X(n) - \text{sign } X(n-1)$, that is, as $X(n) - 1$ if $X(n-1) > 0$, and as $X(n) + 1$ if $X(n+1) < 0$. But the resulting conditional density is then, in either case, just the density $f_0(x) = \frac{1}{2}e^{-|x|}$, and so $V_X(n)$ must be independent of $X(0), \ldots, X(n-1)$. However, $X(0), \ldots, X(n-1)$ are in one-to-one correspondence with $V_X(0), \ldots, V_X(n-1)$, so that $V_X(n)$ is also independent of $V_X(0), \ldots, V_X(n-1)$. Thus, $V_X(n)$, $n = 0, 1, \ldots,$ are i.i.d. with density $f(v) = \frac{1}{2}e^{-|v|}$, $-\infty < v < \infty$.

Now, sign $V_X(0)$ is a Bernoulli random variable taking values ± 1 with equal probability, and since $V_X(1)$ is independent of $V_X(0)$, sign $V_X(1) + V_X(0)$ has a symmetric density and therefore sign(sign $V_X(1) + V_X(0)$) is also a Bernoulli random variable taking values ± 1 with equal probability. It follows by induction that for any $n \geq 1$ the inverse transformations have the form

$$X(n) = \pm 1 + V_X(n),$$

where the random $\pm$ signs are equally likely and independent of $V_X(n)$. Therefore, $X(n)$, $n \geq 1$, has the density

$$f_n(x) = \frac{1}{2}(e^{-|x+1|} + e^{-|x-1|}), \qquad -\infty < x < \infty,$$

and since this density is symmetric, we also have

$$V_X(n) = X(n) \pm 1$$

with the random $\pm$ signs equally likely. (However, here the $\pm$ sign and $X(n)$ are dependent!)

Notice that this example can be easily modified by using any symmetric distribution with finite mean instead of the bilateral exponential. ▲

The initial innovation $V_X(0) = X(0)$ plays a role that sets it slightly apart from subsequent innovations $V_X(n)$, $n = 1, 2, \ldots$. Suppose it turns out that

$$V_X(n) = 0 \qquad \text{for all} \qquad n > 0.$$

Then each $X(n)$ is a deterministic function of $X(0)$ only, and thus, although $X(n)$ is still technically a discrete-time stochastic process, its sample path is completely determined, or predictable with certainty, after observing the initial value $X(0)$. Such a process is called, somewhat awkwardly, a *deterministic* (*random*) *process* or, as we shall say, a *singular process*.

If the initial value $X(0) = x_0$ is not random, then of course, the absence of innovations means that the entire sequence

$$\begin{aligned}
&X(0) = x_0, \\
&X(1) = g_1(X(0)), \\
&X(2) = g_2(X(0), X(1)) = g_2(X(0), g_1(X(0))), \\
&X(3) = g_3(X(0), X(1), X(2)) = \cdots \\
&\ldots.
\end{aligned} \tag{20}$$

is deterministic.

Thus, perhaps the most important aspect of the decomposition discussed above is that by comparing the evolution of a deterministic system (20) with the decomposition (13) of a discrete-time process, that is,

$$\begin{aligned}
&X(0) = x(0) = V_X(0), \\
&X(1) = g_1(X(0)) + V_X(1), \\
&X(2) = g_2(X(0), X(1)) + V_X(2), \\
&X(3) = g_3(X(0), X(1), X(2)) + V_X(3), \\
&\ldots.
\end{aligned} \tag{21}$$

we can see that any discrete-time process could be generated by a deterministic dynamic system $\mathcal{G}$ disturbed by an additive noise process $V_X(n)$. Turning this around, if we wish to model the effect of a noise-type disturbance on a deterministic system $\mathcal{G}$ (whose evolution is described by (20)), then adding a zero mean unpredictable noise process $V_X(n)$, as in (21), provides an adequate model.

Example 2.1.4

Suppose a number x_0 is to be replicated a large number of times in a computer memory, and for some reason it is desired to avoid repeatedly recalling the same memory location. It is, of course, possible to copy x_0 to x_1, then x_1 to x_2, and so forth, but in this way, a large error may eventually be accumulated. It has been proposed instead to use an averaging scheme, that is, to store the nth replica as

$$x_n = \frac{1}{n}(x_0 + \cdots + x_{n-1}).$$

In the complete absence of errors, this would work just as well, that is, we would have $x_n = x_0$ for all n. If, however, each replication is distorted by an additive random error, the nth replica will actually be

$$X(n) = \frac{1}{n}(X(0) + \cdots + X(n-1)) + V(n), \qquad n = 1, 2, \ldots,$$

where $V(n)$, $n = 1, 2, \ldots$, is a zero mean mean square unpredictable random sequence.

Clearly,

$$V_X(0) = V(0) = X(0) = x_0,$$

$$V_X(n) = V(n) = X(n) - \frac{1}{n}(X(0) + \cdots + X(n-1)), \qquad n = 1, 2, \ldots,$$

is an innovation process for $X(n)$, $n = 0, 1, \ldots$. The inverse transformation $X(n) = \mathcal{G}\{V_X(n)\}$ is, in this case,

$$X(0) = V_X(0) = x_0,$$

$$X(n) = V_X(0) + \frac{1}{2}V_X(1) + \frac{1}{3}V_X(2) + \cdots + \frac{1}{n}V_X(n-1) + V_X(n), \tag{22}$$

$n = 1, 2, \ldots$, as can be easily verified by induction on n. Notice that, unlike Example 2.1.1, the effect of past innovations does not "wear off" as n becomes large; just the opposite is true. Of course, neither of the processes here is Markov. Suppose that (except for $V_X(0) = x_0$, which is not random), all the innovations have the same variance: $\mathrm{Var}[V_X(n)] = \sigma^2$. Since an innovation process is mean square unpredictable, the variance of the sum (22) is a sum of variances:

$$\mathrm{Var}[X(n)] = \sigma^2\left(\frac{1}{2^2} + \frac{1}{3^2} + \cdots + \frac{1}{n^2} + 1\right).$$

The infinite series $\sum_{k=1}^{\infty} 1/k^2$ is convergent and has the value $\pi^2/6$ so that

$$\lim_{n\to\infty} \mathrm{Var}[X(n)] = \frac{\sigma^2\pi^2}{6}.$$

Thus the average replication scheme does keep the error from increasing to infinity. ▲

Although the Doob decomposition (7) and Theorem 2.1.1 apply to any finite mean discrete-time process, their usefulness for computational purposes is limited by the fact that the causal transformations $\mathcal{G}$ and $\mathcal{H}$ may be very complicated. This is typically the case when they are nonlinear. On the other hand, if these transformations are linear, then the situation is more promising. The transformations $\mathcal{G}$ and $\mathcal{H}$ will be linear if and only if the conditional expectations $\mathrm{E}[X(n) \mid X(0), \ldots, X(n-1)]$ are linear functions of $X(0), \ldots, X(n-1)$. For this it is sufficient (although not necessary) that the process $X(n)$ be a Gaussian process. Since we may then anticipate further additional useful properties, let us now concentrate on the Gaussian process.

LINEAR INNOVATIONS. From now on, let the discrete-time process (possibly vector valued)

$$X(n), \qquad n = 0, 1, \ldots$$

be *Gaussian*, and assume, for simplicity, that its mean function is zero

$$\mathrm{E}[X(n)] = 0, \qquad n = 0, 1, \ldots \tag{23}$$

Then the conditional expectations in (12) will be homogeneous linear functions of the past history of the process. In particular, the functions g_n in (12) will have the form

$$g_n(x_0, \ldots, x_{n-1}) = a(n, 0)x_0 + a(n, 1)x_1 + \cdots + a(n, n-1)x_{n-1}$$

where $a(n, k)$ are constants (matrices for a vector-valued process). This has two important consequences. First, both the causal transformation $\mathcal{G}$ and its causal inverse $\mathcal{G}^{-1}$ will be linear so that equations (14) and (15) will take the form

$$\begin{aligned}
V_X(0) &= X(0),\\
V_X(1) &= a(1, 0)X(0) + X(1),\\
V_X(2) &= a(2, 0)X(0) + a(2, 1)X(1) + X(2),\\
V_X(3) &= a(3, 0)X(0) + a(3, 1)X(1) + a(3, 2)X(2) + X(3),\\
&\text{etc.}
\end{aligned} \tag{24}$$

$$\begin{aligned}
X(0) &= V_X(0),\\
X(1) &= b(1, 0)V_X(0) + V_X(1),\\
X(2) &= b(2, 0)V_X(0) + b(2, 1)V_X(1) + V_X(2),\\
X(3) &= b(3, 0)V_X(0) + b(3, 1)V_X(1) + b(3, 2)V_X(2) + V_X(3),\\
&\text{etc.}
\end{aligned} \tag{25}$$

Remark: Note that for each $n = 1, 2, \ldots,$ the coefficients $a(n, 0)$, $a(n, 1), \ldots,$ $a(n, n-1)$ can be uniquely determined from the coefficients $b(n, 0), b(n, 1), \ldots, b(n, n-1)$, and conversely. In fact, if we write the systems of linear equations (24) and (25) in matrix notation,† we have

$$\begin{pmatrix} V_X(0) \\ V_X(1) \\ V_X(2) \\ \cdots \\ V_X(n) \end{pmatrix} = \begin{bmatrix} 1, & 0, & 0, & \ldots, & 0, & 0 \\ a(1, 0), & 1, & 0, & \ldots, & 0, & 0 \\ a(2, 0), & a(2, 1), & 1, & \ldots, & 0, & 0 \\ \cdots & \cdots & \cdots & \cdots & \cdots & \cdots \\ a(n, 0), & a(n, 1), & a(n, 2), & \ldots, & a(n, n-1), & 1 \end{bmatrix} \begin{pmatrix} X(0) \\ X(1) \\ X(2) \\ \\ X(n) \end{pmatrix},$$

$$\begin{pmatrix} X(0) \\ X(1) \\ X(2) \\ \cdots \\ X(n) \end{pmatrix} = \begin{bmatrix} 1, & 0, & 0, & \ldots, & 0, & 0 \\ b(1, 0), & 1, & 0, & \ldots, & 0, & 0 \\ b(2, 0), & b(2, 1), & 1, & \ldots, & 0, & 0 \\ \cdots & \cdots & \cdots & \cdots & \cdots & \cdots \\ b(n, 0), & b(n, 1), & b(n, 2), & \ldots, & b(n, n-1), & 1 \end{bmatrix} \begin{pmatrix} V_X(0) \\ V_X(1) \\ V_X(2) \\ \\ V_X(n) \end{pmatrix},$$

Notice that both matrices are lower triangular with 1's along the diagonal and hence are nonsingular.

If we define infinite square matrices

$$\mathscr{A} = [\alpha(i, j)], \qquad \mathscr{B} = [\beta(i, j)] \tag{26}$$

where

$$\alpha(i, j) = \begin{cases} -a(i, j) & \text{for } 0 \le j < i, \\ 0 & \text{for } j \ge i, \end{cases}$$

and

$$\beta(i, j) = \begin{cases} b(i, j) & \text{for } 0 \le j < i, \\ 0 & \text{for } j \ge i, \end{cases}$$

then the linear causal system equations (24) and (25) can be written symbolically as

$$\mathbf{V}_X = (\mathscr{I} - \mathscr{A})\mathbf{X}, \qquad \mathbf{X} = (\mathscr{I} + \mathscr{B})\mathbf{V}_X, \tag{27}$$

where $\mathscr{I}$ is an infinite identity matrix. We can think of these infinite matrices as linear operators that transform infinite sequences into infinite sequences. From (27), we see that the operators $(\mathscr{I} + \mathscr{B})$ and $(\mathscr{I} - \mathscr{A})$ are inverses of each other, that is,

$$(\mathscr{I} - \mathscr{A})(\mathscr{I} + \mathscr{B}) = \mathscr{I}, \tag{28}$$

or

$$\mathscr{B} = \mathscr{A} + \mathscr{A}\mathscr{B}$$

This is known as the Volterra equation. Note that because the operators $(\mathscr{I} + \mathscr{B})$ and $(\mathscr{I} - \mathscr{A})$ are both invertible,

$$\mathscr{A} = \mathscr{B}(\mathscr{I} + \mathscr{B})^{-1}, \qquad \text{and} \qquad \mathscr{B} = (\mathscr{I} - \mathscr{A})^{-1}\mathscr{A},$$

† For vector-valued processes, the coefficients $a(n, k)$ and $b(n, k)$ are themselves square matrices, but let us consider the scalar case for the moment.

which shows symbolically how one of these operators is obtained from the other. Note that the "triangular" form of the pair (26) of Volterra operators is essential here; it guarantees that both linear systems $\mathcal{G}$ and $\mathcal{H}$ are causal and invertible. Я

Example 2.1.5

Since the coefficients $a(i, j)$ and $b(i, j)$ are entries of the infinite matrices $(\mathcal{I} - \mathcal{A})$ and $(\mathcal{I} + \mathcal{B})$, respectively, Equation (28) requires that for all $i = 0, 1, \ldots, j = 0, 1, \ldots$

$$\sum_{l=0}^{\infty} a(i, l) b(l, j) = \delta_{ij}, \tag{29}$$

where δ_{ij} is the Kronecker delta ($\delta_{ij} = 1$ if $i = j$, $\delta_{ij} = 0$ if $i \neq j$). Furthermore, since

$$a(i, j) = b(i, j) = 0 \qquad \text{if} \qquad 0 \leq i < j, \tag{30}$$

and

$$a(i, j) = b(i, j) = 1 \qquad \text{if} \qquad 0 \leq i = j,$$

condition (29) is automatically satisfied for any $0 \leq i \leq j$. Because of (30), we thus only have to make sure that

$$\sum_{l=j}^{i} a(i, l) b(l, j) = 0 \qquad \text{for all } 0 \leq j < i.$$

As an example, take

$$a(i, j) = \frac{(-1)^{i-j}}{(i-j)!} \qquad b(i, j) = \frac{1}{(i-j)!}, \qquad 0 \leq j \leq i.$$

Then

$$\sum_{l=j}^{i} \frac{(-1)^{i-l}}{(i-l)!} \frac{1}{(l-j)!} = \frac{1}{(i-j)!} \sum_{l=j}^{i} \frac{(i-j)!}{(i-l)!\,(l-j)!} (-1)^{i-l}$$
$$= \frac{1}{(i-j)!} \sum_{l=j}^{i} \binom{i-j}{i-l} (-1)^{i-l} = 0$$

since by thc binomial theorem

$$\sum_{l=j}^{i} \binom{i-j}{i-l} (-1)^{i-l}$$
$$= \binom{i-j}{0}(-1)^0 + \binom{i-j}{1}(-1)^1 + \binom{i-j}{2}(-1)^2 + \cdots + \binom{i-j}{i-j}(-1)^{i-j}$$
$$= (1-1)^{i-j} = 0.$$

Thus the equations

$$v_n = \sum_{k=0}^{n} \frac{(-1)^{n-k}}{(n-k)!} x_k, \qquad x_n = \sum_{k=0}^{n} \frac{1}{(n-k)!} v_k, \qquad n = 0, 1, \ldots,$$

provide a simple example of a pair of causal and invertible transformations between sequences $v_0, v_1, \ldots$ and $x_0, x_1, \ldots$. ▲

The innovation process $V_X(n)$, being a linear transformation of a Gaussian process, will also be Gaussian. But since an innovation process is mean square unpredictable, we have from (16)

$$\begin{aligned} &E[V_X(k)V_X(n) \mid V_X(0), \ldots, V_X(n-1)] \\ &= V_X(k)E[V_X(n) \mid V_X(0), \ldots, V_X(n-1)] \\ &= 0 \end{aligned}$$

for any $0 \leq k < n$. Hence, by taking expectations of the above

$$E[V_X(k)V_X(n)] = 0, \qquad 0 \leq k < n,$$

and since by assumption (23) $E[X(0)] = E[V_X(0)] = 0$, we also have

$$E[V_X(n)] = 0 \qquad \text{for all } n = 0, 1, \ldots$$

Thus the random variables $V_X(k)$ and $V_X(n)$, $k \neq n$ are uncorrelated and hence, being Gaussian, are also *independent*.

Now, an innovation process $V_X(n)$ consists of forward increments of the martingale $U(n)$ in the Doob decomposition $X(n) = Y(n) + U(n)$. Since these increments are mean square unpredictable and, in the Gaussian case, even independent (i.e., unpredictable in any sense), it is natural to regard the innovation process $V_X(n)$ as a noisy component of the process $X(n)$. After all, it can be argued that the concept of unpredictability should be a distinctive feature of what we would intuitively call random noise.

Since a Gaussian innovation process, that is, a sequence of independent Gaussian random variables $V_X(n)$, is used very frequently in applications as a model for noise, we shall introduce a special term for such a process.

Definition 2.1.1
A discrete-time Gaussian process (possibly vector valued) with zero mean function and with correlation function such that $R(n_1, n_2) = 0$ whenever $n_1 \neq n_2$ will be called discrete-time *white Gaussian noise*.†

† The reason for calling the noise white will be made clear in Chapter Five. For the time being, imagine that such a process is arriving at the input of a color TV receiver, as is approximately the case when there is no station on the air. You would see the familiar chaotic pattern of red, blue, and green dots, and if a color photograph with sufficiently long exposure were taken of the screen, it would appear uniformly gray, which is a shade of white.

Note that the probability law of discrete-time white Gaussian noise is then completely determined by its variance function

$$\sigma^2(n) = R(n, n).$$

In particular, if

$$\sigma^2(n) = 1 \qquad \text{for all } n$$

or, in the vector-valued case,

$$\boldsymbol{\sigma}^2(n) = \mathbf{I}, \qquad \text{the identity matrix,}$$

we will refer to the white Gaussian noise as *standard*. We will now summarize these findings as:

THEOREM 2.1.2. *Let* X(n), n = 0, 1, ..., *be a discrete-time Gaussian process (possibly vector valued) with mean function*

$$\mu_X(n) = 0, \qquad n = 0, 1, \ldots .$$

Then the innovation process $V_X(n)$, n = 0, 1, ..., *defined by*

$$V_X(0) = X(0),$$

$$V_X(n) = X(n) - \mathrm{E}[X(n) \mid X(0), \ldots, X(n-1)], \qquad n = 1, 2, \ldots,$$

is discrete-time white Gaussian noise. Furthermore, it is related to the process X(n) *by a causal and causally invertible linear transformation*

$$V_X(n) = \sum_{k=0}^{n} a(n, k)X(k), \qquad X(n) = \sum_{k=0}^{n} b(n, k)V_X(k), \qquad (31\text{a, b})$$

where a(n, n) = b(n, n) = 1, n = 0, 1, ...

We express this briefly by saying that the processes X(n) *and* $V_X(n)$ *are causally linearly equivalent.*

Linearity of transformations (31) provides a little extra benefit. It allows us to separate the effect of the initial innovation

$$V_X(0) = X(0)$$

in the form

$$X(n) = b(n, 0)X(0) + \sum_{k=1}^{n} b(n, k)V_X(k), \qquad n = 1, 2, \ldots .$$

This represents a decomposition of the process $X(n)$ into a *singular process*

$$X_s(n) = b(n, 0)X(0), \qquad n = 0, 1, \ldots,$$

which can be predicted with certainty from its initial value $X(0) = x_0$, and a *regular process*

$$X_r(n) = \sum_{k=1}^{n} b(n, k)V(k), \qquad n = 1, 2, \ldots$$

which is entirely built up from innovations after time $n = 0$. It should be noted that in this decomposition

$$X(n) = X_s(n) + X_r(n), \qquad n = 1, 2, \ldots \tag{32}$$

the two component processes are *both Gaussian* and *mutually independent*, since $X(0) = V_X(0)$ is independent of all future innovations $V_X(n)$, $n \geq 1$.

Equation (32) is a particular case of the *Wold-Cramèr decomposition*, which we will discuss again later.

Example 2.1.6

Consider a discrete-time Gaussian process

$$Y(n), \qquad n = 0, 1, \ldots \tag{33}$$

with zero mean function and with correlation function

$$R_Y(n_1, n_2) = \sigma^2 \Bigg[(-1)^{n_1+n_2} + (-1)^{n_2}\rho^{n_1} + (-1)^{n_1}\rho^{n_2} + \rho^{n_2-n_1}\frac{1-\rho^{2(n_1+1)}}{1-\rho^2}\Bigg],$$

$0 \leq n_1 \leq n_2$, where $\sigma^2 > 0$ and $|\rho| < 1$.

Comparing this with the correlation function

$$R_X(n_1, n_2) = \sigma^2\rho^{n_2-n_1}\frac{1-\rho^{2(n_1+1)}}{1-\rho^2}, \qquad 0 \leq n_1 \leq n_2$$

of the Gauss-Markov process $X(n)$, $n = 0, 1, \ldots$, of Example 2.1.1, we can see that the correlation function is

$$R_Y(n_1, n_2) = (-1)^{n_1+n_2}R_X(0, 0) + (-1)^{n_2}R_X(0, n_1) + (-1)^{n_1}R_X(0, n_2) + R(n_1, n_2).$$

But this is exactly what we would get if we defined the original process (33) by

$$Y(n) = (-1)^n X(0) + X(n), \qquad n = 0, 1, \ldots, \tag{34}$$

since both $Y(n)$ and $X(n)$ are zero mean Gaussian processes, and thus their probability laws are completely determined by their correlation functions. It follows that the probability laws of the original process (33) and the one defined by (34) must be identical. Now using (34), $Y(0) = 2X(0)$ and $X(n) = Y(n) - (-1)^n \frac{1}{2} Y(0)$, and so for $n \geq 1$

$$\begin{aligned} E[Y(n) \mid Y(0), \ldots, Y(n-1)] &= E[Y(n) \mid X(0), \ldots, X(n-1)] \\ &= (-1)^n X(0) + E[X(n) \mid X(0), \ldots, X(n-1)] \\ &= (-1)^n X(0) + \rho X(n-1) = (-1)^n \frac{1+\rho}{2} Y(0) + \rho Y(n-1). \end{aligned}$$

Therefore the linear innovation process $V_Y(n)$, $n = 0, 1, \ldots$, is given by

$$V_Y(0) = Y(0),$$

$$V_Y(n) = Y(n) - \rho Y(n-1) - (-1)^n \frac{1+\rho}{2} Y(0), \qquad n = 1, 2, \ldots$$

The inverse transformation is

$$\begin{aligned} Y(0) &= V_Y(0), \\ Y(n) &= \frac{(-1)^n + \rho^n}{2} V_Y(0) + \rho^{n-1} V_Y(1) + \rho^{n-2} V_Y(2) \\ &\quad + \cdots + \rho V_Y(n-1) + V_Y(n), \qquad n = 1, 2, \ldots, \end{aligned} \tag{35}$$

as may be readily verified by repeated substitution, as in Example 2.1.1. It follows that if we define

$$Y_s(n) = \frac{(-1)^n + \rho^n}{2} V_Y(0), \qquad n = 0, 1, \ldots$$

and

$$Y_r(n) = \sum_{k=1}^{n} \rho^{n-k} V_Y(k), \qquad n = 1, 2, \ldots,$$

$$Y_r(0) = 0,$$

we obtain the Wold-Cramér decomposition of the process (33):

$$Y(n) = Y_s(n) + Y_r(n), \qquad n = 0, 1, \ldots.$$

Since the innovation process $V_Y(n)$, $n = 0, 1, \ldots$, is white Gaussian noise, the singular and regular component processes are indeed independent. Also from (35), we see that for $n \geq 1$

$$E[Y(n) \mid Y(0)] = E[Y(n) \mid V_Y(0)] = Y_s(n),$$

and

$$Y_r(n) = Y(n) - E[Y(n) \mid Y(0)],$$

so that the entire singular component process can be predicted without error from the initial innovation $V_Y(0) = Y(0)$, while the regular component process cannot be predicted from the initial innovation at all. ▲

So far in this section we have exclusively treated discrete-time processes with semi-infinite time domain $\mathcal{T} = \{0, 1, \ldots\}$. Let us extend the above results to discrete-time processes with time domain

$$\mathcal{T} = \{\ldots, -1, 0, 1, \ldots\}, \tag{36}$$

the case quite often encountered when dealing with stationary processes. Unfortunately, some difficulty is created by the fact that with time domain (36) we have no "initial" value for the process, so we shall be forced to resort to rather heuristic arguments.

We are going to consider only the linear case, and we therefore begin by assuming that we are given a *discrete-time Gaussian process*

$$X(n), \qquad n = \ldots, -1, 0, 1, \ldots .$$

As before, we define the *innovation process*

$$V_X(n), \qquad n = \ldots, -1, 0, 1, \ldots$$

by

$$V_X(n) = X(n) - \mathrm{E}[X(n) \mid X(n'), n' < n]. \tag{37}$$

However, the conditional expectation now depends on an infinite number of past random variables $\ldots, X(n-2), X(n-1)$ of the process, and thus it is a *functional* of this infinite sequence of past values. Since the process $X(n)$ is, by assumption, Gaussian we know (see Chapter One) that this functional must be linear and that the conditional expectation in (37) will be a Gaussian random variable. Consequently, the innovation process (37) will also be Gaussian. Taking the conditional expectation of both sides of (37)

$$\begin{aligned}\mathrm{E}[V_X(n) \mid X(n'), n' < n] &= \mathrm{E}[X(n) \mid X(n'), n' < n] - \mathrm{E}[X(n) \mid X(n'), n' < n] \\ &= 0,\end{aligned}$$

and since all the past innovations $\ldots, V_X(n-2), V_X(n-1)$ are, according to (37), completely determined by the corresponding past history $\ldots, X(n-2), X(n-1)$ of the process $X(n)$, $n = \ldots, -1, 0, 1, \ldots$, it also follows that

$$\mathrm{E}[V_X(n) \mid V_X(n'), n' < n] = 0 \quad \text{for all } n = \ldots, -1, 0, 1, \ldots .$$

The innovation process being Gaussian thus implies that the random variables $V_X(n)$, $n = \ldots, -1, 0, 1, \ldots$, are all independent and hence have mean zero. In other words, the innovation process is again *discrete-time white Gaussian noise.*

Since, as we have already mentioned, the conditional expectation in (37) is a linear functional of the past history of the process, it follows that for each n the innovation $V_X(n)$ is itself a linear functional of the past and present $\ldots, X(n-2)$, $X(n-1)$, $X(n)$. Thus, we again say that the innovation process $V_X(n)$ can be obtained by passing the process $X(n)$ through a *causal linear system* $\mathcal{G}$; that is, we can write

$$V_X(n) = \mathcal{G}\{X(n)\},$$

just as before.

Now the question is whether this linear transformation is also causally invertible, that is, if there is a causal linear system $\mathcal{H}$ such that

$$X(n) = \mathcal{H}\{V_X(n)\}.$$

This time, however, we run into trouble; such a system need not exist! For example, let

$$Y(n), \qquad n = \ldots, -1, 0, 1, \ldots \tag{38}$$

be a doubly infinite sequence of independent standard Gaussian random variables, and let Z be another standard Gaussian random variable independent of the entire sequence $Y(n)$, $n = \ldots, -1, 0, 1, \ldots$. Define a discrete-time Gaussian process by

$$X(n) = Y(n) + (-1)^n Z, \qquad n = \ldots, -1, 0, 1, \ldots. \tag{39}$$

By the properties of conditional expectation

$$\mathrm{E}[X(n) \mid X(n'), n' < n] = (-1)^n Z$$

so that the innovation process for $X(n)$ is just the sequence (38)

$$V_X(n) = Y(n), \qquad n = \ldots, -1, 0, 1, \ldots.$$

Thus for each n, we have the relation

$$X(n) = V_X(n) + (-1)^n Z$$

with the innovations independent of the random variable Z. This shows that $X(n)$ cannot be obtained from only the innovation process even if we allowed the system $\mathcal{H}$ to be noncausal and nonlinear.

This example also suggests how to remedy the situation. From definition (39) of the process, we will have

$$\mathrm{E}[X(n) \mid X(n'), n' < n_0] = (-1)^n Z \qquad \text{for any } n_0 \leq n,$$

and, hence, also

$$\lim_{n_0 \to -\infty} \mathrm{E}[X(n) \mid X(n'), n' < n_0] = (-1)^n Z.$$

Thus, returning to the original discrete-time Gaussian process,

$$X(n), \qquad n = \ldots, -1, 0, 1, \ldots \tag{40}$$

we would expect that the missing piece of information that prevented us from reconstructing process (40) from its innovation process

$$V_X(n) = X(n) - \mathrm{E}[X(n) \mid X(n'), n' < n], \qquad n = \ldots, -1, 0, 1, \ldots \tag{41}$$

is an additional discrete-time process defined by

$$X_s(n) = \lim_{n_0 \to -\infty} \mathrm{E}[X(n) \mid X(n'), n' < n_0], \qquad n = \ldots, -1, 0, 1, \ldots . \tag{42}$$

This is indeed true, but since we cannot provide a rigorous proof without using rather abstract concepts, we resort to heuristics. Let $n_0 < n$ and consider the difference

$$D = \mathrm{E}[X(n) \mid X(n'), n' < n] - \mathrm{E}[X(n) \mid X(n'), n' < n_0]. \tag{43}$$

We argue that this difference D can only be a function of $X(n_0)$, $X(n_0 + 1), \ldots,$ $X(n-1)$, for if we take the conditional expectation of the difference (43), then

$$\mathrm{E}[D \mid X(n'), n' < n_0] = 0,$$

which would not be true if D also depended on some $X(n')$, $n' < n_0$. Since $X(n)$, $n = \ldots, -1, 0, 1, \ldots$, is a Gaussian process, the difference (43) must further be a *linear* function of $X(n_0)$, $X(n_0 + 1), \ldots, X(n-1)$,

$$D = -\sum_{k=n_0}^{n-1} a(n-1, k) X(k),$$

where the a's are constants (matrices for vector-valued process) and the minus sign was chosen as a matter of convenience. Thus, we can write (41) as

$$\begin{aligned} V_X(n) &= X(n) - D - \mathrm{E}[X(n) \mid X(n'), n' < n_0] \\ &= \sum_{k=n_0}^{n} a(n, k) X(k) - \mathrm{E}[X(n) \mid X(n'), n' < n_0]. \end{aligned}$$

Now, letting $n_0 \to -\infty$ and using (42), we should get the representation for the innovation process

$$V_X(n) = \sum_{k=-\infty}^{n} a(n, k) X(k) - X_s(n) \qquad n = \ldots, -1, 0, 1, \ldots, \tag{44}$$

where $a(n, n) = 1$.

The process $X_s(n)$, $n = \ldots, -1, 0, 1, \ldots$, defined by (42) is called the *singular* component of the process $X(n)$, $n = \ldots, -1, 0, 1, \ldots$, while the difference

$$X_r(n) = X(n) - X_s(n) \qquad n = \ldots, -1, 0, 1, \ldots$$

is called the *regular* or *purely indeterministic* component.

Example 2.1.7

Let $X(n)$, $n = \dots, -1, 0, 1, \dots$, be a discrete-time Gaussian process satisfying the equation

$$X(n) + X(n-2) = 0 \qquad \text{for all } n = \dots, -1, 0, 1, \dots. \tag{45}$$

It should be clear that this is a singular process, for if we know two successive values $X(n_0) = x_1$ and $X(n_0 - 1) = x_2$ for some time n_0, no matter how far back in the past, we have, by (45),

$$X(n_0 \pm 2k) = (-1)^k x_1, \qquad X(n_0 - 1 \pm 2k) = (-1)^k x_2,$$

for all $k = 0, 1, \dots$. In other words, the entire sample path of the process is determined by these two values. It follows that the process is determined by just two Gaussian random variables, say Z_1 and Z_2, and that we can define

$$X(n) = Z_1 \cos\frac{n\pi}{2} + Z_2 \sin\frac{n\pi}{2}, \qquad n = \dots, -1, 0, 1, \dots. \tag{46}$$

The reader is invited to verify that with this definition, Equation (45) is indeed satisfied.

It should also be noted that a process satisfying Equation (45) contains no innovations, since for any $n = \dots, -1, 0, 1, \dots$

$$\begin{aligned} V_X(n) &= X(n) - \mathrm{E}[X(n) \mid X(n'), n' < n] \\ &= -X(n-2) - \mathrm{E}[-X(n-2) \mid X(n'), n' < n] \\ &= -X(n-2) + X(n-2) = 0. \end{aligned}$$

If we further assume that the two Gaussian random variables Z_1 and Z_2 in (46) are independent with zero mean and common variance $\sigma_Z^2 > 0$, we have for the correlation function

$$\begin{aligned} R_X(n_1, n_2) &= \mathrm{E}\left[\left(Z_1 \cos\frac{n_1\pi}{2} + Z_2 \sin\frac{n_1\pi}{2}\right)\left(Z_1 \cos\frac{n_2\pi}{2} + Z_2 \sin\frac{n_2\pi}{2}\right)\right] \\ &= \mathrm{E}[Z_1^2] \cos\frac{n_1\pi}{2} \cos\frac{n_2\pi}{2} + \mathrm{E}[Z_2^2] \sin\frac{n_1\pi}{2} \sin\frac{n_2\pi}{2} \\ &= \sigma_Z^2 \cos(n_1 - n_2)\frac{\pi}{2}, \end{aligned}$$

so that for this case the process is also stationary with correlation function

$$R_X(n) = \sigma_Z^2 \cos\frac{n\pi}{2} \qquad n = 0, \pm 1, \dots. \tag{47}$$

and zero mean function.

The reader should be warned, however, that a singular Gaussian process, even a stationary one, may be far more complicated than this example may suggest; it may, for instance, be defined as a trigonometric series similar to (46) but with an infinite number of random variables Z_j.

As we shall see shortly, however, a singular process will always have zero innovations. ▲

From the definition of the singular component

$$X_s(n) = \lim_{n_0 \to -\infty} \mathrm{E}[X(n) \mid X(n'), n' < n_0] \tag{48}$$

we see that since the conditional expectation $\mathrm{E}[X(n) \mid X(n'), n' < n_0]$ is based on the past $\ldots, X(n_0 - 2), X(n_0 - 1)$, the limit in (48) can be interpreted as the best predictor of $X(n)$ based on the "infinitely remote past" of the process $X(n)$. In other words, $X_s(n)$ is the part of the process $X(n)$ that can be predicted with certainty from the past history $X(n'), n' < n_0$, no matter how distant the past may be from the present time n. For this reason, $X_s(n)$ is sometimes also called a deterministic process, but we prefer the term singular, since "deterministic random process" just does not sound right. The singular component process has no innovations of its own, since from (48)

$$X_s(n) - \mathrm{E}[X_s(n) \mid X(n'), n' < n] = X_s(n) - X_s(n) = 0,$$

and hence also

$$V_{X_s}(n) = X_s(n) - \mathrm{E}[X_s(n) \mid X_s(n'), n' < n] = 0$$

for all $n = \ldots, -1, 0, 1, \ldots$. Consequently, all the innovations $V_X(n)$ in (41) are due to the regular component process:

$$V_X(n) = V_{X_r}(n) = X_r(n) - \mathrm{E}[X_r(n) \mid X_r(n'), n' < n].$$

Now if the original process (40) were itself regular, that is, if its singular components were identically zero, then the discussion leading to the representation (44) would still apply, but the last term on the right-hand side of (44) would be missing. Thus, we would have obtained the representation

$$V_{X_r}(n) = \sum_{k=-\infty}^{n} a(n, k) X_r(k).$$

However, as we have just seen, $V_X(n) = V_{X_r}(n)$ for all n, and so we have, in general,

$$V_X(n) = \sum_{k=-\infty}^{n} a(n, k) X_r(k), \qquad n = \ldots, -1, 0, 1, \ldots, \tag{49a}$$

where $a(n, n) = 1$. This clearly represents a causal linear transformation, and it can be shown that this transformation is also *causally invertible*, that is, that

$$X_r(n) = \sum_{k=-\infty}^{n} b(n, k)V_X(k), \qquad n = \ldots, -1, 0, 1, \ldots, \tag{49b}$$

where again $b(n, n) = 1$.

Remark: Although we cannot prove this statement as we did for the process $X(n)$, $n = 0, 1, \ldots$ (since we now have no first element $X(0)$ to begin with), we give a plausibility argument. The infinite system of linear equations (49a) has the form

$$\begin{aligned} V_X(n) &= \cdots + a(n, n-3)X_r(n-3) + a(n, n-2)X_r(n-2) + a(n, n-1)X_r(n-1) + X_r(n) \\ V_X(n-1) &= \cdots + a(n-1, n-3)X_r(n-3) + a(n-1, n-2)X_r(n-2) + X_r(n-1) \\ V_X(n-2) &= \cdots + a(n-2, n-3)X_r(n-3) + X_r(n-2) \\ &\cdots \end{aligned} \tag{50}$$

By multiplying the second equation above by $a(n, n-1)$ and then subtracting from the first, we eliminate the random variable $X_r(n-1)$. The coefficient of $X_r(n-2)$ becomes $a(n, n-2) - a(n, n-1)a(n-1, n-2)$. By multiplying the third equation by this constant and again subtracting from the new first equation, we eliminate the variable $X_r(n-2)$. A reader familiar with linear algebra may recognize this procedure as the Gauss elimination method. If we continued in this fashion "indefinitely" the first equation in (50) would be turned into an equation of the form

$$V_X(n) + b(n, n-1)V_X(n-1) + b(n, n-2)V_X(n-2) + b(n, n-3)V_x(n-3) + \cdots = X_r(n),$$

the sole "surviving" variable on the right-hand side being $X_r(n)$. Thus, we would obtain the causal inverse (49b). Of course, the catch is that "indefinite" elimination is not a legitimate procedure. We would have to show that after a large but finite number of such eliminations the remaining infinite linear combination of past $X(n-k)$'s is small (in the mean square sense) and will tend to zero as the elimination process continues. Я

Example 2.1.8

Let $V(n)$, $n = \ldots, -1, 0, 1, \ldots$, be discrete-time stationary white Gaussian noise with variance $\sigma_V^2 > 0$. Define a discrete-time Gaussian process by

$$X(n) = V(n) - \gamma V(n-1) - \gamma^2 V(n-2), \qquad n = \ldots, -1, 0, 1, \ldots, \tag{51}$$

where $|\gamma| < 1$ is a constant. Clearly, the process has a zero mean function, and its correlation function is easily evaluated by taking the expectation of the product $E[X(n_1)X(n_2)]$.

$$\begin{aligned} E[X(n_1)X(n_2)] = E[&V(n_1)V(n_2) - \gamma V(n_1)V(n_2-1) - \gamma^2 V(n_1)V(n_2-2) \\ &- \gamma V(n_1-1)V(n_2) + \gamma^2 V(n_1-1)V(n_2-1) + \gamma^3 V(n_1-1)V(n_2-2) \\ &- \gamma^2 V(n_1-2)V(n_2) + \gamma^3 V(n_1-2)V(n_2-1) + \gamma^4 V(n_1-2)V(n_2-2)]. \end{aligned}$$

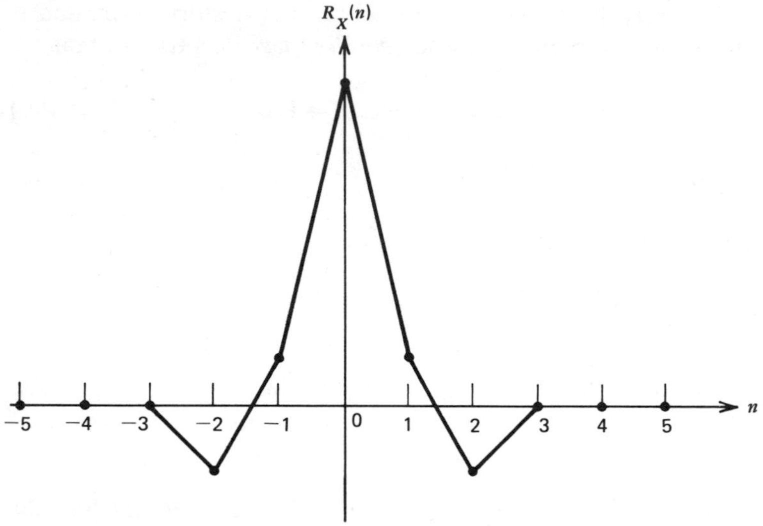

Figure 2.1.4

Since $V(n)$ is white noise

$$E[V(n)V(m)] = \begin{cases} \sigma_V^2 & \text{if} \quad n = m, \\ 0 & \text{if} \quad n \neq m, \end{cases}$$

we find that

$$R_X(n_1, n_2) = \begin{cases} \sigma_V^2(1 + \gamma^2 + \gamma^4) & \text{if} \quad n_1 = n_2, \\ \sigma_V^2 \gamma(\gamma^2 - 1) & \text{if} \quad n_1 = n_2 \pm 1, \\ -\sigma_V^2 \gamma^2 & \text{if} \quad n_1 = n_2 \pm 2, \\ 0 & \text{otherwise.} \end{cases}$$

Since the correlation function depends only on the difference $|n_1 - n_2|$, the process is stationary. The correlation function (with $\sigma_V = 2$ and $\gamma = -\frac{1}{2}$) is depicted in Figure 2.1.4. It turns out, as might have been expected, that the process (51) is regular and that the white Gaussian noise $V(n)$, $n = \ldots, -1, 0, 1, \ldots$, is actually its innovation process. To prove this, we claim that the white Gaussian noise $V(n)$ can be obtained from the process $X(n)$ by a causal linear filter:

$$V(n) = \sum_{k=0}^{\infty} f_k \gamma^k X(n-k), \qquad n = \ldots, -1, 0, 1, \ldots, \tag{53}$$

where $f_0, f_1, f_2, \ldots$ is the Fibonacci sequence 1, 1, 2, 3, 5, 8, ..., that is, a sequence of integers defined by $f_0 = f_1 = 1$ and $f_{k+1} = f_k + f_{k-1}$, $k = 1, 2, \ldots$. To verify this merely note that for any n, we have, from (53),

$$\begin{aligned} &V(n) - \gamma V(n-1) - \gamma^2 V(n-2) \\ &\quad = X(n) + f_1 \gamma X(n-1) + f_2 \gamma^2 X(n-2) + f_3 \gamma^3 X(n-3) + f_4 \gamma^4 X(n-4) + \cdots \\ &\qquad - f_0 \gamma X(n-1) - f_1 \gamma^2 X(n-2) - f_2 \gamma^3 X(n-3) - f_3 \gamma^4 X(n-4) + \cdots \\ &\qquad - f_0 \gamma^2 X(n-2) - f_1 \gamma^3 X(n-3) - f_2 \gamma^4 X(n-4) + \cdots \\ &\quad = X(n) \end{aligned}$$

since $f_1 = f_0 = 1$ and $f_{k+1} = f_k + f_{k-1}$ so that all but the first term cancels.

Since (51) and (53) show that $X(n)$ and $V(n)$ are causally linearly equivalent, we have

$$\begin{aligned} \mathrm{E}[X(n) \mid X(n'), n' < n] &= \mathrm{E}[X(n) \mid V(n'), n' < n] \\ &= \mathrm{E}[V(n)] - \gamma V(n-1) - \gamma^2 V(n-2) \\ &= X(n) - V(n) \end{aligned} \tag{54}$$

by (51), using the fact that $\mathrm{E}[V(n)] = 0$. Consequently,

$$V_X(n) = X(n) - \mathrm{E}[X(n) \mid X(n'), n' < n] = X(n) - X(n) + V(n) = V(n)$$

so that the white Gaussian noise $V(n)$ is indeed the innovation process for $X(n)$. Furthermore, for $n_0 \le n - 2$, we have, by (51),

$$\begin{aligned} \mathrm{E}[X(n) \mid X(n'), n' < n_0] &= \mathrm{E}[X(n) \mid V(n'), n' < n_0] \\ &= \mathrm{E}[V(n) - \gamma V(n-1) - \gamma^2 V(n-2)] = 0 \end{aligned}$$

so that

$$X_s(n) = \lim_{n_0 \to -\infty} \mathrm{E}[X(n) \mid X(n'), n' < n_0] = 0$$

for all n. It follows that the process $X(n)$, $n = \ldots, -1, 0, 1, \ldots$ is regular.

Notice also that from (54) and (53)

$$\mathrm{E}[X(n) \mid X(n'), n' < n] = -\sum_{k=1}^{\infty} f_k \gamma^k X(n-k)$$

so that the one-step best predictor involves the entire past history of the process in spite of the fact that $X(n)$ is independent of the past $\cdots X(n-5)$, $X(n-4)$, $X(n-3)$ as seen from the correlation function R_X. ▲

The pair of causal linear relations (49a,b) is the representation we were seeking. Let us only add that for a Gaussian process $X(n)$, $n = \ldots, -1, 0, 1, \ldots$,

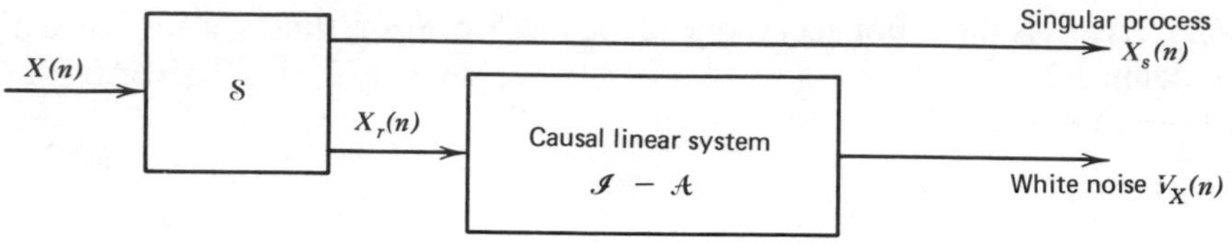

Figure 2.1.5

both the regular and singular components are also Gaussian processes, and furthermore they are *mutually independent*. The decomposition

$$X(n) = X_r(n) + X_s(n), \qquad n = \ldots, -1, 0, 1, \ldots$$

is known as the *Wold-Cramér decomposition.*

As we have seen, the regular component is causally linearly equivalent to the innovation process $V_X(n)$, $n = \ldots, -1, 0, 1, \ldots$, which is white Gaussian noise. Thus the process $X(n)$ can be converted into a singular process $X_s(n)$ and white Gaussian noise as shown in Figure 2.1.5. Here, $\mathcal{S}$ is also a causal linear system that performs the Wold-Cramér decomposition, while $\mathcal{J} - \mathcal{A}$ is a system described by equation (49a).† This conversion results in no loss of information, since the process $X(n)$ can be reconstructed as shown in Figure 2.1.6.

The system $\mathcal{J} - \mathcal{B}$ is the inverse of the system $\mathcal{J} - \mathcal{A}$ and is described by equation (49b).

If the Gaussian process $X(n)$, $n = \ldots, -1, 0, 1, \ldots$, is also *stationary*, then both its singular and regular components $X_s(n)$ and $X_r(n)$, as well as the innovation process $V_X(n)$, $n = \ldots, -1, 0, 1, \ldots$, will all be *stationary processes*. Also, the linear systems $\mathcal{J} - \mathcal{A}$ and $\mathcal{J} + \mathcal{B}$ are, in this case, *time invariant*, that is, the rela-

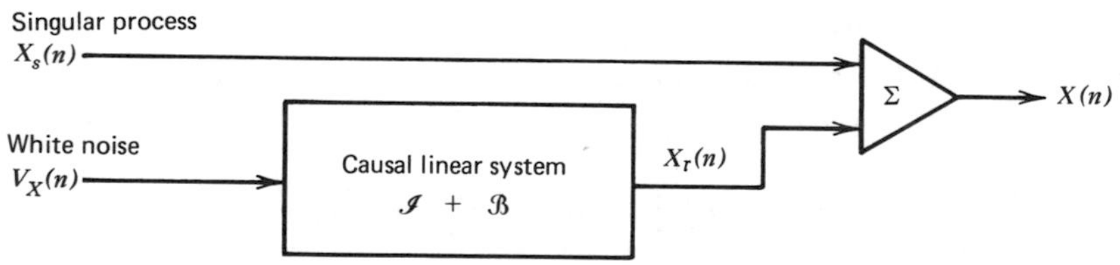

Figure 2.1.6

† We have chosen the symbols $\mathcal{J} - \mathcal{A}$ and $\mathcal{J} + \mathcal{B}$ to remind the reader that $\mathcal{A}$ and $\mathcal{B}$ are really linear Volterra operators as discussed earlier.

tions (49a) and (49b) now become

$$V_X(n) = \sum_{k=-\infty}^{n} a(n-k)X_r(k), \tag{55a}$$

$$X_r(n) = \sum_{k=-\infty}^{n} b(n-k)V_X(k), \qquad n = \ldots, -1, 0, 1, \ldots, \tag{55b}$$

where $a(0) = b(0) = 1$. Notice that (55a,b) can also be rewritten as convolutions

$$V_X(n) = \sum_{k=0}^{\infty} a(k)X_r(n-k), \tag{56a}$$

$$X_r(n) = \sum_{k=0}^{\infty} b(k)V_X(n-k), \qquad n = \ldots, -1, 0, 1, \ldots. \tag{56b}$$

Example 2.1.9

Suppose we take the stationary singular process

$$X_s(n) = Z_1 \cos\frac{n\pi}{2} + Z_2 \sin\frac{n\pi}{2}, \qquad n = \ldots, -1, 0, 1, \ldots, \tag{57}$$

of Example 2.1.7, and the stationary regular process

$$X_r(n) = V(n) - \gamma V(n-1) - \gamma^2 V(n-2), \qquad n = \ldots, -1, 0, 1, \ldots \tag{58}$$

of Example 2.1.8, where the Gaussian random variables Z_1, Z_2, and $V(n), n = \ldots, -1, 0, 1, \ldots$, are all independent.

The discrete-time process defined by

$$X(n) = X_s(n) + X_r(n), \qquad n = \ldots, -1, 0, 1, \ldots \tag{59}$$

will then be a stationary Gaussian process, with $X_s(n)$ and $X_r(n)$ being the singular and regular component processes of its Wold-Cramér decomposition. Since both $X_s(n)$ and $X_r(n)$ have zero mean functions, so will the process $X(n)$, and since the two components are independent, the correlation function of the process (59) will be a sum

$$R_X(n) = R_{X_s}(n) + R_{X_r}(n), \qquad n = 0, \pm 1, \ldots$$

of the correlation functions (47) and (52)

$$R_{X_s}(n) = \sigma_Z^2 \cos\frac{n\pi}{2},$$

$$R_{X_r}(n) = \begin{cases} \sigma_V^2(1+\gamma^2+\gamma^4) & \text{if} \quad n = 0, \\ \sigma_V^2\gamma(\gamma^2-1) & \text{if} \quad n = \pm 1, \\ -\sigma_V^2\gamma^2 & \text{if} \quad n = \pm 2, \\ 0 & \text{if} \quad |n| > 2. \end{cases}$$

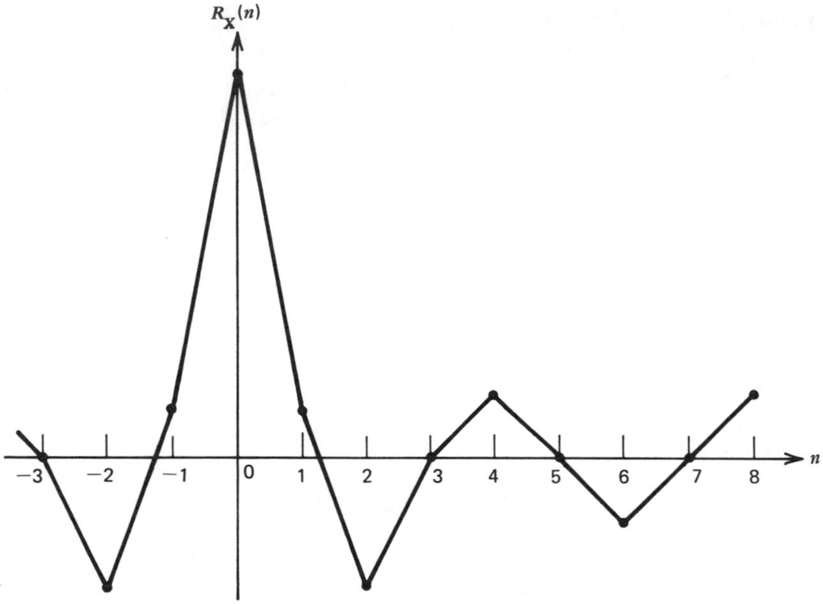

Figure 2.1.7

The correlation function $R_X(n)$ with $\sigma_Z = 1$, $\sigma_V = 2$, and $\gamma = -\frac{1}{2}$ is depicted in Figure 2.1.7.

Generating the process $X(n)$ from the singular process $X_s(n)$ and the white noise $V(n)$ is easy, since the three equations (57), (58), and (59) specify the construction completely. A block scheme of this generation is shown in Figure 2.1.8.

A decomposition of the process $X(n)$ into its singular component $X_s(n)$ and its innovation white noise $V(n)$ is a little bit more complicated. First, we must separate the singular component. This can be accomplished by noticing that if we form the average

$$\frac{1}{m}(X(n) + X(n-4) + X(n-8) + X(n-12) + \cdots + X(n-4(m-1)))$$

or

$$\frac{1}{m}\sum_{k=0}^{m-1} X(n-4k) = \frac{1}{m}\sum_{k=0}^{m-1} X_s(n-4k) + \frac{1}{m}\sum_{k=0}^{m-1} X_r(n-4k)$$

then, since the singular component $X_s(n)$ satisfies the equation $X_s(n) = -X_s(n-2) = X_s(n-4)$, the first average on the right-hand side equals $X_s(n)$. On the other hand, $X_r(n)$, $X_r(n-4)$, $X_r(n-8)$, ... are independent identically dis-

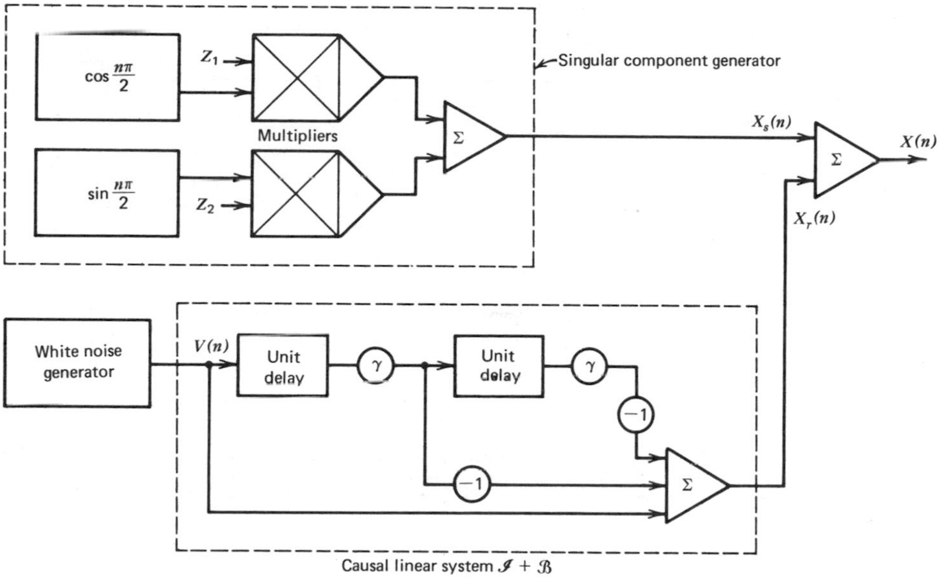

Figure 2.1.8

tributed Gaussian random variables with zero mean, and hence the second average above will approach zero as $m \to \infty$. Thus for large m, we have approximately

$$\frac{1}{m} \sum_{k=0}^{m-1} X(n - 4k) \doteq X_s(n), \qquad n = \ldots, -1, 0, 1, \ldots$$

and the desired separation is accomplished. The regular component $X_r(n)$ can then be obtained as the difference

$$X_r(n) = X(n) - X_s(n).$$

Finally, to convert the regular component into innovation white noise $V(n)$, we employ the equation

$$V(n) = \sum_{k=0}^{\infty} f_k \gamma^k X_r(n - k), \qquad n = \ldots, -1, 0, 1, \ldots$$

established in Example 2.1.8. An overall block scheme of this decomposition is shown in Figure 2.1.9.

The reader should be warned, however, that this example is deceptive in that it is too simple. In general, the determination of these linear systems may be very difficult, although various methods are available in the literature. This is especially true if the Gaussian process $X(n)$ is defined by its correlation function, which is often the case. ▲

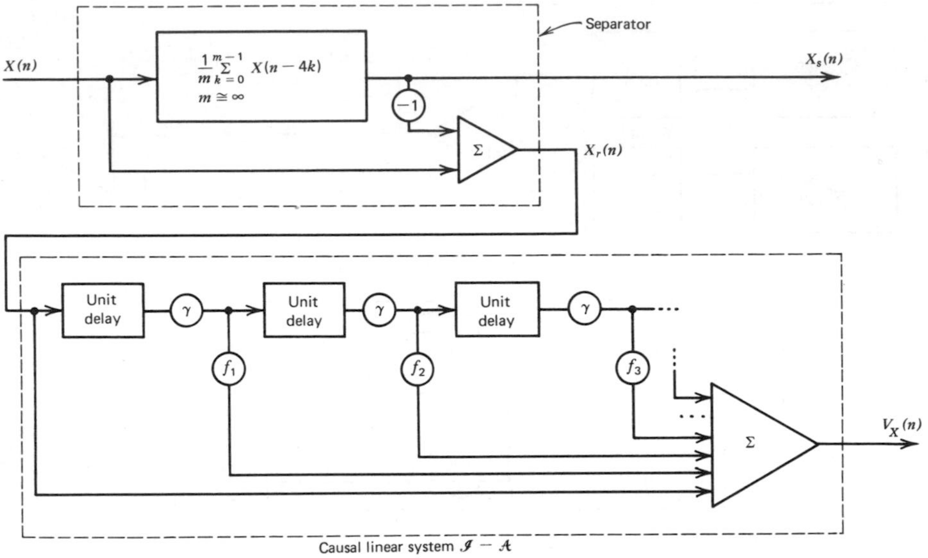

Figure 2.1.9

Remark: If all but a finite number of the coefficients $a(k)$, $k = 0, 1, \ldots$, are zero, then the representation (56a) is often called an mth order *autoregressive representation* of the stationary process $X_r(n)$. It is then usually written in the form

$$X_r(n) = a(1)X_r(n-1) + a(2)X_r(n-2) + \cdots + a(m)X_r(n-m) + V_x(n),$$

$n = \ldots, -1, 0, 1, \ldots$, suggesting that the regular process $X_r(n)$ can be thought of as being generated by an mth order linear difference equation disturbed by additive white Gaussian noise $V_x(n)$.

On the other hand, if all but a finite number of the coefficients $b(k)$, $k = 0, 1, \ldots$, are zero, for instance, if $b(k) = 0$ for $k \geq m$, $(m > 1)$, then the representation (56b) is called the *moving average* of the stationary process $X_r(n)$. In this case the reason for this terminology is obvious, since

$$X_r(n) = b(0)V_x(n) + b(1)V_x(n-1) + \cdots + b(m)V_x(n-m),$$

$n = \ldots, -1, 0, 1, \ldots$, so that for each n, $X_r(n)$ is a weighted average of m i.i.d. random variables which "moves" along with n.

However, in any nontrivial case, it is impossible for *both* (56a) and (56b) to have only a finite number of nonzero coefficients. Я

EXERCISES

Exercise 2.1

1. Suppose that the discrete-time process $X(n)$, $n = 0, 1, \ldots$, is obtained as a sum $X(n) = \sum_{k=1}^{n} Z(k)$, $n > 0$, $X(0) = 0$, of i.i.d. discrete random variables $Z(k)$, $k = 1, 2, \ldots$, such

that $P(Z(k) = 1) = p$, $P(Z(k) = -1) = q = 1 - p$, where $0 < p < 1$. Show that the predictable process $Y(n)$ in the Doob decomposition (7) is just a deterministic sequence, while the martingale $U(n)$ is a linear transformation of the binomial process. (See Chapter One.)

2. A discrete-time binary (i.e., with the state space $\mathcal{S} = \{0, 1\}$) process $X(n)$, $n = 0, 1, \ldots$, is defined as follows:
   (i) $X(0) = 1$
   (ii) If $X(n-2) = X(n-1) = 0$, then $X(n) = 1$; otherwise, $X(n) = 1$ or 0 with equal probability $\frac{1}{2}$.

   Suppose the initial portion of the sample path $X(n)$, $n = 0, \ldots, 20$ is the sequence:

   $$1\,0\,0\,1\,1\,0\,1\,1\,1\,0\,1\,0\,0\,1\,1\,0\,0\,1\,0\,0\,1.$$

   Determine the corresponding portions of the sample paths $y(n)$ and $u(n)$ of the predictable process $Y(n)$ and the martingale $U(n)$ in the Doob decomposition (7).

3. The quadratic variation $Q_X(n)$, $n = 1, 2, \ldots$, of a finite-variance discrete-time process $X(n)$, $n = 0, 1, \ldots$, is defined by $Q_X(n) = \sum_{k=0}^{n-1} (\Delta X(n))^2$. Show that the expectation $\mathrm{E}[Q_X(n)] = \mathrm{E}[Q_Y(n)] + \mathrm{E}[U^2(n)]$, where $Q_Y(n)$ is the quadratic variation of the predictable process $Y(n)$ and $U(n)$, is the martingale in the Doob decomposition (7).

4. A discrete-time process $X(n)$, $n = 1, 2, \ldots$, has kth order densities

   $$f_{1,\ldots,k}(x_1, \ldots, x_k) = 2^{-2k}(x_1 x_2 \cdots x_k)^{-1/2} e^{-\sum_{j=1}^{k} (\sqrt{x_j} - \sqrt{x_{j-1}})}$$

   if $x_k > x_{k-1} > \cdots > x_1 > 0$, and $f_{1,\ldots,k}(x_1, \ldots, x_k) = 0$ otherwise, where $x_0 = 0$. Obtain the innovation process $V_X(n) = \mathcal{G}\{X(n)\}$, $n = 1, 2, \ldots$. Write an expression for the predictable process $Y(n)$, $n = 1, 2, \ldots$, in the Doob decomposition (7) and argue that the sample paths of this process must be nondecreasing, that is, $y(1) \leq y(2) \leq y(3) \leq \cdots$.

5. Let $X(n)$, $n = 0, 1, \ldots$, be the positions of a particle performing a random walk described in Exercise 1.37. Assume the initial position $X(0) = 0$ and $X(1) = \pm 1$ with equal probability. Obtain expressions for the martingale $U(n)$ and the predictable process $Y(n)$, both in terms of the process $X(n)$, in the Doob decomposition (7).

6. Suppose $Z(n)$, $n = 0, 1, \ldots$, is a Gauss-Markov process such that $Z(0)$ is Gaussian with mean zero and variance σ^2 and the conditional density of $Z(n+1)$, given $Z(n) = z$, is Gaussian with mean βz and variance σ^2. A new process $X(n)$, $n = 0, 1, \ldots$, is defined by the memoryless transformation $X(n) = Z^3(n)$ for all n. Obtain the innovation process $V_X(n)$, $n = 0, 1, \ldots$, as a causal transformation of the process $X(n)$. For $n = 0$, 1, 2, 3, obtain the causal inverse $X(n) = \mathcal{H}\{V_X(n)\}$. Is the innovation process Gaussian?

7. Consider a discrete-time dynamic system whose state $x(n)$, $n = 0, 1, \ldots$, evolves according to the equation $\Delta x(n) = \alpha\, \Delta x(n-1)$, $n = 1, 2, \ldots$, $\Delta x(0) = 1$, $x(0) = 0$, where α is a constant. Suppose now that the system is disturbed by a white Gaussian noise $V(n)$, $n = 1, 2, \ldots$, so that we now have $\Delta X(n) = \alpha\, \Delta X(n-1) + V(n+1)$, $n = 1$, 2, $\ldots$, $\Delta X(0) = 1 + V(1)$. Express the innovation process $V_X(n)$, $n = 0, 1, \ldots$, as a causal transformation of the process $X(n)$ and find the causal inverse $X(n) = \mathcal{H}\{V_X(n)\}$.

What happens if $|\alpha| = 1$? How does the constant α influence the behavior of the system as $n \to \infty$?

8. Consider a point source that at each unit of time emits an electron in a random direction and with a random velocity. Let $Z(n)$ be the velocity of the electron emitted at time n, and assume that $Z(n)$, $n = 1, 2, \ldots$, are i.i.d. random variables having the Maxwell distribution with parameter σ (see Appendix Two). Assume further that each emitted electron flies away from the source along a straight line with constant velocity. Define a process $X(n)$, $n = 1, 2, \ldots$, where $X(n)$ is the electrostatic potential at the source just prior to the emission of the nth electron. Obtain an expression for the innovation $V_X(n)$ in terms of $Z(1), \ldots, Z(n)$ and find its mean and variance. Try to derive the causal transformations $X(n) = \mathcal{G}\{V_X(n)\}$ and $V_X(n) = \mathcal{H}\{X(n)\}$, at least for the first few values of n.

9. Let $Z(n)$, $n = 1, 2, \ldots$, be a (scalar-valued) white Gaussian noise, and let the process $X(n)$, $n = 1, 2, \ldots$, be defined by

$$X(n) = (-1)^n \gamma X(n-2) + Z(n),$$

where $\gamma > 0$ is a constant and $X(-1) = Y_1$, $X(0) = Y_2$, with $Y = (Y_1, Y_2)'$ being a Gaussian random vector independent of the noise sequence $Z(n)$. Obtain the linear transformations (31a,b). Decompose the process $X(n)$ into a singular process depending only on $\mathbf{Y}$ and a regular process depending only on the noise $Z(n)$, $n = 1, 2, \ldots$.

10. Suppose $Z(n)$, $n = 1, 2, \ldots$, is a Gauss-Markov process with zero mean function and with correlation function

$$R_Z(n_1, n_2) = \sigma^2 \rho^{n_2 - n_1} \frac{1 - \rho^{2(n_1+1)}}{1 - \rho^2},$$

$1 \le n_1 \le n_2$, $\sigma > 0$, $|\rho| < 1$. A process $X(n)$, $n = 1, 2, \ldots$, is constructed as follows: $X(1) = Z(1)$, $X(n+1) = \gamma X(n) + Z(n+1)$, $n = 1, 2, \ldots$, where γ is a constant. Obtain the innovation process $V_X(n)$, $n = 1, 2, \ldots$ as a linear combination of $X(1), \ldots, X(n)$ and also as a linear combination of $Z(1), \ldots, Z(n)$. Use the latter to obtain the variance function of the innovation process $V_X(n)$.

11. In Example 2.1.7, assume that the two Gaussian random variables Z_1 and Z_2 in equation (46) are no longer independent but have some nonzero correlation coefficient ρ. How does this affect the correlation function $R_X(n_1, n_2)$? Is the process still stationary?

12. Suppose that a stationary Gaussian process $X(n)$, $n = \ldots, -1, 0, 1, \ldots$, satisfies the equation $\Delta X(n+2) = -\Delta X(n)$ for all n. Show that this is a singular process and that it can be expressed as

$$X(n) = Z_0 + Z_1 \cos \frac{n\pi}{2} + Z_2 \sin \frac{n\pi}{2}$$

where Z_0, Z_1, Z_2 are Gaussian random variables. Obtain the correlation function $R_X(n_1, n_2)$. Are Z_0, Z_1 and Z_2 independent?

13. Show that if $X(n)$, $n = \ldots, -1, 0, 1, \ldots$, is a finite-variance process, then the coefficients $a(n, k)$ and $b(n, k)$ in (49a,b) must be such that for all n

$$\sum_{k=-\infty}^{n} a^2(n, k) < \infty \qquad \text{and} \qquad \sum_{k=-\infty}^{n} b^2(n, k) < \infty.$$

Also show that for a stationary process these conditions become

$$\sum_{k=0}^{\infty} a^2(k) < \infty \qquad \text{and} \qquad \sum_{k=0}^{\infty} b^2(k) < \infty.$$

14. Show that for a stationary regular process $X(n)$, $n = \ldots, -1, 0, 1, \ldots$, the coefficients $a(k)$ and $b(k)$ in (56a,b) satisfy the condition

$$\sum_{k=0}^{m} a(k)b(m-k) = 0 \qquad \text{for all } m = 1, 2, \ldots.$$

(*Hint:* Substitute (56b) into (56a), multiply by $V_X(n-m)$, and take expectations.)

15. Use Exercise 2.1.14 to verify that if a regular stationary process $X(n)$, $n = \ldots, -1, 0, 1, \ldots$, has the moving average representation

$$X(n) = V_X(n) - 3V_X(n-1) + 3V_X(n-2) - V_X(n-3)$$

then (56a) has the form

$$V_X(n) = \frac{1}{2} \sum_{k=0}^{\infty} (k+1)(k+2)X(n-k).$$

2.2. LINEAR TRANSFORMATIONS

Consider a linear discrete-time dynamic system and a discrete-time stochastic process $U(n)$ (no longer the martingale of Section 2.1). If the process $U(n)$ is applied to the input of the system, the output will also be a discrete-time stochastic process $Z(n)$, as pictured in Figure 2.2.1. Given the probability law of the input process $U(n)$ and a specification of the system in some form (e.g., its impulse-response function), we would like to determine the probability law of the output process $Z(n)$. In general, that means determining its distributions, for example, the infinite family of joint density functions

$$f_{n_1, \ldots, n_k}(z_1, \ldots, z_k).$$

Since the system is not assumed memoryless, this would be quite a formidable problem, even for a relatively simple input process $U(n)$.

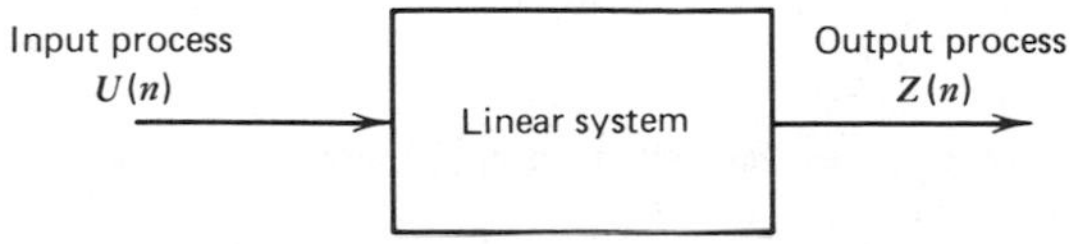

Figure 2.2.1

On the other hand, linearity of the system and linearity of the expectation operator suggests that it should not be too difficult to establish the relationship between moment functions of the input and output processes.

In particular, if we restrict our attention to the first two moment functions of the output process, we only have to find the relationships between the mean functions and correlation functions of the two. Restrictive as it may seem, this will still solve a very important class of problems, namely when the input process $U(n)$ is Gaussian. For in that case, as a consequence of the linearity of the system, we would expect the output process $Z(n)$ to be Gaussian also and hence its probability law to be completely determined by its mean and correlation functions.

Let us now begin to develop the relations between the mean functions and the correlation functions. First, we consider the case when the linear system is specified by its *impulse-response function*

$$h(n, k). \tag{1}$$

Recall that $h(n, k)$ is defined as the output $z(n)$ at time n if the input sequence $u(n)$ is a sequence of zeros except for a unit input $u(k) = 1$ at the time k. Recall also that for multiple input–multiple output linear systems, (1) is a matrix, with its (i, j)th entry $h_{ij}(n, k)$ the response at the ith output terminal to the unit impulse at the jth input terminal.

We shall consider here only linear systems, which are *causal*

$$h(n, k) = 0 \qquad \text{for} \quad n < k, \tag{2}$$

and *stable*

$$\sum_{k=-\infty}^{n} |h(n, k)| < \infty \qquad \text{for all } n, \tag{3}$$

with (2) and (3) being satisfied for each entry h_{ij} if h is a matrix.† If $u(n)$, $n = \ldots, -1, 0, 1, \ldots$, is a (vector-valued) input sequence to such a linear system, then the output sequence $z(n)$, $n = \ldots, -1, 0, 1, \ldots$, is given by

$$z(n) = \sum_{k=-\infty}^{n} h(n, k)u(k), \qquad n = \cdots -1, 0, 1, \ldots. \tag{4}$$

Many practical linear systems are *time-invariant*. In this special case, the impulse-response function $h(n, k)$ depends only on the difference $n - k$, which means that the response to a unit input depends only on the time elapsed since the arrival of this input. For a time-invariant linear system we shall abuse our notation slightly by writing the impulse-response function as

$$h(n, k) = h(n - k), \qquad k \leq n.$$

† The results derived in this section generalize easily to the matrix case. We use scalar notation (no bold face) and primes to indicate necessary transpositions.

The input-output equation (4) then takes the form of a convolution

$$z(n) = \sum_{k=-\infty}^{n} h(n-k)u(k) = \sum_{k=0}^{\infty} h(k)u(n-k),$$

while the stability condition (3) simplifies to

$$\sum_{k=0}^{\infty} |h(k)| < \infty. \tag{5}$$

Formula (4) applies as well to input sequences $u(n)$, $n = 0, 1, \ldots$, which "begin" at time $n = 0$; we merely set $u(n) = 0$ for negative n so that, by linearity, we get in that case

$$z(n) = \sum_{k=0}^{n} h(n, k)u(k), \qquad n = 0, 1, \ldots.$$

Suppose now that the input is a discrete-time Gaussian process. We will assume that it is defined on the time domain $\mathcal{T} = \{\ldots, -1, 0, 1, \ldots\}$,

$$U(n), \qquad n = \ldots, -1, 0, 1, \ldots$$

since the case $n = 0, 1, \ldots$, in view of the above remark, then becomes a special case of the former by setting $U(n) = 0$ for $n < 0$.

In other words, we assume that the output process $Z(n)$, $n = \ldots, -1, 0, 1, \ldots$, of the linear system is defined by

$$Z(n) = \sum_{k=-\infty}^{n} h(n, k)U(k), \qquad n = \ldots, -1, 0, 1, \ldots. \tag{6}$$

Of course, in the case when the summation in (6) is actually an infinite series, $z(n)$ is, for each n, really defined as a limit

$$z(n) = \lim_{l \to -\infty} \sum_{k=l}^{n} h(n, k)u(k),$$

and when we replace the input $u(n)$ by a discrete-time stochastic process $U(n)$, $n = \ldots, -1, 0, 1, \ldots$, we have to decide what kind of stochastic convergence is meant by this limit. Here, we will be satisfied with MS convergence, that is, we define the output process by

$$Z(n) = \underset{l \to -\infty}{\text{l.i.m.}} \sum_{k=l}^{n} h(n, k)U(k), \qquad n = \ldots, -1, 0, 1, \ldots. \tag{7}$$

Before addressing the question of whether such an MS limit actually exists, we proceed formally with equation (6) to derive the relations between the mean functions and correlation functions.

Taking expectations on both sides of equation (6), we have

$$\mathrm{E}[Z(n)] = \mathrm{E}\left[\sum_{k=-\infty}^{n} h(n, k)U(k)\right] = \sum_{k=-\infty}^{n} h(n, k)\mathrm{E}[U(k)]. \tag{8}$$

Similarly, by considering the product

$$Z(n_1)Z'(n_2) = \left(\sum_{k_1=-\infty}^{n_1} h(n_1, k_1)U(k_1)\right)\left(\sum_{k_2=-\infty}^{n_2} h(n_2, k_2)U(k_2)\right)'$$
$$= \sum_{k_1=-\infty}^{n_1} \sum_{k_2=-\infty}^{n_2} h(n_1, k_1)U(k_1)U'(k_2)h'(n_2, k_2),$$

and taking expectations on both sides, we get

$$\begin{aligned}\mathrm{E}[Z(n_1)Z'(n_2)] &= \mathrm{E}\left[\sum_{k_1=-\infty}^{n_1} \sum_{k_2=-\infty}^{n_2} h(n_1, k_1)U(k_1)U'(k_2)h'(n_2, k_2)\right] \\ &= \sum_{k_1=-\infty}^{n_1} \sum_{k_2=-\infty}^{n_2} h(n_1, k_1)\mathrm{E}[U(k_1)U'(k_2)]h'(n_2, k_2).\end{aligned} \tag{9}$$

If $U(n) = 0$ for $n < 0$, all the sums involved in (6), (8), and (9) are finite, and so $Z(n)$ is a Gaussian process, and its mean and correlation functions are related to those of the input process by equations (8) and (9).

With a doubly infinite input process $U(n)$, $n = \ldots, -1, 0, 1, \ldots$, we only have to show that the MS limit in (7) exists. If it does, then the interchange of the expectation operator with the infinite series in (8) and (9) will be justified (see Chapter One, Exercise 1.48), and $Z(n)$ will again be a Gaussian process with mean and correlation function given by (8) and (9). We will show the existence of the MS limit for the scalar case only; for vector-valued processes the idea is the same, only the notation becomes somewhat unwieldy.

By the Loève criterion (see Chapter One), we know that the MS limit (7) will exist if and only if

$$\mathrm{E}\left[\left(\sum_{k_1=l_1}^{n} h(n, k_1)U(k_1)\right)\left(\sum_{k_2=l_2}^{n} h(n, k_2)U(k_2)\right)\right]$$
$$= \sum_{k_1=l_1}^{n} \sum_{k_2=l_2}^{n} h(n, k_1)\mathrm{E}[U(k_1)U(k_2)]h(n, k_2)$$

converges as $l_1 \to -\infty$ and $l_2 \to -\infty$. But this requires the convergence of the double infinite series

$$\sum_{k_1=-\infty}^{n} \sum_{k_2=-\infty}^{n} h(n, k_1)\mathrm{E}[(U(k_1)U(k_2)]h(n, k_2)$$

which will converge if it converges absolutely, that is, if

$$\sum_{k_1=-\infty}^{n} \sum_{k_2=-\infty}^{n} |h(n, k_1)|\, |E[U(k_1)U(k_2)]|\, |h(n, k_2)| < \infty. \tag{10}$$

Since, by the Schwarz inequality,

$$|E[U(k_1)U(k_2)]| \le \sqrt{E[U^2(k_1)]E[U^2(k_2)]}$$

condition (10) will be assured if for all $k \le n$, $E[U^2(k)] \le c$, a finite constant, because then (10) is upper bounded by

$$\sum_{k_1=-\infty}^{n} \sum_{k_2=-\infty}^{n} |h(n, k_1)|\, c\, |h(n, k_2)| = c \sum_{k=-\infty}^{n} |h(n, k)|^2,$$

which is finite by the stability assumption (3).

Let us now summarize all this in the following theorem.

THEOREM 2.2.2. *Let the input process* $U(n)$, $n = \ldots, -1, 0, 1, \ldots$, *be a discrete-time Gaussian process with mean function* $\mu_U(n)$ *and correlation function* $R_U(n_1, n_2)$.

If, for every n *there is a constant* c_n *such that the variance function*

$$\sigma_U^2(k) \le c_n \qquad \text{for all } k \le n \tag{11}$$

then the output process $Z(n)$, $n = \ldots, -1, 0, 1, \ldots$, *defined by equation* (6), *is also a Gaussian process.*

Its mean function $\mu_Z(n)$ *and correlation function* $R_Z(n_1, n_2)$ *are given by*

$$\mu_Z(n) = \sum_{k=-\infty}^{n} h(n, k)\mu_U(k), \qquad n = \ldots, -1, 0, 1, \ldots, \tag{12}$$

$$R_Z(n_1, n_2) = \sum_{k_1=-\infty}^{n_1} \sum_{k_2=-\infty}^{n_2} h(n_1, k_1)R_U(k_1, k_2)h'(n_2, k_2), \tag{13}$$

$$n_1 = \ldots, -1, 0, 1, \ldots, \qquad n_2 = \ldots, -1, 0, 1, \ldots.$$

Notice that condition (11) is quite natural for a physical system; it requires that all the inputs $U(k)$ that entered the system up to the present time n have bounded expected power. Of course, if $U(n) = 0$ for $n < 0$, (11) is satisfied automatically.

It is sometimes also useful to have an expression for the cross-correlation function between the output and the input. It can be easily obtained by multiplying equation (6) with $n = n_1$ by $U'(n_2)$ and taking expectations. The result is the relation

$$R_{Z,U}(n_1, n_2) = \sum_{k=-\infty}^{n_1} h(n_1, k) R_U(k, n_2), \tag{14}$$

$$n_1 = \ldots, -1, 0, 1, \ldots; \qquad n_2 = \ldots, -1, 0, 1, \ldots.$$

Example 2.2.1

Consider the discrete-time linear system shown in Figure 2.2.2. Here the clock produces a sequence of pulses at times $n = 3m$, $m = 0, \pm 1, \ldots$, which control the switch S. Thus the switch closes and discharges the capacitor at each discrete time instant which is a multiple of 3. Therefore, if a unit input arrives at time $n = 3m$, it has no effect whatsoever; if it arrives at time $n = 3m + 1$, the output sequence is 1, 2, 0, 0, ...; if it arrives at time $n = 3m + 2$, the output sequence is 1, 0, 0, ... Consequently, the impulse-response function of this system is given by

$$h(n, k) = \begin{cases} 1 & \text{if} \quad n = k = 3m + 1 \quad \text{or} \quad n = k = 3m + 2, \\ 2 & \text{if} \quad k = 3m + 1,\ n = 3m + 2, \\ 0 & \text{otherwise,} \end{cases}$$

where $m = 0, \pm 1, \ldots$. Clearly, the system is causal and stable.

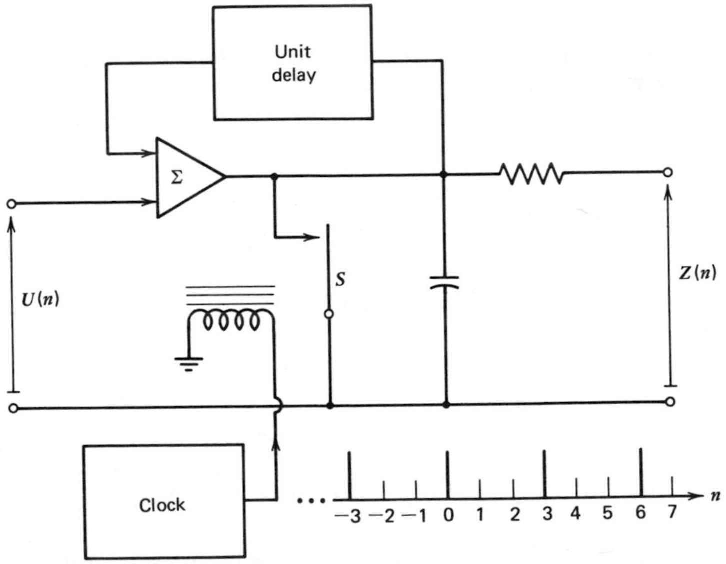

Figure 2.2.2

From (12) the output mean function is

$$\mu_Z(n) = \begin{cases} \mu_U(n) & \text{if } n = 3m + 1, \\ \mu_U(n) + 2\mu_U(n-1) & \text{if } n = 3m + 2, \\ 0 & \text{if } n = 3m, \ m = 0, \pm 1, \ldots. \end{cases}$$

Thus, for instance, if the input mean function is a constant, $\mu_U(n) = \mu$, the output mean function will follow the periodic sequence $\ldots, \mu, 3\mu, 0, \mu, 3\mu, 0, \ldots$, with the zeros corresponding to times that are multiples of 3.

From (13) the output correlation function is, for $n_1 \le n_2$, given by

$$R_Z(n_1, n_2) = \begin{cases} R_U(n_1, n_2) & \text{if } n_1 = n_2 = 3m + 1 \\ R_U(n_1, n_2) + 2R_U(n_1, n_2 - 1) & \text{if } n_1 = 3m + 1 \text{ and } n_2 = 3m + 2, \\ R_U(n_1, n_2) + 2R_U(n_1, n_2 - 1) + 2R_U(n_1 - 1, n_2) \\ \quad + 4R_U(n_1 - 1, n_2 - 1) & \text{if } n_1 = n_2 = 3m + 2 \text{ and} \\ 0 & \text{otherwise,} \end{cases}$$

where again $m = 0, \pm 1, \ldots$.

For instance, if the input process $U(n)$, $n = \ldots, -1, 0, 1, \ldots$, has correlation function

$$R_U(n_1, n_2) = \sigma^2 \gamma^{|n_1 - n_2|}, \qquad \sigma^2 > 0, \qquad |\gamma| < 1,$$

that is, if it is a stationary Gauss-Markov process, the output correlation function will be (for $n_1 \le n_2$)

$$R_Z(n_1, n_2) = \begin{cases} \sigma^2 & \text{if } n_1 = n_2 = 3m + 1, \\ \sigma^2(1 + 2\gamma) & \text{if } n_1 = 3m + 1, \ n_2 = 3m + 2, \\ \sigma^2(5 + 4\gamma) & \text{if } n_1 = n_2 = 3m + 2, \\ 0 & \text{otherwise.} \end{cases}$$

The output-input cross-correlation function can be easily obtained from (14); we leave this to the reader. ▲

Example 2.2.2

Suppose that a linear system operating in discrete time $n = 0, 1, \ldots$ has the impulse-response function

$$h(n, k) = \begin{cases} \dfrac{k+1}{n} \alpha^{n-k-1} & \text{if } 0 \le k < n, \\ 0 & \text{if } 0 \le n \le k, \end{cases} \tag{15}$$

where $|\alpha| < 1$. Let the input to this system be a discrete-time Gaussian process $U(n)$, $n = 0, 1, \ldots$, with mean function

$$\mu_U(n) = \cos n\pi, \qquad n = 0, 1, \ldots,$$

variance function

$$\sigma_U^2(n) = 1, \qquad n = 0, 1, \ldots, \tag{16}$$

and such that $U(n_1)$ and $U(n_2)$ are independent whenever $|n_1 - n_2| > 1$ while

$$\text{Cov}[U(n), U(n+1)] = \frac{(-1)^n}{2}, \qquad n = 0, 1, \ldots. \tag{17}$$

The mean function of the Gaussian output process

$$Z(n) = \sum_{k=0}^{n} h(n, k)U(k), \qquad n = 0, 1, \ldots$$

is readily obtained from (12). First, since by (15), $h(0, 0) = 0$, we have

$$Z(0) = 0 \qquad \text{so that} \qquad \mu_Z(0) = 0.$$

For $n > 0$, we get

$$\mu_Z(n) = \sum_{k=0}^{n-1} \frac{k+1}{n} \alpha^{n-k-1} \cos k\pi = \frac{\alpha^{n-1}}{n} \sum_{k=0}^{n-1} (k+1)\left(\frac{-1}{\alpha}\right)^k \tag{18}$$

since $\cos k\pi = (-1)^k$. The sum can be further simplified by noticing that for any real number x such that $x \neq 1$

$$\sum_{k=0}^{n-1} (k+1)x^k = \frac{d}{dx}\left(\sum_{k=0}^{n-1} x^{k+1}\right) = \frac{d}{dx}\left(x\frac{1-x^n}{1-x}\right) = \frac{1-(n+1)x^n + nx^{n+1}}{(1-x)^2}.$$

Putting $x = -1/\alpha$ and substituting into (18), we obtain, after a little algebra,

$$\mu_Z(n) = \frac{\alpha^{n+1} + (-1)^{n-1}[(n+1)\alpha + n]}{n(1+\alpha)^2}, \qquad n = 1, 2, \ldots$$

Notice that for large n we have approximately

$$\mu_Z(n) \doteq \frac{1}{1+\alpha}(-1)^{n-1} = \frac{1}{1+\alpha}\mu_U(n-1)$$

so that "in the steady state" the system behaves as a unit delay with an attenuator.

Let us next assume that the mean function $\mu_U(n)$ has been subtracted from the input process, so that now the input and hence also the output process both have zero mean.

It then follows from (16) and (17) that the correlation function of the input process equals

$$R_U(n_1, n_2) = \begin{cases} 1 & \text{if } n_2 = n_1, \\ \frac{1}{2}(-1)^{n_1} & \text{if } n_2 = n_1 + 1, \\ \frac{1}{2}(-1)^{n_2} & \text{if } n_2 = n_1 - 1, \\ 0 & \text{if } |n_1 - n_2| > 1, \end{cases}$$

where $n_1 = 0, 1, \ldots$ and $n_2 = 0, 1, \ldots$. Since $Z(0) = 0$, the correlation function $R_Z(n_1, n_2) = 0$ whenever either $n_1 = 0$ or $n_2 = 0$ (or both). For $n_1 > 0$ and $n_2 > 0$, we get from (13) and (15) the expression

$$R_Z(n_1, n_2) = \sum_{k_1=0}^{n_1-1} \sum_{k_2=0}^{n_2-1} \frac{k_1+1}{n_1} \frac{k_2+1}{n_2} \alpha^{n_1+n_2-k_1-k_2-2} R_U(k_1, k_2)$$

$$= \frac{\alpha^{n_1+n_2-2}}{n_1 n_2} \sum_{k_1=0}^{n_1-1} \sum_{k_2=0}^{n_2-1} (k_1+1)(k_2+1) \frac{R_U(k_1, k_2)}{\alpha^{k_1+k_2}}. \tag{19}$$

To evaluate the double sum, let us first consider the case $n_1 = n_2 = n$. Since $R_U(k_1, k_2) = 0$ unless $k_2 = k_1$ or $k_2 = k_1 \pm 1$, we can split the double sum into three single sums by first setting $k_1 = k_2 = k$ and summing over $k = 0, \ldots, n-1$, then setting $k_1 = k$, $k_2 = k-1$ and summing over $k = 1, \ldots, n-1$ and finally setting $k_2 = k$, $k_1 = k-1$ and again summing over $k = 1, \ldots, n-1$. We get

$$\sum_{k_1=0}^{n-1} \sum_{k_2=0}^{n-1} (k_1+1)(k_2+1) \frac{R_U(k_1, k_2)}{\alpha^{k_1+k_2}}$$

$$= \sum_{k=0}^{n-1} \frac{(k+1)^2}{\alpha^{2k}} + \frac{1}{2} \sum_{k=1}^{n-1} \frac{(k+1)k}{\alpha^{2k-1}} (-1)^{k-1} + \frac{1}{2} \sum_{k=1}^{n-1} \frac{k(k+1)}{\alpha^{2k-1}} (-1)^{k-1}, \tag{20}$$

with the last two terms absent if $n = 1$. The sums can again be evaluated by twice differentiating a finite geometric progression with some additional algebraic manipulation. By substituting the result into (19), we obtain an expression for the variance function $\sigma_Z^2(n) = R_Z(n, n)$. For $n = 1$, we have $\sigma_Z^2(1) = 1$, and for $n > 0$, we obtain the rather complicated expression

$$\sigma_Z^2(n) = \frac{1}{n^2} \left[\frac{\alpha^{2n+2}(\alpha^2+1) - (n+1)^2\alpha^4 + (2n^2+2n-1)\alpha^2 - n^2}{(\alpha^2-1)^3} \right.$$

$$\left. + \frac{2\alpha^{2n+3} + (-1)^n(n(n+1)\alpha^5 + 2(n^2-1)\alpha^3 + n(n-1)\alpha)}{(\alpha^2+1)^3} \right]. \tag{21}$$

For $n_1 \neq n_2$, for instance, if $0 < n_1 < n_2$, the double sum in (19) can again be written in the form of (20) with n replaced by n_1 and with an additional term

$$\frac{n_1(n_1+1)}{\alpha^{2n_1-1}} (-1)^{n_1}$$

in the last sum, since with $k_2 = k$ and $k_1 = k - 1$, the summation now ranges over $k = 1, \ldots, n_1$. For $n_1 = 1$, we obtain

$$R_Z(n_1, n_2) = \frac{\alpha^{n_2 - 2}}{n_2}\left(1 - \frac{2}{\alpha}\right), \qquad n_2 > 1.$$

For $n_1 > 1$ the correlation function equals

$$R_Z(n_1, n_2) = \frac{\alpha^{n_2 + 1}}{n_1 n_2}[\cdots] + (-1)^{n_1}\frac{(n_1 + 1)\alpha^{n_2 - n_1 - 1}}{n_2},$$

$n_2 > n_1$, where in the brackets we have the same expression as in (21) with n replaced by n_1.

We can also evaluate the output-input cross-correlation $R_{Z,U}(n_1, n_2)$ from (14), but we shall leave that as an exercise.

The reader should keep in mind, however, that in spite of the formidable expressions obtained here, if a numerical value for the parameter α were given, the calculation of the correlation function would be a relatively simple matter. One would then directly use formula (19) and calculate the double sum, for example, using a programmable calculator. ▲

Example 2.2.3

Consider a causal linear system with two input terminals and one output terminal, as shown in Figure 2.2.3. The impulse-response function $\mathbf{h}(n, k)$ is then a 1×2 matrix, and we assume that it is given by

$$\mathbf{h}(n, k) = \begin{cases} \left[\alpha^{n-k} \sin \dfrac{n\pi}{2} \quad \alpha^{n-k} \cos \dfrac{n\pi}{2}\right] & \text{if } k \le n, \\ [\quad 0 \qquad\qquad 0 \quad] & \text{if } k > n, \end{cases}$$

where α is a constant such that $|\alpha| < 1$. Clearly, this condition guarantees that the system is stable.

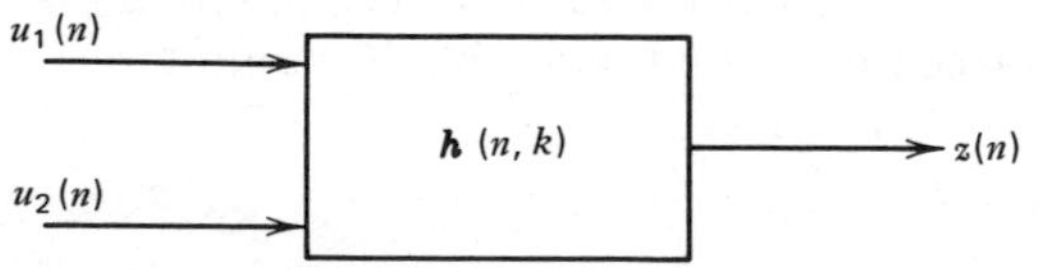

Figure 2.2.3

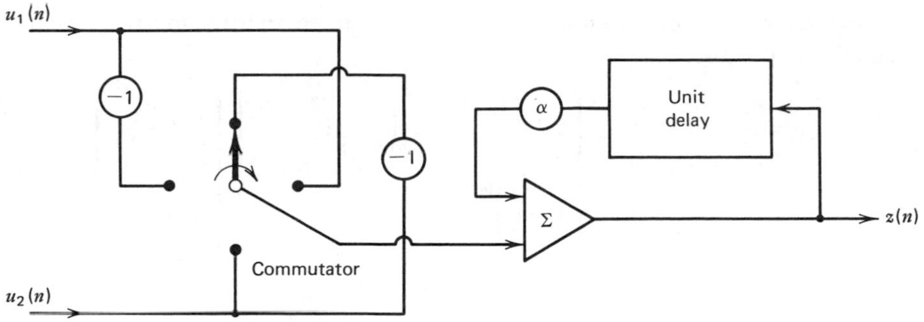

Figure 2.2.4

This system can be imagined to have an internal structure, as shown in Figure 2.2.4, where the commutator rotates in the clockwise direction one-quarter revolution every time unit and its position is shown at time $n = 0$.

Suppose now that the input to this system is a discrete-time stationary Gaussian process

$$\mathbf{U}(n) = \begin{pmatrix} U_1(n) \\ U_2(n) \end{pmatrix}, \qquad n = \ldots, -1, 0, 1, \ldots,$$

with zero mean function and with correlation function

$$\mathbf{R}_{\mathbf{U}}(n_1, n_2) = \begin{bmatrix} 2\gamma^{|n_1 - n_2|} & \left(-\dfrac{\gamma}{3}\right)^{|n_1 - n_2|} \\ \left(-\dfrac{\gamma}{3}\right)^{|n_1 - n_2|} & 2\gamma^{|n_1 - n_2|} \end{bmatrix}$$

$n_1 = \ldots, -1, 0, 1, \ldots$; $n_2 = \ldots, -1, 0, 1, \ldots$, where $|\gamma| < 1$ is a constant. According to (13) the correlation function of the output proccss $Z(n), n = \ldots, -1, 0, 1, \ldots$, will be given by

$$R_Z(n_1, n_2) = \sum_{k_1=-\infty}^{n_1} \sum_{k_2=-\infty}^{n_2} \left[\alpha^{n_1-k_1} \sin \frac{n_1 \pi}{2} \quad \alpha^{n_1-k_1} \cos \frac{n_1 \pi}{2}\right]$$

$$\times \begin{bmatrix} 2\gamma^{|k_1 - k_2|} & \left(-\dfrac{\gamma}{3}\right)^{|k_1 - k_2|} \\ \left(-\dfrac{\gamma}{3}\right)^{|k_1 - k_2|} & 2\gamma^{|k_1 - k_2|} \end{bmatrix} \begin{bmatrix} \alpha^{n_2-k_2} \sin \dfrac{n_2 \pi}{2} \\ \alpha^{n_2-k_2} \cos \dfrac{n_2 \pi}{2} \end{bmatrix}. \tag{22}$$

Noticing that the powers of α and γ can be factored from all three matrices, we first calculate the product

$$\begin{bmatrix} \sin \frac{n_1 \pi}{2} & \cos \frac{n_1 \pi}{2} \end{bmatrix} \begin{bmatrix} 2 & \left(-\frac{1}{3}\right)^{|k_1 - k_2|} \\ \left(-\frac{1}{3}\right)^{|k_1 - k_2|} & 2 \end{bmatrix} \begin{bmatrix} \sin \frac{n_2 \pi}{2} \\ \cos \frac{n_2 \pi}{2} \end{bmatrix}$$

$$= 2 \sin \frac{n_1 \pi}{2} \sin \frac{n_2 \pi}{2} + \left(-\frac{1}{3}\right)^{|k_1 - k_2|} \sin \frac{n_1 \pi}{2} \cos \frac{n_2 \pi}{2}$$
$$+ \left(-\frac{1}{3}\right)^{|k_1 - k_2|} \cos \frac{n_1 \pi}{2} \sin \frac{n_2 \pi}{2} + 2 \cos \frac{n_1 \pi}{2} \cos \frac{n_2 \pi}{2}$$
$$= \left(-\frac{1}{3}\right)^{|k_1 - k_2|} \sin(n_1 + n_2)\frac{\pi}{2} + 2 \cos(n_1 - n_2)\frac{\pi}{2}.$$

Substituting back into (22), we obtain

$$R_Z(n_1, n_2) = \sin(n_1 + n_2)\frac{\pi}{2} \sum_{k_1 = -\infty}^{n_1} \sum_{k_2 = -\infty}^{n_2} \alpha^{n_1 + n_2 - (k_1 + k_2)} \left(-\frac{\gamma}{3}\right)^{|k_1 - k_2|}$$
$$+ 2 \cos(n_1 - n_2)\frac{\pi}{2} \sum_{k_1 = -\infty}^{n_1} \sum_{k_2 = -\infty}^{n_2} \alpha^{n_1 + n_2 - (k_1 + k_2)} \gamma^{|k_1 - k_2|}, \quad (23)$$
$$n_1 = \ldots, -1, 0, 1, \ldots; n_2 = \ldots, -1, 0, 1, \ldots.$$

To calculate the double infinite series, let us take $n_1 \leq n_2$ and begin with the second series in the above expression. We can decompose it into a sum of three series

$$\sum_{k_1 = -\infty}^{n_1} \sum_{k_2 = -\infty}^{n_2} \alpha^{n_1 + n_2 - (k_1 + k_2)} \gamma^{|k_1 - k_2|} = \sum_{k = -\infty}^{n_1} \alpha^{n_1 + n_2 - 2k}$$
$$+ 2 \sum_{k_1 = -\infty}^{n_1 - 1} \sum_{k_2 = k_1 + 1}^{n_1} \alpha^{n_1 + n_2 - (k_1 + k_2)} \gamma^{k_2 - k_1}$$
$$+ \sum_{k_1 = -\infty}^{n_1} \sum_{k_2 = n_1 + 1}^{n_2} \alpha^{n_1 + n_2 - (k_1 + k_2)} \gamma^{k_2 - k_1},$$

where the last one is absent if $n_1 = n_2$. Using the formulas for the sum of a geometric series

$$\sum_{j=0}^{n} x^j = \frac{1 - x^{j+1}}{1 - x}, \qquad |x| \neq 1, \qquad \sum_{j=0}^{\infty} x^j = \frac{1}{1 - x}, \qquad |x| < 1,$$

we can now evaluate all three infinite series above. For the first one, we have

$$\sum_{k=-\infty}^{n_1} \alpha^{n_1+n_2-2k} = \alpha^{n_2-n_1} \sum_{k=-\infty}^{n_1} \alpha^{2(n_1-k_1)} = \frac{\alpha^{n_2-n_1}}{1-\alpha^2}.$$

The second one takes a little bit longer,

$$\begin{aligned}
\sum_{k_1=-\infty}^{n_1-1} \sum_{k_2=k_1+1}^{n_1} \alpha^{n_1+n_2-(k_1+k_2)}\gamma^{k_2-k_1} &= \sum_{k_1=-\infty}^{n_1-1} \alpha^{n_1+n_2-k_1}\gamma^{-k_1} \sum_{k_2=k_1+1}^{n_1} \left(\frac{\gamma}{\alpha}\right)^{k_2} \\
&= \sum_{k_1=-\infty}^{n_1-1} \alpha^{n_1+n_2-k_1}\gamma^{-k_1} \left(\frac{\gamma}{\alpha}\right)^{k_1+1} \frac{1-\left(\dfrac{\gamma}{\alpha}\right)^{n_1-k_1}}{1-\dfrac{\gamma}{\alpha}} \\
&= \frac{\gamma\alpha^{n_2-n_1}}{\alpha-\gamma}\left(\sum_{k_1=-\infty}^{n_1-1} \alpha^{2(n_1-k_1)} - \sum_{k_1=-\infty}^{n_1-1} (\alpha\gamma)^{n_1-k_1}\right) \\
&= \frac{\alpha^3\gamma^2\alpha^{n_2-n_1}}{(1-\alpha^2)(1-\alpha\gamma)(\alpha-\gamma)},
\end{aligned}$$

where we have made the assumption that $|\gamma| \neq |\alpha|$.

Under the same assumption, we obtain similarly for the third series

$$\sum_{k_1=-\infty}^{n_1} \sum_{k_2=n_1+1}^{n_2} \alpha^{n_1+n_2-(k_1+k_2)}\gamma^{k_2-k_1} = \frac{\gamma\alpha^{n_2-n_1}\left(1-\left(\dfrac{\gamma}{\alpha}\right)^{n_2-n_1}\right)}{(1-\alpha\gamma)(\alpha-\gamma)},$$

which gives the correct result (zero) even if $n_1 = n_2$. Putting everything together, we have

$$\sum_{k_1=-\infty}^{n_1} \sum_{k_2=-\infty}^{n_2} \alpha^{n_1+n_2-(k_1+k_2)}\gamma^{|k_1-k_2|} = \frac{\alpha^{n_2-n_1}}{(1-\alpha\gamma)(\alpha-\gamma)}\left[\gamma\left(1-\left(\frac{\alpha}{\gamma}\right)^{n_2-n_1}\right) + \frac{\alpha^3\gamma^2+(1-\alpha\gamma)(\alpha-\gamma)}{1-\alpha^2}\right]$$

$n_1 \leq n_2$, $|\gamma| \neq |\alpha|$, and by just replacing γ by $-\gamma/3$ in the above expression, we also get the sum of the first series in (23). Substituting, we finally obtain the desired output correlation function

$$\begin{aligned}
R_Z(n_1, n_2) = \sin(n_1+n_2)&\left[\frac{\pi}{2}\frac{\alpha^{|n_1-n_2|}}{1+\dfrac{\alpha\gamma}{3}}\frac{1+\dfrac{\alpha\gamma}{3}}{1-\alpha^2} - \frac{\gamma}{3}\left(1-\left(\frac{-\gamma}{3\alpha}\right)^{|n_1-n_2|}\right)\right] \\
&+ 2\cos(n_1-n_2)\frac{\pi}{2}\frac{\alpha^{|n_1-n_2|}}{1-\alpha\gamma}\left[\frac{1+\alpha\gamma}{1-\alpha^2} + \gamma\left(1-\left(\frac{\gamma}{\alpha}\right)^{|n_1-n_2|}\right)\right],
\end{aligned}$$

which holds for all $n_1 = \ldots, -1, 0, 1, \ldots$; $n_2 = \ldots, -1, 0, 1, \ldots$ under the assumptions

$$|\gamma| \neq |\alpha| \qquad \text{and} \qquad |\gamma| \neq 3|\alpha|.$$

We could go on and calculate the correlation function if either of these two assumptions is violated and also evaluate the cross-correlation function $R_{Z,U}$, but we shall leave this as an exercise. ▲

If the Gaussian input process is *stationary*, and the causal and stable linear system is *time-invariant* the relations (12) through (14) are further simplified.

If $U(n)$, $n = \ldots, -1, 0, 1, \ldots$, is a stationary Gaussian process with (constant) mean function $\mu_U(n) = \mu_U$ and with correlation function $R_U(n)$, $n = 0, \pm 1, \ldots$, condition (11) is satisfied automatically, (12) becomes simply

$$\mu_Z(n) = \sum_{k=-\infty}^{n} h(n-k)\mu_U = \mu_U \sum_{k=0}^{\infty} h(k),$$

while (13) can be rearranged as

$$R_Z(n_1, n_2) = \sum_{k_1=-\infty}^{n_1} \sum_{k_2=-\infty}^{n_2} h(n_1 - k_1)R_U(k_1 - k_2)h'(n_2 - k_2)$$

$$= \sum_{l_1=0}^{\infty} \sum_{l_2=0}^{\infty} h(l_1)R_U((n_1 - n_2) - (l_1 - l_2))h'(l_2),$$

where the last expression is obtained by substitution $n_1 - k_1 = l_1, n_2 - k_2 = l_2$. It is seen that the output process $Z(n)$, $n = \ldots, -1, 0, 1, \ldots$, then has a constant mean function and a correlation function depending only on the time difference $n_1 - n_2$, and being Gaussian, it follows that it is then also a stationary process.

We thus have:

Corollary 2.2.1

Let $U(n)$, $n = \ldots, -1, 0, 1, \ldots$, be a discrete-time stationary Gaussian process with mean function $\mu_U(n) = \mu_U$, and with correlation function $R_U(n)$, $n = 0, \pm 1, \ldots$. If this process is applied to the input of a causal, linear, stable, and time-invariant linear system with impulse-response function

$$h(n), \qquad n = 0, 1, \ldots,$$

then the output process $Z(n)$, $n = \ldots, -1, 0, 1, \ldots$, is also a discrete-time stationary Gaussian process with mean function

$$\mu_Z = \mu_U \sum_{k=0}^{\infty} h(k),$$

and with correlation function

$$R_Z(n) = \sum_{k_1=0}^{\infty} \sum_{k_2=0}^{\infty} h(k_1)R_U(n - (k_1 - k_2))h'(k_2), \tag{24}$$

$n = 0, \pm 1, \ldots$. The cross-correlation function between the output and the input is given by

$$R_{Z,U}(n) = \sum_{k=0}^{\infty} h(k)R_U(n - k), \qquad n = 0, \pm 1, \ldots. \tag{25}$$

Example 2.2.4

Suppose our linear system is designed to compute first order differences of the input sequence $u(n)$,

$$z(n) = \Delta u(n-1) = u(n) - u(n-1), \qquad n = \ldots, -1, 0, 1, \ldots$$

as shown in Figure 2.2.5. Clearly, this is a causal, stable, and time-invariant system, and its impulse-response function is

$$h(n) = \begin{cases} 1 & \text{if } n = 0, \\ -1 & \text{if } n = 1, \\ 0 & \text{if } n > 1. \end{cases}$$

Suppose now that the input to this system is a discrete-time stationary Gaussian process

$$U(n), \qquad n = \ldots, -1, 0, 1, \ldots$$

with zero mean function, and with correlation function

$$R_U(n) = \sigma^2 e^{-\alpha n^2}, \qquad n = 0, \pm 1, \ldots,$$

where $\sigma^2 > 0$ and $\alpha > 0$ are constants. Let us first evaluate the output-input cross-correlation function. Substituting into (25), we obtain

$$\begin{aligned} R_{Z,U}(n) = R_U(n) - R_U(n-1) &= \sigma^2 e^{-\alpha n^2} - \sigma^2 e^{-\alpha(n-1)^2} \\ &= \sigma^2 e^{-\alpha n^2}(1 - e^{2\alpha n - \alpha}), \end{aligned}$$

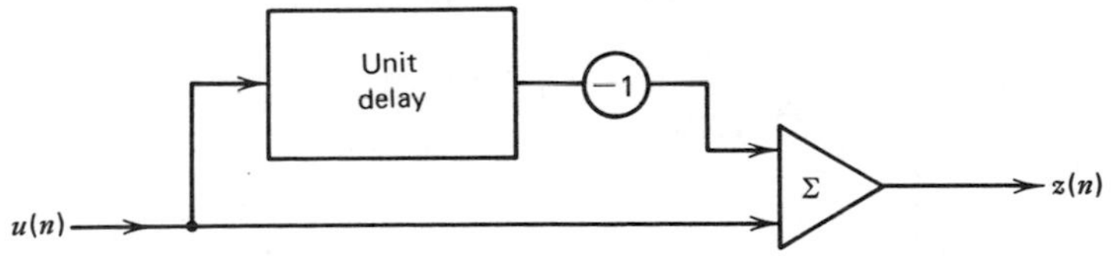

Figure 2.2.5

or

$$R_{Z,U}(n) = (1 - e^{2\alpha n - \alpha})R_U(n), \qquad n = 0, \pm 1, \ldots$$

The output correlation function $R_Z(n)$ can now be evaluated using equation (27).† We get

$$\begin{aligned} R_Z(n) &= R_{Z,U}(n) - R_{Z,U}(n+1) \\ &= \sigma^2 e^{-\alpha n^2} - \sigma^2 e^{-\alpha(n-1)^2} - \sigma^2 e^{-\alpha(n+1)^2} + \sigma^2 e^{-\alpha n^2} \\ &= 2\sigma^2 e^{-\alpha n^2} - \sigma^2 e^{-\alpha n^2}(e^{2\alpha n - \alpha} + e^{-2\alpha n - \alpha}) \\ &= 2\sigma^2 e^{-\alpha n^2} - 2\sigma^2 e^{-\alpha n^2} e^{-\alpha}\left(\frac{e^{2\alpha n} + e^{-2\alpha n}}{2}\right) \\ &= 2\sigma^2 e^{-\alpha n^2}\left(1 - \frac{1}{e^{\alpha}}\cosh 2\alpha n\right), \end{aligned}$$

or

$$R_Z(n) = 2\left(1 - \frac{1}{e^{\alpha}}\cosh 2\alpha n\right)R_U(n), \qquad n = 0, \pm 1, \ldots$$

The correlation functions $R_U(n)$, $R_Z(n)$, and the cross-correlation function $R_{Z,U}(n)$ are plotted (with $\sigma^2 = 1$ and $\alpha = .01$) in Figure 2.2.6. Notice that the cross-correlation is not quite symmetric about $n = 0$. ▲

Example 2.2.5

Suppose we have a linear causal time-invariant system with the impulse-response function

$$h(n) = n\alpha^{-n}, \qquad n = 0, 1, \ldots,$$

where $\alpha > 1$ is a constant. It may, for instance, represent an *RC* smoothing filter, as pictured in Figure 2.2.7, whose output is sampled at discrete, equally spaced time units.

Assume next that the input to this system is stationary discrete-time white Gaussian noise‡ $U(n)$, $n = \ldots, -1, 0, 1, \ldots$, with variance $\sigma^2 > 0$. The mean function of the stationary Gaussian output process $Z(n)$, $n = \ldots, -1, 0, 1, \ldots$, will of course be zero, while the correlation function is, by (24), simply

$$R_Z(n) = \sigma^2 \sum_{k_1=0}^{\infty} \sum_{k_2=0}^{\infty} k_1 \alpha^{-k_1} k_2 \alpha^{-k_2},$$

† See "Remark" following Example 2.2.5.

‡ See Definition 2.1.1.

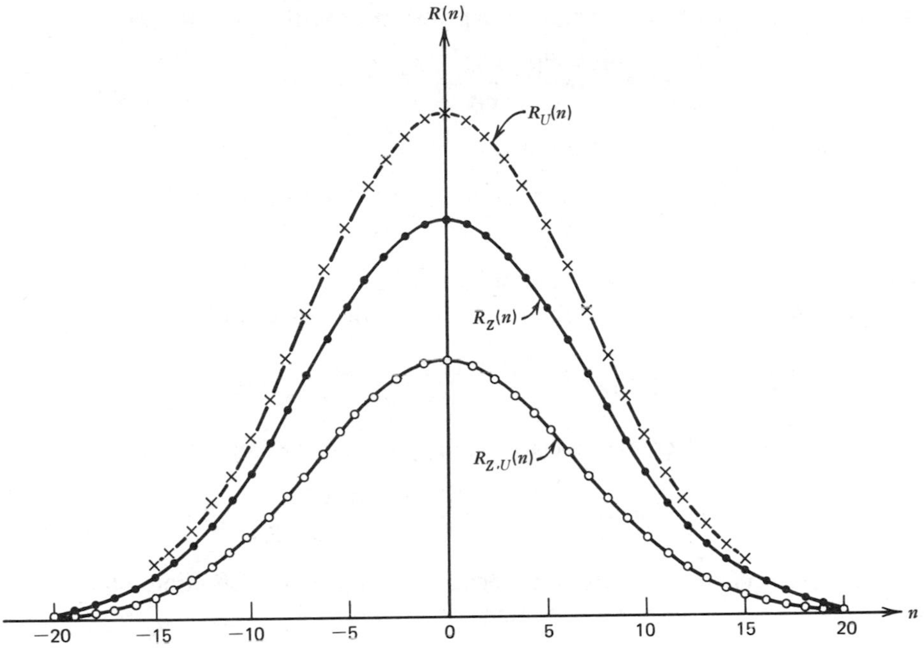

Figure 2.2.6

with the summation restricted to k_1 and k_2 such that $k_1 - k_2 = n$, since the correlation function $R_U(n) = 0$ for $n \neq 0$. For $n \geq 0$, this becomes by setting $k_1 = k$ and $k_2 = k - n$

$$\begin{aligned} R_Z(n) &= \sigma^2 \sum_{k=n}^{\infty} k(k-n)\alpha^{-2k+n} \\ &= \sigma^2\alpha^n\left(\sum_{k=n}^{\infty} k^2\alpha^{-2k} - n\sum_{k=n}^{\infty} k\alpha^{-2k}\right) \end{aligned} \tag{26}$$

To calculate the sums of the two infinite series, we note that for a real number $|x| < 1$ we obtain by differentiating a geometric progression just as in Example 2.2.2,

$$\sum_{k=0}^{\infty} kx^k = \frac{x}{(1-x)^2} \qquad \text{and} \qquad \sum_{k=0}^{n-1} kx^k = \frac{x - nx^n + (n-1)x^{n+1}}{(1-x)^2},$$

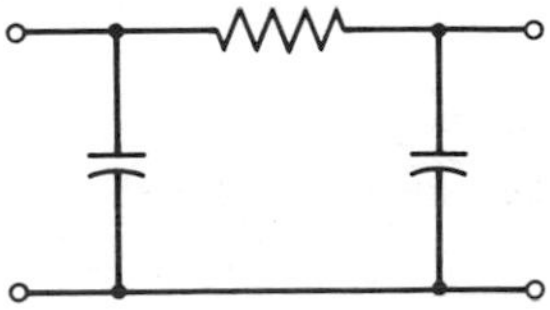

Figure 2.2.7

and hence by subtracting the second expression from the first, we get

$$\sum_{k=n}^{\infty} kx^k = \frac{nx^n - (n-1)x^{n+1}}{(1-x)^2}, \qquad n = 0, 1, \ldots.$$

By a similar technique, we also obtain

$$\sum_{k=n}^{\infty} k^2x^k = \frac{n^2x^n - (2n^2 - 2n - 1)x^{n+1} + (n-1)^2x^{n+2}}{(1-x)^3},$$

$n = 0, 1, \ldots$. Putting $x = \alpha^{-2}$ and substituting into (26), we arrive, after some algebraic manipulation, at the final formula for the correlation function

$$R_Z(n) = \sigma^2\alpha^2(|n| + 1)\frac{\alpha^2 - |n| + 1}{(\alpha^2 - 1)^3}\alpha^{-|n|}, \qquad n = 0, \pm 1, \ldots$$

For the output-input cross-correlation function we have immediately, from (25),

$$R_{Z,U}(n) = \begin{cases} \sigma^2 h(n) & \text{if } n = 0, 1, \ldots, \\ 0 & \text{if } n = -1, -2, \ldots \end{cases}$$

The cross-correlation is zero for negative n, since by definition of a cross-correlation function between stationary processes

$$R_{Z,U}(n_1 - n_2) = \mathrm{E}[Z(n_1)U(n_2)]$$

and if $n = n_1 - n_2$ is negative, that is, if $n_1 < n_2$, then $\mathrm{E}[Z(n_1)U(n_2)] = \mathrm{E}[Z(n_1)]\mathrm{E}[U(n_2)] = 0$. The reason for this is that for a causal system a present output $Z(n_1)$ can only depend on a future input $U(n_2)$ via a possible correlation between future, past, and present input, which is zero if the input is white Gaussian noise.

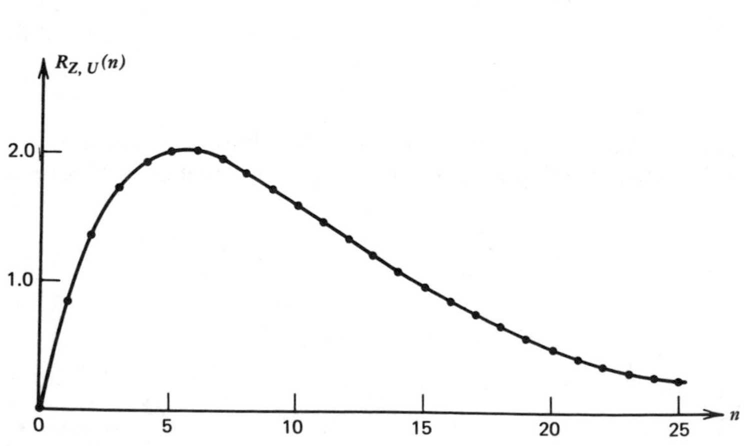

Figure 2.2.8

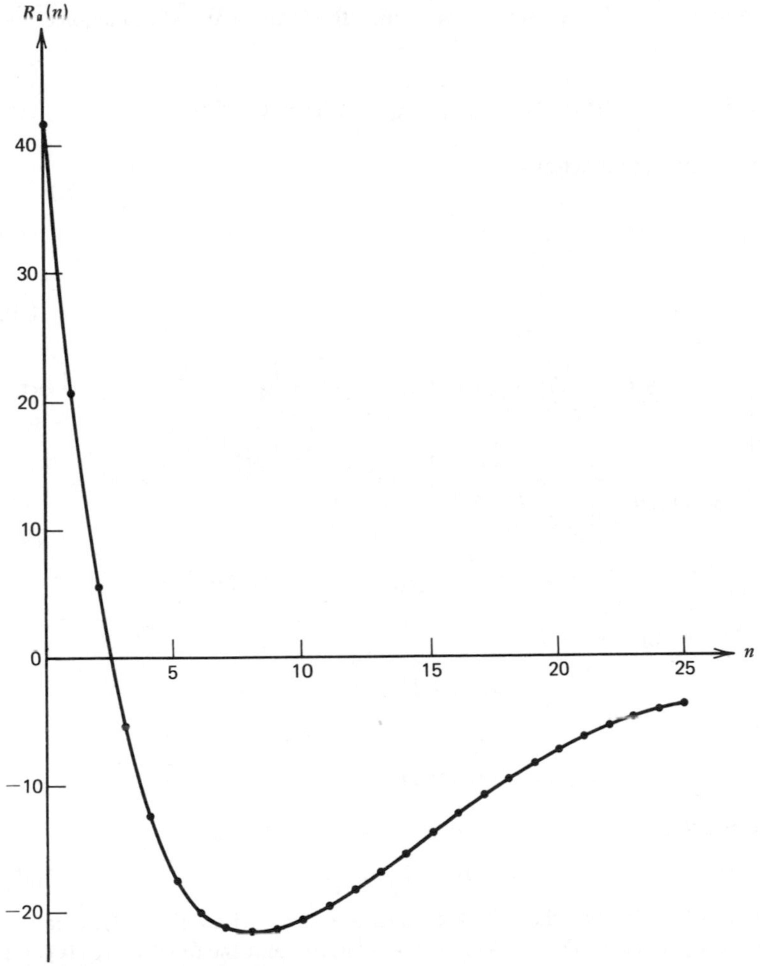

Figure 2.2.9

The functions $R_Z(n)$ and $R_{Z,U}(n)$ are plotted in Figures 2.2.8 and 2.2.9 with $\sigma^2 = 1$ and $\alpha = 1.2$. ▲

Remark: Notice that with the aid of (25) we can write (24) as

$$R_Z(n) = \sum_{k=0}^{\infty} R_{Z,U}(n+k)h'(k), \qquad n = 0, \pm 1, \ldots. \tag{27}$$

This allows us to evaluate the output and the cross-correlation function by employing various transform techniques. For example, if we define the familiar *transfer function* of the linear system by

$$H(\nu) = \sum_{n=0}^{\infty} h(n)e^{-\iota\omega n}, \qquad |\nu| < \frac{1}{2} \qquad \text{where } \omega = 2\pi\nu \tag{28}$$

and if we similarly define the functions

$$S_U(\nu) = \sum_{n=-\infty}^{\infty} R_U(n)e^{-\iota\omega n}, \tag{29}$$

$$S_{Z,U}(\nu) = \sum_{n=-\infty}^{\infty} R_{Z,U}(n)e^{-\iota\omega n}, \tag{30}$$

$$S_Z(\nu) = \sum_{n=-\infty}^{\infty} R_Z(n)e^{-\iota\omega n}, \qquad |\nu| < \frac{1}{2}\dagger, \tag{31}$$

we get, from (25),

$$\begin{aligned} S_{Z,U}(\nu) &= \sum_{n=-\infty}^{\infty} \sum_{k=0}^{\infty} h(k)R_U(n-k)e^{-\iota\omega k}e^{-\iota\omega(n-k)} \\ &= \sum_{k=0}^{\infty} h(k)e^{-\iota\omega k} \sum_{l=-\infty}^{\infty} R_U(l)e^{-\iota\omega l} = H(\nu)S_U(\nu). \end{aligned} \tag{32}$$

Similarly, from (27), we obtain

$$S_Z(\nu) = S_{Z,U}(\nu)H'(-\nu), \tag{33}$$

and combining these

$$S_Z(\nu) = H(\nu)S_U(\nu)H'(-\nu). \tag{34}$$

For scalar-valued processes, the latter relation can also be written as

$$S_Z(\nu) = |H(\nu)|^2 S_U(\nu), \tag{35}$$

because, as seen from (28), $H(-\nu)$ is the complex conjugate of $H(\nu)$. Equations (32) through (35) are known as (discrete-time) *Wiener-Khintchine relations*, and the functions $S_U(\nu)$ and $S_Z(\nu)$ are called *power spectra* of the input and output process, respectively.

A reader familiar with Fourier series will recognize from (31) that the values $R_Z(n)$, $n = 0, \pm 1, \ldots,$ of the correlation function are, in fact, coefficients of the Fourier series expansion of the function $S_Z(\nu)$, $|\nu| < \frac{1}{2}$. Consequently, the correlation function $R_Z(n)$ can be recovered from the power spectrum $S_Z(\nu)$ by the formula

$$R_Z(n) = \int_{-\pi}^{\pi} S_Z(\nu)e^{\iota\omega n}\, d\nu, \qquad n = 0, 1, \ldots. \qquad Я$$

† The existence of the transfer function (28) is assured by the stability condition (5), while the existence of the remaining three functions is a consequence of the properties of correlation functions—a result known as Herglotz lemma. Note that (28) and (30) are generally complex-valued functions. However, (29) and (31) are real valued because of the symmetry of a correlation function.

Example 2.2.6
Let us calculate the transfer function

$$H(\nu), \qquad |\nu| < \frac{1}{2}$$

for the time-invariant system of Example 2.2.5. From (28)

$$H(\nu) = \sum_{n=0}^{\infty} n\alpha^{-n}e^{-\iota\omega n} = \sum_{n=0}^{\infty} n\left(\frac{1}{\alpha e^{\iota\omega}}\right)^n = \frac{\dfrac{1}{\alpha e^{\iota\omega}}}{\left(1 - \dfrac{1}{\alpha e^{\iota\omega}}\right)^2}$$

$$= \frac{\alpha e^{\iota\omega}}{(\alpha e^{\iota\omega} - 1)^2}, \qquad |\nu| < \frac{1}{2},$$

where we have again used the formula

$$\sum_{k=0}^{\infty} kx^k = \frac{x}{(1-x)^2}, \qquad |x| < 1,$$

which is also valid for x a complex number.

For white Gaussian noise $U(n)$, $n = \ldots, -1, 0, 1, \ldots$, with variance $\sigma^2 > 0$, we obtain from (29) the power spectrum

$$S_U(\gamma) = \sum_{n=-\infty}^{\infty} R_U(n)e^{-\iota\omega n} = \sigma^2, \qquad |\nu| < \frac{1}{2}$$

since $R_U(0) = \sigma^2$ and $R_U(n) = 0$ for $n \neq 0$. By (32) then, $S_{Z,U}(\nu) = \sigma^2 H(\nu)$, $|\nu| < \frac{1}{2}$, and by 34), the output power spectrum is

$$S_Z(\nu) = \sigma^2 H(\nu)H(-\nu) = \sigma^2 \frac{\alpha e^{\iota\omega}}{(\alpha e^{\iota\omega} - 1)^2} \frac{\alpha e^{-\iota\omega}}{(\alpha e^{-\iota\omega} - 1)^2}$$

$$= \frac{\sigma^2\alpha^2}{(\alpha^2 - 2\alpha\cos\omega + 1)^2}, \qquad |\nu| < \frac{1}{2},$$

since $\alpha e^{\iota\omega} + \alpha e^{-\iota\omega} = 2\alpha\cos\omega$. Notice that while $H(\nu)$ and $S_{Z,U}(\nu)$ are complex-valued functions, $S_U(\nu)$ and $S_Z(\nu)$ are both real valued. ▲

Another common way to define a linear discrete-time system is by means of the state-space representation:

$$\mathbf{x}(n+1) = \mathbf{a}(n)\mathbf{x}(n) + \mathbf{b}(n)\mathbf{u}(n) \tag{36a}$$

$$\mathbf{z}(n) = \mathbf{c}(n)\mathbf{x}(n), \qquad n = 0, 1, \ldots, \tag{36b}$$

together with an initial state $\mathbf{x}(0)$.

Here again, $\mathbf{u}(n)$ is the input sequence, $\mathbf{z}(n)$ is the output sequence, and $\mathbf{x}(n)$ is a sequence of internal states of the system. Typically, all three sequences are vector valued, so that $\mathbf{a}(n)$, $\mathbf{b}(n)$, and $\mathbf{c}(n)$, $n = 0, 1, \ldots$, in (36) are matrices of appropriate order.

If now the input sequence is a discrete-time random process, so will the sequence of states be, as will the sequence of outputs of the system. In addition, the initial state $\mathbf{x}(0)$ may also be a random variable. Hence, we can again seek the relations between the first and second order moment functions of these processes. Before we do so, a few remarks are in order.

First, the state-output equation (36b) represents a linear memoryless transformation (see Chapter One). Hence, we immediately have

$$\begin{aligned} \boldsymbol{\mu}_{\mathbf{z}}(n) &= \mathbf{c}(n)\boldsymbol{\mu}_{\mathbf{x}}(n), \qquad n = 0, 1, \ldots, \\ \mathbf{R}_{\mathbf{z}}(n_1, n_2) &= \mathbf{c}(n_1)\mathbf{R}_{\mathbf{x}}(n_1, n_2)\mathbf{c}'(n_2), \\ \mathbf{R}_{\mathbf{z},\mathbf{x}}(n_1, n_2) &= \mathbf{c}(n_1)\mathbf{R}_{\mathbf{x}}(n_1, n_2), \qquad n = 0, 1, \ldots;\ n_2 = 0, 1, \ldots \end{aligned} \tag{37}$$

Therefore we now concentrate on the input-state equation (36a). By introducing the system transition matrix $\boldsymbol{\phi}(n, k)$ defined for $0 \leq k \leq n$ by the recurrence

$$\begin{aligned} \boldsymbol{\phi}(n + 1, k) &= \mathbf{a}(n)\boldsymbol{\phi}(n, k), \\ \boldsymbol{\phi}(n, n) &= \mathbf{I}\dagger, \end{aligned} \tag{38}$$

the identity matrix, we get from (36a) for $n = 1, 2, \ldots, k = 0, \ldots, n - 1$,

$$\mathbf{x}(n) = \boldsymbol{\phi}(n, k)\mathbf{x}(k) + \sum_{l=k}^{n-1} \boldsymbol{\phi}(n, l + 1)\mathbf{b}(l)\mathbf{u}(l) \tag{39}$$

and in particular taking $k = 0$

$$\mathbf{x}(n) = \boldsymbol{\phi}(n, 0)\mathbf{x}(0) + \sum_{l=0}^{n-1} \boldsymbol{\phi}(n, l + 1)\mathbf{b}(l)\mathbf{u}(l), \qquad n = 1, 2, \ldots$$

But this expression has the general form

$$\mathbf{x}(n) = \sum_{k=0}^{n} \mathbf{h}(n, k)\tilde{\mathbf{u}}(k), \qquad n = 0, 1, \ldots, \tag{40}$$

where $\tilde{\mathbf{u}}$ is an augmented vector

$$\tilde{\mathbf{u}}(n) = \begin{pmatrix} \mathbf{x}(0) \\ \mathbf{u}(n) \end{pmatrix}, \tag{41}$$

† In other words, $\boldsymbol{\phi}(n + 1, k) = \mathbf{a}(n)\mathbf{a}(n - 1) \cdots \mathbf{a}(k)$ for $0 \leq k \leq n$.

and $\mathbf{h}$ is an augmented matrix

$$\begin{aligned} \mathbf{h}(n, k) &= [\boldsymbol{\phi}(n, 0) \mid \boldsymbol{\phi}(n, k+1)\mathbf{b}(k)], \qquad 0 \le k < n \\ \mathbf{h}(n, n) &= [\boldsymbol{\phi}(n, 0) \mid \mathbf{0}], \qquad n = 0, 1, \ldots. \end{aligned} \tag{42}$$

This way, the state-input equation (36a) is converted into relation (40) with modified input (41) and impulse-response function given by (42). Thus Theorem 2.2.1, equations (12) through (14) and equation (37) can be used to obtain all desired relations between the first and second order moment functions of the three discrete-time processes, $\mathbf{U}(n)$, $\mathbf{X}(n)$, and $\mathbf{Z}(n)$. Note that since, in this case, $\mathbf{U}(n)$ is defined only for $n = 0, 1, \ldots$, the assumption that $\mathbf{U}(n)$, $n = 0, 1, \ldots$, is a discrete-time Gaussian process and that the initial state $\mathbf{X}(0)$ (if random) is also Gaussian implies that both $\mathbf{X}(n)$ and $\mathbf{Z}(n)$, $n = 0, 1, \ldots$, are also Gaussian processes. No further condition on the input process is needed.

Example 2.2.7

Consider a linear system with the input-state equation

$$x(n+1) = \alpha\left(1 - \frac{1}{n+1}\right)x(n) + u(n), \qquad n = 0, 1, \ldots, \tag{43}$$

and initial state $x(0) = 0$.

The system transition function is easily evaluated, since

$$a(n) = \frac{n}{n+1}\alpha, \qquad n = 0, 1, \ldots$$

and we have for $0 \le k \le n$

$$\begin{aligned} \phi(n, k) &= \frac{n-1}{n}\alpha\frac{n-2}{n-1}\alpha \cdots \frac{k+1}{k+2}\alpha\frac{k}{k+1}\alpha \\ &= \frac{k}{n}\alpha^{n-k}. \end{aligned}$$

Thus by (22) we can write

$$\begin{aligned} \mathbf{h}(n, k) &= \left[0 \quad \frac{k+1}{n}\alpha^{n-k-1}\right], \\ \mathbf{h}(n, n) &= [0 \quad 0], \qquad \text{since } \phi(n, 0) = 0. \end{aligned}$$

Consequently, equation (40) becomes

$$\begin{aligned} x(n) &= \sum_{k=0}^{n-1}\left[0 \quad \frac{k+1}{n}\alpha^{n-k-1}\right]\begin{bmatrix} x(0) \\ u(k) \end{bmatrix} \\ &= \sum_{k=0}^{n-1}\frac{k+1}{n}\alpha^{n-k-1}u(k), \qquad n = 1, 2, \ldots \end{aligned}$$

and $x(0) = 0$. Thus the impulse-response function for the linear system (43) is the same as the one in Example 2.2.2. ▲

In applications, we often encounter a special case, namely that the input process $\mathbf{U}(n)$, $n = 0, 1, \ldots$, to the discrete-time linear system given in the state-space representation is *white Gaussian noise*† with variance function $\boldsymbol{\sigma}_{\mathbf{U}}^2(n)$, $n = 0, 1, \ldots$.

As before, we will consider only the state-input equation (36a)

$$\mathbf{X}(n+1) = \mathbf{a}(n)\mathbf{X}(n) + \mathbf{b}(n)\mathbf{U}(n), \qquad n = 0, 1, \ldots, \tag{44}$$

with a Gaussian initial state $\mathbf{X}(0)$, possibly degenerate to a constant vector $\mathbf{X}(0) = \mathbf{x}(0)$.

Since, by definition, the mean function of white Gaussian noise is identically zero, we have, by taking the expectation of (44) and using (38),

$$\boldsymbol{\mu}_{\mathbf{X}}(n+1) = \mathbf{a}(n)\boldsymbol{\mu}_{\mathbf{X}}(n) = \boldsymbol{\phi}(n+1, 0)\boldsymbol{\mu}_{\mathbf{X}}(0), \qquad n = 0, 1, \ldots, \tag{45}$$

where $\boldsymbol{\mu}_{\mathbf{X}}(0) = E[\mathbf{X}(0)]$. This takes care of the mean function, and since the system is linear, we can easily modify the state sequence $\mathbf{X}(n)$ by subtracting from each $\mathbf{X}(n)$ the mean $\boldsymbol{\mu}_{\mathbf{X}}(n)$ as computed in (45). Thus, we will assume from now on that for all $n = 0, 1, \ldots$

$$\boldsymbol{\mu}_{\mathbf{X}}(n) = \boldsymbol{\phi}(n+1, 0)E[\mathbf{X}(0)] = \mathbf{0}.$$

It remains to calculate the correlation function $\mathbf{R}_{\mathbf{X}}(n_1, n_2)$. Before we start, we make two more simplifying assumptions.

First, we assume that the matrices $\mathbf{b}(n)$ in (44) are all identities, so that (44) becomes

$$\mathbf{X}(n+1) = \mathbf{a}(n)\mathbf{X}(n) + \mathbf{U}(n), \qquad n = 0, 1, \ldots \tag{46}$$

There is actually no loss of generality in this assumption, since if $\mathbf{U}(n)$, $n = 0, 1, \ldots$, is white Gaussian noise with variance function $\boldsymbol{\sigma}_{\mathbf{U}}^2(n)$, then $\mathbf{b}(n)\mathbf{U}(n)$, $n = 0, 1, \ldots$, is again white Gaussian noise with variance function

$$\mathbf{b}(n)\boldsymbol{\sigma}_{\mathbf{U}}^2(n)\mathbf{b}'(n), \qquad n = 0, 1, \ldots.$$

Thus, passing from (44) to (46) simply means replacing the original white Gaussian noise by new white Gaussian noise.

Let us therefore work from now on with the simplified equation (46). To avoid introducing new symbols, we will still call the variance function of the white Gaussian noise $\boldsymbol{\sigma}_{\mathbf{U}}^2(n)$, and no confusion should arise.

Second, we assume that the *initial state* $\mathbf{X}(0)$ *is independent of the white*

† See Definition 2.1.1.

Gaussian noise $\mathbf{U}(n)$, $n = 0, 1, \ldots$. (If $\mathbf{X}(0) = \mathbf{x}(0)$ is not random, this is always true.)

As a consequence, the "present" noise input $\mathbf{U}(n)$ will be independent of both the "present and future" state $\mathbf{X}(k)$, $k \geq n$; that is, we will have

$$\mathrm{E}[\mathbf{U}(n)\mathbf{X}'(k)] = \mathbf{0} \qquad \text{for all } 0 \leq k \leq n. \tag{47}$$

This can be easily verified by induction from equation (46). Multiplying $\mathbf{U}(n)$ by $\mathbf{X}'(k)$, $1 \leq k \leq n$, and taking expectations, we get

$$\mathrm{E}[\mathbf{U}(n)\mathbf{X}'(k)] = \mathrm{E}[\mathbf{U}(n)\mathbf{X}'(k-1)]\mathbf{a}'(k-1)$$

since $\mathrm{E}[\mathbf{U}(n)\mathbf{U}'(k-1)] = \mathbf{0}$, and hence the independence of the initial state $\mathrm{E}[\mathbf{U}(n)\mathbf{X}'(0)] = \mathbf{0}$ implies $\mathrm{E}[\mathbf{U}(n)\mathbf{X}'(1)] = \mathbf{0}$, and so on up to $\mathrm{E}[\mathbf{U}(n)\mathbf{X}'(n)] = 0$, when $k = n$. Using (47), we can now obtain a recurrence relation for the variance function

$$\boldsymbol{\sigma}_{\mathbf{X}}^2(n) = \mathrm{E}[\mathbf{X}(n)\mathbf{X}'(n)].$$

From equation (46) we have, for all $n = 0, 1, \ldots$,

$$\begin{aligned}\mathbf{X}(n+1)\mathbf{X}'(n+1) = {} & \mathbf{a}(n)\mathbf{X}(n)\mathbf{X}'(n)\mathbf{a}'(n) + \mathbf{a}(n)\mathbf{X}(n)\mathbf{U}'(n) \\ & + \mathbf{U}(n)\mathbf{X}'(n)\mathbf{a}'(n) + \mathbf{U}(n)\mathbf{U}'(n).\end{aligned}$$

Taking expectations and using (47) with $k = n$, that is,

$$\mathrm{E}[\mathbf{X}(n)\mathbf{U}'(n)] = \mathrm{E}[\mathbf{U}(n)\mathbf{X}'(n)] = \mathbf{0},$$

we obtain

$$\boldsymbol{\sigma}_{\mathbf{X}}^2(n+1) = \mathbf{a}(n)\boldsymbol{\sigma}_{\mathbf{X}}^2(n)\mathbf{a}'(n) + \boldsymbol{\sigma}_{\mathbf{U}}^2(n), \qquad n = 0, 1, \ldots \tag{48}$$

where $\boldsymbol{\sigma}_{\mathbf{X}}^2(0) = \mathrm{E}[\mathbf{X}(0)\mathbf{X}'(0)]$ is the variance matrix of the initial state.

For the correlation function

$$\mathbf{R}_{\mathbf{X}}(n, k) = \mathrm{E}[\mathbf{X}(n)\mathbf{X}'(k)], \qquad 0 \leq k < n,$$

we use equation (39) with $\mathbf{b}(n) = \mathbf{I}$, that is,

$$\mathbf{X}(n) = \boldsymbol{\phi}(n, k)\mathbf{X}(k) + \sum_{l=k}^{n-1} \boldsymbol{\phi}(n, l+1)\mathbf{U}(l),$$

multiply by $\mathbf{X}'(k)$ and take the expectation. We get

$$\mathrm{E}[\mathbf{X}(n)\mathbf{X}'(k)] = \boldsymbol{\phi}(n, k)\mathrm{E}[\mathbf{X}(k)\mathbf{X}'(k)] + \sum_{l=k}^{n-1} \boldsymbol{\phi}(n, l+1)\mathrm{E}[\mathbf{U}(l)\mathbf{X}'(k)]. \tag{49}$$

But by (47) the expectations behind the summation sign are all zero, and so (49) becomes

$$\mathbf{R}_{\mathbf{X}}(n, k) = \boldsymbol{\phi}(n, k)\boldsymbol{\sigma}_{\mathbf{X}}^2(k), \qquad 0 \leq k < n. \tag{50}$$

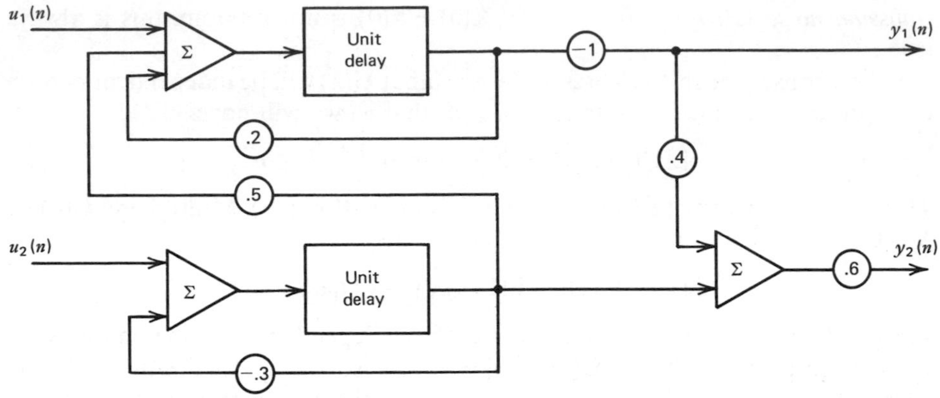

Figure 2.2.10

Since $\mathbf{R}'_{\mathbf{X}}(k, n) = \mathbf{R}_{\mathbf{X}}(n, k)$ and $\mathbf{R}_{\mathbf{X}}(n, n) = \boldsymbol{\sigma}^2_{\mathbf{X}}(n)$ the recurrence relation (48), together with (50), allows us to evaluate the entire correlation function of the process $\mathbf{X}(n)$, $n = 0, 1, \ldots$.

Remark: By repeated substitution into the recurrence (48) it is possible to obtain an explicit formula for $\boldsymbol{\sigma}^2_{\mathbf{X}}(n)$ and hence also for $\mathbf{R}_{\mathbf{X}}(n, k)$ solely in terms of the matrices $\mathbf{a}(n)$, $\boldsymbol{\sigma}^2_{\mathbf{U}}(n)$ $n = 0, 1, \ldots$ and $\boldsymbol{\sigma}^2_{\mathbf{X}}(0)$. However, equations (48) and (50) are better suited for actual calculation, especially on a computer. Я

Example 2.2.8

Consider a linear discrete-type system, as shown in Figure 2.2.10. The state space representation is easily seen to be

$$\begin{pmatrix} X_1(n+1) \\ X_2(n+1) \end{pmatrix} = \begin{bmatrix} .2 & .5 \\ 0 & -.3 \end{bmatrix} \begin{pmatrix} x_1(n) \\ x_2(n) \end{pmatrix} + \begin{pmatrix} u_1(n) \\ u_2(n) \end{pmatrix},$$

$$\begin{pmatrix} y_1(n) \\ y_2(n) \end{pmatrix} = \begin{bmatrix} -1 & 0 \\ -.24 & .6 \end{bmatrix} \begin{pmatrix} x_1(n) \\ x_2(n) \end{pmatrix}, \qquad n = 0, 1, \ldots.$$

Let the initial state

$$\begin{pmatrix} X_1(0) \\ X_2(0) \end{pmatrix}$$

be Gaussian with zero mean vector and with covariance matrix

$$\boldsymbol{\sigma}^2_{\mathbf{X}}(0) = \begin{bmatrix} 0 & 0 \\ 0 & 1 \end{bmatrix},$$

and assume that the input

$$\begin{pmatrix} U_1(n) \\ U_2(n) \end{pmatrix}, \qquad n = 0, 1, \ldots,$$

is discrete-time stationary white Gaussian noise with variance function

$$\boldsymbol{\sigma}_{\mathbf{U}}^2(n) = \begin{bmatrix} 1.2 & .7 \\ .7 & .8 \end{bmatrix}, \qquad n = 0, 1, \ldots.$$

The recurrence relation (48) for the variance function of the state process

$$\begin{pmatrix} X_1(n) \\ X_2(n) \end{pmatrix}, \qquad n = 0, 1, \ldots,$$

takes the form

$$\boldsymbol{\sigma}_{\mathbf{X}}^2(n+1) = \begin{bmatrix} .2 & .5 \\ 0 & -.3 \end{bmatrix} \boldsymbol{\sigma}_{\mathbf{X}}^2(n) \begin{bmatrix} .2 & 0 \\ .5 & -.3 \end{bmatrix} + \begin{bmatrix} 1.2 & .7 \\ .7 & .8 \end{bmatrix}, \qquad n = 0, 1, \ldots,$$

$$\boldsymbol{\sigma}_{\mathbf{X}}^2(0) = \begin{bmatrix} 0 & 0 \\ 0 & 1 \end{bmatrix}.$$

The successive terms of the sequence of matrices $\boldsymbol{\sigma}_{\mathbf{X}}^2(0)$, $\boldsymbol{\sigma}_{\mathbf{X}}^2(1)$, ... can be easily evaluated even with a small calculator. The first five terms are

$$\begin{bmatrix} 0 & 0 \\ 0 & 1 \end{bmatrix}, \begin{bmatrix} 1.45 & .55 \\ .55 & .89 \end{bmatrix}, \begin{bmatrix} 1.59 & .53 \\ .53 & .88 \end{bmatrix}, \begin{bmatrix} 1.59 & .54 \\ .54 & .88 \end{bmatrix}, \begin{bmatrix} 1.59 & .54 \\ .54 & .88 \end{bmatrix}$$

The correlation function of the state process is obtained from equation (50),

$$\mathbf{R}_{\mathbf{X}}(n, k) = \boldsymbol{\phi}(n, k)\boldsymbol{\sigma}_{\mathbf{X}}^2(k), \qquad 0 \le k \le n.$$

The state transition matrix $\boldsymbol{\phi}(n, k)$ for this system is just the $(n-k)$th power of the matrix $\mathbf{a}(n) = \mathbf{a}$

$$\boldsymbol{\phi}(n, k) = \begin{bmatrix} .2 & .5 \\ 0 & -.3 \end{bmatrix}^{n-k}, \qquad 0 \le k \le n.$$

However, the mth power of a 2×2 matrix of the form

$$\mathbf{a} = \begin{bmatrix} \alpha & \beta \\ 0 & \gamma \end{bmatrix}, \qquad \text{where } \alpha \ne \gamma$$

is

$$\mathbf{a}^m = \begin{bmatrix} \alpha^m & \beta \dfrac{\alpha^m - \gamma^m}{\alpha - \gamma} \\ 0 & \gamma^m \end{bmatrix}, \qquad m = 0, 1, \ldots,$$

as can be easily verified by induction. Applying this to our case the state correlation function becomes

$$\mathbf{R_X}(n, k) = \begin{bmatrix} (.2)^{n-k} & (.2)^{n-k} - (-.3)^{n-k} \\ 0 & (-.3)^{n-k} \end{bmatrix} \sigma_{\mathbf{X}}^2(n), \qquad 0 \le k \le n.$$

Thus, for example,

$$\mathbf{R_X}(5, 2) = \begin{bmatrix} .03 & .04 \\ -.01 & -.02 \end{bmatrix}.$$

The correlation function of the output process

$$\mathbf{Y} = \begin{pmatrix} Y_1(n) \\ Y_2(n) \end{pmatrix}, \qquad n = 0, 1, \ldots,$$

is then obtained from the state-output equation of the system (see equation (37)). In our particular case

$$\mathbf{R_Y}(n_1, n_2) = \begin{bmatrix} -1 & 0 \\ -.24 & .6 \end{bmatrix} \mathbf{R_X}(n_1, n_2) \begin{bmatrix} -1 & -.24 \\ 0 & .6 \end{bmatrix}$$

so that, for example,

$$\mathbf{R_Y}(5, 2) = \begin{bmatrix} .031 & -.017 \\ .013 & -.010 \end{bmatrix}.$$ ▲

Since the discrete-time process $\mathbf{X}(n)$, $n = 0, 1, \ldots,$ is Gaussian, we have thus calculated everything that is needed to specify its probability law. However, from equation (46)

$$\mathbf{X}(n + 1) = \mathbf{a}(n)\mathbf{X}(n) + \mathbf{U}(n)$$

and the fact that $\mathbf{U}(n)$ is independent of $\mathbf{X}(0), \ldots, \mathbf{X}(n)$, it follows that

$$E[\mathbf{X}(n + 1) \mid \mathbf{X}(0), \ldots, \mathbf{X}(n)] = \mathbf{a}(n)\mathbf{X}(n) + E[\mathbf{U}(n)] = \mathbf{a}(n)\mathbf{X}(n),$$

and so the process $\mathbf{X}(n)$, $n = 0, 1, \ldots,$ is not only Gaussian but also *Markov.*

Thus a *discrete-time process* $\mathbf{X}(n)$, $n = 0, 1, \ldots,$ *defined by the state-input equation*

$$\mathbf{X}(n + 1) = \mathbf{a}(n)\mathbf{X}(n) + \mathbf{U}(n), \qquad n = 0, 1, \ldots$$

with Gaussian initial state $\mathbf{X}(0)$ *independent of the white Gaussian noise* $\mathbf{U}(n)$, $n = 0, 1, \ldots,$ *must* be a *Gauss-Markov process.*

Surprisingly, perhaps, the converse is also true: *Every discrete-time Gauss-Markov process* $\mathbf{X}(n)$, $n = 0, 1, \ldots,$ *can be obtained as the solution of a state-input equation*

$$\mathbf{X}(n + 1) = \mathbf{a}(n)\mathbf{X}(n) + \mathbf{U}(n), \qquad n = 0, 1, \ldots, \tag{51}$$

driven by a white Gaussian noise input $\mathbf{U}(n)$, $n = 0, 1, \ldots$, *independent of a Gaussian initial state* $\mathbf{X}(0)$.

To show the truth of this statement, suppose that $\mathbf{X}(n)$, $n = 0, 1, \ldots$, is a discrete-time Gauss-Markov process with correlation function $\mathbf{R_X}(n_1, n_2)$. We assume (with no loss of generality) that its mean function is zero and, to avoid the need of working with matrix pseudoinverses, also that $\boldsymbol{\sigma}_\mathbf{X}^2(n) = \mathbf{R}(n, n)$ is a nonsingular matrix for each $n = 0, 1, \ldots$. Under this assumption we know from Chapter One that the correlation function of a Gauss-Markov process must satisfy

$$\mathbf{R_X}(n + 1, k) = \mathbf{R_X}(n + 1, n)\mathbf{R_X}^{-1}(n, n)\mathbf{R_X}(n, k) \tag{52}$$

for all $0 \le k \le n$. Consequently, if we define the matrices $\mathbf{a}(n)$ in the state-input equation (51) as

$$\mathbf{a}(n) = \mathbf{R_X}(n + 1, n)\mathbf{R_X}^{-1}(n, n), \qquad n = 0, 1, \ldots, \tag{53}$$

it follows immediately from (52) that for $0 \le k \le n$

$$\begin{aligned}\mathbf{R_X}(n, k) &= \mathbf{a}(n - 1)\mathbf{R_X}(n - 1, k) = \mathbf{a}(n - 1) \cdots \mathbf{a}(k)\mathbf{R_X}(k, k)\\ &= \boldsymbol{\phi}(n, k)\boldsymbol{\sigma}_\mathbf{X}^2(k),\end{aligned}$$

which is (50). Substitution for $\mathbf{a}(n)$ from (53) into the recurrence (48) determines the variance function

$$\boldsymbol{\sigma}_\mathbf{U}^2(n) = \mathbf{R_X}(n + 1, n + 1) - \mathbf{R_X}(n + 1, n)\mathbf{R_X}^{-1}(n, n)\mathbf{R_X}(n, n + 1), \qquad n = 0, 1, \ldots \tag{54}$$

of the white Gaussian noise input $\mathbf{U}(n)$, $n = 0, 1, \ldots$, in (51), and since the correlation function $\mathbf{R_X}$ now satisfies both (48) and (50), the Gaussian process $\mathbf{X}(n)$, $n = 0, 1, \ldots$, is indeed a solution of equation (51).

Example 2.2.9

Consider a stationary Gauss-Markov process

$$\begin{pmatrix} X_1(n) \\ X_2(n) \end{pmatrix}, \qquad n = 0, 1, \ldots$$

with zero mean function and with correlation function

$$\mathbf{R_X}(n_1, n_2) = \left(\frac{1}{2}\right)^{|n_1 - n_2|/2} \begin{bmatrix} 2 & 1 \\ 1 & 2 \end{bmatrix}$$

$n_1 = 0, 1, \ldots; n_2 = 0, 1, \ldots$. (The reader is invited to verify that this is indeed a Gauss-Markov process.) We wish to construct a linear system

$$\mathbf{X}(n + 1) = \mathbf{a}(n)\mathbf{X}(n) + \mathbf{U}(n), \qquad n = 0, 1, \ldots$$

with Gaussian initial state $\mathbf{X}(0)$ and white Gaussian noise $\mathbf{U}(n)$, $n = 0, 1, \ldots$, such that $\mathbf{X}(n)$, $n = 0, 1, \ldots$, would be the above Gauss-Markov process.

Clearly, the initial state

$$\begin{pmatrix} X_1(0) \\ X_2(0) \end{pmatrix}$$

must have zero mean vector and covariance matrix

$$\boldsymbol{\sigma}_{\mathbf{X}}^2(0) = \mathbf{R}_{\mathbf{X}}(0, 0) = \begin{bmatrix} 2 & 1 \\ 1 & 2 \end{bmatrix}.$$

To determine the system matrix $\mathbf{a}(n)$, we use equation (53). Since

$$\mathbf{R}_{\mathbf{X}}^{-1}(n, n) = \begin{bmatrix} 2 & 1 \\ 1 & 2 \end{bmatrix}^{-1} = \begin{bmatrix} \frac{2}{3} & -\frac{1}{3} \\ -\frac{1}{3} & \frac{2}{3} \end{bmatrix}$$

and

$$\mathbf{R}_{\mathbf{X}}(n + 1, n) = \frac{1}{\sqrt{2}} \begin{bmatrix} 2 & 1 \\ 1 & 2 \end{bmatrix}$$

we obtain

$$\mathbf{a}(n) = \frac{1}{\sqrt{2}} \begin{bmatrix} 2 & 1 \\ 1 & 2 \end{bmatrix} \begin{bmatrix} \dfrac{2}{3} & -\dfrac{1}{3} \\ -\dfrac{1}{3} & \dfrac{2}{3} \end{bmatrix} = \begin{bmatrix} \dfrac{1}{\sqrt{2}} & 0 \\ 0 & \dfrac{1}{\sqrt{2}} \end{bmatrix}$$

for all $n = 0, 1, \ldots$. Thus the system state-input equations are

$$X_1(n + 1) = \frac{1}{\sqrt{2}} X_1(n) + U_1(n),$$

$$X_2(n + 1) = \frac{1}{\sqrt{2}} X_2(n) + U_2(n), \qquad n = 0, 1, \ldots.$$

It remains to determine the variance function $\boldsymbol{\sigma}_{\mathbf{U}}^2(n)$, $n = 0, 1, \ldots$, of the white noise input. From equation (54), we have

$$\begin{aligned} \boldsymbol{\sigma}_{\mathbf{U}}^2(n) &= \begin{bmatrix} 2 & 1 \\ 1 & 2 \end{bmatrix} - \frac{1}{\sqrt{2}} \begin{bmatrix} 2 & 1 \\ 1 & 2 \end{bmatrix} \begin{bmatrix} \frac{2}{3} & -\frac{1}{3} \\ -\frac{1}{3} & \frac{2}{3} \end{bmatrix} \frac{1}{\sqrt{2}} \begin{bmatrix} 2 & 1 \\ 1 & 2 \end{bmatrix} \\ &= \begin{bmatrix} 2 & 1 \\ 1 & 2 \end{bmatrix} - \begin{bmatrix} 1 & \frac{1}{2} \\ \frac{1}{2} & 1 \end{bmatrix} = \begin{bmatrix} 1 & \frac{1}{2} \\ \frac{1}{2} & 1 \end{bmatrix}, \qquad n = 0, , \ldots \end{aligned}$$

The white Gaussian noise is stationary, as might have been expected. ▲

The possibility of representing an arbitrary discrete-time Gauss-Markov process as the state of a linear system (51) driven by white Gaussian noise has great practical significance in filtering and prediction, as is shown in Section 2.3.

Remark: It should be clear that various equations relating the first and second order moment functions of inputs, outputs, and states of a linear system derived throughout this section remain valid even if the input process $U(n)$ is not Gaussian but merely a finite-variance process. Of course, in such a case we can no longer assert that the output process $Z(n)$ or the state process $X(n)$ are Gaussian, and so the knowledge of their mean and correlation functions may be of little value.

However, if the input process $U(n)$ is a sequence of *independent* random variables, we can often use the fact that $Z(n)$ and $X(n)$ are in general linear combinations of the $U(n)$'s and invoke the central limit theorem to conclude that $Z(n)$ and $X(n)$ will be at least *approximately Gaussian*. Of course, whether or not we can make such a conclusion will now depend crucially on the structure of the linear system as well as on the actual distributions of the inputs $U(n)$. The following example illustrates the point. Я

Example 2.2.10
Let $U(n)$, $n = \ldots, -1, 0, 1, \ldots$, be a sequence of i.i.d. random variables taking values + and -1 with equal probability $\frac{1}{2}$.

If this sequence is applied to the input of a linear system with the impulse-response function

$$h(n, k) = \begin{cases} \dfrac{1}{\sqrt{m}} & \text{if } \; n - m + 1 \leq k \leq n, \\ 0 & \text{otherwise} \end{cases}$$

where m is a large positive integer, then the output

$$Z(n) = \sum_{k=-\infty}^{n} h(n, k)U(k), \qquad n = \ldots, -1, 0, 1, \ldots$$

will consist of approximately Gaussian random variables. This is so because by the central limit theorem

$$Z(n) = \frac{1}{\sqrt{m}}(U(n - m + 1) + \cdots + U(n))$$

will be approximately standard Gaussian, the approximation being better, the larger the integer m.

On the other hand, if we consider a linear system with the impulse-response function

$$h(n, k) = \begin{cases} 1 & \text{if } \; k = n, \\ -1 & \text{if } \; k = n - 1, \\ 0 & \text{otherwise,} \end{cases}$$

then with the same input sequence, we have

$$Z(n) = U(n) - U(n-1), \qquad n = \ldots, -1, 0, 1, \ldots$$

so that each $Z(n)$ will take on only the values $+1$, 0, and -1, with probabilities $\frac{1}{4}$, $\frac{1}{2}$, and $\frac{1}{4}$, respectively. This can hardly be considered approximately Gaussian. ▲

EXERCISES

Exercise 2.2

1. Find the output-input cross-correlation functions $R_{Z,U}(n_1, n_2)$ for the linear systems in Example 2.2.1 and in Example 2.2.2.
2. For the linear system and the input process in Example 2.2.3, calculate the output correlation function $R_Z(n_1, n_2)$ if:
   (a) $|\gamma| = |\alpha|$,
   (b) $|\gamma| = 3|\alpha|$.
3. A zero mean process $U(n)$, $n = 0, 1, \ldots$, with correlation function

   $$R_U(n_1, n_2) = 4(3^{-|n_1-n_2|} - 3^{(n_1+n_2)})$$

   is an input to a linear system with impulse-response function

   $$h(n, k) = \begin{cases} \dfrac{1}{\sqrt{n+1}} & \text{if } 0 \le k \le n, \\ 0 & \text{otherwise.} \end{cases}$$

   Find the mean function and the correlation function of the output process $Z(n)$, $n = 0, 1, \ldots$. What can be said about the probability law of the output process for large n?
4. A single-input/single-output linear system is defined as follows: If a unit impulse arrives at the input at an even time n, it is delayed by two time units; if it arrives at an odd time n, it is delayed by three time units. The amplitude of the input is not affected. Obtain an expression for the correlation function of the output process $Z(n)$ if the input process $U(n)$, $n = \ldots, -1, 0, 1, \ldots$, is a zero mean stationary Gauss-Markov process.
5. A linear system calculates successive differences of the input process, that is, $Z(n) = \Delta U(n-1)$, $n = \cdots -1, 0, 1, \ldots$. Show that the correlation functions of the input and output processes are related by the formula

   $$R_Z(n_1, n_2) = \Delta_1 \Delta_2 R_U(n_1 - 1, n_2 - 1),$$

   where Δ_1 and Δ_2 are forward difference operators acting on the first and the second variables, respectively. Also show that if the input process is stationary so is the output process, and the formula then becomes

   $$R_Z(n) = -\Delta^2 R_U(n-1),$$

   where $\Delta^2 = \Delta\Delta$. Assume that the input process has a zero mean function.

6. Use Exercise 2.2.5 to obtain the correlation function of $Z(n) = \Delta U(n-1)$ if the input process $U(n)$, $n = \cdots -1, 0, 1, \ldots$, is a stationary Gaussian process with correlation function

$$R_U(n) = (1 - |n|)\left(\frac{1}{2}\right)^{|n|}.$$

7. Show that if the input process $\mathbf{U}(n)$, $n = \ldots, -1, 0, 1, \ldots$, applied to a time-invariant causal linear filter with impulse response function $\mathbf{h}(n)$ is stationary white Gaussian noise, then the correlation function of the output process $\mathbf{Z}(n)$ is given by

$$\mathbf{R_Z}(n) = \sum_{k=0}^{\infty} \mathbf{h}(k)\mathbf{h}'(n+k), \qquad n = 0, 1, \ldots.$$

8. A binomial filter of order m, a positive integer, is a causal time-invariant linear filter with impulse-response function

$$h(n) = \begin{cases} \binom{m}{n} & \text{if } 0 \le n \le m, \\ 0 & \text{otherwise.} \end{cases}$$

Find the mean function and the correlation function of the output process $Z(n)$ of this filter if the input process $U(n)$, $n = \ldots, -1, 0, 1, \ldots$, is a Gaussian process with mean function

$$\mu_U(n) = (-1)^n$$

and correlation function

$$R_U(n_1, n_2) = \begin{cases} \sigma^2 & \text{if } n_1 = n_2, \\ 0 & \text{if } n_1 \neq n_2. \end{cases}$$

(*Hint:* Use Exercise 2.2.7.)

9. A scalar-valued stationary white Gaussian noise $U(n)$, $n = \ldots, 1, 0, 1, \ldots$, is passed through a cascade of two identical causal time-invariant filters with impulse-response function

$$h(n) = \alpha^n, \qquad n = 0, 1, \ldots,$$

where $|\alpha| < 1$ is a constant. Find the correlation function of the output process $Z(n)$ of the cascade.

10. Consider a linear system in the state-space representation $\mathbf{X}(n+1) = \mathbf{a}(n)\mathbf{X}(n) + \mathbf{U}(n)$, $n = 0, 1, \ldots$, with white Gaussian noise input $\mathbf{U}(n)$ and initial state $\mathbf{X}(0) = \mathbf{0}$. Derive a formula for the cross-correlation function $\mathbf{R}_{\mathbf{X},\mathbf{U}}(n_1, n_2)$ in terms of the system transition matrix $\boldsymbol{\phi}(n, k)$.

11. Use equation (48) to derive an explicit expression for the variance function $\boldsymbol{\sigma}_\mathbf{X}^2(n)$, $n = 1, 2, \ldots$, if $\boldsymbol{\sigma}_\mathbf{X}^2(0) = \mathbf{0}$ and the matrices $\mathbf{a}(n) = \mathbf{a}$, $\boldsymbol{\sigma}_\mathbf{U}^2(n) = \boldsymbol{\sigma}_\mathbf{U}^2$ are independent of n. Can you further simplify your expression if $\boldsymbol{\sigma}_\mathbf{U}^2 = \sigma^2\mathbf{I}$ with $\mathbf{I}$ the identity matrix?

12. Consider the linear system

$$\begin{pmatrix} X_1(n+1) \\ X_2(n+1) \end{pmatrix} = \begin{bmatrix} \alpha & 0 \\ 1 & \beta \end{bmatrix} \begin{pmatrix} X_1(n) \\ X_2(n) \end{pmatrix} + \begin{pmatrix} U(n) \\ 0 \end{pmatrix},$$

$n = 0, 1, \ldots,$ where $\mathbf{X}(0) = \mathbf{0}$, α and β are constants, and $U(n)$, $n = 0, 1, \ldots,$ is a stationary white Gaussian noise. Find conditions under which $\lim_{n\to\infty} \boldsymbol{\sigma}_{\mathbf{X}}^2(n) = \boldsymbol{\sigma}_{\mathbf{X}}^2(\infty)$ exists and evaluate the matrix $\boldsymbol{\sigma}_{\mathbf{X}}^2(\infty)$.

13. Consider a scalar linear system

$$X(n+1) = \sqrt{n}X(n) + U(n), \qquad n = 0, 1, \ldots,$$

with $X(0) = 1$ and $U(n)$, $n = 0, 1, \ldots,$ the standard white Gaussian noise. Calculate the mean function $\mu_X(n)$ and the correlation function $R_X(n_1, n_2)$ for at least the first few values of n.

14. A process $Y(n)$, $n = 0, 1, \ldots,$ satisfies a second order linear difference equation

$$2Y(n+2) + Y(n+1) + Y(n) = 2U(n), \qquad Y(0) = 0,$$

$Y(1) = 1$, with $U(n)$, $n = 0, 1, \ldots,$ a standard white Gaussian noise. Transform this equation into the state-space representation and evaluate the mean function $\mu_Y(n)$ and the correlation function $R_Y(n_1, n_2)$ at least for the first few values of n. (*Hint:* Define the state vector $\mathbf{X}(n) = (Y(n+1), Y(n))'$.)

15. A zero mean Gauss-Markov process $X(n)$, $n = 0, 1, \ldots,$ has correlation function

$$R_X(n_1, n_2) = \frac{(1+n_1)(1+n_2)}{(1+n_1^2)(1+n_2^2)}.$$

Obtain the state-input equation (51) for this process and determine the initial state $X(0)$ and the variance function $\sigma_U^2(n)$.

2.3. FILTERING AND PREDICTION OF DISCRETE-TIME GAUSSIAN PROCESSES

In this section we illustrate the application of the material of Sections 2.1 and 2.2 to the problem of filtering and prediction. We do not intend to study this problem in any great detail, because this is, after all, usually the subject of another course. We merely wish to examine some of the basic ideas from the point of view of stochastic processes, especially the role of the innovation process. Furthermore, we restrict ourselves to discrete-time processes with semi-infinite time domain $\mathcal{T} = 0, 1, \ldots,$ thus completely omitting the classical Wiener filtering theory of stationary processes, in spite of the fact that the innovation approach is applicable there as well.

We begin with a *pure prediction problem.* Suppose that we have a discrete-time Gaussian process

$$X(n), \qquad n = 0, 1, \ldots,$$

whose probability law has been completely specified. By this we mean that either the mean and correlation function of this process are given or that the evolution of the process is known to obey some given relationship, for example, a difference

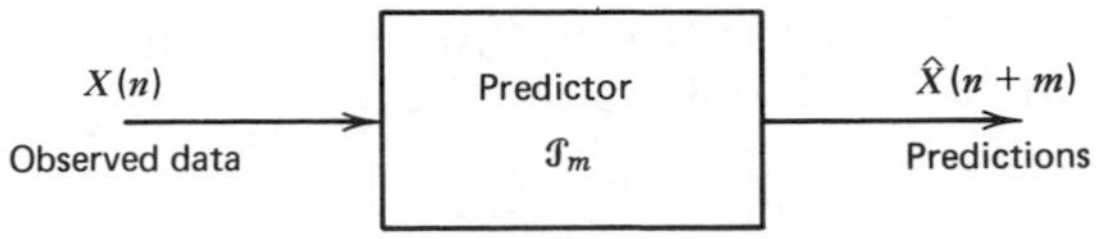

Figure 2.3.1

equation with known initial conditions. We will assume for simplicity that the mean function $\mu_X(n)$ is identically zero.

Suppose now that after observing the process up to some time n, that is, after having observed the values $X(0) = x(0), \ldots, X(n) = x(n)$, we wish to use these observations to predict the value of $X(n+m)$ m time units into the future. Here, m is a fixed positive integer.

Denote the predicted value of $X(n+m)$ by $\hat{X}(n+m)$. It is a function of the observations and thus can be considered a random variable, hence the capital X. We want the prediction to be *optimal* in the sense that the mean square error

$$\varepsilon^2(n) = \mathrm{E}[(\hat{X}(n+m) - X(n+m))^2] \tag{1}$$

is minimized. But we are not satisfied with the prediction $\hat{X}(n+m)$ at a single time instance n only; we wish to keep computing the prediction for *each* n, at least for some period of time. For example, the process $X(n)$ may represent the state of a dynamic system, such as the position and velocity vector of a spacecraft sampled at discrete time units, and we wish to keep predicting its state m time units ahead during its orbit. Thus, we really want to specify a system (computer), called a *predictor* $\mathcal{T}_m$, which would accept the flow of observed data $X(0), X(1), \ldots, X(n), \ldots$ at its input and produce a sequence of predictions $\hat{X}(m), \hat{X}(1+m), \ldots, \hat{X}(n+m), \ldots$ at its output in real time, as indicated in Figure 2.3.1.

Now, we already know what the predictor should really do (see Chapter One). The mean square error (1) is minimized if $\hat{X}(n+m)$ is chosen equal to the conditional expectation

$$\hat{X}(n+m) = \mathrm{E}[X(n+m) \mid X(0), \ldots, X(n)]. \tag{2}$$

Thus the optimal predictor $\mathcal{T}_m$ should simply keep computing the conditional expectation (2), and the problem seems to be solved. The trouble is that the computation of the conditional expectation (2) is no simple matter, even if the process $X(n)$ is zero mean Gaussian as we have assumed. It would generally involve the inversion of an $n \times n$ matrix and some additional matrix multiplication, and for larger n, this would soon become computationally infeasible, especially in real time.

However, in Theorem 2.1.2 we saw that the process $X(n)$ can be represented as a causal linear transformation of its innovation process $V_X(n)$, which is discrete-time white Gaussian noise. Thus, by (2.1.31b) we have for each n

$$X(n+m) = \sum_{k=0}^{n+m} b(n+m, k)V_X(k), \qquad n = 0, 1, \ldots$$

Split the sum into two parts

$$X(n+m) = \sum_{k=0}^{n} b(n+m, k)V_X(k) + \sum_{k=n+1}^{n+m} b(n+m, k)V_X(k)$$

and take the conditional expectation as required by (2):

$$\hat{X}(n+m) = \sum_{k=0}^{n} b(n+m, k)\mathrm{E}[V_X(k) \mid X(0), \ldots, X(n)] + \sum_{k=n+1}^{n+m} b(n+m, k)\mathrm{E}[V_X(k) \mid X(0), \ldots, X(n)]. \tag{3}$$

Now recall from Theorem 2.1.2, that the Gaussian process $X(n)$, $n = 0, 1, \ldots$, and its innovation process $V_X(n)$, $n = 0, 1, \ldots$, are causally linearly equivalent, which means in particular that $X(0), \ldots, X(n)$ are uniquely determined by $V_X(0), \ldots, V_X(n)$, and vice versa. Therefore, conditioning on $X(0), \ldots, X(n)$ is the same as conditioning on $V_X(0), \ldots, V_X(n)$; they carry the same information, and hence

$$\mathrm{E}[V_X(k) \mid X(0), \ldots, X(n)] = \mathrm{E}[V_X(k) \mid V_X(0), \ldots, V_X(n)].$$

But for $0 \le k \le n$, by the properties of conditional expectation,

$$\mathrm{E}[V_X(k) \mid V_X(0), \ldots, V_X(n)] = V_X(k),$$

while for $k > n$

$$\mathrm{E}[V_X(k) \mid V_X(0), \ldots, V_X(n)] = \mathrm{E}[V_X(k)] = 0,$$

since the innovation process is white Gaussian noise, that is, a sequence of independent zero mean Gaussian random variables.

Substituting back into (3), this becomes

$$\hat{X}(n+m) = \sum_{k=0}^{n} b(n+m, k)V_X(k), \qquad n = 0, 1, \ldots, \tag{4}$$

so that we have found an expression for the optimal predictor in terms of the innovation process.

However, the input to the predictor is not the innovation process $V_X(n)$ but the process $X(n)$ itself. This is no problem, since these two processes are causally linearly equivalent, so we can use (2.1.31a) to write

$$V_X(k) = \sum_{l=0}^{k} a(k, l)X(l), \qquad k = 0, 1, \ldots, \tag{5}$$

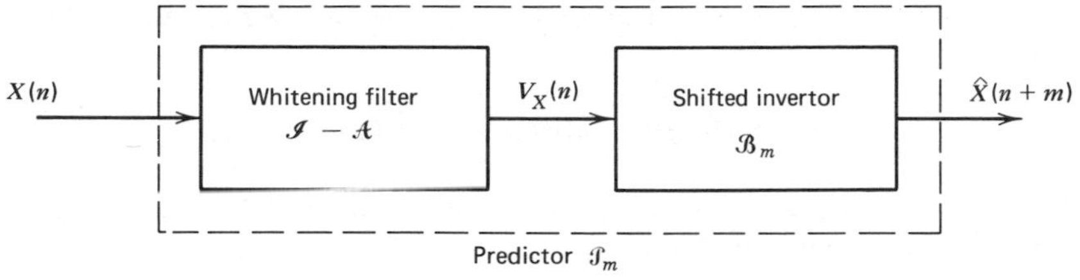

Figure 2.3.2

Substituting into (4), we get the desired form of the predictor:

$$\hat{X}(n+m) = \sum_{k=0}^{n} b(n+m, k) \sum_{l=0}^{k} a(k, l)X(l), \qquad n = 0, 1, \ldots. \tag{6}$$

Note that the predictor is then a cascade of two causal linear systems as portrayed in Figure 2.3.2, the first $\mathcal{I} - \mathcal{A}$ with impulse-response function $a(n, k)$, $0 \le k \le n$, turns the process $X(n)$ into white Gaussian noise (and is therefore often called a *whitening filter*), and the second $\mathcal{B}_m$ with impulse-response function $b_m(n, k) = b(n+m, k)$, $0 \le k \le n$, which changes the white Gaussian noise into the desired output. Since its impulse response is obtained by taking the causal inverse of $a(n, k)$, shifting it by m time units to the left, and chopping off the "noncausal excess," as in Figure 2.3.3, we refer to it here as a shifted invertor.

Remark: We can also easily obtain an expression for the mean square prediction error. Writing

$$X(n+m) = \sum_{k=0}^{n+m} b(n+m, k)V_X(k)$$

and subtracting equation (4), we have

$$X(n+m) - \hat{X}(n+m) = \sum_{k=n+1}^{n+m} b(n+m, k)V_X(k).$$

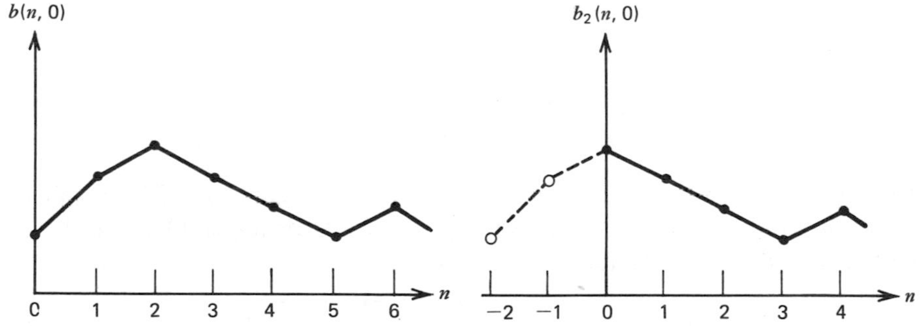

Figure 2.3.3

Now the innovations $V_X(n)$ are independent random variables with zero mean, and hence by the Bienayme equality

$$\varepsilon^2(n) = \mathrm{E}[(X(n+m) - \hat{X}(n+m))^2] = \sum_{k=n+1}^{n+m} b^2(n+m, k)\sigma_V^2(k), \tag{7}$$

where $\sigma_V^2(n) = \mathrm{E}[V_X^2(n)]$ are variances of the innovation process. These can, in turn, be related to the correlation function $R_X(n_1, n_2)$ of the process $X(n)$ (assuming it has a zero mean function) since from (5)

$$\sigma_V^2(n) = \sum_{k_1=0}^{n} \sum_{k_2=0}^{n} a(n, k_1)R_X(k_1, k_2)a(n, k_2).$$ Я

The reader may argue at this point that although we have obtained some insight into the structure of the optimal predictor (and we certainly did), it still remains to actually determine the coefficients $a(n, k)$ and $b(n, k)$ for all $0 \le k \le n$. Now we know from Section 2.1 that the set of coefficients $\mathcal{I} - \mathcal{A} = \{a(n, k), 0 \le k \le n\}$ is related to the set of coefficients $\mathcal{I} + \mathcal{B} = \{b(n, k), 0 \le k \le n\}$ by a system of linear equations and that one set completely determines the other. Furthermore, the systems of linear equations are triangular, which greatly simplifies their computation. Thus we only need to know either one of the two sets of coefficients.

Now everything depends on the way the probability law of the process $X(n)$, $n = 0, 1, \ldots$, is specified. If we are lucky, we may get one of these sets directly from the specification of the process. Fortunately, this is often the case for in many applications the process $X(n)$ represents the state of a dynamic system of known structure disturbed by additive white Gaussian noise. In that case, we can read the coefficients $a(n, k)$ directly from the system equations.

Example 2.3.1

Suppose that a discrete-time Gaussian process

$$Y(n), \qquad n = 0, 1, \ldots$$

is defined by

$$Y(n) = \sum_{k=0}^{n} \frac{\gamma^{n-k}}{(n-k)!} U(k), \qquad n = 0, 1, \ldots, \tag{8}$$

where $\gamma \neq 0$ is a constant and $U(n)$, $n = 0, 1, \ldots$, is a sequence of independent Gaussian random variables with means $\mu_U(n)$ and variances $\sigma_U^2(n)$, $n = 0, 1, \ldots$. Since expression (8) is linear, the expectation

$$\mu_Y(n) = \mathrm{E}[Y(n)] = \sum_{k=0}^{n} \frac{\gamma^{n-k}}{(n-k)!} \mu_U(k), \tag{9}$$

and hence if we define

$$X(n) = Y(n) - \mu_Y(n), \qquad n = 0, 1, \ldots \tag{10}$$

then

$$X(n) = \sum_{k=0}^{n} \frac{\gamma^{n-k}}{(n-k)!} V_X(n), \qquad n = 0, 1, \ldots \tag{11}$$

is a representation of the zero-mean Gaussian process (10) in terms of its innovation process

$$V_X(n) = U(n) - \mu_U(n), \qquad n = 0, 1, \ldots .$$

Suppose now that we wish to obtain the best predictor of $Y(n + m)$, $m \geq 1$ fixed, based on the observations $Y(0), \ldots, Y(n)$. From (10), it is clear that the best predictor $\hat{Y}(n + m)$ will be

$$\hat{Y}(n + m) = \hat{X}(n + m) + \mu_Y(n + m), \tag{12}$$

where $\hat{X}(n + m)$ is the best predictor of $X(n + m)$ based on $X(0) = Y(0) - \mu_Y(0)$, $\ldots$, $X(n) = Y(n) - \mu_Y(n)$. From equation (6), we know that this best predictor will have the form

$$\hat{X}(n + m) = \sum_{k=0}^{n} b(n + m, k) \sum_{l=0}^{k} a(k, l) X(l), \tag{13}$$

where the coefficients are those of the pair of causally equivalent transformations

$$V_X(n) = \sum_{k=0}^{n} a(n, k) X(k), \qquad X(n) = \sum_{k=0}^{n} b(n, k) V_X(k).$$

From (11) we have directly

$$b(n, k) = \frac{\gamma^{n-k}}{(n-k)!}, \qquad 0 \leq k \leq n,$$

and it only remains to find the coefficients $a(n, k)$, $0 \leq k \leq n$, of the inverse transformation. These are given by

$$a(n, k) = \frac{(-\gamma)^{n-k}}{(n-k)!}, \qquad 0 \leq k \leq n,$$

which can be easily verified exactly the same way as used in Example 2.1.4; the presence of the constant γ causes no trouble.

Thus we can write the predictor equation (13) explicitly as

$$\hat{X}(n + m) = \sum_{k=0}^{n} \frac{\gamma^{n+m-k}}{(n + m - k)!} \sum_{l=0}^{k} \frac{(-\gamma)^{k-l}}{(k - l)!} X(l).$$

Substituting for $X(l) = Y(l) - \mu_Y(l)$, and using the relation

$$\sum_{l=0}^{k} \frac{(-\gamma)^{k-l}}{(k-l)!} \mu_Y(l) = \mu_U(k), \qquad k = 0, 1, \ldots$$

due to the fact that with $a(n, k)$ and $b(n, k)$ as above, the causally equivalent transformations hold between any pair of sequences, specifically $\mu_U(n)$, $n = 0, 1, \ldots$, and $\mu_Y(n)$, $n = 0, 1, \ldots$, we obtain

$$\hat{X}(n+m) = \sum_{k=0}^{n} \frac{\gamma^{n+m-k}}{(n+m-k)!} \sum_{l=0}^{k} \frac{(-\gamma)^{k-l}}{(k-l)!} Y(l) - \sum_{k=0}^{n} \frac{\gamma^{n+m-k}}{(n+m-k)!} \mu_U(k).$$

Adding equation (9), with n replaced by $n + m$, and using (12), this becomes

$$\hat{Y}(n+m) = \sum_{k=0}^{n} \frac{\gamma^{n+m-k}}{(n+m-k)!} \sum_{l=0}^{k} \frac{(-\gamma)^{k-l}}{(k-l)!} Y(l) + \sum_{k=n+1}^{n+m} \frac{\gamma^{n+m-k}}{(n+m-k)!} \mu_U(k).$$

This expression may be rewritten in the somewhat more pleasing form

$$\hat{Y}(n+m) = \sum_{l=0}^{n} \sum_{k=l}^{n} \frac{(-1)^{k-l} \gamma^{n+m-k}}{(n+m-k)!\,(k-l)!} Y(l) + \sum_{l=1}^{m} \frac{\gamma^{m-l}}{(m-l)!} \mu_U(n+l), \qquad n = 0, 1, \ldots,$$

which shows directly the coefficients of the linear combination of the observed values $Y(0), \ldots, Y(n)$.

The expression for the mean square prediction error is obtained from equation (7):

$$\varepsilon^2(n) = \mathrm{E}[(\hat{Y}(n+m) - Y(n+m))^2] = \sum_{k=n+1}^{n+m} \left(\frac{\gamma^{n+m-k}}{(n+m-k)!} \right)^2 \sigma_U^2(k) = \sum_{l=1}^{m} \left(\frac{\gamma^{m-l}}{(m-l)!} \right)^2 \sigma_U^2(n+l).$$

It should be noted that if the variances $\sigma_U^2(n)$ do not depend on n, neither in this case does the mean square error. Also, the mean square error then remains bounded for all $m \geq 1$, since the infinite series

$$\sum_{n=0}^{\infty} \left(\frac{\gamma^n}{n!} \right)^2$$

converges for any value of the constant γ. ▲

The case in which the probability law of the process $X(n)$, $n = 0, 1, \ldots,$ is specified by the correlation function R_X is more difficult. From representation (2.1.31b)

$$X(n) = \sum_{k=0}^{n} b(n, k)V_X(k), \qquad n = 0, 1, \ldots$$

and the fact that the innovation process $V_X(n)$ is white Gaussian noise, that is, its correlation function

$$R_V(n_1, n_2) = \begin{cases} \sigma_V^2(n_1) & \text{if} \quad n_1 = n_2 \\ 0 & \text{if} \quad n_1 \neq n_2 \end{cases}$$

we obtain, by using Theorem 2.2.1, the relation

$$R_X(n_1, n_2) = \sum_{k=0}^{n_1} b(n_1, k)\sigma_V^2(k)b(n_2, k), \qquad 0 \leq n_1 \leq n_2, \tag{14}$$

where we have assumed for simplicity that $X(n)$ is a scalar-valued process. Equation (14) represents a system of equations from which the coefficients $b(n, k)$, $0 \leq k \leq n$ (and also the variances $\sigma_V^2(n)$, $n = 0, 1, \ldots$) can be determined. To see how this can be done, recall first that $b(n, n) = 1$, $n = 0, 1, \ldots$ and that $\sigma_V^2(0) = R_X(0, 0)$, since $V_X(0) = X(0)$ by definition. Thus, setting $n_1 = 0$ in (14), we get

$$R_X(0, n_2) = R_X(0, 0)b(n_2, 0), \qquad n_2 = 1, 2, \ldots,$$

so that we have the values $b(n, k)$ for $k = 0$. Next, with $n_1 = n_2 = 1$ in (14)

$$R_X(1, 1) = b(1, 0)R_X(0, 0) + \sigma_V^2(1),$$

and since $b(1, 0)$ is already known, we get $\sigma_V^2(1)$.

Next, setting $n_1 = 1$ in (14), this becomes

$$R_X(1, n_2) = b(1, 0)R_X(0, 0)b(n_2, 0) + \sigma_V^2(1)b(n_2, 1),$$
$$n_2 = 2, 3, \ldots$$

from which we obtain the values $b(n, k)$ for $k = 1$, since $b(n_2, 0)$, $b(1, 0)$, and $\sigma_V^2(1)$ are already known from previous steps. It should now be clear that by continuing this way we can indeed compute the coefficients $b(n, k)$, $0 \leq k \leq n$ for any $n = 0, 1, \ldots,$ at least numerically. This is not to say that the indicated procedure is the only one or even that it is an efficient one. We only wanted to demonstrate the possibility of computing the $b(n, k)$'s from the correlation function R_X. The search for an efficient procedure for computing the coefficients would lead us too deeply into the area of filtering and prediction, and as already mentioned, that is not the goal of this section.

Let us now turn our attention to the *filtering problem*. Actually, we are going to discuss filtering and prediction combined together. To begin again, let

$$X(n), \qquad n = 0, 1, \ldots$$

be a discrete-time Gaussian process with zero mean function. This time, however, we cannot observe the past values of the process directly. Instead, we can observe the values $Y(0) = y(0), \ldots, Y(n) = y(n)$ of another discrete-time Gaussian process

$$Y(n) = X(n) + Z(n), \qquad n = 0, 1, \ldots, \tag{15}$$

where $Z(n)$, $n = 0, 1, \ldots$ is discrete-time white Gaussian noise. Thinking again of the process $X(n)$ as representing the state (position and velocity vector) of a spacecraft in orbit, the process $Y(n)$ would represent the successive measurements of the state by a ground station, the noise $Z(n)$ modeling the random measurement errors. We thus may refer to it as *observation noise*. As before, we want to obtain a sequence of predictions $\hat{X}(n+m)$, $n = 0, 1, \ldots$, m fixed, but this time based only on the sequence of observations $Y(0), \ldots, Y(n)$, $n = 0, 1, \ldots$. Note that this time it would be nontrivial to have $m = 0$, which would correspond to "filtering out" the observation noise to get the best estimate $\hat{X}(n)$ of the current state $X(n)$. This is called a *pure filtering problem*. (It would even make sense to take m negative, which is called a smoothing problem. But we will not discuss this case here.)

In what follows we will consider the case $m = 1$, a *one-step prediction and filtering problem*, for vector-valued processes.† Thus, our problem is to find a filter-predictor $\hat{\mathbf{X}}(n+1)$ based on $\mathbf{Y}(0), \ldots, \mathbf{Y}(n)$, again for each $n = 0, 1, \ldots$ and operating in real time. It should be, of course, also *optimal*, again in the sense of minimizing the mean square error defined by

$$\mathrm{E}[(\hat{\mathbf{X}}(n+1) - \mathbf{X}(n+1))'(\hat{\mathbf{X}}(n+1) - \mathbf{X}(n+1))].$$

As before, we already know the solution; it must be the conditional expectation

$$\hat{\mathbf{X}}(n+1) = \mathrm{E}[\mathbf{X}(n+1) \mid \mathbf{Y}(0), \ldots, \mathbf{Y}(n)], \qquad n = 0, 1, \ldots.$$

However, our model for the observation process (15) is not yet completely specified, we have not said anything about a possible dependence between the observation noise $\mathbf{Z}(n)$, $n = 0, 1, \ldots$, and the process $\mathbf{X}(n)$, $n = 0, 1, \ldots$.

At first, it may seem that in most practical problems they can safely be assumed independent. But this is not quite so. Taking the example of the spacecraft, the ground station may wish to make midcourse corrections by transmitting commands to the spacecraft and thus changing its future states. Now these com-

† All vectors are considered column vectors.

mands are certainly dependent on the present and past observations† $\mathbf{Y}(0), \ldots,$ $\mathbf{Y}(n)$, but these contain the noise values $\mathbf{Z}(0), \ldots, \mathbf{Z}(n)$. Thus, future states may actually depend on the past and present observation noise.

However, in order to keep our illustration simple, we will abandon this more general case and assume that the state and the observation noise are indeed independent. Since we are dealing with zero mean Gaussian processes, this amounts to the condition

$$\mathrm{E}[\mathbf{X}(n)\mathbf{Z}'(k)] = \mathbf{0} \qquad \text{for all} \quad n = 0, 1, \ldots; k = 0, 1, \ldots. \tag{16}$$

We also need to make an assumption about the specification of the probability law of the state process $\mathbf{X}(n)$, $n = 0, 1, \ldots$. This time we will assume that this process is Gauss-Markov, which, as we have seen in Section 2.2, is equivalent to assuming that it can be represented as a state of the dynamic system

$$\mathbf{X}(n + 1) = \mathbf{a}(n)\mathbf{X}(n) + \mathbf{U}(n), \qquad n = 0, 1, \ldots$$

with Gaussian initial state $\mathbf{X}(0)$ and a white Gaussian noise process

$$\mathbf{U}(n), \qquad n = 0, 1, \ldots$$

to be referred to as the *system noise*. We assume that the matrices

$$\mathbf{a}(n), \qquad n = 0, 1, \ldots$$

and the variance function (matrix)

$$\boldsymbol{\sigma}_{\mathbf{U}}^2(n), \qquad n = 0, 1, \ldots$$

of the system noise are known. In view of the fact that we are allowing the state process $\mathbf{X}(n)$, $n = 0, 1, \ldots$, to be vector valued and that in reality it usually represents an actual state of a known physical dynamic system, these assumptions are not as restrictive as they may seem at first. At any rate, they allow us to construct a filter-predictor that is very well suited for computer implementation and has been successfully used for well over a decade—the so-called *Kalman filter*.

Before we begin deriving the filter, let us recapitulate the problem. We have a state process

$$\mathbf{X}(n + 1) = \mathbf{a}(n)\mathbf{X}(n) + \mathbf{U}(n), \qquad n = 0, 1, \ldots, \tag{17a}$$

and an observation process

$$\mathbf{Y}(n) = \mathbf{X}(n) + \mathbf{Z}(n), \qquad n = 0, 1, \ldots. \tag{17b}$$

The initial state $\mathbf{X}(0)$ is Gaussian with zero mean and covariance matrix $\boldsymbol{\sigma}_{\mathbf{X}}^2(0)$. The system noise $\mathbf{U}(n)$ and the observation noise $\mathbf{Z}(n)$ are both white Gaussian noises

† For those familiar with the term, we are talking about closed loop control of the state.

with variance functions $\boldsymbol{\sigma}_{\mathbf{U}}^2(n)$, $n = 0, 1, \ldots$, and $\boldsymbol{\sigma}_{\mathbf{Z}}^2(n)$, $n = 0, 1, \ldots$, respectively.

We will also assume that both these noises are mutually independent, that is, that

$$\mathrm{E}[\mathbf{U}(n)\mathbf{Z}'(k)] = \mathbf{0} \qquad \text{for all} \quad n = 0, 1, \ldots; k = 0, 1, \ldots. \tag{17c}$$

It must be mentioned, however, that had we assumed that (16) holds only for $0 \le n \le k$, this would, in view of equation (17a), force us to assume that (17c) can possibly hold only for $0 \le n < k$ since by multiplying (17a) by $\mathbf{Z}'(n)$

$$\mathbf{X}(n+1)\mathbf{Z}'(n) = \mathbf{a}(n)\mathbf{X}(n)\mathbf{Z}'(n) + \mathbf{U}(n)\mathbf{Z}'(n)$$

and taking expectations we would have

$$\mathrm{E}[\mathbf{X}(n+1)\mathbf{Z}'(n)] = \mathrm{E}[\mathbf{U}(n)\mathbf{Z}'(n)]$$

The subsequent derivation of the Kalman filter can be carried through even under these more general assumptions. However, since the purpose of this section is only to elucidate the techniques rather than to develop a complete theory, we will assume from now on that both conditions (16) and (17c) are satisfied.

Equations (17a,b), assumptions (16) and (17c), together with the collection of matrices

$$\mathbf{a}(n), \qquad \boldsymbol{\sigma}_{\mathbf{X}}^2(0), \qquad \boldsymbol{\sigma}_{\mathbf{U}}^2(n), \qquad \boldsymbol{\sigma}_{\mathbf{Z}}^2(n) \tag{17d}$$

represent the data available.

Our goal is to find an optimal one-step filter-predictor

$$\hat{\mathbf{X}}(n+1) = \mathrm{E}[\mathbf{X}(n+1) \mid \mathbf{Y}(0), \ldots, \mathbf{Y}(n)], \qquad n = 0, 1, \ldots, \tag{18}$$

in a form that can be implemented on a computer in real time.

Let us begin with the innovation process

$$\mathbf{V}_{\mathbf{Y}}(n), \qquad n = 0, 1, \ldots$$

for the observation process

$$\mathbf{Y}(n), \qquad n = 0, 1, \ldots.$$

Since these two processes are causally linearly equivalent, we can replace the conditioning on $\mathbf{Y}(0), \ldots, \mathbf{Y}(n)$ in (18) by conditioning on $\mathbf{V}_{\mathbf{Y}}(0), \ldots, \mathbf{V}_{\mathbf{Y}}(n)$ so that

$$\hat{\mathbf{X}}(n+1) = \mathrm{E}[\mathbf{X}(n+1) \mid \mathbf{V}_{\mathbf{Y}}(0), \ldots, \mathbf{V}_{\mathbf{Y}}(n)], \qquad n = 0, 1, \ldots. \tag{19}$$

Since all the processes involved here are zero mean Gaussian, the conditional expectation (19) must be a homogeneous linear combination

$$\hat{\mathbf{X}}(n+1) = \sum_{k=0}^{n} \mathbf{c}(n, k)\mathbf{V}_{\mathbf{Y}}(k), \qquad n = 0, 1, \ldots, \tag{20}$$

where $\mathbf{c}(n, k)$ are, so far, unknown matrices. To find their values, multiply (20) on the right by $\mathbf{V}'_{\mathbf{Y}}(j)$ and take the expectation. Since the innovation process $\mathbf{V}_{\mathbf{Y}}(n)$ is white Gaussian noise, that is,

$$\mathrm{E}[\mathbf{V}_{\mathbf{Y}}(k)\mathbf{V}'_{\mathbf{Y}}(j)] = \begin{cases} \boldsymbol{\sigma}^2_{\mathbf{V}}(k) & \text{if} \quad j = k, \\ \mathbf{0} & \text{if} \quad j \neq k, \end{cases}$$

we obtain for any $j = 0, 1, \ldots, n$

$$\mathrm{E}[\hat{\mathbf{X}}(n+1)\mathbf{V}'_{\mathbf{Y}}(j)] = \mathbf{c}(n, j)\boldsymbol{\sigma}^2_{\mathbf{V}}(j). \tag{21}$$

However, by (19) for any $j = 0, 1, \ldots, n$

$$\begin{aligned}\hat{\mathbf{X}}(n+1)\mathbf{V}'_{\mathbf{Y}}(j) &= \mathrm{E}[\mathbf{X}(n+1) \mid \mathbf{V}_{\mathbf{Y}}(0), \ldots, \mathbf{V}_{\mathbf{Y}}(n)]\mathbf{V}'_{\mathbf{Y}}(j) \\ &= \mathrm{E}[\mathbf{X}(n+1)\mathbf{V}'_{\mathbf{Y}}(j) \mid \mathbf{V}_{\mathbf{Y}}(0), \ldots, \mathbf{V}_{\mathbf{Y}}(n)],\end{aligned}$$

and hence by taking expectations

$$\mathrm{E}[\hat{\mathbf{X}}(n+1)\mathbf{V}'_{\mathbf{Y}}(j)] = \mathrm{E}[\mathbf{X}(n+1)\mathbf{V}'_{\mathbf{Y}}(j)].$$

Thus (21) becomes

$$\mathrm{E}[\mathbf{X}(n+1)\mathbf{V}'_{\mathbf{Y}}(j)] = \mathbf{c}(n, j)\boldsymbol{\sigma}^2_{\mathbf{V}}(j),$$

and assuming that the matrices $\boldsymbol{\sigma}^2_{\mathbf{V}}(j)$ are nonsingular,† we find that

$$\mathbf{c}(n, j) = \mathrm{E}[\mathbf{X}(n+1)\mathbf{V}'_{\mathbf{Y}}(j)](\boldsymbol{\sigma}^2_{\mathbf{V}}(j))^{-1}, \qquad j = 0, \ldots, n.$$

We can now substitute into (20), thus obtaining

$$\hat{\mathbf{X}}(n+1) = \sum_{k=0}^{n} \mathrm{E}[\mathbf{X}(n+1)\mathbf{V}'_{\mathbf{Y}}(k)](\boldsymbol{\sigma}^2_{\mathbf{V}}(k))^{-1}\mathbf{V}_{\mathbf{Y}}(k), \qquad n = 0, 1, \ldots. \tag{22}$$

Using the state equation (17a)

$$\mathbf{X}(n+1) = \mathbf{a}(n)\mathbf{X}(n) + \mathbf{U}(n)$$

we get for the expectations in (22)

$$\mathrm{E}[\mathbf{X}(n+1)\mathbf{V}'_{\mathbf{Y}}(k)] = \mathbf{a}(n)\mathrm{E}[\mathbf{X}(n)\mathbf{V}'_{\mathbf{Y}}(k)] + \mathrm{E}[\mathbf{U}(n)\mathbf{V}'_{\mathbf{Y}}(k)], \qquad k = 0, \ldots, n. \tag{23}$$

However, for $0 \leq k \leq n$ the state $\mathbf{X}(k)$ is independent of the system noise $\mathbf{U}(n)$, so that (zero mean)

$$\mathrm{E}[\mathbf{U}(n)\mathbf{X}'(k)] = \mathbf{0} \qquad \text{for} \quad 0 \leq k \leq n.$$

Substituting for $\mathbf{X}'(k)$ from the observation equation (17b)

$$\mathbf{Y}(k) = \mathbf{X}(k) + \mathbf{Z}(k)$$

† In the singular case, we would have to use pseudoinverses; we wish to keep things simple here.

this becomes

$$E[U(n)Y'(k)] - E[U(n)Z'(k)] = \mathbf{0},$$

and since by (17c) the system and observation noises are assumed independent the second expectation is zero. Thus

$$E[U(n)Y'(k)] = \mathbf{0}, \qquad k = 0, \ldots, n.$$

and since $\mathbf{Y}(0), \ldots, \mathbf{Y}(n)$ and $\mathbf{V_Y}(0), \ldots, \mathbf{V_Y}(n)$ are causally linearly equivalent, we must also have

$$E[U(n)V'_Y(k)] = \mathbf{0} \qquad \text{for} \quad k = 0, \ldots, n.$$

Therefore the second expectation on the right-hand side of (23) is zero, and if we then substitute from (23) to (22), we obtain

$$\hat{\mathbf{X}}(n+1) = \mathbf{a}(n) \sum_{k=0}^{n} E[\mathbf{X}(n)\mathbf{V}'_{\mathbf{Y}}(k)](\boldsymbol{\sigma}^2_{\mathbf{V}}(k))^{-1}\mathbf{V}_{\mathbf{Y}}(k), \qquad n = 0, 1, \ldots. \tag{24}$$

Now equation (22) holds for all $n = 0, 1, \ldots$; in particular

$$\hat{\mathbf{X}}(n) = \sum_{k=0}^{n-1} E[\mathbf{X}(n)\mathbf{V}'_{\mathbf{Y}}(k)](\boldsymbol{\sigma}^2_{\mathbf{V}}(k))^{-1}\mathbf{V}_{\mathbf{Y}}(k),$$

so that if we separate the last term in the summation in equation (24), we can write

$$\hat{\mathbf{X}}(n+1) = \mathbf{a}(n)[\hat{\mathbf{X}}(n) + E[\mathbf{X}(n)\mathbf{V}'_{\mathbf{Y}}(n)](\boldsymbol{\sigma}^2_{\mathbf{V}}(n))^{-1}\mathbf{V}_{\mathbf{Y}}(n)],$$

or by denoting

$$\mathbf{g}(n) = E[\mathbf{X}(n)\mathbf{V}'_{\mathbf{Y}}(n)](\boldsymbol{\sigma}^2_{\mathbf{V}}(n))^{-1}, \qquad n = 0, 1, \ldots \tag{25}$$

as

$$\hat{\mathbf{X}}(n+1) = \mathbf{a}(n)(\hat{\mathbf{X}}(n) + \mathbf{g}(n)\mathbf{V}_{\mathbf{Y}}(n)), \qquad n = 0, 1, \ldots, \tag{26}$$

where we set $\hat{\mathbf{X}}(0) = \mathbf{0}$ since at time $n = -1$ we have no observations and hence the best estimate of the initial state $\mathbf{X}(0)$ is just its expectation, which was assumed to be zero.

Now $\mathbf{V}_{\mathbf{Y}}(n)$ is an innovation process of the observation process $\mathbf{Y}(n)$ and as such is, according to Theorem 2.1.2, defined by

$$\mathbf{V}_{\mathbf{Y}}(0) = \mathbf{Y}(0),$$

$$\mathbf{V}_{\mathbf{Y}}(n) = \mathbf{Y}(n) - E[\mathbf{Y}(n) \mid \mathbf{Y}(0), \ldots, \mathbf{Y}(n-1)], \qquad n = 1, 2, \ldots. \tag{27}$$

Substituting for $\mathbf{Y}(n) = \mathbf{X}(n) + \mathbf{Z}(n)$ into the conditional expectation above, we get

$$E[\mathbf{Y}(n) \mid \mathbf{Y}(0), \ldots, \mathbf{Y}(n-1)] = E[\mathbf{X}(n) \mid \mathbf{Y}(0), \ldots, \mathbf{Y}(n-1)] + E[\mathbf{Z}(n) \mid \mathbf{Y}(0), \ldots, \mathbf{Y}(n-1)] = \hat{\mathbf{X}}(n)$$

since $\hat{\mathbf{X}}(n) = \mathrm{E}[\mathbf{X}(n) \mid \mathbf{Y}(0), \ldots, \mathbf{Y}(n-1)]$ by definition (18) and $\mathrm{E}[\mathbf{Z}(n) \mid \mathbf{Y}(0), \ldots, \mathbf{Y}(n-1)] = \mathrm{E}[\mathbf{Z}(n)] = 0$ because, by assumption (16), the noise $\mathbf{Z}(n)$ is independent of $\mathbf{X}(0), \ldots, \mathbf{X}(n-1)$ and, being white Gaussian noise, also of its own past $\mathbf{Z}(0), \ldots, \mathbf{Z}(n-1)$.

It follows that

$$\mathbf{V}_{\mathbf{Y}}(0) = \mathbf{Y}(0), \qquad \mathbf{V}_{\mathbf{Y}}(n) = \mathbf{Y}(n) - \hat{\mathbf{X}}(n), \qquad n = 1, 2, \ldots,$$

so that upon substitution into (26), we obtain

$$\hat{\mathbf{X}}(n+1) = \mathbf{a}(n)(\hat{\mathbf{X}}(n) + \mathbf{g}(n)\mathbf{Y}(n) - \mathbf{g}(n)\hat{\mathbf{X}}(n))$$

or, with $\mathbf{I}$ being the identity matrix

$$\hat{\mathbf{X}}(n+1) = \mathbf{a}(n)((\mathbf{I} - \mathbf{g}(n))\hat{\mathbf{X}}(n) + \mathbf{g}(n)\mathbf{Y}(n)), \qquad n = 0, 1, \ldots, \qquad \hat{\mathbf{X}}(0) = \mathbf{0}. \tag{28}$$

This is the *main* equation of the Kalman filter. Note that it is a recurrence relation, the "next" value $\hat{\mathbf{X}}(n+1)$ is obtained as a weighted average of the "present" value $\hat{\mathbf{X}}(n)$ and the "present" observation $\mathbf{Y}(n)$. The weighting factor $\mathbf{g}(n)$ is a matrix defined by equation (25) and is usually called the *Kalman gain.* Notice that this equation is easily implemented on a computer, since the past observations need not be stored; only the "present" value of $\hat{\mathbf{X}}(n)$ is needed at each step. This, of course, is a consequence of the Markovian character of the state process.

Remark: If the initial state $\mathbf{X}(0)$ has a nonzero mean vector, which is usually the case in applications, we initiate the recurrence (28) with $\hat{\mathbf{X}}(0) = \boldsymbol{\mu}_{\mathbf{X}}(0)$. Я

To complete the filter specification, it is necessary to show how the Kalman gain sequence

$$\mathbf{g}(n), \qquad n = 0, 1, \ldots$$

is calculated from the problem data (17d).

Let us denote the prediction error

$$\tilde{\mathbf{X}}(n) = \hat{\mathbf{X}}(n) - \mathbf{X}(n), \qquad n = 0, 1, \ldots,$$

and let

$$\boldsymbol{\varepsilon}^2(n) = \mathrm{E}[\tilde{\mathbf{X}}(n)\tilde{\mathbf{X}}'(n)], \qquad n = 0, 1, \ldots \tag{29}$$

be its covariance matrix. From definition (27) of the innovation process $\mathbf{V}_{\mathbf{Y}}(n)$, $n = 0, 1, \ldots,$ we get upon substitution for $\mathbf{Y}(n)$ from the observation equation $\mathbf{Y}(n) = \mathbf{X}(n) + \mathbf{Z}(n)$ the equation

$$\mathbf{V}_{\mathbf{Y}}(n) = -\tilde{\mathbf{X}}(n) + \mathbf{Z}(n), \qquad n = 0, 1, \ldots . \tag{30}$$

Since $\mathbf{X}(n)$ and $\mathbf{Z}(n)$ are independent, it follows from (30) that

$$\mathrm{E}[\mathbf{X}(n)\mathbf{V}'_{\mathbf{Y}}(n)] = -\mathrm{E}[\mathbf{X}(n)\tilde{\mathbf{X}}'(n)]. \tag{31}$$

But by the orthogonality principle (see Exercise 1.38), the right-hand side of (31) equals the minimum mean square prediction error $\boldsymbol{\varepsilon}(n)$, so that

$$\mathrm{E}[\mathbf{X}(n)\mathbf{V}'_{\mathbf{Y}}(n)] = \boldsymbol{\varepsilon}^2(n), \qquad n = 0, 1, \ldots.$$

However $\tilde{\mathbf{X}}(n)$ and $\mathbf{Z}(n)$ are also independent, and so the covariance matrix (variance function) of the left-hand side of (30) is the sum of the covariance matrices of the two random vectors on the right-hand side. In symbols

$$\boldsymbol{\sigma}^2_{\mathbf{V}}(n) = \boldsymbol{\varepsilon}^2(n) + \boldsymbol{\sigma}^2_{\mathbf{Z}}(n), \qquad n = 0, 1, \ldots.\dagger \tag{32}$$

Thus we can write the Kalman gain (25) as

$$\mathbf{g}(n) = \boldsymbol{\varepsilon}^2(n)[\boldsymbol{\varepsilon}^2(n) + \boldsymbol{\sigma}^2_{\mathbf{Z}}(n)]^{-1}, \qquad n = 0, 1, \ldots \tag{33}$$

To tie things together, it is only necessary to establish a recurrence relation for the covariance matrices $\boldsymbol{\varepsilon}^2(n)$. To this end, recall again from Exercise 1.38 that the orthogonality principle allows us to write the prediction error covariance matrix as

$$\boldsymbol{\varepsilon}^2(n) = \mathrm{E}[\mathbf{X}(n)\mathbf{X}'(n)] - \mathrm{E}[\hat{\mathbf{X}}(n)\hat{\mathbf{X}}'(n)], \tag{34}$$

$n = 0, 1, \ldots$. Using the state equation (17a) and the independence of $X(n)$ and $U(n)$, we get

$$\mathrm{E}[\mathbf{X}(n+1)\mathbf{X}'(n+1)] = \mathbf{a}(n)\mathrm{E}[\mathbf{X}(n)\mathbf{X}'(n)]\mathbf{a}'(n) + \boldsymbol{\sigma}^2_{\mathbf{U}}(n).$$

Similarly, from equation (26) and the independence of $\hat{X}(n)$ and $V_Y(n)$, we obtain

$$\begin{aligned}\mathrm{E}[\hat{\mathbf{X}}(n+1)\hat{\mathbf{X}}'(n+1)] &= \mathbf{a}(n)\mathrm{E}[\hat{X}(n)\hat{X}'(n)]\mathbf{a}'(n) + \mathbf{a}(n)\mathbf{g}(n)\boldsymbol{\sigma}^2_{\mathbf{V}}(n)\mathbf{g}'(n)\mathbf{a}'(n)\\ &= \mathbf{a}(n)\mathrm{E}[\hat{\mathbf{X}}(n)\hat{\mathbf{X}}'(n)]\mathbf{a}'(n)\mathbf{a}(n)\boldsymbol{\varepsilon}^2(n)\mathbf{g}'(n)\mathbf{a}'(n),\end{aligned}$$

since by (32) and (33) $\mathbf{g}(n) = \boldsymbol{\varepsilon}^2(n)(\boldsymbol{\sigma}^2_{\mathbf{V}}(n))^{-1}$ or $\mathbf{g}(n)\boldsymbol{\sigma}^2_{\mathbf{V}}(n) = \boldsymbol{\varepsilon}^2(n)$. Substitution into (34) yields, after minor simplification, the desired recurrence

$$\boldsymbol{\varepsilon}^2(n+1) = \mathbf{a}(n)\boldsymbol{\varepsilon}^2(n)(\mathbf{I} - \mathbf{g}'(n))\mathbf{a}'(n) + \boldsymbol{\sigma}^2_{\mathbf{U}}(n), \qquad n = 0, 1, \ldots. \tag{35a}$$

Coupled with

$$\mathbf{g}(n) = \boldsymbol{\varepsilon}^2(n)(\boldsymbol{\varepsilon}^2(n) + \boldsymbol{\sigma}^2_{\mathbf{Z}}(n))^{-1}, \qquad n = 0, 1, \ldots, \tag{35b}$$

where $\boldsymbol{\varepsilon}^2(0) = \boldsymbol{\sigma}^2_{\mathbf{X}}(0)$ (since $\hat{\mathbf{X}}(0) = 0$, or more generally $\boldsymbol{\mu}_{\mathbf{X}}(0)$) this constitutes the *auxiliary set of Kalman filter equations*, from which the sequence of Kalman gains

† It follows that if the matrices $\boldsymbol{\sigma}^2_{\mathbf{Z}}(n)$ are nonsingular the assumed nonsingularity of $\boldsymbol{\sigma}^2_{\mathbf{V}}(n)$ will be guaranteed.

$\mathbf{g}(n)$, $n = 0, 1, \ldots$, can be computed. Notice that the computation is again recursive; to get $\mathbf{g}(n+1)$ and $\boldsymbol{\varepsilon}^2(n+1)$, we only need $\mathbf{g}(n)$ and $\boldsymbol{\varepsilon}^2(n)$. The computation involves a matrix inversion and is therefore somewhat slower than that of the main equation (28). On the other hand, the auxiliary set (35a) and (35b) only involves the problem data and not the observations, so it can be done prior to receiving the observations, and the resulting sequence of Kalman gains can be stored.

Example 2.3.2

Consider a pair of coupled deterministic difference equations

$$\Delta X_1(n) = \alpha X_2(n),$$
$$\Delta X_2(n) = -\beta X_1(n), \qquad n = 0, 1, \ldots, \tag{36}$$

where $0 < \alpha < 1$ and $0 < \beta < 1$ are constants, and $\Delta X_j(n) = X_j(n+1) - X_j(n)$ denotes a one-step forward increment. These equations represent a discrete version of the so-called Volterra-Lottka predator-prey equations. For their original interpretation, consider a population of two species sharing a common territory, for instance, wolves and caribou in the Alaskan tundra, and the first (predators) prey on the second (prey). Let $X_1(n)$ denote the excess of the number of predators over their average number at the beginning of the nth period, say a year. Similarly, let $X_2(n)$ be the excess in the number of prey. Then the Volterra-Lottka equations (36) express a simple fact of life, namely that the increment of the number of predators is proportional to the number of prey, while the decrement of the number of prey is proportional to the number of predators. We feel obliged to say that a pair of coupled equations of the Volterra-Lottka type can be encountered in a number of physical situations more in line with engineering applications, but we prefer here the former and perhaps more colorful interpretation.

Let us assume now that the changes in the number of both predator and prey populations are further subjected to random changes, which we express by adding noise terms $U_1(n)$ and $U_2(n)$ to the right-hand sides of the two equations (36). The equations can then be rewritten in the standard form of a state-input equation

$$\begin{pmatrix} X_1(n+1) \\ X_2(n+1) \end{pmatrix} = \begin{bmatrix} 1 & \alpha \\ -\beta & 1 \end{bmatrix} \begin{pmatrix} X_1(n) \\ X_2(n) \end{pmatrix} + \begin{pmatrix} U_1(n) \\ U_2(n) \end{pmatrix}, \qquad n = 0, 1, \ldots.$$

Suppose next that we observe the numbers of each population, our observations, however, being subjected to random additive errors. That is, we assume that the observation equation is of the form

$$\begin{pmatrix} Y_1(n) \\ Y_2(n) \end{pmatrix} = \begin{pmatrix} X_1(n) \\ X_2(n) \end{pmatrix} + \begin{pmatrix} Z_1(n) \\ Z_2(n) \end{pmatrix}, \qquad n = 0, 1, \ldots.$$

Our task is now to predict the "next year" populations $\hat{Y}_1(n+1)$ and $\hat{Y}_2(n+1)$ based on the up-to-date observation record $Y_1(0), \ldots, Y_1(n), Y_2(0), \ldots, Y_2(n)$. We wish to employ the Kalman filter technique, and we therefore assume that the conditions under which we derived the filter equations are satisfied. In particular, we assume that $\mathbf{U}(n)$ and $\mathbf{Z}(n)$, $n = 0, 1, \ldots$, are mutually independent white Gaussian noises, the observation noise also being independent of the state process $\mathbf{X}(n)$, $n = 0, 1, \ldots$. Since both $\mathbf{X}(n)$ and $\mathbf{Y}(n)$ actually represent deviations from the average, we will assume that the initial state $\mathbf{X}(0) = \mathbf{0}$. (The Gaussian assumption may seem rather odd in the present context, since both $\mathbf{X}(n)$ and $\mathbf{Y}(n)$ are actually integer valued. However, if we are dealing with reasonably large populations, it may be quite a reasonable approximation.) To complete the description we need to specify the variance functions $\boldsymbol{\sigma}_{\mathbf{U}}^2(n)$ and $\boldsymbol{\sigma}_{\mathbf{Z}}^2(n)$ of the system and observation noises. Let us assume that both these noises are stationary, $\boldsymbol{\sigma}_{\mathbf{U}}^2(n) = \boldsymbol{\sigma}_{\mathbf{U}}^2$, $\boldsymbol{\sigma}_{\mathbf{Z}}^2(n) = \boldsymbol{\sigma}_{\mathbf{Z}}^2$, $n = 0, 1, \ldots$, and that

$$\boldsymbol{\sigma}_{\mathbf{U}}^2 = \begin{bmatrix} 1 & .9 \\ .9 & 1.5 \end{bmatrix}, \qquad \boldsymbol{\sigma}_{\mathbf{Z}}^2 = \begin{bmatrix} 2 & .6 \\ .6 & 1 \end{bmatrix}.$$

Thus the predator is less subject to random changes but harder to observe than the prey and there is a positive correlation between both their random changes and their respective observation errors.

We are now ready to write down the Kalman filter equations. The basic equation (28) is

$$\begin{pmatrix} \hat{Y}_1(n+1) \\ \hat{Y}_2(n+1) \end{pmatrix} = \begin{bmatrix} 1 & \alpha \\ -\beta & 1 \end{bmatrix} \left(\begin{bmatrix} 1 - g_{11}(n) & -g_{12}(n) \\ -g_{21}(n) & 1 - g_{22}(n) \end{bmatrix} \begin{pmatrix} \hat{Y}_1(n) \\ \hat{Y}_2(n) \end{pmatrix} + \begin{bmatrix} g_{11}(n) & g_{12}(n) \\ g_{21}(n) & g_{22}(n) \end{bmatrix} \begin{pmatrix} Y_1(n) \\ Y_2(n) \end{pmatrix} \right),$$

$$n = 0, 1, \ldots, \qquad \begin{pmatrix} \hat{Y}_1(0) \\ \hat{Y}_2(0) \end{pmatrix} = \begin{pmatrix} 0 \\ 0 \end{pmatrix},$$

where $[g_{ij}(n)]$ are the Kalman gain matrices. The auxiliary Kalman filter equations (35a,b) will be

$$\begin{bmatrix} \varepsilon_{11}(n+1) & \varepsilon_{12}(n+1) \\ \varepsilon_{21}(n+1) & \varepsilon_{22}(n+1) \end{bmatrix} = \begin{bmatrix} 1 & \alpha \\ -\beta & 1 \end{bmatrix} \begin{bmatrix} \varepsilon_{11}(n) & \varepsilon_{12}(n) \\ \varepsilon_{21}(n) & \varepsilon_{22}(n) \end{bmatrix} \times \begin{bmatrix} 1 - g_{11}(n) & -g_{21}(n) \\ -g_{12}(n) & 1 - g_{22}(n) \end{bmatrix} \begin{bmatrix} 1 & -\beta \\ \alpha & 1 \end{bmatrix} + \begin{bmatrix} 1 & .9 \\ .9 & 1.5 \end{bmatrix},$$

$$\begin{bmatrix} g_{11}(n) & g_{12}(n) \\ g_{21}(n) & g_{22}(n) \end{bmatrix} = \begin{bmatrix} \varepsilon_{11}(n) & \varepsilon_{12}(n) \\ \varepsilon_{21}(n) & \varepsilon_{22}(n) \end{bmatrix} \begin{bmatrix} 2 + \varepsilon_{11}(n) & .6 + \varepsilon_{12}(n) \\ .6 + \varepsilon_{21}(n) & 1 + \varepsilon_{22}(n) \end{bmatrix}^{-1},$$

$n = 0, 1, \ldots$, where $[\varepsilon_{ij}(n)]$ is the mean square error matrix $\boldsymbol{\varepsilon}^2(n)$ and $\boldsymbol{\varepsilon}^2(0) = \mathbf{0}$.

For a specific example, let us choose $\alpha = 0.5$ and $\beta = 0.7$. The first few mean square error matrices and Kalman gain matrices (for $n = 0, 1, \ldots, 5$) are calculated below.

$$\boldsymbol{\varepsilon}^2(n) = \begin{bmatrix} 0 & 0 \\ 0 & 0 \end{bmatrix}, \begin{bmatrix} 1 & .9 \\ .9 & 1.5 \end{bmatrix}, \begin{bmatrix} 2.085 & 1.031 \\ 1.031 & 1.878 \end{bmatrix}, \begin{bmatrix} 2.540 & .767 \\ .767 & 2.120 \end{bmatrix},$$

$$\begin{bmatrix} 2.624 & .672 \\ .672 & 2.250 \end{bmatrix}, \begin{bmatrix} 2.631 & .659 \\ .659 & 2.281 \end{bmatrix}$$

$$\mathbf{g}(n) = \begin{bmatrix} 0 & 0 \\ 0 & 0 \end{bmatrix}, \begin{bmatrix} .219 & .229 \\ 0 & .6 \end{bmatrix}, \begin{bmatrix} .475 & .089 \\ -.010 & .658 \end{bmatrix}, \begin{bmatrix} .559 & .001 \\ -.041 & .697 \end{bmatrix},$$

$$\begin{bmatrix} .572 & -.017 \\ -.051 & .712 \end{bmatrix}, \begin{bmatrix} .573 & -.019 \\ -.052 & .715 \end{bmatrix}$$

There is no appreciable change in $\boldsymbol{\varepsilon}^2(n)$ or $\mathbf{g}(n)$ for $n > 5$. ▲

The foregoing derivation, long as it may seem, is actually one of the simplest ways to obtain the Kalman filter equations. Not surprisingly, the Kalman filter theory is far more extensive than what has been presented here. Nevertheless, the line of reasoning outlined above is quite universal. For example, in applications the observation equation (17b) more often has the form

$$\mathbf{Y}(n) = \mathbf{c}(n)\mathbf{X}(n) + \mathbf{Z}(n),$$

where $\mathbf{c}(n)$ are not necessarily square matrices. The reader is invited to verify that the Kalman equations for this more complicated case can be derived by the innovation method used here just as easily. For instance, the main equation (28) would have the same form, with the matrices $\mathbf{c}(n)$ replacing the identity matrix there.

As a final remark, let us mention that since we have worked exclusively with the second order moment functions, the derivation of the Kalman filter would apply also to non-Gaussian processes. However, in that case the filter-predictor $\hat{\mathbf{X}}(n + 1)$ need no longer be optimal (in the mean square error sense); it would only be optimal among all *linear* filters. If that is what is desired, fine, but then one may just as well work with the processes as if they were Gaussian in the first place. For the processes, whose probability law is drastically different from a Gaussian law, even the mean square error may be quite an inappropriate criterion of optimality. For instance, if $X(n)$ takes only two values, say $+1$ and -1, the minimum mean square error filter-predictor would typically be predicting values in the interval $(-1, +1)$, which can, in fact, never occur. In such cases, it is better to approach the prediction and filtering problem from scratch.

EXERCISES

Exercise 2.3

1. Find an expression for the best predictor $\hat{X}(n+m)$ if $X(n)$, $n = 0, 1, \ldots$, is the process defined in Example 2.1.1. Also evaluate the mean square prediction error.
2. Repeat Exercise 2.3.1 if $X(n)$, $n = 1, 2, \ldots$, is the process defined in Example 2.1.4.
3. Find the mean square prediction error $\varepsilon^2(n) = \mathrm{E}[(\hat{X}(n+3) - X(n+3))^2]$ if the process $X(n)$, $n = 0, 1, \ldots$, is obtained by passing a stationary white Gaussian noise $U(n)$, $n = 0, 1, \ldots$, through a linear filter with impulse-response function
$$h(n, k) = \begin{cases} \binom{7}{n-k} & \text{if} \quad 0 \le n - k \le 7, \\ 0 & \text{otherwise.} \end{cases}$$
Can you obtain an explicit formula for the predictor $\hat{X}(n+3)$ at least for small n?
4. A Gaussian process $X(n)$, $n = 0, 1, \ldots$, is defined by the equation
$$X(n) = \sum_{k=1}^{n} \binom{k+2}{2} X(n-k) + U(n), \qquad n = 1, 2, \ldots,$$
$X(0) = U(0)$, where $U(n)$, $n = 0, 1, \ldots$, is a sequence of independent standard Gaussian random variables. Show that $U(n)$ is the innovation process for $X(n)$ and that
$$X(n) = U(n) - 3U(n-1) + 3U(n-2) - U(n-3),$$
$n = 0, 1, \ldots$, where $U(-3) = U(-2) = U(-1) = 0$. Use this result to obtain the two-step best predictor of $X(12)$ as a linear combination of $X(0), \ldots, X(10)$. Also calculate the mean square prediction error.
5. Under the assumptions of Exercise 2.3.4 show that except for the first few values of n the mean square error of m-step prediction does not depend on n. Evaluate the mean square prediction error for $m = 1, 2, \ldots$.
6. Suppose that the state $X(n)$, $n = 0, 1, \ldots$, of a linear dynamic system evolves according to the equation $X(n+1) = \alpha X(n)$, where $\alpha \neq 0$ is a constant. The initial state $X(0)$ is a Gaussian random variable with mean zero and variance σ_0^2. The observation process is $Y(n) = X(n) + Z(n)$, $n = 0, 1, \ldots$, where the observation noise $Z(n)$ is stationary white Gaussian noise with variance σ_Z^2. Note that both $X(n)$ and $Y(n)$ are scalars. Write the auxiliary Kalman filter equations for this case and find the Kalman gains $g(n)$ and the mean square errors $\varepsilon^2(n)$ for all n. (*Hint:* Substitute from (35b) into (35a) and obtain an equation for $1/\varepsilon^2(n)$.)
7. Under the assumptions of Exercise 2.3.6 find the limits $g(\infty) = \lim_{n\to\infty} g(n)$ and $\varepsilon^2(\infty) = \lim_{n\to\infty} \varepsilon^2(n)$ in terms of $\alpha, \sigma_0^2, \sigma_Z^2$. Is there a value or values of α for which $\varepsilon^2(\infty) = 0$?
8. Suppose that $\mathbf{X}(n)$, $n = 0, 1, \ldots$, is a zero mean Gauss-Markov process with correlation function $\mathbf{R}_{\mathbf{X}}(n_1, n_2)$. The observation process is defined by $\mathbf{Y}(n) = \mathbf{X}(n) + \mathbf{Z}(n)$, $n = 0, 1, \ldots$, with the white Gaussian observation noise $\mathbf{Z}(n)$ independent of the process $\mathbf{X}(n)$. Show that the MMSE one-step recursive filter-predictor $\hat{\mathbf{X}}(n+1)$ for

$\mathbf{X}(n+1)$ based on $\mathbf{Y}(0), \ldots, \mathbf{Y}(n)$ is the Kalman filter. Write the Kalman filter equations (28) and (35a,b) in terms of the correlation function $\mathbf{R_X}(n_1, n_2)$ and the variance function $\boldsymbol{\sigma}_\mathbf{Z}^2(n)$. (*Hint:* Recall Example 2.2.9.)

9. Suppose that equation (7b) for the observation process is replaced by

$$\mathbf{Y}(n) = \mathbf{c}(n)\mathbf{X}(n) + \mathbf{Z}(n), \qquad n = 0, 1, \ldots,$$

where the $\mathbf{c}(n)$'s are generally rectangular matrices. Repeat the derivation of the Kalman filter equations to show that the main equation (28) now becomes

$$\hat{\mathbf{X}}(n+1) = \mathbf{a}(n)((\mathbf{I} - \mathbf{g}(n)\mathbf{c}(n))\hat{\mathbf{X}}(n) + \mathbf{g}(n)\mathbf{Y}(n))$$

with Kalman gain

$$\mathbf{g}(n) = \boldsymbol{\varepsilon}^2(n)\mathbf{c}'(n)(\mathbf{c}(n)\boldsymbol{\varepsilon}^2(n)\mathbf{c}'(n) + \boldsymbol{\sigma}_\mathbf{Z}^2(n))^{-1}.$$

What happens to the auxiliary equation (35a)?

10. For the situation as described in Example 2.3.2, suppose that $N_1(n)$ and $N_2(n)$ are the numbers (rather than excess over average) of predators and prey, respectively, at the beginning of the nth year. These are assumed to obey the equations

$$\Delta N_1(n) = 0.5N_2(n) - 8 + U_1(n),$$
$$\Delta N_2(n) = -0.7N_1(n) + 7 + U_2(n),$$

with the initial numbers established prior to year zero to be $N_1(0) = 10$, $N_2(0) = 16$ (in appropriate units). Show that the means $\mu_1(n) = \mathrm{E}[N_1(n)] = 10$ and $\mu_2(n) = \mathrm{E}[N_2(n)] = 16$ for all $n = 0, 1, \ldots$ so that the excesses are $X_1(n) = N_1(n) - 10$, $X_2(n) = N_2(n) - 16$.

11. (Exercise 2.3.10 continued.) Suppose that the observations of the numbers (rather than excesses over average) of predators and prey up to the beginning of the fifth year were as in the table below:

Year n	1	2	3	4	5
Predators	11	9	9	12	11
Prey	13	17	16	14	17

Predict the numbers of predators and prey for the beginning of the sixth year. On graph paper, plot the ellipse with the property that the point with coordinates $N_1(6)$ and $N_2(6)$ will be covered by the ellipse with probability at least .95. The noise covariance matrices are as in Example 2.3.2. (*Hint:* Consult Exercise 1.47.)

CHAPTER THREE

MARKOV CHAINS

3.1. DISCRETE-TIME MARKOV CHAINS—INTRODUCTION

In this section we will study discrete-time stochastic processes

$$X(n), \qquad n = 0, 1, \ldots, \tag{1}$$

that are Markov, that is, whose probability law satisfies the Markov property (Chapter One). For a discrete-time process (1), this property can be written as

$$f_{n+1|0,\ldots,n}(x_{n+1} \mid x_0, \ldots, x_n) = f_{n+1|n}(x_{n+1} \mid x_n) \qquad \text{for all } n = 0, 1, \ldots. \tag{2}$$

The importance of this class of stochastic processes stems from the fact that most discrete-time dynamic systems in the state-space representation have the form

$$x_{n+1} = g_n(x_n, u_n), \qquad n = 0, 1, \ldots, \tag{3}$$

where g_n is an input-state transition function (and generally nonlinear, vector valued). If the input sequence u_n, $n = 0, 1, \ldots$, represents a random disturbance (noise) of the system, or more specifically, if

$$U(n), \qquad n = 0, 1, \ldots, \tag{4}$$

is a sequence of independent random variables such that its present and future values

$$U(n), U(n+1), \ldots$$

are also independent of past and present states

$$X(0), \ldots, X(n),$$

of the system, then it is clear from (3) that the state sequence

$$X(n), \qquad n = 0, 1, \ldots,$$

will be a discrete-time Markov process. The probability law of this process will then be completely determined by the initial density $f_{X(0)}(x)$ of $X(0)$, and by the probability law of the sequence (4), since the transition probability density is

$$f_{n+1|n}(x_{n+1} \mid x_n)\,\Delta x \doteq P(x_{n+1} \le g_n(x_n, U(n)) \le x_{n+1} + \Delta x),$$

where the probability on the right-hand side can be determined from the first order density

$$f_n(u)\,\Delta u \doteq P(u \le U(n) \le u + \Delta u)$$

of the noise sequence (4).

Example 3.1.1
Almost any kind of digital data processing unit can be regarded as a finite automaton, a device with a finite number of internal states that transforms strings of input symbols into strings of output symbols, the symbols belonging to some finite alphabet of symbols. Formally, a finite automaton can be defined as a pair of transformations

$$x = g(x', u),$$
$$y = h(x'),$$

where x' is the present state, u is the present input symbol, y is the present output, and x is the next state. Thus, if x_0 is an initial state and $u_0, u_1, \ldots$ is a string of input symbols, then the automaton undergoes a sequence of state transitions

$$x_1 = g(x_0, u_0), \qquad x_2 = g(x_1, u_1), \qquad x_3 = g(x_2, u_2), \ldots,$$

and produces an output string

$$y_0 = h(x_0), \qquad y_1 = h(x_1), \qquad y_2 = h(x_2), \ldots .$$

As a specific example, consider a serial binary adder designed to accept two input strings of 0's and 1's producing at its output a string of 0's and 1's corresponding to the sum of the two numbers, again in binary notation. Such an automaton can most easily be constructed with four internal states denoted by

$$0/0, \; 1/0, \; 0/1, \; 1/1,$$

and operating according to the following table giving the next state x for each of the four possible present states x' and each of the four possible present inputs

$$\mathbf{u} = \begin{bmatrix} 0 \\ 0 \end{bmatrix}, \begin{bmatrix} 0 \\ 1 \end{bmatrix}, \begin{bmatrix} 1 \\ 0 \end{bmatrix}, \begin{bmatrix} 1 \\ 1 \end{bmatrix}.$$

<table>
<tr><td colspan="2" rowspan="2">NEXT STATE
$g(\mathbf{x}', \mathbf{u})$</td><td colspan="3">PRESENT INPUT $\mathbf{u}$</td></tr>
<tr><td>$\begin{bmatrix} 0 \\ 0 \end{bmatrix}$</td><td>$\begin{bmatrix} 0 \\ 1 \end{bmatrix}$ or $\begin{bmatrix} 1 \\ 0 \end{bmatrix}$</td><td>$\begin{bmatrix} 1 \\ 1 \end{bmatrix}$</td></tr>
<tr><td rowspan="4">PRESENT STATE $\mathbf{x}'$</td><td>0/0</td><td>0/0</td><td>1/0</td><td>0/1</td></tr>
<tr><td>1/0</td><td>0/0</td><td>1/0</td><td>0/1</td></tr>
<tr><td>0/1</td><td>1/0</td><td>0/1</td><td>1/1</td></tr>
<tr><td>1/1</td><td>1/0</td><td>0/1</td><td>1/1</td></tr>
</table>

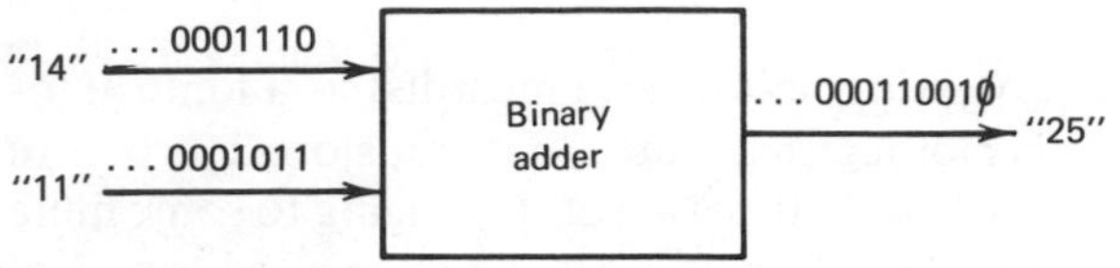

Figure 3.1.1

The state-output transformation $y = h(\mathbf{x}')$ is simple; y is just the upper digit in the symbol for $\mathbf{x}$:

State $\mathbf{x}'$	0/0 or 0/1	1/0 or 1/1
Output y	0	1

The initial state is always $\mathbf{x}_0 = 0/0$. For instance, in adding the numbers 14 and 11 as in Figure 3.1.1 the automaton goes through a sequence of states

$$0/0,\ 1/0,\ 0/1,\ 0/1,\ 1/1,\ 1/0,\ 0/0,\ 0/0,\ \ldots$$

thus producing the sequence of outputs

$$\cdots 00110010,$$

written from right to left as the symbols are emerging from the machine. If the first (i.e., rightmost) symbol (which is always 0) is thrown out, the resulting binary number 11001 is $25 = 14 + 11$.

If the inputs were noisy, that is, if some of the input symbols were randomly changed, or if the automaton were imperfect, occasionally making a random incorrect transition, then provided these random changes were independent in time, the sequence of states $X(n)$, $n = 0, 1, \ldots$ would form a discrete-time Markov process.† Thus, Markov processes are clearly important in modeling effects of random noise and errors in digital systems. Note that not only the time domain but also the state space of such a Markov process is discrete. Actually, as long as finite automata are involved, the state space is, in fact, finite. ▲

Example 3.1.2

Given a discrete-time stochastic process $Y(n)$, $n = 0, 1, \ldots$, representing a time evolution of some physical quantity, the assumption that the process is Markov may seem too restrictive for many applications. The Markov property says, in effect, that the memory of the sequence $Y(n)$, $n = 0, 1, \ldots$, extends only one step to

† The output sequence $Y(n)$, $n = 0, 1, \ldots$, would not necessarily be Markov, but its probability law would nevertheless be completely determined by the Markov process $X(n)$.

the past, while in practice we often deal with systems where the memory extends m steps into the past with $m > 1$. That is, we may frequently be dealing with a discrete-time stochastic process $Y(n)$, $n = 0, 1, \ldots$, whose conditional density functions satisfy the condition

$$f_{n+1|0,\ldots,n}(y_{n+1} \mid y_0, \ldots, y_n) = f_{n+1|n-m+1,\ldots,n}(y_{n+1} \mid y_{n-m+1}, \ldots, y_n),$$

where $m > 1$ is a fixed integer rather than the condition (2). Such processes are sometimes referred to as having the m-Markov property, but there is really no need to introduce a new class of process. All we have to do to convert such an m-Markov process into an ordinary (1-Markov) process is to define a vector-valued discrete-time process

$$\mathbf{X}(n) = \begin{pmatrix} Y(n) \\ Y(n-1) \\ \vdots \\ Y(n-m+1) \end{pmatrix}, \qquad n = m-1, m, m+1, \ldots$$

which is then clearly Markov, since by (1) the conditional distribution of $\mathbf{X}(n+1)$, given the past $\mathbf{X}(n), \ldots, \mathbf{X}(m-1)$, now depends only on the vector $\mathbf{X}(n)$. This, of course, is nothing but the familiar device used to convert a given dynamic system into a canonical state-space representation (3) by suitably defining the state of the system.

As an example, consider one of the fundamental problems in communication theory—the problem of obtaining a suitable probabilistic description of a source of messages to be transmitted. Consider, for instance, a teletype with the messages being ordinary sentences in English. Each message is then a string of symbols—each symbol one of the 26 letters or a blank space in the simplest case—and thus if $Y(n)$ stands for the symbol emitted by the source at the time n, we have a discrete-time stochastic process with state space $\mathcal{S}_Y = \{A, B, C, \ldots, Z, -\}$. But the process $Y(n)$ is definitely not Markov, since for example, the conditional probability of $Y(n+1) = E$, given $Y(n) = H$ and $Y(n-1) = T$, that is, that TH is followed by E, is hardly the same as that of $Y(n+1) = E$, given $Y(n) = H$ and $Y(n-1) = Q$, which is almost certainly zero. However, if instead of single symbols we consider a process $\mathbf{X}(n)$, with each $\mathbf{X}(n)$ representing a block of, say, five consecutive letters, the resulting sequence will now be quite close to being Markov. Of course, each $\mathbf{X}(n)$ is now a 5-tuple of symbols, so that instead of 26 original symbols we now have 26^5 new symbols and a message such as

A-RANDOM-VARIABLE-IS- . . . ,

now transforms into a string

A-RAN, -RAND, RANDO, ANDOM, NDOM-, DOM-V, OM-VA, M-VAR,

-VARI, VARIA, ARIAB, ...

However, this increased complexity is compensated for by the advantage of having an almost Markov source. ▲

In the following discussion, we consider only discrete-time Markov processes (1) with a *discrete state space*, that is, such that the state space $\mathcal{S}$ of the process (1) is a finite or countably infinite set.† Since elements of such a set can always be indexed by integers, we will almost always regard $\mathcal{S}$ as a subset of the set of all integers. Markov processes with a discrete state space $\mathcal{S}$ are commonly called *Markov chains*. The reason for this terminology comes from the fact that if the random variables $X(n)$ are all discrete, the entire probability law of the process $X(n)$ can be described in terms of joint probability mass distributions

$$p_{0,\ldots,n}(x_0, \ldots, x_k) = P(X(0) = x_0, \ldots, X(n) = x_n), \qquad n = 0, 1, \ldots,$$

which, as we have seen in Chapter One, can always be expressed as a product of conditional probabilities

$$\begin{aligned} &P(X(0) = x_0, \ldots, X(n) = x_n) \\ &= P(X(n) = x_n \mid X(0) = x_0, \ldots, X(n-1) = x_{n-1}) \\ &\quad \cdot P(X(n-1) = x_{n-1} \mid X(0) = x_0, \ldots, X(n-2) = x_{n-2}) \\ &\quad \cdots P(X(2) = x_2 \mid X(0) = x_0, X(1) = x_1)P(X(1) = x_1 \mid X(0) = x_0)P(X(0) = x_0). \end{aligned}$$

However, by the Markov property,

$$P(X(k) = x_k \mid X(0) = x_0, \ldots, X(k-1) = x_k) = P(X(k) = x_k \mid X(k-1) = x_{k-1})$$

for all $k = 1, 2, \ldots$ and any $x_j \in \mathcal{S}$, $j = 0, \ldots, k$. Thus the joint probability mass distribution can be expressed as the product of the conditional probabilities of $X(k) = x_k$, given the immediate predecessor $X(k-1)$, the so-called *one-step transition probabilities*

$$P(X(k) = x_k \mid X(k-1) = x_{k-1}), \qquad k = 1, 2, \ldots, \tag{5}$$

and the initial probability $P(X(0) = x_0)$, that is,

$$\begin{aligned} &P(X(0) = x_0, \ldots, X(n) = x_n) \\ &\quad = P(X(n) = x_n \mid X(n-1) = x_{n-1})P(X(n-1) = x_{n-1} \mid X(n-2) = x_{n-2}) \\ &\qquad\qquad \cdots P(X(2) = x_2 \mid X(1) = x_1)P(X(1) = x_1 \mid X(0) = x_0)P(X(0) = x_0). \end{aligned}$$

It follows that the one-step transition probabilities (5), together with the initial

† In view of Example 3.1.1, it may seem that there is no need to consider an infinite state space. However, if we equip a finite automaton with an external memory, for example, a tape, the number of states may be arbitrarily increased and it is then more convenient to regard the state space as infinite.

probability distribution of $X(0)$, completely specify the entire probability law of the discrete-time stochastic process $X(n)$, $n = 0, 1, \ldots$.

In what follows we will deal exclusively with Markov chains that are *homogeneous in time*. This means that the one-step transition probabilities (5) are the same for any time $k = 1, 2, \ldots$

$$P(X(k) = x \mid X(k-1) = x') = p(x \mid x'), \tag{6}$$

for each $k = 1, 2, \ldots$ and any $x \in \mathcal{S}$, $x' \in \mathcal{S}$. This assumption considerably simplifies the specification of the probability law of the process $X(n)$, $n = 0, 1, \ldots$, for now instead of an infinite family of one-step transition probabilities (5), we need only a single function $p(x \mid x')$ of the two state variables x and x', defined by (6).

Remark: A homogeneous Markov chain is sometimes called (more appropriately) a Markov chain with stationary transition probabilities. We will, however, use the shorter term, which also avoids the possible confusion with stationary processes, since a homogeneous Markov chain is *not*, in general, a stationary process. Я

A *discrete-time homogeneous Markov chain* $X(n)$, $n = 0, 1, \ldots$, is thus completely characterized by the initial probability mass distribution

$$p_0(x), \qquad x \in \mathcal{S}, \tag{7}$$

and single set of (one-step) transition probabilities

$$p(x \mid x'), \qquad x \in \mathcal{S},\ x' \in \mathcal{S}. \tag{8}$$

Both (7) and (8) are clearly nonnegative functions satisfying

$$\sum_{x \in \mathcal{S}} p_0(x) = 1, \qquad \sum_{x \in \mathcal{S}} p(x \mid x') = 1 \qquad \text{for all } x' \in \mathcal{S}. \tag{9a, b}$$

The transition probabilities (8) of a homogeneous Markov chain can be displayed in the form of a graph, which is a very convenient device to visualize the operation of the chain. Represent each state $x \in \mathcal{S}$ as a small circle with the state label x inscribed. Then look at each pair of states x and x' for which the transition probability $p(x \mid x')$ is not zero and draw an arrow from x' to x, as in Figure 3.1.2.

Figure 3.1.3 is an example of such a graph for a chain with ten states $\mathcal{S} = \{1, 2, \ldots, 10\}$. The numerical values of the transition probabilities are written alongside the corresponding arrows. Note that (9b) requires that the numbers

Figure 3.1.2

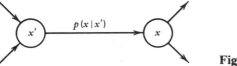

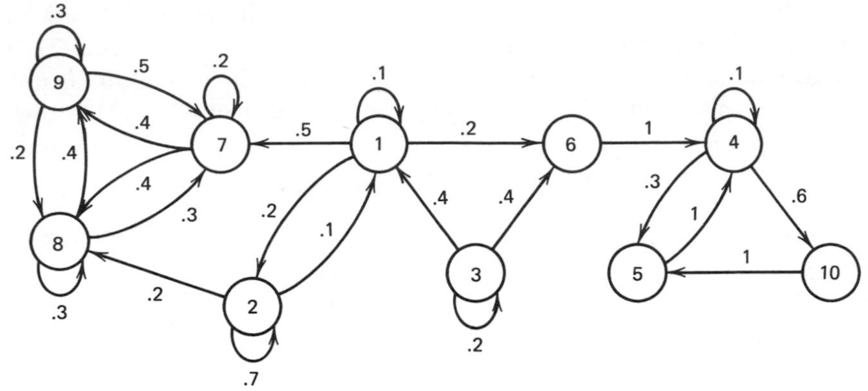

Figure 3.1.3

associated with all arrows emanating from a state must add up to 1. Of course, if the number of states is infinite, or even very large, it is not possible to actually draw such a graph. Nevertheless, if its structure is sufficiently regular, it may still be useful to at least imagine how it would look.

This graph of transition probabilities suggests that one can think of a homogeneous Markov chain as being a sequence of states visited by a traveler walking through the graph along the arrows, selecting at each state an arrow emanating from his current position with probability equal to the number written alongside the arrow. In this context, the Markov property simply means that the traveler at each state chooses his next arrow independently of all the choices made previously in his voyage. In somewhat picturesque language, a Markov traveler starts afresh at each state.

If the number of states of a homogeneous Markov chain is finite, say m, the transition probabilities (8) can also be arranged in an $m \times m$ matrix $\mathbf{P}$, called the transition probability matrix of the chain. If the states are labeled by integers $1, \ldots, m$ then the (i, j)th entry of the matrix $\mathbf{P}$ is the transition probability $p(x \mid x')$ with x' the ith and x the jth state. Thus the ith row contains the transition probabilities *from* the ith state, while the jth column contains the transition probabilities *into* the jth state. Note that (9b) requires that the sums of entries in each row be equal to 1. Of course, all the entries, being probabilities, must be nonnegative. A matrix with these two properties is called *stochastic*.

Thus the transition probability matrix $\mathbf{P}$ of a homogeneous Markov chain is a stochastic matrix, and conversely, any stochastic matrix can serve as a transition probability matrix of a homogeneous Markov chain.

The following is the matrix $\mathbf{P}$ for the chain with transition graph in Figure 3.1.3:

$$\mathbf{P} = \begin{bmatrix} .1 & .2 & 0 & 0 & 0 & .2 & .5 & 0 & 0 & 0 \\ .1 & .7 & 0 & 0 & 0 & 0 & 0 & .2 & 0 & 0 \\ .4 & 0 & .2 & 0 & 0 & .4 & 0 & 0 & 0 & 0 \\ 0 & 0 & 0 & .1 & .3 & 0 & 0 & 0 & 0 & .6 \\ 0 & 0 & 0 & 1 & 0 & 0 & 0 & 0 & 0 & 0 \\ 0 & 0 & 0 & 1 & 0 & 0 & 0 & 0 & 0 & 0 \\ 0 & 0 & 0 & 0 & 0 & 0 & .2 & .4 & .4 & 0 \\ 0 & 0 & 0 & 0 & 0 & 0 & .3 & .3 & .4 & 0 \\ 0 & 0 & 0 & 0 & 0 & 0 & .5 & .2 & .3 & 0 \\ 0 & 0 & 0 & 0 & 1 & 0 & 0 & 0 & 0 & 0 \end{bmatrix}$$

Having described the general structure of a discrete-time homogeneous Markov chain, we now present a few examples.

Example 3.1.3

A relatively simple, but quite important, class of discrete-time homogeneous Markov chains is the random walk (see Chapter One). Generally, a random walk on the state space $\mathcal{S}$ is a discrete-time Markov chain $X(n)$, $n = 0, 1, \ldots$, that is homogeneous not only in time but also in space. This means that the transition probabilities $p(x \mid x')$ are allowed to depend only on the (appropriately defined) *distance* between the states x and x'; that is, $p(x \mid x')$ is the same for all pairs of states x and x' that are the same distance apart. This property can be clearly seen from the transition diagrams of the already familiar Bernoulli random walk (Fig. 3.1.4) and cyclical random walk (Fig. 3.1.5). A slightly more general random walk on the set of all integers is shown in Figure 3.1.6. Random walks defined originally on the set of all integers may also be restricted to subsets such as $\mathcal{S} = \{0, 1, \ldots\}$ or $\mathcal{S} = \{0, \ldots, m\}$, in which case the state $x = 0$ or states $x = 0$ and $x = m$ are referred to as boundaries. If x' is a boundary state, the transition probability

$$p(x \mid x')$$

has to be modified. Such a boundary state x' can be classified according to the value of

$$p(x' \mid x');$$

the two extreme cases in which $p(x' \mid x') = 0$ and $p(x' \mid x') = 1$ are called reflecting and absorbing boundaries, respectively, and the intermediate case $0 < p(x' \mid x') < 1$ is often called an elastic boundary. The terminology is self-

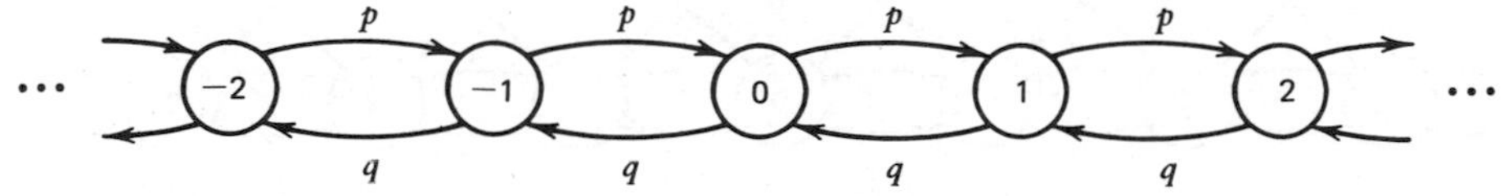

Figure 3.1.4

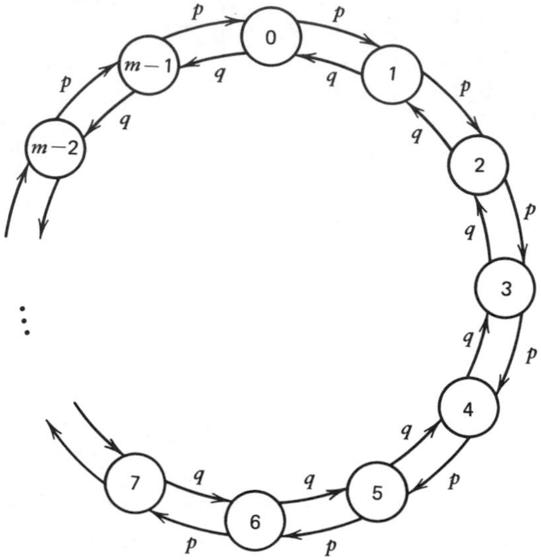

Figure 3.1.5

explanatory (Fig. 3.1.7). Random walks can also be defined on more complicated state spaces; for instance, taking $\mathcal{S}$ to be the set of all ordered pairs $x = (i, j)$ of integers i and j, we may define a simple symmetric random walk on the planar lattice as one with the transition diagram shown in Figure 3.1.8, with all allowed transitions having a probability $\frac{1}{4}$. Of course, there are numerous other possibilities, and if boundaries are also included, the situation may become quite complicated.

Random walks not only serve as basic building blocks for more complex processes (e.g. the Wiener process) but are themselves often used as models for a large variety of physical processes. For instance, a somewhat more sophisticated version of the above random walk on a planar lattice could be used to describe the random motion of an electron or a hole in a thin layer of a semiconductor. To give another example of a rather different nature, consider a line of customers waiting for some service. At each unit of time, there is a fixed probability a, $0 < a < 1$, that a new customer joins the line and that no more than one customer can do so

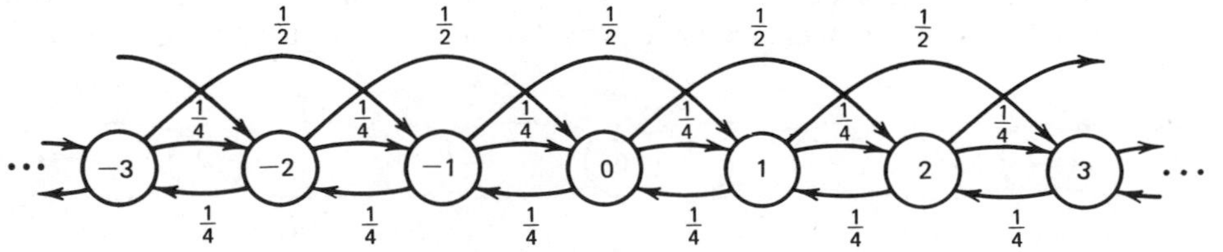

Figure 3.1.6

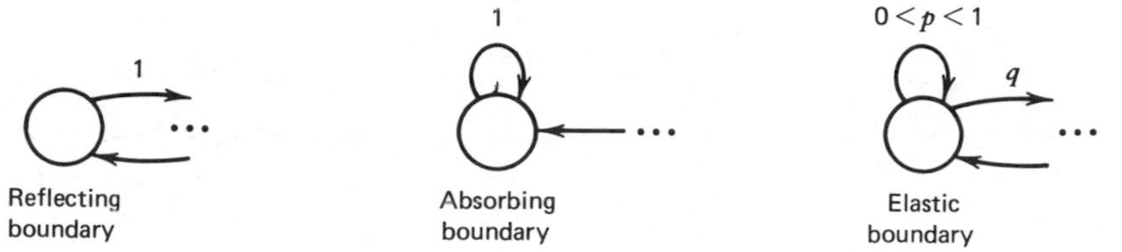

Figure 3.1.7

during the time unit. Also, there is a fixed probability b, $0 < b < 1$, that a customer currently being serviced finishes his or her service during the time unit. Denoting by $X(n)$ the number of customers waiting in line or in service at the beginning of the nth time unit, we see that $X(n)$ can differ from $X(n+1)$ by, at most, one. Assuming that arrivals as well as completions of service are independent, $X(n)$, $n = 0, 1, \ldots$, is a random walk with a transition diagram as pictured in Figure 3.1.9, with the unlabeled transitions from each state except $x = 0$ into itself having the probability

$$1 - a(1-b) - b(1-a) = ab + (1-a)(1-b).$$

Note that $x = 0$ is an elastic boundary corresponding to the server being idle.

▲

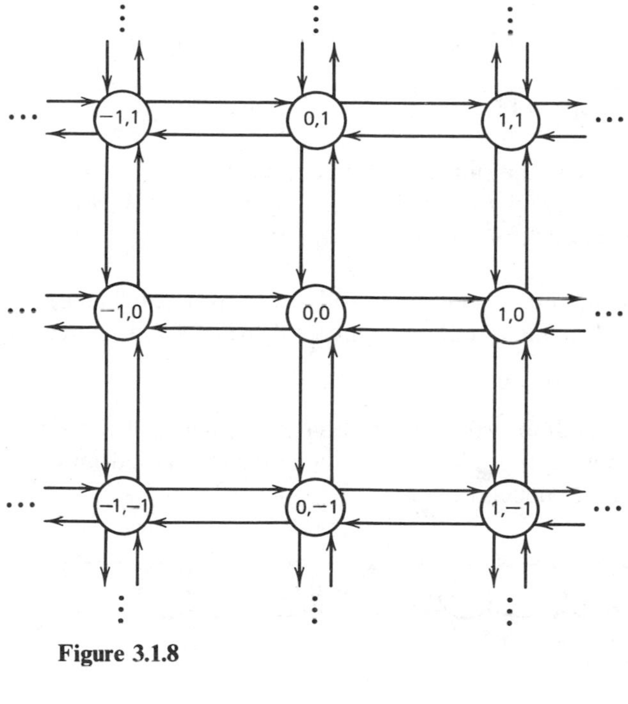

Figure 3.1.8

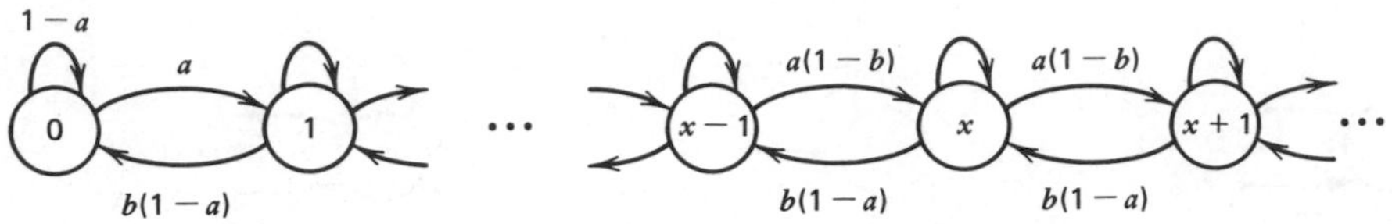

Figure 3.1.9

Example 3.1.4

An example of a Markov chain other than a random walk is provided by the Ehrenfests' model of diffusion. In this model, two containers of the same volume are assumed to contain m molecules of gas. At each time instant $n = 0, 1, \ldots,$ a molecule is selected at random from the entire collection of m molecules, and the diffusion process is simulated by transferring the molecule to the other container. Suppose the containers are labeled A and B, and let $X(n)$ be the number of molecules in A at the time n just before the transfer. Then $X(n)$, $n = 0, 1, \ldots,$ is a discrete-time homogeneous Markov chain with state space $\mathcal{S} = \{0, 1, \ldots, m\}$ and can be represented by the transition diagram in Figure 3.1.10. For $0 < x < m$, the transition probabilities are

$$p(x - 1 \mid x) = \frac{x}{m},$$

and

$$p(x + 1 \mid x) = \frac{m - x}{m},$$

since, for instance, if $X(n) = x$ is the number of molecules in A before the nth transfer, then the number of molecules there decreases by 1, that is, $X(n + 1) = x - 1$, only if a molecule is selected from container A, the probability of which is x/m.

For $x = 0$ and $x = m$, container A is empty or full, respectively, and hence the subsequent transfer necessarily adds or removes a molecule from A. Consequently

$$p(1 \mid 0) = p(m - 1 \mid m) = 1,$$

and so if we borrowed the random walk terminology, we might call the states $x = 0$ and $x = m$ reflecting boundaries. However, this chain is not a random walk, since the transition probabilities depend on x. ▲

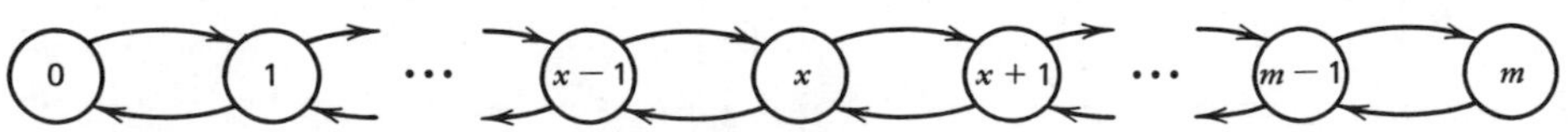

Figure 3.1.10

Example 3.1.5

Consider a light beam shining through a scattering medium, for instance, coming from a star surrounded by a cloud of interstellar dust. Suppose that upon each encounter with a dust particle the light loses a fixed fraction γ, $0 < \gamma < 1$, of its intensity. Let the intensity of the light just after emission from the star be unity, and suppose that during each time unit it travels a unit distance away from the source. Thus, if $Y(n)$, $n = 0, 1, \ldots$, is the intensity of the light at the beginning of the nth time unit, then the intensity $Y(n+1)$ at the beginning of a subsequent time unit is either $Y(n+1) = Y(n)$ or $Y(n+1) = \gamma Y(n)$, depending on whether or not the beam encountered a dust particle during the time unit. (We assume that at most one such collision can occur in each time unit.) Suppose further that the density of the dust cloud decreases with increasing distance from the star, so that after traveling n time units the probability of a collision with a dust particle is $p_n = (n+2)^{-1}$. Under these assumptions, we can write for each $n = 0, 1, \ldots$

$$Y(n+1) = \begin{cases} \gamma Y(n) & \text{with probability } \dfrac{1}{n+2}, \\ Y(n) & \text{with probability } 1 - \dfrac{1}{n+2}, \end{cases}$$

and $Y(0) = 1$. Hence $Y(n)$, $n = 0, 1, \ldots$, is a discrete-time Markov chain, but it is not homogeneous, since the transition probabilities clearly depend on the time n.

We can, however, easily remedy the situation by redefining the state space to consist of pairs of integers

$$\mathbf{x} = (n, k),$$

where $n = 0, 1, \ldots$ is the time and $k = 0, \ldots, n$ is the number of collisions up to time n. The discrete-time process

$$\mathbf{X}(n), \qquad n = 0, 1, \ldots$$

is now a homogeneous Markov chain with initial state $\mathbf{X}(0) = (0, 0)$ and with transition probabilities

$$p(\mathbf{x} \mid \mathbf{x}') = \begin{cases} \dfrac{1}{n+2} & \text{if } \mathbf{x}' = (n, k) \text{ and } \mathbf{x} = (n+1, k+1), \\ \dfrac{n+1}{n+2} & \text{if } \mathbf{x}' = (n, k) \text{ and } \mathbf{x} = (n+1, k) \\ 0 & \text{otherwise.} \end{cases}$$

Furthermore, the original intensity $Y(n)$ is simply

$$Y(n) = \gamma^k \qquad \text{whenever } \mathbf{X}(n) = (n, k)$$

so that the probability law of the process $Y(n)$, $n = 0, 1, \ldots$, is completely determined by the probability law of the discrete-time homogeneous Markov chain $\mathbf{X}(n)$.

It should be clear from this example that by including the time n in the description of the state, we can always convert a discrete-time Markov chain into one that is homogeneous. ▲

In what follows, we shall also need a notation for n-step transition probabilities defined for $n = 1, 2, \ldots$ by

$$p^{(n)}(x \mid x') = P(X(k + n) = x \mid X(k) = x'). \tag{10}$$

It is easily seen that the right-hand side of (10) is independent of the time k, since, for example,

$$p^{(2)}(x \mid x') = \frac{P(X(k + 2) = x, X(k) = x')}{P(X(k) = x')}, \tag{11}$$

and

$$\begin{aligned}
&P(X(k + 2) = x, X(k) = x') \\
&\quad = \sum_{y \in \mathcal{S}} P(X(k + 2) = x \mid X(k + 1) = y, X(k) = x') \\
&\qquad \times P(X(k + 1) = y \mid X(k) = x')P(X(k) = x') \\
&\quad = \sum_{y \in \mathcal{S}} P(X(k + 2) = x \mid X(k + 1) = y)P(X(k + 1) = y \mid X(k) = x')P(X(k) = x') \\
&\quad = \sum_{y \in \mathcal{S}} p(x \mid y)p(y \mid x')P(X(k) = x'),
\end{aligned}$$

by the Markov property and homogeneity assumption (6), so that substituting into (11)

$$p^{(2)}(x \mid x') = \sum_{y \in \mathcal{S}} p(x \mid y)p(y \mid x').$$

Hence, by induction always using $X(k + 1) = y$ as the intermediate step),

$$p^{(n)}(x \mid x') = \sum_{y \in \mathcal{S}} p^{(n-1)}(x \mid y)p(y \mid x') \qquad n = 2, 3, \ldots$$

where $p^{(1)}(\ \mid\ ) = p(\ \mid\ )$. In fact, we have for any $n = 1, 2, \ldots, m = 1, 2, \ldots$

$$p^{(n+m)}(x \mid x') = \sum_{y \in \mathcal{S}} p^{(n)}(x \mid y)p^{(m)}(y \mid x'), \tag{12}$$

which is nothing but the Chapman-Kolmogorov equation for homogeneous Markov chains.

Recall the highly intuitive appeal of this equation: the set of all sample paths leading from state x' to the state x in $n + m$ steps is partitioned into subsets of paths that after m (or n, if you prefer) steps reach an intermediate state y. The probability of any such path decomposes into a product of m- and n-step transition probabilities, since, by virtue of the Markov property, the probability of a transition $y \to x$ is independent of the history of the path prior to reaching the state y.

For a chain with a finite number of states, the n-step transition probabilities can again be arranged into a stochastic matrix $\mathbf{P}^{(n)}$ in exactly the same way as was done for the one-step transition probabilities. The Chapman-Kolmogorov equation (12) can then be written as

$$\mathbf{P}^{(n+m)} = \mathbf{P}^{(n)}\mathbf{P}^{(m)} \tag{13}$$

since the right-hand side of (12) is readily recognized as the definition of matrix multiplication. But since $\mathbf{P}^{(1)} = \mathbf{P}$, the transition probability matrix, it follows from (13) by induction that

$$\mathbf{P}^{(n)} = \mathbf{P}^n, \qquad n = 1, 2, \ldots$$

that is, the matrix of n-step transition probabilities is just the nth power of the transition probability matrix $\mathbf{P}$.

The n-step transition probabilities also allow us to compute the first order probability mass distributions of the process

$$p_n(x) = P(X(n) = x), \qquad n = 1, 2, \ldots$$

from the initial probability mass distribution $p_0(x)$. By the rule for conditional probabilities, we have immediately

$$p_n(x) = \sum_{x' \in \mathcal{S}} p^{(n)}(x \mid x')p_0(x'), \qquad n = 1, 2, \ldots$$

or, more generally, also for any $m = 1, 2, \ldots$

$$p_{n+m}(x) = \sum_{x \in \mathcal{S}} p^{(m)}(x \mid x')p_n(x'). \tag{14}$$

For a finite chain, these relations can also be written in matrix form

$$\mathbf{p}_n = \mathbf{p}_0\,\mathbf{P}^n, \qquad \mathbf{p}_{n+m} = \mathbf{p}_n\,\mathbf{P}^m$$

where $\mathbf{p}_n$ is a row vector with components $p_n(x)$.

Example 3.1.6

The n-step transition probabilities are easy to find for a two-state Markov chain with transition diagram given in Figure 3.1.11. This can be done by writing down the Chapman-Kolmogorov equation

$$p^{(n+1)}(1 \mid x') = p(x \mid 1)p^{(n)}(1 \mid x') + p(x \mid 2)p^{(n)}(2 \mid x'), \tag{15}$$

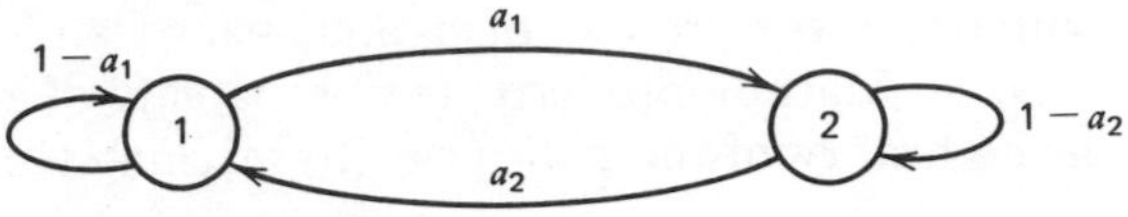

Figure 3.1.11

and using the fact that for a two-state chain $p^{(n)}(2\,|\,x') = 1 - p^{(n)}(1\,|\,x')$ we obtain a simple recurrence relation

$$p^{(n+1)}(1\,|\,x') = (1 - a_1 - a_2)p^{(n)}(1\,|\,x') + a_2, \qquad n = 1, 2, \ldots, x' \in \{1, 2\}.$$

By repeated substitution, we then find the general form of the n-step transition probabilities

$$p^{(n)}(1\,|\,1) = \frac{a_2}{a_1 + a_2} + \frac{a_1}{a_1 + a_2}(1 - a_1 - a_2)^n,$$

$$p^{(n)}(1\,|\,2) = \frac{a_2}{a_1 + a_2} - \frac{a_2}{a_1 + a_2}(1 - a_1 - a_2)^n,$$

and

$$p^{(n)}(2\,|\,1) = \frac{a_1}{a_1 + a_2} - \frac{a_1}{a_1 + a_2}(1 - a_1 - a_2)^n,$$

$$p^{(n)}(2\,|\,2) = \frac{a_1}{a_1 + a_2} + \frac{a_2}{a_1 + a_2}(1 - a_1 - a_2)^n.$$

Equivalently, we could have calculated the nth power of the 2×2 transition matrix

$$\mathbf{P} = \begin{bmatrix} 1 - a_1 & a_1 \\ a_2 & 1 - a_2 \end{bmatrix}$$

to obtain the same result, namely

$$\mathbf{P}^n = \frac{1}{a_1 + a_2}\begin{bmatrix} a_2 & a_1 \\ a_2 & a_1 \end{bmatrix} + \frac{(1 - a_1 - a_2)^n}{a_1 + a_2}\begin{bmatrix} a_1 & -a_1 \\ -a_2 & a_2 \end{bmatrix}.$$

With some effort, we might also obtain an explicit expression for the n-step transition probabilities of a general three-state chain, but beyond that the calculations would become extremely difficult. Of course, if numerical values for the one-step transition probabilities were given, we could calculate the powers $\mathbf{P}^n$ on a computer, but even that would hardly be feasible if the number of states were large.

However, it is sometimes possible to exploit a special structure of a particular Markov chain and obtain explicit expressions for the n-step transition probabilities even if the state space $\mathcal{S}$ of the chain is infinite. Such is the case, for instance, for a Bernoulli random walk where the conditional distribution of $X(n)$, given $X(0) = x'$, that is, $p^{(n)}(x \mid x')$, can be obtained from the fact that $X(n) - x'$ is the same as the number of successes minus the number of failures in n i.i.d. Bernoulli trials.

To illustrate a more complicated case, consider the chain of Example 3.1.5. There, by the very nature of the chain, a transition from a state $\mathbf{x}' = (n_1, k_1)$ to a state $\mathbf{x} = (n_2, k_2)$ can occur in n steps only if $n = n_2 - n_1$ and if exactly $l = k_2 - k_1$ encounters occur in these n time units. The probability of an encounter at time $n_1 + j$ being $1/(n_1 + j + 2)$, the n-step transition probability $p^{(n)}(\mathbf{x} \mid \mathbf{x}')$, with $\mathbf{x}'$ and $\mathbf{x}$ as above, is a sum of $\binom{n}{l}$ terms, each term being a product of the probabilities of an encounter at l specific time units selected from $n_1 + 1, \ldots, n_2$ and no encounters occurring at the remaining $n - l$ time units. In particular, with $\mathbf{x}' = (0, 0)$, the initial state, and $\mathbf{x} = (n, l)$, we have

$$p^{(n)}(\mathbf{x} \mid \mathbf{x}') = \frac{1}{(n+1)!} \sum j_1 j_2 \cdots j_{n-l}$$

the summation being taken over all the $\binom{n}{l}$ subsets $\{j_1, \ldots, j_{n-l}\}$ of the set of integers $\{1, \ldots, n\}$. This is the same as the distribution of a sum of n independent Bernoulli random variables $Z_1, \ldots, Z_n$ with $P(Z_j = 1) = 1/(1 + j)$. The central limit theorem applies to this case (see Chapter One), so, for large n, the transition probabilities $p^{(n)}(\mathbf{x} \mid \mathbf{x}')$, $\mathbf{x}' = (0, 0)$, can be approximated by the Gaussian distribution. We leave the details to the reader as an exercise. ▲

We would like to point out that a homogeneous Markov chain is not necessarily a strictly stationary process. However, it is quite easy to establish a criterion for strict stationarity.

THEOREM 3.1.1 *A homogeneous Markov chain is a strictly stationary process if and only if its initial probability mass distribution* $\mathrm{p}_0(\mathrm{x})$ *satisfies the equation*

$$\sum_{y \in \mathcal{S}} p(y \mid x) p_0(x) = p_0(y) \qquad \text{for all } y \in \mathcal{S}. \tag{16}$$

The initial distribution $\mathrm{p}_0(\mathrm{x})$ *is then called a* stationary distribution *of the chain.*

Proof. The proof is quite simple. By (14), with $m = 1$, we have

$$p_1(x) = p_0(x)$$

if and only if (16) holds. Hence, by induction, for any $n = 1, 2, \ldots$

$$p_n(x) = p_0(x) \tag{17}$$

if and only if (16) is true. Now take a k-tuple of nonnegative integers

$$n_1 < n_2 < \cdots < n_k$$

and consider the kth order probability mass distribution

$$p_{n_1, \ldots, n_k}(x_1, \ldots, x_k) = P(X(n_1) = x_1, \ldots, X(n_k) = x_k).$$

Then, by the Markov property and homogeneity of the process,

$$p_{n_1, \ldots, n_k}(x_1, \ldots, x_k) = p^{(m_{k-1})}(x_k \mid x_{k-1}) p^{(m_{k-2})}(x_{k-1} \mid x_{k-2}) \cdots p^{(m_1)}(x_2 \mid x_1) p_{n_1}(x_1), \quad (18)$$

where $m_1 = n_2 - n_1$, $m_2 = n_3 - n_2$, $\ldots$, $m_{k-1} = n_k - n_{k-1}$. But by (17), this depends only on the time differences $m_1, \ldots, m_{k-1}$ which amounts to strict stationarity according to the definition. ●

Example 3.1.7

Consider a discrete-time homogeneous Markov chain with a finite state space $\mathcal{S} = \{1, \ldots, m\}$, and suppose that the transition probabilities $p(x \mid x')$, in addition to being nonnegative and satisfying

$$\sum_{x \in \mathcal{S}} p(x \mid x') = 1 \qquad \text{for all } x' \in \mathcal{S},$$

also satisfy

$$\sum_{x' \in \mathcal{S}} p(x \mid x') = 1 \qquad \text{for all } x \in \mathcal{S}.$$

That is, if $\mathbf{P} = [p(x \mid x')]$ is the $m \times m$ transition matrix of this chain, then not only the rows but also the columns of the matrix add up to 1. Such a matrix is called doubly stochastic. Then, with the initial probability distribution

$$p_0(x) = \frac{1}{m} \qquad \text{for all } x \in \mathcal{S}$$

the chain becomes a strictly stationary process, since according to Theorem 3.1.1, in this case

$$\sum_{x' \in \mathcal{S}} p(x \mid x') p_0(x') = \frac{1}{m} \sum_{x' \in \mathcal{S}} p(x \mid x') = \frac{1}{m} = p_0(x) \qquad \text{for all } x \in \mathcal{S} \qquad \blacktriangle$$

CLASSIFICATION OF STATES. Let us return for a while to the graph of transition probabilities in Figure 3.1.3, and let us examine the voyages of a Markov traveler. Suppose that the Markov traveler enters state ⑦ or ⑧. Then it is immediately seen that he is doomed to travel forever among the three states ⑦, ⑧, ⑨, since there is no arrow leading away from any of these states.

We express this by saying that the set $\mathcal{A} = \{7, 8, 9\}$ of these three states is *closed*. In general, a subset $\mathcal{A}$ of states is called *closed* if the Markov traveler cannot leave it, that is, if

$$\sum_{x \in \mathcal{A}} p(x \mid x') = 1 \qquad \text{for all } x' \in \mathcal{A}. \tag{19}$$

Note that a chain can have several closed sets, for example, the set $\mathcal{B} = \{4, 5, 10\}$ in Figure 3.1.3 is also closed. Of course, according to the definition (19) the set $\mathcal{S}$ of all states is closed too, so that it is possible for a closed set to contain a smaller closed set. However, the closed sets $\mathcal{A} = \{7, 8, 9\}$ and $\mathcal{B} = \{4, 5, 10\}$ clearly do not contain any smaller closed subset. In fact, any proper subset of the set $\mathcal{A}$ has the property that the Markov traveler not only can but actually will leave such a subset, for if a Markov traveler visits some state infinitely often, then he will infinitely often leave that state along each of the arrows emanating from that state. This is so because the choice of a particular exit arrow is a Bernoulli trial, and these trials are independent, since at each visit to the state the Markov traveler starts afresh. We know (see Exercise 1.3) that in a sequence of independent Bernoulli trials, each of the two outcomes must occur infinitely often.

Consequently, if a closed set $\mathcal{C}$ of states contains no proper closed subset and *if the number of states in* $\mathcal{C}$ *is finite*, then the Markov traveler, upon entering $\mathcal{C}$, will visit each state in $\mathcal{C}$ infinitely often. In other words, once he enters such a minimal (i.e., containing no proper closed subset) finite closed set $\mathcal{C}$, he is doomed to commute forever between *all* states in $\mathcal{C}$.

Furthermore, suppose that $\mathcal{C}$ is a minimal closed set of states (this time not necessarily with a finite number of states), and suppose that some particular state $x \in \mathcal{C}$, once visited, will be *revisited infinitely often*. Call such a state $x \in \mathcal{C}$ *recurrent*. Consider now some other state $y \in \mathcal{C}$. Since $\mathcal{C}$ is, by assumption, a minimal closed set in the transition diagram, there must be a path going from x to y and also a path going from y to x. Hence, there is a positive probability of getting from x to y, and the choice of this entire path can be regarded as a "success" in a sequence of Bernoulli trials, each trial commencing upon each visit to the state x. Since the Markov traveler is assumed to visit x infinitely often, the sequence of Bernoulli trials is infinite, and hence the Markov traveler must sooner or later take this path and find himself in the state y. But he must later return back to the state x, for otherwise he would not be able to keep visiting x infinitely often. Once back in x the entire reasoning can be repeated (a *Markov* traveler starts afresh in each state). It follows that he will also be revisiting the state y infinitely often; that is, y is also a recurrent state. This proves the following important fact:

THEOREM 3.1.2. *If a state* x *in a minimal closed set* $\mathcal{C}$ *is recurrent, then all states in* $\mathcal{C}$ *are recurrent, and* $\mathcal{C}$ *is then called a* recurrent class of states.

If a minimal closed set $\mathcal{C}$ *has only a* finite *number of states, then it must be a recurrent class.*

A Markov chain may contain several recurrent classes, even infinitely many. For example, in Figure 3.1.3, both $\mathcal{A}$ and $\mathcal{B}$ are recurrent classes; in fact, they are the only recurrent classes there. Moreover, recurrent classes must clearly be *disjoint*, for if two recurrent classes had a state in common, they would not be closed sets of states.

A recurrent class sometimes consists of a single state only. In that case, the state is appropriately called *absorbing*, since the Markov traveler, upon visiting such a state, must remain there forever.

A state $x \in \mathcal{S}$, which is not a recurrent state, is called *transient*. By the definition of a recurrent state, this implies that a state is transient if, once visited, it can be revisited only a finite number of times. In other words, a transient state, if ever visited, must eventually be abandoned, never to be entered again.

Since every recurrent state must belong to a minimal closed set of states, the states that do not belong to any such set are necessarily transient.

This raises some interesting questions.

1. Could all states be transient? Or, alternatively, does every Markov chain contain at least one recurrent class?
2. Can a transient state belong to a minimal closed set of states?
3. How do we recognize whether a given state is recurrent or transient?

If the number of states in the state space $\mathcal{S}$ is *finite*, that is, if we deal with a finite Markov chain, the answers are quite easy.

A finite Markov chain must have at least one minimal closed set of states, and every such set, being finite, *is a recurrent class*. Thus the answer to both questions 1 and 2 is no. Since recurrent states are, in this case, exactly those belonging to minimal closed sets of states and since these can be found either by examining the transition diagram or by finding smallest sets $\mathcal{A}$ satisfying the condition (19), question 3 can be answered by merely checking whether or not the state in question belongs to a minimal closed set of states.

On the other hand, if the number of states in $\mathcal{S}$ is *infinite*, the answer to both questions 1 and 2 is yes.

First, an infinite Markov chain need not have any minimal closed set of states. The chain with state space $\mathcal{S} = \{1, 2, \ldots\}$ and transition diagram in Figure 3.1.12 provides a simple example. Since there is no minimal closed set, there are no recurrent states, and thus all states are transient.

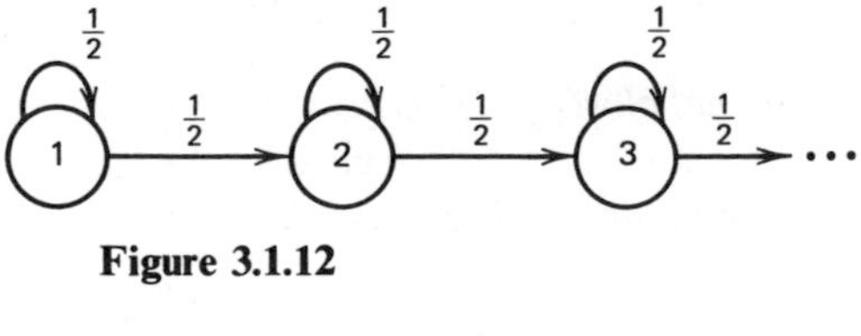

Figure 3.1.12

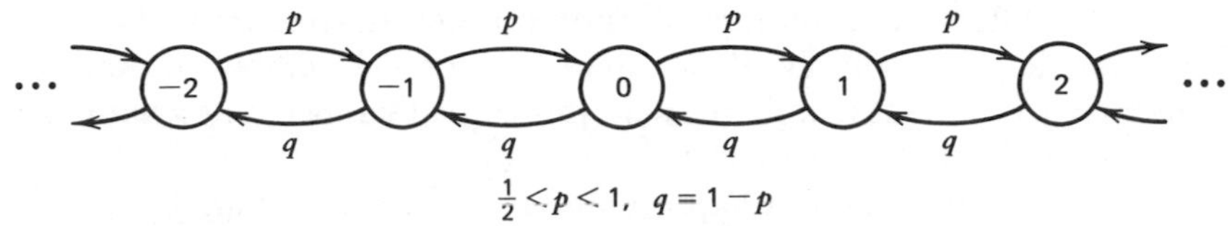

Figure 3.1.13

Second, it is indeed possible that a transient state belongs to a minimal closed set $\mathcal{C}$. Furthermore, in such a case, *all* states in $\mathcal{C}$ must be transient (if one were recurrent, all would have to be recurrent by Theorem 3.1.2), and so each minimal closed set is either a recurrent class or a *transient class*. An example of such a case is provided by a simple Bernoulli random walk with $p \neq q$, that is, a Markov chain with state space $\mathcal{S} = \{\ldots, -1, 0, 1, \ldots\}$ and transition diagram as in Figure 3.1.13. It is easy to see that this chain has a single minimal closed set of states, namely the set $\mathcal{S}$ of all states.

To verify that all states are transient requires a little more work, however. Suppose that the random walk starts at the initial state $x_0 = 0$. Then, from Exercise 1.51, we see that the probability that the random walk will ever enter the state $x_l = -1$ equals q/p, while the probability that it will ever visit a state $x_r > 0$ is 1. By symmetry and the Markov property, it follows that upon visiting any state x, the probability of any future visit to its felt neighbor $x_l = x - 1$ equals q/p, and the probability of any future visit to a state $x_r > x$ is 1. Thus, if at some time the Markov traveler is at state x and exits to the left (probability q), he will return to x with probability 1, while if he exits to the right (probability p), he will return to x with probability q/p. (See Fig. 3.1.14) Consequently his total probability of returning to x equals $q \times 1 + p \times (q/p) = 2q < 1$, since $q < \frac{1}{2}$. Thus the probability of n consecutive returns to x equals $(2q)^n$, which goes to zero as $n \to \infty$. Therefore, the probability of returning to x infinitely often is zero, and x is a transient state.

This example also shows that for an infinite Markov chain, it is no longer a simple matter to tell which state is recurrent and which is transient. It also raises yet another question, which may have already bothered some readers: Is it true that each state is necessarily either transient or recurrent?

It is, after all, quite conceivable that, in our picturesque language, one Markov traveler starting from some state x in the transition diagram will return to

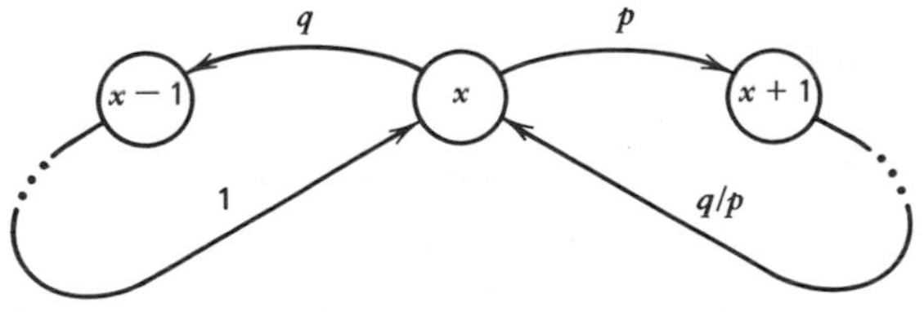

Figure 3.1.14

this state infinitely often, while another Markov traveler, starting from the same state in the same diagram, will revisit this state only a finite number of times.

It is intuitively clear that in a finite chain this cannot happen, but what if the number of states is infinite?

To show that it indeed cannot happen and also to obtain an answer to the previous question 3, it is clearly necessary to have a criterion for a state being recurrent or transient in terms of transition probabilities.

Let us first denote by

$$r^{(n)}(x \mid x'), \qquad x \in \mathcal{S},\ x' \in \mathcal{S}, \qquad n = 1, 2, \ldots, \tag{20}$$

the probability that it takes the Markov traveler, starting from the state x', exactly n steps to enter the state x *for the first time*. Formally,

$$r^{(n)}(x \mid x') = P(X(n) = x,\ X(n-1) \neq x,\ X(n-2) \neq x,\ \ldots,\ X(1) \neq x \mid X(0) = x'),$$

or more generally (since the chain is homogeneous)

$$r^{(n)}(x \mid x') = P(X(m+n) = x,\ X(m+n-1) \neq x,\ X(m+n-2) \neq x,$$
$$\ldots,\ X(m+1) \neq x \mid X(m) = x') \qquad \text{for any } m = 0, 1, \ldots.$$

The probabilities (20) are called *first entrance probabilities* (for $x \neq x'$) or *first return probabilities* (for $x = x'$).

Note that for fixed x and x' the probabilities $r^{(n)}(x \mid x')$ are for different n probabilities of mutually exclusive events. Hence, the sum

$$r(x \mid x') = \sum_{n=1}^{\infty} r^{(n)}(x \mid x'),$$

is the probability that the Markov traveler, starting from x', ever visits the state x. In particular for $x = x'$,

$$r(x \mid x)$$

is the probability that he ever returns to x. Since upon each return to x he begins afresh, the probability that he returns to x at least k times is the kth power

$$(r(x \mid x))^k.$$

Consequently, the probability that the Markov traveler, upon a visit to x, returns to x infinitely many times is the limit

$$\lim_{k \to \infty} (r(x \mid x))^k.$$

Since $r(x \mid x)$ is a probability, that is, $0 \leq r(x \mid x) \leq 1$, this gives us the following criterion:

$$r(x \mid x) < 1 \qquad \text{if and only if } x \text{ is transient}, \tag{21a}$$

$$r(x \mid x) = 1 \qquad \text{if and only if } x \text{ is recurrent}. \tag{21b}$$

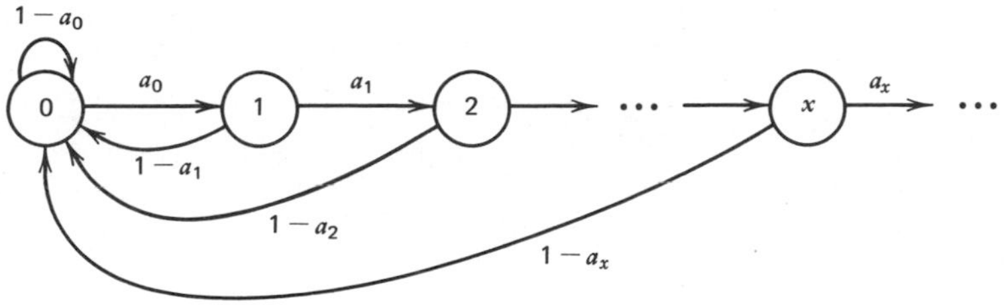

Figure 3.1.15

Example 3.1.8

Consider a discrete-time homogeneous Markov chain with the state space $\mathcal{S} = \{0, 1, \ldots\}$ and transition diagram of Figure 3.1.15, where a_x, $x = 0, 1, \ldots$, is such that $0 < a_x \leq 1$. For the state $x = 0$ the first return probabilities

$$r^{(n)}(0 \mid 0)$$

are easily seen to be

$$r^{(1)}(0 \mid 0) = 1 - a_0$$

and

$$r^{(n)}(0 \mid 0) = a_0 a_1 \cdots a_{n-2}(1 - a_{n-1}), \qquad n = 2, 3, \ldots .$$

Writing the latter product as

$$r^{(n)}(0 \mid 0) = a_0 a_1 \cdots a_{n-2} - a_0 a_1 \cdots a_{n-1},$$

we find that the sum

$$\sum_{k=1}^{n} r^{(k)}(0 \mid 0) = 1 - a_0 a_1 \cdots a_{n-1}.$$

It follows that

$$r(0 \mid 0) = \sum_{k=1}^{\infty} r^{(k)}(0 \mid 0) = 1 - \lim_{n \to \infty} a_0 a_1 \cdots a_{n-1}$$

and hence the chain is recurrent if and only if

$$\lim_{n \to \infty} a_0 a_1 \cdots a_{n-1} = 0.$$

This is equivalent to the infinite series with nonnegative terms

$$\sum_{n=0}^{\infty} \ln \frac{1}{a_n} = +\infty,$$

so that if the above series diverges the chain is recurrent, while if it converges the chain is transient. For example, if

$$a_n = e^{-1/(n+1)}, \qquad n = 0, 1, \ldots,$$

we have

$$\sum_{n=0}^{\infty} \ln \frac{1}{a_n} = \sum_{n=0}^{\infty} \frac{1}{n+1} = +\infty,$$

a recurrent chain, while if

$$a_n = e^{-(1/(n+1))^2}, \qquad n = 0, 1, \ldots,$$

then

$$\sum_{n=0}^{\infty} \ln \frac{1}{a_n} = \sum_{n=0}^{\infty} \left(\frac{1}{n+1}\right)^2 < +\infty$$

and we now have a transient chain. The probability $r(0 \mid 0)$ of ever returning to the state $x = 0$ is then

$$r(0 \mid 0) = 1 - \exp\left(-\sum_{n=0}^{\infty} \ln \frac{1}{a_n}\right),$$

or, upon using the formula

$$\sum_{k=1}^{\infty} \frac{1}{k^2} = \frac{\pi^2}{6},$$

$r(0 \mid 0) = 1 - e^{-\pi^2/6}$, which is 0.8070 to four decimal places. ▲

The first entrance (first return) probabilities are related to n-step transition probabilities by the following basic equation (called the first entrance equation)

$$p^{(n)}(x \mid x') = \sum_{k=1}^{n} p^{(n-k)}(x \mid x) r^{(k)}(x \mid x') \tag{22}$$

valid for any pair of states x, x' and also any $n = 1, 2, \ldots$, provided we define

$$p^{(1)}(x \mid x') = p(x \mid x'), \qquad p^{(0)}(x \mid x) = 1.$$

Although we will not prove (22) formally, the proof can be easily obtained from the following idea. Consider the Markov traveler starting at the state x'. In order to find himself at the state x after n steps, he must have entered (reentered if $x = x'$) the state x for the first time at some step k $(1 \le k \le n)$. The probability of this is $r^{(k)}(x \mid x')$. After that, he starts afresh and, having $n - k$ steps left, the probability that at n he will again be at the state x is $p^{(n-k)}(x \mid x)$. The result then follows from the fact that first visits to x at different times k are mutually exclusive events.

Equation (22) can be used to yield yet another criterion for a state to be recurrent or transient:

$$\sum_{n=1}^{\infty} p^{(n)}(x \mid x) < \infty \qquad \text{if and only if } x \text{ is transient,} \tag{23a}$$

$$\sum_{n=1}^{\infty} p^{(n)}(x \mid x) = \infty \qquad \text{if and only if } x \text{ is recurrent.} \tag{23b}$$

The proof is short. Summing (22) with $x' = x$ over $n = 1, 2, \ldots, m$ we get

$$\begin{aligned}\sum_{n=1}^{m} p^{(n)}(x \mid x) &= \sum_{n=1}^{m} \sum_{k=1}^{n} r^{(k)}(x \mid x) p^{(n-k)}(x \mid x) \\ &= \sum_{k=1}^{m} r^{(k)}(x \mid x) \sum_{n=k}^{m} p^{(n-k)}(x \mid x) = \sum_{k=1}^{m} r^{(k)}(x \mid x) \sum_{l=0}^{m-k} p^{(l)}(x \mid x) \\ &= \sum_{k=1}^{m} r^{(k)}(x \mid x)\left(1 + \sum_{l=1}^{m-k} p^{(l)}(x \mid x)\right)\end{aligned}$$

since $p^{(0)}(x \mid x) = 1$. Dividing by the left-hand side yields

$$1 = \sum_{k=1}^{m} r^{(k)}(x \mid x) \frac{1 + \sum_{l=1}^{m-k} p^{(l)}(x \mid x)}{\sum_{n=1}^{m} p^{(n)}(x \mid x)}$$

Now, let $m \to \infty$. If $\sum_{n=1}^{\infty} p^{(n)}(x \mid x) = p < \infty$, the fraction above converges to $(1 + p)/p > 1$, so that $r(x \mid x) = \sum_{k=1}^{\infty} r^{(k)}(x \mid x) < 1$; that is, x is transient by (21a). If the infinite series $\sum_{n=1}^{\infty} p^{(n)}(x \mid x) = \infty$ then the fraction approaches 1, and hence $r(x \mid x) = 1$; that is, x is recurrent by (21b).

Example 3.1.9

Earlier, we argued that a simple Bernoulli random walk on the set of all integers, that is, a discrete-time homogeneous Markov chain with transition diagram in Figure 3.1.13, is transient if $\frac{1}{2} < p < 1$. We can now use the criterion (23) to verify this statement. Take, for simplicity, $x = 0$. The n-step transition probability $p^{(n)}(0 \mid 0)$ is clearly zero for n odd, since a return to $x = 0$ in an odd number of steps is impossible. Thus, it remains to find only whether the infinite series

$$\sum_{n=1}^{\infty} p^{(2n)}(0 \mid 0) \tag{24}$$

converges or diverges. But for any $n = 1, 2, \ldots$, $p^{(2n)}(0 \mid 0)$ is just the probability of exactly n moves to the right (probability p) and exactly n moves to the left (probability $q = 1 - p$), or equivalently n successes in $2n$ i.i.d. Bernoulli trials with p the probability of success. Thus

$$p^{(2n)}(0\,|\,0) = \binom{2n}{n} p^n q^n, \qquad n = 1, 2, \ldots,$$

are the terms of the infinite series above. A simple sufficient condition for convergence of an infinite series $\sum a_n$ with positive terms is provided by the ratio test

$$\lim_{n\to\infty} \frac{a_{n+1}}{a_n} < 1. \tag{25}$$

Applied to the series (24), the ratios are

$$\frac{p^{(2(n+1))}(0\,|\,0)}{p^{(2n)}(0\,|\,0)} = \frac{\binom{2n+2}{n+1} p^{n+1} q^{n+1}}{\binom{2n}{n} p^n q^n} = \frac{\dfrac{(2n+2)!}{(n+1)!\,(n+1)!}}{\dfrac{(2n)!}{n!\,n!}} pq$$

$$= \frac{(2n+2)(2n+1)}{(n+1)(n+1)} pq = 4pq \frac{1 + \dfrac{1}{2n}}{1 + \dfrac{1}{n}} \to 4pq \quad \text{as} \quad n \to \infty.$$

However, $pq = p(1-p) \leq \frac{1}{4}$ with equality if and only if $p = q = \frac{1}{2}$. Consequently, as long as $p \neq q$, the ratio test shows that the series (24) converges, and the random walk is transient.

The case $p = q = \frac{1}{2}$ (symmetric walk) has to be handled differently, since the limit (25) is 1 and the quotient criterion does not apply. We can, however, use Stirling's formula† for the asymptotic behavior of the factorial

$$n! \sim \sqrt{2\pi}\, n^{n+1/2} e^{-n}$$

to conclude that for large n

$$\binom{2n}{n}\left(\frac{1}{4}\right)^n \sim \frac{1}{\sqrt{\pi n}}.$$

Thus, with $p = q = \frac{1}{2}$, the terms of the infinite series (24) decrease with n as $1/\sqrt{n}$, and so the series diverges to $+\infty$. It follows that the simple symmetric Bernoulli random walk is recurrent, a fact already mentioned several times.

Using the criterion (23) and Stirling's formula, it can be shown that a simple random walk on a planar lattice exhibits similar behavior. It is recurrent if and only if its transition probabilities in the north-south as well as the east-west

† See Appendix Three.

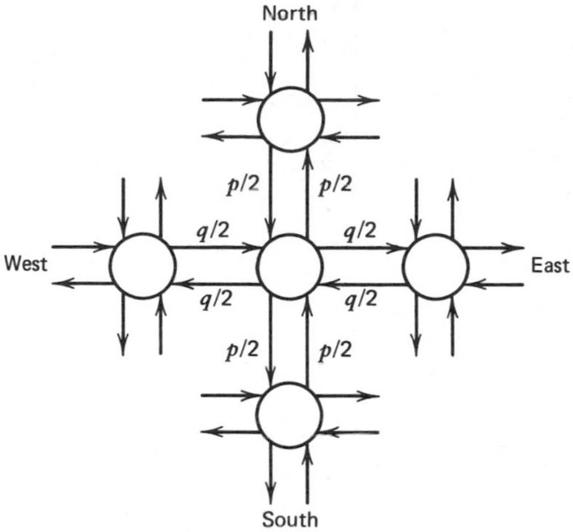

Figure 3.1.16

directions are symmetric, that is, if the transition diagram has the structure shown in Fig. 3.1.16 with $p + q = 1$. The fact that simple random walks on a linear or planar lattice are recurrent as long as they are symmetric may at first thought be attributed to the absence of drift, that is, the fact that the expected displacement of the particle performing a random walk starting at the origin is always zero. It may then come as a surprise to discover that a random walk on a lattice in a three-dimensional space (where each state has six neighbors) is transient even if it is symmetric. In other words, a particle performing such a random walk has a positive probability of never returning to its initial position. Thus, intuition may sometimes be of little value in guessing whether an infinite state Markov chain is transient or recurrent. ▲

The preceding discussion can now be summarized to give an overall picture of the state space of a homogeneous Markov chain. Each state is either transient or recurrent. Recurrent states, if any, can be further grouped into recurrent classes that are minimal closed subsets of recurrent states. Unless all states are transient (which is possible only if $\mathcal{S}$ is infinite), the Markov traveler starting from any state must eventually enter one of the recurrent classes and remain inside this class forever. (See Figure 3.1.17.)

We can thus distinguish two main areas of interest concerning the chain:

1. *Transient behavior*, which deals with problems related to behavior of the Markov traveler before entrance into a recurrence class. In this category, one may be interested in finding the probability of entering a particular recurrence class, the time when such an entrance occurs, and so forth.

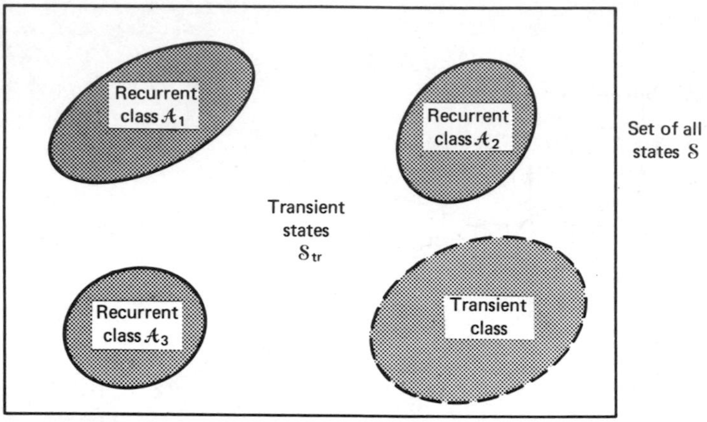

Figure 3.1.17

2. *Steady-state behavior*, that is, questions related to voyages of the Markov traveler after he has entered a recurrence class.

We shall study some of these topics in the following section.

EXERCISES

Exercise 3.1

1. A shift register is a digital logic device that can at any time hold an m-bit binary number. At each clock impulse, each of the m binary digits is shifted one place to the right. The leftmost digit is replaced by the current input digit, while the rightmost digit is released as an output of the register.

   Assuming that the input to the register is a sequence $U(n)$, $n = 0, 1, \ldots,$ of i.i.d. Bernoulli random variables with $P(U(n) = 1) = p$, show that the sequence $\mathbf{X}(n)$, $n = 0, 1, \ldots,$ of successive contents of the register is a homogeneous Markov chain. Find the transition probabilities of this chain.
2. Consider a random digit generator that generates a sequence of independent digits $0, \pm 1, \pm 2, \ldots, \pm 9$ each with equal probability. This sequence is fed into a memory device that operates according to the following rules:

   (i) If the current input digit is zero or if it is of the same sign but no larger in magnitude than the digit in the memory, the memory remains unchanged.

   (ii) If the digit in the memory is zero or if the input digit is of the same sign and if its magnitude exceeds that of the digit in the memory, the memory is erased and replaced by the input digit.

   (iii) If none of the above occurs, the memory digit is erased and replaced by zero.

Show that the sequence of consecutive memory states (digits) is a homogeneous Markov chain and find the transition probabilities.

3. Suppose $X(n)$, $n = 1, 2, \ldots$, is a Bernoulli process (see Chapter One). Define a process $Y(n)$, $n = 1, 2, \ldots$, by $Y(n) = 0$ if $X(n) = 0$, and $Y(n) = k$ if $X(n) = X(n-1) = \cdots = X(n-k+1) = 1$ but $X(n-k) = 0$, where $k = 1, \ldots, n$. ($Y(n)$ is called the length of a run of 1's at the time n.) Show that $Y(n)$, $n = 1, 2, \ldots$, is a homogeneous Markov chain and find its transition probabilities.

4. Consider a homogeneous Markov chain with state space $\mathcal{S} = \{1, 2, 3, 4, 5\}$ and transition probabilities:

$$p(x\,|\,1) = \begin{cases} \frac{1}{2} & \text{if} \quad x = 1 \text{ or } x = 3, \\ 0 & \text{otherwise,} \end{cases} \qquad p(x\,|\,2) = \begin{cases} \frac{1}{4} & \text{if} \quad x = 2, \\ \frac{3}{4} & \text{if} \quad x = 3, \\ 0 & \text{otherwise,} \end{cases}$$

$$p(x\,|\,3) = \begin{cases} \frac{1}{3} & \text{if} \quad x = 3, \\ \frac{2}{3} & \text{if} \quad x = 5, \\ 0 & \text{otherwise,} \end{cases} \qquad p(x\,|\,4) = \begin{cases} \frac{1}{4} & \text{if} \quad x = 1 \text{ or } x = 4, \\ \frac{1}{2} & \text{if} \quad x = 2, \\ 0 & \text{otherwise,} \end{cases}$$

$$p(x\,|\,5) = \begin{cases} \frac{1}{3} & \text{if} \quad x = 1 \text{ or } x = 3 \text{ or } x = 5, \\ 0 & \text{otherwise.} \end{cases}$$

Draw the transition diagram and write the transition probability matrix for this chain. Find the n-step transition probabilities $p^{(n)}(x\,|\,x')$ for $x' = 4$ and $n = 2, 3, 4, 5$.

5. For the chain in Exercise 3.1.4, find which states, if any, are transient. How many minimal closed sets of states are there?

6. A homogeneous Markov chain has a transition diagram as shown in Figure 3.1.E.1, where all indicated transitions from each particular state are equally likely. Verify that if the initial probability mass distribution is

$$p_0(x) = \begin{cases} \frac{1}{4} & \text{if} \quad x = 1 \quad \text{or} \quad x = 4, \\ \frac{1}{6} & \text{if} \quad x = 2 \quad \text{or} \quad x = 3 \quad \text{or} \quad x = 5, \end{cases}$$

then the chain is strictly stationary. Is there any other such initial probability mass distribution?

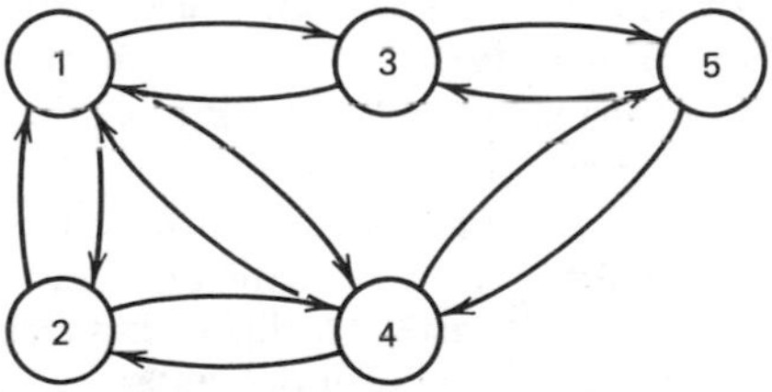

Figure 3.1.E.1

7. A power source contains an overvoltage safety device that protects the attached circuit against voltage surges. However, with each voltage surge, there is a positive probability α that the device fails to reduce the surge and a probability β ($\alpha + \beta < 1$) that the safety device burns, leaving the circuit unprotected against further surges.

With each voltage surge arriving at the circuit, there is a probability γ that the circuit will be damaged. Consider the three following states of the system: system fully operational, safety device burned, circuit damaged. Assuming independence between surges, argue that the sequence of the states of the system after each surge is a homogeneous Markov chain and find its transition probabilities. If at time $n = 0$ the system is fully operational, calculate the probability of each of its three states after n voltage surges.

8. Consider a homogeneous Markov chain with a transition diagram as drawn in Figure 3.1.E.2. Fill in the missing transition probabilities, arrange them in the transition probability matrix **P**, and classify each state as transient, recurrent, or absorbing. Write down the two-step transition probabilities $p^{(2)}(x \mid x')$, and verify that they coincide with entries of the second power of the matrix **P**. Find an initial probability mass distribution $p_0(x)$, $x = 0, \ldots, 5$, for which this chain becomes strictly stationary. Is there only one such distribution?

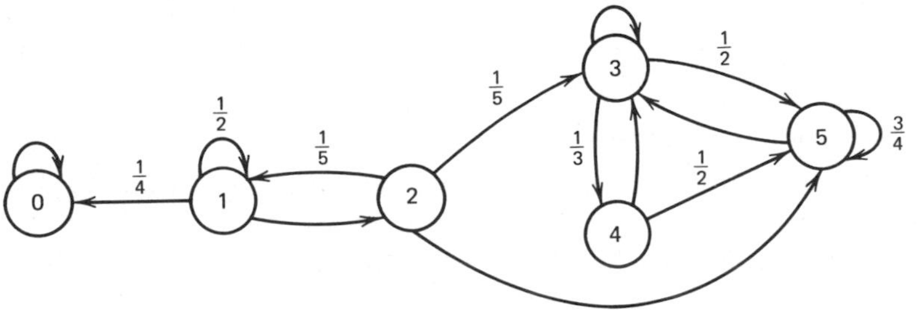

Figure 3.1.E.2

9. For the Markov chain $Y(n)$, $n = 1, 2, \ldots$, defined in Exercise 3.1.3, calculate the first return probabilities $r^{(n)}(0 \mid 0)$ and the transition probabilities $p^{(n)}(0 \mid 0)$ for all $n = 1, 2, \ldots$. Is the chain recurrent or transient? Call the (random) time T between two consective returns to the state $y = 0$ via a nonzero state a cycle of the runs. Find the expectation and the variance of the cycle T.

10. Consider a particle performing a random walk on the set of all integers. Suppose, however, that after each move, there is a probability p that the next move of the particle will be in the same direction and a probability $q = 1 - p$ that it will be in the opposite direction. (See Exercise 1.3.7 for the special case $p = \frac{3}{4}$.) Show that by a suitable choice of the state space $\mathcal{S}$, the motion of the particle can be described as a homogeneous Markov chain. Calculate the first return probability $r^{(n)}(x \mid x)$ for a few values of n. What is your opinion about the nature of this chain—is it recurrent or transient? (*Hint:* For each position x of the particle, consider two states, say, x_l and x_r, depending on the direction the position x is entered.)

11. Consider the chain of Example 3.1.8 with $a_n = 1/(n + 2)$, $n = 0, 1, \ldots$. Determine whether the chain is recurrent or transient. Do the same if $a_n = (n + 1)/(n + 2)$, $n = 0, 1, \ldots$.

12. A gene in a living cell of a primitive organism consists of r units, m of which are mutants and the rest normal. Prior to each division of the cell the gene duplicates. The genes in each of the two daughter cells are assumed to consist of r units chosen at random from the $2m$ mutant units and the $2(r-m)$ normal units of the two duplicate genes. Let $X(n)$ be the number of mutant units in the nth generation daughter cell in a particular line of descent. Argue that $X(n)$, $n = 0, 1, \ldots$, is a homogeneous Markov chain, find its transition probabilities, and classify the states.
13. Consider the branching process $N(n)$, $n = 1, 2, \ldots$, described in Exercise 1.48, where the random variables $X_{j,n}$ all have a Poisson distribution with parameter λ. Show that $N(n)$, $n = 1, 2, \ldots$, is a homogeneous Markov chain and find its transition probabilities. Use the Chapman-Kolmogrov equation (12) with $m = 1$ to obtain $p^{(n)}(0 \mid x')$, $n = 1, 2, \ldots$. Is this chain recurrent or transient?
14. Suppose that the transition probability matrix $\mathbf{P}$ of a finite-state homogeneous Markov chain has all its rows identical. Find the initial probability mass distribution for which this chain is strictly stationary. Is there any simpler way to characterize the probability law of such a chain?
15. For the simple random walk on a planar lattice (Fig. 3.1.16), use the multinomial probability mass distribution (see Appendix Two) to obtain an expression for the n-step transition probabilities $p^{(n)}(0 \mid 0)$. If m particles perform this random walk independently from one another, all starting at the origin, find the probability that at time n there will be exactly k of them back at the origin.

3.2. FURTHER ANALYSIS OF DISCRETE-TIME MARKOV CHAINS

TRANSIENT BEHAVIOR. Here our two main interests are the probability of entering a particular recurrent class and the random time (= number of steps) before entering any recurrent class (the duration of transition period) or before entering a particular recurrent class.

Let $X(n)$, $n = 0, 1, \ldots$, be a homogeneous Markov chain with state space $\mathcal{S}$; let $\mathcal{S}_{\text{tr}}$ denote the class of all transient states, and let $\mathcal{A}_k$, $k = 1, 2, \ldots$, be recurrent classes. That is, we have partitioned the set of all states (see Fig. 3.1.17).

$$\mathcal{S} = \mathcal{S}_{\text{tr}} \cup \mathcal{S}_{\text{rec}}, \tag{1a}$$

where

$$\mathcal{S}_{\text{rec}} = \mathcal{A}_1 \cup \mathcal{A}_2 \cup \mathcal{A}_3 \cup \cdots. \tag{1b}$$

Recall that all the unions appearing on the right-hand side of (1) are disjoint. First, we calculate the probability that the Markov traveler starting at $x \in \mathcal{S}$ enters a particular class $\mathcal{A}_k$. We denote this probability by

$$P(\mathcal{A}_k \mid x),$$

and call it the *absorption probability for the class* $\mathcal{A}_k$. Clearly, since the $\mathcal{A}$'s are disjoint recurrent classes,

$$P(\mathcal{A}_k|x) = 1 \qquad \text{if} \quad x \in \mathcal{A}_k,$$
$$P(\mathcal{A}_k|x) = 0 \qquad \text{if} \quad x \in \mathcal{A}_l, \qquad l \neq k. \tag{2}$$

This leaves the case $x \in \mathcal{S}_{tr}$. Suppose that the first step takes the Markov traveler to the state y. The probability of this is $p(y|x)$. But he starts afresh after visiting y, which means that the probability of entering $\mathcal{A}_k$ after visiting y is the same as if he had started at y, in other words, $P(\mathcal{A}_k|y)$. Summing over all y, we thus get the equation

$$P(\mathcal{A}_k|x) = \sum_{y \in \mathcal{S}} P(\mathcal{A}_k|y)p(y|x).\dagger \tag{3}$$

Using (2), we can write this as

$$P(\mathcal{A}_k|x) - \sum_{y \in \mathcal{S}_{tr}} P(\mathcal{A}_k|y)p(y|x) = \sum_{y \in \mathcal{A}_k} p(y|x), \qquad x \in \mathcal{S}_{tr}. \tag{4}$$

Now (4) is a system of linear equations in the unknowns $P(\mathcal{A}_k|x)$, $x \in \mathcal{S}_{tr}$, and it can be shown that it always has a solution. Thus the calculation of the absorption probabilities amounts to solving the system (4).

Example 3.2.1

Consider a particle performing a simple symmetric planar random walk between two horizontal boundaries, one at level 0 and the other at the level $m > 1$. The nature of the boundaries is such that once the particle reaches either of them, it is forever confined to travel back and forth along that boundary. A skeleton transition diagram of this chain is depicted in Figure 3.2.1. Since there is no lateral restriction the state space, $\mathcal{S}$ is infinite, consisting of all pairs $\mathbf{x} = (j, k)$ of integers $j = \ldots, -1, 0, 1, \ldots, k = 0, 1, \ldots, m$. The boundaries

$$\mathcal{A}_0 = \{(j, k) \in \mathcal{S}: k = 0\} \qquad \text{and} \qquad \mathcal{A}_1 = \{(j, k) \in \mathcal{S}: k = m\}$$

are assumed to be two recurrent classes; the rest of the states are all transient

$$\mathcal{S}_{tr} = \{(j, k) \in \mathcal{S}: 0 < k < m\}.$$

The transition probabilities for $\mathbf{x} = (j, k) \in \mathcal{S}_{tr}$ are

$$p(\mathbf{y}|\mathbf{x}) = \begin{cases} \frac{1}{4} & \text{if} \quad \mathbf{y} = (j \pm 1, k) \quad \text{or} \quad \mathbf{y} = (j, k \pm 1), \\ 0 & \text{otherwise,} \end{cases}$$

and, of course, $p(\mathbf{y}|\mathbf{x}) = 0$ if $\mathbf{y} \in \mathcal{S}_{tr}$ and $\mathbf{x} \in \mathcal{A}_0$ or $\mathbf{x} \in \mathcal{A}_1$. The transition probabi-

† We would like the reader to keep in mind the line of reasoning used to derive (3). It was, in fact, the same idea used to derive the relation (3.1.22). We will refer to it as the *first entrance argument*.

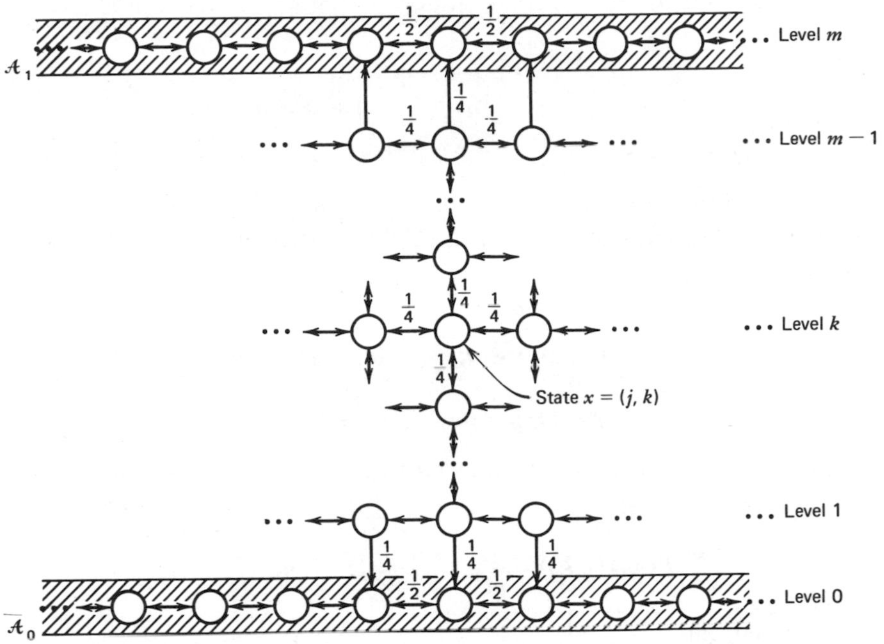

Figure 3.2.1

lities between states within $\mathcal{A}_0$ and $\mathcal{A}_1$ are left unspecified; we require only that both $\mathcal{A}_0$ and $\mathcal{A}_1$ each form a minimal recurrent class.

Such a chain may, for instance, serve as a crude model for the diffusion of charge carriers in a shallow channel of semiconductor material between two metallic boundaries or semiconductors of opposite type.

Assuming that the initial position $\mathbf{x}(0)$ of such a charge carrier is one of the transient states $\mathbf{x} = (j, k) \in \mathcal{S}_{tr}$, it is then of obvious interest to find the absorption probabilities $P(\mathcal{A}_0 \mid \mathbf{x})$ and $P(\mathcal{A}_1 \mid \mathbf{x})$. From the lateral symmetry, it should be clear that these absorption probabilities can only depend on the level k of the initial state $\mathbf{x} = (j, k)$ and not on the lateral j-coordinate. Let us temporarily denote

$$P(\mathcal{A}_1 \mid (j, k)) = a_k, \qquad k = 1, \ldots, m - 1.$$

The system of linear equations (4) then becomes

$$a_1 - \frac{1}{4}a_1 - \frac{1}{4}a_1 - \frac{1}{4}a_2 = 0,$$

$$a_k - \frac{1}{4}a_k - \frac{1}{4}a_k - \frac{1}{4}a_{k-1} - \frac{1}{4}a_{k+1} = 0, \qquad 1 < k < m - 1,$$

$$a_{m-1} - \frac{1}{4}a_{m-1} - \frac{1}{4}a_{m-1} - \frac{1}{4}a_{m-2} = \frac{1}{4},$$

or

$$a_2 = 2a_1,$$
$$a_{k+1} = 2a_k - a_{k-1}, \qquad 1 < k < m-1,$$
$$2a_{m-1} - a_{m-2} = 1.$$

By repeated substitution, we obtain $a_3 = 2a_2 - a_1 = 4a_1 - a_1 = 3a_1$, $a_4 = 2a_3 - a_2 = 6a_1 - 2a_1 = 4a_1, \ldots$, from which it is easily guessed and verified by induction that for $k = 1, \ldots, m-1$

$$a_k = ka_1.$$

Substitution into the last equation for $a_{m-2} = (m-2)a_1$ and $a_{m-1} = (m-1)a_1$ yields

$$2(m-1)a_1 - (m-2)a_1 = 1,$$

from which $a_1 = 1/m$. Hence, $a_k = k/m$, or returning to our original notation for the absorption probabilities,

$$P(\mathcal{A}_1 \mid (j, k)) = \frac{k}{m}, \qquad k = 1, \ldots, m.$$

Similarly, we find that

$$P(\mathcal{A}_0 \mid (j, k)) = \frac{m-k}{m}, \qquad k = 1, \ldots, m,$$

so that for any $x \in \mathcal{S}_{tr}$, $P(\mathcal{A}_0 \mid \mathbf{x}) + P(\mathcal{A}_1 \mid \mathbf{x}) = 1$; that is, eventual absorption in one of the two recurrent classes is certain. ▲

Example 3.2.2

The system of linear equations (4) is also useful for calculating various first entrance probabilities even if there are originally no transient states. To illustrate this kind of application, consider, for instance, a cyclic random walk with the transition diagram shown in Figure 3.1.5, where $\frac{1}{2} < p < 1$, $q = 1 - p$. Suppose that the Markov traveler begins at the initial state $X(0) = 0$, and we wish to find the probability that his first return to $x = 0$ occurs only after traveling the full circle, or equivalently the probability that he has visited all the states $x = 1, \ldots, m$ before returning to zero. Let us denote the event described above by R_{00}. Suppose the first transition the traveler makes is to the state $x = 1$. The probability of this is p. Upon reaching the state $x = 1$, the event R_{00} can occur if and only if he will return to zero *without ever making use of the transition* ⓪ ← ①. Let this event be denoted by B_{10}. Then the conditional probability of R_{00} given that the first transition is to the state $x = 1$ is, by the Markov property, just the probability $P(B_{10})$. Similarly, denoting by B_{m0} the probability that the Markov traveler returns to zero from the state m without ever making use of the transition ⓜ → ⓪,

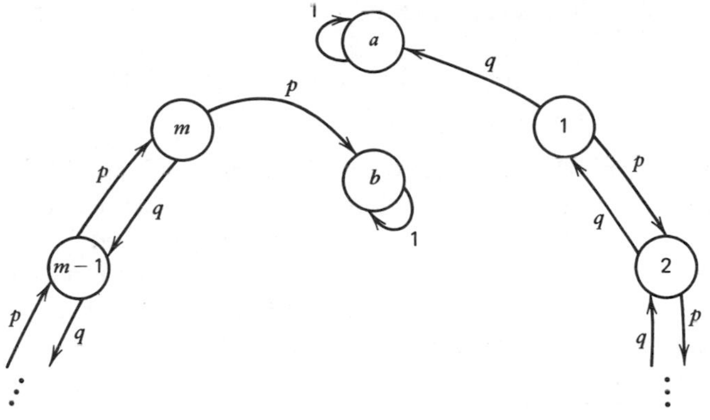

Figure 3.2.2

we find that $P(B_{m0})$ is the conditional probability of R_{00} given that the first transition is to the state $x = m$. Consequently,

$$P(R_{00}) = pP(B_{10}) + qP(B_{m0}). \tag{5}$$

To find the probability $P(B_{10})$, we modify the transition diagram in Figure 3.1.5 by disconnecting the link ⓜ → ⓪ and splitting the state $x = 0$ into two absorbing states a and b, as shown in Figure 3.2.2. From the definition of the event B_{10}, it is clear that the probability $P(B_{10})$ is then the same as the absorption probability $P(b \mid x = 1)$ starting from $x = 1$ in this modified chain. But the modified chain is just a Bernoulli random walk with $m + 2$ states and absorbing boundaries at each of the end states a and b, and thus $P(B_{10})$ is the probability of absorption at b starting from $x = 1$. Equivalently, it is the probability of an unrestricted Bernoulli random walk reaching level m before reaching level -1 starting from level 0. In Exercise 1.51 we found that this probability equals

$$\frac{\left(\dfrac{p}{q}\right)^{m+1} - \dfrac{p}{q}}{\left(\dfrac{p}{q}\right)^{m+1} - 1},$$

which is then also $P(B_{10})$. To find the remaining probability $P(B_{m0})$, we similarly modify the original chain by disconnecting the link ⓜ → ⓪, and exactly the same reasoning shows that

$$P(B_{m0}) = \frac{\left(\dfrac{q}{p}\right)^{m+1} - \dfrac{q}{p}}{\left(\dfrac{q}{p}\right)^{m+1} - 1}.$$

Substitution into (5) then yields the desired probability

$$P(R_{00}) = (p^2 + q^2)\frac{p^m - q^m}{p^{m+1} - q^{m+1}}. \qquad \blacktriangle$$

Next we calculate the distribution of the duration of the transient period. Let x be an initial state, and let T be the time at which the Markov traveler enters a recurrent class. We call T an *absorption time*, meaning absorption in the set $\mathcal{S}_{\text{rec}}$ of recurrent states. If there is only a finite number of transient states, T is a random variable with values 0, 1, ..., and $T = 0$ means that the initial state is recurrent, that is, the Markov traveler is in a recurrent class at time $n = 0$. Let

$$f_T(n \mid x) = P(T = n \mid x), \qquad n = 0, 1, \ldots \tag{6}$$

be the probability mass distribution of T, with x being the initial state.

Remark: If the number of transient states is infinite, it may happen that the Markov traveler will remain in the set of transient states forever. The definition (6) still makes sense, but

$$\sum_{n=0}^{\infty} f_T(n \mid x) = P(T < \infty \mid X(0) = x)$$

may now be strictly less than 1. Should this be the case, we refer to the "random variable" T and its "probability mass distribution" $f_T(n \mid x)$ as *defective* and $1 - \sum_{n=0}^{\infty} f_T(n \mid x)$ as its defect. The defect is just the probability of remaining in transient states forever starting from x.

Я

If $x \in \mathcal{S}_{\text{rec}}$, we have, according to the above,

$$f_T(n \mid x) = \begin{cases} 1 & \text{for } n = 0, \\ 0 & \text{for } n > 0. \end{cases} \tag{7}$$

Furthermore, it is not difficult to see that for every $x \in \mathcal{S}$, (6) satisfies the equation

$$f_T(n + 1 \mid x) = \sum_{y \in \mathcal{S}} f_T(n \mid y)p(y \mid x), \qquad n = 0, 1, \ldots \tag{8}$$

The reasoning is exactly the same as the one that led to (3), that is, the first entrance argument, and we won't repeat it. Again, (8) can be simplified somewhat; using (7), we have from (8) with $n = 0$

$$f_T(1 \mid x) = \sum_{y \in \mathcal{S}_{\text{rec}}} p(y \mid x) \qquad \text{for all } x \in \mathcal{S}_{\text{tr}} \tag{9}$$

which is, after all, obvious. Hence, for transient states (8) becomes

$$f_T(n + 1 \mid x) = \sum_{y \in \mathcal{S}_{\text{tr}}} f_T(n \mid y)p(y \mid x), \qquad n = 1, 2, \ldots, \quad x \in \mathcal{S}_{\text{tr}}. \tag{10}$$

We see that the distribution of the absorption time for transient initial states can be computed recursively from (10), using (9) as the initial condition. Of course, the probabilities (6) can also be calculated from the first entrance probabilities (3.1.20), since by the definition of the latter

$$f_T(n \mid x) = \sum_y r^{(n)}(y \mid x), \qquad n = 1, 2, \ldots, x \in \mathcal{S}_{tr}$$

where the summation is over the set of all recurrent states y such that $p(y \mid x) \neq 0$, and obviously $f_T(0 \mid x) = 0$ for $x \in \mathcal{S}_{tr}$. However, it is usually simpler to use equations (10).

Example 3.2.3

Although the recurrence relation (10) looks simple, it may still be quite difficult to obtain an explicit expression for the distributions $f_T(n \mid x)$. The difficulty, of course, arises from the fact that before we can compute $f_T(n + 1 \mid x)$ for some particular $x \in \mathcal{S}$, we must have computed $f_T(n \mid y)$ for *all* states $y \in \mathcal{S}_{tr}$ such that $p(y \mid x) \neq 0$. Even in simple cases such as a Bernoulli random walk with absorbing boundaries, a direct calculation may soon become rather formidable, and various analytical techniques have to be employed. On the other hand, a numerical computation of $f_T(n \mid x)$ is a simpler matter, especially if the number of transient states is finite. The reason for this is that equations (10) or, even more so, (7) and (8) are especially well suited for development of a computer algorithm. To illustrate this, suppose that for some n, all the numbers $f_T(n \mid x)$, $x \in \mathcal{S}$, have been computed and written next to the node representing state x in the transition diagram. For instance, a portion of a transition diagram involving states x, a, b, c may at the nth computational stage look like that shown in Figure 3.2.3, where only the arrows emanating from the state x have been drawn. Then from equation (8)

$$f_T(n + 1 \mid x) = 0.1 f_T(n \mid a) + 0.2 f_T(n \mid b) + 0.5 f_T(n \mid c) + 0.2 f_T(n \mid x),$$

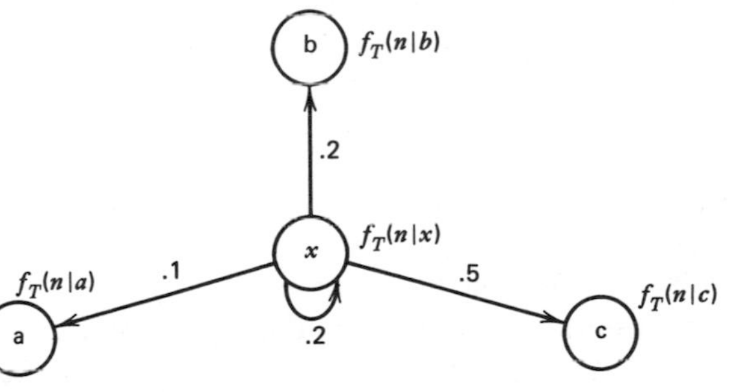

Figure 3.2.3

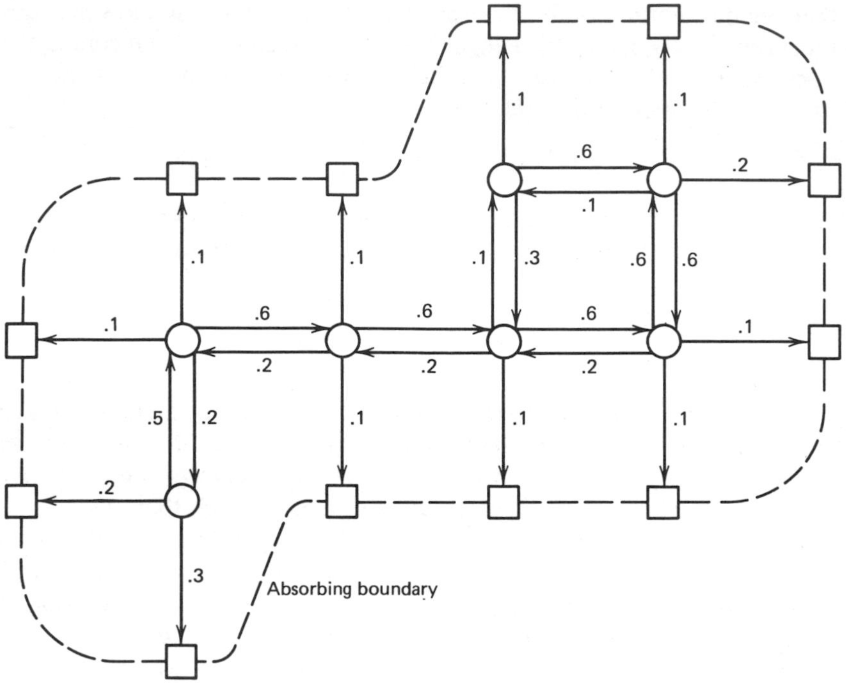

Figure 3.2.4

so that $f_T(n+1 \mid x)$ is calculated by simply summing the products of the numbers $f_T(n \mid y)$ at the points of the arrows times the transition probabilities associated with the arrows. These numbers are the new values $f_T(n+1 \mid x)$ for each state x. The initial values $f_T(0 \mid x)$ are obtained from (7) by assigning 1 to each absorbing state and 0 to each transient state.

As an illustration of this procedure, consider a particle moving on a planar lattice with rather irregular absorbing boundaries defined in the transition diagram depicted in Figure 3.2.4; the boundary states are drawn as little squares. The probabilities $f_T(n \mid x)$ evaluated by the first three steps of this algorithm are shown in Figure 3.2.5, with the values of the transition probabilities omitted for better clarity.

Clearly, this iterative procedure is relatively easy to implement on a computer, especially if a parallel data processing mode is available. Naturally, only a finite number of steps can be performed, but since for each $x \in \mathcal{S}$, $f_T(n \mid x)$, $n = 1, 2, \ldots,$ is a probability distribution, a record can be kept of the sum $f_T(1 \mid x) + f_T(2 \mid x) + \cdots + f_T(n \mid x)$ and the algorithm stopped as soon as the sum is as close to 1 as desired. ▲

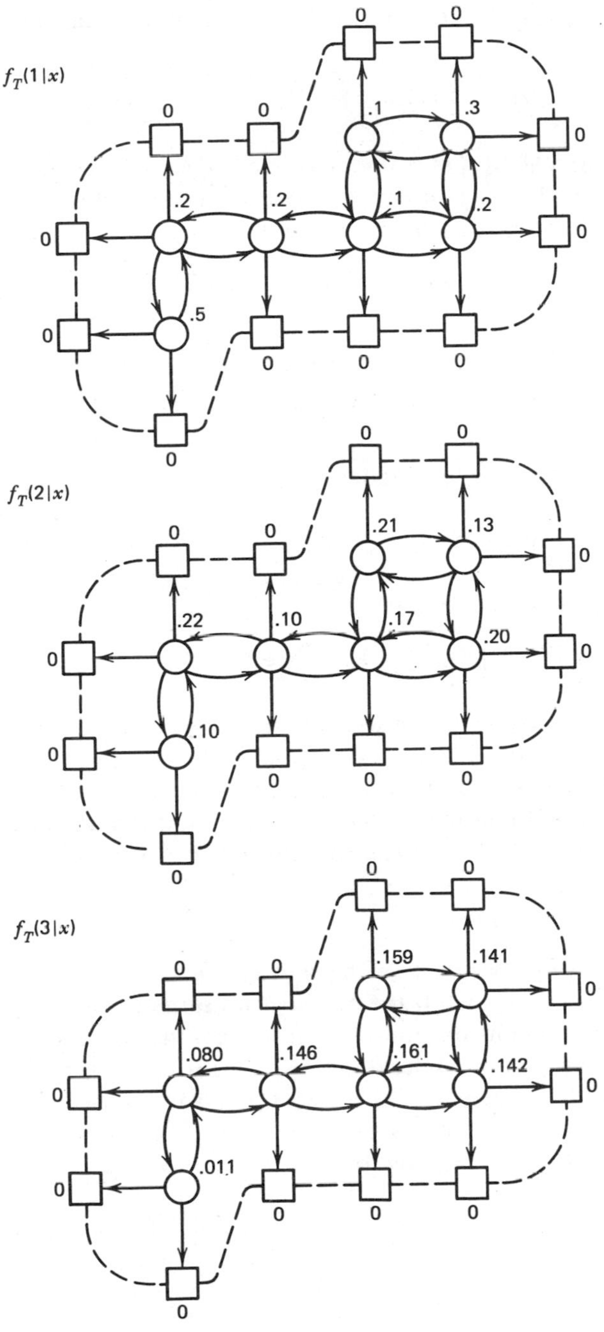
$f_T(1|x)$
.1
.3
.2
.2
.1
.2
.5
$f_T(2|x)$
.21
.13
.22
.10
.17
.20
.10
$f_T(3|x)$
.159
.141
.080
.146
.161
.142
.011

Figure 3.2.5

Sometimes, one may be interested in the expected duration of the transition period, that is, in the expected value

$$\mathrm{E}[T \mid x] = \sum_{n=0}^{\infty} n f_T(n \mid x).$$

(Of course, this makes sense only if $f_T(\ \mid x)$ is not defective.) Although it can be calculated from known $f_T(n \mid x)$, it is again simpler to obtain it directly from equation (8). Multiplying (8) on both sides by n, we can write

$$(n + 1)f_T(n + 1 \mid x) - f_T(n + 1 \mid x) = \sum_{y \in \mathcal{S}} n f_T(n \mid y)p(y \mid x).$$

Summing over $n = 0, 1, \ldots$, this gives

$$\mathrm{E}[T \mid x] - \sum_{n=1}^{\infty} f_T(n \mid x) = \sum_{y \in \mathcal{S}} \mathrm{E}(T \mid y)p(y \mid x)$$

and since $\mathrm{E}[T \mid x] = 0$ for $x \in \mathcal{S}_{\text{rec}}$, while for nondefective $f_T(n \mid x)$

$$\sum_{n=1}^{\infty} f_T(n \mid x) = 1 \qquad \text{for} \quad x \in \mathcal{S}_{\text{tr}},$$

we obtain a system of linear equations

$$\mathrm{E}[T \mid x] - \sum_{y \in \mathcal{S}_{\text{tr}}} \mathrm{E}[T \mid y]p(y \mid x) = 1, \qquad x \in \mathcal{S}_{\text{tr}}. \tag{11}$$

Note that this system of equations could have also been derived by again appealing to the first entrance argument, for if y is the state visited by the traveler at the time $n = 1$, then the expected time to absorption starting from x at time $n = 0$ and being at y at the time $n = 1$ equals 1 plus the expected time to absorption starting from y at the time $n = 1$. Multiplying by $p(y \mid x)$ and adding over $y \in \mathcal{S}$ leads again to (11).

Example 3.2.4

Consider a simplified model of a control system with states $x = 0, 1, \ldots$ operating in discrete time steps. From any state $x' > 0$ the system tries to reach the state zero, but because of some imperfections or random disturbances the next state x of the system could actually be any of the states $0, 1, \ldots, x' - 1$ with equal probability. Thus, if we assume that the sequence $X(n)$, $n = 0, 1, \ldots$, of the consecutive states of the system forms a discrete-time homogeneous Markov chain, the transition probabilities will, for $x' = 1, 2, \ldots$, be given by

$$p(x \mid x') = \begin{cases} \dfrac{1}{x'} & \text{if} \quad x = 0, 1, \ldots, x' - 1, \\[2ex] 0 & \text{if} \quad x \geq x'. \end{cases}$$

The state $x = 0$ is considered absorbing, $p(0 \mid 0) = 1$.

We now wish to find the expected time to absorption given that the initial state of the system $X(0) = k > 0$. For brevity, let

$$\mathrm{E}[T \mid X(0) = k] = t_k, \qquad k = 1, 2, \ldots.$$

Then the system of linear equations (11) takes the form

$$t_k - \frac{1}{k}(t_1 + t_2 + \cdots + t_{k-1}) = 1, \qquad k = 1, 2, \ldots, \tag{12}$$

or, in multiplying the kth equation in (12) by k, we have

$$\begin{aligned} t_1 \qquad\qquad\qquad &= 1, \\ -t_1 + 2t_2 \qquad\quad &= 2, \\ -t_1 - t_2 + 3t_3 \quad\; &= 3, \\ -t_1 - t_2 - t_3 + 4t_4 &= 4. \end{aligned}$$

Solving successively starting from the first equation, we get

$$t_1 = 1,$$

$$t_2 = 1 + \frac{1}{2},$$

$$t_3 = 1 + \frac{1}{3}\left(1 + 1 + \frac{1}{2}\right) = 1 + \frac{1}{3}\left(1 + \frac{3}{2}\right) = 1 + \frac{1}{2} + \frac{1}{3},$$

$$t_4 = 1 + \frac{1}{4}\left(1 + 1 + \frac{1}{2} + 1 + \frac{1}{2} + \frac{1}{3}\right) = 1 + \frac{1}{4}\left(1 + \frac{4}{2} + \frac{4}{3}\right) = 1 + \frac{1}{2} + \frac{1}{3} + \frac{1}{4},$$

which suggests that generally for $k = 1, 2, \ldots,$

$$t_k = 1 + \frac{1}{2} + \frac{1}{3} + \cdots + \frac{1}{k}.$$

Indeed, if we write

$$t_1 = 1,$$

$$t_2 = 1 + \frac{1}{2},$$

$$t_3 = 1 + \frac{1}{2} + \frac{1}{3},$$

$$\cdots$$

$$t_{k-1} = 1 + \frac{1}{2} + \frac{1}{3} + \cdots + \frac{1}{k-1}$$

and add these equations, we find that

$$t_1 + \cdots + t_{k-1} = k - 1 + (k-2)\frac{1}{2} + (k-3)\frac{1}{3} + \cdots + [(k-(k-1)]\frac{1}{k-1}$$

$$= k - 1 + \left(\frac{k}{2} - 1\right) + \left(\frac{k}{3} - 1\right) + \cdots + \left(\frac{k}{k-1} - 1\right)$$

$$= \frac{k}{1} + \frac{k}{2} + \frac{k}{3} + \cdots + \frac{k}{k-1} - (k-1) = k\left(1 + \frac{1}{2} + \frac{1}{3} + \cdots + \frac{1}{k} - 1\right)$$

$$= k(t_k - 1)$$

so that equation (12) is satisfied. Thus the expected time to absorption equals

$$\mathrm{E}[T \mid X(0) = k] = 1 + \frac{1}{2} + \frac{1}{3} + \cdots + \frac{1}{k}$$

for any $k = 1, 2, \ldots$. ▲

The absorption probabilities and the absorption time are by no means all the quantities of interest associated with the transition behavior of the chain. One may, for instance, wish to find the distribution of the time T_k to enter a particular recurrent class $\mathcal{A}_k$ (this distribution may again be defective, since $\mathcal{A}_k$ need not be entered at all), the number of visits to some transient states before absorption, and many others. We will not dwell upon these questions any longer. We would only like to emphasize that practically all these problems can be solved by some modification of the first entrance argument.

STEADY-STATE BEHAVIOR. We will now study the voyages of the Markov traveler inside a recurrent class. Since an escape from a recurrent class is impossible or, which is the same thing, since the sum over all x in a recurrence class

$$\sum_x p(x \mid x') = 1$$

for all x' in the same recurrence class, we may without any loss of generality consider a homogeneous Markov chain with all states forming a single recurrent class. We call such a chain *recurrent.*

We already know that the Markov traveler must visit each state of a recurrent chain infinitely often regardless of the initial state. This indicates that the system whose states form a recurrent Markov chain should eventually reach a dynamic equilibrium. More precisely, we expect that all quantities, such as the frequency of visits to a state, the times between returns to a state, the times to enter one state from another state, and so forth, should converge to a limiting value as the time n goes to infinity and that their limits should be independent of the initial state. In other words, the effect of the initial state should eventually wear off and have no influence on the equilibrium values. We will see shortly that this is indeed the case.

Let us begin by considering a particular state $x \in \mathcal{S}$. Since x is recurrent, the Markov traveler must eventually visit x and then keep returning to x infinitely often.

Define a sequence of random variables

$$T_x(1),\ T_x(2),\ T_x(3),\ \ldots,$$

where $T_x(1)$ is the time (i.e., number of steps) it takes to return to x for the first time, $T_x(2)$ is the time to return to x for the second time, and so on. In general, $T_x(n)$ is the time between the nth and $(n+1)$st visit to the state x. We refer to these random variables as *recurrence times* (for the state x).

We make the following important observation:

For any given state, the sequence of recurrence times is a sequence of independent, identically distributed random variables with values 1, 2, ...

The truth of this statement follows immediately from the Markov property: Since the Markov traveler starts afresh each time he visits x the random variables $T_x(1)$, $T_x(2)$, ... must be independent. Since the chain is homogeneous in time, the distribution of $T_x(1)$ must be the same as that of $T_x(2)$ and so forth.

Consider now the average recurrence time for n returns,

$$\bar{T}_x(n) = \frac{1}{n}\,(T_x(1) + \cdots + T_x(n)).$$

Since the recurrence times $T_x(1)$, $T_x(2)$, ... are i.i.d. positive random variables, their common expected value

$$\mu_x = \sum_{k=1}^{\infty} kP(T_x(1) = k) = \sum_{k=1}^{\infty} kr^{(k)}(x\,|\,x) \tag{13}$$

is either positive and finite or $+\infty$ if the above series diverges. In either case, by the law of large numbers,

$$\bar{T}_x(n) \to \mu_x(x) \qquad \text{as} \qquad n \to \infty. \tag{14}$$

The quantity μ_x given by (13) is called the *mean recurrence time of the state* x and allows us to further distinguish between two distinct categories of recurrent states.

A recurrent state $x \in \mathcal{S}$ is called *positive* if $\mu_x < \infty$ and *null* if $\mu_x = \infty$. Now, for *any recurrent chain, either all states are positive or all states are null, the latter case being possible only if the number of states is infinite.*

To see this, note that for any pair x and y of states in a recurrent chain, there is a nonzero probability that the Markov traveler will visit y between successive returns to x. Hence, if the average recurrence time to y tends to infinity, the average recurrence time to x must also tend to infinity. This clearly cannot happen if there is only a finite number, say m, of states in $\mathcal{S}$, since the Markov traveler must, during each $m + 1$ steps, visit some state at least twice.

Next, let us consider a recurrent chain $X(n)$ with initial state $X(0) = x'$, and another state $x \in \mathcal{S}$. For each $n = 1, 2, \ldots$, let $K_x(n)$ be the number of visits to the state x up to and including the time n. Using the recurrence time $T_x(k)$ defined above, we can write

$$n = N_1 + T_x(1) + \cdots + T_x(K_x(n)) + N_2, \tag{15}$$

where N_1 is the time needed to enter x for the first time, and N_2 is the time from the last (i.e., $K_x(n)$th) visit to x up to n. Since the chain is recurrent, $K_x(n) \to \infty$ as $n \to \infty$, and hence both

$$\frac{N_1}{K_x(n)} \quad \text{and} \quad \frac{N_2}{K_x(n)}$$

go to zero. Consequently, by dividing (15) by $K_x(n)$, we see that both

$$\frac{n}{K_x(n)} \quad \text{and} \quad \frac{T_x(1) + \cdots + T_x(K_x(n))}{K_x(n)}$$

converge to the same limit (possibly $+\infty$). But by (14) the latter expression converges to μ_x and hence also

$$\frac{n}{K_x(n)} \to \mu_x \quad \text{as} \quad n \to \infty.$$

This also implies that as $n \to \infty$,

$$\frac{K_x(n)}{n} \to \pi(x) \quad \text{where} \quad \pi(x) = \frac{1}{\mu_x}. \tag{16}$$

Note that $0 \le \pi(x) \le 1$, since $1 \le \mu(x) \le +\infty$, and that the numbers $\pi(x)$, $x \in \mathcal{S}$, are independent of the initial state $x' \in \mathcal{S}$. Now $K_x(n)$ can be written as a sum of Bernoulli random variables

$$K_x(n) = I_x(1) + \cdots + I_x(n), \tag{17}$$

where

$$I_x(k) = \begin{cases} 1 & \text{if} \quad X(k) = x, \\ 0 & \text{if} \quad X(k) \neq x. \end{cases}$$

But

$$\mathrm{E}[I_x(k)] = P(I_x(k) = 1) = P(X(k) = x) = p^{(k)}(x \mid x'),$$

so that by (17) and (16)

$$\pi(x) = \lim_{n\to\infty} \frac{1}{n} \sum_{k=1}^{n} p^{(k)}(x\,|\,x') \tag{18}$$

for every $x \in \mathcal{S}$ and any initial state $x' \in \mathcal{S}$.

Multiplying both sides of this equation by the transition probability $p(y\,|\,x)$ and summing over all $x \in \mathcal{S}$, we get

$$\sum_{x\in\mathcal{S}} p(y\,|\,x)\pi(x) = \lim_{n\to\infty} \frac{1}{n} \sum_{k=1}^{n} \sum_{x\in\mathcal{S}} p(y\,|\,x)p^{(k)}(x\,|\,x')$$

$$= \lim_{n\to\infty} \frac{1}{n} \sum_{k=1}^{n} p^{(k+1)}(y\,|\,x') = \lim_{n\to\infty} \frac{1}{n} \sum_{k=1}^{n} p^{(k)}(y\,|\,x') \tag{19}$$

since

$$\sum_{k=1}^{n} p^{(k+1)}(y\,|\,x') = \sum_{k=1}^{n} p^{(k)}(y\,|\,x') - p^{(1)}(y\,|\,x') + p^{(n+1)}(y\,|\,x),$$

the last two terms going to zero when divided by $n \to \infty$. Applying (18) to the last term in (19), we obtain the equation

$$\sum_{x\in\mathcal{S}} p(y\,|\,x)\pi(x) = \pi(y), \tag{20}$$

which the numbers $\pi(y)$ must satisfy for each $y \in \mathcal{S}$.

If the recurrent chain is null, then by definition $\mu_x = \infty$, that is, $\pi(x) = 0$ for all $x \in \mathcal{S}$, in which case (20) is trivially satisfied.

If the chain is positive recurrent, however, then the sum

$$\sum_{x\in\mathcal{S}} \pi(x)$$

is not zero and, in fact, must be equal to 1. To see this, note that upon summing both sides of (18) over a finite subset of states, $\mathcal{A} \subset \mathcal{S}$, and interchanging the summation with the limit (which is legitimate for a finite sum), we obtain

$$\sum_{x\in\mathcal{A}} \pi(x) = \lim_{n\to\infty} \frac{1}{n} \sum_{k=1}^{n} \sum_{x\in\mathcal{A}} p^{(k)}(x\,|\,x') \le 1$$

Thus, if $\mathcal{S}$ is itself finite, we have immediately $\sum_{x\in\mathcal{S}} \pi(x) = 1$, while if $\mathcal{S}$ is infinite, we can only conclude that

$$0 < \sum_{x\in\mathcal{S}} \pi(x) \le 1$$

for a positive recurrent chain. But then

$$\pi'(x) = \frac{\pi(x)}{\sum_{x\in\mathcal{S}} \pi(x)}, \qquad x \in \mathcal{S} \tag{21}$$

is a probability distribution on the set of states $\mathcal{S}$, so that we can use it as the initial distribution

$$\pi'(x) = p_0(x) = P(X(0) = x)$$

of the chain $X(n)$. For this choice, however, we will have

$$p_n(x) = p_0(x) \qquad \text{for all } n = 0, 1, \ldots \text{ and all } x \in \mathcal{S}$$

since the chain will be a strictly stationary process according to Theorem 3.1.1.

Now for such an initial distribution

$$\mathrm{E}[I_x(k)] = p_k(x) = \pi'(x),$$

and thus by (16), $\pi'(x) = \pi(x)$, since the $\pi(x)$ were independent of the initial state and hence also the initial distribution. Thus, by (21)

$$\sum_{x \in \mathcal{S}} \pi(x) = 1$$

as asserted.

THEOREM 3.2.1. *Let* X(n), n = 0, 1, ... *be a homogeneous recurrent Markov chain with state space* $\mathcal{S}$. *Then for any initial state* x′ ∈ $\mathcal{S}$ *the limits*

$$\lim_{n \to \infty} \frac{1}{n} \sum_{k=1}^{n} p^{(k)}(x \mid x') = \pi(x), \qquad x \in \mathcal{S} \tag{22}$$

always exist and satisfy the linear system of equations

$$\sum_{x \in \mathcal{S}} p(y \mid x)\pi(x) = \pi(y), \qquad y \in \mathcal{S}. \tag{23}$$

The chain is positive recurrent if and only if the system (23) *has a solution such that*

$$\pi(x) \geq 0 \quad \text{for all } x \in \mathcal{S} \quad \text{and} \quad \sum_{x \in \mathcal{S}} \pi(x) = 1, \tag{24a,b}$$

Otherwise the chain is null recurrent and

$$\pi(x) = 0 \qquad \text{for all } x \in \mathcal{S}.$$

In either case the mean recurrence time of a state x *equals*

$$\mu_x = \frac{1}{\pi(x)} \qquad \text{for all } x \in \mathcal{S}.$$

Example 3.2.5

Let us examine a Bernoulli random walk with the state space $\mathcal{S} = \{0, 1, \ldots\}$, where the state 0 is an elastic boundary. The transition diagram is reproduced in Figure

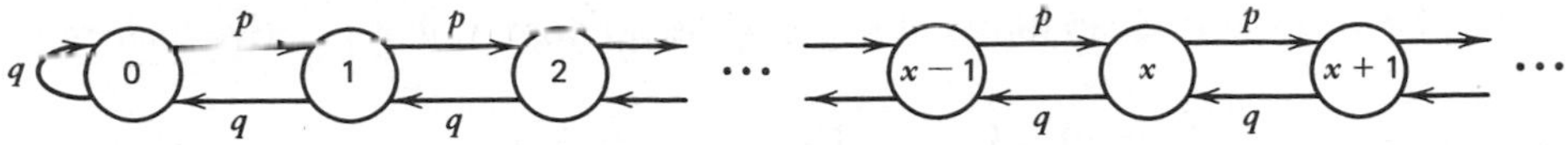

Figure 3.2.6

3.2.6, where $0 < p \le \frac{1}{2}$, $q = 1 - p$, since we already know that with $p > \frac{1}{2}$ the chain would be transient.

From the transition diagram we can readily write the system of equations (23), and we have

$$q\pi(0) + q\pi(1) = \pi(0), \tag{25}$$

and for $y = 1, 2, \ldots,$

$$p\pi(y-1) + q\pi(y+1) = \pi(y). \tag{26}$$

From (25)

$$\pi(1) = \frac{p}{q}\pi(0),$$

and rewriting (26) as

$$\pi(y+1) = \frac{1}{q}\pi(y) - \frac{p}{q}\pi(y-1),$$

we find by repeated substitution that

$$\pi(2) = \left(\frac{p}{q}\right)^2 \pi(0), \qquad \pi(3) = \left(\frac{p}{q}\right)^3 \pi(0), \qquad \text{etc.}$$

In general

$$\pi(x) = \left(\frac{p}{q}\right)^x \pi(0), \qquad x = 0, 1, \ldots,$$

as is easily proved by induction. Summing over all $x = 0, 1, \ldots,$ we find that

$$\sum_{x \in S} \pi(x) = \sum_{x=0}^{\infty} \left(\frac{p}{q}\right)^x \pi(0) \tag{27}$$

so that a solution satisfying the conditions (24) will exist if and only if the infinite series

$$\sum_{x=0}^{\infty} \left(\frac{p}{q}\right)^x < \infty. \tag{28}$$

If $0 < p < \frac{1}{2}$, this is indeed a convergent geometric series with the sum $q/(q-p)$, and hence, setting the right-hand side of (27) equal to 1, we have $\pi(0) =$

$(q-p)/q = 1 - p/q$. Therefore, if $0 < p < \frac{1}{2}$ the random walk is positive recurrent,

$$\pi(x) = \left(1 - \frac{p}{q}\right)\left(\frac{p}{q}\right)^x, \qquad x = 0, 1, \ldots,$$

is its probability mass distribution (this is in fact the modified geometric distribution), and the mean recurrence times

$$\mu_X = \frac{q-p}{q}\left(\frac{q}{p}\right)^x, \qquad x = 0, 1, \ldots$$

are seen to increase geometrically with the distance of the state x from the elastic boundary.

However, if $p = \frac{1}{2}$, $p/q = 1$, the series (28) diverges, and thus no solution of the system (25, 26) that would satisfy the conditions (24) can possibly exist. It follows that with $p = \frac{1}{2}$ the random walk is null recurrent, and $\pi(x) = 0$, $\mu_x = +\infty$ for all $x \in \mathcal{S}$. Recalling that for $\frac{1}{2} < p \leq 1$ the walk is transient we see how its character changes as p increases from zero to one. ▲

We have seen that in a recurrent chain the fraction of the time the Markov traveler spends in a particular state x converges to a limit $\pi(x)$, which does not depend on the initial state $X(0) = x'$ or, more generally, on the initial distribution $p_0(x')$.

The question remains whether the first-order distributions $p_n(x) = P(X(n) = x)$ also converge to a limit regardless of the initial distribution $p_0(x')$. We will shortly see that, except for some rather special cases, this is indeed so. But first we will discuss the significance of such a result.

Thus, suppose that for every state $x \in \mathcal{S}$ the limit

$$\lim_{n\to\infty} p_n(x)$$

exists and is independent of the initial distribution $p_0(x')$. Then we must have

$$\lim_{n\to\infty} p_n(x) = \lim_{n\to\infty} p^{(n)}(x \mid x') = \pi(x). \tag{29}$$

The first equality above follows from the fact that if $p_n(x)$ converges for *any* initial distribution $p_0(x')$, then it converges for any degenerate initial distribution $p_0(x) = 1$ if $x = x'$, in which case $p_n(x) = p^{(n)}(x \mid x')$; conversely, if $p^{(n)}(x \mid x')$ converges to the same limit for any $x \in \mathcal{S}$, then

$$p_n(x) = \sum_{x' \in \mathcal{S}} p^{(n)}(x \mid x') p_0(x')$$

converges to that limit for any $p_0(x')$. The second equality in (29) is a consequence of the fact that whenever a sequence of numbers converges to a limit, then the corresponding sequence of averages, that is,

$$\frac{1}{n}\sum_{k=1}^{n} p^{(k)}(x \mid x')$$

in our case, must converge to the same limit, that is, $\pi(x)$ according to (22).

Now $p_n(x) = P(X(n) = x)$ is the probability of the event that the Markov traveler at time n occupies the state x. Using the frequency interpretation of probability, we can equivalently imagine a large population of Markov travelers (or particles if you prefer) who travel simultaneously through the chain. Then $p_n(x)$ can be viewed as the *fraction of this population that at time n occupy the state x.*

Loosely speaking (29) means that for all n sufficiently large, we have $p_n(x) \doteq \pi(x)$ for each $x \in \mathcal{S}$, that is, the fraction of the population of Markov travelers occupying any given state remains unchanged. Of course, each Markov traveler keeps wandering through the set of states, visiting each state over and over again. The macroscopic properties of the system have stabilized, since the distribution of Markov travelers over states no longer changes with time. In other words, the system has reached a dynamic equilibrium.

Remark: This dynamic equilibrium can also be characterized by a Kirchhoff type of conservation law governing the flow of probability, that is, a current where Markov travelers play the role of electrons.

This law may be stated by saying that in the equilibrium state the *flow of probability out of any subset of states equals the flow of probability into this subset.*

Formally, let $\mathcal{S}_1 \subset \mathcal{S}$ be a subset of states and let $\mathcal{S}_2 = \mathcal{S} - \mathcal{S}_1$ be its complement. Then the flow of probability out of $\mathcal{S}_1$, that is, into $\mathcal{S}_2$, equals

$$P(\mathcal{S}_1 \to \mathcal{S}_2) = \sum_{y \in \mathcal{S}_2} \sum_{x \in \mathcal{S}_1} p(y \mid x)\pi(x),$$

while the flow into $\mathcal{S}_1$, that is, out of $\mathcal{S}_2$, equals

$$P(\mathcal{S}_2 \to \mathcal{S}_1) = \sum_{y \in \mathcal{S}_1} \sum_{x \in \mathcal{S}_2} p(y \mid x)\pi(x),$$

since at equilibrium we have $p_n(x) \doteq \pi(x)$ for large n. The law now says that *for any* $\mathcal{S}_1 \subset \mathcal{S}$ *and* $\mathcal{S}_2 = \mathcal{S} - \mathcal{S}_1$, *or equivalently that for any partition of* $\mathcal{S}$ *into* $\mathcal{S}_1$ *and* $\mathcal{S}_2$ *we have*

$$P(\mathcal{S}_1 \to \mathcal{S}_2) = P(\mathcal{S}_2 \to \mathcal{S}_1). \tag{30}$$

To see this, write the equations

$$\sum_{x \in \mathcal{S}} p(y \mid x)\pi(x) = \pi(y), \qquad y \in \mathcal{S}, \tag{31}$$

as

$$\sum_{x \in \mathcal{S}_1} p(y \mid x)\pi(x) + \sum_{x \in \mathcal{S}_2} p(y \mid x)\pi(x) = \pi(y), \qquad y \in \mathcal{S}. \tag{32}$$

Now add (32) first over $y \in \mathcal{S}_2$, second over $y \in \mathcal{S}_1$, and then subtract this first equation so obtained from the second. This gives

$$P(\mathcal{S}_1 \to \mathcal{S}_2) + \sum_{y \in \mathcal{S}_2} \sum_{x \in \mathcal{S}_2} p(y \mid x)\pi(x) - P(\mathcal{S}_2 \to \mathcal{S}_1) - \sum_{y \in \mathcal{S}_1} \sum_{x \in \mathcal{S}_1} p(y \mid x)\pi(x) = \sum_{y \in \mathcal{S}_2} \pi(y) - \sum_{y \in \mathcal{S}_1} \pi(y). \tag{33}$$

However,

$$\sum_{y \in \mathcal{S}_2} p(y \mid x) = 1 - \sum_{y \in \mathcal{S}_1} p(y \mid x),$$

so that

$$\begin{aligned}
\sum_{y \in \mathcal{S}_2} \sum_{x \in \mathcal{S}_2} p(y \mid x)\pi(x) - \sum_{y \in \mathcal{S}_1} \sum_{x \in \mathcal{S}_1} p(y \mid x)\pi(x)
&= \sum_{x \in \mathcal{S}_1} \pi(x) - \sum_{y \in \mathcal{S}_1} \sum_{x \in \mathcal{S}_2} p(y \mid x)\pi(x) - \sum_{y \in \mathcal{S}_1} \sum_{x \in \mathcal{S}_1} p(y \mid x)\pi(x) \\
&= \sum_{x \in \mathcal{S}_2} \pi(x) - \sum_{y \in \mathcal{S}_1} \sum_{x \in \mathcal{S}} p(y \mid x)\pi(x) = \sum_{y \in \mathcal{S}_2} \pi(y) - \sum_{y \in \mathcal{S}_1} \pi(y)
\end{aligned}$$

by replacing x by y in the first sum and using (31) in the second sum. Substitution into (33) yields (30).

In fact, the law (30) is equivalent to the system of equations (31) and may also be used to solve this system for $\pi(n)$. ∎

Example 3.2.6

Let us take the Ehrenfests' chain described in Example 3.1.4 and let us partition the state space $\mathcal{S} = \{0, \ldots, m\}$ into subsets $\mathcal{S}_1 = \{0, \ldots, x - 1\}$, $\mathcal{S}_2 = \{x, \ldots, m\}$, where x is an arbitrary but fixed state, $x = 1, \ldots, m$. From the transition diagram in Fig. 3.2.7, it is seen that the probability flows are

$$P(\mathcal{S}_1 \to \mathcal{S}_2) = \frac{m - (x - 1)}{m} \pi(x - 1),$$

$$P(\mathcal{S}_2 \to \mathcal{S}_1) = \frac{x}{m} \pi(x),$$

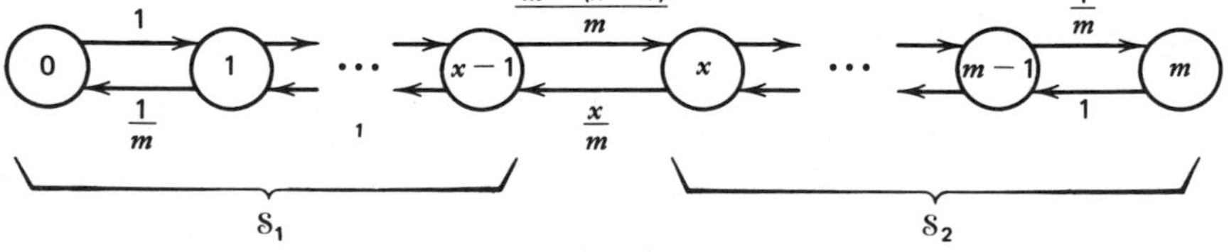

Figure 3.2.7

Equating the flows gives us a simple recurrence formula

$$\pi(x) = \frac{m - (x - 1)}{x} \pi(x - 1),$$

valid for all $x = 1, \ldots, m$. Hence

$$\pi(1) = m\pi(0), \qquad \pi(2) = \frac{m - 1}{2} \pi(1) = \frac{m(m - 1)}{2} \pi(0),$$

$$\pi(3) = \frac{m - 2}{3} \pi(2) = \frac{m(m - 1)(m - 2)}{2 \cdot 3} \pi(0),$$

and in general for $x = 1, \ldots, m$,

$$\pi(x) = \frac{m(m - 1) \cdots (m - x + 1)}{2 \cdot 3 \cdots x} \pi(0) = \binom{m}{x} \pi(0).$$

Summing over all x, we find that

$$1 = \sum_{x=0}^{m} \pi(x) = \pi(0) \sum_{x=0}^{m} \binom{m}{x} = \pi(0)2^m,$$

and hence the equilibrium distribution

$$\pi(x) = \binom{m}{x} \frac{1}{2^m}, \qquad x = 0, \ldots, m,$$

is binomial. Notice that the mean recurrence time for the extreme states $x = 0$ or $x = m$ equals 2^m and thus increases exponentially with m, while for the middle states, for example, $x = m/2$ for m even, we find by Stirling's formula that the mean recurrence time increases only as $\sqrt{m}$. Also by approximating the binomial distribution by the Gaussian distribution with mean $m/2$ and standard deviation $\sigma = \sqrt{m}/2$, we see that for large m almost all of the equilibrium distribution will be concentrated within the $6\sigma/m = 3/\sqrt{m}$ fraction of states around the middle of the chain. Recalling the interpretation of the Ehrenfests' chain as a model for diffusion, this means that the number of molecules in containers A and B will be almost the same most of the time and that the mean time it takes for the chain to return to its initial state (all molecules in A) is extremely large as compared with the time for which the numbers of molecules in A and B are about the same. This very fact was actually the purpose of the Ehrenfest model, since in a sense it reconciles the second law of thermodynamics with the time reversibility of classical mechanics. ▲

Of course, if the recurrent chain is null, that is, $\pi(x) = 0$ for all $x \in \mathcal{S}$, then the dynamic equilibrium is of a degenerate nature. For in this case the population of Markov travelers will eventually spread so thin over the infinite state space $\mathcal{S}$ that there will be virtually nobody occupying any state. (If this sounds weird, remember that we are talking about a limit as the time $n \to \infty$ so that the equilibrium state is not actually reached at any finite time.) Yet, since the chain is recurrent, each of the infinite number of states will be visited infinitely often by each of the Markov travelers. We would like to mention that for a null recurrent chain, (29) is indeed true (with $\pi(x) = 0$), so that the above discussion is valid.

It may be interesting to compare this situation with the chain where all states are transient. Since by (3.1.23a) we have, in this case,

$$\sum_{n=1}^{\infty} p^{(n)}(x \mid x') < \infty,$$

the limit in (29) obviously exists and is zero for all $x \in \mathcal{S}$. Thus the population of Markov travelers is again behaving macroscopically in such a manner that none of the infinite number of states is essentially occupied. This time, however, each state is visited only a finite number of times by each traveler, so that their population is not only spreading thin but also drifting over the infinite set $\mathcal{S}$ away from their initial locations.

Thus the only nondegenerate case of dynamic equilibrium occurs if the chain is positive recurrent. Then the limit (29) is a genuine probability mass distribution, and in fact

$$\pi(x) > 0 \qquad \text{for all} \quad x \in \mathcal{S}$$

since $\pi(x) = 0$ would mean $\mu(x) = \infty$, that is, x is null, and we know that in a recurrent chain either all states are null or all states are positive recurrent, that is, no states are null. In this case, the distribution $\pi(x)$, $x \in \mathcal{S}$, is called the *steady-state distribution.*

Note that, in this case, (29) means that for every $x \in \mathcal{S}$ the first-order distributions $p_n(x)$ of the process converge to the steady-state distribution $\pi(x)$ as $n \to \infty$. Actually, the kth order distributions of the process converge also, for if, as in the proof of Theorem 3.1.1,

$$n_1 < n_2 < \cdots < n_k$$

are nonnegative integers, then by (3.1.18) the kth order probability mass distribution

$$p_{n_1, \ldots, n_k}(x_1, \ldots, x_k) = p^{(m_{k-1})}(x_k \mid x_{k-1}) \cdots p^{(m_2)}(x_3 \mid x_2) p^{(m_1)}(x_2 \mid x_1) p_{n_1}(x_1),$$

where $m_1 = n_2 - n_1$, $m_2 = n_3 - n_2, \ldots, m_{k-1} = m_k = m_{k-1}$. Hence, if as $n_1 \to \infty$, $p_{n_1}(x_1) \to \pi(x_1)$, then also

$$p_{n_1, \ldots, n_k}(x_1, \ldots, x_k) \to p^{(m_{k-1})}(x_n \mid x_{k-1}) \cdots p^{(m_1)}(x_2 \mid x_1)\pi(x_1) \tag{34}$$

However, the right-hand side of (34) is a kth order distribution of a strictly stationary Markov chain with initial distribution $p_0(x) = \pi(x)$, $x \in \mathcal{S}$.

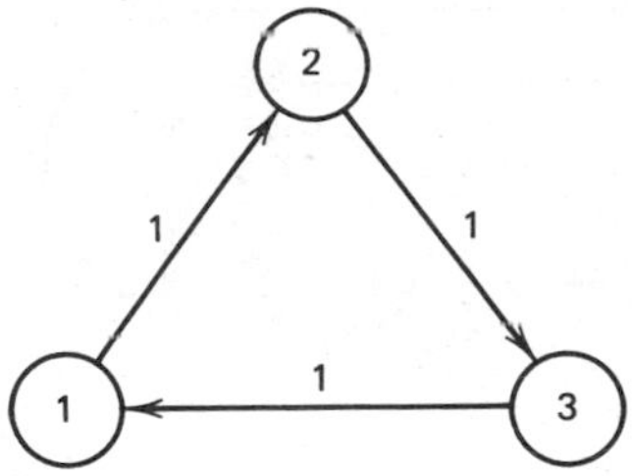

Figure 3.2.8

Hence we can conclude that *if a steady-state distribution* $\pi(x)$ *exists*, that is, when (29) is true, then the probability law of the chain $X(n)$ converges to the probability law of a strictly stationary Markov chain with initial distribution $p_0(x) = \pi(x)$ and the same transition probabilities. We express this by saying that the chain is *asymptotically (strictly) stationary*.

Furthermore, since the steady-state distribution $\pi(x)$ must be unique when it exists, there will be *only one initial distribution* $p_0(x) = \pi(x)$ *for which the process* $X(n)$ *will be strictly stationary*.

It remains to determine when (29) is true for a positive recurrent chain, that is, whether the sequence of n-step transition probabilities $p^{(n)}(x \mid x')$ converges to $\pi(x)$ regardless of the initial state x'.

First, however, we have to dispose of a somewhat special type of regular recurrent chain where the above-mentioned sequence clearly fails to converge. Look at the chain with three states and transition diagram in Figure 3.2.8. This is clearly a recurrent chain, since each state will be visited infinitely often. However, the effect of the initial state will be discernible forever in the behavior of the chain, for the initial state will be visited only at times that are multiples of three and at no other times. That means we will have, for example,

$$p^{(n)}(1 \mid 1) = \begin{cases} 1 & \text{if} \quad n \text{ is a multiple of 3} \\ 0 & \text{otherwise} \end{cases}$$

so that the $\lim_{n\to\infty} p^{(n)}(1 \mid 1)$ does not exist. The reader may object that this is a trivial case of a deterministic process, since the sample path $x(n)$, $n = 0, 1, \ldots$, of the process is clearly determined by the initial state, for example, for $X(0) = 1$, we have the sample path 1, 2, 3, 1, 2, 3, 1, 2, 3, ... But the chain need not be a deterministic process to exhibit periodic behavior. To see this, consider the seven-state recurrent chain with transition diagram given in Figure 3.2.9.

The process $X(n)$, $n = 0, 1, \ldots$, is genuinely random, since at each state the Markov traveler has a choice between two arrows, so that there is a random choice at each time n. However, the initial state is again influencing the chain

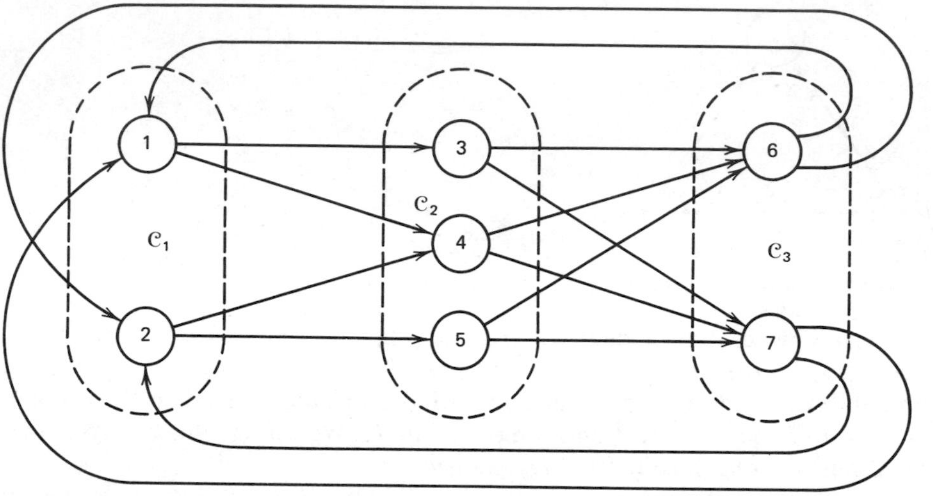

Figure 3.2.9

forever. Just consider the three sets of states $\mathcal{C}_1$, $\mathcal{C}_2$, and $\mathcal{C}_3$ indicated by dotted lines in Figure 3.2.9. It is readily seen that the Markov traveler will forever cycle between these three sets going from $\mathcal{C}_1$ to $\mathcal{C}_2$ to $\mathcal{C}_3$ to $\mathcal{C}_1$, and so on, so that if the initial state is in $\mathcal{C}_1$, say, then the states 1 and 2 can only be visited at times that are multiples of three and at no other times.

Recurrent chains with this property are, for obvious reasons, called *periodic.* It can be shown in general that the set of states $\mathcal{S}$ of a recurrent periodic chain can be decomposed into a finite number $d \geq 2$ of disjoint subsets (called cyclic subclasses)

$$\mathcal{S} = \mathcal{C}_1 \cup \mathcal{C}_2 \cup \cdots \cup \mathcal{C}_d$$

with the property that upon entering a state in $\mathcal{C}_1$, the Markov traveler must go to a state in $\mathcal{C}_2$ and so on in a cyclical fashion, revisiting each subset $\mathcal{C}_k$ after exactly d steps. The number d is called the *period* of the chain and is formally defined as the largest integer with the property that $p^{(n)}(x \mid x)$ is not zero if and only if n is divisible by d.†

Recurrent chains that are not periodic are called *aperiodic.*

It is not difficult to see that for a recurrent chain to be periodic it is necessary that

$$p(x \mid x) = 0 \qquad \text{for all } x \in \mathcal{S} \tag{35}$$

† This definition seems to imply that the period d may depend on x. It can be shown, however, that it does not, so that all states in a periodic recurrent class have the same period.

that is, the Markov traveler must not be allowed to revisit any state in a single step. Thus the converse of (35)

$$p(x \mid x) > 0 \qquad \text{for at least one } x \in \mathcal{S}$$

is a sufficient condition for a recurrent chain to be aperiodic.

We are now ready for a theorem.

THEOREM 3.2.2. *If a positive recurrent chain is aperiodic then it is asymptotically strictly stationary, that is, for any initial state* $x' \in \mathcal{S}$,

$$\lim_{n\to\infty} p^{(n)}(x \mid x') = \pi(x), \qquad x \in \mathcal{S}$$

where $\pi(x)$ *is the unique steady-state distribution.*

Proof. We briefly sketch the idea behind the proof of this important result. The crucial point is the fact that for an aperiodic recurrent chain, one must have for each $x \in \mathcal{S}$ and all n sufficiently large

$$p^{(n)}(x \mid x) > 0. \tag{36}$$

That is, the Markov traveler, starting at a state x, can return to x at any time larger than some n_0. Inequality (36) is, as we have seen, false for periodic chains. To show that it can be false only for periodic chains is a little bit more difficult. It follows from the fact that if $p^{(n_1)}(x \mid x) > 0$ and $p^{(n_2)}(x \mid x) > 0$, then by the Chapman-Kolmogorov equation (3.1.12), also $p^{(n_1+n_2)}(x \mid x) > 0$. Therefore, if $p^{(n)}(x \mid x)$ were zero for an infinite number of n's, it could be nonzero only if n is a multiple of some $d \geq 2$, that is, if the recurrent chain were periodic.

Now consider two independent regular recurrent aperiodic chains $X(n)$ and $Y(n)$ with the same state space $\mathcal{S}$ and identical transition probabilities. These two chains can be regarded as a single chain with state space $\mathcal{S} = \{(x, y): x \in \mathcal{S}_X, y \in \mathcal{S}_Y\}$ and transition probabilities

$$\tilde{p}((x, y) \mid (x', y')) = p(x \mid x')p(y \mid y').$$

Because of (36), this implies that for any pair (x, y) and (x', y') of states of this combined chain

$$\tilde{p}^{(n)}((x, y) \mid (x', y')) > 0 \qquad \text{and} \qquad \tilde{p}^{(n)}((x', y') \mid (x, y)) > 0$$

for some n. But this means that any state in this chain can be visited from any other state, that is, the states of the combined chain form a single recurrent class.

However, this combined chain can be looked upon as describing the voyages of two independent Markov travelers through the transition graph of the original chain. Since the combined chain is recurrent, then regardless of the initial states $X(0) = x'$ and $Y(0) = y'$, there must be a finite time instant when a state $(x, y) \in \mathcal{S} \times \mathcal{S}$ such that $x = y$ is visited for the first time, that is, a time T such that $P(T < \infty) = 1$ when the two Markov travelers meet.

Suppose now that Markov traveler X starts at some arbitrary initial state

$X(0) = x'$, while Markov traveler Y chooses his initial state with the distribution $\pi(y)$,

$$P(Y(0) = y) = \pi(y), \qquad y \in \mathcal{S}$$

where $\pi(y)$ is the unique solution of the system (23). We know from Theorem 3.2.1 that such a solution exists, since the chain is positive recurrent. We now introduce a third Markov traveler Z who begins her journey together with the traveler X and keeps him company until they meet traveler Y. At that moment Z abandons X and joins Y to accompany him forever after. Formally $Z(n)$, $n = 0, 1, \ldots$, is again a Markov chain defined by

$$Z(n) = \begin{cases} X(n) & \text{if} \quad T \geq n, \\ Y(n) & \text{if} \quad T < n, \end{cases}$$

where T is the time when X and Y first meet. However, since $P(T < \infty) = 1$, X will be abandoned after a finite number of steps, and hence

$$\lim_{n\to\infty} P(Z(n) = x) = \lim_{n\to\infty} P(Y(n) = x) = \pi(x),$$

since $P(Y(n) = x) = \pi(x)$ for all $n = 0, 1, \ldots$, according to Theorem 3.1.1. On the other hand, the chain $Z(n)$ has identical transition probabilities with both $X(n)$ and $Y(n)$, and since $Z(0) = X(0) = x'$, this means that

$$P(X(n) = x) = p^{(n)}(x \mid x') \qquad \text{for all } n = 1, 2, \ldots$$

Thus, $\lim_{n\to\infty} p^{(n)}(x \mid x') = \pi(x)$, and since $x' \in \mathcal{S}$ was arbitrary, the theorem is proved. ●

Example 3.2.7

A classical example originally introduced by S. Chandrasekhar as a model for certain processes in astrophysics goes as follows. Consider a fixed volume of space and assume that at time n there are $X(n) = x'$ particles present. During the next time unit, each of these particles may leave the volume with some fixed probability α, $0 < \alpha < 1$, independently of the other particles. Thus the number $Z(n)$ of particles leaving the volume will be a binomial random variable with parameters x' and α. During the same time unit, however, new particles may enter the volume, their number $Y(n)$ being a Poisson random variable with parameter $\lambda > 0$, independent of $X(n)$. Then the number $X(n + 1)$ of particles present in the volume at time n equals

$$X(n + 1) = X(n) + Y(n) - Z(n),$$

where $X(n)$ and $Y(n)$ are independent, while the distribution of $Z(n)$ depends only on the number $X(n) = x'$. It follows that $X(n)$, $n = 0, 1, \ldots$, is a discrete-time homogeneous Markov chain with state space $\mathcal{S} = \{0, 1, \ldots\}$. To find the transition probabilities $p(x \mid x')$, argue as follows. The conditional probability of the number $X(n) - Z(n)$ of old particles remaining in the volume equals

$$P(X(n) - Z(n) = k \mid X(n) = x') = \binom{x'}{k}(1-\alpha)^k \alpha^{x'-k},$$

$k = 0, 1, \ldots, x'$, since each of the x' old particles has probability $1 - \alpha$ of not leaving the volume.

At the same time the probability of $Y(n) = x - k \geq 0$ new arrivals equals

$$P(Y(n) = x - k) = e^{-\lambda} \frac{\lambda^{x-k}}{(x-k)!}, \qquad x = k, k+1, \ldots$$

and since $Y(n)$ is independent of both $X(n)$ and $Z(n)$

$$P(X(n+1) = x \mid X(n) = x') = \sum_k P(Y(n) = x - k) P(X(n) - Z(n) = k \mid X(n) = x'),$$

where the sum is over all $k = 0, 1, \ldots, \min(x', x)$, since $0 \leq k \leq x'$ and $x - k \geq 0$. Hence, if we substitute into the above equation, we find that

$$p(x \mid x') = e^{-\lambda} \sum_{k=0}^{\min(x', x)} \binom{x'}{k}(1-\alpha)^k \alpha^{x'-k} \frac{\lambda^{x-k}}{(x-k)!}, \tag{37}$$

for all $x' \in \mathcal{S}$, $x \in \mathcal{S}$, where $\binom{0}{0} = 1$.

Since clearly $p(x \mid x) > 0$ for any pair of states, the chain cannot have more than one recurrent class and cannot be periodic. It will be shown shortly (Theorem 3.2.4) that under these circumstances the existence of a probability distribution $\pi(x)$, $x \in \mathcal{S}$ satisfying the linear system

$$\sum_{x \in \mathcal{S}} p(y \mid x)\pi(x) = \pi(y), \qquad y \in \mathcal{S}$$

guarantees that the chain is positive recurrent. Such a distribution does exist in our case and it happens to be a Poisson distribution

$$\pi(x) = e^{-\lambda/\alpha} \frac{(\lambda/\alpha)^x}{x!}, \qquad x \in \mathcal{S}. \tag{38}$$

To verify this statement, note that with (37) and (38), we have

$$\sum_{x \in \mathcal{S}} p(y \mid x)\pi(x) = \sum_{x=0}^{\infty} e^{-\lambda/\alpha} \frac{(\lambda/\alpha)^x}{x!} e^{-\lambda} \sum_{k=0}^{\min(x, y)} \binom{x}{k}(1-\alpha)^k \alpha^{x-k} \frac{\lambda^{y-k}}{(y-k)!}$$

Replacing the double sum

$$\sum_{x=0}^{\infty} \sum_{k=0}^{\min(x, y)}$$

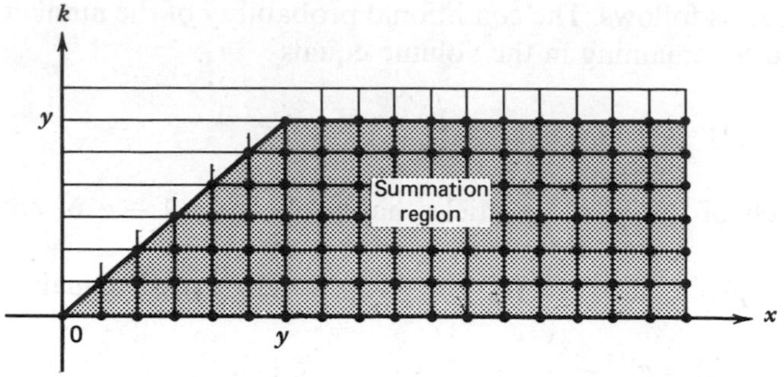

Figure 3.2.10

by the double sum

$$\sum_{k=0}^{y} \sum_{x=k}^{\infty}$$

(see Fig. 3.2.10), we obtain after some cancellation

$$\sum_{x \in \mathcal{S}} p(y \mid x)\pi(x) = e^{-\lambda/\alpha} \sum_{k=0}^{y} \left(\frac{1-\alpha}{\alpha}\right)^k \frac{\lambda^{y-k}}{(y-k)!} e^{-\lambda} \sum_{x=k}^{\infty} \binom{x}{k} \frac{\lambda^x}{x!}$$

$$= e^{-\lambda/\alpha} \lambda^y \sum_{k=0}^{y} \frac{1}{(y-k)!}\frac{1}{k!}\left(\frac{1-\alpha}{\alpha}\right)^k e^{-\lambda} \sum_{x=k}^{\infty} \frac{\lambda^{x-k}}{(x-k)!}$$

$$= e^{-\lambda/\alpha} \frac{\lambda^y}{y!} \sum_{k=0}^{y} \binom{y}{k}\left(\frac{1-\alpha}{\alpha}\right)^k = e^{-\lambda/\alpha} \frac{(\lambda/\alpha)^y}{y!} = \pi(y),$$

since the last sum equals

$$\left(\frac{1-\alpha}{\alpha} + 1\right)^y = \frac{1}{\alpha^y}$$

by the binomial theorem. Thus, according to Theorem 3.2.2, the chain is asymptotically strictly stationary and

$$\lim_{n \to \infty} p^{(n)}(x \mid x') = e^{-\lambda/\alpha} \frac{(\lambda/\alpha)^x}{x!}, \qquad x \in \mathcal{S}.$$

In other words, regardless of the initial number of particles in the volume, the system approaches a steady state when the number of entering particles has a Poisson distribution and the average number of particles approaches λ/α. ▲

DISCUSSION OF THE GENERAL HOMOGENEOUS MARKOV CHAIN—STATIONARITY AND ERGODICITY. As we have seen, the state space $\mathcal{S}$ of a general Markov chain can be decomposed into a set of transient states $\mathcal{S}_{tr}$ and the set of recurrent states $\mathcal{S}_{rec}$. The latter set can, in turn, be partitioned into recurrent classes, each of which may be either null or positive recurrent. Each positive recurrent class can be either periodic (in which case it is decomposable into $d > 2$ cyclic subclasses) or aperiodic. If the state space $\mathcal{S}$ is *finite*, there must be at least one recurrent class, and no recurrent class can be null. Also, if $\mathcal{S}$ is finite, then the entrance into a recurrence class is certain, that is, if $T_{x'}$ denotes the time to enter the set $\mathcal{S}_{\text{rec}}$ starting from a state $x' \in \mathcal{S}$, then

$$P(T_{x'} < \infty) = 1 \qquad \text{for all } x' \in \mathcal{S}.$$

Our first result concerns the existence of a stationary initial distribution.

THEOREM 3.2.3. *A homogeneous Markov chain has a stationary initial distribution* $p_0(x)$ *if and only if there is at least one positive recurrent class. This distribution is unique if and only if there is only one such class.*

Combining this with Theorem 3.2.1, we also have:

THEOREM 3.2.4. *A homogeneous Markov chain has at least one positive recurrent class if and only if the system of equations*

$$\sum_{x \in \mathcal{S}} p(y \mid x)\pi(x) = \pi(y), \qquad y \in \mathcal{S} \tag{39}$$

has a solution such that $\pi(y) \geq 0$, $y \in \mathcal{S}$ *and* $\sum_{y \subset \mathcal{S}} \pi(y) = 1$. *Every state* $x \in \mathcal{S}$ *for which* $\pi(x) > 0$ *for some solution of* (39) *is positive recurrent. If the system* (39) *has only one solution, then there is only one positive recurrent class, namely* $\{x\colon \pi(x) > 0\}$. *If the state space* $\mathcal{S}$ *is finite the system* (39) *always has a solution.*

Proof. The proof for both of these theorems is immediate. We already know that if there is only one positive recurrent class $\mathcal{C} \subset \mathcal{S}$, the system

$$\sum_{x \in \mathcal{C}} p(y \mid x)\pi(x) = \pi(y), \qquad y \in \mathcal{C} \tag{40}$$

has a unique solution such that $\pi(y) > 0$ for $y \in \mathcal{C}$ and $\sum_{y \in \mathcal{C}} \pi(y) = 1$. Setting $p_0(y) = \pi(y)$ for $y \in \mathcal{C}$ and $p_0(y) = 0$ for $y \in \mathcal{S} - \mathcal{C}$, we see that $p_0(x)$ is also a solution of (39). This solution of (39) must be unique, since any other solution $p_0'(x)$ of (39) would have to have $p_0(y) > 0$ for at least one $y \in \mathcal{S} - \mathcal{C}$. But every state in $\mathcal{S} - \mathcal{C}$ is either transient or null recurrent, that is, $p_n(y) \to 0$, which is impossible if $p_0(y) > 0$ is to be a stationary distribution. On the other hand, if there are several regular recurrent classes, say $\mathcal{C}_1$ and $\mathcal{C}_2$, then (40) with $\mathcal{C} = \mathcal{C}_1$ has a unique solution π_1, and (40) with $\mathcal{C} = \mathcal{C}_2$ has a unique solution π_2. Since $\mathcal{C}_1$ and $\mathcal{C}_2$ must be disjoint subsets of $\mathcal{S}$, we can define for any $0 \leq \alpha \leq 1$

$$p_0(y) = \begin{cases} \alpha\pi_1(y) & \text{if } \ y \in \mathcal{C}_1, \\ (1-\alpha)\pi_2(y) & \text{if } \ y \in \mathcal{C}_2, \\ 0 & \text{if } \ y \in \mathcal{S} - \mathcal{C}_1 - \mathcal{C}_1. \end{cases}$$

Then $p_0(y)$ satisfies (40) and hence is a stationary distribution that is not unique, since α can be chosen arbitrarily from the interval [0, 1]. ●

Example 3.2.8

Consider again the chain of Example 3.1.8. We found there that a necessary and sufficient condition for this chain to be recurrent is that the positive probabilities $a_0, a_1, \ldots$ satisfy the condition

$$\lim_{n\to\infty} a_0 a_1 \cdots a_{n-1} = 0. \tag{41}$$

We can now use Theorem 3.2.4 to find a necessary and sufficient condition for the chain to be positive recurrent. The system of equations (39) is, in this case, quite simple; we have

$$a_{y-1}\pi(y-1) = \pi(y), \qquad y = 1, 2, \ldots,$$

and hence by repeated substitution

$$\pi(y) = a_0 a_1 \cdots a_{y-1}\pi(0), \qquad y = 1, 2, \ldots$$

It follows that a solution $\pi(y) \geq 0$, $\sum_{y=0}^{\infty} \pi(y) = 1$ of this system of equations will exist if and only if the infinite series of products

$$1 + a_0 + a_0 a_1 + a_0 a_1 a_2 + \cdots \tag{42}$$

converges to a finite sum. Thus the convergence of this series is necessary and sufficient for the chain to be positive recurrent. If the series (42) diverges and (41) is still satisfied the chain is null recurrent, while if the products (41) fail to converge to zero the chain is transient. Thus we have obtained a complete classification of this chain in terms of the probabilities $a_0, a_1, \ldots$. ▲

Earlier we saw that if we define a sequence of random variables

$$I_x(n) = \begin{cases} 1 & \text{if } X(n) = x, \\ 0 & \text{if } X(n) \neq x, \end{cases} \qquad n = 0, 1, \ldots,$$

where x is a state in a recurrent class, then the average

$$\bar{I}_x(n) = \frac{1}{n}\sum_{k=1}^{n} I_x(k) \to \pi(x) \qquad \text{as } n \to \infty, \tag{43}$$

where

$$\pi(x) = \lim_{n\to\infty} \frac{1}{n}\sum_{k=1}^{n} p^{(k)}(x \mid x'). \tag{44}$$

Since transient states are visited only a finite number of times and since

$$\pi(x) = \lim_{n\to\infty} p^{(n)}(x \mid x') = 0, \tag{45}$$

for x transient, (43) and (44) are true for any state $x \in \mathcal{S}$.

Now suppose that the chain has a single class $\mathcal{C}$ that is positive recurrent and aperiodic and that the entrance to this class from any initial state is certain, that is, $P(T_x < \infty) = 1$ for all $x \in \mathcal{S}$. Then, by Theorem 3.2.2

$$\lim_{n\to\infty} p^{(n)}(x \mid x') = \pi(x) \qquad \text{for all} \quad x \in \mathcal{C},\ x' \in \mathcal{C},$$

and since entrance to $\mathcal{C}$ is certain for every $x' \in \mathcal{S}$, we conclude from (45) and from the fact that if $x \notin \mathcal{C}$ then it is necessarily transient that actually

$$\lim_{n\to\infty} p^{(n)}(x \mid x') = \pi(x) \qquad \textit{for any} \quad x \in \mathcal{S},\ x' \in \mathcal{S}.$$

However, by Theorem 3.2.3 $\pi(x) = p_0(x)$ is the unique stationary distribution, so that

$$\mathrm{E}[I_x(n)] = P(X(n) = x) = p_n(x) = p_0(x) = \pi(x). \tag{46}$$

Thus $\bar{I}_x(n)$, which is the average number of visits to x during the first n steps (time average), converges *regardless of the initial state* $X(0) = x'$ to the expectation (46), that is, the ensemble average. But this is the ergodic property (see Chapter One), since the time average $\bar{I}_x(n)$ over *any sample path of the process*, that is, no matter what the initial state is, converges to the ensemble average.

We state this as a theorem.

THEOREM 3.2.5. *A homogeneous Markov chain is ergodic if and only if it contains a single positive recurrent class that is aperiodic and such that the entrance to this class from any initial state is certain.*

Example 3.2.9

Ergodicity of a Markov chain is extremely useful in applications, since it allows us to actually estimate the transition probabilities $p(x \mid x')$ or other parameters by observing a single sample path of the process for a sufficiently long time. For instance, consider the two-state chain of Example 3.1.6. As long as $0 < a_1 < 1$ and $0 < a_2 < 1$ the chain is clearly aperiodic and positive recurrent and hence is ergodic by Theorem 3.2.5. It follows that we can estimate the parameters a_1 and a_2 by the relative frequencies of transitions, for example, the estimate $\hat{a}_1$ of the parameter a_1 can be obtained as the ratio:

$$\hat{a}_1 = \frac{\text{number of transitions from ① to ②}}{\text{number of transitions from ①}},$$

after having observed a large number of such transitions. Of course, if the state space is infinite, we cannot estimate infinitely many transition probabilities from a finite number of observations. However, if there are only a few unknown parameters on which the transition probabilities are assumed to depend in some way, we can estimate these. For instance, in the case described in Example 3.2.7, we can use the ergodicity of the chain to estimate the ratio λ/α as the number of particles in the volume averaged over a sufficiently large period of time, assuming only that the process has already been running for a long time so that it can be considered stationary. ▲

EXERCISES

Exercise 3.2

1. For the Markov chain with transition diagram in Figure 3.1.3 evaluate the absorption probabilities $P(\mathcal{A}\,|\,x)$ and $P(\mathcal{B}\,|\,x)$ for each of the two recurrent classes $\mathcal{A}$, $\mathcal{B}$ and for each initial state $x \in \mathcal{A}_{\text{tr}}$.
2. For the chain in Exercise 3.2.1, find the probability mass distribution $f_T(n\,|\,x)$ of the absorption time for each initial state $x \in \mathcal{A}_{\text{tr}}$. Also calculate the expected duration of the transition period both directly from the distribution $f_T(n\,|\,x)$ and from the equations (11).
3. Repeat Exercises 3.2.1 and 3.2.2 for the Markov chain with the transition diagram drawn in Figure 3.1.E.2.
4. Suppose that the gene in Exercise 3.1.12 consists originally of $r = 12$ units, one of them a mutant. Suppose further that if the number of mutant units in a cell exceeds two, the cell can no longer reproduce and so dies. For a particular line of descent, find the probability that the line becomes extinct after the nth generation.
5. Consider a homogeneous Markov chain with state space $\mathcal{S} = \{0, 1, \ldots\}$ and transition probabilities

$$p(x\,|\,x') = \begin{cases} \dbinom{2x'}{x} 4^{-x'} & \text{if} \quad x = 0, \ldots, 2x'; \\ 0 & \text{otherwise,} \end{cases}$$

   $x' = 1, 2, \ldots,$ with the state $x = 0$ absorbing. Find the probability of absorption for an arbitrary initial state $x_0 > 0$.
6. In the cyclic random walk of Example 3.2.2, starting at the state zero, find the expected number of steps the Markov traveler makes before completing a full circle.
7. Consider the Markov chain with the transition diagram drawn in Figure 3.2.E.1. Show that the ratio of absorption probabilities $p(\{m_1\}\,|\,0)$ to $p(\{-m_2\}\,|\,0)$ equals the ratio of the products $a_1 a_2 \cdots a_{m_1}$ to $b_1 b_2 \cdots b_{m_2}$ and use this to evaluate the absorption probabilities. Also evaluate the expected time to absorption from the initial state zero assuming $m_1 = m_2$. (*Hint:* The Markov traveler starts afresh upon each return to zero.)

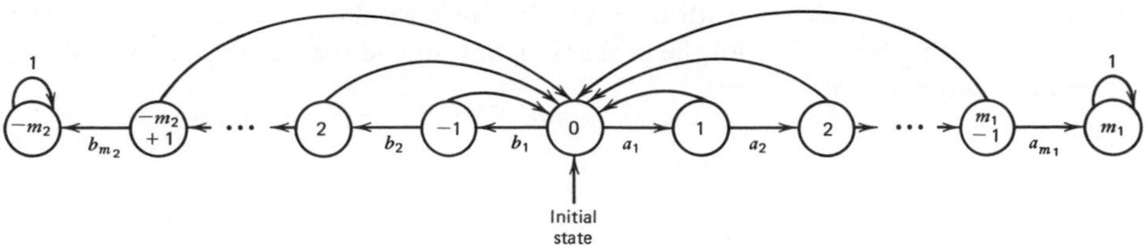

Figure 3.2.E.1

8. Consider a random walk in a planar lattice as depicted in Figure 3.2.E.2. Show that the absorption in one of the states of the set $\mathcal{A} = \{(i, j): i = j = 1, 2, \ldots\}$ is certain. For an arbitrary initial state x, evaluate the expected time T to absorption in the set $\mathcal{A}$. Also evaluate the distribution $f_T(n \mid x)$, at least for a few values of n.

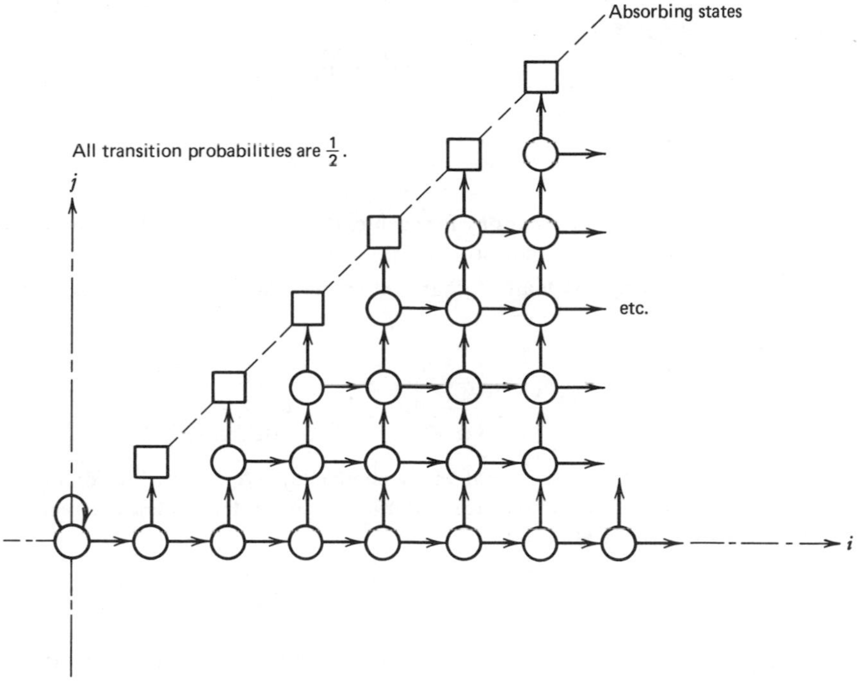

Figure 3.2.E.2

9. Show that the chain of Exercise 3.1.2 is positive recurrent, and find the mean recurrence time μ_x for the initial state $x = 0$. Show further that this chain is aperiodic and thus has a unique steady-state distribution.

10. Consider a Markov chain with the transition diagram shown in Figure 3.2.E.3. Use the Kirchoff's type of law for the probability flow to find the steady-state probability mass distribution $\pi(x)$, $x = 0, 1, 2, 3$.

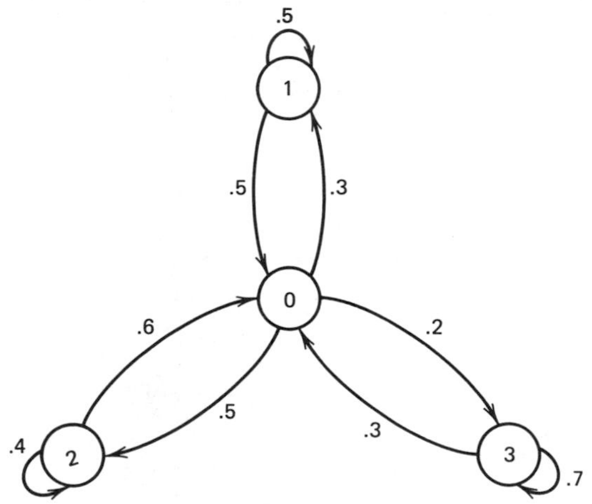

Figure 3.2.E.3

11. Find the mean recurrence times μ_X for each of the seven states of the periodic chain in Figure 3.2.9 with all transition probabilities equal to $\frac{1}{2}$.

12. Consider a homogeneous Markov chain with state space $\mathcal{S} = \{0, 1, \ldots\}$ and transition probabilities

$$p(x \mid x') = \begin{cases} a_{x'} & \text{if } x = x' + 1 \\ b_{x'} & \text{if } x = x' - 1 \\ 1 - a_{x'} - b_{x'} & \text{if } x = x', \\ 0 & \text{otherwise,} \end{cases}$$

where a_j and b_j are positive numbers such that $a_j + b_j < 1$ except for $j = 0$ when $b_0 = 0$. Show that every nonnegative solution of the system of linear equations (39) must be such that for all $x = 1, 2, \ldots,$

$$\pi(x) = \frac{a_0 a_1 \cdots a_{x-1}}{b_0 b_1 \cdots b_x} \pi(0).$$

Hence, establish a criterion for this chain to be positive recurrent and find the steady-state distribution.

13. Find the steady-state distribution for the states of the shift register described in Exercise 3.1.1.

14. Consider a symmetric Bernoulli random walk with a reflecting boundary at the state $m > 0$ and an elastic boundary ($p(0 \mid 0) = \frac{1}{2}$) at the origin. Find the steady-state distribution $\pi(x)$, $x = 0, \ldots, m$, for this Markov chain.

15. Consider a random walk with "directional memory" as described in Exercise 3.1.10. Place a reflecting boundary m steps at each side of the origin, and calculate the steady-state distribution of the position of the particle. What can be said about the motion of the particle if the boundaries were removed? (*Hint*: Use "Kirchoff's law" to show that $\pi(x_l) = \pi(x_r)$ for all except the boundary positions $x = \pm m$.)

3.3. CONTINUOUS-TIME MARKOV CHAINS

Consider a discrete-state dynamic system, such as a relay, a counter, or an arbitrary digital circuit, that changes its state randomly with time. Assume that the system begins its operation (or we begin to observe it) at time $t = 0$, and let $X(t)$ be its state at time $t \geq 0$. Then

$$X(t), \qquad t \geq 0$$

is a continuous time stochastic process with time domain $\mathcal{T} = [0, \infty)$, and since its state space $\mathcal{S}$ is assumed to be discrete, we may as before label the individual states $x \in \mathcal{S}$ by integers, that is, assume that

$$\mathcal{S} = \{\ldots, -1, 0, 1, \ldots\}.$$

The ensemble of such a process must then consist of sample paths $x(t)$, $t \geq 0$, which all look like the one shown in Figure 3.3.1; it consists of flat pieces of integral height interrupted by jumps occurring at random time instances, with each jump having a random integer-valued magnitude.

The values that the sample paths take at the instances of jumps are not particularly important from the point of view of the system. For the sake of convenience we take them to be continuous from the right so that at the instant t of a jump, the value of $x(t)$ indicates the state into which the jump is made.

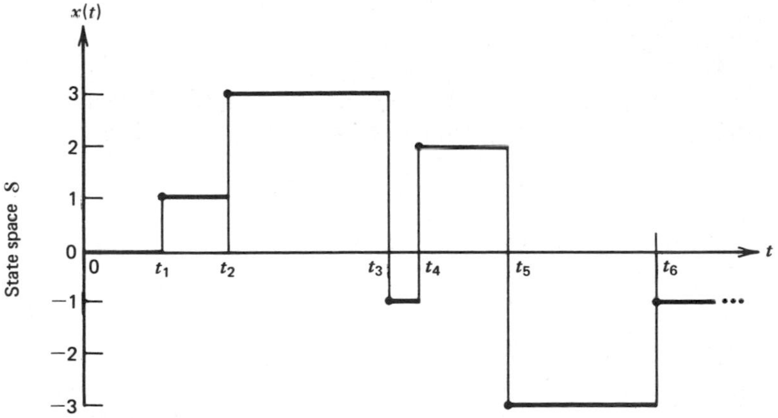

Figure 3.3.1

Since $X(t)$ is to represent the state of the dynamic system at the time t, we would like to assume, as the notion of a state of a dynamic system requires, that the future evolution of the system depends only on the present state $X(t) = x(t)$ and not on the past history $X(t') = x(t')$, $t' < t$. For a stochastic process

$$X(t), \qquad t \geq 0$$

this amounts to the Markov property:

$$P(X(t_2) = x \mid X(t'), 0 \leq t' \leq t_1)\dagger = P(X(t_2) = x \mid X(t_1)) \qquad \text{for all } x \in \mathcal{S} \quad (1)$$

and any $0 \leq t_1 \leq t_2$. The probability law of the process $X(t)$ is then completely specified by the distribution of the initial state

$$P(X(0) = x_0), \qquad x_0 \in \mathcal{S} \tag{2}$$

and by the family of conditional distributions, the so-called *transition probabilities*

$$P(X(t_2) = x \mid X(t_1) = x') \tag{3}$$

for all $x \in \mathcal{S}$, $x' \in \mathcal{S}$, and all $0 \leq t_1 \leq t_2$. As we did for discrete-time Markov chains, we will consider only the case in which the transition probabilities (3) are stationary:

$$P(X(t_2 + \tau) = x \mid X(t_1 + \tau) = x') = P(X(t_2) = x \mid X(t_1) = x') \qquad \text{for any } \tau \geq 0.$$

Since under this assumption the transition probabilities (3) now depend only on the difference $t = t_2 - t_1$, it becomes convenient to introduce a new symbol:

$$p(t, x \mid x') = P(X(t_1 + t) = x \mid X(t_1) = x').$$

If we also denote the initial distribution (2) by

$$p_0(x) = P(X(0) = x),$$

the probability law of the process $X(t)$, $t \geq 0$, is now completely specified by the initial distribution $p_0(x)$, $x \in \mathcal{S}$, and the family of transition probabilities

$$p(t, x \mid x'), \qquad x \in \mathcal{S}, \qquad x' \in \mathcal{S}, \qquad t \geq 0. \tag{4}$$

Obviously, both $p_0(x)$ and $p(t, x \mid x')$ must always be nonnegative and also satisfy

$$\sum_{x \in \mathcal{S}} p_0(x) = 1 \qquad \text{and} \qquad \sum_{x \in \mathcal{S}} p(t, x \mid x') = 1 \qquad \text{(5a,b)}$$

for all $x' \in \mathcal{S}$ and $t \geq 0$.

† Defined as the conditional expectation of the random variable $I_{t_2, x} = 1$ if and only if $X(t_2) = x$.

We will refer to the process $X(t)$, $t \geq 0$, as a *continuous-time homogenous Markov chain*, where "homogeneous" implies that the transition probabilities are assumed stationary, while "chain" reminds us that the state space $\mathcal{S}$ is discrete.

Example 3.3.1
As a simplified model of an electron multiplier, suppose that each primary electron instantaneously releases a random number K of secondary electrons, these numbers being independent random variables for each primary electron and having a common distribution with probability mass distribution

$$P(K = k), \qquad k = 0, 1, \ldots. \tag{6}$$

Assume further that the number N_I of primary electrons hitting the anode during a time interval I has a Poisson distribution with parameter $\lambda = 1$ and that these numbers N_I are, for nonoverlapping time intervals I, independent random variables. (This amounts to assuming that primary electrons are hitting the anode according to a Poisson process.)

Suppose the multiplier was switched on at the time $t = 0$, and let $X(t)$ be the total number of secondary electrons released by the time t.

Now it is easy to see that $X(t)$, $t \geq 0$, is a continuous-time homogeneous Markov chain. $X(t)$ is clearly an integer-valued random variable, so the state space is $\mathcal{S} = \{0, 1, \ldots\}$. To check the Markov property (1), take $0 < t_1 < t_2$, and let $K = X(t_2) - X(t_1)$ be the number of secondary electrons released during the time interval $I_2 = (t_1, t_2]$. According to the assumptions made, this depends only on the number N_{I_2} of primary electrons hitting the anode during I_2; in fact,

$$\begin{aligned} P(K = k) &= \sum_{n=0}^{\infty} P(K = k \mid N_{I_2} = n) P(N_{I_2} = n) \\ &= e^{-(t_2 - t_1)} \sum_{n=0}^{\infty} P(K = k \mid N_{I_2} = n) \frac{(t_2 - t_1)^n}{n!}, \qquad k = 0, 1, \ldots. \end{aligned} \tag{7}$$

Similarly $X(t_1)$ depends only on N_{I_1}, the number of primary electrons hitting the anode during the time interval $I_1 = [0, t_1]$, and since I_1 and I_2 do not overlap, N_{I_1} and N_{I_2} are independent and consequently so are $X(t_1)$ and $X(t_2) - X(t_1)$. Hence the conditional probability $P(X(t_2) = x_2 \mid X(t), 0 \leq t \leq t_1)$ must be the same as the conditional probability $P(X(t_2) = x_2 \mid X(t_1))$, and the Markov property is established.

Furthermore, according to (7)

$$\begin{aligned} P(X(t_2) = x_2 \mid X(t_1) = x_1) &= P(K = x_2 - x_1 \mid X(t_1) = x_1) = P(K = k) \\ &= e^{-(t_2 - t_1)} \sum_{n=0}^{\infty} P(K = k \mid N_{I_2} = n) \frac{(t_2 - t_1)^n}{n!} \qquad \text{for } k = x_2 - x_1 \geq 0, \end{aligned} \tag{8}$$

while clearly

$$P(X(t_2) = x_2 \mid X(t_1) = x_1) = 0 \qquad \text{for } x_2 < x_1, \tag{9}$$

so that the chain is also homogeneous.

Since $X(0) = 0$ the initial distribution is degenerate

$$p_0(x) = \begin{cases} 1 & \text{if } x = 0, \\ 0 & \text{if } x = 1, 2, \ldots. \end{cases}$$

To find the transition probabilities, note that the conditional probability distribution of the number K of secondary electrons released during I_2, given that $N_{I_2} = n$ primary electrons hit the anode, will be the same as the sum of n i.i.d. discrete random variables each with probability mass distribution (6). Denoting the probability mass distribution of such a sum by

$$p_n(k), \qquad k = 0, 1, \ldots,$$

we have

$$P(K = k \mid N_{I_2} = n) = p_n(k), \qquad \text{for } n = 1, 2, \ldots,$$

while trivially

$$P(K = k \mid N_{I_2} = 0) = p_0(k) = \begin{cases} 1 & \text{if } k = 0, \\ 0 & \text{if } k = 1, 2, \ldots. \end{cases}$$

Hence by (8) and (9) for $t > 0$

$$p(t, x \mid x') = e^{-t} \sum_{n=0}^{\infty} p_n(x - x') \frac{t^n}{n!} \qquad \text{if } x' = 0, 1, \ldots, \qquad x = x', x' + 1, \ldots, \tag{10}$$

while

$$p(t, x \mid x') = 0 \quad \text{if } x < x'.$$

Specifically suppose that the probability mass distribution (6) is modified geometric with parameter $0 < \gamma < 1$, that is,

$$p(k) = \gamma(1 - \gamma)^k, \qquad k = 0, 1, \ldots$$

Since the sum of n i.i.d. modified geometric random variables has the modified negative binomial distribution, we have in this case

$$p_n(k) = \binom{n + k - 1}{k} \gamma^n (1 - \gamma)^k, \qquad k = 0, 1, \ldots, \qquad n = 1, 2, \ldots,$$

and, of course,

$$p_0(k) = \begin{cases} 1 & \text{if } k = 0, \\ 0 & \text{if } k = 1, 2, \ldots. \end{cases}$$

Hence the transition probability (10) becomes

$$p(t, x \mid x') = \begin{cases} e^{-t}(1-\gamma)^{x-x'} \sum_{n=1}^{\infty} \binom{n+x-x'-1}{x-x'} \frac{(\gamma t)^n}{n!} & \text{if } x > x', \\ e^{-t}\left(1 + \sum_{n=1}^{\infty} \frac{(\gamma t)^n}{n!}\right) = e^{-(1-\gamma)t} & \text{if } x = x', \\ 0 & \text{if } x < x', \end{cases} \tag{11}$$

where both x and x' are nonnegative integers. ▲

Remark: The infinite series appearing in (11), that is,

$$\sum_{n=1}^{\infty} \binom{n+k-1}{k} \frac{z^n}{n!}$$

can be shown to converge for every $k = 1, 2, \ldots$ and $-\infty < z < \infty$ to the sum

$$\frac{1}{k!} \frac{d^{k-1}}{dz^{k-1}} (z^k e^z) = \frac{e^z}{k!} Q_k(z),$$

where $Q_k(z)$ is a polynomial in z of degree k. Defining $Q_0(z) = 1$, we can thus write for $x = x' + k$, $k = 0, 1, \ldots,$

$$p(t, x' + k \mid x') = e^{-(1-\gamma)t} \frac{(1-\gamma)^k}{k!} Q_k(\gamma t), \qquad t \geq 0, \qquad x' \in \mathcal{S}.$$ Я

We would like to impose one more condition on the process $X(t)$, $t > 0$. We will require that no sample path $x(t)$, $t \geq 0$, from its ensemble can undergo an infinite number of jumps during a finite time interval. We will refer to the process $X(t)$, $t \geq 0$, satisfying this condition as *nonexploding*. This is certainly quite a reasonable condition for a dynamic system to satify. After all, from the physical point of view, an infinite number of jumps would require an infinite expenditure of energy on the part of the system, and this certainly cannot happen during any finite interval of time.† It can be shown that for this condition to be satisfied, it is necessary that for all $x' \in \mathcal{S}$ the transition probabilities (4) be such that

$$p(t, x' \mid x') \to 1 \qquad \text{as} \quad t \to 0. \tag{12}$$

Notice that because of (5b), this is equivalent to the condition

$$\sum_{x \neq x'} p(t, x \mid x') \to 0 \qquad \text{as} \quad t \to 0. \tag{13}$$

and since this sum is just the probability of the system leaving the state x' after time t, our condition amounts to insisting that the system cannot leave a state

† There are, nevertheless, some important models where it is exactly this phenomenon that is of prime interest.

immediately after reaching it. If the total number of states is finite, this actually prevents the system from undergoing an infinite number of jumps during a finite time interval, so that the condition (12) is also sufficient for the chain to be nonexploding. For a chain with an infinite number of states, however, it is still possible to have an infinite number of jumps in a finite time interval even if instantaneous jumps are excluded by (12). Later on, we will give an example of such a case. But for the time being, we will just assume that we are dealing with nonexploding chains only.

Even with this additional condition (12), it seems that the specification of the probability law of a continuous-time homogeneous Markov chain requires much more information than in the discrete-time case. The transition probabilities (4) are now functions of time, so that even for a two-state chain we now need to specify two functions rather than just two numbers as in the discrete case. But, perhaps surprisingly, this is not so.

The reason is that as for any Markov process the transition probabilities (4) must satisfy the Chapman-Komogorov equation. For our continuous-time homogeneous Markov chain this equation takes on the form

$$p(t+\tau, x \mid x') = \sum_{y \in S} p(\tau, x \mid y) p(t, y \mid x'), \tag{14}$$

where $t \geq 0$ and $\tau \geq 0$ are any two times. Suppose now that we are given the transition probabilities $p(t, x \mid x')$ for t ranging only over some small time interval $0 \leq t \leq dt$, where $dt > 0$. Then we can use (14) with both $0 \leq t \leq dt$ and $0 \leq \tau \leq dt$ to evaluate the transition probabilities at $t + \tau$, that is, for times ranging over the time interval twice as long $0 \leq t \leq 2dt$. But once this is done we can use (14) again and get the transition probabilities for all times $0 \leq t \leq 4dt$. Clearly, continuing in this way, we can actually determine the transition probabilities for any time interval, no matter how long.

What this means is that knowledge of the transition probabilities in an arbitrary small time interval near zero is sufficient to specify them for all times $t \geq 0$.

Let us assume, then, that the transition probabilities $p(t, x \mid x')$ as functions of $t \geq 0$ are sufficiently well behaved so that we can write for small $dt > 0$

$$p(dt, x \mid x') \doteq p(0, x \mid x') + \alpha(x \mid x')\, dt. \tag{15}$$

Since

$$p(0, x \mid x') = \begin{cases} 1 & \text{if} \quad x = x', \\ 0 & \text{if} \quad x \neq x', \end{cases} \tag{16}$$

and according to (12) and (13)

$$p(dt, x \mid x') \to p(0, x \mid x') \qquad \text{as} \quad dt \to 0,$$

(15) actually boils down to the assumption that $p(t, x \mid x')$ has a one-sided (right) derivative $\alpha(x \mid x')$ at $t = 0$. We shall assume from now on that this is indeed the case.

In view of (16), we can write (15) as

$$p(dt, x \mid x') \doteq \alpha(x \mid x')\, dt \qquad \text{if} \qquad x \neq x', \tag{17a}$$

and

$$p(dt, x' \mid x') \doteq 1 - \lambda(x')\, dt, \tag{17b}$$

where we have denoted $\lambda(x') = -\alpha(x' \mid x')$. Since we must have, for any $dt > 0$,

$$\sum_{x \in \mathcal{S}} p(dt, x \mid x') = 1,$$

it follows that

$$\lambda(x') = \sum_{x \neq x'} \alpha(x \mid x'), \qquad x' \in \mathcal{S}.$$

Thus we see that the transition probabilities can actually be characterized by the quantities

$$\alpha(x \mid x'), \qquad x \in \mathcal{S}, \qquad x' \in \mathcal{S}, \qquad x \neq x', \tag{18}$$

which must be nonnegative and such that

$$\sum_{x \neq x'} \alpha(x \mid x') < \infty \qquad \text{for all } x' \in \mathcal{S},$$

but otherwise need not satisfy any further condition. (It is perfectly all right to have $\alpha(x \mid x') > 1$, for example.) These quantities are called *infinitesimal transition rates* (or sometimes intensities) of the continuous-time homogeneous Markov chain, and their physical interpretation is seen from (17a);

$$\alpha(x \mid x') \doteq \frac{p(dt, x \mid x')}{dt}, \qquad x \neq x'.$$

In words, for small $dt > 0$, $\alpha(x \mid x')$ is the probability of the system moving from state x' to another state x, divided by dt. The collection (18) of infinitesimal transition rates $\alpha(x \mid x')$ for all distinct pairs $x \neq x'$ of states is called the *infinitesimal generator* of the process, since together with the initial probability distribution $p_0(x)$, $x \in \mathcal{S}$, they can be used to completely specify the probability law and thus "generate" the process. In modeling physical discrete-state dynamic systems it is usually this infinitesimal generator, rather than the collection of the transition probabilities, that is easiest to get from some physical assumptions about the nature of the system.

Example 3.3.2

Consider a closed volume of a saturated solution of sodium chloride that contains both charged ions of sodium and chlorine and undissolved neutral molecules of NaCl. The particles are in constant random motion as a result of heat and possibly also as a result of stirring of the liquid, which is violent enough so that the long range electrical forces between the ions can be disregarded. Thus, a positive and a negative ion recombine to form a neutral molecule only if they are brought sufficiently close to each other. We can therefore assume that the probability of such a recombination occurring during a small time interval $(t, t + dt]$ is proportional to the product of the number of positive and negative ions present in the volume at the time t. On the other hand, a neutral molecule of NaCl can at any time split into positive and negative ions, and the probability of that occurring during $(t, t + dt]$ is again proportional to the concentration of NaCl molecules at the time t.

Call $X(t)$ the total number of neutral NaCl molecules present in the volume at time t, $t \geq 0$. Let m be the total number of sodium atoms in the solution, ionized or not. Then the total number of chlorine atoms must also be m, and thus the number of positive ions present in the solution at some time t is the same as the number of negative ions, and equals

$$m - X(t).$$

From the assumptions made, it follows that the probability that a sodium and a chlorine ion recombine during $(t, t + dt]$, that is, that the number $X(t)$ of neutral NaCl molecules increases by one, is proportional to

$$(m - X(t))^2 \, dt.$$

Similarly, the probability of ionization occurring during $(t, t + dt]$, that is, the probability that $X(t)$ decreases by one, is proportional to

$$X(t) \, dt.$$

Thus if we regard $X(t)$, $t \geq 0$, as a continuous-time homogeneous Markov chain with the state space $\mathcal{S} = \{0, 1, \ldots, m\}$, we are naturally led to the specification of the infinitesimal generator for the process, namely

$$\alpha(x \mid x') = \begin{cases} (m - x')^2 & \text{if } x = x' + 1, \\ x' & \text{if } x = x' - 1, \\ 0 & \text{otherwise} \end{cases}$$

where $x' = 1, \ldots, m - 1$, while for $x' = 0$

$$\alpha(x \mid 0) = \begin{cases} m^2 & \text{if } x = 1, \\ 0 & \text{otherwise,} \end{cases}$$

and for $x' = m$

$$\alpha(x \mid m) = \begin{cases} m & \text{if } x = m - 1, \\ 0 & \text{otherwise} \end{cases}$$

Of course, this model is rather crude, since more physical details would have to be taken into account to properly describe the behavior of the solution. Nevertheless, the example does illustrate that it is usually much simpler to specify the infinitesimal generator than to try to derive the transition probabilities. ▲

As indicated above, we should be able to recover the transition probabilities $p(t, x \mid x')$ for all $t \geq 0$ from the infinitesimal generator of the process. Since we expect the Chapman-Kolmogorov equation (14) to be the key, let us set $\tau = dt$ in (14) so that

$$p(t + dt, x \mid x') = \sum_{y \in \mathcal{S}} p(dt, x \mid y) p(t, y \mid x'). \tag{19}$$

Separating the term with $y = x$ from the rest in the summation, this becomes

$$p(t + dt, x \mid x') = p(dt, x \mid x) p(t, x \mid x') + \sum_{y \neq x} p(dt, x \mid y) p(t, y \mid x'). \tag{20}$$

Assuming now that $dt > 0$ is small, we have from (17)

$$p(dt, x \mid x) \doteq 1 - \lambda(x)\, dt, \tag{21a}$$

$$p(dt, x \mid y) \doteq \alpha(x \mid y)\, dt, \qquad x \neq y, \tag{21b}$$

and substituting back into (20), we obtain after a slight rearrangement

$$p(t + dt, x \mid x') - p(t, x \mid x') = -\lambda(x) p(t, x \mid x')\, dt + \sum_{y \neq x} \alpha(x \mid y) p(t, y \mid x')\, dt.$$

Dividing both sides by dt and letting $dt \to 0$, we end up with the differential equation.

$$\frac{dp(t, x \mid x')}{dt} = -\lambda(x) p(t, x \mid x') + \sum_{y \neq x} \alpha(x \mid y) p(t, y \mid x') \tag{22}$$

defined for $t \geq 0$ and $x \in \mathcal{S}$, $x' \in \mathcal{S}$. Since from (16)

$$p(0, x \mid x') = \begin{cases} 1 & \text{if } x = x', \\ 0 & \text{if } x \neq x', \end{cases}$$

we have the initial condition needed for the first order equation. Recall that

$$\lambda(x) = \sum_{y \neq x} \alpha(y \mid x), \qquad x \in \mathcal{S},$$

so that the infinitesimal generator is all we need to set up the equation.

This equation is known as the *Kolmogorov-Feller forward equation*.

Taking $x' = x_0$, the initial state, multiplying both sides of (22) by the initial distribution

$$p_0(x_0), \qquad x_0 \in \mathcal{S}$$

and summing over all $x_0 \in \mathcal{S}$, we obtain the equation

$$\frac{dp(t, x)}{dt} = -\lambda(x)p(t, x) + \sum_{y \neq x} \alpha(x \mid y)p(t, y), \qquad t \geq 0, \qquad x \in \mathcal{S}, \tag{23}$$

where

$$p(t, x) = \sum_{x_0 \in \mathcal{S}} p(t, x \mid x_0)p_0(x_0) = P(X(t) = x) \tag{24}$$

are now the first order distributions of the process $X(t)$, $t \geq 0$. The equation (23) with the initial condition

$$p(0, x) = p_0(x), \qquad x \in \mathcal{S}$$

is then referred to as the *Fokker-Planck equation.*

Actually, both are systems of differential equations, one for each state $x \in \mathcal{S}$, and as one might expect, it is quite difficult to obtain an explicit solution. In spite of that, equation (22) remains one of the main tools in studying the behavior of continuous-time homogeneous Markov chains.

Example 3.3.3

Consider a continuous-time homogeneous Markov chain with finite state space $\mathcal{S} = \{1, \ldots, m\}$ and infinitesimal generator

$$\alpha(x \mid x') = 1 \qquad \text{for all} \quad x \in \mathcal{S},\ x' \in \mathcal{S},\ x \neq x'.$$

Here $\lambda(x) = m - 1$ for all $x \in \mathcal{S}$, and the Kolmogorov-Feller equations have a particularly simple form, namely

$$\frac{dp(t, x \mid x')}{dt} = -(m - 1)p(t, x \mid x') + \sum_{y \neq x} p(t, y \mid x').$$

However $\sum_{y \neq x} p(t, y \mid x') = 1 - p(t, x \mid x')$, and hence upon substitution we have for all $x \in \mathcal{S}$, $x' \in \mathcal{S}$,

$$\frac{dp(t, x \mid x')}{dt} = -mp(t, x \mid x') + 1, \qquad t \geq 0.$$

A general solution of this differential equation is

$$p(t, x \mid x') = Ce^{-mt} + \frac{1}{m},$$

as can be readily verified by substitution. The constant C is determined from the initial condition: $p(0, x \mid x') = 1$ if $x = x'$ and $= 0$ if $x \neq x'$, that is, $C = 1 - 1/m$ if $x = x'$ and $C = -1/m$ if $x \neq x'$.

Thus, we find that the transition probabilities are for any $t \geq 0$ and $x \in \mathcal{S}$, $x' \in \mathcal{S}$,

$$p(t, x \mid x') = \begin{cases} \left(1 - \dfrac{1}{m}\right)e^{-mt} + \dfrac{1}{m} & \text{if} \quad x = x', \\ \dfrac{1}{m}(1 - e^{-mt}) & \text{if} \quad x \neq x'. \end{cases}$$

Of course, this example is very simple, and one can hardly expect it to be representative of even a slightly more complicated case. ▲

Remark: If, in going from (19) to (20), we had separated the term with $y = x'$, from the summation we would have obtained the *Kolmogorov-Feller backward* equation

$$\frac{dp(t, x \mid x')}{dt} = -\lambda(x')p(t, x \mid x') + \sum_{y \neq x'} p(t, x \mid y)\alpha(y \mid x'), \qquad t \geq 0, \qquad x \in \mathcal{S}, \qquad x' \in \mathcal{S}, \tag{25}$$

which has certain theoretical advantages over the forward equation. Я

Remark: If the number of states in the state space $\mathcal{S}$ is finite, and if we arrange the transition probabilities into a square matrix

$$\mathbf{P}(t) = [p(t, x \mid x')], \qquad t \geq 0$$

with x a row index and x' a column index, the forward equation (22) can be written neatly as

$$\frac{d\mathbf{P}(t)}{dt} = \mathbf{aP}(t), \qquad t \geq 0,$$

$$\mathbf{P}(0) = \mathbf{I}, \qquad \text{(identity matrix)}, \tag{26}$$

where $\mathbf{a} = [\alpha(x \mid x')]$ is a square matrix with row index x, column index x', and diagonal entries $\alpha(x \mid x) = -\lambda(x)$. (Note that the rows of $\mathbf{a}$ add to zero.) A reader with some familiarity with system theory can immediately write down the solution of (26)

$$\mathbf{P}(t) = e^{\mathbf{a}t}, \qquad t \geq 0,$$

which looks simple but actually is not, since the evaluation of a matrix exponential is generally not an easy matter.

Notice also that the backward equation (25) can also be written in matrix notation as

$$\frac{d\mathbf{P}(t)}{dt} = \mathbf{P}(t)\mathbf{a}. \qquad t \geq 0,$$

so that the matrices $\mathbf{P}(t)$ and $\mathbf{a}$ commute. Я

Example 3.3.4
Consider a continuous-time homogeneous Markov chain with two states only, $\mathcal{S} = \{1, 2\}$. Let the infinitesimal transition rates be

$$\alpha(2 \mid 1) = a_1, \qquad \alpha(1, 2) = a_2,$$

where $a_1 > 0$ and $a_2 > 0$. The 2×2 matrix $\mathbf{a}$ in equation (26) is then

$$\mathbf{a} = \begin{bmatrix} -a_1 & a_1 \\ a_2 & -a_2 \end{bmatrix}$$

and we see that this matrix equation (with dots standing for the time derivatives) is

$$\begin{bmatrix} \dot{p}(t, 1 \mid 1) & \dot{p}(t, 1 \mid 2) \\ \dot{p}(t, 2 \mid 1) & \dot{p}(t, 2 \mid 2) \end{bmatrix} = \begin{bmatrix} -a_1 & a_1 \\ a_2 & -a_2 \end{bmatrix} \begin{bmatrix} p(t, 1 \mid 1) & p(t, 1 \mid 2) \\ p(t, 2 \mid 1) & p(t, 2 \mid 2) \end{bmatrix}$$

It now follows that the unique solution of this system satisfying the initial condition $\mathbf{P}(0) = \mathbf{I}$ equals

$$\mathbf{P}(t) = e^{\mathbf{a}t},$$

where the matrix exponential function is

$$\exp\begin{bmatrix} -a_1 & a_1 \\ a_2 & -a_2 \end{bmatrix} t = \begin{bmatrix} \dfrac{a_2}{a_1 + a_2}\left(1 + \dfrac{a_1}{a_2} e^{-(a_1 + a_2)t}\right), & \dfrac{a_2}{a_1 + a_2}(1 - e^{-(a_1 + a_2)t}) \\ \dfrac{a_1}{a_1 + a_2}(1 - e^{-(a_1 + a_2)t}), & \dfrac{a_1}{a_1 + a_2}\left(1 + \dfrac{a_2}{a_1} e^{-(a_1 + a_2)t}\right) \end{bmatrix}.$$

Of course, in this simple case, the solution can be easily obtained from the Kolmogorov-Feller equations just as in Example 3.3.3. For instance, with $x = 2$, $x' = 1$, we obtain, using $p(t, 1 \mid 1) = 1 - p(t, 2 \mid 1)$,

$$\begin{aligned} \frac{dp(t, 2 \mid 1)}{dt} &= -a_2 p(t, 2 \mid 1) + a_1 p(t, 1 \mid 1) \\ &= -(a_1 + a_2)p(t, 2 \mid 1) + a_1, \qquad t \geq 0, \end{aligned}$$

with the initial condition $p(0, 2 \mid 1) = 0$. The solution of this first order linear differential equation is readily found to be

$$p(t, 2 \mid 1) = \frac{a_1}{a_1 + a_2}(1 - e^{-(a_1 + a_2)t}),$$

as asserted. However, a reader familiar with the matrix exponential may find the general formula (26) easier to use, especially when the eigenvalues and eigenvectors of the matrix $\mathbf{A}$ are easy to obtain. ▲

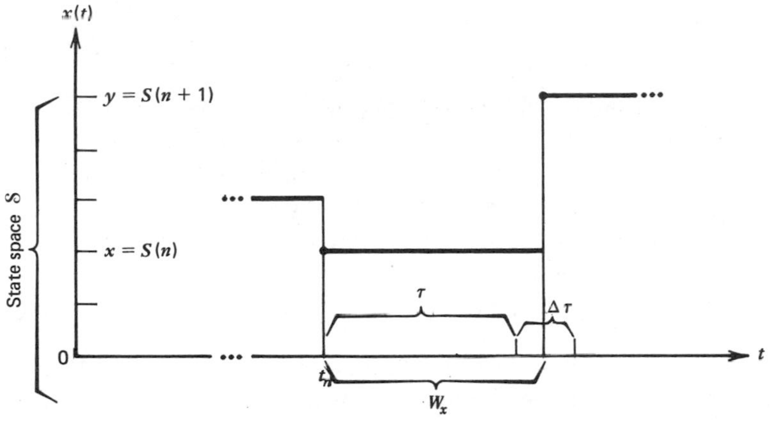

Figure 3.3.2

As was mentioned earlier, since the state space $\mathcal{S}$ is assumed to be the set of all integers, a continuous-time homogeneous Markov chain $X(t)$, $t \geq 0$, begins at some initial state $X(0) \in \mathcal{S}$ and then proceeds with random jumps from state to state, staying in each state for a random amount of time. Each sample path then passes through a sequence of states, and hence such a sequence of consecutive states

$$S(n), \qquad n = 0, 1, \ldots$$

is a discrete-time stochastic process with the same state space $\mathcal{S}$. Thus, $S(0)$ is simply the initial state of the process $X(t)$, $t \geq 0$,

$$S(0) = X(0),$$

while $S(n)$, $n > 0$, is the value of $X(t)$ at the random time of the nth jump.

Suppose now that the nth jump occurred at some time $t_n > 0$ and that the process jumped into the state $x \in \mathcal{S}$ so that

$$X(t_n) = S(n) = x. \tag{27}$$

Denote† the time the process stays in this state by W_x. In other words, W_x is the time between the nth jump and the $(n + 1)$st jump, assuming the nth jump was made into the state x. But the state into which the $(n + 1)$st jump is made is $S(n + 1)$, so that (see Fig. 3.3.2)

$$S(n + 1) = X(t_n + W_x).$$

† This is the only place where we use the letter W for a purpose other than to denote the Wiener process. Since the latter does not appear in this chapter, hopefully no confusion will arise.

Since the process

$$X(t), \qquad t \geq 0,$$

is Markov, the Markov property (1) indicates that the conditional distribution of $S(n+1)$ and W_x, given that $S(n) = x$, that is, that $X(t_n) = x$, should not depend on what happened prior to the time t_n of the nth jump, in particular on the past states $S(0), \ldots, S(n-1)$, as well as on the times spent in these states.†

Clearly, W_x is a nonnegative continuous random variable. To learn more about it, consider the conditional probability

$$P(W_x > \tau + d\tau \,|\, W_x > \tau,\ X(t_n) = x) \tag{28}$$

where $\tau > 0$ and $d\tau > 0$. Since W_x is the time spent in the state x, if $W_x > \tau$ then at time $t_n + \tau$ the process must still be in state x. Thus (28) is equal to

$$P(W_x > \tau + d\tau \,|\, X(t_n + \tau) = x,\ X(t_n) = x) = P(W_x > \tau + d\tau \,|\, X(t_n + \tau) = x), \tag{29}$$

the last equality being again a consequence of the Markov property. The same reasoning as above also implies that

$$W_x > \tau + d\tau \qquad \text{if and only if} \qquad X(t_n + \tau + d\tau) = x$$

and so (29) is also equal to the transition probability

$$P(X(t_n + \tau + d\tau) = x \,|\, X(t_n + \tau) = x) = p(d\tau, x \,|\, x) \doteq 1 - \lambda(x)\, d\tau,$$

for small $d\tau$ by (21a). Therefore, putting all this together

$$P(W_x > \tau + d\tau \,|\, W_x > \tau,\ X(t_n) = x) \doteq 1 - \lambda(x)\, d\tau.$$

But by definition of conditional probability, we also have

$$P(W_x > \tau + d\tau \,|\, W_x > \tau,\ X(t_n) = x) = \frac{P(W_x > \tau + d\tau \,|\, X(t_n) = x)}{P(W_x > \tau \,|\, X(t_n) = x)}.$$

Equating these two expressions we have

$$P(W_x > \tau + d\tau \,|\, X(t_n) = x) = (1 - \lambda(x)\, d\tau) P(W_x > \tau \,|\, X(t_n) = x),$$

which yields a differential equation

$$\frac{d}{d\tau} P(W_x > \tau \,|\, X(t_n) = x) = -\lambda(x) P(W_x > \tau \,|\, X(t_n) = x), \tag{30}$$

† Under closer scrutiny, this argument is rather shaky because the time t_n of the nth jump is random, in fact a stopping time, while the Markov property (1) has been assumed only for nonrandom "present" times t. However, for the Markov chains considered here, it can be shown that the Markov property (1) does extend to the so-called *strong Markov property*, where t is allowed to be a stopping time.

defined for $\tau \geq 0$, with initial condition $P(W_x > 0 \mid X(t) = x) = 1$. The solution

$$P(W_x > \tau \mid X(t_n) = x) = e^{-\lambda(x)\tau}, \qquad \tau > 0$$

shows that W_x *has an exponential distribution with parameter* $\lambda(x)$. Note that by (27)

$$X(t_n) = S(n) = x$$

and so we can also write

$$P(W_x > \tau \mid S(n) = x) = e^{-\lambda(x)\tau}, \qquad \tau > 0. \tag{31}$$

Let us next look at the conditional probability

$$P(S(n+1) = y, \tau \leq W_x < \tau + d\tau \mid S(n) = x), \tag{32}$$

where $y \in \mathcal{S}$, $y \neq x$, $\tau \geq 0$, and $d\tau > 0$ and small. From Figure. 3.3.2, this should be the same as

$$P(X(t_n + \tau + d\tau) = y, W_x \geq \tau \mid X(t_n) = x), \tag{33}$$

since for sufficiently small $d\tau$ the probability of having more than one jump during a time interval of length $d\tau$ is negligible, and if $X(t_n + \tau + d\tau) = y \neq x$, then the time W_x spent in state x must be less than $\tau + d\tau$. Now (33) can also be written as the product

$$P(X(t + \tau + d\tau) = y \mid W_x \geq \tau, X(t_n) = x)P(W_x \geq \tau \mid X(t_n) - x)$$

But again, $W_x > \tau$ if and only if $X(t_n + \tau) = x$, and applying the Markov property and (21b), the first term in this product becomes for small $d\tau$

$$P(X(t_n + \tau + d\tau) = y \mid W_x \geq \tau, X(t_n) = x) = p(d\tau, y \mid x) \doteq \alpha(y \mid x)\, d\tau.$$

Substituting for the second term from (30), we obtain for the conditional probability (32) the expression

$$P(S(n+1) = y, \tau \leq W_x \leq \tau + d\tau \mid S(n) = x) \doteq \alpha(y \mid x) e^{-\lambda(x)\tau}\, d\tau \tag{34}$$

and since from (31) for small $d\tau$

$$P(\tau \leq W_x < \tau + d\tau \mid S(n) = x) \doteq \lambda(x) e^{-\lambda(x)\tau}\, d\tau \tag{35}$$

we conclude from (34) that, given $S(n) = x$, the random variables $S(n+1)$ and W_x must be independent. Also we see from (34) and (35) that

$$P(S(n+1) = y \mid S(n) = x) = \frac{\alpha(y \mid x)}{\lambda(x)}, \qquad y \neq x,$$

and since, as mentioned earlier, the distribution of $S(n+1)$, given that $S(n) = x$, can no longer depend on the previous states $S(0), \ldots, S(n-1)$, it follows that $S(n)$,

$n = 0, 1, \ldots$, is in fact a discrete-time homogeneous Markov chain with transition probabilities given above. For $y = x$ we have by definition

$$P(S(n+1) = y \mid X(n) = x) = 0.$$

Remark: An alert reader may ask what happens if $\lambda(x) = 0$. But this is actually quite simple; $\lambda(x) = 0$ means

$$p(t, x \mid x) = 1 \quad \text{for all } t \geq 0;$$

in other words, that the state x is *absorbing*; once the process enters such a state, it remains there forever. Of course, for an absorbing state, the time $W_x = \infty$, and $P(S(n+1) = y \mid S(n) = x) = 0$ if $y \neq x$ and equals 1 if $y = x$. Я

Let us now summarize these findings.

THEOREM 3.3.1. *Let* $\mathrm{X}(\mathrm{t}), \mathrm{t} \geq 0$ *be a continuous-time homogeneous Markov chain with infinitesimal transition rates* $\alpha(\mathrm{x} \mid \mathrm{x}')$, *where*

$$\lambda(x) = \sum_{y \neq x} \alpha(y \mid x).$$

Then the sequence of consecutive states

$$S(n), \qquad n = 0, 1, \ldots$$

is a discrete-time homogeneous Markov chain, called an embedded *chain of the process* $\mathrm{X}(\mathrm{t})$. *Its transition probabilities are*

$$q(x \mid x') = \frac{\alpha(x \mid x')}{\lambda(x')}, \text{ if } x \neq x' \text{ and } q(x \mid x') = 0 \text{ if } x = x'. \tag{36}$$

for $\lambda(x') > 0$, $x \in \mathcal{S}$, $x' \in \mathcal{S}$. *The states with* $\lambda(x') = 0$ *are absorbing.*

The random times the process $\mathrm{X}(\mathrm{t}), \mathrm{t} \geq 0$, *spends in nonabsorbing states are independent random variables and are also independent of the embedded chain. For any nonabsorbing state* x, *the time* W_x *is called a* sojourn time *of the state and is an exponentially distributed random variable with parameter* $\lambda(\mathrm{x})$.

Example 3.3.5

Consider a piece of equipment subject to random failures. A failure, when it occurs, can be suspected to be either major or minor. If a major failure is suspected the equipment is sent to a repair shop. If a minor failure is suspected, repair is attempted on the spot. If it is then found that the failure is actually major, the equipment is then sent to the repair shop. Once there, the equipment is either repaired and returned or discarded.

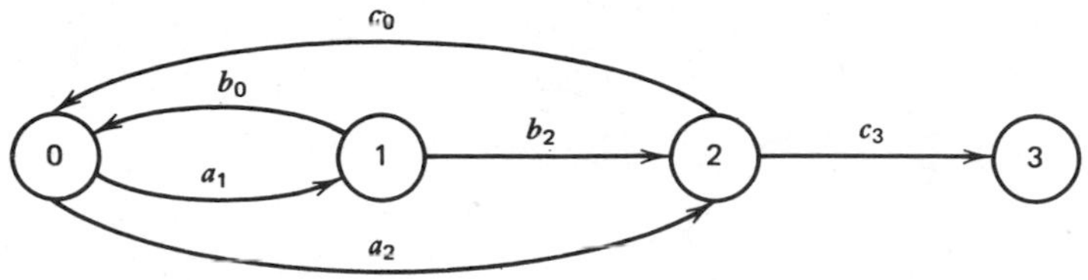

Figure 3.3.3

This entire procedure can be modeled as a four-state dynamic system with states labeled as follows:

0 ... equipment operational
1 ... repair on the spot being attempted
2 ... equipment in a repair shop
3 ... equipment discarded.

The transitions between states are as shown in Figure 3.3.3, where the letters alongside the arrows stand for the infinitesimal transition rates under the assumption that the entire system operates as a continuous-time homogeneous Markov chain. Thus, for instance, $c_0\,dt$ is the probability that the equipment is at the repair shop at time t and will be returned to operation within the time interval $(t, t + dt]$. Notice that this amounts to assuming that all the sojourn time (i.e., the time between failures, the time spent in the repair shop, and the time for which a repair on the spot is attempted) is exponentially distributed. The parameters $\lambda(x)$ of these three times are respectively:

$$\lambda(0) = a_1 + a_2, \qquad \lambda(1) = b_0 + b_2, \qquad \lambda(2) = c_0 + c_3 .$$

State $x = 3$ is absorbing. The transition probabilities $q(x \mid x')$ of the discrete-time embedded Markov chain, calculated according to equation (36) are indicated in the transition diagram depicted in Figure 3.3.4. Clearly, unless $c_3 = 0$, the entire chain is absorbing, and the equipment will eventually be discarded. ▲

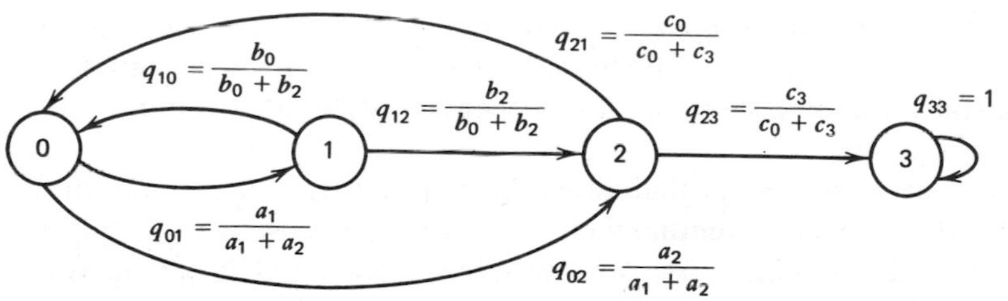

Figure 3.3.4

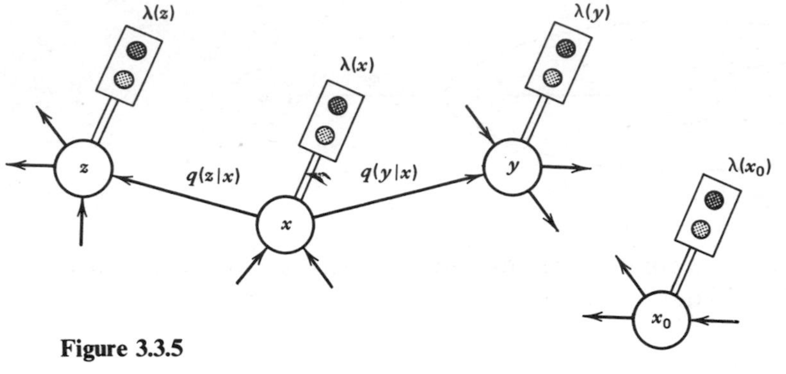

Figure 3.3.5

Theorem 3.3.1 provides us with a good picture of the operation of a continuous-time homogeneous Markov chain. We can look first at the sequence of consecutive states $S(n)$. Since this is a discrete-time homogeneous Markov chain with the same state space $\mathcal{S}$, it is completely described by the transition probabilities (36). We can therefore draw, or at least imagine, its transition probability diagram just as we did when we studied discrete-time Markov chains. This time, however, there will be no arrows looping from a state x back to itself, the absorbing states, if any, being the only exception. Instead, each nonabsorbing state will be equipped with a "traffic light" or "go signal" controlled by a random timer (see Figure 3.3.5).

If we again imagine our Markov traveler and his voyages through the transition diagram, the situation is similar to the discrete-time case: at each state x', the traveler chooses an exit arrow according to the transition probabilities $q(x \mid x')$ with complete disregard to the path taken to state x'. However, the novel feature here is that upon his arrival at each nonabsorbing state x, the traffic light there is activated, showing red. The traveler must then wait at this state for a green light to show before he can exit. The time he must wait at the state x is the random sojourn time W_x and is independent both of the way the traveler arrived at state x and of his choice of the exit arrow. In fact, all the traffic lights in the transition diagram operate independently of one another and are activated only by entry into the state whose sojourn time they control.

In terms of the transition diagram, or equivalently in terms of the transition probabilities $q(x \mid x')$, we can again define a closed set of states, recurrent states and transient states, and in general obtain a general classification of states just as in the discrete-time case.

We can also calculate the probabilities $P(\mathcal{A} \mid x)$ that starting from an initial state x, the Markov traveler eventually enters a recurrent class $\mathcal{A}$ in exactly the same way as in the discrete-time case. We again have a system of linear equations

$$P(\mathcal{A} \mid x) - \sum_{j \in \mathcal{S}_{tr}} P(\mathcal{A} \mid y) q(y \mid x) = \sum_{y \in \mathcal{A}} q(y \mid x), \tag{37}$$

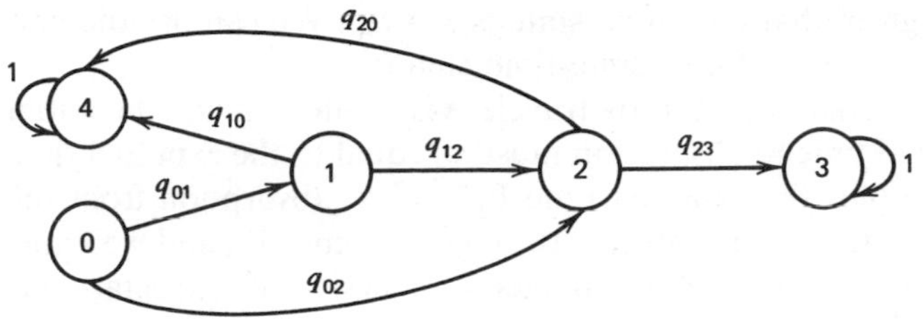

Figure 3.3.6

with $\mathcal{S}_{tr}$ being the set of all transient states. In particular, if each recurrent class consists of a single absorbing state, we can calculate the probabilities of absorption for each of the absorbing states. Note that in these calculations, the quantities $\lambda(x)$ do not appear at all.

Example 3.3.6

Suppose that for the situation described in Example 3.3.5 we wish to find the probability that the equipment, having just failed, will be successfully repaired. This can be accomplished by introducing another absorbing state, labeled, say, 4, signifying a successful return of the equipment to the state 0 and modifying the transition diagram as shown in Figure 3.3.6. With $\mathcal{A} = \{4\}$ and $\mathcal{S}_{tr} = \{0, 1, 2\}$ the desired probability is $P(\mathcal{A}|0)$, and it can be found from the system of linear equations (37):

$$P(\mathcal{A}|0) - (P(\mathcal{A}|0)0 + P(\mathcal{A}|1)q_{01} + P(\mathcal{A}|2)q_{02}) = 0,$$
$$P(\mathcal{A}|1) - (P(\mathcal{A}|0)0 + P(\mathcal{A}|1)0 + P(\mathcal{A}|2)q_{12}) = q_{10},$$
$$P(\mathcal{A}|2) - (P(\mathcal{A}|0)0 + P(\mathcal{A}|1)0 + P(\mathcal{A}|2)0) = q_{20}.$$

The solution is easily found by repeated substitution:

$$P(\mathcal{A}|0) = q_{01}(q_{10} - q_{20}q_{12}) + q_{20}q_{02},$$

which can now be expressed in terms of the original infinitesimal transition rates, the *a*'s, *b*'s, and *c*'s of Example 3.3.5. ▲

The λ's do appear, however, in calculations of probabilities when the event in question depends on the time the Markov traveler spends in each state.

Suppose, for example, that we wish to compute the expected value of the absorption time, that is, the total time T our traveler spends in transient states before entering a recurrent class. Let us assume that this must always happen; such will be the case if the number of transient states is finite, in particular, if the number of all states in $\mathcal{S}$ is finite. We want to compute the expectation

$$\mathrm{E}[T|x] = \mathrm{E}[T|X(0) = x]$$

of the absorption time given that the initial state is $x \in \mathcal{S}_{tr}$. We can use the first entrance argument just as we did for discrete-time chains.

Let $y \in \mathcal{S}$ be the first state the Markov traveler visits after leaving the initial state x. Then the expected time to absorption must be equal to the expected time spent in the initial state plus the expected time $\mathrm{E}[T \mid y]$ to absorption from the state y. The time spent in the initial state x is the sojourn time W_x, and we know that its distribution is exponential with parameter $\lambda(x)$, and so its expectation is

$$\mathrm{E}[W_x] = \frac{1}{\lambda(x)},$$

where $\lambda(x) > 0$, since x is by assumption transient, that is, not absorbing. Consequently, *conditional* upon y being the first state visited after leaving x, we have the equation

$$\mathrm{E}[T \mid S(0) = x, S(1) = y] = \frac{1}{\lambda(x)} + \mathrm{E}[T \mid y].$$

Multiplying by the transition probabilities

$$q(y \mid x) = P(S(1) = y \mid S(0) = x)$$

and summing over all $y \in \mathcal{S}$, this becomes

$$\mathrm{E}[T \mid S(0) = x] = \frac{1}{\lambda(x)} + \sum_{y \in \mathcal{S}} \mathrm{E}[T \mid y] q(y \mid x),$$

and since $S(0) = X(0)$ and $\mathrm{E}[T \mid y] = 0$ unless $y \in \mathcal{S}_{tr}$, we can write this as

$$\mathrm{E}[T \mid x] - \sum_{y \in \mathcal{S}_{tr}} \mathrm{E}[T \mid y] q(y \mid x) = \frac{1}{\lambda(x)}, \qquad x \in \mathcal{S}_{tr} \tag{38}$$

which is a system of linear equations for the unknowns $\mathrm{E}[T \mid x]$, $x \in \mathcal{S}_{tr}$. The reader is invited to compare the system (38) with the system of linear equations for the expected absorption time for discrete-time chains.

Example 3.3.7

In a simple model for the spread of an epidemic, it is assumed that there is a fixed probability $a\,dt$ that two individuals come into contact during the time interval $(t, t + dt]$. If one of them is infected and the other one is not, then the latter becomes infected with probability γ, $0 < \gamma < 1$. The entire population consists of m individuals and contacts are made "at random." This means that if, at time t, there are exactly $x(t)$ individuals infected, the probability of only one of the two individuals in contact being infected equals

$$\frac{x(t)(m - x(t))}{\binom{m}{2}}.$$

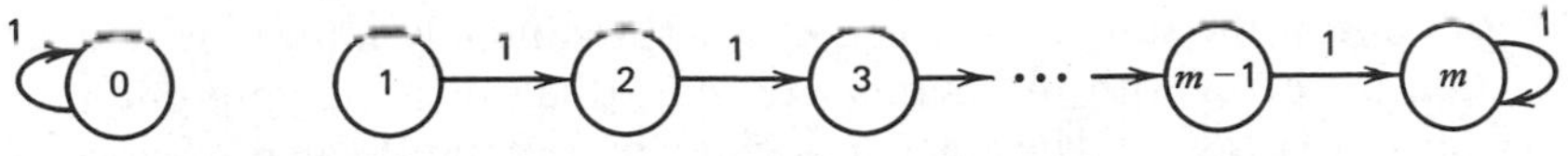

Figure 3.3.7

Now let $X(t)$, $t \geq 0$, the random number of individuals infected at the time t, be a continuous-time homogeneous Markov process. The state space $\mathcal{S} = \{0, 1, \ldots, m\}$ and the infinitesimal generator is, from the above description,

$$\alpha(x \mid x') = \begin{cases} \dfrac{2x'(m - x')}{m(m-1)} a\gamma & \text{if} \quad x = x' + 1, \\ \\ 0 & \text{otherwise,} \end{cases}$$

where $x \in \mathcal{S}$, $x' \in \mathcal{S}$, $x \neq x$. Suppose now that $X(0) = 1$, that is, there is one infected individual at time $t = 0$. Since the number of infected individuals can only increase with time, the state $x = m$ (entire population infected) is absorbing. We wish to find the expected time for this to occur, that is,

$$\mathrm{E}[T \mid 0].$$

The embedded discrete-time chain has a very simple structure, given in Figure 3.3.7, whence $\lambda(x) = \alpha(x + 1 \mid x)$, $x = 1, \ldots, m - 1$. Thus the system of linear equations (38) becomes

$$\mathrm{E}[T \mid x] - \mathrm{E}[T \mid x + 1] = \frac{m(m-1)}{2a\gamma x(n - x)},$$

$x = 1, \ldots, m - 1$, with $\mathrm{E}[T \mid m] = 0$. Adding these $m - 1$ equations, all terms except $\mathrm{E}[T \mid 1]$ on the left-hand side cancel, and we get

$$\mathrm{E}[T \mid 1] = \frac{m(m-1)}{2a\gamma} \sum_{x=1}^{m-1} \frac{1}{x(m - x)}.$$

This can be further simplified by noticing that

$$\frac{1}{x(m-x)} = \frac{1}{m}\left(\frac{1}{x} + \frac{1}{m - x}\right),$$

so that

$$\sum_{x=1}^{m-1} \frac{1}{x(m-x)} = \frac{2}{m} \sum_{x=1}^{m-1} \frac{1}{x},$$

and consequently

$$\mathrm{E}[T \mid 1] = \frac{m-1}{a\gamma}\left(1 + \frac{1}{2} + \frac{1}{3} + \cdots + \frac{1}{m-1}\right).$$

Note that for large m the sum in parenthesis is approximately $\ln(m-1)$, and so the expectation increases with the number of individuals at the asymptotic rate $(1/a)(m-1)\ln(m-1)$. It should be noted that many more sophisticated versions of this simple model can be designed and analyzed and can also be applied to a large number of physical phenomena other than epidemics. ▲

The steady-state behavior of a continuous-time homogeneous Markov chain will also be affected by the random sojourn times. Since a steady-state regime can begin to manifest itself only after an entrance into a recurrence class, we will assume, as we did in the discrete-time case, that the state space $\mathcal{S}$ is now a single recurrence class. Excluding the trivial case when $\mathcal{S}$ consists of a single absorbing state, we may also assume that

$$\lambda(x) > 0 \qquad \text{for all} \qquad x \in \mathcal{S}.$$

In the case of a discrete-time homogeneous Markov chain the steady-state behavior was expressed in terms of the *steady-state distribution* $\pi(x)$, $x \in \mathcal{S}$. We saw there that for an aperiodic regular recurrent chain, the n-step transition probabilities $p^{(n)}(x \mid x')$ converge as $n \to \infty$ to the limit $\pi(x)$ regardless of the initial state x'. For a continuous-time homogeneous Markov chain the corresponding statement would be that the limits

$$\lim_{t \to \infty} p(t, x \mid x') = \pi(x), \qquad x \in \mathcal{S} \tag{39}$$

exist and do not depend on the initial states $x' \in \mathcal{S}$. But if $p(t, x \mid x')$ as a function of $t \geq 0$ converges to a constant, then we would expect that its time derivative should approach zero:

$$\frac{dp(t, x \mid x')}{dt} \to 0 \qquad \text{as} \qquad t \to \infty. \tag{40}$$

On the other hand we know that the transition probabilities satisfy the Kolmogorov-Feller forward equations

$$\frac{dp(t, x \mid x')}{dt} = -\lambda(x)p(t, x \mid x) + \sum_{y \neq x} \alpha(x \mid y)p(t, y \mid x'),\ t \geq 0,\ x \in \mathcal{S},\ x' \in \mathcal{S}, \tag{41}$$

and so if (39) and (40) are true, then by letting $t \to \infty$ the equations in (41) reduce to

$$0 = -\lambda(x)\pi(x) + \sum_{y \neq x} \alpha(x \mid y)\pi(y), \qquad x \in \mathcal{S}$$

or

$$\sum_{y \neq x} \alpha(x \mid y)\pi(y) = \lambda(x)\pi(x), \qquad x \in \mathcal{S}. \tag{42}$$

But this is just a system of linear equations for the unknowns $\pi(x)$, $x \in \mathcal{S}$.

So far, however, we have defined the quantities $\pi(x)$, $x \in \mathcal{S}$, as limits (39) of the transition probabilities without really knowing if they actually exist and are independent of x'. The following theorem, which we present without proof, sets things straight.

THEOREM 3.3.2. *Let* $X(t)$, $t \geq 0$, *be a continuous-time homogeneous Markov chain*† *such that the state space* $\mathcal{S}$ *is a minimal closed set of states. Then for every* $x \in \mathcal{S}$ *and* $x' \in \mathcal{S}$, *the transition probability* $p(t, x \mid x')$ *converges as* $t \to \infty$ *to a limit* $\pi(x)$ *independent of the initial state* x'. *The nonnegative numbers* $\pi(x)$, $x \in \mathcal{S}$, *always satisfy the system of linear equations (42) and either this system has exactly one solution such that*

$$\pi(x) > 0 \quad \text{for all} \quad x \in \mathcal{S} \quad \text{and} \quad \sum_{x \in \mathcal{S}} \pi(x) = 1,$$

or it has no solution satisfying this condition, in which case

$$\pi(x) = 0 \qquad \text{for all } x \in \mathcal{S}.$$

In the former case we call the probability mass distribution $\pi(x)$, $x \in \mathcal{S}$, *a* steady-state distribution *of the continuous-time chain* $X(t)$, $t \geq 0$.

Example 3.3.8

Consider the now classic problem of a telephone exchange with m available trunk-lines, and let $X(t)$ be the number of lines that are busy (i.e., carrying a telephone call) at time $t \geq 0$. Suppose that telephone calls are arriving at the exchange at a constant rate and, more precisely, that the probability that a call arrives during an interval $(t, t + dt]$ is approximately $a\,dt$, where $a > 0$ is a constant. Thus, if $X(t) = x' < m$ is the number of busy lines at time t, an incoming call in $(t, t + dt]$ increases the number of busy lines by one. If, however, $x' = m$, the incoming call is lost, since there is no free trunkline available at the time. On the other hand, assume that each call currently in progress has a fixed probability $b\,dt$, ($b > 0$), of terminating during $(t, t + dt]$. Thus if $X(t) = x' > 0$, the probability that a busy line will become free during $(t, t + dt]$ is approximately x' times $b\,dt$.

If we assume that the process $X(t)$, $t \geq 0$, is a continuous-time homogeneous Markov chain with the state space $\mathcal{S} = \{0, 1, \ldots, m\}$ with the above specifications, then its infinitesimal generator is defined by

$$\alpha(x \mid x') = \begin{cases} a & \text{if } \; x = x' + 1, \\ x'b & \text{if } \; x = x' - 1, \\ 0 & \text{otherwise}, \end{cases} \tag{43}$$

† Recall that we are dealing only with nonexploding chains with no instantaneous jumps.

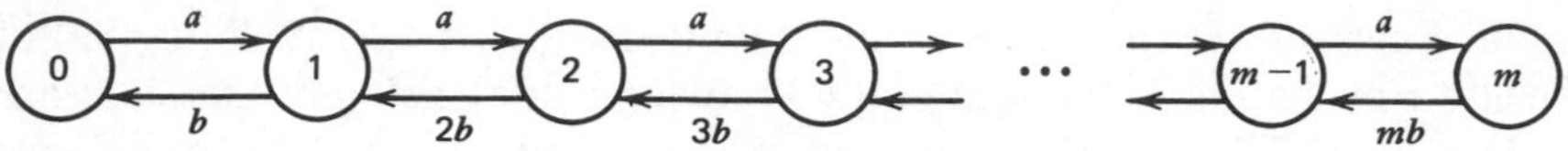

Figure 3.3.8

provided $x \in \mathcal{S}$, $x' \in \mathcal{S}$, $x \neq x'$. From the transition diagram of this chain in Figure 3.3.8, it is clear that $\mathcal{S}$ is a single finite recurrence class, and so the steady-state distribution $\pi(x)$, $x \in \mathcal{S}$, must exist.

From (43) we have

$$\lambda(x) = \begin{cases} a + xb & \text{if} \quad x = 0, 1, \ldots, m-1 \\ mb & \text{if} \quad x = m, \end{cases}$$

and hence the system of linear equations (42) is

$$a\pi(x-1) + (x+1)b\pi(x+1) = (a+bx)\pi(x), \qquad x = 1, \ldots, m-1, \quad \text{(44a)}$$

$$b\pi(1) = a\pi(0), \qquad a\pi(m-1) = mb\pi(m). \quad \text{(44b)}$$

As shown in the remark following this example, the solution of a system of linear equations like (44a) above always satisfies the relation

$$(x+1)b\pi(x+1) = a\pi(x), \qquad x = 0, 1, \ldots,$$

which also satisfies (44b). Hence by repeated substitution, we obtain

$$\pi(x) = \frac{1}{x!}\left(\frac{a}{b}\right)^x \pi(0), \qquad x = 0, 1, \ldots, m.$$

Summing over all $x = 0, \ldots, m$, we get

$$1 = \sum_{x=0}^{m} \pi(x) = \pi(0) \sum_{x=0}^{m} \frac{1}{x!}\left(\frac{a}{b}\right)^x$$

that is,

$$\pi(0) = \left(\sum_{x=0}^{m} \frac{1}{x!}\left(\frac{a}{b}\right)^x\right)^{-1},$$

so that the steady-state distribution is given by

$$\pi(x) = \frac{\dfrac{1}{x!}\left(\dfrac{a}{b}\right)^x}{\displaystyle\sum_{k=0}^{m} \frac{1}{k!}\left(\frac{a}{b}\right)^k}, \qquad x = 0, \ldots, m.$$

In particular, the steady-state probability that all m available lines are busy, or equivalently that incoming calls will be lost, equals

$$\pi(m) = \frac{\frac{1}{m!}\left(\frac{a}{b}\right)^m}{\sum_{k=0}^{m} \frac{1}{k!}\left(\frac{a}{b}\right)^k}$$

which is known as Erlang's loss formula. Although we have formulated this example in terms of a classic telephone exchange problem, it should be noted that it would also apply to a variety of problems of similar nature. For instance, $X(t)$ could stand for the number of computer terminals busy at time t, the number of atoms in an excited state caused by impinging radiation, and many others. However, the assumption that $X(t)$, $t \geq 0$, is a continuous-time homogeneous Markov chain implies that the incoming calls (or equivalent concept, e.g., the radiation quanta) form a Poisson process, while the duration of each call (or, e.g., the time each atom remains excited) are i.i.d. exponential random variables independent of the Poisson process. ▲

Remark: Continuous-time homogeneous Markov chains, with state space $\mathcal{S} = \{0, \ldots, n\}$ or $\mathcal{S} = \{0, 1, \ldots\}$ such that a transition from any given state $x' \in \mathcal{S}$ can only be made into the neighboring states $x = x' \pm 1$ (provided $x \in \mathcal{S}$), are also called *birth and death processes*. The infinitesimal generator of such a process is necessarily of the form

$$\alpha(x \mid x') = \begin{cases} a(x') & \text{if} \quad x = x' + 1, \\ b(x') & \text{if} \quad x = x' - 1, \\ 0 & \text{otherwise} \end{cases}$$

where the nonnegative constants $a(x)$, $x \in \mathcal{S}$, and $b(x)$, $x \in \mathcal{S}$, are called birth and death parameters respectively.†

For a birth and death process the system of linear equations (42) will be

$$a(x-1)\pi(x-1) + b(x+1)\pi(x+1) = (a(x) + b(x))\pi(x)$$

Rewriting this as

$$b(x+1)\pi(x+1) - a(x)\pi(x) = b(x)\pi(x) - a(x-1)\pi(x-1)$$

we see that these equations will be satisfied if both sides are always equal to the same constant. Since for $x = 0$ we have necessarily $b(0) = 0$, and so

$$b(1)\pi(1) = a(0)\pi(0)$$

† In some literature, it has become customary to denote the birth and death parameters by λ_x and μ_x, respectively. We have chosen a different notation, since we use λ and μ for different purposes.

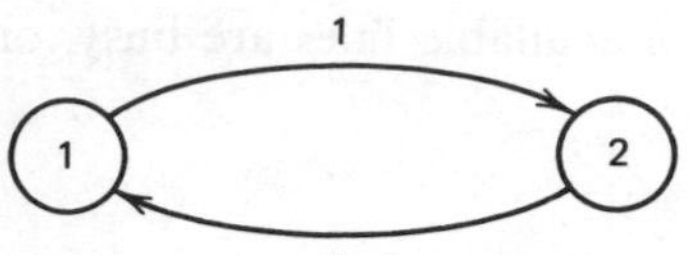

Figure 3.3.9

the constant must be zero. Thus, for all $x \in \mathcal{S}$

$$b(x+1)\pi(x+1) - a(x)\pi(x) = 0,$$

from which $\pi(x)$, $x \in \mathcal{S}$, can be easily evaluated by repeated substitution. Я

Remark: The reader may recall that in the discrete-time case we had to exclude periodic behavior before we could assert that the n-step transition probabilities converge. No such thing is needed in the continuous-time case. A continuous-time homogeneous Markov chain cannot exhibit any periodic behavior, since any such periodicity would soon be obliterated by the random and independent sojourn times the Markov traveler spends in each state. Я

Example 3.3.9
An almost trivial example can be obtained by setting $a_1 = a_2 = 1$ in the two-state chain of Example 3.3.4, that is, considering a continuous-time homogeneous Markov chain $X(t)$, $t \geq 0$, with the transition diagram in Figure 3.3.9. From Example 3.3.4 the transition probabilities are

$$p(t, 1 \mid 1) = \frac{1}{2}(1 + e^{-2t}),\ p(t, 2 \mid 1) = \frac{1}{2}(1 - e^{-2t}),$$

$$p(t, 1 \mid 2) = \frac{1}{2}(1 - e^{-2t}),\ p(t, 2 \mid 2) = \frac{1}{2}(1 + e^{-2t}), \qquad t \geq 0,$$

which clearly do not show any kind of periodic behavior. This is, after all, quite understandable, since although by Theorem 3.3.1 the embedded discrete-time chain $S(n)$, $n = 0, 1, \ldots$, would clearly be periodic, forever alternating between the two states, the time between two consecutive visits to state 1, say, is random, in this particular case having the Erlang density

$$f(t) = te^{-t}, \qquad t > 0. \qquad \blacktriangle$$

For the discrete-time homogeneous Markov chain with all states recurrent, we originally defined the quantities $\pi(x)$ as *the limit* (as $n \to \infty$) *of the average number of visits to the state* x. By definition, a continuous-time homogeneous Markov chain $X(t)$, $t \geq 0$, is recurrent if the discrete-time embedded chain

$$S(n), \qquad n = 0, 1, \ldots$$

formed by the consecutive sequence of states is recurrent. Let us denote by

$$\pi_s(x), \qquad x \in \mathcal{S}$$

the limit of the average number of visits to x by the embedded chain $S(n)$. Recall from Section 3.1 that the numbers $\pi_s(x)$ must satisfy the linear system of equations

$$\sum_{y \in \mathcal{S}} q(x \mid y)\pi_s(y) = \pi_s(x), \qquad x \in \mathcal{S} \tag{45}$$

where $q(y \mid x)$ are the transition probabilities of the embedded chain. But from (36), since $\lambda(y) > 0$,

$$q(x \mid y) = \frac{\alpha(x \mid y)}{\lambda(y)} \qquad \text{for } x \neq y$$

while $q(x \mid x) = 0$ for all $x \in \mathcal{S}$. Thus if we substitute the transition probabilities into (45) we obtain a system of linear equations

$$\sum_{y \neq x} \alpha(x \mid y)\frac{\pi_s(y)}{\lambda(y)} = \pi_s(x), \qquad x \in \mathcal{S}.$$

But the right-hand side can be written as

$$\lambda(x)\frac{\pi_s(x)}{\lambda(x)},$$

and comparison with the linear equations (42)

$$\sum_{y \neq x} \alpha(x - y)\pi(y) = \lambda(x)\pi(x), \qquad x \in \mathcal{S}$$

reveals that we must have

$$\pi(x) = \frac{c\pi_s(x)}{\lambda(x)} \qquad \text{for all } x \in \mathcal{S} \tag{46}$$

with $c > 0$ a proportionality constant to ensure $\sum_{x \in \mathcal{S}} \pi(x) = 1$.

Now as $\pi_s(x)$ is, for large n, approximately the average number of times the recurrent state x has been visited, while $(\lambda(x))^{-1}$ is the expected time spent in this state upon each visit, we conclude that $\pi(x)$ *should be the fraction of time spent in the recurrent state x in the steady-state regime.*

Equivalently, its reciprocal

$$\mu_x = \frac{1}{\lambda(x)\pi(x)}$$

is then the mean time between consecutive returns to x, the *mean* recurrent time, just as in the discrete-time case.

We also know that for the discrete-time homogeneous Markov chain $S(n)$, $n = 0, 1, \ldots$, we either have $\pi_s(x) > 0$ for all states x in the recurrence class $\mathcal{S}$ or $\pi_s(x) = 0$ for all these states, depending on whether the class $\mathcal{S}$ is positive recurrent or null recurrent.

It follows that for our continuous-time homogeneous Markov chain, the condition

$$\lim_{t \to \infty} p(t, x \mid x') = \pi(x) > 0, \qquad x \in \mathcal{S}$$

implies that all states $x \in \mathcal{S}$ are *positive recurrent*, that the finite numbers

$$\mu_x < \infty$$

are *mean recurrence times*, and that the recurrence times themselves are independent, identically distributed random variables.

The condition

$$\lim_{t \to \infty} p(t, x \mid x') = 0, \qquad x \in \mathcal{S},$$

which is possible only if the number of states in $\mathcal{S}$ is infinite, implies that either all states $x \in \mathcal{S}$ are *null recurrent* or all states $x \in \mathcal{S}$ are *transient*. This can be determined by checking whether the minimal closed set of states $\mathcal{S}$ is recurrent or transient for the embedded discrete-time chain $S(n)$, $n = 0, 1, \ldots$. However, the embedded chain being null recurrent or positive recurrent *does not imply* that the continuous-time Markov chain, although necessarily recurrent, must be accordingly null or positive recurrent. It may be positive recurrent while its embedded chain is null, or vice versa. The following example illustrates this case.

Example 3.3.10

Consider a continuous-time homogenous Markov chain $X(t)$, $t \geq 0$, with state space $\mathcal{S} = \{0, 1, \ldots\}$, and with the transition diagram as shown in Figure 3.3.10. Thus, from any state $x \in \mathcal{S}$ the chain can only either move to the next state $x + 1$

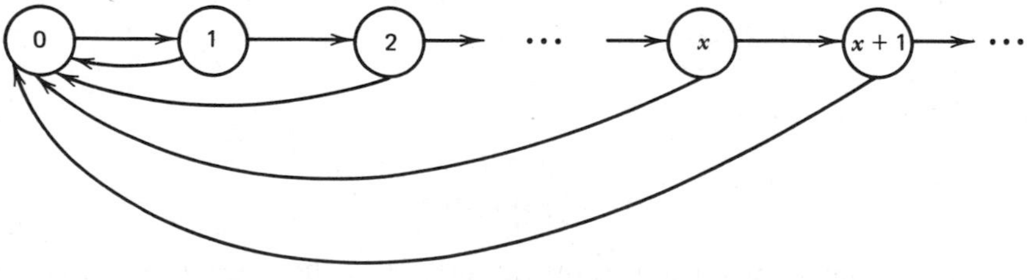

Figure 3.3.10

or (if $x > 0$) return to the state $x = 0$. In other words, the infinitesimal generator is of the form

$$\alpha(x \mid x') = \begin{cases} a(x') & \text{if } \; x = x' + 1, \\ b(x') & \text{if } \; x = 0, \; x' > 0, \\ 0 & \text{otherwise} \end{cases} \tag{47}$$

with $x \neq x'$, $x \in \mathcal{S}$, $x' \in \mathcal{S}$, as usual. Such a chain may serve, for instance, as the first step toward a model for an electron climbing up the successive energy levels by absorbing quanta of radiation with the possibility of dropping at any time to the ground level by releasing the appropriate amount of energy. Of course, without also allowing the electron to drop to intermediate energy levels, the model is not very realistic. However, to keep the forthcoming calculations simple, we are willing to sacrifice some reality. The reader who feels unhappy about this may consider instead a capacitor being charged by a random stream of unit charges with the possibility of spontaneous random discharge due perhaps to some defects in its dielectric layer.

Let us now return to mathematics and note that as long as all the constants $a(x)$ and $b(x)$ are positive, the entire state space $\mathcal{S}$ is a single recurrent class. Let us write down the system of linear equations (42). In view of (47), we obtain quite a simple set of equations, namely

$$a(x-1)\pi(x-1) = (a(x) + b(x))\pi(x), \qquad x = 1, 2, \ldots, \tag{48}$$

and

$$\sum_{x=1}^{\infty} b(x)\pi(x) = a(0)\pi(0) \tag{49}$$

From (48) for each $x = 1, 2, \ldots,$

$$\pi(x) = \frac{a(x-1)}{a(x) + b(x)} \pi(x-1),$$

and hence

$$\pi(x) = \frac{a(x-1)a(x-2) \cdots a(0)}{(a(x) + b(x))(a(x-1) + b(x-1)) \cdots (a(1) + b(1))} \pi(0). \tag{50}$$

Suppose now that we choose

$$a(x) = 1 \qquad \text{for all } x = 0, 1, \ldots,$$

and

$$b(x) = \frac{1}{x+1} \qquad \text{for } \; x = 1, 2, \ldots . \tag{51}$$

Substitution into (50) yields

$$\pi(x) = \frac{1}{1+\dfrac{1}{x+1}} \frac{1}{1+\dfrac{1}{x}} \cdots \frac{1}{1+\dfrac{1}{2}} \pi(0)$$

$$= \frac{x+1}{x+2} \frac{x}{x+1} \cdots \frac{2}{3} \pi(0) = \frac{2}{x+2} \pi(0), \qquad x = 0, 1, 2, \ldots$$

However, the harmonic series $\sum_{x=0}^{\infty} (x+2)^{-1}$ diverges to infinity and thus there is no solution of the linear system (48, 49) that would satisfy

$$\sum_{x=0}^{\infty} \pi(x) = 1.$$

In other words, $\pi(x) = 0$ for all $x \in \mathcal{S}$ is the only solution, and hence the chain is null recurrent. Thus, although the chain will return to the state $x = 0$ infinitely often, the mean recurrence time μ_0 is infinite.

On the other hand, suppose we choose instead

$$a(x) = x + 1 \qquad \text{for } x = 0, 1, \ldots,$$

and (52)

$$b(x) = 1 \qquad \text{for all } x = 1, 2, \ldots .$$

Notice that the ratio $a(x)/b(x) = x + 1$, which may be thought of as the tendency of $X(t)$ to increase, is for all $x = 1, 2, \ldots$ the same as before, that is, with the choice (51), so that one may think that the chain should behave at least asymptotically as the previous one. However, if we substitute (52) into (50), we find that

$$\pi(x) = \frac{x}{x+2} \frac{x-1}{x+1} \frac{x-2}{x} \cdots \frac{3}{5} \frac{2}{4} \frac{1}{3} \pi(0) = \frac{2}{(x+2)(x+1)} \pi(0), \qquad x = 0, 1, \ldots .$$

Using the identity

$$\frac{1}{(x+2)(x+1)} = \frac{1}{x+1} - \frac{1}{x+2}$$

we find that this time

$$\sum_{x=0}^{\infty} \pi(x) = 2\pi(0) \sum_{x=0}^{\infty} \left(\frac{1}{x+1} - \frac{1}{x+2} \right) = 2\pi(0),$$

and hence

$$\pi(x) = \frac{1}{(x+2)(x+1)}, \qquad x = 0, 1, \ldots,$$

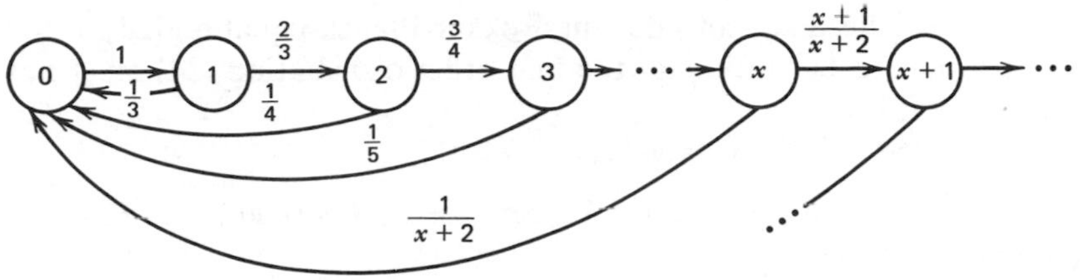

Figure 3.3.11

is now a genuine probability distribution on $\mathcal{S}$. Thus, our chain is positive recurrent, and the mean recurrence time for the state $x = 0$ is $\mu_0 = 2$. Interestingly enough, the embedded discrete-time chain is the same in both cases, as may be easily verified by calculating its transition probabilities

$$q(x \mid x') = \frac{\alpha(x \mid x')}{\lambda(x)}.$$

We find that for both the choice (51) and (52) the resulting transition diagram is as depicted in Figure 3.3.11. The difference between the two continuous-time chains rests entirely in the distributions of the sojourn times W_x, $x = 0, 1, \ldots$. They are, of course, exponentially distributed with parameter $\lambda(x)$, but in case (51) we have

$$\lambda(x) = \frac{x+2}{x+1}, \qquad x = 0, 1, \ldots,$$

while in case (52)

$$\lambda(x) = x + 2, \qquad x = 0, 1, \ldots.$$

Thus, in the former case the expected time spent in a state x is (for larger x) almost one, while in the latter case, this expected time decreases to zero as $x \to \infty$. Consequently, although the embedded chain is null recurrent, and thus the expected number of transitions between two consecutive returns to the state $x = 0$ is infinite, the time between the returns has finite expectation, since, roughly, the farther from $x = 0$ the Markov traveler gets, the faster he travels.

Finally, let us mention that either choice (51) or choice (52) would hardly make sense for the physical models mentioned earlier. Especially in the capacitor model, one would expect $b(x)$ to increase rather than decrease or remain constant with increasing x. We leave this case to the reader as an exercise. ▲

The last question we shall address is whether the steady-state distribution

$$\pi(x), \qquad x \in \mathcal{S},$$

is also a stationary distribution for the continuous-time homogeneous Markov chain $X(t)$, $t \geq 0$, as was the case in discrete time.

To see that it is indeed so, consider the Fokker-Planck equation (23). If the process $X(t)$, $t \geq 0$, is to be stationary the first order distribution (24) must not depend on the time t,

$$p(t, x) = p_0(x), \qquad t \geq 0.$$

Then the time derivative of $p(t, x)$ must be zero for all $t > 0$, and the Fokker-Planck equation (23) becomes

$$\sum_{y \neq x} \alpha(x \mid y) p_0(y) = \lambda(x) p_0(x), \qquad x \in \mathcal{S}.$$

But this system of linear equations is identical with the system (42), which for a positive recurrent chain has a unique solution

$$p_0(x) = \pi(x), \qquad x \in \mathcal{S},$$

the steady-state distribution. Since for a *homogeneous* Markov process first order stationarity implies strict stationarity, it follows that a continuous-time homogeneous Markov chain $X(t)$, $t \geq 0$, is a *strictly stationary process if and only if its initial distribution* $p_0(x)$ *coincides with the steady state distribution.*

Remark: Just as in the discrete-time case, if there are several positive recurrent classes the process $X(t)$, $t \geq 0$, is strictly stationary if and only if the initial distribution is taken to be any mixture of the steady-state distributions corresponding to these positive recurrent classes.

For the strictly stationary process $X(t)$, $t \geq 0$, to be also *ergodic*, it is necessary (and sufficient) that there be only one positive recurrent class. Я

Example 3.3.11

Imagine a particle performing a simple continuous-time random walk on the set $\mathcal{S}$ of all integers. Suppose also that the particle is subjected to a central force that increases linearly with the distance of the particle from the state $x = 0$. (You may imagine that the particle is tied to an infinitely stretchable spring anchored at the origin.) Let $X(t)$ be the position of the particle at the time $t \geq 0$, and then $X(t)$ is a continuous-time homogeneous Markov chain with state space $\mathcal{S} = \{\ldots, -1, 0, 1, \ldots\}$, and transition diagram as in Figure 3.3.12 with $a > 0$ and $b > 0$ constants. The infinitesimal generator is

$$\alpha(x \mid x') = \begin{cases} a & \text{if } \ |x| = |x'| + 1, \\ |x'| b & \text{if } \ |x| = |x'| - 1, \\ 0 & \text{otherwise} \end{cases}$$

where as usual $x \neq x'$, $x \in \mathcal{S}$, $x' \in \mathcal{S}$. Because of the obvious symmetry about $x = 0$, $\mathcal{S}$ is a single recurrent class, and hence if the steady-state distribution $\pi(x)$, $x \in \mathcal{S}$, exists, it will also be symmetric, that is,

$$\pi(-x) = \pi(x) \qquad \text{for all } x \in \mathcal{S}.$$

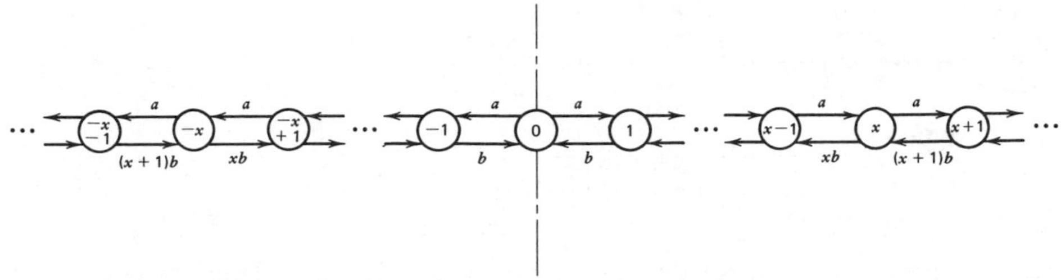

Figure 3.3.12

For $x = 1, 2, \ldots$, the linear equations (42) will be

$$a\pi(x-1) + (x+1)b\pi(x+1) = (a + xb)\pi(x),$$

which, as we have seen in Example 3.3.8, are satisfied by

$$\pi(x) = \frac{1}{x!}\left(\frac{a}{b}\right)^x \pi(0).$$

Now by symmetry

$$\sum_{x \in \mathcal{S}} \pi(x) = \pi(0) + 2\sum_{x=1}^{\infty} \pi(x) = \pi(0)\left(1 + 2\sum_{k=1}^{\infty} \frac{1}{k!}\left(\frac{a}{b}\right)^k\right)$$
$$= \pi(0)(1 + 2(e^{a/b} - 1)),$$

so that if we set

$$\pi(0) = \frac{1}{2e^{a/b} - 1},$$

we have

$$\sum_{x \in \mathcal{S}} \pi(x) = 1$$

and the process is positive recurrent. Thus, if we choose the initial distribution $p_0(x)$ of the particle to be

$$p_0(x) = \frac{\dfrac{1}{|x|!}\left(\dfrac{a}{b}\right)^{|x|}}{2e^{a/b} - 1} \qquad x \in \mathcal{S}$$

the process $X(t)$, $t \geq 0$, will be strictly stationary and ergodic.

Notice also that the steady-state distribution $\pi_s(x)$, the embedded discrete-time Markov chain $S(n)$, $n = 0, 1, \ldots$, is not necessarily the same as that of the continuous-time chain $X(t)$, $t \geq 0$. In fact, by (46), we have for the former

$$\pi_s(x) = c\lambda(x)\pi(x), \qquad x \in \mathcal{S}$$

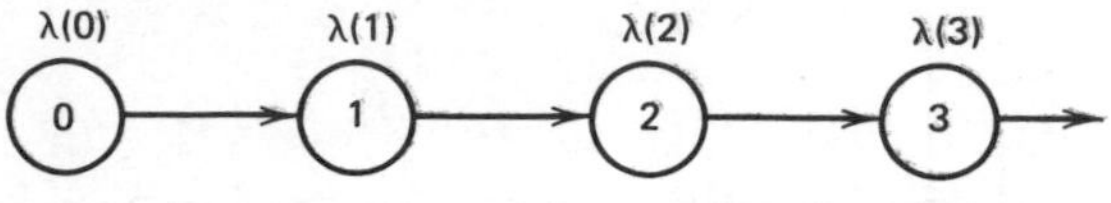

Figure 3.3.13

so that unless $\lambda(x) = 1/c$ for all $x \in \mathcal{S}$, π, and π_s are different. Thus the embedded discrete-time chain of our strictly stationary process $X(t)$, $t \geq 0$, will not be stationary. Again, the sojourn times make all the difference. ▲

Remark: At the beginning of this section we made the assumption that the continuous-time homogeneous Markov chain $X(t)$, $t \geq 0$, is *nonexploding*. We have mentioned that this is always the case if the number of states is finite, but we have given no condition for nonexplosion of an infinite state process. A simple sufficient condition of this kind is furnished by the following example, which also illustrates the phenomenon of explosion.

Consider a continuous-time homogeneous Markov chain $X(t)$, $t \geq 0$, with the state space $\mathcal{S} = \{0, 1, \ldots\}$ consisting of all nonnegative integers and the transition diagram as in Figure 3.3.13. Starting with the initial state $x_0 = 0$, each transition from a state x can only be made to the next state $x + 1$, and so all sample paths of this process will be step functions with all jumps of magnitude $+1$. Clearly, all states are transient, and the discrete-time chain $S(n)$, $n = 0, 1, \ldots$, is degenerate, $S(n) = n$, $n = 0, 1, \ldots$.

Suppose now that the parameters $\lambda(x)$ are

$$\lambda(x) = 2^x, \qquad x = 0, 1, \ldots.$$

Then the time $T^{(n)}$ needed to reach state $x = n$ is just the sum of the first $n + 1$ sojourn times $W_0 + W_1 + \cdots + W_n$, and the expectation of this time equals

$$\mathrm{E}[T^{(n)}] = \sum_{x=0}^{n} \frac{1}{2^x} = 2 - \frac{1}{2^n}$$

which approaches 2 as $n \to \infty$. But that means that the expected value of the time $T^{(\infty)}$, $\mathrm{E}[T^{(\infty)}] = 2$, which is only possible if the time $T^{(\infty)}$, is *always finite*. It follows that the process will pass through *all* of its states in a finite time, and since the number of states is infinite, each sample path is literally blown to infinity after a finite (random) time $T^{(\infty)}$. A typical sample path is shown in Figure 3.3.14. We say that the process $X(t)$, $t \geq 0$, has exploded at the *explosion time* $T^{(\infty)}$.

It is also clear from this example how to prevent explosions; if we require that for all states $x \in \mathcal{S}$

$$\lambda(x) \leq C, \tag{53}$$

where C is a finite constant, then the sum of sojourn times for any sequence of states will always increase to infinity, and no explosion can occur in a finite time. Thus, (53) is, in

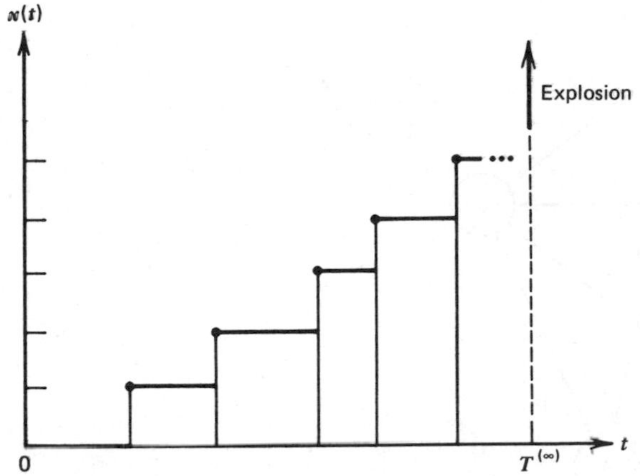

Figure 3.3.14

general, a sufficient condition for nonexplosion. Note that if the number of states is finite, (53) is automatically satisfied, so that a continuous-time homogeneous Markov chain with a finite state space cannot explode. Я

EXERCISES

Exercise 3.3

1. Verify by direct calculation that the transition probabilities $p(t, x \mid x')$ found in Example 3.3.3 satisfy the Chapman-Kolmogorov equation (14).
2. Consider a Markov chain with state space $\mathcal{S} = \{1, 2, 3\}$ and with the infinitesimal transition rates $\alpha(2 \mid 1) = 0.5$, $\alpha(3 \mid 1) = 0$, $\alpha(1 \mid 2) = 1.2$, $\alpha(3 \mid 2) = 0$, $\alpha(1 \mid 3) = 0.6$, and $\alpha(2 \mid 3) = 0.2$. For the initial state $x_0 = 3$, use the Fokker-Planck equation (23) to find the first order distributions $p(t, x)$ of the process.
3. There are m parallel input lines each carrying an analog signal measuring some parameter of a complex system. At any time interval of length dt, there is a probability $b\,dt$ that the signal at any of these lines may change, requiring reevaluation of the state of the system. This is done by a central processor, which, however, has only $r \leq m$ parallel units. If a signal changes, any currently unoccupied unit acts immediately upon it, but if all units are busy the signal is stored until some processing unit is free again. Each of the busy processing units has probability $a\,dt$ of finishing its computation in a time interval dt.

   Calling $X(t)$ the number of system parameters requiring reevaluation at time t, write the infinitesimal generator for this Markov chain. Also write the Kolmogorov-Feller differential equations and solve for the case $m = r = 2$.
4. Consider a continuous-time homogeneous Markov chain with the transition diagram depicted in Figure 3.3.E.1. The numbers along the arrows are the infinitesimal transition rates. Write the Fokker-Planck differential equation for this chain and try to solve the system if the initial distribution is uniform.

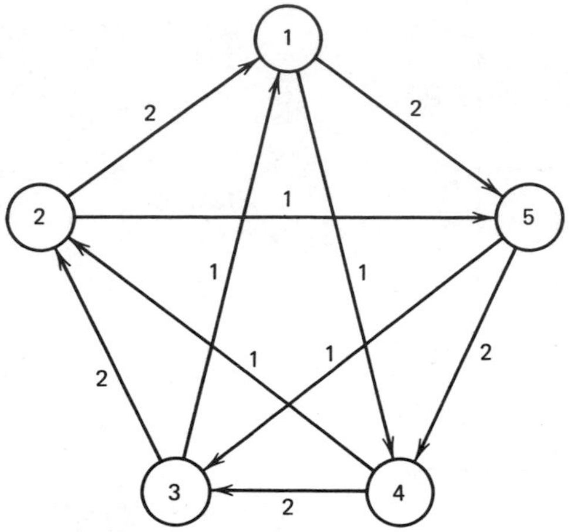

Figure 3.3.E.1

5. For the model described in Example 3.3.5, find the expected lifetime of the equipment, that is, the expected time to absorption in the state $x = 3$ if the initial state is $x = 0$ (new equipment).
6. Data are transmitted from an earth station to a space probe by a forward link, and their reception is acknowledged by a separate feedback link from the probe to the earth station. During any time interval of length dt, there is a probability $a\,dt$ that any of the two links may fail. There is also a probability $b\,dt$ that a failed forward link will become operational again. For the failed feedback link the probability that it will become operational during dt is also $b\,dt$, provided, however, that the forward link is operational during this time interval. Otherwise the feedback link remains inoperational.

   Construct a model for this system as a continuous-time Markov chain. Assuming that $a = b$ and that at $t = 0$ both links are operational, find the probability that at least one will remain operational for at least time $t > 0$.
7. In Exercise 3.3.6, calculate the steady-state distribution for the chain. Use it to find the expected fraction of time that the system is not fully operational and the expected fraction of time for which only the forward link works.
8. Consider a birth and death process with state space $\mathcal{S} = \{0, 1, \ldots\}$ and infinitesimal generator

$$\alpha(x \mid x') = \begin{cases} \alpha x' + \gamma & \text{if} \quad x = x' + 1, \\ \beta x' & \text{if} \quad x = x' - 1, \\ 0 & \text{otherwise}, \end{cases}$$

   where $\alpha < \beta$ and γ are positive parameters. (γ is called an immigration parameter.) Find the steady-state distribution for this chain. What happens if $\alpha \geq \beta$?

9. For the process in Exercise 3.3.8, write the Fokker-Planck equations (23). Assuming that the initial state $X(0) = x_0$, use this equation to show that the expectation $\mu(t) = \mathrm{E}[X(t)]$ satisfies the differential equation

$$\frac{d\mu(t)}{dt} = \gamma + (\alpha - \beta)\mu(t), \qquad t \geq 0.$$

Solve this equation and find the limit $\mu(\infty) = \lim_{t \to \infty} \mu(t)$.

10. Consider a four-state continuous time homogeneous Markov chain with infinitesimal generator $\alpha(2|1) = \alpha(1|3) = \alpha(4|3) = 1$, $\alpha(3|2) = \alpha(2|3) = 2$, $\alpha(1|4) = \alpha(4|2) = 3$, and $\alpha(x|x') = 0$ for all other $x \neq x'$. Find the steady-state distribution for this chain and the expected sojourn times for each state. Also find the transition probabilities and the steady-state distribution of the corresponding embedded chain.

11. A DC power source supplies m identical circuits. Each circuit may at any time be either off, drawing 1 milliwatt of power, or on, drawing 10 milliwatts of power. During each time interval dt, there is probability $a\,dt$ that an "off" circuit will go on and probability $b\,dt$ that an "on" circuit will go off independently of all other circuits. Construct a continuous-time Markov chain model for this system and find the asymptotic distribution of the power supplied by the source. If $a = 1.2$, $b = 1$, $m = 100$, and the source is overloaded when the power exceeds 0.5 watt, find the expected time between successive overloadings.

12. In a simplified model for a neural synaptic junction, there are several receptor sites at which molecules of an appropriate neurotransmitter may become permanently attached. This process can be temporarily slowed down by injecting into the bloodstream a synthetic chemical whose molecules somewhat resemble those of the neurotransmitter. These molecules become attached to the receptor sites only for a random period of time, assumed exponentially distributed with parameter λ, but while attached, they prevent attachment of the neurotransmitter at that particular site. Assume that during any small interval of length dt there is a probability $a\,dt$ that a neurotransmitter molecule becomes attached at a particular site and probability $b\,dt$ that the synthetic molecule becomes attached there. For a particular receptor site, evaluate the probability that no molecule is attached there at time $t > 0$, assuming the site is still free at time $t = 0$.

13. In a certain ionization process the number $X(t)$ of heavy ions per unit volume fluctuates between m_1 and m_2 (where $0 < m_1 < m_2$ are integers). During a time interval $[t, t + dt]$, each heavy ion present produces a new heavy ion with probability $a(m_2 - X(t))\,dt$ or recombines with a free electron with probability $b(X(t) - m_1)\,dt$, thus turning into a neutral atom. Individual ions are assumed to act independently of one another. Model this ionization process as a finite-state birth and death process and find the steady-state distribution of ions.

14. In Example 3.3.10, try some other choices of the parameters $a(x)$ and $b(x)$. For instance, try $a(x) = e^{-x}$, $b(x) = e^x$ or $a(x) = x^2 + 1$, $b(x) = x^2$ or any other choice you may find interesting.

15. Consider a continuous-time homogeneous Markov chain with state space $\mathcal{S} = \{1, \ldots, 5\}$ and sojourn time parameters $\lambda(x) = x$, $x = 1, \ldots, 5$. Suppose further

that the corresponding embedded discrete-time chain is identical with the one in Exercise 3.1.6. Determine the infinitesimal generator and find the steady-state distribution for this continuous-time chain.

16. Consider once more the random walk with "directional memory" and reflecting boundaries at $\pm m$ as described in Exercise 3.2.15. Suppose now that the particle stays in each position $x = -m, \ldots, m$ for an exponentially distributed length of time, with the parameter $\lambda(x) = (1 + |x|)^{-1}$ if the particle arrived at x in the direction away from the origin and $\lambda(x) = 1 + |x|$ if it arrived at the position x while moving toward the origin. Find the steady-state distribution of the corresponding continuous-time Markov chain, and find the expected time the particle stays at each position x (not state) regardless of the direction from which it arrived.

CHAPTER FOUR

CONTINUOUS-TIME PROCESSES

4.1. STOCHASTIC CALCULUS—CONTINUITY, DIFFERENTIATION, AND INTEGRATION

As in ordinary calculus, we begin with questions concerning continuity of sample functions. Recall that a deterministic function $x(t)$ is, by definition, continuous at some t_0 if

$$\lim_{t \to t_0} x(t) = x(t_0)$$

or, in other words, if for every $\varepsilon > 0$ there exists a $\delta > 0$ such that

$$|x(t) - x(t_0)| < \varepsilon \qquad \text{whenever} \qquad |t - t_0| < \delta.$$

Since sample functions $x(t)$ of a stochastic process $X(t)$ are deterministic functions, the above definition applies to sample functions as well.

However, a stochastic process is not just a single sample function but a whole ensemble of sample functions, and *any property of members of this ensemble can only be defined in terms of its probability of possessing this property* or, loosely speaking, in terms of the fraction of the members of this ensemble that have this property.

With $X(t)$ being the representative sample function, the proper question to ask is then "What is the probability that $X(t)$ is continuous at some (fixed) point t_0?"

From this point of view, it is then natural to say that a stochastic process $X(t)$ is continuous at t_0 if the probability of the event $[X(t)$ continuous at $t_0]$ is unity; in symbols,

$$P(X(t) \text{ continuous at } t_0) = 1. \tag{1}$$

Now a deterministic function $x(t)$ defined on an interval $\mathcal{T}$ is called continuous if it is continuous at every point $t_0 \in \mathcal{T}$.

Similarly a stochastic process $X(t)$, $t \in \mathcal{T}$, is called *continuous* if (1) is true for every $t_0 \in \mathcal{T}$, that is, if

$$t_0 \in \mathcal{T} \Rightarrow P(X(t) \text{ is continuous at } t_0) = 1. \tag{2}$$

However, this *does not mean* that the sample functions of such a process must be continuous! In fact, as the following example illustrates, a process $X(t)$, $t \in \mathcal{T}$, may be continuous according to the above definition, and at the same time, none of its sample functions $x(t)$ need be continuous.

Example 4.1.1

Let $X(t)$, $t \geq 0$, be a stochastic process defined by

$$X(t) = \begin{cases} 0 & \text{if} \quad t < T, \\ 1 & \text{if} \quad t \geq T, \end{cases}$$

where T itself is a continuous random variable with density $f_T(t) = 0$ for $t < 0$. Then every sample function $x(t)$ is a unit step function with a jump at the random time T and is hence discontinuous. On the other hand, for *any* fixed $t_0 \geq 0$

$$P(X(t) \text{ is continuous at } t_0) = P(T \neq t_0) = 1 - P(T = t_0) = 1$$

since $P(T = t_0) = 0$ for a continuous random variable T. Thus the process $X(t)$, $t \geq 0$, is continuous according to the definition (2). ▲

This example shows that continuity of $X(t)$ in fact means that the sample functions have no *fixed discontinuity*, that is, the probability of obtaining a sample function with discontinuity at any fixed t_0 is zero.

To define the notion of continuity for a stochastic process $X(t)$ that would entail ordinary continuity of its sample functions, we need to strengthen the definition (2).

Definition 4.1.1

A stochastic process $X(t)$, $t \in \mathcal{T}$, is called *sample continuous* if

$$P(X(t) \text{ is continuous at } t_0 \text{ for every } t_0 \in \mathcal{T}) = 1. \tag{3}$$

Carefully note the distinction between (2) and (3). Clearly, sample continuity implies continuity, but as the preceding example shows, the converse need not be true.

The notion of continuity involves that of a limit, which in turn rests upon the notion of convergence. Thus the statement "$X(t)$ is continuous at t_0" is the same as "$X(t)$ converges to $X(t_0)$ as $t \to t_0$," which is further equivalent to "$x(t_n)$ converges to $x(t_0)$ for any sequence t_n converging to t_0 as $n \to \infty$." When dealing with a stochastic process $X(t)$ rather than with a deterministic function, we have, as seen in Chapter One, several types of convergence that can be applied to the sequence $X(t_n)$. In equation (2), we have in fact used the strongest type, namely almost sure convergence.

We now give another definition of continuity of the process $X(\cdot)$, which, although weaker than (2), is most useful because it depends only on the first two moments of the process.

Definition 4.1.2
A finite-variance stochastic process $X(t)$, $t \in \mathcal{T}$, is said to be *continuous in mean square* (abbreviated MS continuous) at $t_0 \in \mathcal{T}$ if

$$\lim_{t \to t_0} \mathrm{E}[(X(t) - X(t_0))^2] = 0. \tag{4}$$

It is said to be MS continuous (on $\mathcal{T}$) if (4) holds for every $t_0 \in \mathcal{T}$.

Remark: The statement (4) is frequently written as

$$\underset{t \to t_0}{\text{l.i.m.}}\ X(t) = X(t_0)$$

where "l.i.m." is suggested by the initials of limit in mean (square). We shall occasionally use this symbol. It must be emphasized that *MS continuity does not imply sample continuity* of the process. The reader is invited to verify this by showing that the process in Example 4.1.1 is MS continuous. Я

Example 4.1.2
Let $N(t)$, $t \geq 0$, be a Poisson process with parameter $\lambda > 0$ (see Chapter One). We know that the increments $N(t_2) - N(t_1)$, $0 \leq t_1 \leq t_2$, have the Poisson distribution

$$P(N(t_2) - N(t_1) = n) = e^{-\lambda(t_2 - t_1)} \frac{\lambda^n (t_2 - t_1)^n}{n!}, \tag{5}$$

$n = 0, 1, \ldots,$ and hence for any $t_0 \geq 0$, $t \geq 0$, we also have

$$\mathrm{E}[(N(t) - N(t_0))^2] = \lambda|t - t_0| + \lambda^2 |t - t_0|^2,$$

the right-hand side being the second moment (= variance + mean^2) of the Poisson distribution (5). Thus

$$\lim_{t \to t_0} \mathrm{E}[(N(t) - N(t_0))^2] = 0,$$

that is, the Poisson process is MS continuous on its time domain $t \geq 0$.

And yet we know that all sample functions of $N(t)$, $t \geq 0$, are step functions with integral jumps so that the Poisson process is not sample continuous at any $t_0 \geq 0$. ▲

Of course, for a finite-variance process, sample continuity does imply MS continuity, so that MS continuity is a necessary condition for sample continuity.

An advantage of the MS continuity concept is that there is a simple criterion for determining whether it holds in terms of the correlation function of the process.

Before stating this criterion, let us mention the obvious fact that for any type of continuity discussed above, it is necessary that the *mean function* $\mu_X(t)$ *of the process be a continuous function of t.* If, for instance $\mu_X(t)$ were discontinuous at some t_0, then the inequality

$$(\mu_X(t) - \mu_X(t_0))^2 = (\mathrm{E}[X(t) - X(t_0)])^2 \leq \mathrm{E}[(X(t) - X(t_0))^2]$$

would imply that $X(t)$ is not MS continuous at t_0 and hence is also not sample continuous.

We will therefore assume from now on that $X(t)$ is a finite-variance process with

$$\mu_X(t) = 0 \qquad \text{for all} \quad t \in \mathcal{T}.$$

The reader should be reminded that there is really nothing restrictive about this assumption, since one can always consider the process $X(t) - \mu_X(t)$ in place of $X(t)$. Now let us state and prove the MS continuity requirement.

THEOREM 4.1.1 (*MS Continuity Criterion*)
A finite variance process $X(t)$, $t \in \mathcal{T}$ *is MS continuous at* $t_0 \in \mathcal{T}$ *if and only if its correlation function* $R_X(t_1, t_2)$ *is continuous at* (t_0, t_0). *Hence the process* $X(t)$ *is MS continuous on* $\mathcal{T}$ *if and only if* $R_X(t_1, t_2)$ *is continuous at every diagonal point* $t_1 = t_2$.

Proof. By the definition of the correlation function

$$\mathrm{E}[(X(t) - X(t_0))^2] = R_X(t, t) - R_X(t, t_0) - R_X(t_0, t) + R_X(t_0, t_0). \tag{6}$$

Hence, if $R_X(\cdot, \cdot)$ is continuous at (t_0, t_0), then by letting $t \to t_0$ the right-hand side of (6) approaches zero and $X(t)$ is MS continuous at t_0 by Definition 4.1.2.

Conversely, we have

$$\begin{aligned} R_X(t_1, t_2) - R_X(t_0, t_0) &= \mathrm{E}[(X(t_1) - X(t_0))X(t_2)] \\ &\quad + \mathrm{E}[X(t_0)(X(t_2) - X(t_0))] \end{aligned} \tag{7}$$

and by the Schwarz inequality

$$|\mathrm{E}[(X(t_1) - X(t_0))X(t_2)]|^2 \leq \mathrm{E}[(X(t_1) - X(t_0))^2]\mathrm{E}[X^2(t_2)]$$

and (8)

$$|\mathrm{E}[X(t_0)(X(t_2) - X(t_0))]|^2 \leq \mathrm{E}[X^2(t_0)]\mathrm{E}[(X(t_2) - X(t_0))^2].$$

Now if $X(t)$ is MS continuous at t_0, then as $t_1 \to t_0, t_2 \to t_0$, the right-hand sides of (8) go to zero, and so by (7)

$$\lim_{t_1 \to t_0} \lim_{t_2 \to t_0} R_X(t_1, t_2) = R_X(t_0, t_0)$$

that is, $R_X(t_1, t_2)$ is continuous at (t_0, t_0). ●

Example 4.1.3
Consider a finite-variance process $X(t)$, $t \geq 0$, with zero mean function $\mu_X(t) = 0$, $t \geq 0$, and with correlation function

$$R_X(t_1, t_2) = \sigma^2(e^{-\alpha|t_1 - t_2|} - e^{-\alpha(t_1 + t_2)})$$

$t_1 \geq 0$, $t_2 \geq 0$, where $\sigma^2 > 0$ and $\alpha > 0$ are constants. It is easily seen that this correlation function is continuous at any $t_1 = t_2 = t_0 \geq 0$, and hence the process $X(t)$ is MS continuous on $t \geq 0$. ▲

Example 4.1.4
Let $X(t)$, $t \geq 0$, be a stochastic process such that for every $t \geq 0$, $X(t)$ is a random variable with zero mean and finite variance σ^2, and for any $t_1 \neq t_2$ the random variables $X(t_1)$ and $X(t_2)$ are uncorrelated. Then

$$\mu_X(t) = 0, \qquad t \geq 0,$$

while the correlation function

$$R_X(t_1, t_2) = \begin{cases} \sigma^2 & \text{if } t_1 = t_2, \\ 0 & \text{if } t_1 \neq t_2. \end{cases}$$

Thus the correlation function is discontinuous at every diagonal point $t_1 = t_2$, and hence the process is not MS continuous at any $t_0 \geq 0$. Notice that this process is very strange, so that the result is hardly a surprise. ▲

Example 4.1.5
Let $Z(n)$, $n = 0, 1, \ldots$, be a sequence of independent random variables with zero means and finite nonzero variances $E[Z^2(n)] = \sigma_n^2$. Define a continuous-time process

$$X(t), \qquad t \geq 0$$

by setting

$$X(t) = Z(n) \qquad \text{if} \qquad n \leq t < n + 1, \qquad n = 0, 1, \ldots.$$

A typical sample path of this process will look like the one pictured in Figure 4.1.1. Clearly, this process is not sample continuous; in fact, the sample paths are discontinuous at every $t = n$, $n = 1, 2, \ldots$, with probability 1, while they are continuous for all other times $t > 0$. We might say that the process has *fixed discontinuities* at integral time instances.

Now the mean function of this process

$$\mu_X(t) = 0 \qquad \text{for all} \qquad t \geq 0$$

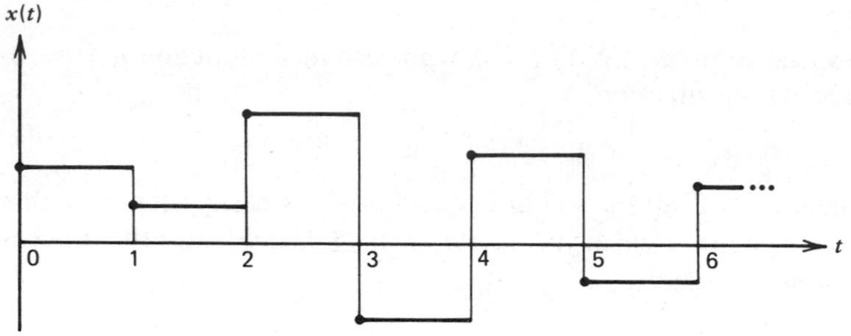

Figure 4.1.1

since all the random variables $Z(n)$ have zero means. The correlation function

$$R_X(t_1, t_2) = \begin{cases} \sigma_n^2 & \text{if } n \le t_1 < n+1 \text{ and } \\ & \quad n \le t_2 < n+1 \text{ for some } n = 0, 1, \ldots, \\ 0 & \text{otherwise,} \end{cases}$$

since in the first case, $R_X(t_1, t_2) = E[X(t_1)X(t_2)] = E[Z^2(n)] = \sigma_n^2$, while in the second case, $X(t_1)X(t_2)$ is equal to a product of two different $Z(n)$'s, and since the latter are independent and have zero mean, the expectation is zero. In Figure 4.1.2,

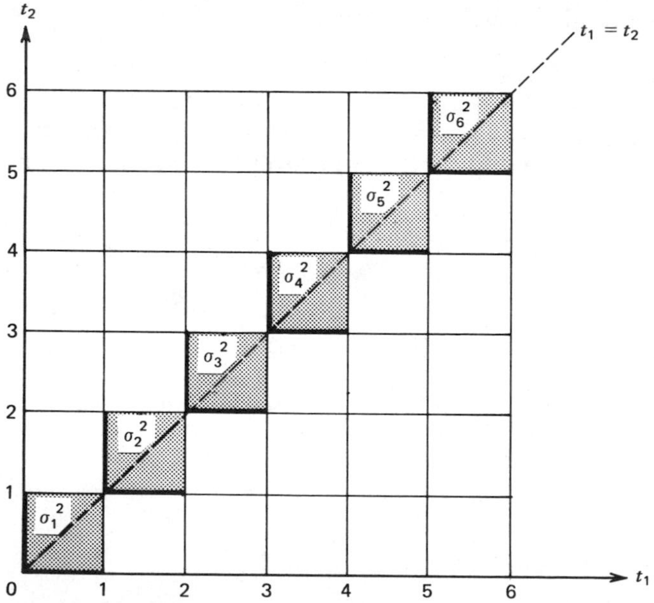

Figure 4.1.2

which shows the domain of the correlation function, note that it has value zero everywhere except on the string of squares along the diagonal $t_1 = t_2$. It follows that the correlation function $R_X(t_1, t_2)$ is continuous at every diagonal point $t_1 = t_2 \neq n$, $n = 1, 2, \ldots$, while it will be discontinuous whenever $t_1 = t_2 = n$, a positive integer. It follows that the process $X(t)$, $t \geq 0$, will be MS continuous everywhere except at the integral time instances $t = 1, 2, \ldots$. Thus MS discontinuities correspond exactly to the fixed discontinuities of the sample paths of its ensemble.

This is quite a universal feature—whenever an MS discontinuity at some $t_0 \in \mathcal{T}$ is found, one may typically suspect a fixed discontinuity of the ensemble at this time instant. (Or at least a positive probability of such a fixed discontinuity.)

▲

An easy by-product of the proof of Theorem 4.1.1 is the following theorem.

THEOREM 4.1.2. *If a correlation function* $R_X(t_1, t_2)$ *is continuous at every diagonal point* $t_1 = t_2$, *then it is continuous everywhere.*

Proof. Equations (7) and (8) are valid if the point (t_0, t_0) is replaced by (t_{01}, t_{02}), with t_{01} and t_{02} not necessarily equal. If $R_X(t_1, t_2)$ is continuous at the diagonal, then according to the previous theorem the process $X(t)$ is MS continuous at every point, in particular both at t_{01} and at t_{02}. But then the right-hand sides of (8) again go to zero, and so

$$\lim_{t_1 \to t_{01}} \lim_{t_2 \to t_{02}} R_X(t_1, t_2) = R_X(t_{01}, t_{02})$$

by (7). ●

It is no big surprise that the continuity of correlation functions implies MS continuity of the process. Let

$$\rho_{X(t), X(t_0)} = \frac{R_X(t, t_0)}{\sqrt{\sigma_X^2(t)\sigma_X^2(t_0)}}$$

where $\sigma_X^2(t) = R_X(t, t)$, be the correlation coefficient of $X(t)$ and $X(t_0)$. If R_X is continuous, then

$$|\rho_{X(t), X(t_0)}| \to 1 \qquad \text{as} \quad t \to t_0. \tag{9}$$

But a correlation coefficient is ± 1 if and only if the two random variables are linearly dependent. Thus by (9), $X(t)$ and $X(t_0)$ are, for t approaching t_0, close to being linearly dependent, that is, $X(t) \doteq aX(t_0)$, so that $X(t)$ can hardly be discontinuous at t_0.

It is therefore not unexpected that by imposing a somewhat more strict condition on the correlation function one can also obtain a criterion for sample continuity of the process.

Sample-Continuity Criterion

Let $X(t)$, $t \in \mathcal{T}$, be a finite variance process. If there exist constants C and $\gamma > 1$ such that for all $t_1 < t_2$ in some finite time interval $I \subset \mathcal{T}$

$$E[(X(t_2) - X(t_1))^2] \leq C(t_2 - t_1)^\gamma \tag{10}$$

then the process is sample continuous in the interval I.

If the process $X(t)$ is Gaussian, then it is sufficient that the exponent $\gamma > 0$.

Example 4.1.6

Consider the standard Wiener process $W_0(t)$, $t \geq 0$. By definition, its increments $W_0(t_2) - W_0(t_1)$, $t_1 < t_2$, have the Gaussian density with mean zero and variance $t_2 - t_1$. Thus for any $0 \leq t_1 < t_2$

$$E[(W_0(t_2) - W_0(t_1))^2] = t_2 - t_1$$

and since the process is Gaussian, the sample-continuity criterion is satisfied with $C = 1$ and $\gamma = 1$. It follows that the standard Wiener process is sample continuous. Clearly, the same is also true for the nonstandard Wiener process

$$W(t) = \sigma W_0(t) + \mu t, \qquad t \geq 0, \qquad \sigma > 0$$

as well as an inhomogeneous Wiener process defined by

$$W(t) = \sigma(t) W_0(t) + \mu(t), \qquad t \geq 0$$

provided that $\sigma(t) > 0$ and $\mu(t)$ are continuous functions of $t \geq 0$.

Notice that for a Poisson process $N(t)$, $t \geq 0$, with parameter $\lambda = 1$, we would also have

$$E[(N(t_2) - N(t_1))^2] = t_2 - t_1, \qquad 0 \leq t_1 < t_2$$

but since the Poisson process is not Gaussian, the sample-continuity criterion would require $\gamma > 1$. In fact, the Wiener process is essentially the only process with independent increments that is sample continuous. ▲

The above criterion is not the most general one that can be given for the sample continuity of the process. However, it has the advantage that condition (10) can be expressed entirely in terms of the correlation function of the process, since

$$E[(X(t_2) - X(t_1))^2] = R_X(t_2, t_2) - R_X(t_2, t_1) - R_X(t_1, t_2) + R_X(t_1, t_1), \tag{11}$$

which is, in fact, the second mixed difference of $R_X(t_1, t_2)$ at a diagonal point. Thus, if we denote

$$\begin{aligned}\Delta^2_{\tau_1, \tau_2} R_X(t_1, t_2) = R_X(t_1 + \tau_1, t_2 + \tau_2) &- R_X(t_1 + \tau_2, t_2) \\ &- R_X(t_1, t_2 + \tau_2) \\ &+ R_X(t_1, t_2), \qquad \tau_1 > 0, \qquad \tau_2 > 0,\end{aligned} \tag{12}$$

then condition (10) can be written as

$$\Delta^2_{\tau,\tau} R_X(t, t) \le C\tau^\gamma.$$

The proof of the criterion, although not particularly difficult, will not be given here.

Example 4.1.7
Consider a finite-variance process

$$X(t), \qquad 0 \le t < 1,$$

with zero mean function and with correlation function

$$R_X(t_1, t_2) = \frac{\sigma^2}{1 - t_1 t_2}, \qquad 0 \le t_1 < 1, \qquad 0 \le t_2 < 1.$$

Noticing that the correlation function is the sum of a convergent geometric series,

$$R_X(t_1, t_2) = \sigma^2(1 + t_1 t_2 + (t_1 t_2)^2 + (t_1 t_2)^3 + \cdots)$$

we can write the second difference (12) as

$$\begin{aligned}\Delta^2_{\tau,\tau} R_X(t, t) = {} & \sigma^2(1 + (t + \tau)^2 + (t + \tau)^4 + (t + \tau)^6 + \cdots) \\ & - 2\sigma^2(1 + t(t + \tau) + t^2(t + \tau)^2 + t^3(t + \tau)^3 + \cdots) \\ & + \sigma^2(1 + t^2 + t^4 + t^6 + \cdots),\end{aligned}$$

where as long as $0 \le t \le t + \tau < 1$, all three series are convergent. Adding corresponding terms, we see that the first terms cancel, the second terms add to τ^2, while τ^2 can be factored from all remaining terms, leaving an infinite series that converges and hence is bounded by some constant C. Therefore, we can write for all $0 \le t \le t + \tau < 1$,

$$\Delta^2_{\tau,\tau} R_X(t, t) \le \sigma^2(\tau^2 + C\tau^2) = \sigma^2(1 + C)\tau^2,$$

and it follows that the process is sample continuous on its time domain $\mathcal{T} = [0, 1)$. ▲

Example 4.1.8
Let $X(t)$, $t \ge 0$, be the Ornstein-Uhlenbeck process, that is, a Gauss-Markov process with zero mean function and with correlation function

$$R_X(t_1, t_2) = \sigma^2(e^{-\alpha|t_1 - t_2|} + e^{-\alpha(t_1 + t_2)}),$$

$t_1 \geq 0$, $t_2 \geq 0$, with σ^2 and α positive constants. For $0 \leq t \leq t + \tau$, the second difference becomes

$$\begin{aligned}\Delta^2_{\tau,\tau} R_X(t, t) &= \sigma^2(1 + e^{-\alpha(2t+2\tau)}) - 2\sigma^2(e^{-\alpha\tau} + e^{-\alpha(2t+\tau)}) + \sigma^2(1 + e^{-\alpha 2t}) \\ &= 2\sigma^2(1 - e^{-\alpha\tau}) + \sigma^2 e^{-2\alpha t}(e^{-2\alpha\tau} - 2e^{-\alpha\tau} + 1) \\ &= \sigma^2(1 - e^{-\alpha\tau})[2 + e^{-2\alpha\tau}(1 - e^{-\alpha\tau})].\end{aligned}$$

The term in brackets is clearly bounded by some constant C ($C < 3$), while for $\tau > 0$

$$1 - e^{-\alpha\tau} \leq \alpha\tau.$$

Consequently, we have for all $t \geq 0$, $\tau \geq 0$, $\Delta^2_{\tau,\tau} R_X(t, t) \leq \sigma^2 C\tau$, and since the process is Gaussian, it is therefore sample continuous on any finite interval in its time domain.

This should come as no surprise, since we know that the Ornstein-Uhlenbeck process can be obtained by a continuous-time transformation of the standard Wiener process, and since we established sample continuity of the latter in Example 4.1.6, the sample continuity of the former follows automatically by the above-mentioned fact. ▲

Both the MS-continuity and the sample-continuity criteria take on particularly simple forms if the process $X(t)$ is wide-sense stationary, for then, as we know, the correlation function depends only on the time difference

$$R_X(t_1, t_2) = R_X(t_2 - t_1)$$

so that

$$R_X(t, t) = R_X(0)$$

and (see (11))

$$\mathrm{E}[(X(t_2) - X(t_1))^2] = 2R_X(0) - 2R_X(t_2 - t_1)$$

since (for a real-valued process) $R_X(t_1, t_2) = R_X(t_2, t_1)$. Thus, we have the following *continuity criteria*:

> Let $X(t)$, $t \in \mathcal{T}$, be a wide-sense stationary process. Then $X(t)$ is MS continuous if and only if its correlation function $R_X(\tau)$ is continuous at $\tau = 0$, in which case it is continuous everywhere. For the process to be sample continuous on any finite interval $I \subset \mathcal{T}$ of length l, it is sufficient that there are constants C and $\gamma > 1$ such that
>
> $$R_X(0) - R_X(\tau) \leq C\tau^\gamma, \qquad 0 < \tau \leq l. \tag{13}$$
>
> If the process is also Gaussian, it is sufficient that $\gamma > 0$.

Example 4.1.9
Let $X(t)$, $-\infty < t < \infty$, be a wide-sense stationary process with zero mean function $\mu_X(t) = 0$ and correlation function

$$R_X(\tau) = \sigma^2 e^{-\alpha|\tau|}, \qquad -\infty < \tau < \infty,$$

where $\sigma^2 > 0$ and $\alpha > 0$ are constants. We have

$$R_X(0) - R_X(\tau) = \sigma^2(1 - e^{-\alpha\tau}) \leq \sigma^2 \alpha \tau^\gamma, \qquad \tau \geq 0$$

where the constant γ can be one but not larger. Thus, we cannot conclude that the process is sample continuous.

If, however, the process $X(t)$, $-\infty < t < \infty$, is known to be Gaussian, then by choosing $\gamma = 1$, criterion (13) applies and the process is sample continuous on any finite interval in its time domain. Of course, in this case we recognize the process as a stationary Ornstein-Uhlenbeck process. ▲

Remark: The entire discussion above applies without change to vector-valued processes, since a vector-valued process is continuous (in whatever sense) if each of the component processes is continuous. For complex-valued processes the only change is that the left-hand side of (13) must be replaced by $R_X(0) - \mathcal{Re}\{R_X(\tau)\}$ where $\mathcal{Re}$ denotes the real part. Я

DIFFERENTIATION. If $X(t)$, $t \in \mathcal{T}$, is a continuous-time stochastic process, then it would be natural to define the time derivative of the process

$$\dot{X}(t) = \frac{d}{dt} X(t), \qquad t \in \mathcal{T} \tag{14}$$

as a continuous-time process whose sample paths $\dot{x}(t)$ are time derivatives of the sample paths $x(t)$ of the process $X(t)$, that is $\dot{x}(t) = dx(t)/dt$, $t \in \mathcal{T}$. Of course, for this definition to be meaningful the sample paths $x(t)$ would all have to be differentiable over $\mathcal{T}$.

In practice, when we talk about the derivative of the process, we clearly assume this to be the case. However, even if all sample paths of $X(t)$ are indeed differentiable, the determination of the probability law of the process $\dot{X}(t)$ defined by (14) could be extremely complicated, since the derivative is a limit

$$\dot{X}(t) = \lim_{\tau \to 0} \frac{X(t + \tau) - X(t)}{\tau}. \tag{15}$$

We therefore prefer to replace definition (15) by a weaker version by using the limit in mean square instead.

Definition 4.1.3
Let $X(t)$, $t \in \mathcal{T}$, be a continuous-time finite-variance stochastic process. Then the process $\dot{X}(t)$, $t \in \mathcal{T}$, defined by

$$\dot{X}(t) = \underset{\tau \to 0}{\text{l.i.m.}} \frac{X(t+\tau) - X(t)}{\tau} \tag{16}$$

provided the MS limit exists, is called the mean square derivative (MS derivative) of the process $X(t)$.

Recall that the symbol l.i.m. in (16) means

$$\lim_{\tau \to 0} \mathrm{E}\left[\left(\frac{X(t+\tau) - X(t)}{\tau} - \dot{X}(t)\right)^2\right] = 0 \tag{17}$$

that is, that (for each fixed t) there is a random variable $\dot{X}(t)$ such that for every sequence $\tau_1, \tau_2, \ldots$ converging to zero

$$\mathrm{E}\left[\left(\frac{X(t+\tau_n) - X(t)}{\tau_n} - \dot{X}(t)\right)^2\right] \text{ converges to zero as } n \to \infty. \tag{18}$$

Now if $X(t)$ is a finite-variance process with all sample paths differentiable, that is, such that the *sample derivative* (15) exists, then (17) is true with $\dot{X}(t)$ the sample derivative.

Hence if a *finite-variance process* $X(t)$ *has a sample derivative* $\dot{X}(t)$, *then it has an MS derivative, and these two derivatives are equivalent* (i.e., equal with probability one for each $t \in \mathcal{T}$).

Example 4.1.10
Suppose that a process $X(t)$, $t \geq 0$, is obtained as a finite linear combination of sine waves with random amplitudes $Z_1, \ldots, Z_n$, radial frequencies $\Omega_1, \ldots, \Omega_n$, and phases $\Theta_1, \ldots, \Theta_n$:

$$X(t) = \sum_{k=1}^{n} Z_k \sin(\Omega_k t + \Theta_k), \qquad t \geq 0.$$

Clearly, all sample functions of this process being of the form

$$x(t) = \sum_{k=1}^{n} z_k \sin(\omega_k t + \theta_k), \qquad t \geq 0$$

are differentiable, and so this process has both a sample derivative and an MS derivative

$$\dot{X}(t) = \sum_{k=1}^{n} Z_k \Omega_k \cos(\Omega_k t + \Theta_k), \qquad t \geq 0. \qquad \blacktriangle$$

Remark: The converse of the above statement is not true. The existence of the MS derivative does not imply existence of the sample derivative. The situation is quite similar to that in the previous subsection. Also, sufficient conditions for the existence of sample derivatives can be obtained in terms of the correlation function R_X. The advantage of working with the MS derivative is that its second order properties can be related to those of the process $X(t)$. Я

According to the Loève criterion (see Chapter One), the sequence (18) converges to zero if and only if

$$E\left[\frac{X(t+\tau_n)-X(t)}{\tau_n}\frac{X(t+\tau_m)-X(t)}{\tau_m}\right] \tag{19}$$

converges as both $n \to \infty$ and $m \to \infty$. By multiplying the terms inside the expectation and writing the result as the corresponding linear combination of expectations, we see that (19) equals

$$\frac{1}{\tau_n \tau_m}(R_X(t+\tau_n, t+\tau_m) - R_X(t+\tau_n, t) - R_X(t, t+\tau_m) + R_X(t, t)) \tag{20}$$

where R_X is the correlation function of $X(t)$. But the limit of (20) as $n \to \infty$ and $m \to \infty$ is either of the mixed partial derivatives

$$\frac{\partial^2}{\partial t_1\, \partial t_2} R_X(t_1, t_2), \qquad \frac{\partial^2}{\partial t_2\, \partial t_1} R_X(t_1, t_2)$$

at the diagonal point $t_1 = t$, $t_2 = t$. Thus (using a well-known calculus fact), the limit of (20) is guaranteed to exist for any $\tau_n \to 0$, $\tau_m \to 0$ if the partial derivatives exist and are continuous at the diagonal points $t_1 = t$, $t_2 = t$.

Once the existence of the MS derivative is established, it is easy to obtain its mean and correlation functions, for then (17) implies that

$$\lim_{\tau \to 0} E\left[\frac{X(t+\tau)-X(t)}{\tau}\right] = E[\dot{X}(t)]$$

so that

$$\mu_{\dot{X}}(t) = \frac{d}{dt}\mu_X(t). \tag{21}$$

Similarly

$$\lim_{\tau_1 \to 0}\lim_{\tau_2 \to 0} E\left[\frac{X(t_1+\tau_1)-X(t_1)}{\tau_1}\frac{X(t_2+\tau_2)-X(t_2)}{\tau_2}\right] = E[\dot{X}(t_1)\dot{X}(t_2)]$$

whence

$$R_{\dot{X}}(t_1, t_2) = \frac{\partial^2}{\partial t_1\, \partial t_2} R_X(t_1, t_2). \tag{22}$$

The cross-correlation function between $\dot{X}(t)$ and $X(t)$ is obtained in the same way:

$$R_{\dot{X},X}(t_1, t_2) = \frac{\partial}{\partial t_1} R_X(t_1, t_2). \tag{23}$$

Finally, if $X(t)$ is a Gaussian process and its MS derivative $\dot{X}(t)$ exists, then it is also a Gaussian process, since then $[X(t+\tau) - X(t)]/\tau$ is Gaussian and convergence in a mean square implies convergence in distribution.

To summarize, we state the following theorem.

THEOREM 4.1.3. *Let* $X(t)$, $t \in \mathcal{T}$, *be a finite-variance process with mean function* $\mu_X(t)$ *and correlation function* $R_X(t_1, t_2)$. *If the mean function is differentiable and if the correlation function is such that the mixed partial derivatives*

$$\frac{\partial^2 R_X(t_1, t_2)}{\partial t_1\, \partial t_2}, \qquad \frac{\partial^2 R_X(t_1, t_2)}{\partial t_2, \partial t_1}$$

exist and are continuous at every point $t_1 = t_2 = t \in \mathcal{T}$, *then the MS derivative* $\dot{X}(t)$, $t \in \mathcal{T}$, *exists. Its mean function, correlation function, and cross-correlation function with* $X(t)$ *are given by (21), (22), and (23). If* $X(t)$ *is a Gaussian process, then* $\dot{X}(t)$ *is also a Gaussian process.*

Example 4.1.11

Consider the process $X(t)$, $0 \le t < 1$, of Example 4.1.7. The mean function, being identically zero, is trivially differentiable. As for the correlation function we have for all $0 \le t_1 < 1$ and $0 \le t_2 < 1$,

$$\frac{\partial R_X(t_1, t_2)}{\partial t_1} = \sigma^2 \frac{t_2}{(1 - t_1 t_2)^2}$$

and

$$\frac{\partial^2 R_X(t_1, t_2)}{\partial t_1\, \partial t_2} = \sigma^2 \frac{1 + t_1 t_2}{(1 - t_1 t_2)^3} = \frac{\partial^2 R_X(t_1, t_2)}{\partial t_2\, \partial t_1}.$$

Clearly, this is a continuous function in both variables t_1 and t_2, and hence by Theorem 4.1.3, the MS derivative

$$\dot{X}(t), \qquad 0 \le t < 1,$$

exists. The mean function of this new process is again zero, and the correlation function is

$$R_{\dot{X}}(t_1, t_2) = \sigma^2 \frac{1 + t_1 t_2}{(1 - t_1 t_2)^3}, \qquad 0 \le t_1 < 1, \qquad 0 \le t_2 < 1.$$

The cross-correlation function is given by

$$R_{\dot{X},X}(t_1, t_2) = \sigma^2 \frac{t_2}{(1 - t_1 t_2)^2}, \qquad 0 \le t_1 < 1, \qquad 0 \le t_2 < 1.$$

Applying Theorem 4.1.3 once more to the process $\dot{X}(t)$, $0 \le t < 1$, it can be shown that this process is also MS differentiable, that is, the second MS derivative $\ddot{X}(t)$, $0 \le t < 1$, of the original process also exists. In fact, this process has MS derivatives of all orders. ▲

Example 4.1.12

Consider again the standard Wiener process $W_0(t)$, $t \ge 0$. From its correlation function

$$R_{W_0}(t_1, t_2) = \begin{cases} t_1 & \text{if } 0 \le t_1 \le t_2 \\ t_2 & \text{if } 0 \le t_2 \le t_1, \end{cases}$$

it is plain that even the first partial derivatives do not exist anywhere along the diagonal $t_1 = t_2$. It follows that the standard Wiener process $W_0(t)$, $t \ge 0$, does not have an MS derivative at any time instant $t \ge 0$. Consequently, it cannot be sample differentiable. Actually, it can be shown by more advanced methods that almost all of its sample functions are nowhere differentiable; in symbols,

$$P(\dot{W}_0(t) \text{ exists at some } t \ge 0) = 0.$$

Notice that this is a stronger assertion than $W_0(t)$ not being sample differentiable, for this would only mean that the above probability is less than 1.

This is also true for a nonstandard and inhomogeneous Wiener process $W(t) = \sigma(t)W_0(t) + \mu(t)$, for the nondifferentiability of sample paths cannot be undone by any choice of $\mu(t)$ and $\sigma(t) > 0$.

The Ornstein-Uhlenbeck process shares this nondifferentiability of sample paths with the Wiener process, since likewise, the nondifferentiability cannot be remedied by a continuous-time transformation. ▲

In case the process $X(t)$, $t \in \mathcal{T}$, is wide-sense stationary, that is,

$$\mu_X(t) = \mu \qquad \text{and} \qquad R_X(t_1, t_2) = R_X(t_2 - t_1)$$

the above theorem simplifies to the following corollary.

Corollary 4.1.3

If $X(t)$, $t \in \mathcal{T}$, is a wide-sense stationary process with correlation function $R_X(\tau)$, then the MS derivative $\dot{X}(t)$, $t \in \mathcal{T}$, exists whenever the second derivative

$$\frac{d^2}{dt^2} R_X(\tau)$$

exists and is continuous at $t = 0$. Then $\dot{X}(t)$ is also a wide-sense stationary process with mean function

$$\mu_{\dot{X}}(t) = 0,$$

correlation function

$$R_{\dot{X}}(\tau) = -\frac{d^2}{dt^2} R_X(\tau),$$

and cross-correlation function between $X(t)$ and $\dot{X}(t)$

$$R_{X,\dot{X}}(\tau) = \frac{d}{dt} R_X(\tau) = R_{\dot{X},X}(-\tau).$$

If $X(t)$ is Gaussian, so is $\dot{X}(t)$.

Example 4.1.13

Consider a stationary Gaussian process $X(t)$, $-\infty < t < \infty$, with zero mean function and with correlation function

$$R_X(\tau) = \frac{\sigma^2\lambda^2}{\lambda^2 + \tau^2}, \qquad -\infty < \tau < \infty,$$

where $\sigma > 0$ and $\lambda > 0$ are constants. We see immediately that the second derivative

$$\frac{d^2 R_X(\tau)}{d\tau^2} = 2\sigma^2\lambda^2 \frac{3\tau^2 - \lambda^2}{(\lambda^2 + \tau^2)^3}$$

is continuous at $\tau = 0$. Consequently, the MS derivative $\dot{X}(t)$, $-\infty < t < \infty$, exists and is a stationary Gaussian process with zero mean function and with correlation function

$$R_{\dot{X}}(\tau) = 2\sigma^2\lambda^2 \frac{\lambda^2 - 3\tau^2}{(\lambda^2 + \tau^2)^3}, \qquad -\infty < \tau < \infty.$$

The cross-correlation function is

$$R_{X,\dot{X}}(\tau) = \frac{d}{d\tau}\left(\frac{\sigma^2\lambda^2}{\lambda^2 + \tau^2}\right) = \frac{-2\sigma^2\lambda^2\tau}{(\lambda^2 + \tau^2)^2}, \qquad -\infty < \tau < \infty. \qquad \blacktriangle$$

INTEGRATION. Let $X(t)$, $t \in \mathcal{T}$, be a continuous time stochastic process and let $t_0 \in \mathcal{T}$ be fixed. We would like to define a new process $Y(t)$ by

$$Y(t) = \int_{t_0}^{t} X(\tau)\, d\tau. \tag{24}$$

As before, a natural way to define $Y(t)$ would be to assume that its sample functions $y(t)$ are integrals

$$y(t) = \int_{t_0}^{t} x(\tau)\, d\tau \tag{25}$$

of sample functions $x(t)$ of the process $X(t)$, provided that the latter process has all sample functions integrable over the time domain of interest. A sufficient condition for this or, more precisely, for the probability that sample functions are integrable to be 1, is that $X(t)$ be sample continuous in the interval $[t_0, t]$. We then refer to the process $Y(t)$ as the *sample integral* of the process $X(t)$.

However, as with continuity and differentiation, we prefer to have a mean square version of the integral (24). To arrive at such a version, recall the definition of an ordinary integral (25) of the deterministic function $x(t)$. This (Riemann) integral is defined as a limit of approximating Riemann sums

$$s_n(t) = \sum_{k=0}^{n-1} x(\tau_k)\, \Delta t_k$$

where $t_0 < t_1 < t_2 < \cdots < t_n = t$ is a partition $\mathcal{T}$ of the interval $[t_0, t]$ into n subintervals of lengths $\Delta t_k = t_{k+1} - t_k$ and τ_k are points such that $t_k \le \tau_k \le t_{k+1}$, $k = 0, \ldots, n-1$. Then

$$\int_{t_0}^{t} x(\tau)\, d\tau = \lim s_n(t)$$

where the limit is taken over all partitions $\mathcal{T}$ such that

$$\max_{k=0, \ldots, n-1} \Delta t_k \to 0 \quad \text{as} \quad n \to \infty. \tag{26}$$

Now if $X(t)$ is a stochastic process, then the sum

$$S_n(t) = \sum_{k=0}^{n-1} X(\tau_k)\, \Delta t_k \tag{27}$$

is, for every fixed partition $\mathcal{T}$ and fixed choice of points τ_k, a random variable. Hence, to define a mean square limit we are led to the following definition:

Definition 4.1.4
Let $X(t)$, $t \in \mathcal{T}$, be a finite-variance process. Then the process $Y(t)$, $t \ge t_0 \in \mathcal{T}$, defined by

$$Y(t) = \text{l.i.m. } S_n(t),$$

where the l.i.m. is taken over all partitions $\mathcal{T}$ satisfying (26), is called an MS integral of the process $X(t)$.

If a finite-variance process $X(t)$ has a sample integral, then it also has an MS integral and they are equivalent. It can be shown that MS continuity of the process $X(\tau)$, $\tau \in [t_0, t]$ is, in fact, a sufficient condition for this. Since we will only deal with MS-continuous processes, we will use the notation

$$Y(t) = \int_{t_0}^{t} X(\tau)\, d\tau \tag{28}$$

for both MS and sample integrals. The integral (28) can be manipulated in exactly the same way as an ordinary Riemann integral of a deterministic function. Also, if $Y(t)$ is a finite-variance process such that its MS derivative

$$\dot{Y}(t) = X(t) \qquad \text{exists for all} \quad t \geq t_0,$$

then

$$\int_{t_0}^{t} X(\tau)\, d\tau = Y(t) - Y(t_0).$$

Furthermore, for a finite-variance MS-continuous process $X(t)$

$$\mathrm{E}\left[\int_{t_0}^{t} X(\tau)\, d\tau\right] = \int_{t_0}^{t} \mathrm{E}[X(\tau)]\, d\tau \tag{29}$$

and

$$\mathrm{E}\left[\int_{t_0}^{t} X(\tau)\, d\tau \int_{t_0'}^{t'} X(\tau)\, d\tau\right] = \int_{t_0}^{t}\int_{t_0'}^{t'} \mathrm{E}[X(\tau)X(\tau')]\, d\tau\, d\tau'. \tag{30}$$

For a process $X(t)$ defined on $t \in (-\infty, \infty)$, we will also need the improper integral

$$Y(t) = \int_{-\infty}^{t} X(\tau)\, d\tau. \tag{31}$$

This is defined as the limit of integrals (28) with $t_0 \to -\infty$, the limit being the MS limit for the MS integral. By the Loève criterion, such an MS limit will exist if and only if

$$\mathrm{E}\left[\int_{t_0}^{t} X(\tau)\, d\tau \int_{t_0'}^{t} X(\tau)\, d\tau\right] \tag{32}$$

converges as $t_0 \to -\infty$ and $t_0' \to -\infty$. However, according to (29) and (30), (32) equals

$$\int_{t_0}^{t}\int_{t_0'}^{t} R_X(\tau, \tau')\, d\tau\, d\tau',$$

which will converge if and only if the ordinary improper integral

$$\int_{-\infty}^{t}\int_{-\infty}^{t} R_X(\tau, \tau')\, d\tau\, d\tau' \tag{33}$$

exists. It can be shown that the existence of the integral (33) is also sufficient for the existence of the improper sample integral (31).

We restate some of the above as a theorem:

THEOREM 4.1.4. *Let* X(t), $t \in \mathcal{T}$, *be an MS-continuous process, and let* $t_0 \in \mathcal{T}$, *possibly* $t_0 = -\infty$ *if* $\mathcal{T} = (-\infty, \infty)$. *If the double integral*

$$\int_{t_0}^{t}\int_{t_0}^{t} R_X(\tau, \tau')\, d\tau\, d\tau', \qquad t \in \mathcal{T}$$

exists (which is always the case if $-\infty < t_0 < t < \infty$), *then the process*

$$Y(t) = \int_{t_0}^{t} X(\tau)\, d\tau, \qquad t \geq t_0, \qquad t \in \mathcal{T} \tag{34}$$

exists as both the sample integral and MS integral of X(t). *The process (34) is a sample-continuous finite-variance process with mean function*

$$\mu_Y(t) = \int_{t_0}^{t} \mu_X(\tau)\, d\tau, \tag{35}$$

correlation function

$$R_Y(t_1, t_2) = \int_{t_0}^{t_1}\int_{t_0}^{t_2} R_X(\tau_1, \tau_2)\, d\tau_1\, d\tau_2 \tag{36}$$

and cross-correlation function between Y(t) *and* X(t)

$$R_{Y,X}(t_1, t_2) = \int_{t_0}^{t_1} R_X(\tau, t_2)\, d\tau. \tag{37}$$

It is a Gaussian process whenever X(t) *is a Gaussian process.*

Formulas (35) through (37) follow immediately from (29) and (30). $Y(t)$ is sample continuous because it is an integral with t as the upper limit. If $X(t)$ is a Gaussian process, then the approximating sums (27), being linear combinations of Gaussian random variables, are also Gaussian. Hence, their MS limit, which exists under the assumptions of the theorem, must also be Gaussian.

Example 4.1.14

Let us define a process $Y(t)$, $t \geq 0$, as the integral of the standard Wiener process

$$Y(t) = \int_{0}^{t} W_0(\tau)\, d\tau, \qquad t \geq 0.$$

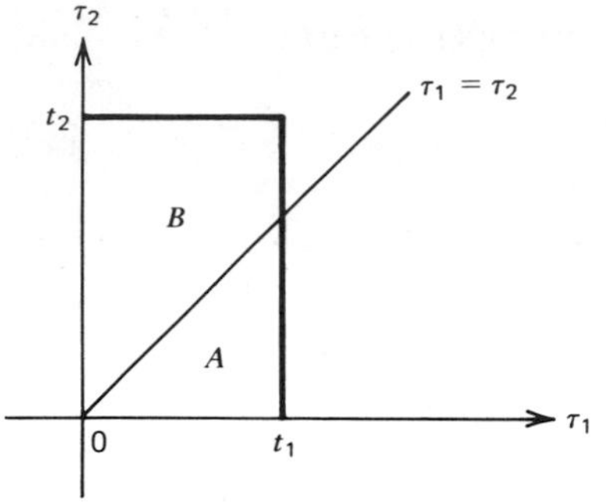

Figure 4.1.3

By Theorem 4.1.4, the process will clearly exist, and it will be a Gaussian process with mean function

$$\mu_Y(t) = \int_0^t 0\, d\tau = 0, \qquad t \geq 0.$$

To calculate its correlation function, let us choose $0 \leq t_1 \leq t_2$ and consider the integration region of the double integral (36) as shown in Figure 4.1.3. Since, for the standard Wiener process, $R_{W_0}(\tau_1, \tau_2) = \tau_2$ if $0 \leq \tau_2 \leq \tau_1$, and $R_{W_0}(\tau_1, \tau_2) = \tau_1$ if $0 \leq \tau_1 \leq \tau_2$, we can write the double integral (36) as a sum

$$\int_0^{t_1} \int_0^{t_2} R_{W_0}(\tau_1, \tau_2)\, d\tau_1\, d\tau_2 = \iint_A \tau_2\, d\tau_1\, d\tau_2 + \iint_B \tau_1\, d\tau_1\, d\tau_2,$$

where A and B are the regions in Figure 4.1.3. Now

$$\iint_A \tau_2\, d\tau_1\, d\tau_2 = \int_0^{t_1} \left(\int_0^{\tau_1} \tau_2\, d\tau_2 \right) d\tau_1 = \int_0^{t_1} \frac{\tau_1^2}{2}\, d\tau_1 = \frac{t_1^3}{6},$$

and

$$\iint_B \tau_1\, d\tau_1\, d\tau_2 = \int_0^{t_1} \left(\int_{\tau_1}^{t_2} \tau_1\, d\tau_2 \right) d\tau_1 = \int_0^{t_1} \tau_1(t_2 - \tau_1)\, d\tau_1$$

$$= \frac{t_1^2 t_2}{2} - \frac{t_1^3}{3}.$$

Hence for $0 \leq t_1 \leq t_2$, we have

$$\int_0^{t_1} \int_0^{t_2} R_{W_0}(\tau_1, \tau_2)\, d\tau_1\, d\tau_2 = \frac{t_1^2}{6}(3t_2 - t_1),$$

and so

$$R_Y(t_1, t_2) = \begin{cases} \dfrac{t_1^2}{6}(3t_2 - t_1) & \text{if } 0 \le t_1 \le t_2, \\ \dfrac{t_2^2}{6}(3t_1 - t_2) & \text{if } 0 \le t_2 \le t_1, \end{cases}$$

is the desired correlation function. Notice that the variance $\sigma_Y^2(t) = R_Y(t, t)$ increases as the cube of the time t.

For the cross-correlation, we have from (37)

$$R_{Y, W_0}(t_1, t_2) = \begin{cases} \dfrac{t_1^2}{2} & \text{if } 0 \le t_1 \le t_2, \\ t_2\left(t_1 - \dfrac{t_2}{2}\right) & \text{if } 0 \le t_2 \le t_1. \end{cases}$$

▲

Example 4.1.15

Let $X(t)$, $-\infty < t < \infty$, be a finite-variance process with zero mean function and with correlation function

$$R_X(t_1, t_2) = \begin{cases} \sigma^2 e^{(1+\alpha)t_1 + (1-\alpha)t_2} & \text{if } -\infty < t_1 \le t_2 < \infty, \\ \sigma^2 e^{(1-\alpha)t_1 + (1+\alpha)t_2} & \text{if } -\infty < t_2 \le t_1 < \infty, \end{cases}$$

where $\sigma > 0$ and $\alpha > 0$ are constants. Since the double integral

$$\int_{-\infty}^{t} \int_{-\infty}^{t} R_X(\tau_1, \tau_2)\, d\tau_1\, d\tau_2 = 2\sigma^2 \int_{-\infty}^{t} e^{(1-\alpha)\tau_1} \left(\int_{-\infty}^{\tau_1} e^{(1+\alpha)\tau_2}\, d\tau_2 \right) d\tau_1$$

$$= \frac{2\sigma^2}{1+\alpha} \int_{-\infty}^{t} e^{2\tau_1}\, d\tau_1 = \frac{\sigma^2}{1+\alpha} e^{2t} < \infty$$

for all $-\infty < t < \infty$, the process defined as an improper integral

$$Y(t) = \int_{-\infty}^{t} X(\tau)\, d\tau, \qquad -\infty < t < \infty$$

exists. The process clearly has its mean function $\mu_Y(t)$ identically zero. Its correlation function $R_Y(t_1, t_2)$, as well as the cross-correlation function $R_{Y, X}(t_1, t_2)$, can be easily evaluated from (36) and (37). We leave this to the reader as an exercise.

It may be noted that the process $X(t)$, $-\infty < t < \infty$, with the correlation function as above, can for instance be obtained by multiplying a stationary Ornstein-Uhlenbeck process by the deterministic time function e^t. ▲

Remark: For an improper double integral

$$\int_{-\infty}^{t}\int_{-\infty}^{t} R_X(\tau_1, \tau_2)\, d\tau_1\, d\tau_2$$

to exist, it is clearly necessary that $R_X(t_1, t_2) \to 0$ as $t_1 \to -\infty$ and $t_2 \to -\infty$. This can never be the case if $R_X(t_1, t_2) = R_X(t_2 - t_1)$, for by choosing $t_1 = t_2 = t$, we would have $R_X(t, t) = \sigma_X^2 > 0$ for all such t_1 and t_2. It follows that if

$$X(t), \qquad -\infty < t < \infty$$

is a wide-sense stationary process, then the integral

$$\int_{-\infty}^{t} X(\tau)\, d\tau$$

cannot exist except in the trivial case $X(t) = 0$ for all $-\infty < t < \infty$. Я

EXERCISES

Exercise 4.1

1. Recall from your calculus course that a sum or a product of two continuous functions is again a continuous function. Show that the same is true for MS continuity of two finite-variance stochastic processes $X(t)$ and $Y(t)$, independent in the case of the product.
2. Suppose $X(t)$, $t \geq 0$, is a process with independent increments such that the increment $X(t_2) - X(t_1)$, $0 \leq t_1 < t_2$, has the gamma density:

$$f_{X(t_2)-X(t_1)}(x) = \frac{\lambda^{t_2 - t_1}}{\Gamma(t_2 - t_1)} x^{t_2 - t_1 - 1} e^{-\lambda x}, \qquad x \geq 0.$$

   Show that the process is MS continuous but fails the sample-continuity criterion. (Indeed, it is not a sample-continuous process.)
3. Consider a phase-modulated sine wave $X(t) = A\cos(\omega t + \pi N(t))$, $t > 0$, where $\omega > 0$ is a constant, $N(t)$, $t \geq 0$, is a Poisson process, and A is a symmetric Bernoulli random variable with $P(A = 1) = P(A = -1) = \frac{1}{2}$ independent of $N(t)$, $t \geq 0$. Sketch a typical sample path of this process to convince yourself that the process is not sample continuous. Show, however, that the process is MS continuous. (*Hint:* The correlation function $R_X(t_1, t_2)$ can be derived by noticing that $X(t_2) = \pm X(t_1)$, depending on whether $N(t_2) - N(t_1)$ is even or odd.)
4. Show that if a finite-variance process $X(t)$, $t \in \mathcal{T}$, is MS differentiable, then it is necessarily MS continuous. Show that if $X(t)$ is, in addition, a Gaussian process, it is also sample continuous. (*Hint:* Verify that the sample continuity criterion is satisfied with $\gamma = 1$.)
5. By applying Theorem 4.1.3 to the MS derivative of a process $X(t)$, $t \in \mathcal{T}$, obtain a condition for existence and a formula for the correlation function of the second MS derivative $\ddot{X}(t)$, $t \in \mathcal{T}$, of the process $X(t)$. Generalize to the nth MS derivative and specialize to wide-sense stationary processes.

6. Derive a formula for the cross-correlation function between the nth and the mth MS derivative of a process $X(t)$, $t \in \mathcal{T}$.
7. If $X(t)$, $t \in \mathcal{T}$, is an MS differentiable wide-sense stationary process, show that $\text{Cov}[X(t), \dot{X}(t)] = 0$ for all $t \in \mathcal{T}$. (*Hint:* Differentiate the identity $R_X(\tau) = R_X(-\tau)$ and set $\tau = 0$.)
8. Check whether the process $X(t)$ of Example 4.1.1 is MS differentiable, and if so, find the correlation function of $\dot{X}(t)$.
9. Let Z be a random variable with finite variance, and let $X(t) = Z$ for all $t \in \mathcal{T}$, in the sense that $\text{E}[(X(t) - Z)^2] = 0$. Show that this is the case if and only if $R_X(t, t) = 0$ for all $t \in \mathcal{T}$.
10. Use the result of Exercise 4.1.9 to show that $R(t_1, t_2) = e^{-(t_1 - t_2)^4}$ cannot be a correlation function.
11. Show that the process $X(t)$, $-\infty < t < \infty$, defined in Example 4.1.13 has MS derivatives of all orders. Obtain the expression for the variance function of its nth MS derivative for all $n = 1, 2, \ldots$. (*Hint:* $R_X(\tau)$ is a sum of a geometric series—the Taylor series of $R_X(\tau)$ at $\tau = 0$.)
12. Use the result of Exercise 4.1.4 to argue that if a Gaussian process has MS derivatives of all orders, its sample functions are infinitely differentiable as well. (Such processes are called analytic.)
13. Below are various correlation functions of a zero mean wide-sense stationary process $X(t)$, $-\infty < t < \infty$. In each case, find the correlation function of the MS derivative $X(t)$, $-\infty < t < \infty$, and the cross-correlation function between $X(t)$ and $\dot{X}(t)$. Which of these processes also have the second MS derivative $\ddot{X}(t)$, $-\infty < t < \infty$?

    (a) $R_X(\tau) = \sigma^2 e^{-\alpha|\tau|}(1 + \alpha|\tau|), \qquad \alpha > 0$

    (b) $R_X(\tau) = \sigma^2 e^{-\alpha|\tau|}\left(\cos \omega\tau + \dfrac{\alpha}{\omega} \sin \omega|\tau|\right), \qquad \alpha > 0, \qquad \omega > 0$

    (c) $R_X(\tau) = \sigma^2 e^{-\alpha|\tau|}(1 + \alpha|\tau| + \frac{1}{3}\alpha^2\tau^2), \qquad \alpha > 0$

    (d) $R_X(\tau) = \sigma^2 e^{-\alpha\tau^2}, \qquad \alpha > 0$

    (e) $R_X(\tau) = \dfrac{\sin \omega\tau}{\tau}, \qquad \omega > 0.$

14. The current $X(t)$, $-\infty < t < \infty$, through a 1 henry inductance is assumed to be a zero mean stationary Gaussian process with the correlation function

    $$R_X(\tau) = 4e^{-|\tau|}(\cos 2\tau + \tfrac{1}{2} \sin 2|\tau|), \qquad \text{amperes}^2.$$

    Find the probability that the voltage $Y(t)$ across the inductance at a fixed time t has magnitude greater than 5 volts.
15. Verify from the definitions that both MS differentiation and MS integration obey the usual rules of ordinary calculus, in particular the rule for differentiating and integrating linear combinations, the rule for differentiating a product and the rule for integration by parts for independent processes, the chain rule for differentiation, and the substitution rule for integrals, the last two for deterministic time transformations only.

16. For the process of Example 4.1.15, find the correlation function $R_Y(t_1, t_2)$ and the cross-correlation function $R_{X,Y}(t_1, t_2)$.
17. Let $X(t)$, $-\infty < t < \infty$, be a stationary Ornstein-Uhlenbeck process. Show that the process $Y(t)$, $-\infty < t < \infty$, defined by

$$Y(t) = \frac{1}{\Delta}\int_{t-\Delta}^{t} X(\tau)\,d\tau,$$

with $\Delta > 0$ a positive constant, is a stationary Gaussian process and find its correlation function.
18. Find the correlation function of the process

$$Y(t) = \int_0^t (W_0^2(\tau) - \tau)\,d\tau, \qquad t \geq 0,$$

where $W_0(t)$, $t \geq 0$, is the standard Wiener process. (*Hint:* Use the result of Exercise 1.20.)
19. Let $Y(t) = \int_0^t X(\tau)\,d\tau$, $t \geq 0$, where $X(t)$, $-\infty < t < \infty$, is a zero mean wide-sense stationary process. Calculate the variance function $R_Y(t, t)$ when the correlation function $R_X(\tau)$ is as in Exercise 4.1.13, a, b, c.
20. If $W(t)$, $-\infty < t < \infty$, is a zero mean Wiener process with intensity parameter σ, what is the correlation function of the process

$$Y(t) = \int_{t-1}^{t} (W(t) - W(\tau))\,d\tau, \qquad 1 \leq t < \infty?$$

Is it a strictly stationary process?
21. The velocity of an aircraft is estimated by means of an inertial navigational device. The error $Y(t)$ of the estimate after t seconds of flight is given by

$$Y(t) = g\int_0^t \sin X(\tau)\,d\tau,$$

where $g = 980$ meter/second^2 is the gravitational acceleration and $X(t)$ is the angular error of the gyro axis. Assuming that $X(t)$, $-\infty < t < \infty$, is a zero mean stationary Ornstein-Uhlenbeck process with correlation function $R_X(\tau) = 4.10^{-8}e^{-.008|\tau|}$ radians^2, find the mean square error of the estimate after 10 hours of flight. (*Hint:* Approximate $\sin x$ by x and justify.)

4.2. LINEAR TRANSFORMATIONS AND DIFFERENTIAL EQUATIONS

A large majority of dynamic systems encountered in engineering are linear or at least can be considered linear as a first approximation. Given a function of time $u(t)$, representing, for example, an input to a system, the action of the system can be represented by an operator $\mathcal{O}$ which transforms the input function $u(t)$ into an output time function $x(t)$, as pictured in Figure 4.2.1. This transformation is

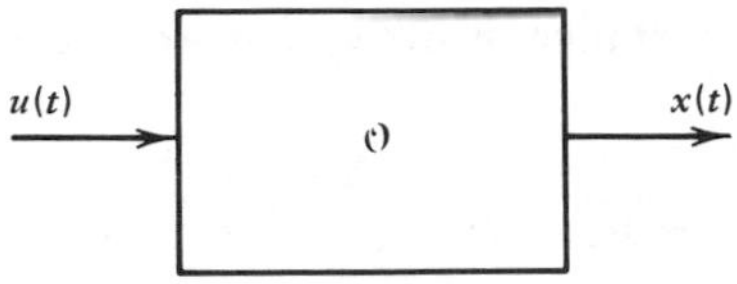

Figure 4.2.1

generally not memoryless, that is, the value of the output at some fixed time t may depend on all values of the input during some interval of time.

We express this symbolically as

$$X(\ ) = \mathcal{O}\{u(\ )\}$$

to indicate that $\mathcal{O}$ operates on the entire input as a function.

The system $\mathcal{O}$ is called *linear* if, when applied to a linear combination of inputs, it produces an output that is the same linear combination of the inputs. In symbols,

$$\mathcal{O}\left\{\sum_k \alpha_k u_k(\ )\right\} = \sum_k \alpha_k \mathcal{O}\{u_k(\ )\}$$

for any constants α_k.

This linearity property is commonly referred to as the *superposition principle* and is used to justify the representation of the linear system by an integral

$$x(t) = \int_{-\infty}^{\infty} h(t, \tau)u(\tau)\, d\tau, \qquad -\infty < t < +\infty, \tag{1}$$

where the time domain of both the input and the output functions is the entire time axis $(-\infty, \infty)$. (If $u(t)$ is defined only for $t \geq t_0$, we set $u(t) = 0$ for $t < t_0$.)

The function $h(t, \tau)$ is called the *impulse-response function*, and it represents the output of the system at the time t if a unit impulse $\delta(\tau)$ (Dirac delta function) is applied to its input at the time τ.

The heuristic argument behind formula (1) is that an input $u(t)$ can be approximated by a linear combination of successive unit impulses $\Delta(t)$ of width $d\tau$

$$u(t) \doteq \sum_k u(\tau_k)\, \Delta(t - \tau_k)\, d\tau$$

and thus by linearity

$$x(t) \doteq \sum_k u(\tau_k)\mathcal{O}\{\Delta(t - \tau_k)\}\, d\tau. \tag{2}$$

Hence, calling $h(t, \tau) = \mathcal{O}\{\delta(t - \tau)\}$ with δ the Dirac delta function, the sum (2) is seen to approximate the integral in (1).

It should be noted that the impulse-response function $h(t, \tau)$ may involve the

Dirac delta function or even its derivatives. For instance, if the system is just a delay by a fixed time θ, then

$$x(t) = \int_{-\infty}^{\infty} \delta(t - \tau - \theta)u(\tau)\, d\tau = u(t - \theta).$$

Similarly, a differentiator $x(t) = (d/dt)u(t)$ can be expressed in the integral form (1) by choosing $h(t, \tau) = \dot{\delta}(t - \tau)$, where $\dot{\delta} = (d/dt)\,\delta$ is the derivative of the delta function.

Now, a linear system is called *stable*† if bounded inputs produce a finite output at any time $-\infty < t < \infty$. A necessary and sufficient condition for this to be so is that the impulse response function satisfy the condition

$$\int_{-\infty}^{\infty} |h(t, \tau)|\, d\tau < \infty \qquad \text{for all} \qquad t \in (-\infty, \infty). \tag{3}$$

A linear system is called *causal* (or sometimes nonanticipating) if the output $x(t)$ at any fixed time t can only depend on the past and present input $u(t')$, $t' \le t$, that is, not on any of the future values $u(t')$, $t' > t$. Since the impulse-response function $h(t, \tau)$ is the output at time t resulting from the unit impulse input arriving at time τ, we must have

$$h(t, \tau) = 0 \qquad \text{for} \qquad -\infty < t < \tau < \infty$$

for a system to be causal. Hence, for a causal system equation (1) becomes

$$x(t) = \int_{-\infty}^{t} h(t, \tau)u(\tau)\, d\tau, \qquad -\infty < t < \infty. \tag{4}$$

In the rest of this section, unless otherwise stated, we will deal with stable and causal linear systems whose impulse-response function does not contain a delta function.

We will now study transformations of a continuous-time stochastic process by a linear system; that is, we will consider a linear system whose input is a finite-variance stochastic process $U(t)$, $t \in \mathcal{T}$. Then the output will also be a stochastic process $X(t)$, $t \in \mathcal{T}$ and

$$X(t) = \int_{-\infty}^{t} h(t, \tau)U(\tau)\, d\tau. \tag{5}$$

The time domain $\mathcal{T}$ will be considered the entire time axis $\mathcal{T} = (-\infty, \infty)$, since if $\mathcal{T}$ is an interval, say $t \ge t_0$, we can always set $U(t) = 0$ for $t \le t_0$ or, equivalently,

† In the literature, the definition of stability is often strengthened by requiring that the output be not only finite but also bounded for all $-\infty < t < \infty$. For our purposes, the weaker definition given above will suffice.

replace the lower limit in the integral (5) by t_0. Of course, the output process will in such a case have the same time domain $t \geq t_0$.

The output process $X(t)$ is related to the input process $U(t)$ by the integral formula (5), and we would like to interpret it as a sample integral. That is, the sample functions $x(t)$ of the process $X(t)$ should be integrals (4) of the sample functions $u(t)$ of the input process.

Now $X(t)$ is an integral, as defined in the previous section, of the process

$$h(t, \tau)U(\tau), \qquad \tau < t \tag{6}$$

and if we assume that $U(t)$ is an MS-continuous process and that $h(t, \tau)$ is, for each fixed t, a continuous function of $\tau < t$, the process (6) will also be MS continuous.

Furthermore, if the variance function $\sigma_U^2(t) = R_U(t, t)$ of the input process is bounded on every time interval $(-\infty, t]$, that is, if for every $-\infty < t < \infty$ there is a constant $C(t)$ such that

$$\sigma_U^2(t') \leq C(t) \qquad \text{for all} \qquad -\infty < t' \leq t \tag{7}$$

then the double integral

$$\int_{-\infty}^{t}\int_{-\infty}^{t} h(t, \tau_1)R_U(\tau_1, \tau_2)h(t, \tau_2)\, d\tau_1\, d\tau_2 \tag{8}$$

will exist for all $-\infty < t < \infty$. This follows immediately from the Schwarz inequality (assuming, with no loss of generality, $\mu_U(t) = 0$)

$$|R_U(\tau_1, \tau_2)| \leq \sigma_U(\tau_1)\sigma_U(\tau_2)$$

since

$$\left|\int_{-\infty}^{t}\int_{-\infty}^{t} h(t, \tau_1)R_U(\tau_1, \tau_2)h(t, \tau_2)\, d\tau_1\, d\tau_2\right| \leq \left(\int_{-\infty}^{t} |h(t, \tau)|\sigma_U(\tau)\, d\tau\right)^2$$

$$\leq C^2(t)\left(\int_{-\infty}^{t} |h(t, \tau)|\, d\tau\right)^2,$$

which must be finite by the stability condition (3).

However, the integrand in (8) is the correlation function of the process (6). Thus, according to Theorem 4.1.3, the integral (5) will exist as a sample integral, and the output process $X(t)$ will be sample continuous. In fact, if we apply Theorem 4.1.3 in its entirety, we obtain a new theorem.

THEOREM 4.2.1. *Let* $U(t)$, $t \in \mathcal{T}$, *be an MS-continuous process with mean function* μ_U *and correlation function* R_U *whose variance function satisfies* (7). *Let* $h(t, \tau)$ *be an impulse-response function of a stable and* *causal linear system that is continuous in τ for every* t ∈ $\mathcal{T}$. *Then the output process* X(t), t ∈ $\mathcal{T}$, *defined as a sample integral (or MS integral)*

$$X(t) = \int_{-\infty}^{t} h(t, \tau)U(\tau)\, d\tau, \qquad t \in \mathcal{T} \tag{9}$$

is a sample-continuous process with mean function

$$\mu_X(t) = \int_{-\infty}^{t} h(t, \tau)\mu_U(\tau)\, d\tau, \tag{10}$$

correlation function

$$R_X(t_1, t_2) = \int_{-\infty}^{t_1} \int_{-\infty}^{t_2} h(t_1, \tau_1)R_U(\tau_1, \tau_2)h(t_2, \tau_2)\, d\tau_1\, d\tau_2 \tag{11}$$

and cross-correlation function between X(t) *and* U(t)

$$R_{X,U}(t_1, t_2) = \int_{-\infty}^{t_1} h(t_1, \tau)R_U(\tau, t_2)\, d\tau \tag{12}$$

If U(t) *is a Gaussian process, then the output process* X(t) *is also a Gaussian process.*

Remark: The assumption that the impulse-response function $h(t, \tau)$ be continuous is not really necessary. It is sufficient if $h(t, \tau)$ is only piecewise continuous in τ. Я

Comparing (11) and (12), it is seen that the expression for the correlation function can also be written as

$$R_X(t_1, t_2) = \int_{-\infty}^{t_2} R_{X,U}(t_1, \tau)h(t_2, \tau)\, d\tau, \tag{13}$$

so that the correlation function can be evaluated in two steps yielding the cross-correlation function after the first step.

Example 4.2.1

Consider a linear system defined for $t \geq 0$ and having the impulse-response function

$$h(t, \tau) = \begin{cases} \dfrac{1 + \tau}{1 + t} & \text{if } 0 \leq \tau \leq t, \\ 0 & \text{otherwise.} \end{cases}$$

For example, this may represent a simple *R-C* circuit, as shown in Figure 4.2.2, where the resistance of the variable resistor increases linearly from 0 at $t = 0$ with

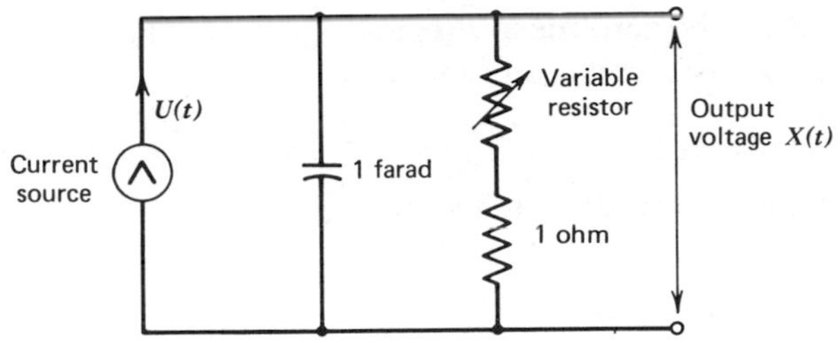

Figure 4.2.2

the rate of 1 ohm/second (i.e., its resistance at the time $t \geq 0$ is t ohms) and it can be shown that this circuit has the above impulse-response function. Clearly, the system is causal and stable according to our definition (3) of stability.

Assume now that the current source produces a finite-variance stochastic process $U(t)$, $t \geq 0$, with mean function

$$\mu_U(t) = \sin t, \qquad t \geq 0$$

and with correlation function

$$R_U(t_1, t_2) = e^{-|t_1 - t_2|} + \mu_U(t_1)\mu_U(t_2), \qquad t_1 \geq 0, \qquad t_2 \geq 0.$$

The output voltage

$$X(t) = \int_0^t \frac{1 + \tau}{1 + t} U(\tau)\, d\tau, \qquad t \geq 0$$

will then also be a finite-variance process, and we wish to evaluate its mean function and its correlation function. Since $\sigma_U^2(t) = 1 + \sin^2 t$, condition (7) is trivially satisfied (as it must be for any finite-variance process with a semi-infinite time domain $t \geq 0$). Hence, by Theorem 4.2.1, we have for the mean function

$$\mu_X(t) = \int_0^t \frac{1 + \tau}{1 + t} \sin \tau\, d\tau, \qquad t \geq 0.$$

This integral is easily evaluated, and we get

$$\mu_X(t) = \frac{1 + \sin t}{1 + t} - \cos t, \qquad t \geq 0.$$

Note that for large t, we have approximately $\mu_X(t) \doteq -\cos t$, which results from the fact that as the resistance increases the system tends to behave as a perfect integrator.

Let us next compute the correlation function of the output process. Since the system is linear, the computation may be simplified by first assuming that the input process has been replaced by the zero mean process $U_0(t) = U(t) - \mu_U(t)$, $t \geq 0$. The new input process now has correlation function

$$R_{U_0}(t_1, t_2) = e^{-|t_1 - t_2|}$$

and we can proceed to compute the correlation function $R_{X_0}(t_1, t_2)$ of the corresponding output process

$$X_0(t) = X(t) - \mu_X(t), \qquad t \geq 0.$$

Instead of directly evaluating the double integral (11), let us first calculate the cross-correlation function $R_{X_0, U_0}(t_1, t_2)$ from equation (12). We have the integral

$$R_{X_0, U_0}(t_1, t_2) = \int_0^{t_1} \frac{1 + \tau}{1 + t_1} e^{-|\tau - t_2|} \, d\tau,$$

and because of the absolute value in the exponent, we have to distinguish two separate cases:

Case 1: For $0 \leq t_1 \leq t_2$ the integral becomes

$$\begin{aligned} R_{X_0, U_0}(t_1, t_2) &= \int_0^{t_1} \frac{1 + \tau}{1 + t_1} e^{-(t_2 - \tau)} \, d\tau \\ &= \frac{e^{-t_2}}{1 + t_1} \left(\int_0^{t_1} e^{\tau} \, d\tau + \int_0^{t_1} \tau e^{\tau} \, d\tau \right) = \frac{t_1}{1 + t_1} e^{-(t_2 - t_1)}. \end{aligned}$$

Case 2: For $0 \leq t_2 \leq t_1$, we have to split the integral into two integrals:

$$R_{X_0, U_0}(t_1, t_2) = \int_0^{t_2} \frac{1 + \tau}{1 + t_1} e^{-(t_2 - \tau)} \, d\tau + \int_{t_2}^{t_1} \frac{1 + \tau}{1 + t_1} e^{-(\tau - t_2)} \, d\tau.$$

The first integral equals

$$\frac{t_2}{1 + t_1}$$

while the second integral is

$$\frac{-1}{1 + t_1} ((2 + t_1) e^{-(t_1 - t_2)} - 2 - t_2).$$

Adding and simplifying, we obtain for Case 2,

$$R_{X_0, U_0}(t_1, t_2) = -\frac{2 + t_1}{1 + t_1} e^{-(t_1 - t_2)} + \frac{2(1 + t_2)}{1 + t_1}.$$

Thus we have computed the cross-correlation function between $X_0(t)$ and $U_0(t)$, from which we can immediately obtain the cross-correlation between the original output $X(t)$ and input $U(t)$, since clearly

$$R_{X,U}(t_1, t_2) = R_{X_0,U_0}(t_1, t_2) + \mu_X(t_1)\mu_U(t_2).$$

Finally, let us begin calculating the correlation function $R_{X_0}(t_1, t_2)$. Since we already have the cross-correlation function, we can use equation (13). Furthermore, since any correlation function of a scalar-valued process is symmetric, it is sufficient to consider only the case $0 \le t_2 \le t_1$. This has the advantage that we can substitute for the cross-correlation function in (13) only the expression obtained in Case 2. Thus for $0 \le t_2 \le t_1$

$$\begin{aligned} R_{X_0}(t_1, t_2) &= \int_0^{t_2} \left(\frac{2(1+\tau)}{1+t_1} - \frac{2+t_1}{1+t_1} e^{-(t_1-\tau)} \right) \frac{1+\tau}{1+t_2}\, d\tau \\ &= \frac{2}{(1+t_1)(1+t_2)} \int_0^{t_2} (1+\tau)^2\, d\tau \\ &\quad - \frac{2+t_1}{(1+t_1)(1+t_2)} e^{-t_1} \int_0^{t_2} (1+\tau)e^{\tau}\, d\tau \\ &= \frac{t_2}{(1+t_1)(1+t_2)} \left(2t_2\left(1 + \frac{1}{3}t_2\right) - t_1 e^{-(t_1-t_2)} \right). \end{aligned}$$

For $0 \le t_1 \le t_2$, we merely interchange t_1 and t_2 in the above expression.

The variance function $\sigma_{X_0}^2(t)$ is obtained by setting $t_1 = t_2 = t$:

$$\sigma_{X_0}^2(t) = \left(\frac{t}{1+t}\right)^2 \left(1 + \frac{2}{3}t\right), \qquad t \ge 0$$

Notice that for large t, we have approximately

$$\sigma_{X_0}^2(t) \doteq \frac{2}{3}t.$$

The correlation function of the original process is then

$$R_X(t_1, t_2) = R_{X_0}(t_1, t_2) + \mu_X(t_1)\mu_X(t_2).$$

This again shows that when dealing with linear systems, there is no loss of generality in assuming that the mean functions are zero. ▲

If the impulse-response function contains a Dirac delta function or its derivatives, the above theorem no longer applies. Nevertheless, formulas (10), (11), (12), and (13) can still be used, provided their left-hand sides define functions that can qualify as mean and correlation functions.

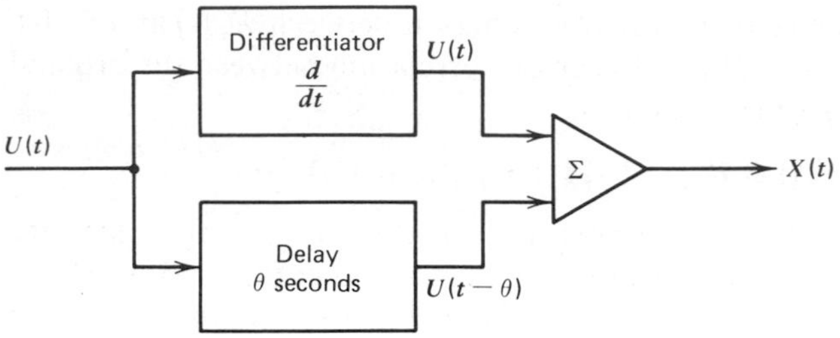

Figure 4.2.3

Example 4.2.2

Consider a linear system with the block scheme depicted in Figure 4.2.3, where the input is an MS-differentiable process with zero mean function and correlation function $R_U(t_1, t_2)$.

Since the output process

$$X(t) = \dot{U}(t) + U(t - \theta)$$

we can immediately obtain the cross-correlation function

$$\begin{aligned} R_{X,U}(t_1, t_2) &= \mathrm{E}[X(t_1)U(t_2)] \\ &= \mathrm{E}[\dot{U}(t_1)U(t_2)] + \mathrm{E}[U(t_1 - \theta)U(t_2)] \\ &= \frac{\partial}{\partial t_1} R_U(t_1, t_2) + R_U(t_1 - \theta, t_2), \end{aligned}$$

where we have used the fact established in Theorem 4.1.3, namely that

$$R_{\dot{U},U}(t_1, t_2) = \frac{\partial}{\partial t_1} R_U(t_1, t_2).$$

Similarly, the correlation function is easily evaluated from

$$\begin{aligned} R_X(t_1, t_2) &= \mathrm{E}[X(t_1)X(t_2)] \\ &= \mathrm{E}[\dot{U}(t_1)\dot{U}(t_2)] + \mathrm{E}[U(t_1 - \theta)\dot{U}(t_2)] \\ &\quad + \mathrm{E}[\dot{U}(t_1)U(t_2 - \theta)] + \mathrm{E}[U(t_1 - \theta)U(t_2 - \theta)]. \end{aligned}$$

$$\begin{aligned} R_X(t_1, t_2) &= \frac{\partial^2}{\partial t_1\, \partial t_2} R_U(t_1, t_2) + \frac{\partial}{\partial t_2} R_U(t_1 - \theta, t_2) \\ &\quad + \frac{\partial}{\partial t_1} R_U(t_1, t_2 - \theta) + R_U(t_1 - \theta, t_2 - \theta). \end{aligned}$$

On the other hand, the equation

$$X(t) = \dot{U}(t) + U(t - \theta)$$

can be written symbolically with the aid of the Dirac delta function as an integral

$$X(t) = \int_{-\infty}^{t} (\dot{\delta}(t - \tau) + \delta(t - \theta - \tau))U(\tau)\, d\tau$$

once we recall that if $f(t)$ is a function continuous at some t_0, then

$$\int_{-\infty}^{\infty} \delta(t - t_0)f(t)\, dt = f(t_0) \qquad \text{and} \qquad \int_{-\infty}^{\infty} \dot{\delta}(t - t_0)f(t)\, dt = \dot{f}(t_0).$$

This shows that we can regard the system as having the impulse-response function

$$h(t, \tau) = \dot{\delta}(t - \tau) + \delta(t - \theta - \tau).$$

And, indeed, if we formally evaluate the cross-correlation function $R_{X,U}(t_1, t_2)$ from equation (12) we obtain, using the above-mentioned properties of the delta function,

$$\begin{aligned} R_{X,U}(t_1, t_2) &= \int_{-\infty}^{t_1} (\dot{\delta}(t_1 - \tau) + \delta(t_1 - \theta - \tau))R_U(\tau, t_2)\, d\tau \\ &= \frac{\partial}{\partial t_1} R_U(t_1, t_2) + R_U(t_1 - \theta, t_2), \end{aligned}$$

and if we then use (13) for the correlation function, we get

$$\begin{aligned} R_X(t_1, t_2) &= \int_{-\infty}^{t_2} \left(\frac{\partial}{\partial t_1} R_U(t_1, \tau) + R_U(t_1 - \theta, \tau)\right)(\dot{\delta}(t_2 - \tau) + \delta(t_2 - \theta - \tau))\, d\tau \\ &= \frac{\partial^2}{\partial t_1\, \partial t_2} R_U(t_1, t_2) + \frac{\partial}{\partial t_2} R_U(t_1 - \theta, t_2) \\ &\quad + \frac{\partial}{\partial t_1} R_U(t_1, t_2 - \theta) + R_U(t_1 - \theta, t_2 - \theta). \end{aligned}$$

Thus, in spite of the fact that the impulse-response function above is not a legitimate function, the formal application of the expressions for the correlation functions yield the correct result. However, caution must be exercised when employing the delta function. For instance, in the present example, we would have to make sure that the input process is indeed MS differentiable. ▲

TIME-INVARIANT LINEAR SYSTEMS. A linear system is called time-invariant if its impulse-response function depends only on the difference $t - \tau$, that is, with the usual abuse of notation,

$$h(t, \tau) = h(t - \tau).$$

Physically, this means that the response of the system to a unit impulse depends only on the time elapsed since the arrival of this impulse.

For a stable, causal and time-invariant system, equation (9) becomes

$$X(t) = \int_{-\infty}^{t} h(t-\tau)U(\tau)\,d\tau.$$

If the input process to such a system is a wide-sense stationary process with mean function $\mu_U(t) = \mu_U$ and a correlation function $R_U(t_1, t_2) = R_U(t_1 - t_2)$, then the relations (10), (11), and (12) can be rewritten by simple substitution as

$$\mu_X = \int_{-\infty}^{t} h(t-\tau)\mu_U\,d\tau = \mu_U \int_0^{\infty} h(\tau)\,d\tau,$$

$$R_X(t_1, t_2) = \int_{-\infty}^{t_1} \int_{-\infty}^{t_2} h(t_1-\tau_1)R_U(\tau_1-\tau_2)h(t_2-\tau_2)\,d\tau_1\,d\tau_2$$

$$= \int_0^{\infty} \int_0^{\infty} h(\tau_1)R_U(t_1 - t_2 - (\tau_1 - \tau_2))h(\tau_2)\,d\tau_1\,d\tau_2,$$

$$R_{X,U}(t_1, t_2) = \int_{-\infty}^{t_1} h(t_1-\tau)R_U(\tau - t_2)\,d\tau = \int_0^{\infty} h(\tau)R_U(t_1 - t_2 - \tau)\,d\tau.$$

But then μ_X is a constant, and both R_X and $R_{X,U}$ depend only on the difference $t_1 - t_2$, so that the *output process is wide-sense stationary*, and the output and input process are jointly wide-sense stationary.

These expressions can be written in more elegant form:

$$\mu_X = C\mu_U, \qquad C = \int_0^{\infty} h(\tau)\,d\tau, \tag{14}$$

$$R_{X,U}(-\tau) = \int_0^{\infty} h(\tau_1)R_U(\tau - \tau_1)\,d\tau_1 = R_{U,X}(\tau), \tag{15}$$

$$R_X(\tau) = \int_0^{\infty} R_{X,U}(\tau - \tau_2)h(\tau_2)\,d\tau_2. \tag{16}$$

The reader may recognize that the last two integrals are both convolutions. Thus, their evaluation may frequently be facilitated by means of Laplace or Fourier transforms. This aspect is discussed in more detail in Section 5.2.

It should also be mentioned that if the input process to a time-invariant linear system is *strictly stationary* the *output process is also strictly stationary.*

This can be seen by noticing that for any $\tau' > 0$

$$X(t+\tau') = \int_{-\infty}^{t+\tau'} h(t+\tau'-\tau)U(\tau)\,d\tau = \int_{-\infty}^{t} h(t-\tau)U(\tau+\tau')\,d\tau$$

and using the fact that for a strictly stationary process the probability law of $U(\tau + \tau')$ is the same as that of $U(\tau)$.

Example 4.2.3
Consider a time-invariant linear system with the impulse-response function

$$h(\tau) = \begin{cases} \tau e^{-\gamma\tau} & \text{if } \tau \geq 0, \\ 0 & \text{if } \tau > 0, \end{cases}$$

where $\gamma > 0$ is a constant. Clearly, the system is causal and stable; in fact, the impulse-response function is recognized as that of a critically damped resonant circuit. Let us assume that the input process $U(t)$, $-\infty < t < \infty$, is a zero mean stationary Ornstein-Uhlenbeck process with correlation function

$$R_U(\tau) = e^{-|\tau|}, \qquad -\infty < \tau < \infty.$$

We also assume that $\gamma \neq 1$; the case $\gamma = 1$ requires special treatment and is left to the reader as an exercise.

The output process

$$X(t) = \int_{-\infty}^{t} (t - \tau)e^{-\gamma(t-\tau)}U(\tau)\, d\tau, \qquad -\infty < t < \infty$$

is then a zero mean stationary Gaussian process. Let us first compute the input-output cross-correlation function. From Equation (15) we have

$$R_{U,X}(\tau) = \int_0^\infty \tau_1 e^{-\gamma\tau_1} e^{-|\tau - \tau_1|}\, d\tau_1, \qquad -\infty < \tau < \infty.$$

Because of the absolute value in the exponent, we have to distinguish two cases.

First, for $\tau \leq 0$, we have $|\tau - \tau_1| = \tau_1 - \tau$ over the integration region $0 < \tau_1 < \infty$, and so

$$R_{U,X}(\tau) = \int_0^\infty \tau_1 e^{-\gamma\tau_1} e^{-(\tau_1 - \tau)}\, d\tau_1 = \frac{e^\tau}{(1+\gamma)^2}.$$

Second, for $\tau > 0$, we have $|\tau - \tau_1| = \tau - \tau_1$ over $0 < \tau_1 \leq \tau$ and $|\tau - \tau_1| = \tau_1 - \tau$ over $\tau < \tau_1 < \infty$. Hence

$$\begin{aligned} R_{U,X}(\tau) &= \int_0^\tau \tau_1 e^{-\gamma\tau_1} e^{-(\tau - \tau_1)}\, d\tau_1 + \int_\tau^\infty \tau_1 e^{\gamma\tau_1} e^{-(\tau_1 - \tau)}\, d\tau_1 \\ &= \frac{1}{(1-\gamma)^2}[e^{-\tau} - e^{-\gamma\tau}(1 - (1-\gamma)\tau)] \\ &\quad + \frac{1}{(1+\gamma)^2} e^{-\gamma\tau}(1 + (1+\gamma)\tau). \end{aligned}$$

This expression may be somewhat simplified by factoring $e^{-\gamma\tau}$ from the last two terms. The resulting expression for the cross-correlation function is then

$$R_{U,X}(\tau) = \begin{cases} \dfrac{e^{-\tau}}{(1-\gamma)^2} + \dfrac{2e^{-\gamma\tau}}{1-\gamma^2}\left(\tau - \dfrac{2\gamma}{1-\gamma^2}\right) & \text{if } \tau > 0, \\ \dfrac{e^{\tau}}{(1+\gamma)^2} & \text{if } \tau \le 0. \end{cases}$$

The function is continuous at $\tau = 0$, being equal to $(1+\gamma)^{-2}$ there.

Next we wish to evaluate the correlation function of the output process from equation (16):

$$R_X(\tau) = \int_0^{\infty} R_{X,U}(\tau - \tau_1)\tau_1 e^{-\gamma\tau_1}\, d\tau_1, \qquad -\infty < \tau_1 < \infty.$$

Since a correlation function is symmetric, it is enough to take $\tau \ge 0$. But the integral still splits in two, depending on the sign of the argument $\tau - \tau_1$ of the cross-correlation function. In particular, we get

$$\begin{aligned} R_X(\tau) &= \int_0^{\tau} R_{X,U}(\tau - \tau_1)\tau_1 e^{-\gamma\tau_1}\, d\tau_1 + \int_{\tau}^{\infty} R_{X,U}(\tau - \tau_1)\tau_1 e^{-\gamma\tau_1}\, d\tau_1 \\ &= \int_0^{\tau} \frac{e^{-(\tau-\tau_1)}}{(1+\gamma)^2}\tau_1 e^{-\gamma\tau_1}\, d\tau_1 \\ &\quad + \int_{\tau}^{\infty} \left(\frac{e^{-(\tau_1-\tau)}}{(1-\gamma)^2} - \frac{2e^{-\gamma(\tau_1-\tau)}}{1-\gamma^2}\left(\tau - \tau_1 + \frac{2\gamma}{1-\gamma^2}\right)\right)\tau_1 e^{-\gamma\tau_1}\, d\tau_1 \end{aligned}$$

After some tedious but otherwise straightforward integration, we obtain

$$R_X(\tau) = \frac{2}{(1-\gamma^2)^2} e^{-\gamma|\tau|} + \frac{4(1-2\gamma) + 2(1-3\gamma)(1+\gamma)|\tau|}{(1-\gamma^2)^2(1+\gamma)^2} e^{-|\tau|}$$

$-\infty < \tau < \infty$. Notice that for $\tau = 0$, we obtain the variance function

$$\sigma_X^2(\tau) = 2\frac{3 - 2\gamma + \gamma^2}{(1-\gamma^2)^2(1+\gamma)^2}$$

a constant, since the process is stationary.

We see that even in this rather simple case, the evaluation of the correlation function of the output process is a tiresome procedure. ▲

Remark: The entire discussion above also applies to linear transformations of vector-valued processes. In such a case the linear system may be visualized as having m input and n output terminals, as pictured in Figure 4.2.4, and its operation is then described

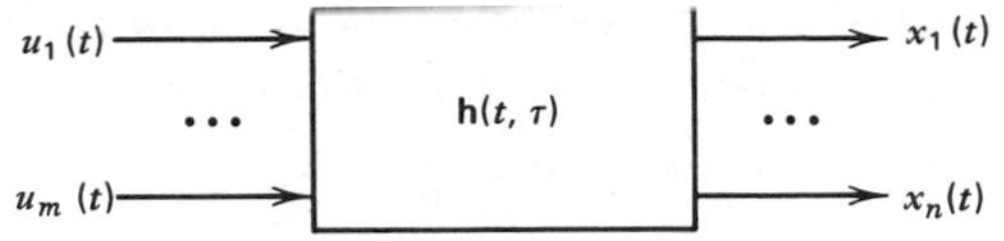

Figure 4.2.4

by the matrix equation

$$\mathbf{x}(t) = \int_{-\infty}^{t} \mathbf{h}(t, \tau)\mathbf{u}(\tau)\, d\tau.$$

Here, $\mathbf{u}(t)$ and $\mathbf{x}(t)$ are column vectors with m and n components, respectively, and the impulse-response function is an $n \times m$ matrix

$$\mathbf{h}(t, \tau) = [h_{ij}(t, \tau)]$$

with the (i, j)th entry being the response on the ith output at time t to a unit impulse applied to the jth input at time τ, with zero on the remaining input terminals.

The results derived above remain valid with the exception of equations (11), (13) and (16), where transposition must be used to ensure an appropriate order of the resulting matrices. In particular,

$$\mathbf{R_X}(t_1, t_2) = \int_{-\infty}^{t_1} \int_{-\infty}^{t_2} \mathbf{h}(t_1, \tau_1)\mathbf{R_U}(\tau_1, \tau_2)\mathbf{h}'(t_2, \tau_2)\, d\tau_1\, d\tau_2,$$

$$\mathbf{R_X}(t_1, t_2) = \int_{-\infty}^{t_2} \mathbf{R_{X,U}}(t_1, \tau)\mathbf{h}'(t_2, \tau)\, d\tau,$$

$$\mathbf{R_X}(\tau) = \int_{0}^{\infty} \mathbf{R_{X,U}}(\tau - \tau_2)\mathbf{h}'(\tau_2)\, d\tau.$$

It also applies with minor change to complex-valued processes and complex-valued impulse-response functions $h(t, \tau)$. In accordance with the definition of the correlation function of a complex-valued process

$$R_U(t_1, t_2) = \mathrm{E}[U(t_1)U^*(t_2)],$$

with the asterisk denoting the complex conjugate, equations (11), (13), and (16) are now

$$R_X(t_1, t_2) = \int_{-\infty}^{t_1} \int_{-\infty}^{t_2} h(t_1, \tau_1)R_U(t_1, t_2)h^*(t_2, \tau_2)\, d\tau_1\, d\tau_1,$$

$$R_X(t_1, t_2) = \int_{-\infty}^{t_2} R_{X,U}(t_1, \tau)h^*(t_2, \tau)\, d\tau$$

$$R_X(\tau) = \int_{0}^{\infty} R_{X,U}(\tau - \tau_2)h^*(\tau_2)\, d\tau_2.$$

Equations (10), (12), (14), and (15) remain unchanged. Я

LINEAR DIFFERENTIAL EQUATIONS. In practice, the operation of a linear system is usually described by one or several linear differential equations

$$a_0(t)x^{(n)}(t) + a_1(t)x^{(n-1)}(t) + \cdots + a_n(t)x(t) = u(t) \tag{17}$$

defined on a time interval† $t_0 \le t$ with prescribed initial conditions

$$x(t_0),\ x^{(1)}(t_0),\ \ldots,\ x^{(n-1)}(t_0) \tag{18}$$

and a given forcing function $u(t)$ defined on the same time interval $t_0 \le t$.

A solution of equation (17) is defined to be an n times differentiable function $x(t)$ defined for $t_0 \le t$ satisfying equation (17) and such that at the time t_0 its value and the values of its first $n - 1$ derivatives are equal to the initial conditions (18).

In terms of linear systems the forcing function $u(t)$ is the input, and the solution $x(t)$ is the output of the system. It is easy to see that the system defined by equation (17) is indeed linear according to the definition given in the previous section.

Linear differential equations have been extensively studied, and a great deal is known about their properties, such as conditions for existence and uniqueness of a solution and many other results. We will not dwell on these topics (the reader can consult any text on differential equations) but will simply assume that the appropriate conditions for existence and uniqueness of the solutions are satisfied.

We will, however, need the following well-known results.

Consider the homogeneous equation

$$a_0(t)x^{(n)}(t) + a_1(t)x^{(n-1)}(t) + \cdots + a_n(t)x(t) = 0 \tag{19}$$

and let $x_0(t)$, $t_0 \le t$, be the solution of this equation satisfying the initial conditions (18).

Next, for each $t_0 \le \tau$, τ fixed, let $h(t, \tau)$ be a function of t defined on the interval $\tau \le t$ and such that it is, on this interval, a solution of equation (19) with initial conditions

$$\frac{\partial^k}{\partial t^k} h(t, \tau)\bigg|_{t=\tau} = \begin{cases} 0 & \text{for } k = 0, \ldots, n-2 \\ \dfrac{1}{a_0(t)} & \text{for } k = n-1. \end{cases} \tag{20}$$

Then

$$x(t) = x_0(t) + \int_0^t h(t, \tau)u(\tau)\, d\tau, \qquad t_0 \le t, \tag{21}$$

† Strictly speaking, we should consider only bounded time intervals $t_0 \le t \le t_1$. However, in what follows, we shall leave out the right endpoint t_1, since we are mainly concerned with a solution for prescribed initial conditions.

is the solution of equation (17) satisfying the initial conditions (18). Furthermore, the function $x_0(t)$ is a linear combination

$$x_0(t) = x(t_0)\xi_0(t) + x^{(1)}(t_0)\xi_1(t) + \cdots + x^{(n-1)}(t_0)\xi_{n-1}(t), \tag{22}$$

where the function $\xi_k(t)$ $(k = 0, \ldots, n-1)$ is a solution of (19) obtained by setting the initial condition $x^{(k)}(t_0) = 1$ and the rest $= 0$.

Remark: The functions $\xi_0(t), \ldots, \xi_{n-1}(t)$ form the so-called fundamental system of solutions of the homogeneous equation (19). The function $h(t, \tau)$ is known as a (one-sided) Green function. In the language of linear systems $x_0(t)$ is called the *zero-input response*, while the integral $\int_{t_0}^{t} h(t, \tau)u(\tau)\,d\tau$ is called the *zero-state response*. The function $h(t, \tau)$ is just the impulse-response function encountered earlier. Я

We will not prove the above result (apart from questions of existence and uniqueness of the solution, the proof is quite easy: just substitute (21) into (17) and verify that it results in an identity and that the initial conditions are met.)

What we are really interested in is the case where the forcing function (input) $u(t)$ is a stochastic process and the initial conditions (18) are random variables. With a suitable interpretation of the derivatives (e.g., as MS derivatives) the solution $X(t)$ will then be a stochastic process.

In fact, using the above result, we can write this process as

$$X(t) = \sum_{k=0}^{n-1} X^{(k)}(t_0)\xi_k(t) + \int_{t_0}^{t} h(t, \tau)U(\tau)\,d\tau, \qquad t_0 \le t.$$

Observe that the solution process $X(t)$ is a sum of two processes, the *singular component*

$$X_s(t) = \sum_{k=0}^{n-1} X^{(k)}(t_0)\xi_k(t), \tag{23}$$

and the *regular component*

$$X_r(t) = \int_{t_0}^{t} h(t, \tau)U(\tau)\,d\tau, \tag{24}$$

of the process $X(t)$.

The singular component $X_s(t)$ is a stochastic process and is often said to be of a deterministic type, since its sample functions $x_s(t)$ are determined by the value of the random vector of initial conditions

$$(X(t_0), X^{(1)}(t_0), \ldots, X^{(n-1)}(t_0))'. \tag{25}$$

Suppose now that the process $U(t)$ is a finite-variance process with mean function $\mu_U(t)$ and that the random vector of initial conditions (25) has mean vector

$$\boldsymbol{\mu}_X(t_0) = \begin{pmatrix} \mu_X^{(0)} \\ \vdots \\ \mu_X^{(n-1)} \end{pmatrix},$$

where $\mu_X^{(k)} = \mathrm{E}[X^{(k)}(t_0)]$, $k = 0, \ldots, n-1$. Since the differential equation (17) is linear, the mean function $\mu_X(t)$ of the process $X(t)$ will be a solution of the same equation, with the right-hand side (forcing function) replaced by $\mu_U(t)$ and with initial conditions $\mu_X^{(0)}, \ldots, \mu_X^{(n-1)}$, that is,

$$\mu_X(t) = \sum_{k=0}^{n-1} \mu_X^{(k)} \xi_k(t) + \int_{t_0}^{t} h(t, \tau)\mu_U(\tau)\, d\tau, \qquad t_0 \le t. \tag{26}$$

Consequently we will, with no loss of generality, assume that the mean vector of initial conditions $\boldsymbol{\mu}_X(t_0)$ and the mean function $\mu_U(t)$ are zero, for if $X(t)$ is the solution obtained under these assumptions, then $X(t) + \mu_X(t)$ is the solution of the original system.

Assume further that $R_U(t_1, t_2)$ is the correlation function of the process $U(t)$ (with $\mu_U(t) = 0$) and that

$$\boldsymbol{\sigma}_X^2(t_0) = \mathrm{E}[X^{(k)}(t_0)X^{(l)}(t_0)]$$

is the $n \times n$ covariance matrix of the random vector of initial conditions (25) (with means assumed now zero).

Then by (23) the correlation function of the singular component $X_s(t)$ is

$$R_{X_s}(t_1, t_2) = \boldsymbol{\xi}'(t_1)\boldsymbol{\sigma}_X^2(t_0)\boldsymbol{\xi}(t_2), \qquad t_0 \le t_1, \qquad t_0 \le t_2, \tag{27}$$

with the vector

$$\boldsymbol{\xi}(t) = \begin{pmatrix} \xi_0(t) \\ \vdots \\ \xi_{n-1}(t) \end{pmatrix}, \tag{28}$$

and the correlation function of the regular component $X_r(t)$ is, according to Theorem 4.2.1, applied to (24)

$$R_{X_r}(t_1, t_2) = \int_{t_0}^{t_1} \int_{t_0}^{t_2} h(t_1, \tau_1)R_U(\tau_1, \tau_2)h(t_2, \tau_2)\, d\tau_1\, d\tau_2, \qquad t_0 \le t_1, \qquad t_0 \le t_2. \tag{29}$$

If, further, the random vector of initial conditions (25) is independent of the process $U(t)$, then the correlation functions of the process $X(t)$ will be the sum

$$R_X(t_1, t_2) = R_{X_s}(t_1, t_2) + R_{X_r}(t_1, t_2) \tag{30}$$

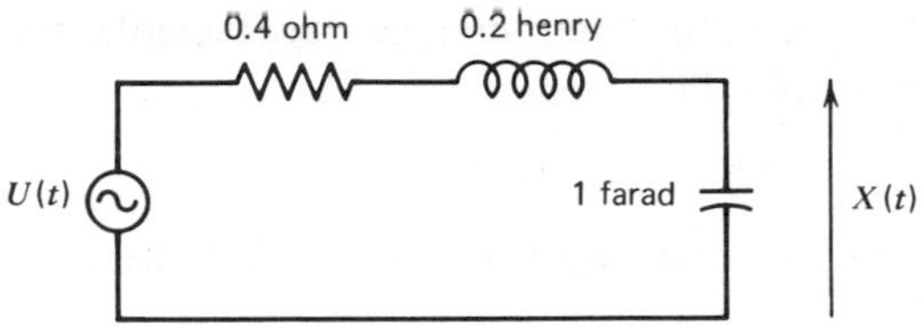

Figure 4.2.5

of the correlations functions (27) and (29). Needless to say, if the initial conditions and the process $U(t)$ are both Gaussian the solution process $X(t)$ will be a Gaussian process, and thus its probability law will be completely determined.

Example 4.2.4

Consider a resonant circuit as shown in Figure 4.2.5, where $X(t)$ is the voltage across the capacitor and $U(t)$ is a random voltage source, representing, for instance, the thermal noise of the circuit resistance.

As is well known, the voltage $X(t)$ will obey a second order linear differential equation, which, with the values of the circuit components as shown, takes the form

$$\ddot{X}(t) + 2\dot{X}(t) + 5X(t) = 5U(t).$$

Suppose we begin to observe the circuit at time $t_0 = 0$, and let the voltage $X(0)$ and its derivative $\dot{X}(0)$ (i.e., the charge in the capacitor and the current in the circuit) be Gaussian random variables with mean vector

$$\boldsymbol{\mu}_{\mathbf{X}}(t_0) = \begin{pmatrix} 1 \\ 0 \end{pmatrix},$$

and with the covariance matrix

$$\boldsymbol{\sigma}^2_{\mathbf{X}}(t_0) = \begin{bmatrix} 1 & -\frac{1}{2} \\ -\frac{1}{2} & 1 \end{bmatrix}$$

The source voltage $U(t)$, $t \geq 0$, is assumed to be a Gaussian process with zero mean function and with correlation function

$$R_U(t_1, t_2) = \begin{cases} \sigma^2\left(1 - \dfrac{1}{\tau_0}|t_1 - t_2|\right) & \text{if } |t_1 - t_2| \leq \tau_0, \\ 0 & \text{if } |t_1 - t_2| > \tau_0, \end{cases} \tag{31}$$

where $\sigma > 0$ and $\tau_0 > 0$ are constants.

First, we consider the homogeneous differential equation

$$\ddot{x}(t) + 2\dot{x}(t) + 5x(t) = 0, \qquad t \geq 0$$

This is a second order linear differential equation with constant coefficients. Its general solution is of the form (see Appendix Three)

$$x(t) = c_1 e^{\lambda_1 t} + c_2 e^{\lambda_2 t}, \qquad t \geq 0$$

where $\lambda_{1,2} = -1 \pm 2\imath$ are the two complex conjugate roots of the characteristic equation $\lambda^2 + 2\lambda + 5 = 0$.

It follows that the functions $\xi_0(t)$ and $\xi_1(t)$, $t \geq 0$, are both of the form $c_1 e^{\lambda_1 t} + c_2 e^{\lambda_2 t}$, where the constants c_1 and c_2 are determined from the conditions

$$\xi_0(0) = c_1 + c_2 = 1,$$

$$\dot{\xi}_0(0) = \lambda_1 c_1 + \lambda_2 c_2 = 0,$$

for the function $\xi_0(t)$, and from the conditions

$$\xi_1(0) = c_1 + c_2 = 0,$$

$$\dot{\xi}_1(0) = \lambda_1 c_1 + \lambda_2 c_2 = 1,$$

for the function $\xi_1(t)$. Thus, in the first case

$$c_1 = \frac{-\lambda_2}{\lambda_1 - \lambda_2} = \frac{1}{2} + \frac{1}{4\imath}, \qquad c_2 = \frac{\lambda_1}{\lambda_1 - \lambda_2} = \frac{1}{2} - \frac{1}{4\imath},$$

and so

$$\xi_0(t) = \left(\frac{1}{2} + \frac{1}{4\imath}\right)e^{(-1+2\imath)t} + \left(\frac{1}{2} - \frac{1}{4\imath}\right)e^{(-1-2\imath)t}$$

$$= e^{-t}\left(\cos 2t + \frac{1}{2}\sin 2t\right), \qquad t \geq 0.$$

Similarly for $\xi_1(t)$, we have

$$c_1 = \frac{1}{\lambda_1 - \lambda_2} = \frac{1}{4\imath}, \qquad c_2 = \frac{-1}{\lambda_1 - \lambda_2} = -\frac{1}{4\imath},$$

which gives

$$\xi_1(t) = \frac{1}{2}e^{-t}\sin 2t, \qquad t \geq 0.$$

By (20) the impulse-response function $h(t, \tau)$, $0 \leq \tau \leq t$, as a function of t also satisfies the homogeneous equation with the initial conditions

$$h(\tau, \tau) = 0, \qquad \left.\frac{\partial}{\partial t} h(t, \tau)\right|_{t=\tau} = 1.$$

This means that in the present case we have simply

$$h(t, \tau) = \xi_1(t - \tau), \qquad 0 \leq \tau \leq t,$$

or

$$h(t, \tau) = \begin{cases} \frac{1}{2}e^{-(t-\tau)} \sin 2(t-\tau) & \text{if } 0 \le \tau \le t, \\ 0 & \text{if } 0 \le t < \tau. \end{cases}$$

According to (21) and (22) the process $X(t)$, $t \ge 0$, is then given by the equation

$$X(t) = X(0)e^{-t}\left(\cos 2t + \frac{1}{2} \sin t\right) + \dot{X}(0)\frac{1}{2}e^{-t} \sin 2t$$

$$+ \int_0^t \frac{1}{2} e^{-(t-\tau)} \sin 2(t-\tau) 5U(\tau)\, d\tau, \qquad t \ge 0.$$

Since $\mu_U(t) = 0$, $t \ge 0$, the mean function $\mu_X(t)$, $t \ge 0$, is by (26)

$$\mu_X(t) = 1\xi_0(t) + 0\xi_1(t) = e^{-t}\left(\cos 2t + \frac{1}{2} \sin 2t\right).$$

After subtracting the mean function, the correlation function $R_{X_s}(t_1, t_2)$ of the singular component equals

$$R_{X_s}(t_1, t_2) = (\xi_0(t_1), \xi_1(t_1))\begin{bmatrix} 1 & -\frac{1}{2} \\ -\frac{1}{2} & 1 \end{bmatrix}\begin{pmatrix} \xi_0(t_2) \\ \xi_1(t_2) \end{pmatrix}$$

$$= \xi_0(t_1)\xi_0(t_2) - \frac{1}{2}\xi_0(t_1)\xi_1(t_2)$$

$$- \frac{1}{2}\xi_1(t_1)\xi_0(t_2) + \xi_1(t_1)\xi_1(t_2).$$

After substituting for $\xi_0(t)$ and $\xi_1(t)$ and applying some trigonometric identities, we can write it as

$$R_{X_s}(t_1, t_2) = e^{-(t_1+t_2)}\left(\frac{3}{4} \cos 2t_1 \cos 2t_2 + \frac{1}{2} \sin\left(2t_1 + \frac{\pi}{4}\right)\sin\left(2t_2 + \frac{\pi}{4}\right)\right),$$

$$t_1 \ge 0, \qquad t_2 \ge 0.$$

In particular with $t_1 = t_2 = t \ge 0$, we have the variance function

$$\sigma_{X_s}^2(t) = e^{-2t}\left(\frac{3}{4} \cos^2 2t + \frac{1}{2} \sin^2\left(2t + \frac{\pi}{4}\right)\right).$$

Notice that the singular component

$$X_s(t) = X(0)e^{-t}\left(\cos 2t + \frac{1}{2} \sin 2t\right) + \dot{X}(0)\frac{1}{2}e^{-t} \sin 2t$$

simply describes the voltage due to the random initial condition only, that is, with $U(t)$ being identically zero.

The regular component

$$X_r(t) = \int_0^t \frac{1}{2} e^{-(t-\tau)} \sin 2(t-\tau) 5U(\tau)\, d\tau, \qquad t \geq 0,$$

is, on the other hand, the voltage component entirely due to the random source $U(t)$. Its correlation function can be computed from (29):

$$R_{X_r}(t_1, t_2) =$$
$$25 \int_0^{t_1} \int_0^{t_2} \frac{1}{2} e^{-(t_1-\tau_2)} \sin 2(t_1-\tau_1) R_U(\tau_1, \tau_2) \frac{1}{2} e^{-(t_2-\tau_2)} \sin 2(t_2-\tau_2)\, d\tau_1\, d\tau_2.$$

Although, with $R_U(t_1, t_2)$ as in (31), this integral can be evaluated by elementary integration, the procedure is rather lengthy and the resulting expression a somewhat formidable mixture of trigonometric functions and exponentials. Therefore, we will calculate only the variance function

$$\sigma_{X_r}^2(t) = 25 \int_0^t \int_0^t \frac{1}{4} e^{-2t+(\tau_1+\tau_2)} \sin 2(t-\tau_1) \sin 2(t-\tau_2) R_U(\tau_1, \tau_2)\, d\tau_1\, d\tau_2,$$

under the additional assumptions that the constant τ_0 in the correlation function (31) is much smaller than 1. The integration region in the double integral above is then a narrow strip along the line $\tau_1 = \tau_2$ (see Fig. 4.2.6), which for $0 < \tau_0 \ll 1 < t$ can be replaced by a rectangle, as indicated by the dotted lines. Rotating the coordinate system by 45 degrees, that is, making the substitution $y_1 = (1/\sqrt{2}) \times (\tau_1 + \tau_2)$, $y_2 = (1/\sqrt{2})(\tau_1 - \tau_2)$, the double integral above is approximately

$$25 \frac{\sigma^2}{4} e^{-2t} \int_0^{t\sqrt{2}} \left(\int_{-\tau_0/\sqrt{2}}^{\tau_0/\sqrt{2}} e^{y_1\sqrt{2}} \sin 2\left(t - \frac{1}{\sqrt{2}}(y_1 + y_2)\right) \right.$$
$$\left. \times \sin 2\left(t - \frac{1}{\sqrt{2}}(y_1 - y_2)\right)\left(1 - \frac{\sqrt{2}|y_2|}{\tau_0}\right) dy_2 \right) dy_1.$$

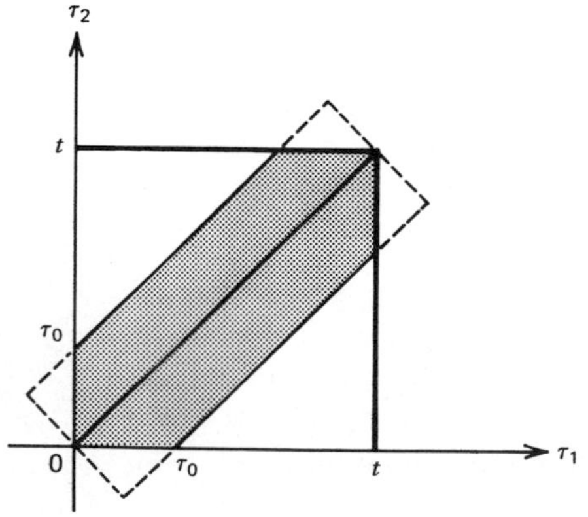

Figure 4.2.6

For $0 < \tau_0 \ll 1$ we can further neglect the variable y_2 in the argument of the sine functions so that the integral factors into a product of two integrals

$$25\,\frac{\sigma^2}{4}\,e^{-2t}\int_0^{t\sqrt{2}} e^{y_1\sqrt{2}}\sin^2 2\left(t-\frac{y_1}{\sqrt{2}}\right)dy_1\int_{-\tau_0/\sqrt{2}}^{\tau_0/\sqrt{2}}\left(1-\frac{\sqrt{2}\,|y_2|}{\tau_0}\right)dy_2 .$$

The second integral is just $\tau_0\sqrt{2}$; the first takes a little longer:

$$\int_0^{t\sqrt{2}} e^{y_1\sqrt{2}}\sin^2 2\left(t-\frac{y_1}{\sqrt{2}}\right)dy_1=\frac{e^{2t}}{\sqrt{2}}\int_0^{2t} e^{-y}\sin^2 y\,dy$$

$$=\frac{e^{2t}}{2\sqrt{2}}\left(\frac{4}{5}-e^{-2t}\left(1-\frac{1}{5}\cos 4t+\frac{2}{5}\sin 4t\right)\right).$$

Putting everything together we have approximately

$$\sigma_{X_r}^2(t) \doteq 25\,\frac{\sigma^2\tau_0}{8}\left(\frac{4}{5}-e^{-2t}\left(1-\frac{1}{5}\cos 4t+\frac{2}{5}\sin 4t\right)\right)$$

for the variance function of the regular component $X_r(t)$. ▲

In engineering applications, it is common practice to describe differential linear systems (i.e., linear systems whose output or outputs are solutions of a system of linear differential equations with inputs as forcing functions) in the so-called *state-space representation*

$$\dot{\mathbf{x}}(t)=\mathbf{a}(t)\mathbf{x}(t)+\mathbf{b}(t)\mathbf{u}(t), \tag{32a}$$

$$\mathbf{y}(t)=\mathbf{c}(t)\mathbf{x}(t), \qquad t_0 \le t, \tag{32b}$$

where $\mathbf{x}(t)$, $\mathbf{y}(t)$, and $\mathbf{u}(t)$ are vector-valued functions and $\mathbf{a}(t)$, $\mathbf{b}(t)$, and $\mathbf{c}(t)$ are matrix-valued functions. Here again, $\mathbf{u}(t)$ and $\mathbf{y}(t)$ are the input and output functions, while $\mathbf{x}(t)$ is the state of the system.

To complete the specification, the initial state

$$\mathbf{x}(t_0)$$

must also be given.

The advantage of this representation is that any linear differential system can be expressed in form (32) by appropriately defining the state vector $\mathbf{x}(t)$ so that this representation is canonical. We assume that the reader is familiar with at least the basics of the theory behind this representation.

Under appropriate conditions on the functions $\mathbf{a}(t)$, $\mathbf{b}(t)$, and $\mathbf{c}(t)$, the system (32) has a unique solution

$$\mathbf{x}(t)=\boldsymbol{\phi}(t,t_0)\mathbf{x}(t_0)+\int_{t_0}^{t}\boldsymbol{\phi}(t,\tau)\mathbf{b}(\tau)\mathbf{u}(\tau)\,d\tau, \tag{33a}$$

$$\mathbf{y}(t)=\mathbf{c}(t)\mathbf{x}(t), \qquad t_0 \le t, \tag{33b}$$

where $\boldsymbol{\phi}(t_1, t_2)$ is the (zero-input) *state-transition function* (matrix valued) satisfying for every fixed t_2, $t_0 \le t_2$, the differential equation

$$\frac{\partial}{\partial t_1}\boldsymbol{\phi}(t_1, t_2) = \mathbf{a}(t_1)\boldsymbol{\phi}(t_1, t_2), \qquad t_2 \le t_1, \tag{34}$$

with initial condition

$$\boldsymbol{\phi}(t_2, t_2) = \mathbf{I}, \qquad \text{the identity matrix.} \tag{35}$$

If now the input $\mathbf{U}(t)$ is a (vector-valued) stochastic process and the initial state is a random vector $\mathbf{X}(t_0)$, then both the state $\mathbf{X}(t)$ and the output $\mathbf{Y}(t)$ will be stochastic processes satisfying (33), that is,

$$\begin{aligned}
&\mathbf{X}(t) = \mathbf{X}_s(t) + \mathbf{X}_r(t),\\
&\mathbf{X}_s(t) = \boldsymbol{\phi}(t, t_0)\mathbf{X}(t_0), \qquad \mathbf{X}_r(t) = \int_{t_0}^{t} \boldsymbol{\phi}(t, \tau)\mathbf{b}(\tau)\mathbf{U}(\tau)\, d\tau, \qquad (36)\\
&\mathbf{Y}(t) = \mathbf{Y}_s(t) + \mathbf{Y}_r(t),\\
&\mathbf{Y}_s(t) = \mathbf{c}(t)\mathbf{X}_s(t), \qquad \mathbf{Y}_r(t) = \mathbf{c}(t)\mathbf{X}_r(t),
\end{aligned}$$

$t_0 \le t$, where the subscripts s and r again denote the singular and regular components. If $\boldsymbol{\mu}_{\mathbf{U}}(t)$ is the mean function of the process $\mathbf{U}(t)$, then by linearity the mean functions of the state and output process satisfy the same system of equations

$$\begin{aligned}
&\dot{\boldsymbol{\mu}}_{\mathbf{X}}(t) = \mathbf{a}(t)\boldsymbol{\mu}_{\mathbf{X}}(t) + \mathbf{b}(t)\boldsymbol{\mu}_{\mathbf{U}}(t)\\
&\boldsymbol{\mu}_{\mathbf{Y}}(t) = \mathbf{c}(t)\boldsymbol{\mu}_{\mathbf{X}}(t), \qquad \boldsymbol{\mu}_{\mathbf{X}}(t_0) = \mathrm{E}[\mathbf{X}(t_0)],
\end{aligned} \tag{37}$$

so that by (33)

$$\begin{aligned}
&\boldsymbol{\mu}_{\mathbf{X}}(t) = \boldsymbol{\phi}(t, t_0)\boldsymbol{\mu}_{\mathbf{X}}(t_0) + \int_{t_0}^{t} \boldsymbol{\phi}(t, \tau)\mathbf{b}(\tau)\boldsymbol{\mu}_{\mathbf{U}}(\tau)\, d\tau\\
&\boldsymbol{\mu}_{\mathbf{Y}}(t) = \mathbf{c}(t)\boldsymbol{\mu}_{\mathbf{X}}(t), \qquad t_0 \le t.
\end{aligned}$$

Hence, assuming now that $\mathrm{E}[\mathbf{X}(t_0)] = 0$, that $\boldsymbol{\mu}_{\mathbf{U}}(t) = 0$, and that $\mathbf{X}(t_0)$ and the input process $\mathbf{U}(t)$ are independent, we obtain expressions for the correlation functions analogous to those in (27) through (30), namely

$$\mathbf{R}_{\mathbf{X}}(t_1, t_2) = \mathbf{R}_{\mathbf{X}_s}(t_1, t_2) + \mathbf{R}_{\mathbf{X}_r}(t_1, t_2), \tag{38a}$$

$$\mathbf{R}_{\mathbf{X}_s}(t_1, t_2) = \boldsymbol{\phi}(t_1, t_0)\boldsymbol{\sigma}_{\mathbf{X}}^2(t_0)\boldsymbol{\phi}'(t_2, t_0), \tag{38b}$$

with

$$\boldsymbol{\sigma}_{\mathbf{X}}^2(t_0) = \mathrm{E}[\mathbf{X}(t_0)\mathbf{X}'(t_0)],$$

$$\mathbf{R}_{\mathbf{X}_r}(t_1, t_2) = \int_{t_0}^{t_1}\int_{t_0}^{t_2} \boldsymbol{\phi}(t_1, \tau_1)\mathbf{b}(\tau_1)\mathbf{R}_{\mathbf{U}}(\tau_1, \tau_2)\mathbf{b}'(\tau_2)\boldsymbol{\phi}'(t_2, \tau_2)\, d\tau_1\, d\tau_2 \tag{38c}$$

and since (33b) is a linear memoryless transformation

$$\mathbf{R}_{\mathbf{Y}}(t_1, t_2) = \mathbf{c}(t_1)\mathbf{R}_{\mathbf{X}}(t_1, t_2)\mathbf{c}'(t_2),$$

all for $t_0 \le t_1$, $t_0 \le t_2$. If $\mathbf{X}(t_0)$ and the input process $\mathbf{U}(t)$ are Gaussian, both $\mathbf{X}(t)$ and $\mathbf{Y}(t)$ will be Gaussian.

The expressions for cross-correlation between the processes involved may be easily obtained from (36) by multiplying the appropriate equations by $\mathbf{U}(t_2)$, $\mathbf{X}(t_2)$, or $\mathbf{Y}(t_2)$ and taking expectations.

Example 4.2.5

A charged particle moving in an electromagnetic field with velocity $\mathbf{v}$ experiences a Lorentz force

$$\mathbf{f} = q\mathbf{E} + q(\mathbf{v} \times \mathbf{B}),$$

where $\mathbf{E}$ and $\mathbf{B}$ are the electric intensity magnetic induction vectors of the field, and q is the charge of the particle. With the aid of Newton's law

$$\mathbf{f} = m\dot{\mathbf{v}},$$

where m is the mass of the particle, we obtain the differential equation for the velocity vector

$$m\dot{\mathbf{v}} = q\mathbf{E} + q(\mathbf{v} \times \mathbf{B}).$$

Suppose now that the magnetic field is homogeneous and constant, that is, the vector $\mathbf{B}$ does not change with either space or time, and let both the electric intensity vector $\mathbf{E}$ and the initial velocity vector $\mathbf{v}(t_0)$ both be perpendicular to the vector $\mathbf{B}$. Then the particle remains confined to the plane determined by the vectors $\mathbf{E}$ and $\mathbf{v}(t_0)$, and if we choose a Cartesian coordinate system in this plane, we get the pair of differential equations

$$\begin{aligned} m\dot{v}_1 &= qE_1 + qBv_2, \\ m\dot{v}_2 &= qE_2 - qBv_1, \end{aligned} \tag{39}$$

with (v_1, v_2) and (E_1, E_2) being the Cartesian components of the vectors $\mathbf{v}$ and $\mathbf{E}$.

As is customary in physics, let the state $\mathbf{x}(t)$ of the particle at a time t be represented by its position and momentum, that is, by a column vector

$$\mathbf{x}(t) = \begin{pmatrix} x_1(t) \\ x_2(t) \\ x_3(t) \\ x_4(t) \end{pmatrix}$$

where $x_1(t)$ and $x_2(t)$ are the Cartesian coordinates of its position and $x_3(t) = mv_1(t)$, $x_4(t) = mv_2(t)$ are the corresponding coordinates of the momentum. The differential equations (39), together with the obvious relations $m\dot{x}_1 = x_3$, $m\dot{x}_2 = x_4$ between the positions and momenta, allows us to express the system for the state of the particle in state-space representation (32a):

$$\begin{pmatrix} \dot{x}_1(t) \\ \dot{x}_2(t) \\ \dot{x}_3(t) \\ \dot{x}_4(t) \end{pmatrix} = \begin{bmatrix} 0 & 0 & \frac{1}{m} & 0 \\ 0 & 0 & 0 & \frac{1}{m} \\ 0 & 0 & 0 & \frac{qB}{m} \\ 0 & 0 & \frac{-qB}{m} & 0 \end{bmatrix} \begin{pmatrix} x_1(t) \\ x_2(t) \\ x_3(t) \\ x_4(t) \end{pmatrix} + \begin{bmatrix} 0 & 0 \\ 0 & 0 \\ q & 0 \\ 0 & q \end{bmatrix} \begin{pmatrix} E_1 \\ E_2 \end{pmatrix} \tag{40}$$

If the "output" of this system is to be the trajectory of the particle, then by calling $y_1(t) = x_1(t)$ and $y_2(t) = x_2(t)$, we could also write the second (state-output) equation of the representation (32b) as

$$\begin{pmatrix} y_1(t) \\ y_2(t) \end{pmatrix} = \begin{bmatrix} 1 & 0 & 0 & 0 \\ 0 & 1 & 0 & 0 \end{bmatrix} \begin{pmatrix} x_1(t) \\ x_2(t) \\ x_3(t) \\ x_4(t) \end{pmatrix}.$$

However, since this equation represents just the trivial identity $y_1(t) = x_1(t)$, $y_2(t) = x_2(t)$, we will consider only the input-state equation (40).

Suppose now that the particle is injected into the field at the origin of the coordinate system with unit velocity vector $v(0)$ in the direction of the second coordinate axis. This specifies the initial state $\mathbf{x}(0)$ as

$$\mathbf{x}(0) = \begin{pmatrix} 0 \\ 0 \\ 0 \\ m \end{pmatrix} \tag{41}$$

Suppose further that the electric field vector $\mathbf{E}$ has the magnitude $e_0 = B/m$ and that it rotates in the clockwise direction with the radial frequency $\omega = qB/m$. The arrangement is shown in Figure 4.2.7. We can now write the input-state equation (40) in standard form (32a):

$$\dot{\mathbf{x}}(t) = \mathbf{a}(t)\mathbf{x}(t) + \mathbf{b}(t)\mathbf{u}(t), \qquad t \geq 0$$

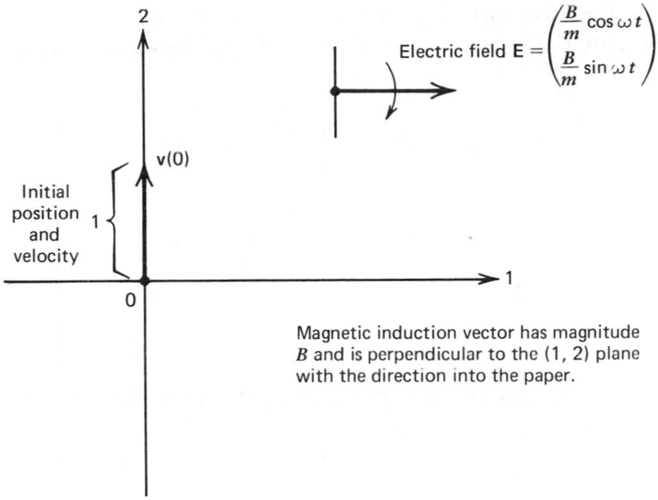

Figure 4.2.7

where

$$\mathbf{a}(t) = \begin{bmatrix} 0 & 0 & \frac{1}{m} & 0 \\ 0 & 0 & 0 & \frac{1}{m} \\ 0 & 0 & 0 & \omega \\ 0 & 0 & -\omega & 0 \end{bmatrix}, \qquad \mathbf{b}(t) = \begin{bmatrix} 0 & 0 \\ 0 & 0 \\ q & 0 \\ 0 & q \end{bmatrix},$$

$$\mathbf{u}(t) = \begin{pmatrix} \frac{B}{m}\cos\omega t \\ -\frac{B}{m}\sin\omega t \end{pmatrix}, \qquad \omega = \frac{qB}{m},$$

and the initial state given by (41).

According to (33a) the solution $\mathbf{x}(t)$, $t \geq 0$, of this system will have the form

$$\mathbf{x}(t) = \boldsymbol{\phi}(t, 0)\mathbf{x}(0) + \int_0^t \boldsymbol{\phi}(t, \tau)\mathbf{b}(\tau)\mathbf{u}(\tau)\, d\tau, \tag{42}$$

where the state-transition function $\boldsymbol{\phi}(t_1, t_2)$, $0 \le t_2 \le t_1$, is obtained from the differential equation (34) with the initial condition (35). In the present case, we have the matrix

$$\boldsymbol{\phi}(t_1, t_2) = \begin{bmatrix} 1 & 0 & \dfrac{1}{m\omega} \sin \omega(t_1 - t_2) & \dfrac{1}{m\omega}(1 - \cos \omega(t_1 - t_2)) \\ 0 & 1 & \dfrac{-1}{m\omega}(1 - \cos(t_1 - t_2)) & \dfrac{1}{m\omega} \sin \omega(t_1 - t_2) \\ 0 & 0 & \cos \omega(t_1 - t_2) & \sin \omega(t_1 - t_2) \\ 0 & 0 & -\sin \omega(t_1 - t_2) & \cos \omega(t_1 - t_2) \end{bmatrix}$$

as is readily verified. Substituting into (42) and performing the integration (which is quite simple in this case), we obtain the solution

$$\mathbf{x}(t) = \begin{pmatrix} t \sin \omega t \\ t \cos \omega t \\ m \sin \omega t + m\omega t \cos \omega t \\ m \cos \omega t - m\omega t \sin \omega t \end{pmatrix}, \qquad t \ge 0. \tag{43}$$

It is seen that the trajectory of the particle will be an Archimedean spiral† unwinding from the origin in the clockwise direction. The particle will complete each revolution in $2\pi/\omega$ seconds, so that its angular velocity will remain a constant ω. The magnitude of its velocity and hence also its kinetic energy will, however, increase with time; in fact, at times $t = n\pi/2\omega$, $n = 1, 2, \ldots$, when it passes through the coordinate axes, its kinetic energy will equal $\frac{1}{2}m(1 + \omega^2 t^2)$. In other words, the arrangement will act as a particle accelerator, actually a crude version of the classical cyclotron. For instance, if a target is placed at a distance d from the origin, as shown in Figure 4.2.8, the kinetic energy of the particle upon hitting the target will be $(m/2)(1 + \omega^2 d^2)$.

It is clearly of interest to investigate the effect of some random disturbances of the system parameters. Specifically, assume that the initial velocity vector $\mathbf{v}(0)$ and the magnitude e_0 of the electric field are subjected to random errors

$$\mathbf{V}(0) = \mathbf{v}(0) + \mathbf{V}, \qquad E_0 = e_0 + Z(t), \qquad t \ge 0,$$

where $\mathbf{V}$ is a random vector with zero mean and covariance matrix

$$\begin{bmatrix} \sigma_1^2 & 0 \\ 0 & \sigma_2^2 \end{bmatrix}$$

† A spiral with polar equation such that the distance from the origin increases linearly with the angle.

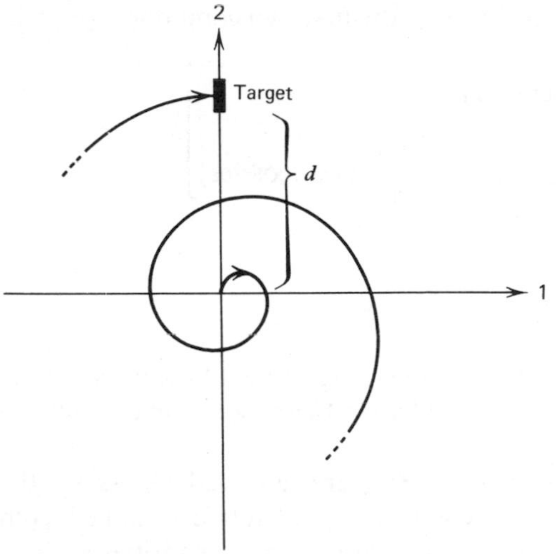

Figure 4.2.8

while $Z(t)$ is a stationary stochastic process with zero mean function and with correlation function $R_Z(\tau)$. Then

$$\mathbf{X}(t) = \boldsymbol{\phi}(t, 0)\mathbf{X}(0) + \int_0^t \boldsymbol{\phi}(t, \tau)\mathbf{b}(\tau)\mathbf{U}(\tau)\, d\tau, \qquad t \geq 0,$$

with $\boldsymbol{\mu}_{\mathbf{X}}(0) = 0$, $\boldsymbol{\mu}_{\mathbf{U}}(t) = 0$, $t \geq 0$,

$$\boldsymbol{\sigma}_{\mathbf{X}}^2(0) = \begin{bmatrix} 0 & 0 & 0 & 0 \\ 0 & 0 & 0 & 0 \\ 0 & 0 & \sigma_1^2 & 0 \\ 0 & 0 & 0 & \sigma_2^2 \end{bmatrix}$$

$$\mathbf{R}_{\mathbf{U}}(t_1, t_2) = R_Z(t_1 - t_2)\begin{bmatrix} \cos \omega t_1 \cos \omega t_2 & \cos \omega t_1 \sin \omega t_2 \\ \sin \omega t_1 \cos \omega t_2 & \sin \omega t_1 \sin \omega t_2 \end{bmatrix}$$

$t_1 \geq 0$, $t_2 \geq 0$, will be a zero mean stochastic process representing the random deviation from the nominal state $\mathbf{x}(t)$ as given by (43). The singular component $\mathbf{X}_s(t) = \boldsymbol{\phi}(t, 0)\mathbf{X}(0)$ represents the effect of errors due to the random initial velocity, and its correlation function is easily evaluated from (38), that is,

$$\mathbf{R}_{\mathbf{X}_s}(t_1, t_2) = \boldsymbol{\phi}(t_1, 0)\boldsymbol{\sigma}_{\mathbf{X}}^2(0)\boldsymbol{\phi}'(t_2, 0).$$

For instance, we find that the variance function of the first two components equals

$$\boldsymbol{\sigma}^2_{\binom{X_{s1}}{X_{s2}}}(t) = \begin{bmatrix} \dfrac{2\sigma_1^2}{m\omega^2}(1 - \cos \omega t) & 0 \\ 0 & \dfrac{2\sigma_2^2}{m\omega^2}(1 - \cos \omega t) \end{bmatrix}.$$

The regular component

$$\mathbf{X}_r(t) = \int_0^t \boldsymbol{\phi}(t, \tau)\mathbf{b}(\tau)\mathbf{U}(\tau)\, d\tau$$

represents the effect of the errors due to the random magnitude of the electrostatic field, and its correlation function can be evaluated with some effort but no fundamental difficulties from (38).

Furthermore, if both $\mathbf{V}$ and the process $Z(t)$ are assumed Gaussian, the process $\mathbf{X}(t)$ will be also. Its probability law is then completely determined from its correlation function, and it is possible to calculate various quantities such as the probability distribution of the kinetic energy of the particle upon hitting the target.

Since our intention was to illustrate the application of the state-space representation to a specific physical situation, not get involved in a long algebraic calculation, we end the discussion at this point. ▲

EXERCISES

Exercise 4.2

1. A linear system has impulse-response function

$$h(t, \tau) = \begin{cases} \left(\cos \dfrac{t}{2\pi}\right) e^{-\beta(t-\tau)} & \text{if } \tau \le t, \\ 0 & \text{if } \tau > t, \end{cases}$$

with $\beta > 0$ a constant. The input $U(t)$, $t \ge 0$, to this system is the Ornstein-Uhlenbeck process with mean function

$$\mu_U(t) = 1 - e^{-\beta t}, \qquad t \ge 0,$$

and covariance function

$$\gamma(t_1, t_2) = \sigma^2(e^{-\beta|t_1 - t_2|} - e^{-\beta(t_1 + t_2)}), \qquad t_1 \ge 0, \qquad t_2 \ge 0.$$

Find the mean and the covariance function of the output process $X(t)$, $t \ge 0$.

2. A unit impulse arriving at the input of a linear system at time τ produces at its output a rectangular pulse of unit width. The height of the pulse is determined by sampling the signal of the local oscillator of frequency 1 hertz and thus equals $a \sin(2\pi\tau)$ with

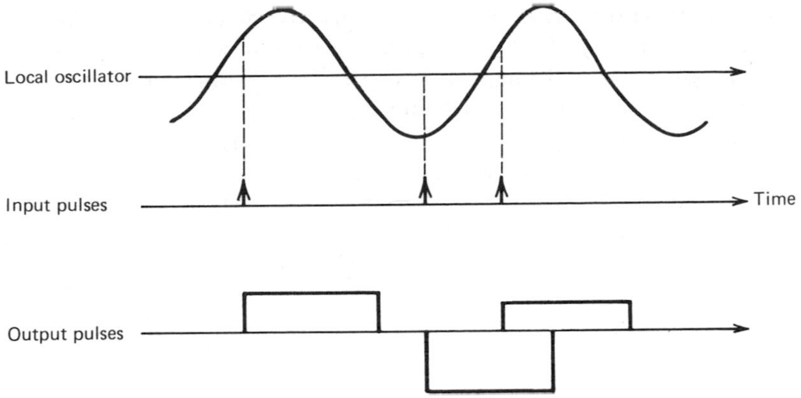

Figure 4.2.E.1

$a > 0$ a constant. (See Fig. 4.2.E.1.) The input to this system is a zero mean wide-sense stationary process $U(t)$, $-\infty < t < \infty$, with the triangular correlation function

$$R_U(\tau) = \begin{cases} \sigma^2(1 - |\tau|) & \text{if } |\tau| \le 1, \\ 0 & \text{if } |\tau| > 1. \end{cases}$$

Find the impulse-response function of this system and determine the variance function of the output process $X(t)$, $-\infty < t < \infty$.

3. A time-varying *R-C* circuit has impulse-response function

$$h(t, \tau) = \begin{cases} ae^{-b(t)(t-\tau)} & \text{if } \tau \le t, \\ 0 & \text{if } \tau > t, \end{cases}$$

where $a > 0$ is a constant and $b(t) > 0$ is a function of time. At time $t = 0$, an input $U(t)$ is applied to the circuit. Assuming $U(t)$, $t \ge 0$, is a zero mean Gaussian process with correlation function

$$R_U(t_1, t_2) = \sigma^2 e^{-\alpha|t_1 - t_2|}, \qquad \alpha > 0,$$

where α is much larger than $b(t)$ find approximately the correlation function R_X of the output process $X(t)$, $t \ge 0$.

4. Suppose that a stationary Ornstein-Uhlenbeck process $U(t)$, $-\infty < t < \infty$, is applied to the input of a time-invariant linear system with impulse-response function:

(a) $h(\tau) = \begin{cases} \tau & \text{if } 0 \le \tau \le 1, \\ 0 & \text{otherwise,} \end{cases}$

(b) $h(\tau) = \begin{cases} 1 - \tau & \text{if } 0 \le \tau \le 1, \\ 0 & \text{otherwise,} \end{cases}$

(c) $h(\tau) = \begin{cases} 1 & \text{if } 0 \le \tau \le 1, \\ 0 & \text{otherwise.} \end{cases}$

For each of these three cases evaluate the cross-correlation function $R_{X,U}$ and the correlation function R_X of the output process $X(t)$, $-\infty < t < \infty$. Assume $\alpha = 1$.

5. Under the assumptions of Example 4.2.3, find the cross-correlation function $R_{X,U}$ and the correlation function R_X if the parameter $\gamma = 1$.
6. A three-input/three-output linear system is constructed using threee single-input/single-output linear systems and three differential amplifiers of unit gain as shown in Figure 4.2.E.2. The impulse-response functions h_a, h_b, and h_c are as in Exercise 4.2.4, the components $U_1(t)$, $U_2(t)$, $U_3(t)$ of the input vector $\mathbf{U}(t)$ are independent zero mean stationary Ornstein-Uhlenbeck processes with identical parameters. Write the integral formula for the matrix correlation function of the output vector $\mathbf{X}(t)$. Find the covariance matrix $\boldsymbol{\sigma}^2 = \mathbf{R_X}(t, t)$ of the stationary output process.

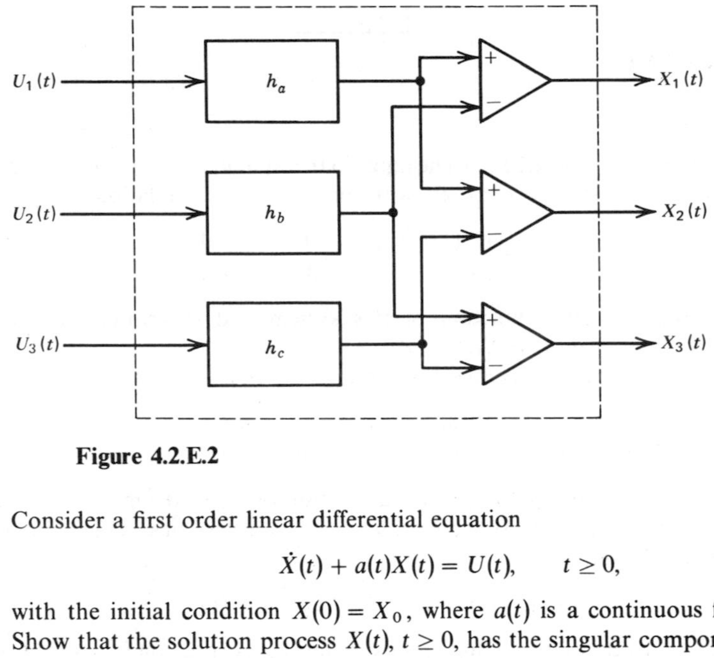

Figure 4.2.E.2

7. Consider a first order linear differential equation

$$\dot{X}(t) + a(t)X(t) = U(t), \qquad t \geq 0,$$

with the initial condition $X(0) = X_0$, where $a(t)$ is a continuous function of time. Show that the solution process $X(t)$, $t \geq 0$, has the singular component

$$X_s(t) = X_0 e^{-\int_\tau^t a(\tau)\,d\tau},$$

and the regular component

$$X_r(t) = \int_0^t U(\tau)e^{-\int_0^t a(\tau')\,d\tau'}\,d\tau.$$

8. Suppose a process $X(t)$, $t \geq 0$, satisfies the linear differential equation

$$\dot{X}(t) + tX(t) = U(t), \qquad t \geq 0,$$

where $U(t)$ is a zero mean stationary Ornstein-Uhlenbeck process. The initial condition $X(0) = X_0$ is a standard Gaussian random variable independent of the process

$U(t)$. Show that the singular and regular component processes are independent Gaussian processes and find their mean and correlation functions. (For the latter leave the answer in an integral form but find the asymptotic behavior of the variance function.

9. Repeat Exercise 4.2.8 if $U(t)$, $t \geq 0$, is the standard Wiener process.
10. Find the mean and correlation functions of the process $X(t)$, $t \geq 1$, that satisfies the differential equation

$$t\dot{X}(t) + X(t) = U(t)$$

with initial condition $X(1) = 0$. Assume that $U(t)$ is a wide-sense stationary process with mean $\mu_U = 1$ and correlation function $R_U(\tau) = e^{-\alpha|\tau|} + 1$, $\alpha > 0$.

11. Consider an ideal resonant system whose output $X(t)$ satisfies the second order linear differential equation

$$\ddot{X}(t) + \omega^2 X(t) = U(t).$$

Suppose that $X(0) = 0$ and that the input $U(t)$, $t \geq 0$, is a zero mean process with correlation function

$$R_U(t_1, t_2) = \sigma^2(e^{-\alpha|t_1 - t_2|} - e^{-\alpha(t_1 + t_2)}).$$

Show that for large t the variance function $\sigma_X^2(t)$ increases asymptotically as a linear function of t, that is, that $\lim_{t\to\infty} (1/t)\sigma_X^2(t) = c$, and determine the constant c in terms of the parameters ω, σ^2, and α.

12. Find the first order densities $f_t(x)$ of the process $X(t)$, $t \geq 0$, satisfying the differential equation

$$3\dot{X}(t) + 2X(t) = W_0(t), \qquad t \geq 0,$$

where $X(0) = x_0$, a constant, and $W_0(t)$ is the standard Wiener process.

13. Suppose that the process $X(t)$, $t \geq 0$, satisfies the differential equation

$$\dot{X}(t) + AX(t) = U(t), \qquad t \geq 0, \qquad X(0) = 0,$$

where A is a random variable with $\mathrm{E}[e^{-tA}] < \infty$ and is independent of the input process $U(t)$. Show that the mean function of $X(t)$ is given by

$$\mu_X(t) = \int_0^t \psi_A(\imath(t - \tau))\mu_U(\tau)\, d\tau, \qquad t \geq 0,$$

where $\psi_A(\zeta) = \mathrm{E}[e^{\imath\zeta A}]$ is the characteristic function of A. Derive an analogous formula for the correlation function $R_X(t_1, t_2)$. (*Hint:* Condition on A.)

14. The process $X(t)$, $t \geq 0$, satisfies the differential equation

$$\dot{X}(t) = -X(t)W_0(t), \qquad t \geq 0,$$

with $X(0)$ a Gaussian random variable with mean μ_0 and variance σ_0^2 and is independent of the standard Wiener process $W_0(t)$, $t \geq 0$. Find the mean function and the correlation function of the process $X(t)$. Is this process Gaussian? (*Hint:* Recall from Example 4.1.14 that an integrated Wiener process is Gaussian with mean and covariance functions derived there.)

15. Consider a linear system with the state-space representation

$$\begin{pmatrix} \dot{x}_1(t) \\ \dot{x}_2(t) \end{pmatrix} = \begin{bmatrix} 1 & t \\ 0 & -1 \end{bmatrix} \begin{pmatrix} x_1(t) \\ x_2(t) \end{pmatrix} + \begin{pmatrix} 1 \\ 0 \end{pmatrix} u(t),$$

$$y(t) = [e^{-t2} \quad 1] \begin{pmatrix} x_1(t) \\ x_2(t) \end{pmatrix}, \qquad t \geq 0.$$

Assume that the initial state vector is a pair of standard Gaussian random variables and that the input process is stationary and Gaussian with zero mean function and correlation function

$$R_U(\tau) = \begin{cases} 1 - |\tau| & \text{if } \quad |\tau| \leq 1, \\ 0 & \text{if } \quad |\tau| > 1. \end{cases}$$

Find the state-transition function $\boldsymbol{\phi}(t_1, t_2)$ of the system and use it to calculate the variance functions of the singular and regular state processes $\mathbf{X}_s(t)$ and $\mathbf{X}_r(t)$, $t \geq 0$. (*Hint*: $\phi_{21}(t_1, t_2) = 0$.) Also calculate the variance function $\sigma_Y^2(t)$ of the output process $Y(t)$, $t \geq 0$. Can you tell without all the calculation what the output is for large t?

4.3. RANDOM SIGNALS, NOISE, AND THE WIENER INTEGRAL

Among the many reasons for introducing randomness into the description and study of engineering systems, two categories are the most prominent. One, which is quite common in almost any design, is that the system is to operate under a variety of conditions that can, at best, be characterized by the probability that particular conditions will be encountered. For example, a voice communications system must be designed to be able to transmit not one particular acoustic signal but the entire class of acoustic signals corresponding to various messages to be transmitted. Such a class of signals may then be described by the relative frequency with which these signals are being used. Thus, from the designer's point of view, the input signal constitutes a stochastic process.

The other, no less important, reason has its origin in our inability to describe complex physical systems in sufficient detail. Historically, this situation was first encountered in the development of classical statistical mechanics by Maxwell and Boltzman. However, we consider here a somewhat simpler example. Suppose we have a single communication link utilizing a laser beam with carrier frequency amplitude modulated by an acoustic signal. Since the carrier frequency is many orders of magnitude higher than the modulating frequency, we may regard the laser beam as a ray of light whose intensity varies according to the modulating signal. If the laser beam travels in a perfect vacuum, then the signal at the receiver site should be a perfect replica of the transmitted signal. However, if the beam travels in the earth's atmosphere, for instance, if the transmitter and receiver site are on mountains with a desert in between them, the beam will be partially

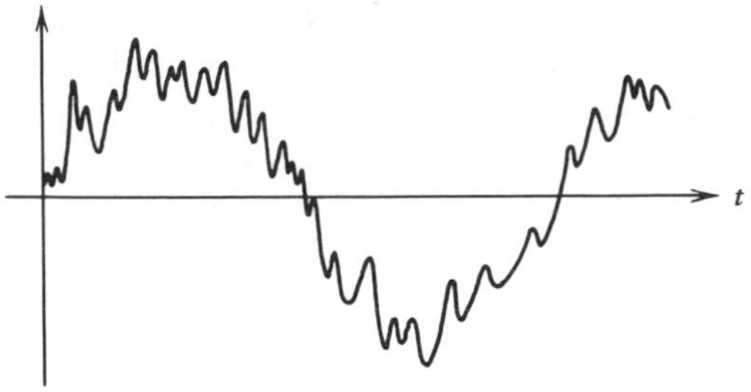

Figure 4.3.1

obstructed and scattered by dust particles, which will cause small but persistent variations of its intensity at the receiving site.

If the transmitted signal is a 1000 hertz sinusoid, the receiver will register a signal which on the oscilloscope may look like Figure 4.3.1. If we could account exactly for the effect of each individual dust particle in the path of the beam, that is, if we knew exactly when and where the particle crosses the beam's path and the change in intensity it will cause, we would, at least in principle, be able to calculate the resulting intensity at the receiver at any time t. Clearly, this is practically impossible. Thus, instead, we regard the effect of the medium between the transmitter and receiver as a source of *noise*, that is, a random signal of a very irregular nature that is superimposed on the transmitted signal.

As another example to illustrate the randomness introduced into the model as a substitute for a complex description of an otherwise deterministic situation, consider a satellite orbiting the moon. Such a satellite can be regarded as a dynamic system whose state (i.e., position and momentum) obeys a set of differential equations of celestial mechanics, and hence if its state is known at some initial time, its trajectory (state as a function of time) can be exactly predicted indefinitely into the future.

But that would require that the gravitational pull of the moon's mass be known exactly for any point in space around it. Not only may such exact knowledge not be available but also it would make the set of equations of the satellite motion so complicated as to render any exact solution impossible. Again, we would prefer to describe the irregularities in the gravitational pull as a random noise acting as a disturbance to an otherwise smooth trajectory of the satellite.

It should be noted that in both of the above examples, randomness of the first category will also be present. In the communication example the transmitted signal can only be known as belonging to a class of signals representing messages to be sent. In the satellite example the initial state is hardly ever known exactly. Rather, we only have a probability distribution over a class of initial conditions.

Nevertheless, there is an important difference between these two categories of randomness, which is perhaps best illustrated by the satellite example. In the case of an ideal homogeneous moon, that is, in the complete absence of the noise component in the gravitational force, the randomness of the satellite trajectory will only be the result of the randomness in the initial state. But for any initial state the trajectory would obey a deterministic system of differential equations. Hence, if the satellite state were observed for some time, it should be possible to continuously improve the estimate of the initial state and thus to predict with increasing accuracy the trajectory indefinitely into the future.

On the other hand, the presence of irregularities in the gravitational force will cause the trajectory to be disturbed all the time and thus will make the actual trajectory deviate more from the calculated one the further we go into the future. Briefly the presence of noise will make exact prediction of the trajectory impossible even if the initial state were known exactly.

We will refer to random processes resulting from the two categories of randomness in the model as being *random signals* and *random noise*, respectively.

In the following discussion, we attempt to decompose a random process, for example, the one modeling the actual trajectory of the satellite, into these two components, trying at the same time to characterize them. The main distinctions will be based on the idea that the noise component should be unpredictable over arbitrarily small increments of time, while the signal component should be sufficiently smooth so that its infinitesimal increments can be well approximated by differentials.

Throughout this section we shall assume that the processes we will be dealing with have *continuous sample paths*. That is, we assume that the noise does not cause a discontinuity in the sample path of the process and, in fact, that the sample paths of the noise component are themselves continuous, although otherwise highly irregular. This is clearly the case in the satellite example and also in the laser example as long as it is regarded as a continuous ray of light.†

Let

$$X(t), \qquad 0 \leq t$$

be a *sample-continuous finite-variance* stochastic process. We may think of $X(t)$ as a component of the trajectory of a dynamic system, for example, the altitude of the satellite over the moon's "sea level."

We would like now to decompose this process into a "smooth" part due to system dynamics and an "irregular" part due to the noise.

To this end, consider the forward increment at time t,

$$dX(t) = X(t + dt) - X(t)$$

of the process over a small increment of time $dt > 0$.

† However, if the beam is regarded as a stream of photons, this assumption would not apply. This case will be treated in Chapter 7.

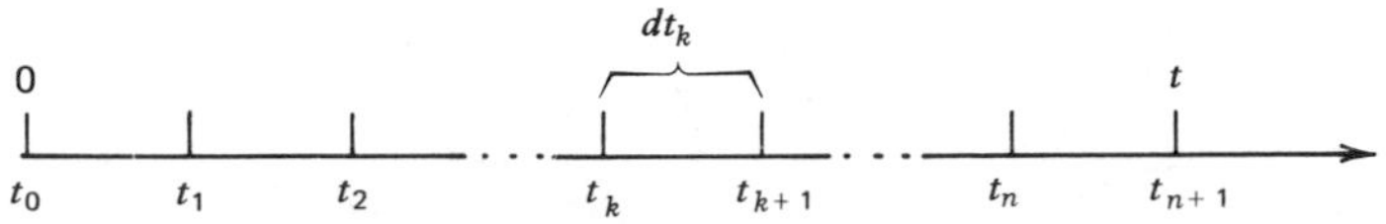

Figure 4.3.2

If we wanted to predict the value of this increment after observing the process up to time t, then the best (in the sense of mean square error) prediction would be the conditional expectation

$$E[dX(t) \mid X(t'),\ 0 \le t' \le t].$$

Recall that this conditional expectation, being a functional of the path $x(t')$, $0 \le t' \le t$, can also be regarded as a random variable dependent on the evolution of the process $X(t')$ over the time interval $0 \le t' \le t$ (see Chapter One).

Let us denote this random variable by $dY(t)$. Then the difference

$$dU(t) = dX(t) - dY(t) \tag{1}$$

represents the prediction error, or the (mean square) *unpredictable part* of the increment $dX(t)$.

Now partition the time interval $[0, t]$ into a large number of small subintervals, and denote the partition points in increasing order by $t_0 = 0, t_1, \ldots, t_n, t_{n+1} = t$ as in Figure 4.3.2. Writing equation (1) for each of these subintervals as

$$dX(t_k) = dY(t_k) + dU(t_k),$$

and adding over $k = 0, \ldots, n$, we obtain

$$X(t) - X(0) = \sum_{k=0}^{n} dY(t_k) + \sum_{k=0}^{n} dU(t_k). \tag{2}$$

What we would like to do now is to let the partition become finer and finer and thus obtain on the right-hand side of (2) two processes

$$X(t) - X(0) = Y(t) + U(t). \tag{3}$$

The process $Y(t)$ would then consist of the infinitesimal increments $dY(t_k)$, each of which would be predictable from the past $0 \le t' \le t_k$. Hence, it would be a natural candidate for the "smooth" part of $X(t)$, that is, for the random signal.

The process $U(t)$, on the other hand, would have all infinitesimal increments $dU(t_k)$ unpredictable from the past and would then be the irregular or noisy part of $X(t)$.

To see how this could be accomplished, let's look first at an increment

$$dY(t_k) = E[dX(t_k) \mid X(t'),\ 0 \le t' \le t_k] \tag{4}$$

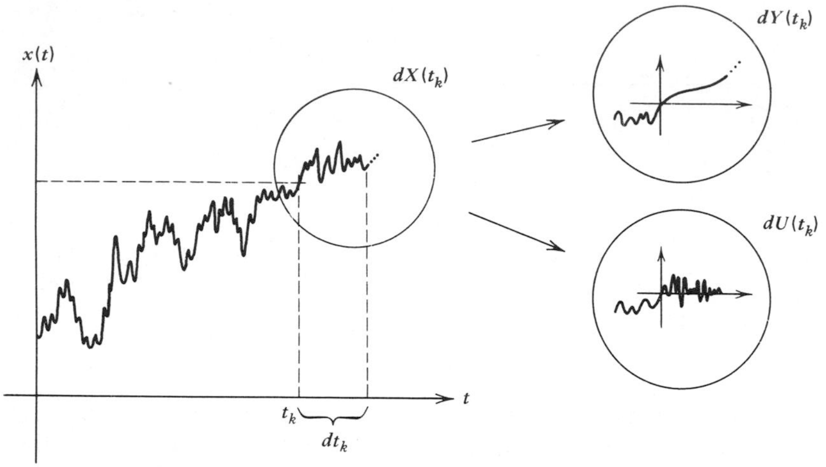

Figure 4.3.3

as pictured in Figure 4.3.3. Since we expect the predictable part to be smooth, the predicted increment (4) should, for small dt_k, be of the form

$$dY(t_k) \doteq A(t_k)\, dt_k, \tag{5}$$

where

$$A(t_k) = a(X(t'), 0 \leq t' \leq t_k) \tag{6}$$

is a functional of the past up to time t_k, that is, a random variable whose value is determined by the past $X(t')$, $0 \leq t' \leq t_k$, of the process $X(t)$. This means that the value of the process $Y(t)$ at time $t + dt$ could be approximated by

$$Y(t) + A(t)\, dt$$

with increasing accuracy as $dt \to 0$. However, $Y(t) + A(t)\, dt$ is a functional of the past $X(t')$, $0 \leq t' \leq t$, and thus is determined by the portion of this process up to time t. In other words, the process $Y(t)$ has the distinctive property that its sample path in an infinitesimal interval $[t, t + dt)$ sticking into the future can be predicted from the portion of the sample path of the process $X(t)$ running only to the present time instant t. For this reason, the process $Y(t)$ with such a property is often called *predictable* or *previsible* with respect to the process $X(t)$.

Unfortunately, another terminology is frequently used. Instead of predictable, the process $Y(t)$ is sometimes called *nonanticipating* with respect to the process $X(t)$. This sounds rather contradictory, but the reason for its use becomes clear if one considers the above argument in reverse. That is, knowing the sample path of *the process* $Y(t)$ up to the present time instant t, it is impossible to predict or anticipate the sample path of *the process* $X(t)$ in any interval $[t, t + dt)$ sticking into the future, no matter how small it is.

Remark: Remember, however, that a process can only be called predictable or nonanticipating *with respect to another process* ($X(t)$ in our discussion here). Here is where the real confusion with this terminology may arise, for if a process were to be *nonanticipating with respect to itself*, its future increments would have to be unpredictable from its own past. This would mean that the process would be a martingale or, if you prefer, pure noise. On the other hand, a process *predictable with respect to itself* would have just the opposite property—its sample paths would have to be reasonably smooth functions of time.

Thus, although the term "nonanticipating" is somewhat more frequently used, especially in the American literature, we feel that it may be less confusing to use the former term, and thus we will refer to the process $Y(t)$ *as predictable with respect to the process* $X(t)$.† Я

Let us now return to equation (5). After substitution into (2) the first sum on the right-hand side there becomes

$$\sum_{k=0}^{n} A(t_k)\, dt_k \tag{7}$$

so that, as $n \to \infty$, this sum would approximate an integral

$$Y(t) = \int_0^t A(\tau)\, d\tau. \tag{8}$$

Notice, however, that such an integral is not the same as the MS integral defined in Section 4.1, since the approximating sums (7) are more special than those in (4.1.27) of Section 4.1. This is so because the sums (4.1.27) if applied to a process called $A(t)$ would be

$$\sum_{k=0}^{n-1} A(\tau_k)\, dt_k,$$

with τ_k an arbitrary point inside the kth partition interval, while in (7) $\tau_k = t_k$ is always the *left endpoint* of the kth partition interval. A little thought reveals that this is exactly what is needed to make the process $Y(t)$ predictable with respect to the process $X(t)$.

Remark: A process $Y(t)$ predictable with respect to a process $X(t)$ need not be always expressible in the integral form (8) even if both $Y(t)$ and $X(t)$ are sample continuous. Such a case, however, is not likely to be encountered in a "nice" applied problem. Я

† With a little of the confusion thus removed, we will occasionally refer to a process as predictable as long as it is clear from the context what the reference process $X(t)$ is.

Next, let's deal with the second process on the right-hand side of equation (3), the process $U(t)$. Since this process is obtained by patching together the unpredictable increments $dU(t_k)$, we expect its sample paths to be highly irregular. Thus, we certainly cannot express its increments in the form (5) as we did for the process $Y(t)$. Nevertheless, since both $X(t)$ and $Y(t)$ are sample-continuous processes, equation (3) implies that the sample paths of $U(t)$ should also be *continuous*. From the same equation, it must be a finite-variance process, and its value at $t = 0$ must be

$$U(0) = 0.$$

Furthermore, by taking the conditional expectation

$$\mathrm{E}[dU(t) \,|\, X(t'), 0 \le t' \le t], \tag{9}$$

we see from equation (1) and the fact that

$$dY(t) = \mathrm{E}[dX(t) \,|\, X(t'), 0 \le t' \le t],$$

these conditional expectations (9) must equal zero. Consequently, applying the smoothing property of conditional expectation, we also have

$$\mathrm{E}[dU(t_k) \,|\, X(t'), 0 \le t' \le t_0] = 0, \tag{10}$$

for any $t_k \ge t_0 \ge 0$. But then if we consider a partition $t_0 < t_1 < \cdots < t_{n+1} = t$ of an interval $[t_0, t]$ and write

$$U(t) = \sum_{k=0}^{n} dU(t_k) + U(t_0) \tag{11}$$

we have by (10)

$$\mathrm{E}[U(t) \,|\, X(t'), 0 \le t' \le t_0] = \mathrm{E}[U(t_0) \,|\, X(t'), 0 \le t' \le t_0] = U(t_0), \tag{12}$$

where the last equality follows from the fact that the value of $U(t_0)$ is completely determined by the process $X(t')$ over the interval $0 \le t' \le t_0$.

To summarize, the process $U(t)$ in equation (3) must be a sample-continuous finite-variance process such that

$$U(0) = 0, \tag{13}$$

and for any $0 \le t_0 < t$

$$\mathrm{E}[U(t) | X(t'), 0 \le t' \le t_0] = U(t_0). \tag{14}$$

Note that by taking unconditional expectations on both sides of (14), we have

$$\mathrm{E}[U(t)] = \mathrm{E}[U(t_0)] \qquad \text{for any } 0 \le t_0 < t$$

and consequently by choosing $t_0 = 0$ and using (13)

$$\mathrm{E}[U(t)] = 0 \qquad \text{for all } 0 \le t.$$

Hence, $U(t)$ is a zero mean process.

Since, as already mentioned, the portion $U(t'), 0 \le t' \le t_0$ of the process $U(t)$ must be completely determined by the portion $X(t'), 0 \le t' \le t$, of the process $X(t)$, the smoothing property of conditional expectation applied to (14) gives

$$E[U(t) \mid U(t'), 0 \le t' \le t_0] = U(t_0) \tag{15}$$

for any $0 \le t_0 < t$.

Recalling now Definition 1.14 of Chapter One, we see that $U(t), t \ge 0$, is a *finite-variance zero mean martingale.* Since the martingale concepts are arrived at by generalizing the concept of independence, this result agrees quite nicely with the intuitive notion of $U(t)$ as the "noisy component" of the process $X(t)$.

Returning to equation (3)

$$X(t) - X(0) = Y(t) + U(t), \qquad t \ge 0,$$

we have obtained a decomposition of a finite-variance sample-continuous process $X(t)$ into two finite-variance processes:

$Y(t), 0 \le t$ — a sample-continuous process predictable with respect to $X(t)$,

and

$U(t), 0 \le t$ — a sample-continuous martingale with $U(0) = 0$.

Such a decomposition is known as the *Doob-Meyer-Fisk decomposition.* We will not state this result in the form of a theorem, since our derivation was mostly heuristic and we do not discuss conditions that the process $X(t)$ must satisfy for such a decomposition to exist.

Our main reason for going through this is the interpretation of such a decomposition if $X(t)$ is the state of a continuous dynamic system disturbed by noise. In such a case, the predictable part $Y(t)$ represents the smooth behavior usually determined by a set of differential equations describing the system.

The martingale $U(t)$, on the other hand, represents the noise component. Note that $U(t)$ indeed has all the properties we would intuitively expect noise to have:

1. It cannot be predicted, since the best (mean square error) prediction of its value $U(t)$ at some future time t based on the observation of its values $U(t')$ up to the present time $t_0 < t$ would be

$$E[U(t) \mid U(t'), 0 \le t' \le t_0] = U(t_0)$$

that is, just its present value $U(t_0)$.

2. Its expectation is $E[U(t)] = 0$ and, in fact, its average value over any time interval of length $\tau > 0$, no matter how short, is zero with probability one:

$$\frac{1}{\tau}\int_t^{t+\tau} U(t')\,dt' = 0 \tag{16}$$

(This follows by dividing the interval $[t, t+\tau]$ into n equal subintervals, of length τ/n, approximating the integral (16) by the average

$$\frac{1}{n}\sum_{k=1}^{n} U\left(t + \frac{k-1}{n}\tau\right)$$

and applying the (strong) law of large numbers to the martingale sequence $U(k) = U(t + (k - 1/n)\tau)$, $k = 1, 2, \ldots$.)

However, a finite-variance sample-continuous martingale has further properties that make it even more appealing as a representation of continuous noise. As we are now going to show, under moderate assumptions every such martingale is, in fact, a transformation of the standard Wiener process.

To this end, we make use of a famous theorem of Lévy and Doob:

THEOREM 4.3.1. *Let* $W(t)$, $t \geq 0$, *be a finite-variance sample-continuous martingale with* $W(0) = 0$, *such that for any* $t \geq 0$ *and* $dt > 0$

$$E[dW(t)\,|\,W(t'), 0 \leq t' \leq t] = 0 \tag{17}$$

and

$$E[(dW(t))^2\,|\,W(t'), 0 \leq t' \leq t] \doteq dt, \tag{18}$$

where $dW(t) = W(t + dt) - W(t)$ *denotes a forward increment. Then* $W(t)$, $t \geq 0$ *is the standard Wiener process.*

In order to preserve the continuity of our discussion, a sketch of the proof of this theorem is postponed to the appendix at the end of the section.

Right now, we will proceed to show how our martingale noise $U(t)$ in the Doob-Meyer-Fisk decomposition can be expressed as a transformed Wiener process.

Consider again a forward increment

$$dU(t) = U(t + dt) - U(t), \qquad dt > 0.$$

Taking the conditional expectation of its square, we have

$$E[(dU(t))^2\,|\,X(t'), 0 \leq t' \leq t] = E[U^2(t + dt)\,|\,X(t'), 0 \leq t' \leq t]$$
$$- 2E[U(t + dt)U(t)\,|\,X(t'), 0 \leq t' \leq t] + E[U^2(t)\,|\,X(t'), 0 \leq t' \leq t]. \tag{19}$$

Since $U(t)$ is completely determined by $X(t')$, $0 \le t' \le t$, we have, by using the martingale property, namely

$$\mathrm{E}[U(t+dt) \,|\, X(t'), 0 \le t' \le t] = U(t),$$

the identity

$$\begin{aligned}\mathrm{E}[U(t+dt)U(t) \,|\, X(t'), 0 \le t' \le t] &= U(t)\mathrm{E}[U(t+dt) \,|\, X(t'), 0 \le t' \le t] \\ &= U^2(t) = \mathrm{E}[U^2(t) \,|\, X(t'), 0 \le t' \le t].\end{aligned}$$

Substituting into (19), this becomes

$$\mathrm{E}[(dU(t))^2 \,|\, X(t'), 0 \le t' \le t] = \mathrm{E}[dU^2(t) \,|\, X(t'), 0 \le t' \le t], \tag{20}$$

where $dU^2(t) = U^2(t+dt) - U^2(t)$ is the forward increment of the square of the process $U(t)$.

But since $U(t)$ represents the component of the process $X(t)$ that is due to the noise, $U^2(t)$ is the energy of the noise component, and thus the increment $dU^2(t)$ is the energy increment caused by the noise during the time increment dt.

For most physical systems it is not unreasonable to assume that the average energy increment $\mathrm{E}[dU^2(t) \,|\, \cdots]$ will either be constant or at most a smoothly varying function of time.

Thus, we will assume as we did with the increments $dA(t)$ that for small dt

$$\mathrm{E}[dU^2(t) \,|\, X(t'), 0 \le t' \le t] \doteq B^2(t)\, dt, \tag{21}$$

where

$$B(t) = b(X(t'), 0 \le t' \le t) \tag{22}$$

is a nonnegative random variable that is determined by the portion of the process $X(t)$ up to the time t.

If we assume that $B(t)$ is actually strictly positive for all $t > 0$ (this assumption is not really necessary; we are making it merely for convenience) we can define a new process $W(t)$, $t \ge 0$, by

$$W(0) = U(0) = 0 \tag{23}$$

and such that its forward increments are

$$dW(t) = \frac{dU(t)}{B(t)}. \tag{24}$$

But since $B(t)$ is completely determined by the portion $X(t')$, $0 \le t' \le t$, we have for any $t_k \ge t_0 \ge 0$,

$$\mathrm{E}[dW(t_k) \,|\, X(t'), 0 \le t' \le t_k] = \frac{1}{B(t_k)} \mathrm{E}[dU(t_k) \,|\, X(t'), 0 \le t' \le t_k] = 0,$$

and hence by the smoothing property of conditional expectation we also have

$$E[dW(t_k) \mid W(t'), 0 \le t' \le t_0] = 0. \tag{25}$$

Thus, we see that the process $W(t)$, $t \ge 0$, is a martingale.

Since by setting $t_k = t_0 = t$ we get from (25)

$$E[dW(t) \mid W(t'), 0 \le t' \le t] = 0$$

and from (24), (20), and (21) for small $dt > 0$

$$\begin{aligned} E[(dW(t))^2 \mid X(t'), 0 \le t' \le t] &= \frac{1}{B^2(t)} E[(dU(t))^2 \mid X(t'), 0 \le t \le t] \\ &= \frac{1}{B^2(t)} E[dU^2(t) \mid X(t'), 0 \le t' \le t] \\ &\doteq dt \end{aligned}$$

and hence also (the smoothing property again)

$$E[(dW(t))^2 \mid W(t'), 0 \le t' \le t] \doteq dt. \tag{26}$$

Thus from (23), (25), and (26) and the fact that $W(t)$ is a martingale, we conclude by using Lévy-Doob's theorem that $W(t)$ must be a standard Wiener process $W_0(t)$, $t \ge 0$.

If we again consider a partition of the interval $[0, t]$ with partition points $t_0 = 0 < t_1 < \cdots < t_{n+1} = t$ (Figure 4.3.2) and write (24) for each subinterval as

$$dU(t_k) = B(t_k)\, dW_0(t_k) = B(t_k)(W_0(t_{k+1}) - W_0(t_k)),$$

$k = 0, \ldots, n$, then by summing these increments over k and using the fact that $U(0) = 0$, we would expect that $U(t)$ can be written in integral form

$$U(t) = \int_0^t B(\tau)\, dW_0(\tau), \tag{27}$$

the integral being defined as an MS limit of the approximating sums

$$S_n(t) = \sum_{k=0}^{n} B(t_k)(W_0(t_{k+1}) - W_0(t_k)) \tag{28}$$

as $n \to \infty$, that is, as the partition becomes finer and finer.

It should be emphasized that the integral in (27) is *not* an integral we studied in Section 4.1. It is a new type of integral, called the *stochastic Itō integral*, and it would therefore be necessary to first establish its existence, that is, to show that the approximating sums (28) do converge to a limit that is independent of the choice of the partition points $0 < t_1 < \cdots < t_n$. This program was successfully carried out by Ito and we will not repeat it here. We only wish to point out that this Ito integral

$$\int_0^t B(\tau)\,dW_0(\tau) = \text{l.i.m.} \sum_{k=0}^{n} B(t_k)(W_0(t_{k+1}) - W_0(t_k))$$

owes its existence to three crucial facts:

1. The process $B(t)$ is *predictable* with respect to the process $X(t)$.
2. The increment $dW_0(\tau)$ of the standard Wiener process is *independent* of the portion $X(t')$, $0 \le t' \le \tau$, of the process $X(t)$ and hence by (22) also of the portion $B(t')$, $0 \le t' \le \tau$, of the process $B(t)$.
3. The original process $X(t)$ is a finite-variance sample-continuous process and the portion $W_0(t')$, $0 \le t' \le t$, of *the Wiener process is determined* by the corresponding portion $X(t')$, $0 \le t' \le t$, of the process $X(t)$.

These facts are easily verified from the preceding discussion. The predictability of the process $B(t)$ follows from (21) and (22) in the same way that the predictability of the process $Y(t)$ followed from (5) and (6). Facts 2 and 3 follow from the definition of the Wiener process (23), (24), and the fact that the Wiener process has independent increments.

We will investigate some properties of the Itō integral in Section 6.5.

Remark: For readers familiar with the notion of a Stieltjes integral, we would like to emphasize that the Itō integral is *not* a stochastic version of a Stieltjes integral. In fact, the Stieltjes integral of the sample paths $b(t)$ of the process $B(t)$ with respect to the sample paths $w_0(t)$ of the standard Wiener process, that is, $\int_0^t b(\tau)\,dw_0(\tau)$, does not exist at all, since the Wiener sample paths are just too irregular. Я

Let us now return to the Doob-Meyer-Fisk decomposition (3) of the original process $X(t)$. Using (8) and (27), we rewrite it as

$$X(t) = X(0) + \int_0^t A(\tau)\,d\tau + \int_0^t B(\tau)\,dW_0(\tau) \tag{29}$$

Recalling the satellite example mentioned in the introduction, for example, if $X(t)$ is the satellite altitude (or in general a state component of a dynamic system) the terms on the right-hand side of (29) have a nice interpretation: The term $X(0)$ is just the initial altitude. The first integral describes the evolution of the component of altitude with time that is due to the smooth dynamics of satellite motion. The third term, on the other hand, represents the irregular component, which is entirely due to noise. Furthermore, we see that *any continuous noise component can be regarded as a smooth transformation* (recall that $B(t)$ behaves smoothly) *of a standard Wiener process.*

Thus the *Wiener process is, in fact, a standard form of any continuous type noise.* (We will see in Chapter Seven that the standard Poisson process plays the same central role for discontinuous, that is, impulsive type, noise.)

THE WIENER INTEGRAL. In the preceding discussion, we have shown that the "noisy component" of a finite-variance sample-continuous process can usually be represented by a zero mean finite-variance martingale

$$U(t), \qquad t \geq 0,$$

which in turn can be obtained as a transformation of a standard Wiener process $W_0(t)$, $t \geq 0$, by means of an Itō integral

$$U(t) = \int_0^t B(\tau)\, dW_0(\tau). \tag{30}$$

Let us now assume further that the noise, that is, the *martingale $U(t)$ is itself a Gaussian process.* Then the forward increments

$$dU(t) = U(t + dt) - U(t)$$

will all have Gaussian distributions, and $U(t)$, $t \geq 0$, being a martingale, will then result in these increments actually being independent.

We have used the fact that a *Gaussian martingale must have independent increments* before. Let us just remind the reader that this simply follows from the fact that by the definition of a martingale, its increments are mean square unpredictable,

$$\mathrm{E}[dU(t) \,|\, U(t'), 0 \leq t' \leq t] = 0,$$

and mean square unpredictability of a sequence of Gaussian random variables implies their independence (see Chapter One).

Let us now examine the consequences of this Gaussian assumption. Recall first from the preceding section that the unpredictable process $B(t)$ in (30) was defined by

$$\mathrm{E}[(dU(t))^2 \,|\, U(t'), 0 \leq t' \leq t] \doteq B^2(t)\, dt.$$

Then noticing that since $dU(t)$ and hence also $(dU(t))^2$ is now independent of the past $U(t')$, $0 \leq t' \leq t$, that is,

$$\mathrm{E}[(dU(t))^2 \,|\, U(t'), 0 \leq t' \leq t] = \mathrm{E}[(dU(t))^2],$$

we can conclude that in the Gaussian case $B(t)$ is no longer a random process but an ordinary deterministic function of time

$$b^2(t)\, dt \doteq \mathrm{E}[(dU(t))^2].$$

Thus (30) becomes

$$U(t) = \int_0^t b(\tau)\, dW_0(\tau), \tag{31}$$

which is a special case of the Itō integral, often called the *Wiener integral.*

Unlike the Itō integral (30) the existence and properties of the Wiener integral are relatively easy to establish.

The integral (31) has been defined as an MS limit of approximating sums

$$S_n = \sum_{k=0}^{n-1} b(t'_k)(W_0(t'_{k+1}) - W_0(t'_k)), \tag{32}$$

where $t'_0 = 0 < t'_1 < \cdots < t'_{n-1} < t'_n = t$ is a partition of the interval $[0, t]$ into n subintervals.

Consider now another partition

$$t''_0 = 0 < t''_1 < \cdots < t''_{m-1} < t''_m = t$$

of the same interval $[0, t]$ into m subintervals, and denote the corresponding approximating sum

$$S_m = \sum_{j=0}^{m-1} b(t''_j)(W_0(t''_{j+1}) - W_0(t''_j)).$$

Then the expectation of the product $S_n S_m$ equals

$$\mathrm{E}[S_n S_m] = \sum_{k=0}^{n-1} \sum_{j=0}^{m-1} b(t'_k)b(t''_j)\mathrm{E}[(W_0(t'_{k+1}) - W_0(t'_k))(W_0(t''_{j+1}) - W_0(t''_j))] \tag{33}$$

But since the standard Wiener process is a zero mean process with independent increments, the expectations behind the double summation in (33) will be zero unless the intervals†

$$I'_k = (t'_k, t'_{k+1}) \qquad \text{and} \qquad I''_j = (t''_j, t''_{j+1})$$

overlap. If they do overlap, that is, if their intersection is a nonempty interval,

$$I'_k \cap I''_j = (t_l, t_{l+1}),$$

then by the properties of the standard Wiener process

$$\mathrm{E}[(W_0(t'_{k+1}) - W_0(t'_k))(W_0(t''_{j+1}) - W_0(t''_j))] = \mathrm{E}[(W_0(t_{l+1}) - W_0(t_l))^2] = t_{l+1} - t_l.$$

Hence if

$$t_0 = 0 < t_1 < \cdots < t_{r-1} < t_r = t$$

† There is no restriction on how these intervals are constructed (they could be open, closed, etc.). Parentheses are not meant to imply open intervals here.

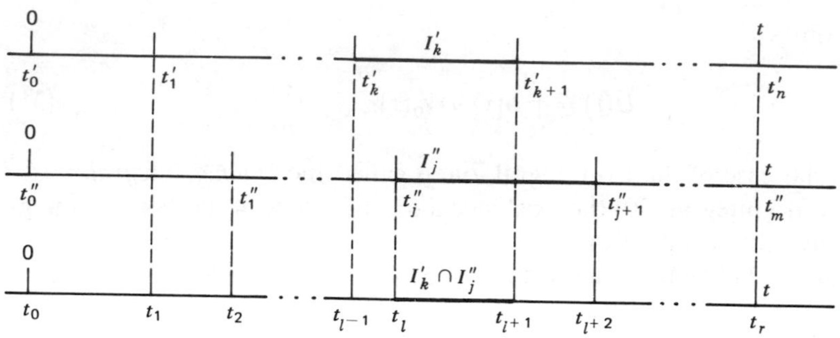

Figure 4.3.4

is a partition of the interval $[0, t]$ obtained by intersecting the two partitions as pictured in Figure 4.3.4, then (33) becomes

$$E[S_n S_m] = \sum_{l=0}^{r-1} b^2(t_l)(t_{l+1} - t_l). \tag{34}$$

Now the right-hand side of this expression is easily recognized as an approximating Riemann sum of the ordinary integral

$$\int_0^t b^2(\tau)\, d\tau.$$

Thus, if this integral exists as n and m tend to infinity, the right-hand side and so also the left-hand side of equation (34) converge, and therefore by the Loève criterion the random sums S_n converge in MS, that is, the Wiener integral

$$\int_0^t b(\tau)\, dW_0(\tau) = \text{l.i.m. } S_n \tag{35}$$

exists. Furthermore, from (35) it is easy to see that

$$E\left[\int_0^t b(\tau)\, dW_0(\tau)\right] = 0$$

and

$$E\left[\left(\int_0^{t_1} b(\tau)\, dW_0(\tau)\right)\left(\int_0^{t_2} b(\tau)\, dW_0(\tau)\right)\right] = \int_0^{\min(t_1, t_2)} b^2(\tau)\, d\tau.$$

We thus have a theorem:

THEOREM 4.3.2. *Let* $W_0(t)$, $t \geq 0$, *be a standard Wiener process, and let* $b(t)$, $t \geq 0$, *be a function of time such that for all* $t \geq 0$ *the integral*

$$\int_0^t |b(\tau)|^2\, d\tau < \infty. \tag{36}$$

Then the Wiener integral

$$U(t) = \int_0^t b(\tau)\, dW_0(\tau), \tag{37}$$

defined as an MS limit of the approximating sums (32) exists. Furthermore, the stochastic process (37)

$$U(t), \qquad t \geq 0,$$

is a sample-continuous Gaussian process with independent increments, zero mean function, and correlation function

$$R_U(t_1, t_2) = \int_0^{\min(t_1, t_2)} |b(\tau)|^2\, d\tau, \qquad 0 \leq t_1, \qquad 0 \leq t_2. \tag{38}$$

Example 4.3.1
Consider the Wiener integral

$$U(t) = \int_0^t \sin \omega\tau\, dW_0(\tau), \qquad t \geq 0.$$

Since $b(t) = \sin \omega t$ is in absolute value bounded by one, condition (36) of Theorem 4.3.2 is trivially satisfied. Hence, $U(t)$, $t \geq 0$, is a zero mean sample-continuous Gaussian process with independent increments. Its correlation function is easily obtained from (38).

$$\begin{aligned} R_U(t_1, t_2) &= \int_0^{t_m} |\sin \omega\tau|^2\, d\tau = \int_0^{t_m} \left(\frac{1}{2} - \frac{1}{2}\cos 2\omega\tau\right) d\tau \\ &= \frac{1}{2} t_m - \frac{1}{4\omega} \sin 2\omega t_m, \end{aligned}$$

where

$$t_m = \min(t_1, t_2), \qquad t_1 \geq 0, \qquad t_2 \geq 0.$$

It follows that the increments $U(t_2) - U(t_1)$, $0 \leq t_1 < t_2$, will have Gaussian distributions with zero mean and with variance

$$\begin{aligned} \sigma^2 &= R_U(t_2, t_2) - 2R_U(t_1, t_2) + R_U(t_1, t_1) \\ &= \frac{1}{2}(t_2 - t_1) - \frac{1}{4\omega}(\sin 2\omega t_2 - \sin 2\omega t_1) \\ &= \frac{1}{2}(t_2 - t_1) - \frac{1}{2\omega} \cos \omega(t_1 + t_2) \sin \omega(t_2 - t_1). \end{aligned}$$ ▲

Remark: The Wiener integral shares many properties with the ordinary integral, for instance linearity:

$$\int_0^t (\alpha_1 b_1(\tau) + \alpha_2 b_2(\tau))\, dW_0(\tau) = \alpha_1 \int_0^t b_1(\tau)\, dW_0(\tau) + \alpha_2 \int_0^t b_2(\tau)\, dW_0(\tau),$$

where α_1, α_2 are constants, and $b_1(t)$ and $b_2(t)$ are functions satisfying (36);

additivity:

$$\int_0^{t_1} b(\tau)\, dW_0(\tau) + \int_{t_1}^{t_2} b(\tau)\, dW_0(\tau) = \int_0^{t_2} b(\tau)\, dW_0(\tau),$$

$0 \le t_1 \le t_2$; and some other properties, as can be readily verified from its definition.

However, caution must be exercised in extending some other properties of an ordinary integral to the Wiener integral. For instance, $b(t) > 0$ for all $t > 0$ does not imply that $\int_0^t b(\tau)\, dW_0(\tau) > 0$, as may be seen by choosing $b(t) = 1$, that is,

$$\int_0^t 1\, dW_0(\tau)\, d\tau = W_0(t),$$

which may be negative. Я

Looking at the Gaussian process

$$U(t) = \int_0^t b(\tau)\, dW_0(\tau), \qquad t \ge 0 \tag{39}$$

we notice that it very much resembles the Wiener process itself: we have $U(0) = 0$, the process has Gaussian and independent increments, and it is also a martingale. In fact, the only difference is that the square of its *intensity*, defined in analogy with the Wiener process as the time derivative of its variance function

$$\frac{d}{dt}\sigma_U^2(t) = \frac{d}{dt}\int_0^t b^2(\tau)\, d\tau = b^2(t), \qquad t \ge 0$$

is not a constant but a function of time. For this reason, we call the process (39) defined by means of the Wiener integral an *inhomogeneous Wiener process*, and the function

$$b(t), \qquad t \ge 0$$

its *intensity function*.

Remark: More generally, an inhomogeneous Wiener process with intensity function $b(t)$, $t \ge 0$, and *drift function* $a(t)$, $t \ge 0$, would be defined by

$$U(t) = \int_0^t a(\tau)\, d\tau + \int_0^t b(\tau)\, dW_0(\tau), \qquad t \ge 0.$$

This process would have mean function

$$\mu_U(t) = \int_0^t a(\tau)\, d\tau, \qquad t \geq 0,$$

so that the drift function $a(t)$, $t \geq 0$, is assumed integrable.

In particular if both $a(t) = \mu$ and $b(t) = \sigma > 0$ are constants, we get the familiar Wiener process with drift parameter μ and intensity parameter σ. Я

The Wiener integral allows us to define a still more general class of Gaussian processes, namely

$$X(t) = \int_0^t h(t, \tau)\, dW_0(\tau), \qquad 0 \leq t, \tag{40}$$

where $h(t, \tau)$ is a function such that

$$\int_0^t |h(t, \tau)|^2\, d\tau < \infty \qquad \text{for all } 0 \leq t. \tag{41}$$

The condition (41) guarantees the existence of the integral in (40), since for each fixed t we can set $h(t, \tau) = b(\tau)$. The process (40) will clearly be a Gaussian process with $X(0) = 0$, with zero mean function, and with correlation function

$$R_X(t_1, t_2) = \int_0^{\min(t_1, t_2)} h(t_1, \tau) h(t_2, \tau)\, d\tau.$$

Furthermore, if $h(t, \tau)$ is a continuous function of the first variable t the process $X(t)$ will also be sample continuous. However, unlike the inhomogeneous Wiener process (39), the process $X(t)$ will no longer be a martingale, since for $t_1 < t_2$

$$\mathrm{E}[X(t_2) \mid X(t'), 0 \leq t' \leq t_1] = \int_0^{t_1} h(t_2, \tau)\, dW_0(\tau) \neq X(t_1)$$

so that its increments can no longer be independent. (In fact, the process being Gaussian, its increments will not even be uncorrelated.)

Example 4.3.2

Consider the Wiener integral

$$X(t) = \int_0^t h(t, \tau)\, dW_0(\tau), \qquad t \geq 0,$$

where

$$h(t, \tau) = \begin{cases} 1 & \text{if } 0 < t - \tau < 1, \\ 0 & \text{otherwise.} \end{cases}$$

Clearly, $X(t)$, $t \geq 0$, will be a Gaussian process with zero mean function and with correlation function

$$R_X(t_1, t_2) = \int_0^{\min(t_1, t_2)} h(t_1, \tau)h(t_2, \tau)\, d\tau$$

$$= \begin{cases} 1 - |t_1 - t_2| & \text{if } |t_1 - t_2| \leq 1, \\ 0 & \text{if } |t_1 - t_2| > 1, \end{cases}$$

$t_1 \geq 0$, $t_2 \geq 0$.

However, with the function $h(t, \tau)$ as above, we can also write directly

$$X(t) = \int_0^t dW_0(\tau) = W_0(t) \qquad \text{for } 0 \leq t < 1,$$

and

$$X(t) = \int_{t-1}^t dW_0(\tau) = W_0(t) - W_0(t-1) \qquad \text{for } t \geq 1.$$

In other words, setting $W_0(t) = 0$ for $t < 0$,

$$X(t) = W_0(t) - W_0(t-1), \qquad t \geq 0.$$

The reader is invited to verify that the correlation function of this process is indeed the same as the one obtained above. ▲

Example 4.3.3

If $b(t)$, $t \geq 0$, is a square-integrable function such that its time derivative $\dot{b}(t)$, $t \geq 0$, exists and is also square integrable, then the Wiener integral may also be integrated by parts:

$$\int_0^t b(\tau)\, dW_0(\tau) = b(t)W_0(t) - \int_0^t \dot{b}(\tau)W_0(\tau)\, d\tau,$$

or more generally for $0 < t_1 < t_2$,

$$\int_{t_1}^{t_2} b(\tau)\, dW_0(\tau) = b(t_2)W_0(t_2) - b(t_1)W_0(t_1) - \int_{t_1}^{t_2} \dot{b}(\tau)W_0(\tau)\, d\tau.$$

This can be proved directly from the definition of the Wiener integral. Notice that the integral on the right-hand side of these equations is an ordinary integral dealt with in Section 4.1.

As an example, let $b(t) = t$, $t \geq 0$. Then, on one hand, the process

$$U(t) = \int_0^t \tau\, dW_0(\tau), \qquad t \geq 0 \tag{42}$$

is a Gaussian process with independent increments, zero mean function, and correlation function

$$R_U(t_1, t_2) = \int_0^{\min(t_1, t_2)} \tau^2 \, d\tau = \frac{1}{3} \min(t_1^3, t_2^3).$$

On the other hand, the by-parts integration formula gives

$$U(t) = tW_0(t) - \int_0^t W_0(\tau) \, d\tau, \qquad t \geq 0. \tag{43}$$

Clearly, the right-hand side is a zero mean Gaussian process with correlation function

$$R_U(t_1, t_2) = R_X(t_1, t_2) - R_{X,Y}(t_1, t_2) - R_{Y,X}(t_1, t_2) + R_Y(t_1, t_2), \tag{44}$$

where

$$X(t) = tW_0(t) \qquad \text{and} \qquad Y(t) = \int_0^t W_0(\tau) \, d\tau.$$

To see that for, say, $0 \leq t_1 \leq t_2$, this again yields the correlation function

$$R_U(t_1, t_2) = \frac{1}{3} t_1^3,$$

note that with $0 \leq t_1 \leq t_2$

$$R_X(t_1, t_2) = t_1^2 t_2,$$

while from Example 4.1.14,

$$R_Y(t_1, t_2) = \frac{t_1^2 t_2}{2} - \frac{t_1^3}{6}, \qquad R_{Y,X}(t_1, t_2) = \frac{t_1^2 t_2}{2},$$

and

$$R_{X,Y}(t_1, t_2) = R_{Y,X}(t_2, t_1) = t_1^2\left(t_2 - \frac{t_1}{2}\right).$$

Substitution into (44) shows that indeed $R_U(t_1, t_2) = \frac{1}{3}t_1^3$, $0 \leq t_1 \leq t_2$, so that both (42) and (43) define the same Gaussian process. ▲

Example 4.3.4

It is also possible to define an improper Wiener integral, that is, one where the region of integration is an unbounded interval. In particular, we are interested in Wiener integrals of the type

$$\int_{-\infty}^t h(t, \tau) \, dW_0(\tau). \tag{45}$$

First, we need to extend the standard Wiener process $W_0(t)$ to the time domain $\mathcal{T} = (-\infty, \infty)$. This is easy, since we just take two independent standard Wiener processes $W_{0,1}(t)$, $t \geq 0$, and $W_{0,2}(t)$, $t \geq 0$, and define

$$W_0(t) = \begin{cases} W_{0,1}(t) & \text{if } t \geq 0. \\ W_{0,2}(-t) & \text{if } t < 0. \end{cases}$$

Then, provided that

$$\int_{-\infty}^{t} |h(t, \tau)|^2 \, d\tau < \infty, \tag{46}$$

we define the integral (45) as usual

$$\int_{-\infty}^{t} h(t, \tau) \, dW_0(\tau) = \underset{t_0 \to -\infty}{\text{l.i.m.}} \int_{t_0}^{t} h(t, \tau) \, dW_0(\tau).$$

If condition (46) is satisfied, this integral will define a Gaussian process

$$X(t) = \int_{-\infty}^{t} h(t, \tau) \, dW_0(\tau), \qquad -\infty < t < \infty$$

with zero mean function and with correlation function

$$R_X(t_1, t_2) = \int_{-\infty}^{\min(t_1, t_2)} h(t_1, \tau) h(t_2, \tau) \, d\tau, \qquad -\infty < t_1 < \infty, \qquad -\infty < t_2 < \infty.$$

For example, with

$$h(t, \tau) = \begin{cases} \beta e^{-\alpha(t-\tau)} & \text{if } -\infty < \tau \leq t < \infty, \\ 0 & \text{if } -\infty < t < \tau < \infty, \end{cases}$$

$\alpha > 0$, $\beta > 0$, which obviously satisfy (46), the zero mean Gaussian process

$$X(t) = \beta \int_{-\infty}^{t} e^{-\alpha(t-\tau)} \, dW_0(\tau), \qquad -\infty < t < \infty,$$

will have correlation function

$$R_X(t_1, t_2) = \beta^2 \int_{-\infty}^{\min(t_1, t_2)} e^{-\alpha(t_1 - \tau)} e^{-\alpha(t_2 - \tau)} \, d\tau = \frac{\beta^2}{2\alpha} e^{-\alpha|t_1 - t_2|}.$$

That is, $X(t)$, $-\infty < t < \infty$, will be a stationary Ornstein-Uhlenbeck process with parameters $\sigma^2 = \beta^2/2\alpha$ and α.

Thus we have obtained an important representation of a stationary zero mean Ornstein-Uhlenbeck process $X(t)$, $-\infty < t < \infty$, as a Wiener integral

$$X(t) = \sqrt{2\alpha\sigma^2} \int_{-\infty}^{t} e^{-\alpha(t-\tau)} \, dW_0(\tau). \tag{47}$$

▲

APPENDIX

SKETCH OF THE PROOF OF LÉVY-DOOB THEOREM. Let $W(t)$, $t \geq 0$, be a finite-variance sample-continuous martingale with $W(0) = 0$. Consider a time interval with endpoints t and $t + \tau$ and partition it into n equal subintervals of length τ/n, as shown in Figure 4.3.5.

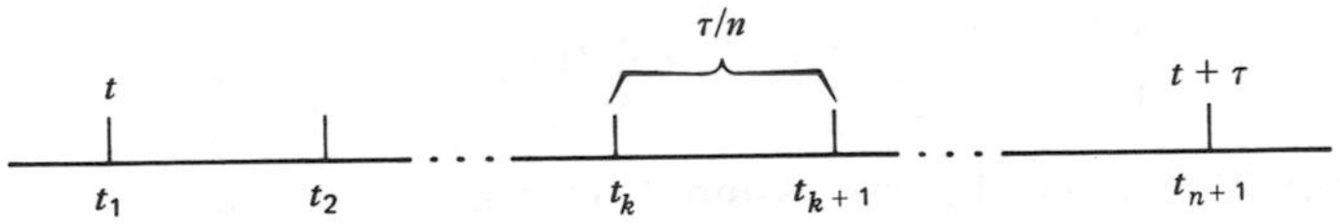

Figure 4.3.5

Consider then the conditional characteristic functions

$$\psi_k(\zeta) = \mathrm{E}[e^{\iota\zeta(W(t_k) - W(t_1))} \mid W(t'), 0 \leq t' \leq t]. \tag{A1}$$

Clearly $\psi_1(\zeta) = 1$, and upon denoting the forward increments

$$dW(t_k) = W(t_{k+1}) - W(t_k),$$

we can write, using the properties of conditional expectation,

$$\begin{aligned}\psi_{k+1}(\zeta) &= \mathrm{E}[e^{\iota\zeta(W(t_k) - W(t_1))} e^{\iota\zeta\, dW(t_k)} \mid W(t'), 0 \leq t' \leq t] \\ &= \mathrm{E}[e^{\iota\zeta(W(t_k) - W(t_1))} \mathrm{E}[e^{\iota\zeta\, dW(t_k)} \mid W(t'), 0 \leq t' \leq t_k] \mid W(t'), 0 \leq t' \leq t].\end{aligned} \tag{A2}$$

Expand the complex exponential into a Taylor series

$$e^{\iota\zeta\, dW(t_k)} = 1 + \iota\zeta\, dW(t_k) - \frac{1}{2}\zeta^2 (dW(t_k))^2 + R_k,$$

where R_k denotes the remainder of the series.

Next we want to substitute this expansion into the inner conditional expectation in the last term of (A2). Since $W(t)$, $t \geq 0$, is a martingale (cf. 17)

$$\mathrm{E}[dW(t_k) \mid W(t'), 0 \leq t' \leq t_k] = 0.$$

Further, by assumption (18) of the Lévy-Doob theorem for n large or τ/n sufficiently small

$$\mathrm{E}[(dW(t_k))^2 \mid W(t'), 0 \leq t' \leq t_k] \doteq \frac{\tau}{n}. \tag{A3}$$

Hence the inner conditional expectation in (A2) becomes

$$\mathrm{E}[e^{\iota\zeta\, dW(t_k)} \mid W(t'), 0 \leq t' \leq t_k] \doteq 1 - \frac{1}{2}\zeta^2 \frac{\tau}{n} + \mathrm{E}[R_k \mid W(t'), 0 \leq t' \leq t_k].$$

Upon substitution into (A2), we obtain using (A1)

$$\psi_{k+1}(\zeta) \doteq \left(1 - \frac{1}{2}\zeta^2 \frac{\tau}{n}\right)\psi_k(\zeta) + Q_k, \tag{A4}$$

where

$$Q_k = \mathrm{E}[e^{\iota\zeta(W(t_k) - W(t_1))}\mathrm{E}[R_k \mid W(t'), 0 \le t' \le t_k] \mid W(t'), 0 \le t' \le t]. \qquad \text{(A5)}$$

However, as is well known, the remainder of a Taylor expansion of a complex exponential is in absolute value not greater than the magnitude of the first omitted term:

$$|R_k| \le \frac{1}{3!}|\iota\zeta\, dW(t_k)|^3 = \frac{1}{3!}\zeta^3 |dW(t_k)|^3.$$

Since the process $W(t)$, $t \ge 0$, has by assumption continuous sample paths, condition (A3) seems to suggest that for small enough τ/n we should have

$$\mathrm{E}[|dW(t_k)|^3 \mid W(t'), 0 \le t' \le t] \doteq \left(\frac{\tau}{n}\right)^{3/2}.$$

This, of course, would require a rigorous proof, but taking it for granted it implies that

$$|R_k| \le \frac{1}{3!}\zeta^3\left(\frac{\tau}{n}\right)^{3/2},$$

and hence by (A5) also

$$|Q_k| \le |\psi_k(\zeta)|\frac{1}{3!}\zeta^3\left(\frac{\tau}{n}\right)^{3/2} \le \frac{1}{3!}\zeta^3\left(\frac{\tau}{n}\right)^{3/2} \qquad \text{(A6)}$$

since the absolute value of any characteristic function cannot exceed 1. Returning now to (A4), we obtain by repeated substitution for $k = 1, 2, \ldots, n$

$$\psi_{n+1}(\zeta) = \left(1 - \frac{1}{2}\zeta^2\frac{\tau}{n}\right)^n \psi_1(\zeta) + \sum_{l=0}^{n-1}\left(1 - \frac{1}{2}\zeta^2\frac{\tau}{n}\right)^l Q_{n-l}.$$

Upon letting $n \to \infty$ and using $\psi_1(\zeta) = 1$, the first term in the above expression converges to

$$e^{-\tau\zeta^2/2},$$

while the second term approaches zero, since by (A6), for $\zeta \ne 0$ and τ sufficiently small

$$\left|\sum_{l=0}^{n-1}\left(1 - \frac{1}{2}\zeta^2\frac{\tau}{n}\right)^l Q_{n-l}\right| \le \frac{1}{3!}\zeta^3\left(\frac{\tau}{n}\right)^{3/2}\sum_{l=0}^{n-1}\left(1 - \frac{1}{2}\zeta^2\frac{\tau}{n}\right)^l$$

$$= \frac{1}{3!}\zeta^3\left(\frac{\tau}{n}\right)^{3/2}\frac{1 - \left(1 - \frac{1}{2}\zeta^2\frac{\tau}{n}\right)^n}{\frac{1}{2}\zeta^2\frac{\tau}{n}} \to 0$$

as $n \to \infty$. However, for any $n = 1, 2, \ldots$, by (A1)

$$\psi_{n+1}(\zeta) = \mathrm{E}[e^{\iota\zeta(W(t+\tau)-W(t))} \mid W(t'),\, 0 \le t' \le t],$$

and so

$$\mathrm{E}[e^{\iota\zeta(W(t+\tau)-W(t))} \mid W(t'),\, 0 \le t' \le t] = e^{-\tau\zeta^2/2},$$

But this means that the conditional distribution of the increment $W(t+\tau) - W(t)$ is Gaussian with zero mean and variance τ. Thus the process $W(t)$, $t \ge 0$, being a martingale with Gaussian increments, the increments must, in fact, be independent, which together with the assumption $W(0) = 0$ implies that $W(t)$, $t \ge 0$, is indeed a standard Wiener process.

EXERCISES

Exercise 4.3

In Exercises 4.3.1 through 4.3.5, assume that $X(t)$, $t \ge 0$, $X(0) = 0$, is a finite-variance, sample-continuous process with the Doob-Meyer-Fisk decomposition

$$X(t) = U(t) + Y(t), \qquad U(t) = \int_0^t B(\tau)\, dW_0(\tau), \qquad Y(t) = \int_0^t A(\tau)\, d\tau.$$

1. Show that the mean function of the process $X(t)$ equals

$$\mu_X(t) = \int_0^t \mathrm{E}[A(\tau)]\, d\tau, \qquad t \ge 0.$$

2. Use the orthogonality principle to show that

$$\mathrm{E}[Y(t)\, dU(t)] = \mathrm{E}[A(t)\, dU(t)] = 0.$$

Use this result to argue that for small $dt > 0$

$$\mathrm{E}[(dX(t))^2] \doteq \mathrm{E}[B^2(t)]\, dt.$$

3. Establish the identity

$$\mathrm{E}\left[\left(\int_0^t A(\tau)\, d\tau\right)^2\right] = 2\int_0^t \mathrm{E}[A(\tau)(X(t) - X(\tau))]\, d\tau.$$

(*Hint:* Write $\left(\int_0^t A(\tau)\, d\tau\right)^2 = 2\int_0^t A(\tau)\left(\int_\tau^t A(\tau')\, d\tau'\right) d\tau$.)

4. Show that

$$\mathrm{E}[X^2(t)] = \int_0^t \mathrm{E}[B^2(\tau)]\, d\tau + 2\int_0^t \mathrm{E}[A(\tau)X(\tau)]\, d\tau.$$

(*Hint:* Write $X^2(t) \doteq \left(\sum_k dX(t_k)\right)^2 = \sum_k dX(t_k)^2 + 2\sum_k X(t_k)\, dX(t_k)$

and use Exercise 2.)

5. Show that the process $X(t)$, $t \geq 0$, cannot have an MS derivative unless $B(t) = 0$ for all $t \geq 0$.
6. Suppose that a finite-variance process $X(t)$, $t \geq 0$, is a linear combination of sine waves,

$$X(t) = \sum_{k=1}^{n} Z_k \sin(\Omega_k t + \Theta_k)$$

with random amplitudes, frequencies, and phases. Give a heuristic argument in favor of the proposition that such a process should be predictable with respect to itself and hence represent a pure random signal.
7. For each of the functions $b(t)$, $t \in \mathcal{T}$, below, check whether the Wiener integral

$$U(t) = \int_0^t b(\tau)\, dW_0(\tau)$$

exists and if so find the correlation function R_U of the process $U(t)$, $t \in \mathcal{T}$.

(a) $b(t) = \dfrac{1}{(1+t)\ln(1+t)}, \qquad t \geq 0,$

(b) $b(t) = \dfrac{1}{\sqrt{1-t^2}}, \qquad 0 \leq t < 1,$

(c) $b(t) = 2\cos^2 t, \qquad t \geq 0,$

(d) $b(t) = \dfrac{1}{\sqrt{t+1}}, \qquad t \geq 0$

(e) $b(t) = e^{-(2+i)t}, \qquad t \geq 0$ (complex-valued process)

8. Find the covariance matrix of the Gaussian random vector $\mathbf{X}' = (X_1, X_2, X_3)$ whose components are defined as Wiener integrals

$$X_j = \int_0^1 \tau^j\, dW_0(\tau), \qquad j = 1, 2, 3.$$

9. Find the mean function and the correlation function for each of the Gaussian processes $X(t)$, $t \in \mathcal{T}$, below:

(a) $X(t) = \displaystyle\int_0^t (t-\tau)e^{-\alpha(t-\tau)}\, dW_0(\tau), \qquad t \geq 0, \qquad \alpha > 0.$

(b) $X(t) = \displaystyle\int_0^t \cos\tau\, d\tau + \int_0^t \frac{t-\tau}{t+\tau}\, dW_0(\tau), \qquad t \geq 0.$

(c) $X(t) = \displaystyle\int_{t-1}^t (t-\tau)\, dW_0(\tau), \qquad -\infty < t < \infty.$

(d) $X(t) = \displaystyle\int_0^t \cos(t\tau)\, dW_0(\tau), \qquad t \geq 0.$

(e) $X(t) = \displaystyle\int_{-\infty}^t \frac{1}{1+\tau^2}\, d\tau - \int_{-\infty}^{\infty} e^{-|t-\tau|}\, dW_0(\tau), \qquad -\infty < t < \infty.$

4.4. WHITE GAUSSIAN NOISE

Let us return for a moment to the basic result of Section 4.3, namely that under certain regularity conditions a finite-variance sample-continuous process $X(t)$, $t \geq 0$, can be decomposed into a "smooth" signal component and a continuous but highly erratic "noisy" component, in particular

$$X(t) - X(0) = \int_0^t A(\tau)\, d\tau + \int_0^t B(\tau)\, dW_0(\tau), \tag{1}$$

the continuous noise component being a transformation of the standard Wiener process $W_0(t)$.

Looking at equation (1) or at its approximate form written in terms of forward increments

$$dX(t) = A(t)\, dt + B(t)\, dW_0(t)$$

we are naturally tempted to rewrite it as a differential equation

$$\dot{X}(t) = A(t) + B(t)\dot{W}_0(t) \tag{2}$$

the dots indicating time derivatives.

Such a differential equation would have particular appeal if the predictable processes $A(t)$ and $B(t)$ were actually memoryless transformations of the process $X(t)$,

$$A(t) = a(t, X(t)), \qquad B(t) = b(t, X(t)),$$

where $a(t, x)$ and $b(t, x)$ were reasonably smooth functions of the time t and the state x. Then the differential equation (2) would have the form of the state-input equation

$$\dot{X}(t) = a(t, X(t)) + b(t, X(t))\dot{W}_0(t) \tag{3}$$

of a state-space representation of a general dynamic system with a "noise $\dot{W}_0(t)$" acting as the input or a random disturbance.

Unfortunately, as we have seen, the standard Wiener process is not differentiable with respect to time, and hence no such "noise $\dot{W}_0(t)$" can exist.

From a strictly mathematical point of view we would thus be left with two possibilities, either to stick to the integral version, that is, in terms of dynamic systems, to work with integral equations

$$X(t) - X(0) = \int_0^t a(\tau, X(\tau))\, d\tau + \int_0^t b(\tau, X(\tau))\, dW_0(\tau)$$

instead of the differential equation (3), or go back to the drawing board and redefine the concept of a random variable and a stochastic process in a more abstract setting.†

† This, indeed, has been done, and it leads to stochastic counterparts of generalized functions like the Dirac delta function.

But we need not be discouraged so easily. After all, we commonly work with the Dirac delta function as a representation of a unit impulse; in fact, we even define the impulse-response function of a linear system as a response of the system to a delta function input. And yet, the delta function does not exist in the strict mathematical sense as a function nor could there be a physical impulse of infinite height, zero width, and unit area.

Thus we introduce a symbolic process $\dot{W}_0(t)$, called *white Gaussian noise*† as a stochastic counterpart of the Dirac delta function. Since we cannot define it (at least not as a stochastic process in the sense used in this book), we instead specify the rules for its manipulation. We do this as follows:

> White Gaussian noise $\dot{W}_0(t)$ on the time domain $\mathcal{T} = [0, \infty)$ or $(-\infty, \infty)$ is a symbolic process that, as far as linear operations are concerned, can be manipulated as if it were a stationary Gaussian process with mean function
>
> $$\mu_{\dot{W}_0}(t) = 0, \qquad t \in \mathcal{T}$$
>
> and correlation function
>
> $$R_{\dot{W}_0}(t_1, t_2) = \delta(t_1 - t_2), \qquad t_1 \in \mathcal{T}, \qquad t_2 \in \mathcal{T}, \tag{4}$$
>
> where δ is the Dirac delta function.

Example 4.4.1

Notice that the "correlation function" (4) is exactly what would be obtained if we calculated the correlation function of the "MS derivative" of a standard Wiener process

$$W_0(t), \qquad 0 \leq t$$

with the aid of the Dirac delta function. Differentiating the function

$$R_{W_0}(t_1, t_2) = \begin{cases} t_1 & \text{if } 0 \leq t_1 \leq t_2 \\ t_2 & \text{if } 0 \leq t_2 \leq t_1 \end{cases}$$

with respect to t_1, we would have

$$\frac{\partial R_{W_0}(t_1, t_2)}{\partial t_1} = \begin{cases} 1 & \text{if } 0 \leq t_1 < t_2, \\ 0 & \text{if } 0 \leq t_2 < t_1, \end{cases} \tag{5}$$

except at $t_1 = t_2$, where the derivative does not exist. That is, for t_1 fixed, the function (5) as a function of t_2 would be a unit step function as pictured in Figure

† The reason for calling the noise white will become apparent in Section 5.2.

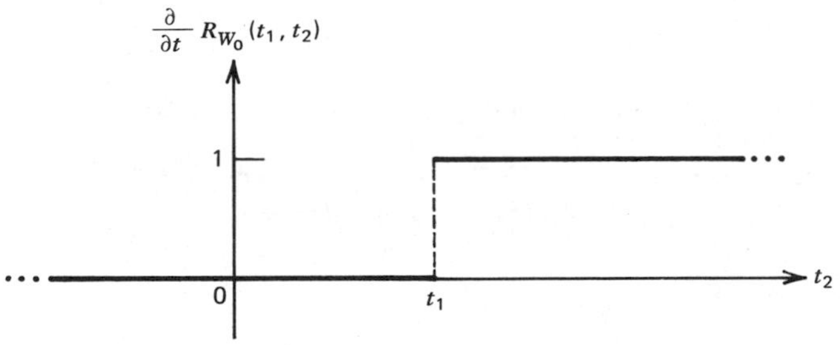

Figure 4.4.1

4.4.1. Hence, formally differentiating this function with respect to t_2, we would get

$$\frac{\partial^2}{\partial t_1\, \partial t_2} R_{W_0}(t_1, t_2) = \delta(t_1 - t_2),$$

that is, the function (4).

We also see that the "cross-correlation function" between $\dot{W}_0(t)$ and $W_0(t)$ should then be given by (5), that is,

$$R_{\dot{W}_0, W_0}(t_1, t_2) = \begin{cases} 1 & \text{if } 0 \le t_1 < t_2, \\ 0 & \text{if } 0 \le t_2 \le t_1, \end{cases}$$

where we have chosen its value to be zero at the discontinuity point $t_1 = t_2$. (This choice is dictated by the requirement that since $W_0(0) = 0$, we wish to have $R_{\dot{W}_0}(0, 0) = \mathrm{E}[\dot{W}_0(0) W_0(0)] = 0$.)

It follows that white Gaussian noise $\dot{W}_0(t)$ at the time $t \ge t_0$ behaves as if it were independent of the past and present of $W_0(t')$, $0 \le t' \le t_0$. ▲

Now we shall investigate these rules for white Gaussian noise. Consider, for instance, the formal integral

$$W_0(t) = \int_0^t \dot{W}_0(\tau)\, d\tau, \qquad t \ge 0,$$

which, if we are to be successful, should give us back the standard Wiener process. Since an integral is a linear operation, $W_0(t)$ should be a Gaussian process. Clearly

$$W_0(0) = \int_0^0 \dot{W}_0(\tau)\, d\tau = 0, \tag{6}$$

and using the formulas from Section 4.1, we have

$$\mu_{W_0}(t) = \int_0^t \mu_{\dot{W}_0}(\tau)\, d\tau = \int_0^t 0\, d\tau = 0, \tag{7}$$

and for $0 \le t_1 < t_2$

$$R_{W_0}(t_1, t_2) = \int_0^{t_1} \int_0^{t_2} R_{\dot{W}_0}(\tau_1 - \tau_2)\, d\tau_1\, d\tau_2$$

$$= \int_0^{t_1} \int_0^{t_2} \delta(\tau_1 - \tau_2)\, d\tau_1\, d\tau_2 = \int_0^{t_1} d\tau_1 = t_1,$$

since for $0 \le \tau_1 \le t_1$

$$\int_0^{t_2} \delta(\tau_1 - \tau_2)\, d\tau_2 = \int_{\tau_1 - t_2}^{\tau_2} \delta(\tau)\, d\tau = 1 \qquad \text{since } \tau_1 - t_2 < 0 < \tau_2 .$$

Similarly for $0 \le t_2 < t_1$, we get $R_{W_0}(t_1, t_2) = t_2$, so that

$$R_{W_0}(t_1, t_2) = \min(t_1, t_2). \tag{8}$$

But a Gaussian process satisfying (6), (7), and (8) is, by definition, indeed a standard Wiener process.

More generally, if we take

$$U(t) = \int_0^t b(\tau)\dot{W}_0(\tau)\, d\tau, \qquad 0 \le t$$

where

$$\int_0^t b^2(\tau)\, d\tau < \infty$$

we get an inhomogeneous Wiener process with intensity function $b(t)$.

In fact, white Gaussian noise allows us to write the Wiener integral

$$X(t) = \int_0^t h(t, \tau)\, dW_0(\tau) = \int_0^t h(t, \tau)\dot{W}_0(\tau)\, d\tau, \tag{9}$$

and manipulate it as an ordinary integral. (Recall, however, that for the Wiener integral to exist the function $h(t, \tau)$ must be square integrable.)

Thus the process $X(t)$ will be a Gaussian process with zero mean function and correlation function

$$R_X(t_1, t_2) = \int_0^{t_1} \int_0^{t_2} h(t_1, \tau_1)\, \delta(\tau_1 - \tau_2) h(t_2, \tau_2)\, d\tau_1\, d\tau_2 \tag{10}$$

$$= \int_0^{\min(t_1, t_2)} h(t_1, \tau) h(t_2, \tau)\, d\tau,$$

as it should.

So far, all three processes, $W_0(t)$, $U(t)$, and $X(t)$, in the above examples were ordinary processes. We may, however, also use white Gaussian noise $\dot{W}_0(t)$ even in such expressions as

$$Y(t) = X(t) + \dot{W}_0(t), \tag{11}$$

where $X(t)$ is an ordinary finite-variance process.

If $X(t)$ is a Gaussian process with mean function and correlation function $\mu_X(t)$ and $R_X(t_1, t_2)$, respectively, and is also considered independent of the white Gaussian noise $\dot{W}_0(t)$, then according to our rule for manipulating the white Gaussian noise, the process $Y(t)$ should again be Gaussian with mean function

$$\mu_Y(t) = \mu_X(t)$$

and correlation function

$$R_Y(t_1, t_2) = R_X(t_1, t_2) + \delta(t_1 - t_2).$$

But the presence of the delta function in this expression shows that $Y(t)$ cannot be an ordinary process.

We will refer to "processes" that are written as linear combinations of ordinary finite-variance (Gaussian or not) processes and white Gaussian noises as *generalized processes.*

In particular, the process

$$\dot{U}(t) = b(t)\dot{W}_0(t),$$

where $b(t)$ is positive and square integrable over the time domain considered, is the simplest example of such a generalized process. Since it appears to be a formal derivative of the inhomogeneous Wiener process $U(t) = \int_0^t b(\tau)\, dW_0(\tau)$, we may refer to this process as an *inhomogeneous white Gaussian noise with intensity* $b(t)$. Note that the correlation function of a generalized process will always contain delta functions. If the ordinary processes involved in the linear combination are all Gaussian, we will regard the resulting generalized process as Gaussian. If a finite-variance process other than Gaussian is involved, for example, if $X(t)$ in (11) is not Gaussian, then the notion of a probability law for such a generalized process becomes meaningless, and only the mean and the correlation function may be used. Of course, we can always consider an integrated version of a generalized process, which, if the integrals exist, will be an ordinary process. For instance, the integrated version of the generalized process (11)

$$\int_0^t Y(\tau)\, d\tau = \int_0^t X(\tau)\, d\tau + W_0(t)$$

is a sum of the process $\int_0^t X(\tau)\, d\tau$ and a standard Wiener process.

Example 4.4.2

Consider a generalized process

$$X(t) = W_0(t) + \alpha\dot{W}_0(t), \qquad t \geq 0,$$

with α a constant. Since it is defined by means of linear operations, it could be regarded as if it were a Gaussian process with zero mean function and with the "correlation function"

$$R_X(t_1, t_2) = R_{W_0}(t_1, t_2) + \alpha R_{W_0, \dot{W}_0}(t_1, t_2) + \alpha R_{\dot{W}_0, W_0}(t_1, t_2) + \alpha^2 R_{\dot{W}_0}(t_1, t_2).$$

Using the results of Example 4.4.1, this expression equals

$$R_X(t_1, t_2) = \min(t_1, t_2) + \alpha, \qquad 0 \le t_1, \qquad 0 \le t_2, \qquad t_1 \ne t_2 \tag{12}$$

and

$$R_X(t, t) = t + \alpha^2 \, \delta(0), \qquad 0 \le t.$$

The reader should verify that by formally differentiating the ordinary process

$$Y(t) = \int_0^t X(\tau)\, d\tau = \int_0^t W_0(\tau)\, d\tau + \alpha W_0(t), \qquad t \ge 0$$

the "correlation function"

$$R_{\dot{Y}}(t_1, t_2) = \frac{\partial^2}{\partial t_1 \, \partial t_2} R_Y(t_1, t_2)$$

again yields the expression (12). ▲

Remark: In replacing $dW_0(\tau)$ by $\dot{W}_0(\tau)\, d\tau$ in a Wiener integral and treating the result as an ordinary stochastic integral, caution must be exercised with the limits of integration. For instance, if we write

$$W_0(t) = \int_0^t dW_0(\tau) = \int_0^t \dot{W}_0(\tau)\, d\tau$$

and formally calculate the covariance

$$\mathrm{E}[W_0(t)\dot{W}_0(t)] = \int_0^t \mathrm{E}[\dot{W}_0(\tau)\dot{W}_0(t)]\, d\tau = \int_0^t \delta(\tau - t)\, d\tau = \frac{1}{2},$$

which is incorrect, since $\mathrm{E}[W_0(t)\, dW_0(t)] = 0$ by independence of increments of the Wiener process. Strictly speaking, we should write the Wiener integral

$$\int_0^t b(\tau)\, dW_0(\tau) \qquad \text{as} \qquad \int_0^{t-} b(\tau)\dot{W}_0(\tau)\, d\tau,$$

rather than just $\int_0^t b(\tau)\dot{W}_0(\tau)\, d\tau$. Я

As another example involving the concept of white Gaussian noise, consider a first order linear differential equation, the so-called *Langevin equation*

$$\dot{X}(t) + \alpha X(t) = \beta \dot{W}_0(t), \qquad t \ge 0, \qquad X(0) = 0, \tag{13}$$

where $\alpha > 0$ and $\beta > 0$ are constants.

Now, a deterministic first order linear differential equation

$$\dot{x}(t) + \alpha x(t) = g(t), \qquad t \geq 0$$

with initial condition $x(0) = 0$ has the solution

$$x(t) = \int_0^t e^{-\alpha(t-\tau)} g(\tau)\, d\tau, \qquad t \geq 0,$$

and hence formally we should also have

$$X(t) = \int_0^t e^{-\alpha(t-\tau)} \beta \dot{W}_0(\tau)\, d\tau, \qquad t \geq 0 \tag{14}$$

for our differential equation (13) with a forcing function $g(t) = \beta \dot{W}_0(t)$.

However, the above integral that has the form (9)

$$X(t) = \int_0^t h(t, \tau) \dot{W}_0(\tau)\, d\tau,$$

with $h(t, \tau) = \beta e^{-\alpha(t-\tau)}$, in other words, that of the Wiener integral

$$X(t) = \int_0^t h(t, \tau)\, dW_0(\tau),$$

since the function $\beta e^{-\alpha(t-\tau)}$ is clearly square-integrable,

$$\int_0^t (\beta e^{-\alpha(t-\tau)})^2\, d\tau < \infty.$$

It then follows that the solution $X(t)$, $t \geq 0$, of our differential equation (13) should be a Gaussian process with zero mean function and with correlation function given by (10):

$$\begin{aligned} R_X(t_1, t_2) &= \int_0^{\min(t_1, t_2)} \beta^2 e^{-\alpha(t_1-\tau)} e^{-\alpha(t_2-\tau)}\, d\tau \\ &= \frac{\beta^2}{2\alpha}(e^{-\alpha|t_1-t_2|} - e^{-\alpha(t_1+t_2)}), \end{aligned} \tag{15}$$

$t_1 \geq 0$, $t_2 \geq 0$. In other words, the process $X(t)$, $t \geq 0$, is an Ornstein-Uhlenbeck process with parameters $\sigma^2 = \beta^2/2\alpha$ and 2α.

We could also calculate the cross-correlation function between $X(t)$ and $\dot{W}_0(t)$. Using (14) and our rule for manipulating white Gaussian noise, we get

$$
\begin{aligned}
&R_{X,\dot{W}_0}(t_1, t_2) \\
&= E[X(t_1)\dot{W}_0(t_2)] = E\left[\int_0^{t_1} \beta e^{-\alpha(t_1-\tau)}\dot{W}_0(\tau)\,d\tau\,\dot{W}_0(t_2)\right] = \int_0^{t_1} \beta e^{-\alpha(t_1-\tau)} R_{\dot{W}_0}(\tau, t_2)\,d\tau \\
&= \int_0^{t_1} \beta e^{-\alpha(t_1-\tau)}\,\delta(\tau - t_2)\,d\tau = \begin{cases} \beta e^{-\alpha(t_1-t_2)} & \text{if } 0 \le t_2 < t_1, \\ 0 & \text{if } 0 \le t_1 \le t_2. \end{cases}
\end{aligned}
$$

Since both $X(t)$ and $\dot{W}(t)$ are Gaussian, it follows that for $0 \le t_1 \le t_2$, $X(t_1)$ and $\dot{W}_0(t_2)$ should be independent.

To verify that we have indeed obtained a correct solution of the differential equation (13), recall that a rigorous version of this equation, that is, one that would not contain any generalized process, would be an integral equation

$$X(t) + \int_0^t \alpha X(\tau)\,d\tau = \beta W_0(t), \qquad t \ge 0. \tag{16}$$

If now $X(t)$, $t \ge 0$, is a zero-mean Ornstein-Uhlenbeck process with correlation function (15) then the left-hand side of (16) is a zero-mean Gaussian process with correlation function

$$R_X(t_1, t_2) + \alpha\int_0^{t_1} R_X(\tau, t_2)\,d\tau + \alpha\int_0^{t_2} R_X(t_1, \tau)\,d\tau + \alpha^2\int_0^{t_1}\int_0^{t_2} R_X(\tau_1, \tau_2)\,d\tau_1\,d\tau_2.$$

The reader may verify for himself that this expression equals

$$\beta^2 \min(t_1, t_2),$$

which is the correlation function of $\beta W_0(t)$, $t \ge 0$. Thus both sides of equation (16) represent the same Gaussian process and so the Ornstein-Uhlenbeck process $X(t)$, $t \ge 0$, is indeed the solution of this equation.

In other words, with our rule for white Gaussian noise, the illegitimate differentiation of the integral equation (16) and subsequent formal solution of the differential equation with a symbolic process $\dot{W}_0(t)$ on the right-hand side yields at the end a completely rigorous solution. Clearly, if we were to solve the integral equation (16) without differentiating it, the solution would be harder to find.

It should be emphasized that in all the examples above, the operations performed on white Gaussian noise were *linear*, and our rule specifically stated that it *applies only to linear operations*. We will, however, consider some nonlinear operations involving white Gaussian noise in Chapter Six.

Although we have apparently been successful in introducing the symbolic process $\dot{W}_0(t)$ such that the formal manipulation with it does not lead to a contradiction, the big question remains as to whether it is meaningful to model

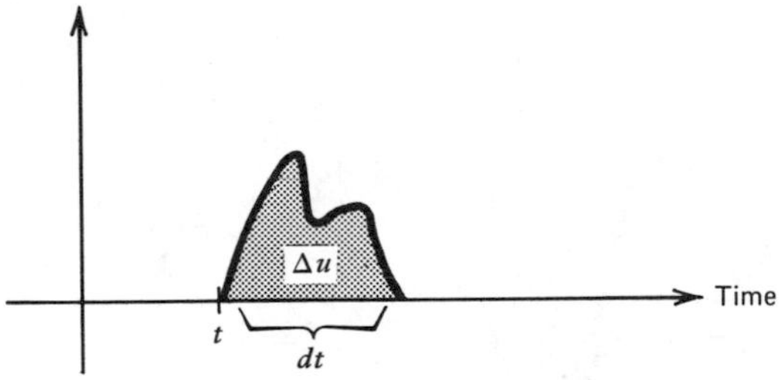

Figure 4.4.2

an actual physical noise by means of such a symbolic process, at least as a limiting or idealized case.

Since white Gaussian noise $\dot{W}_0(t)$ is formally a "derivative" of a standard Wiener process, which can be derived as a limit of a symmetric Bernoulli random walk, let us begin with a similar situation in a little more physical context.

Consider a source that is producing a continuous stream of electric charges. Suppose that the amount of charge Δu delivered during a time interval $(t, t + \Delta t)$ can be represented as an area of an impulse as depicted in Figure 4.4.2. During the delivery of this charge the average current is

$$\frac{\Delta u}{\Delta t}$$

and thus the energy dissipated in a 1-ohm resistor by this charge would be

$$\left(\frac{\Delta u}{\Delta t}\right)^2 \Delta t. \tag{17}$$

Let us now take a fixed time interval, say $[0, t]$, and divide this interval into n subintervals

$$I_k = [t_k, t_{k+1}), \qquad t_k = k\,\Delta t, \quad k = 0, \ldots, n-1,$$

so that each subinterval has length,

$$\Delta t = \frac{t}{n}$$

and let Δu_k be the amount of charge delivered during the interval I_k. We thus have a stream of consecutive charges

$$\Delta u_0, \Delta u_1, \ldots, \Delta u_{n-1}$$

and let us assume that the polarities (signs) of these charges are random and independent and that each charge is equally likely to be positive or negative.

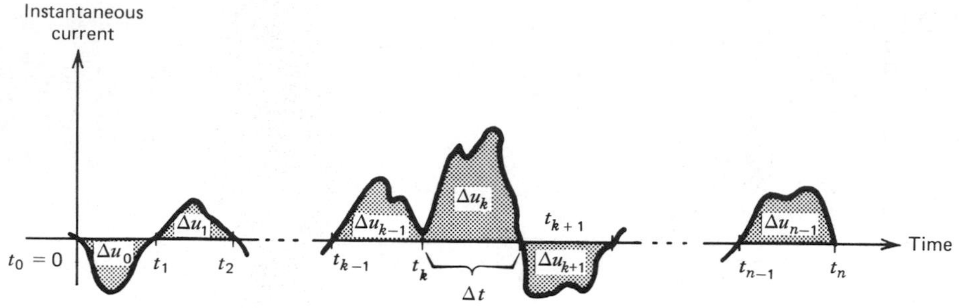

Figure 4.4.3

This can be conveniently expressed by multiplying each amount of charge Δu_k by a random variable X_k with values ± 1, the random variables

$$X_0, X_1, \ldots, X_{n-1}$$

being independent identically distributed with

$$P(X_k = +1) = P(X_k = -1) = \frac{1}{2}. \tag{18}$$

The resulting stream of charges with random polarities can now be depicted as in Figure 4.4.3. The total charge delivered during the time interval $[0, t]$ will then be given by the sum

$$U_n(t) = \sum_{k=0}^{n-1} X_k \Delta u_k \tag{19}$$

where the summands $X_k \Delta u_k$ are now independent random variables with means

$$\mathrm{E}[X_k \Delta u_k] = \Delta u_k \mathrm{E}[X_k] = 0$$

and variances

$$\mathrm{Var}[X_k \Delta u_k] = (\Delta u_k)^2 \mathrm{Var}[X_k] = (\Delta u_k)^2.$$

Suppose now that we let $n \to \infty$, that is, $\Delta t \to 0$. To keep the total charge finite, we must then allow the individual charges also to decrease to zero

$$\Delta u_k \to 0, \qquad k = 0, \ldots, n$$

but we may insist that the energies (17) dissipated by the individual charges remain nonzero and finite,

$$\left(\frac{\Delta u_k}{\Delta t}\right)^2 \Delta t = b^2(t_k),$$

where $b(t)$ is some fixed function such that

$$\int_0^t b^2(\tau)\, d\tau < \infty. \tag{20}$$

Then the variance of the total charge $U_n(t)$ in (19) will be

$$\sum_{k=0}^{n-1} (\Delta u_k)^2 = \sum_{k=0}^{n-1} b^2(t_k)\, \Delta t,$$

which in the limit $n \to \infty$ would become the integral (20). But then the total charge $U_n(t)$ will converge in MS to a random variable $U(t)$, which, according to the Central limit theorem, will be Gaussian with mean $\mu(t) = 0$ and variance

$$\sigma^2(t) = \int_0^t b^2(\tau)\, d\tau.$$

If we now let the time t vary over some interval $0 \leq t$, then $U(t)$ will be a stochastic process. Clearly this process will have

$$U(0) = 0$$

and by the construction above its forward increments

$$U(t_1) - U(0),\ U(t_2) - U(t_1), \ldots, \qquad 0 < t_1 < t_2 < \cdots$$

will be independent Gaussian random variables with zero mean and variances

$$\int_0^{t_1} b^2(\tau)\, d\tau,\ \int_{t_1}^{t_2} b^2(\tau)\, d\tau, \ldots,$$

respectively. In other words, the process

$$U(t), \qquad 0 \leq t,$$

will be an inhomogeneous Wiener process with intensity function $b(t)$.

Now, from the physical point of view, we can regard this process as representing the cumulative charge as a function of time, the square of the intensity $b^2(t)$ being the energy of the infinitesimal charge at time t. If, for instance, the source of these infinitesimal charges were connected to a capacitor, then the sample paths of $U(t)$ would be proportional to the variation of voltage on the capacitor's terminals, as pictured in Figure 4.4.4. But as the source of charges in the above picture is thought of as a source of noise, it follows that the voltage

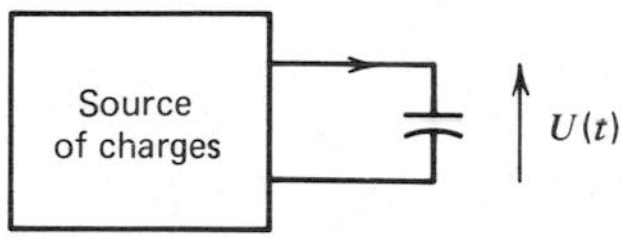

Figure 4.4.4

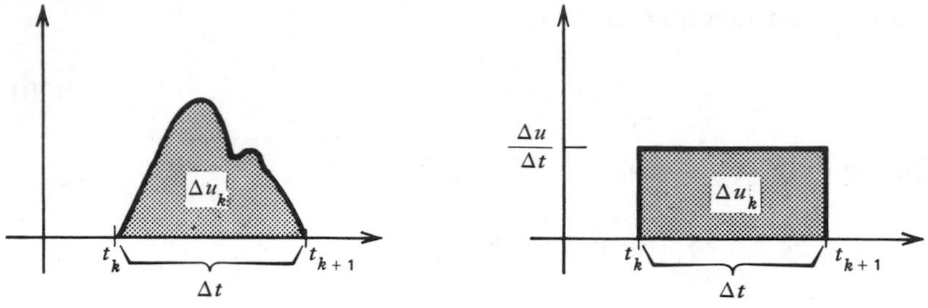

Figure 4.4.5

$U(t)$, that is, the Wiener process, should be more properly called an integral of the noise process rather than noise process itself. (Recall that the capacitor in the above circuit acts as an ideal integrator, since the source is assumed to have zero internal resistance.)

Thus, we are again led to our original temptation to define the noise itself as a time derivative $\dot{W}_0(t)$ of a Wiener process. In our context, however, the noise should then be the *current* rather than the charge. Going back to our stream of charges in Figure 4.4.3 the plot of the current as a function of time is, of course, the heavy line drawn there, and as the polarities X_k of the pulses are random, the current process, say $\dot{U}_n(t)$, is a genuine stochastic process with probability law depending on the shape of the pulses. The mean function of this process will be identically zero because of (18), and the correlation function $R_{\dot{U}_n}(t_1, t_2)$ will be zero as long as t_1 and t_2 do not fall into the same interval I_k.

For t_1, t_2 both in the same interval I_k, the correlation function depends on the shape of the pulse, but if we replace the pulse by a rectangular pulse with the same area, that is, representing the same amount of charge Δu_k as pictured in Figure 4.4.5, the correlation function will be equal to

$$\left(\frac{\Delta u_k}{\Delta t_k}\right)^2$$

If now $n \to \infty$, that is, $\Delta t \to 0$ and $\Delta u_k \to 0$ such that (17) remains constant, then

$$\left(\frac{\Delta u_k}{\Delta t}\right) = \frac{b^2(t_k)}{\Delta t} \to \infty.$$

Thus, in the limit the correlation function would become

$$\lim_{n\to\infty} R_{\dot{U}_n}(t_1, t_2) = \begin{cases} \infty & \text{if} \quad t_1 = t_2, \\ 0 & \text{if} \quad t_1 \neq t_2. \end{cases}$$

However, note that for $t_1 \in I_k$ and any n the integral

$$\int_0^t R_{\dot{U}_n}(t_1, t_2)\, dt_2 = \left(\frac{\Delta u_k}{\Delta t}\right)^2 \Delta t = b^2(t_1),$$

so that in the limit

$$\lim_{n\to\infty} \int_0^t R_{\dot{U}_n}(t_1, t_2)\, dt_2 = b^2(t_1) \qquad \text{for all } 0 \le t_1 \le t.$$

This suggests that we might define the correlation function of the "process" $\dot{U}(t)$ as

$$R_{\dot{U}}(t_1, t_2) = b^2(t_1)\, \delta(t_1 - t_2),$$

where δ is the Dirac delta function.

Of course, no stochastic process as we defined it can have a delta function as a correlation function. The reason for the appearance of the delta function in the above expression is due to the fact that the process $\dot{U}_n(t)$ consists of independent current impulses of width Δt and average height

$$\frac{\Delta u_k}{\Delta t} = \frac{b(t_k)}{\sqrt{\Delta t}}$$

so that as $n \to \infty$, that is, $\Delta t \to 0$ and $\Delta u_k \to 0$, the pulses are becoming narrower, their areas decrease to zero, but their average heights increase to infinity.

Thus, "in the limit" the "process" $\dot{U}(t)$ would consist of infinitely densely packed current pulses of zero width, infinite height, zero area, and random polarity, and yet would dissipate a finite energy during any finite interval of time.

Clearly, such a physical process cannot possibly exist. There is, however, no need to despair. As we have seen, if we do not literally go to the limit, the current process $\dot{U}_n(t)$ is, even for very large n, quite legitimate and a physically meaningful stochastic process. We can therefore regard the nonexistent process $\dot{U}(t)$ as an *idealization* of the physical process $\dot{U}_n(t)$, that is, of a stream of independent current impulses of random polarity provided only that the rate with which these pulses are occurring is much greater than the time scale of the physical model under consideration.

For example, suppose we wish to model a voltage across a capacitor caused by the thermal noise of a resistor in the circuit (Fig. 4.4.6).

As is customary, we first represent the noisy resistor by an ideal resistor in parallel with a current noise source. In reality, the noise source generates a stream of electrons, but the rate with which they are being generated may be so high that the noise current appears continuous.

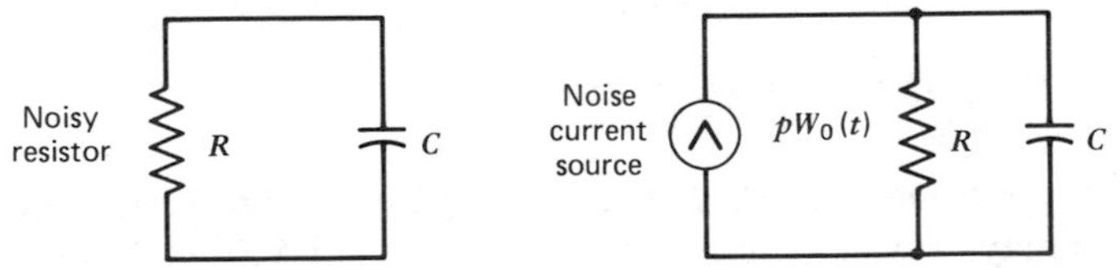

Figure 4.4.6

In that case, we may replace the actual noise current, which may be something like the process $U_n(t)$, by the idealized symbolic process $\dot{U}(t)$. In particular, if the resistor is held at constant temperature or, more precisely, if we assume that the power of the noise source does not change with time, we would model the noise current by a constant p times white Gaussian noise $\dot{W}_0(t)$. Denoting by $X(t)$ the voltage across the capacitor, we would then obtain from Kirchhoff's current law the differential equation

$$C\dot{X}(t) + RX(t) = p\dot{W}_0(t), \qquad t \geq 0.$$

Dividing by the capacitance C and assuming that the initial voltage $X(0) = 0$, the equation would become

$$\dot{X}(t) + \alpha X(t) = \beta \dot{W}_0(t), \qquad t \geq 0, \qquad X(0) = 0,$$

where $\alpha = 1/RC$ and $\beta = p/C$. But this is just the Langevin equation we met before, and so the voltage $X(t)$, $t \geq 0$, would be an Ornstein-Uhlenbeck process with parameters $\sigma^2 = p^2(R/2C)$ and $1/RC$. Of course, since we have used an idealized process for the noise current, the Ornstein-Uhlenbeck voltage process $X(t)$, $t \geq 0$, is also an idealization of the actual voltage waveforms in a sense that it describes the time changes of the voltage only over time intervals considerably larger than the average times between successive emission of electrons. But this may generally be quite satisfactory, at least over the range of radio frequencies.

To illustrate the nature of this idealization more clearly, consider yet another famous physical phenomenon, Brownian motion. As you probably know, Brownian motion refers to the erratic motion of a heavy particle suspended in a liquid caused by impacts of the molecules of the liquid on the particle. Denoting by $X(t)$ a coordinate of the velocity of the particle at time t, the force acting upon the particle equals

$$-\alpha X(t) + \beta U_n(t). \tag{21}$$

Here, the first term represents friction or drag, while the second represents the push imparted upon the particle by the colliding molecules. In reality, the process $U_n(t)$ consists of a stream of impulses of random magnitude and random signs (since we are considering one Cartesian coordinate only). However, we may again attempt an idealization and replace the process $U_n(t)$ by white Gaussian noise $\dot{W}_0(t)$. Then equating the force (21) with the acceleration $\dot{X}(t)$ times the mass m of the particle, we arrive once more at the Langevin equation

$$\dot{X}(t) + \frac{\alpha}{m} X(t) = \frac{\beta}{m} \dot{W}_0(t), \qquad t \geq 0.$$

Thus the velocity coordinate $X(t)$, $t \geq 0$, will again be an Ornstein-Uhlenbeck process. (We may as well assume the initial condition $X(0) = 0$, since the effect of the initial condition wears off for larger t rather quickly.) Let us now

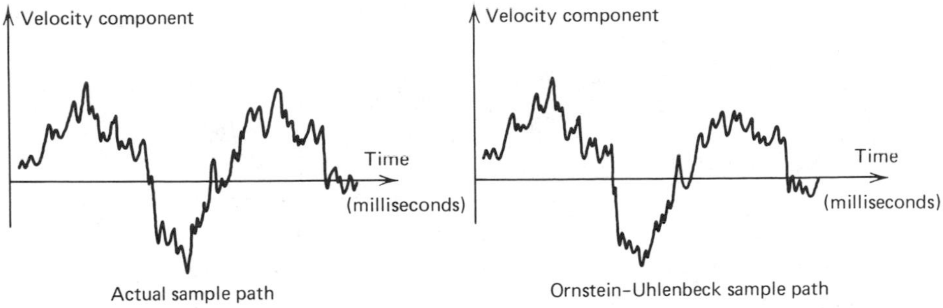

Figure 4.4.7

try to compare the typical sample path of the actual velocity component with a path of the Ornstein-Uhlenbeck process, that is, the one obtained by replacing the impulsive force $\beta U_n(t)$ by the white Gaussian noise idealization $\beta \dot{W}_0(t)$. A typical response of the differential system

$$m\dot{x}(t) + \alpha x(t) = \beta u_n(t)$$

to an impulsive function $\beta u_n(t)$ will be a superposition of responses to each individual impulse (impact force).

On a time scale that is much larger than the average time between consecutive collisions, the result will be a continuous but extremely erratic function of time, for all practical purposes indistinguishable from a typical sample path of the Ornstein-Uhlenbeck process, as in Figure 4.4.7. However, on a time scale comparable with the average intercollision time the typical sample path of the actual velocity process will be more or less a smooth function, while as we know, the sample paths of the Ornstein-Uhlenbeck process, just like those of the Wiener process, look alike under any time scale magnification—a continuous but nowhere differentiable function. Thus, on the time scale of, say, picoseconds, the effect of idealization will begin to show up as in Figure 4.4.8.

We see here a situation characteristic of almost any mathematical model of a physical process—any mathematical model involves a degree of idealization that produces a good match with reality only within certain ranges of the parameter (the time scale in our example) involved. If we wanted to obtain a mathematical model for the velocity component on the time scale comparable with the average intercollision time, we would have to use a different mathematical model for the collision force term.†

To conclude our discussion, let us summarize what we have gained by introducing the concept of a white Gaussian noise. For one thing, it allows us to apply

† For instance, that of a point process dealt with in Chapter Seven.

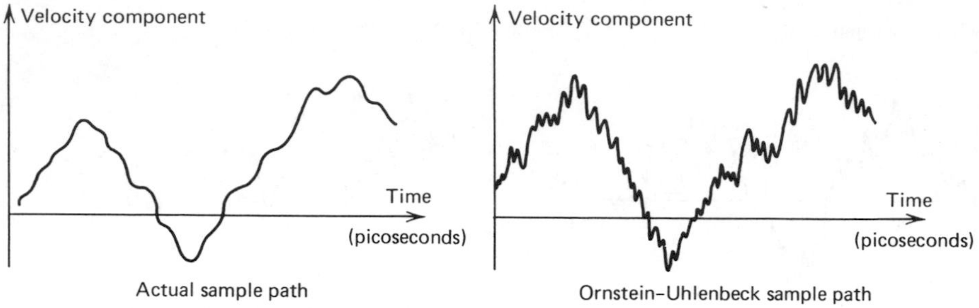

Figure 4.4.8

the rules of MS calculus even to processes that are not MS differentiable, and it greatly simplifies calculations involving the Wiener integral. We shall study this further in Section 5.1. More importantly, however, the white Gaussian noise represents an idealized form of a *continuous physical noise* just like the Dirac delta function is an idealized form of a unit impulse. Thus, whenever we wish to model a physical noise that in reality may consist of densely packed narrow impulses of constant energy and random polarity, we may reach for a white Gaussian noise as a suitable mathematically tractable idealization.

VECTOR-VALUED WHITE GAUSSIAN NOISE. On many occasions, there is a need for a process that would be a model for vector-valued noise. The simplest way to accomplish this is to consider a generalized vector-valued process

$$\dot{W}_0(t) = (\dot{W}_{01}(t), \ldots, \dot{W}_{0m}(t))'$$

each component $\dot{W}_{0k}(t)$ of which is a white Gaussian noise and the component processes $\dot{W}_{01}(t), \ldots, \dot{W}_{0m}(t)$ behave as if they were *mutually independent.*

In other words $\dot{\mathbf{W}}_0(t)$ acts upon linear systems as a stationary vector-valued Gaussian process with zero mean vector and correlation function

$$\mathbf{R}_{\dot{\mathbf{W}}_0}(\tau) = \delta(\tau)\mathbf{I},$$

where $\mathbf{I}$ is the $m \times m$ identity matrix.

Interdependence between components and nonstationarity may, if needed, be introduced by the transformation

$$\dot{\mathbf{U}}(t) = \mathbf{b}(t)\dot{\mathbf{W}}_0(t)$$

where the intensity $\mathbf{b}(t)$ is now an $m \times m$ nonsingular square integrable matrix-valued function. The correlation function is then

$$\mathbf{R}_{\dot{\mathbf{U}}}(t_1, t_2) = \delta(t_1 - t_2)\mathbf{b}(t_1)\mathbf{b}'(t_1).$$

Note that $\dot{\mathbf{U}}(t)$ is then the formal time derivative of the vector-valued Gaussian martingale

$$\mathbf{U}(t) = \int_0^t \mathbf{b}(\tau)\, d\mathbf{W}_0(\tau),$$

where $\mathbf{W}_0(t)$ is now a vector-valued process whose components are independent standard Wiener processes.

LINEAR SYSTEMS WITH WHITE GAUSSIAN NOISE INPUT. In Section 4.2, we studied differential linear systems in the state-space representation

$$\dot{\mathbf{x}}(t) = \mathbf{a}(t)\mathbf{x}(t) + \mathbf{b}(t)\mathbf{u}(t), \tag{22a}$$

$$\mathbf{y}(t) = \mathbf{c}(t)\mathbf{x}(t), \qquad t_0 \le t, \tag{22b}$$

with the input being a stochastic process and the initial state $x(t_0)$ a random vector.

We now wish to extend this to the case where the input process is (vector-valued) white Gaussian noise. In other words, we want to consider the system

$$\dot{\mathbf{X}}(t) = \mathbf{a}(t)\mathbf{X}(t) + \mathbf{b}(t)\dot{\mathbf{W}}_0(t), \tag{23a}$$

$$\mathbf{Y}(t) = \mathbf{c}(t)\mathbf{X}(t), \qquad t_0 \le t \tag{23b}$$

with initial condition $\mathbf{X}(t_0)$, where $\dot{\mathbf{W}}_0(t)$ is (vector-valued) white Gaussian noise; $\mathbf{X}(t_0)$ is a Gaussian random vector; and $\mathbf{a}(t)$, $\mathbf{b}(t)$, and $\mathbf{c}(t)$ are matrices of appropriate dimension. We also assume that $\mathbf{X}(t_0)$ and $\dot{\mathbf{W}}_0(t)$ are independent for all $t_0 \le t$.

According to Section 4.2 the general solution of the system (23) is

$$\mathbf{X}(t) = \boldsymbol{\phi}(t, t_0)\mathbf{X}(t_0) + \int_{t_0}^t \boldsymbol{\phi}(t, \tau)\mathbf{b}(\tau)\dot{\mathbf{W}}_0(\tau)\, d\tau, \tag{24a}$$

$$\mathbf{Y}(t) = \mathbf{c}(t)\mathbf{X}(t), \qquad t_0 \le t, \tag{24b}$$

where the state-transition function $\boldsymbol{\phi}(t_1, t_2)$ is determined from (4.2.34) and (4.2.35). Now the integral on the right-hand side of (24a) is, in fact, a Wiener integral

$$\int_{t_0}^t \boldsymbol{\phi}(t, \tau)\mathbf{b}(\tau)\, d\mathbf{W}_0(\tau),$$

which exists provided the matrix-valued function $\boldsymbol{\phi}(t, \tau)\mathbf{b}(\tau)$ is a square-integrable function of τ. Then, both $\mathbf{X}(t)$ and $\mathbf{Y}(t)$ will be ordinary Gaussian processes, and their mean and correlation functions can be determined as in Section 4.2, with $\mathbf{U}(t) = \dot{\mathbf{W}}_0(t)$ considered, according to our rule for white Gaussian noise, a Gaussian process with

$$\boldsymbol{\mu}_{\dot{\mathbf{W}}_0}(t) = 0 \qquad \text{and} \qquad \mathbf{R}_{\dot{\mathbf{W}}_0}(t_1, t_2) = \delta(t_1 - t_2)\mathbf{I}. \tag{25}$$

In particular, the differential equation for the state mean function (4.2.37) will become

$$\dot{\boldsymbol{\mu}}_{\mathbf{X}}(t) = \mathbf{a}(t)\boldsymbol{\mu}_X(t), \qquad t_0 \le t,$$
$$\boldsymbol{\mu}_{\mathbf{X}}(t_0) = \mathrm{E}[\mathbf{X}(t_0)]$$

with the solution

$$\boldsymbol{\mu}_{\mathbf{X}}(t) = \boldsymbol{\phi}(t, t_0)\boldsymbol{\mu}_{\mathbf{X}}(t_0), \qquad t \ge t_0 .$$

Because of (25) the expressions (4.2.38) for the state correlation functions can also be simplified; in particular, we will have instead of (4.2.38c)

$$\mathbf{R}_{\mathbf{X}_r}(t_1, t_2) = \int_{t_0}^{\min(t_1, t_2)} \boldsymbol{\phi}(t_1, \tau)\mathbf{b}(\tau)\mathbf{b}'(\tau)\boldsymbol{\phi}'(t_2, \tau)\, d\tau. \tag{26}$$

More importantly, however, it is now possible to obtain a differential equation for the state correlation functions, which is frequently more tractable than integral expressions.

First, note that by multiplying (24a) from the right by $\dot{\mathbf{W}}_0'(\tau_1)$ with $t \le \tau_1$, we get

$$\mathbf{X}(t)\dot{\mathbf{W}}_0'(\tau_1) = \boldsymbol{\phi}(t, t_0)\mathbf{X}(t_0)\dot{\mathbf{W}}_0'(\tau_1) + \int_0^t \boldsymbol{\phi}(t, \tau)\mathbf{b}(\tau)\dot{\mathbf{W}}_0(\tau)\dot{\mathbf{W}}_0'(\tau_1)\, d\tau, \tag{27}$$

and since $\mathrm{E}[\mathbf{X}(t_0)\dot{\mathbf{W}}_0'(\tau_1)] = 0$, $\mathbf{X}(t_0)$ and $\dot{\mathbf{W}}_0(t)$ being independent by assumption, we get, by taking the expectation of (27),

$$\mathrm{E}[\mathbf{X}(t)\dot{\mathbf{W}}_0'(\tau_1)] = \int_0^t \boldsymbol{\phi}(t, \tau)\mathbf{b}(\tau)\, \delta(\tau - \tau_1)\, d\tau = 0 \tag{28}$$

for $t_0 \le t \le t_1$. Hence the *state* $\mathbf{X}(t)$ *is independent of present and future noise* $\dot{\mathbf{W}}_0(\tau_1)$.

Now upon taking $t_0 \le t_1 \le t_2$, multiplying the equation

$$\mathbf{X}(t_2) = \boldsymbol{\phi}(t_2, t_1)\mathbf{X}(t_1) + \int_{t_1}^{t_2} \boldsymbol{\phi}(t_2, \tau)\mathbf{b}(\tau)\dot{\mathbf{W}}_0(\tau)\, d\tau$$

from the right by $\mathbf{X}'(t_1)$, and taking the expectation, we get, by using the transpose of equation (28)

$$\mathbf{R}_{\mathbf{X}}(t_2, t_1) = \boldsymbol{\phi}(t_2, t_1)\mathbf{R}_{\mathbf{X}}(t_1, t_2)$$

or using the fact that $\mathbf{R}_{\mathbf{X}}(t_1, t_2) = \mathbf{R}_{\mathbf{X}}'(t_2, t_1)$ and denoting the variance function of the state

$$\boldsymbol{\sigma}_{\mathbf{X}}^2(t) = \mathbf{R}_{\mathbf{X}}(t, t) \tag{29}$$

we have the relation

$$\mathbf{R}_{\mathbf{X}}(t_1, t_2) = \begin{cases} \boldsymbol{\sigma}_{\mathbf{X}}^2(t_1)\boldsymbol{\phi}'(t_2, t_1) & \text{if} \quad t_1 \le t_2, \\ \boldsymbol{\phi}(t_1, t_2)\boldsymbol{\sigma}_{\mathbf{X}}^2(t_2) & \text{if} \quad t_1 \ge t_2. \end{cases} \tag{30}$$

In other words, the correlation function can be determined from the variance function (29). To get the differential equation for the variance function, we note that by setting $t_1 = t_2 = t$ in equation (26) and using the fact that (cf. 4.2.34)

$$\frac{\partial}{\partial t}\boldsymbol{\phi}(t, \tau) = \mathbf{a}(t)\boldsymbol{\phi}(t, \tau), \qquad \tau \le t, \qquad \boldsymbol{\phi}(\tau, \tau) = \mathbf{I},$$

we get after differentiation

$$\frac{d}{dt}\mathbf{R}_{\mathbf{X}_r}(t, t) = \mathbf{a}(t)\mathbf{R}_{\mathbf{X}_r}(t, t) + \mathbf{R}_{\mathbf{X}_r}(t, t)\mathbf{a}'(t) + \mathbf{b}(t)\mathbf{b}'(t). \tag{31}$$

Since the correlation function of the singular component $\mathbf{X}_s(t)$ becomes, from (4.2.38b) with $t_1 = t_2 = t$,

$$\mathbf{R}_{\mathbf{X}_s}(t, t) = \boldsymbol{\phi}(t, t_0)\boldsymbol{\sigma}_{\mathbf{X}}^2(t_0)\boldsymbol{\phi}'(t, t_0)$$

we obtain similarly

$$\frac{d}{dt}\mathbf{R}_{\mathbf{X}_s}(t, t) = \mathbf{a}(t)\mathbf{R}_{\mathbf{X}_s}(t, t) + \mathbf{R}_{\mathbf{X}_s}(t, t)\mathbf{a}'(t). \tag{32}$$

Hence, using (4.2.38a), that is,

$$\boldsymbol{\sigma}_{\mathbf{X}}^2(t) = \mathbf{R}_{\mathbf{X}_s}(t, t) + \mathbf{R}_{\mathbf{X}_r}(t, t)$$

we have, by adding (31) and (32), the differential equation

$$\dot{\boldsymbol{\sigma}}_{\mathbf{X}}^2(t) = \mathbf{a}(t)\boldsymbol{\sigma}_{\mathbf{X}}^2(t) + \boldsymbol{\sigma}_{\mathbf{X}}^2(t)\mathbf{a}'(t) + \mathbf{b}(t)\mathbf{b}'(t), \qquad t_0 \le t, \tag{33}$$

with the initial condition

$$\boldsymbol{\sigma}_{\mathbf{X}}^2(t_0) = \mathrm{E}[\mathbf{X}(t_0)\mathbf{X}'(t_0)]. \tag{34}$$

The system of differential equations (33) and (34), together with (30), allows us to evaluate the state correlation function. These equations are frequently referred to as *propagation equations* for the correlation function.

It should also be noted that since the state-transition function of a linear differential system (22) generally satisfies the relation

$$\boldsymbol{\phi}(t_3, t_1) = \boldsymbol{\phi}(t_3, t_2)\boldsymbol{\phi}(t_2, t_1),$$

we get from (30), with $t_1 \le t_2 \le t_3$,

$$\begin{aligned} \mathbf{R}_{\mathbf{X}}(t_1, t_3) &= \boldsymbol{\sigma}_{\mathbf{X}}^2(t_1)\boldsymbol{\phi}'(t_3, t_1) \\ &= \boldsymbol{\sigma}_{\mathbf{X}}^2(t_1)\boldsymbol{\phi}'(t_2, t_1)(\boldsymbol{\sigma}_{\mathbf{X}}^2(t_2))^{-1}\boldsymbol{\sigma}_{\mathbf{X}}^2(t_2)\boldsymbol{\phi}'(t_3, t_2) \\ &= \mathbf{R}_{\mathbf{X}}(t_1, t_2)(\boldsymbol{\sigma}_{\mathbf{X}}^2(t_2))^{-1}\mathbf{R}_{\mathbf{X}}(t_2, t_3), \end{aligned}$$

assuming that $\boldsymbol{\sigma}_{\mathbf{X}}^2(t_2)$ is nonsingular.

But this means that the state process $X(t)$, being Gaussian, is also a Markov process (see Chapter One). This result is not really surprising; after all, the future of the system whose state obeys a first order differential equation

$$\dot{\mathbf{X}}(t) = \mathbf{a}(t)\mathbf{X}(t) + \mathbf{b}(t)\mathbf{U}(t)$$

could possibly depend on the past $\mathbf{X}(t')$, $t' < t$, only through the input $\mathbf{U}(t)$. If this input is white Gaussian noise $\mathbf{U}(t) = \dot{\mathbf{W}}_0(t)$, then its present $\dot{\mathbf{W}}_0(t)$ is also independent of its past $\dot{\mathbf{W}}_0(t')$, $t' < t$, and thus the future of $\mathbf{X}(t)$ can only depend on its present value. In fact, we can state a general theorem:

THEOREM 4.4.1. *A (vector-valued) finite-variance MS differentiable process* $\mathbf{X}(t)$, $t_0 \le t$, *is Gauss-Markov if and only if it satisfies the system of differential equations*

$$\dot{\mathbf{X}}(t) = \mathbf{a}(t)\mathbf{X}(t) + \mathbf{b}(t)\dot{\mathbf{W}}_0(t)$$

with Gaussian initial condition $\mathbf{X}(t_0)$ *independent of the white Gaussian noise* $\dot{\mathbf{W}}_0(t)$, $t_0 \le t$.

Example 4.4.3

Consider a parallel resonant circuit as shown in Figure 4.4.9, where i_N represents a noisy current. From the well-known relations between the currents i_C, i_R, and i_L and the voltage v,

$$v = L\frac{di_L}{dt} = Ri_R, \qquad i_C = C\frac{dv}{dt},$$

and Kirchhoff's current law $i_N = i_C + i_R + i_L$, we obtain a pair of first order differential equations

$$\frac{di_L}{dt} = \frac{R}{L}i_R, \qquad \frac{di_R}{dt} = \frac{1}{R}\frac{dv}{dt} = \frac{1}{RC}(i_N - i_L + i_R).$$

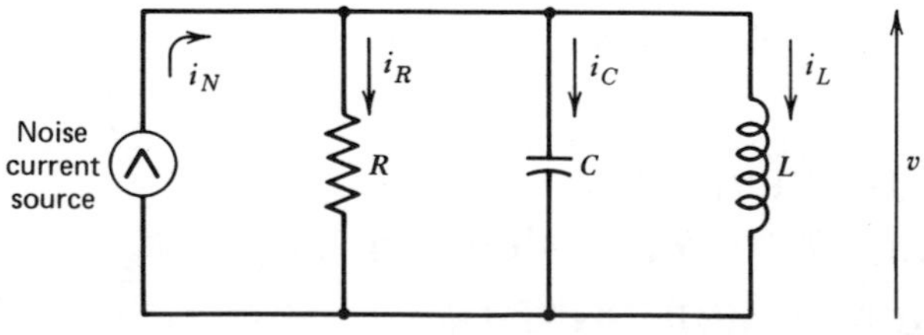

Figure 4.4.9

Calling $x_1 = i_L$ and $x_2 = i_R$ the components of the state x of the system, renaming the input current $u = i_N$, and abbreviating

$$\alpha = \frac{R}{L}, \qquad \beta = \frac{1}{RC},$$

we can write the state-input equation for this system in standard form:

$$\begin{pmatrix} \dot{X}_1(t) \\ \dot{X}_2(t) \end{pmatrix} = \begin{bmatrix} 0 & \alpha \\ -\beta & -\beta \end{bmatrix} \begin{pmatrix} x_1(t) \\ x_2(t) \end{pmatrix} + \begin{pmatrix} 0 \\ \beta \end{pmatrix} u(t)$$

Assuming further that the input is "switched on" at time $t = 0$ and that the initial states $x_1(0) = x_2(0) = 0$, that is, that the initial charge in the capacitor as well as the initial current through the inductance are both zero, the state of the system at time $t \geq 0$ is given by

$$\mathbf{X}(t) = \int_0^t \boldsymbol{\phi}(t, \tau)\beta u(\tau)\, d\tau.$$

Assuming that $4\alpha > \beta$, that is, that the resistance $R > \frac{1}{2}\sqrt{L/C}$ (subcritical damping), the state-transition matrix is found to be

$$\boldsymbol{\phi}(t, \tau) = \begin{bmatrix} \cos \omega(t - \tau) + \dfrac{\beta}{2\omega} \sin \omega(t - \tau) & \dfrac{\alpha}{\omega} \sin \omega(t - \tau) \\ -\dfrac{\beta}{\omega} \sin \omega(t - \tau) & \cos \omega(t - \tau) - \dfrac{\beta}{2\omega} \sin \omega(t - \tau) \end{bmatrix} e^{-\beta(t-\tau)/2},$$

$0 \leq \tau \leq t$, where $\omega = (\beta/2)\sqrt{(4\alpha/\beta) - 1}$. Suppose now that the noise current is modeled as a white Gaussian noise process

$$U(t) = \sigma \dot{W}_0(t), \qquad t \geq 0$$

with $\sigma > 0$ a constant. Then the state of the system $X(t)$, $t \geq 0$, that is, the currents through the resistance and inductance are, respectively,

$$X_1(t) = \int_0^t \frac{\alpha\beta\sigma}{\omega} e^{-\beta/2(t-\tau)} \sin \omega(t - \tau)\dot{W}_0(\tau)\, d\tau,$$

$$X_2(t) = \int_0^t \beta\sigma e^{-\beta/2(t-\tau)} \left(\cos \omega(t - \tau) - \frac{\beta}{2\omega} \sin \omega(t - \tau) \right) \dot{W}_0(\tau)\, d\tau.$$

We could now determine the correlation functions of these processes using the propagation equations. For instance, for the variance function of the process $X_1(t)$, $t \geq 0$, we would calculate

$$\sigma_{X_1}^2(t) = \int_0^t \left(\frac{\alpha\beta\sigma}{\omega}\right)^2 e^{-\beta(t-\tau)} \sin^2 \omega(t-\tau)\, d\tau.$$

After some rather straightforward integration and a little algebra, we get the expression

$$\sigma_{X_1}^2(t) = \frac{\alpha\sigma^2}{2} - \frac{1}{2}\left(\frac{\alpha\beta\sigma}{\omega}\right)^2 \left(\frac{1}{\beta} + \frac{\beta}{\beta^2 + 4\omega^2} \cos 2\omega t + \frac{2\omega}{\beta^2 + 4\omega^2} \sin 2\omega t\right) e^{-\beta t}.$$

On the other hand, we could also compute the variance matrix function

$$\boldsymbol{\sigma}_{\mathbf{X}}^2(t) = \begin{bmatrix} \sigma_{11}(t) & \sigma_{12}(t) \\ \sigma_{21}(t) & \sigma_{22}(t) \end{bmatrix}, \qquad t \geq 0$$

of the state vector

$$\mathbf{X}(t) = \begin{pmatrix} X_1(t) \\ X_2(t) \end{pmatrix}$$

from the differential equation (33). Since, in our case,

$$\mathbf{a}(t) = \begin{bmatrix} 0 & \alpha \\ -\beta & -\beta \end{bmatrix}, \qquad \mathbf{b}(t) = \begin{bmatrix} 0 \\ \beta \end{bmatrix} \sigma,$$

the equation would take the form

$$\begin{bmatrix} \dot\sigma_{11}(t) & \dot\sigma_{12}(t) \\ \dot\sigma_{21}(t) & \dot\sigma_{22}(t) \end{bmatrix} = \begin{bmatrix} 0 & \alpha \\ -\beta & -\beta \end{bmatrix} \begin{bmatrix} \sigma_{11}(t) & \sigma_{12}(t) \\ \sigma_{21}(t) & \sigma_{22}(t) \end{bmatrix}$$
$$+ \begin{bmatrix} \sigma_{11}(t) & \sigma_{12}(t) \\ \sigma_{21}(t) & \sigma_{22}(t) \end{bmatrix} \begin{bmatrix} 0 & -\beta \\ \alpha & -\beta \end{bmatrix} + \sigma^2 \begin{bmatrix} 0 \\ \beta \end{bmatrix} [0 \quad \beta],$$

$t \geq 0$, with the initial condition $\boldsymbol{\sigma}_{\mathbf{X}}^2(0) = \mathbf{0}$. Using the symmetry of the variance matrix,

$$\sigma_{12}(t) = \sigma_{21}(t)$$

this would reduce to a system of three linear first order equations

$$\dot\sigma_{11}(t) = 2\alpha\sigma_{12}(t),$$
$$\dot\sigma_{12}(t) = \alpha\sigma_{22}(t) - \beta\sigma_{11}(t) - \beta\sigma_{12}(t),$$
$$\dot\sigma_{22}(t) = -2\beta\sigma_{12}(t) - 2\beta\sigma_{22}(t) + \sigma^2\beta^2,$$

$t \geq 0$, with initial conditions $\sigma_{11}(0) = \sigma_{12}(0) = \sigma_{22}(0) = 0$.

Although an explicit solution of this system of equations is feasible, given sufficient time and effort, in practice we would be more interested in the steady state of the system than in its transient behavior. From physical considerations, as well as from the integral expressions for the variance functions above, it is clear that as $t \to \infty$ the variance matrix $\boldsymbol{\sigma}_{\mathbf{X}}^2(t)$ tends to a constant. Its entries can then be calculated quite easily by setting the time derivatives in the above system all equal to zero. Denoting the steady-state variance matrix $\boldsymbol{\sigma}_{\mathbf{X}}^2(\infty)$, we thus have

$$\boldsymbol{\sigma}_{\mathbf{X}}^2(\infty) = \begin{bmatrix} \dfrac{\sigma^2\alpha}{2} & 0 \\ 0 & \dfrac{\sigma^2\beta}{2} \end{bmatrix}.$$

It follows that for large t or, if you prefer, in the steady state, the currents $X_1(t)$ and $X_2(t)$ become independent (at the same t).

We can also quite easily evaluate the correlation function $\mathbf{R}_{\mathbf{X}}(t_1, t_2)$ for large t_1 and t_2. Using (30) we see that in the steady state

$$\mathbf{R}_{\mathbf{X}}(\tau) = \begin{bmatrix} \alpha \cos \omega\tau + \dfrac{\alpha\beta}{2\omega} \sin \omega|\tau| & \dfrac{\alpha\beta}{\omega} \sin \omega\tau \\ -\dfrac{\alpha\beta}{\omega} \sin \omega\tau & \beta \cos \omega\tau - \dfrac{\beta^2}{2\omega} \sin \omega|\tau| \end{bmatrix} \frac{\sigma^2}{2} e^{-\beta|\tau|/2},$$

$\tau = t_1 - t_2$. Thus, in the steady state, the vector-valued process $\mathbf{X}(t)$ becomes a zero mean stationary Gauss-Markov process. Note, however, that the component processes $\mathbf{X}_1(t)$ and $\mathbf{X}_2(t)$ will not be Markov, because each above satisfies a linear second order rather than first order differential equation. For instance, the current through the resistor $X_2(t)$ satisfies the equation

$$\ddot{X}_2(t) + \beta\dot{X}_2(t) + \alpha\beta X_2(t) = \sigma\ddot{W}_0(t),$$

where $\ddot{W}_0(t)$ is a symbolic process acting, as far as linear operations are concerned, as a stationary zero mean Gaussian process with correlation function

$$R_{\ddot{W}_0}(\tau) = \mathring{\delta}(\tau).$$

If we take the time domain $-\infty < t < \infty$, then in spite of the presence of this symbolic process, the process $X_2(t)$, $-\infty < t < \infty$, satisfying the above second

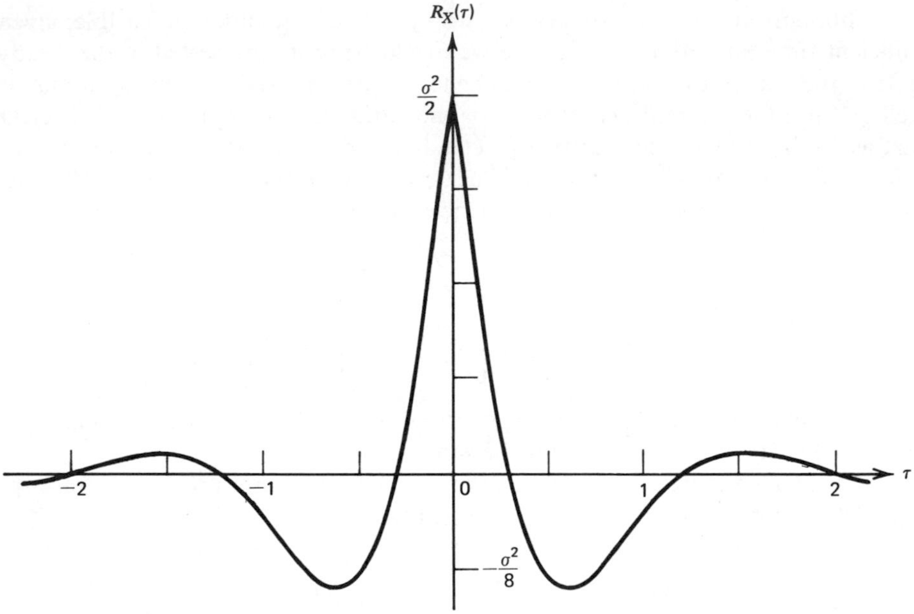

Figure 4.4.10

order differential equation will be a genuine stationary zero mean Gaussian process with correlation function

$$R_{X_2}(\tau) = \frac{\beta\sigma^2}{2} e^{-\beta|\tau|/2}\left(\cos \omega\tau - \frac{\beta}{2\omega} \sin \omega|\tau|\right),$$

$-\infty < \tau < \infty$. Notice the oscillatory behavior of this function, as seen from an example (with $\alpha = \beta = 1$) in Figure 4.4.10. ▲

EXERCISES

Exercise 4.4

1. Let $X(t)$, $t \geq 0$, be an Ornstein-Uhlenbeck process with zero mean function and with the correlation function (15). Verify by directly computing the conditional expectation $E[dX(t) \mid X(t'), 0 \leq t' \leq t]$ that equation (16), is, in fact, the Doob-Meyer-Fisk decomposition of $X(t)$. What are the functionals $a(x(t'), 0 \leq t' \leq t)$ and $b(x(t'), 0 \leq t' \leq t)$ in this case?
2. If $X(t)$, $t \geq 0$, is a zero mean Ornstein-Uhlenbeck process with correlation function (15), show that for all $t \geq 0$

$$\int_0^t X(\tau)\, d\tau = \beta \int_0^t \frac{1 - e^{-\alpha(t-\tau)}}{\alpha} \dot{W}_0(\tau)\, d\tau.$$

3. The Langevin equation (13) suggests that the generalized process $Y(t) = \beta \dot{W}_0(t) - \alpha X(t)$, $t \geq 0$, should have the "correlation function"

$$R_Y(t_1, t_2) = \beta^2 R_{\dot{W}_0}(t_1, t_2) - \alpha\beta R_{\dot{W}_0, X}(t_1, t_2) - \alpha\beta R_{X, \dot{W}_0}(t_1, t_2) + \alpha^2 R_X(t_1, t_2).$$

On the other hand, since $Y(t) = \dot{X}(t)$ is the formal derivative of the Ornstein-Uhlenbeck process $X(t)$, $t \geq 0$, we should also have

$$R_Y(t_1, t_2) = \frac{\partial^2}{\partial t_1 \, \partial t_2} R_X(t_1, t_2),$$

where R_X is as in (15). Verify that these two formulas indeed yield the same "correlation function" R_Y.

4. Find the mean function and the correlation function of the process $X(t)$, $t \geq 0$, satisfying the differential equation

$$\alpha_0 \dot{X}(t) + \alpha_1 X(t) = \gamma + \sigma \dot{W}_0(t), \qquad X(0) = 0,$$

where α_0, α_1, γ, and σ are positive constants. Express the process $X(t)$, $t \geq 0$, in the integral form

$$X(t) = \int_0^t a(\tau)\, d\tau + \int_0^t h(t, \tau)\, dW_0(\tau).$$

5. For each of the differential equations below, find the correlation function of the process $X(t)$, $t \geq 0$, and express the process in the form of a Wiener integral and/or a stochastic integral of the Wiener process $W_0(t)$. Assume the initial condition $X(0) = \dot{X}(0) = 0$.
   (a) $\ddot{X}(t) + 2\dot{X}(t) = 3\dot{W}_0(t)$
   (b) $4\ddot{X}(t) + 8\dot{X}(t) + 5X(t) = \dot{W}_0(t)$
   (c) $\ddot{X}(t) + 3\dot{X}(t) + 2X(t) = -\dot{W}_0(t)$
   (d) $\ddot{X}(t) + 2\dot{X}(t) + X(t) = 2\dot{W}_0(t)$
   (e) $\ddot{X}(t) + 2X(t) = \dot{W}_0(t)$

6. Derive a general formula for the solution of the Langevin equation with variable coefficients

$$\dot{X}(t) + \alpha(t) X(t) = \beta(t) \dot{W}_0(t), \qquad t \geq 0, \qquad X(0) = x_0,$$

as a Wiener integral. Assume that $\alpha(t)$ and $\beta(t)$ are both integrable as needed.

7. Let $X(t)$, $t \geq 0$, $X(0) = 0$, be a scalar-valued Gauss-Markov process defined by the differential equation

$$(1 + t^2)\dot{X}(t) + X(t) = \sqrt{2} \arctan t \, \dot{W}_0(t).$$

Find the variance function $\sigma_X^2(t)$, $t \geq 0$, of this process. (*Hint:* Use Equation (33).)

8. Consider a linear differential system

$$\begin{pmatrix} \dot{x}_1(t) \\ \dot{x}_2(t) \end{pmatrix} = \begin{bmatrix} 0 & t \\ t & 0 \end{bmatrix} \begin{pmatrix} x_1(t) \\ x_2(t) \end{pmatrix} + \begin{pmatrix} 1 \\ 0 \end{pmatrix} u(t),$$

$$y(t) = \Big| [0, \; 1] \begin{pmatrix} x_1(t) \\ x_2(t) \end{pmatrix}, \qquad t \geq 0,$$

with initial condition $\mathbf{x}(0) = (0, 1)'$. Assuming that the input $u(t)$ is standard white Gaussian noise, find the mean function $\mu_Y(t)$ and the variance function $\sigma_Y^2(t)$ of the output process $Y(t)$, $t \geq 0$. Is the output process Gauss-Markov?

9. Show that the process $Y(t)$, $t \geq 0$, satisfying the differential equation

$$\dot{Y}(t) + \gamma(t)Y(t) = U(t),$$

where $U(t)$, $t \geq 0$, is an Ornstein-Uhlenbeck process, can always be regarded as the output of a linear dynamic system of the form (23a,b). Identify the matrices $\mathbf{a}(t)$, $\mathbf{b}(t)$, and $\mathbf{c}(t)$, and set up the differential equation (33) for the variance function $\boldsymbol{\sigma}_{\mathbf{X}}^2(t)$. What are the entries of this matrix function in relation to the processes $Y(t)$ and $U(t)$? (*Hint:* Set $X_1(t) = Y(t)$, $X_2(t) = U(t)$, and recall the Langevin equation (13).)

4.5. INNOVATION PROCESSES AND SOME APPLICATIONS

A considerable number of problems in communication engineering and statistical signal processing are concerned with estimating, predicting, detecting, or in general obtaining some information about a signal that has been contaminated by noise. The model most commonly used for this kind of situation is to assume that the noise $N(t)$ is simply superimposed on the signal $X(t)$, that is, that the contaminated signal has the form,

$$Y(t) = X(t) + N(t), \qquad t \in \mathcal{T}. \tag{1}$$

The noise is assumed to be white Gaussian noise possibly multiplied by a deterministic function of time

$$N(t) = b(t)\dot{W}_0(t), \qquad t \in \mathcal{T}, \tag{2}$$

to account for variations in its intensity. Since, in practice, every signal has a finite duration, the time domain $\mathcal{T}$ can be taken as a time interval of finite length, in particular $\mathcal{T} = [0, t_0]$. However, to simplify notation, we leave the upper endpoint unspecified and from now on write just $t \geq 0$. The process $Y(t)$, $t \geq 0$, is then referred to as an *observation process*, a term that arises quite naturally if one considers the general block scheme (Fig. 4.5.1) of almost any kind of signal processing.

The process $Y(t)$ is the process seen or observed by the receiver, and hence it is called an observation process.

It may be rightfully argued that the block diagram in Figure 4.5.1 represents a more general situation than the model of equation (1), where the noise is simply added to the signal. Indeed, in many cases the actual physical noise (thermal noise, atmospheric noise, noise due to roundoffs and quantization, etc.) will distort the signal in a far more complicated fashion than by a simple superposition on the signal waveform. However, the reader may recall from Section 4.3 that as long as

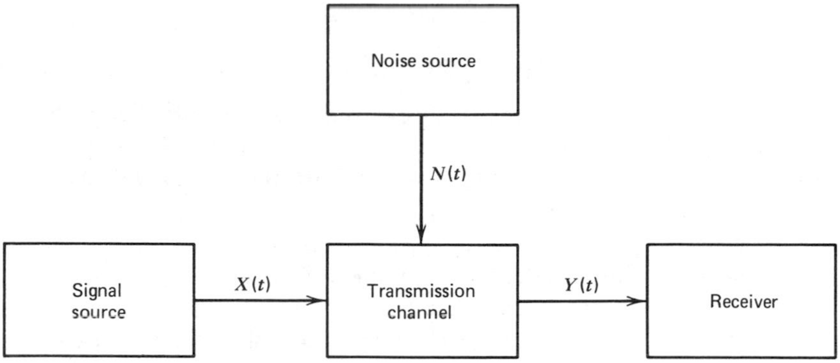

Figure 4.5.1

we interpret the dichotomy between signal and noise as that between a process with relatively smooth sample paths versus a process whose sample paths are highly irregular, it is not unreasonable to assume that the observation process can actually be decomposed into a sum of two such processes. Thus, as long as we are willing to assume that we are dealing only with sample-continuous finite-variance processes, as we do exclusively in this chapter, model (1) is not as restrictive as it may seem. True, there are physical noises of impulsive character, and even sample-continuous noises need not be Gaussian as a result of nonlinearities.† However, in this section, we will restrict our attention to the model of equation (1), with the interpretation that the noise component there need not necessarily be the original noise entering the transmission channel but rather the *noisy component of the observation process as perceived by the receiver.*

However, this immediately raises another problem. In the spirit of our general discussion in Section 4.3, the noise as perceived by the receiver should be the mean square unpredictable part of the observation process. In other words, the receiver sees the signal as the conditional expectation

$$\hat{X}(t) = \mathrm{E}[X(t) \mid Y(t'), 0 \leq t' \leq t], \tag{3}$$

and the noise as the unpredictable residue

$$V(t) = Y(t) - \hat{X}(t).$$

Thus, from the receiver's point of view, the decomposition of the observation process into a signal and noise is

$$Y(t) = \hat{X}(t) + V(t), \qquad t \geq 0, \tag{4}$$

† These models are treated in Chapters Six and Seven.

rather than

$$Y(t) = X(t) + b(t)\dot{W}_0(t), \qquad t \geq 0, \tag{5}$$

as in (1) and (2). This means that unless we can show that despite their different appearance the models (4) and (5) are probabilistically indistinguishable, the interpretation offered above could not be justified. Fortunately, equivalence of (4) and (5) can indeed be established.

But before we attempt to convince the reader that this is so, we need to state in more detail the assumption underlying the model in equation (1).

First, equation (1) involves white Gaussian noise, that is, a symbolic or generalized process, and it must therefore be regarded as the symbolic derivative of an equation

$$Z(t) = \int_0^t X(\tau)\, d\tau + W(t), \qquad t \geq 0, \tag{6}$$

where $Y(t) = \dot{Z}(t)$, $t \geq 0$, and $W(t)$ is an inhomogeneous Wiener process

$$W(t) = \int_0^t b(\tau)\, dW_0(\tau), \qquad t \geq 0. \tag{7}$$

Now $X(t)$, $t \geq 0$, is assumed to be a finite-variance process such that the stochastic integral in (6) exists for all $t \geq 0$. As we know from Section 4.1, this is guaranteed if

$$\int_0^t \mathrm{E}[X^2(\tau)]\, d\tau < \infty \qquad \text{for all } t \geq 0, \tag{8}$$

For the existence of the Wiener integral (7) the deterministic function $b(t)$ must also be square integrable

$$\int_0^t b^2(\tau)\, d\tau < \infty \qquad \text{for all } t \geq 0. \tag{9}$$

To avoid degenerate cases, we will also assume that $b(t) \neq 0$ for all $t > 0$. Finally, some assumption must be made about the type of dependence between the signal $X(t)$ and the Wiener process $W(t)$ in (6). The easiest thing would be to assume, as is often done, that $X(t)$ and $W(t)$ are mutually independent. But it is not really necessary to go that far, and besides, such an assumption would preclude the possibility that the signal may be influenced by past noise, for instance, owing to the presence of a feedback link. We will therefore assume only that the *future noise is independent of the past signal*, or, more precisely, that:

For every $t > 0$ the forward increment $dW(t)$ is independent of the past $X(t')$, $0 \leq t' \leq t$. (10)

Since $W(t)$ is a Wiener process, $dW(t)$ is also independent of its own past and thus also of $X(t')$, $W(t')$, $0 \le t' \le t$. Let us now define a sample-continuous, finite-variance process

$$U(t), \qquad t \ge 0,$$

such that $U(0) = 0$ and such that its forward increments for every $t > 0$ satisfy the equation

$$dU(t) = dZ(t) - \mathrm{E}[dZ(t) \mid Z(t'), 0 \le t' \le t]. \tag{11}$$

Since by (11) $U(t')$, $0 \le t' \le t$, is determined by $Z(t')$, $0 \le t' \le t$, we obtain by the smoothing property of conditional expectation

$$\mathrm{E}[dU(t) \mid U(t'), 0 \le t' \le t] = 0 \qquad \text{for all } t > 0.$$

Thus the process $U(t)$, $t \ge 0$, is a sample-continuous martingale. Next, from (6),

$$dZ(t) = X(t)\,dt + dW(t) \tag{12}$$

and hence upon substitution into the conditional expectation above

$$\begin{aligned} dU(t) = dZ(t) &- \mathrm{E}[X(t) \mid Z(t'), 0 \le t' \le t]\,dt \\ &- \mathrm{E}[dW(t) \mid Z(t'), 0 \le t' \le t]. \end{aligned} \tag{13}$$

However, since formally $Y(t) = \dot{Z}(t)$, $t \ge 0$, conditioning on $Z(t')$, $0 \le t' \le t$, is equivalent to conditioning on $Y(t')$, $0 \le t' \le t$, and so

$$\mathrm{E}[X(t) \mid Z(t'), 0 \le t' \le t] = \mathrm{E}[X(t) \mid Y(t'), 0 \le t' \le t] = \hat{X}(t) \tag{14}$$

by (3). On the other hand, by assumption (10) and the fact that $W(t)$ is a Wiener process

$$\mathrm{E}[dW(t) \mid X(t'), W(t'), 0 \le t' \le t] = 0, \tag{15}$$

and since $Z(t')$, $0 \le t' \le t$ is completely determined by $X(t')$, $W(t')$, $0 \le t' \le t$, we also have

$$\mathrm{E}[dW(t) \mid Z(t'), 0 \le t' \le t] = 0. \tag{16}$$

Thus using (12), (15), and (16), we can rewrite (13) as

$$dU(t) = (X(t) - \hat{X}(t))\,dt + dW(t).$$

Then

$$\begin{aligned} &\mathrm{E}[(dU(t))^2 \mid X(t'), W(t'), 0 \le t' \le t] \\ &\qquad = \mathrm{E}[(X(t) - \hat{X}(t))^2 \mid X(t'), W(t'), 0 \le t' \le t](dt)^2 \\ &\qquad\quad + 2\mathrm{E}[(X(t) - \hat{X}(t))\,dW(t) \mid X(t'), W(t'), 0 \le t' \le t]\,dt \\ &\qquad\quad + \mathrm{E}[(dW(t))^2 \mid X(t'), W(t'), 0 \le t' \le t]. \end{aligned} \tag{17}$$

However,

$$E[(X(t) - \hat{X}(t))\, dW(t) \mid X(t'), W(t'), 0 \le t' \le t]$$
$$= dW(t) E[X(t) - \hat{X}(t) \mid X(t'), W(t'), 0 \le t' \le t]$$

and by (10)

$$E[(dW(t))^2 \mid X(t'), W(t'), 0 \le t' \le t] = E[(dW(t))^2] \doteq b^2(t)\, dt.$$

Hence, taking the conditional expectation of (17), given $U(t'), 0 \le t' \le t$, we get for small dt

$$E[(dU(t))^2 \mid U(t'), 0 \le t' \le t] \doteq b^2(t)\, dt \tag{18}$$

since $(dt)^2$ can be neglected as compared to dt and by (14)

$$E[X(t) - \hat{X}(t) \mid U(t'), 0 \le t' \le t] = E[X(t) - \hat{X}(t) \mid Z(t'), 0 \le t' \le t]$$
$$= \hat{X}(t) - \hat{X}(t) = 0.$$

Since the process $U(t)$, $t \ge 0$, is a sample-continuous martingale with $U(0) = 0$ and satisfies (18), it follows from the Lévy-Doob Theorem 4.3.1 that it must be an inhomogeneous Wiener process:

$$U(t) = \int_0^t b(t)\, dW_0(t) \qquad \text{or} \qquad dU(t) = b(t)\, dW_0(t).$$

At the same time, substitution from (14) and (15) into (13) shows that

$$dU(t) = dZ(t) - \hat{X}(t)\, dt, \tag{19}$$

or equivalently

$$Z(t) = \int_0^t \hat{X}(\tau)\, dt + U(t), \qquad t \ge 0. \tag{20}$$

But if we formally differentiate this equation and call

$$V(t) = \dot{U}(t) = b(t)\dot{W}_0(t)$$

we get equation (4), with $V(t)$ now identified as white Gaussian noise multiplied by the deterministic function $b(t)$. It follows that (4) and (5) are indeed equivalent, in other words, that the noise component $V(t)$ as perceived by the receiver is probabilistically equivalent to the original noise component $b(t)\dot{W}_0(t)$. This result, which was first obtained by Kailath and Frost, will now be summarized as a theorem.

THEOREM 4.5.1. *Let*

$$Y(t) = X(t) + b(t)\dot{W}_0(t), \qquad t \ge 0 \tag{21a}$$

where X(t), t $\ge$ 0, *is a finite-variance mean square integrable process;*

$b(t) \neq 0$, $t \geq 0$, *is a square integrable function; and the white Gaussian noise* $\dot{W}_0(t)$ *is for every* $t > 0$ *independent of the past* $X(t'), 0 \leq t' \leq t$. *Then there is an equivalent representation*

$$Y(t) = \hat{X}(t) + V(t), \qquad t \geq 0, \tag{21b}$$

where

$$\hat{X}(t) = \mathrm{E}[X(t) \mid Y(t'), 0 \leq t' \leq t], \qquad t \geq 0,$$

and $V(t)$, $t \geq 0$, *is again* $b(t)$ *times white Gaussian noise independent of the past* $X(t')$, $0 \leq t' \leq t$, *for every* $t > 0$.

Remark: We have stated the theorem in its original form with generalized processes $\dot{W}_0(t)$, $Y(t)$, and $V(t)$. An equivalent statement avoiding generalized processes can be obtained by replacing (21a) and (21b) by their respective integrated versions as in (6) and (20). Я

The generalized process $V(t)$ defined by

$$V(t) = Y(t) - \hat{X}(t), \qquad t \geq 0$$

is called an *innovation process of the process* $Y(t)$. Since $V(t)$ is the formal derivative of an inhomogeneous Wiener process, the martingale $U(t)$, $t \geq 0$, defined by (11), the latter may be called an *innovation martingale* of the process $Z(t) = \int_0^t Y(\tau)\, d\tau$.

The origin of this terminology can be seen from equation (19) written fully as

$$dU(t) = dZ(t) - \mathrm{E}[X(t) \mid Z(t'), 0 \leq t' \leq t]\, dt.$$

The forward increment $dU(t)$ can be interpreted as new information or a random innovation brought about by the increment $dZ(t)$ of the integrated observation process during the time dt after all the information contained in the past $Z(t')$, $0 \leq t' \leq t$, has been extracted. Thus the formal derivative $V(t) = \dot{U}(t)$ can be interpreted as an instantaneous innovation brought about by $Y(t)$ after extracting the past—hence the term "innovation process."

Theorem 4.5.1 and the concept of the innovation process is an extremely useful theoretical tool in deriving important results in signal estimation (both linear and nonlinear), detection, and similar areas. As an illustration, we will briefly describe its use in deriving a general formula for signal detection first established by Kailath.

Example 4.5.1

A common problem in communication engineering is that of detecting the presence of a random signal waveform in a noisy observation. For example, we may have observed a radar or sonar echo for some period of time, and we wish to determine whether that particular echo indicates the presence of some type of target or whether it is merely due to a random noise such as a background noise, transmission noise, atmospheric noise, or some other noise.

Denoting by $Y(t)$ the output of the radar or sonar receiver at some time t during the observation interval $\mathcal{T}_0$, we thus have to decide between two possibilities:

$$(\mathrm{H}_0) \qquad Y(t) = N(t), \qquad t \in \mathcal{T}_0,$$

that is, the observation is just noise $N(t)$ or

$$(\mathrm{H}_1) \qquad Y(t) = X(t) + N(t), \qquad t \in \mathcal{T}_0,$$

that is, the observation consists of a signal $X(t)$ as a result of the presence of the target masked, however, by the everpresent noise $N(t)$.

The signal $X(t)$, $t \geq 0$, is assumed to be a stochastic process with known probability law and satisfying the assumptions of Theorem 4.5.1. The noise is assumed to have the form

$$N(t) = b(t)\dot{W}_0(t), \qquad t \geq 0,$$

and to keep things simple, we shall take $b(t) = 1$ so that $N(t) = \dot{W}_0(t)$ is just white Gaussian noise. Also for the sake of simplicity, we take the observation interval $\tau_0 = (0, 1)$ and assume that the signal and noise are mutually independent.†

If the observations were made only at a finite number of time instances, say

$$0 < t_1 < \cdots < t_n < 1$$

that is, if we only had the values of n random variables

$$Y(t_1) = y_1, \ldots, Y(t_n) = y_n$$

as a basis of our decision, then it is known from the classical theory of signal detection that the optimum solution of the problem would be as follows: Form the likelihood ratio

$$L(y_1, \ldots, y_n) = \frac{f_0(y_1, \ldots, y_n)}{f_1(y_1, \ldots, y_n)} \tag{22}$$

with f_0 and f_1 being the joint densities of the random variables $Y(t_1) = N(t_1), \ldots,$ $Y(t_n) = N(t_n)$ or $Y(t_1) = X(t_1) + N(t_1), \ldots, Y(t_n) = X(t_n) + N(t_n)$, respectively, and decide that the signal was absent or present depending on whether the likelihood ratio were greater or smaller than some fixed threshold c.

In the present case, we have a continuous observation

$$Y(t) = y(t), \qquad 0 < t < 1$$

† In the subsequent discussion, nothing would change if we admitted as in Theorem 4.5.1 that the signal may depend on past noise.

rather than just n samples, but it should at least be plausible that the likelihood ratio for a continuous observation

$$L(y(t), 0 < t < 1) \tag{23}$$

can be obtained as a limit of likelihood ratios (22), with the samples $Y(t_k) = y_k$ taken at increasing numbers n of time instances $0 < t_1 < \cdots < t_n < 1$, forming finer and finer partitions of the observation interval (0, 1).

However, we must not forget that $Y(t)$ is a generalized process, the formal derivative of a process $Z(t)$, and so before we can talk about densities, we need express the processes involved in the form of increments of their integrals, that is, rewrite (H_0) and (H_1) as

(H_0') $\qquad dZ(t) = dW_0(t), \qquad 0 < t < 1,$

(H_1') $\qquad dZ(t) = X(t)\, dt + dW_0(t), \qquad 0 < t < 1.$

The likelihood ratio of a sample

$$dZ(t_1), \ldots, dZ(t_n),$$

which we now denote by

$$L_n = \frac{f_0(z_1, \ldots, z_n)}{f_1(z_1, \ldots, z_n)} \tag{24}$$

is, however, still difficult to calculate. True, the density f_0 is just the joint density of independent Gaussian random variables

$$dW_0(t_1), \ldots, dW_0(t_n),$$

and hence it easily factors into the product of marginal Gaussian densities with zero means and variances

$$dt_k = t_{k+1} - t_k, \qquad k = 1, \ldots, n, \qquad t_n < t_{n+1} < 1.$$

However, nothing of this sort happens with the joint density f_1 of the random variables

$$X(t_1)\, dt_1 + dW_0(t_1), \ldots, X(t_n)\, dt_n + dW_0(t_n)$$

since it would be completely unrealistic to assume that the samples of the signal $X(t_1), \ldots, X(t_n)$ are independent, especially when the differences dt_k between the time instances become small.

Nevertheless, we can write the density f_1 as the product of conditional densities

$$f_1(z_1, \ldots, z_n) = f_1(z_n \mid z_1, \ldots, z_{n-1}) f_1(z_{n-1} \mid z_1, \ldots, z_{n-2}) \cdots f_1(z_2 \mid z_1) f_1(z_1),$$

and upon taking the logarithm of the likelihood ratio L_n in (24), we now have

$$\ln L_n = \sum_{k=1}^{n} (\ln f_0(z_k) - \ln f_1(z_k \mid z_1, \ldots, z_{k-1})). \tag{25}$$

Of course, we know that for each $k = 1, \ldots, n$

$$f_0(z_k) = \frac{1}{\sqrt{2\pi\, dt_k}} e^{-z_k^2/2\, dt_k} \tag{26}$$

but how do we find the conditional densities in the second term? This is where Theorem 4.5.1 enters the arena. According to this theorem, we can replace expression (H_1'), that is,

$$dZ(t) = X(t)\, dt + dW_0(t), \qquad 0 < t < 1,$$

by the equivalent expression

$$dZ(t) = \hat{X}(t)\, dt + dU(t), \qquad 0 < t < 1,$$

without affecting the problem in any way. Now if we examine the conditional distribution of

$$dZ(t_k) = \hat{X}(t_k)\, dt_k + dU(t_k)$$

we realize that $\hat{X}(t_k)$, being defined as a conditional expectation given the past $Z(t')$, $0 \le t' \le t_k$, will act as a constant, while $dU(t_k)$, being independent of the past, will have the same distribution as $dW_0(t_k)$. It follows that the conditional distribution of $dZ(t_k)$ will then be Gaussian with mean $\hat{X}(t_k)\, dt_k$ and variance dt_k, so that the conditional density is

$$f_1(z_k \mid z_1, \ldots, z_{k-1}) = \frac{1}{\sqrt{2\pi\, dt_k}} e^{-(z_k - \hat{x}(t_k)\, dt_k)^2/2\, dt_k}. \tag{27}$$

If we now substitute from (26) and (27) into (25), we find after taking logarithms that a number of terms cancel, and we end up with the expression

$$\ln L_n = \sum_{k=1}^{n} \left(-\hat{x}(t_k) z_k + \frac{1}{2} \hat{x}^2(t_k)\, dt_k \right)$$

The rest is easy; recalling that z_k represents the value of an increment $dZ(t_k)$, as the partition of the observation interval (0, 1) becomes finer and finer, the above sum will approximate an integral

$$-\int_0^1 \hat{x}(t)\, dZ(t) + \frac{1}{2} \int_0^1 \hat{x}^2(t)\, dt,$$

which is then the logarithm of the likelihood ratio (23). Removing the logarithm and reverting to the original generalized process $Y(t)$, that is, using the symbolic notation $dZ(t) = Y(t)\, dt$, we obtain Kailath's general likelihood ratio formula

$$L(Y(t), 0 < t < 1) = e^{-\int_0^1 \hat{X}(t) Y(t)\, dt + \frac{1}{2} \int_0^1 \hat{X}^2(t)\, dt} \tag{28}$$

From the point of view of applications the apparent simplicity of this formula is somewhat misleading, since $\hat{X}(t) = \mathrm{E}[X(t) \mid Y(t'), 0 \le t' \le t]$ is a functional of the entire past $Y(t')$, $0 \le t' \le t$, and the first integral in the exponent is actually an Itō integral mentioned in Section 4.3. Nevertheless, the generality of the formula (28) allows one to further derive many useful results concerning the likelihood ratio, and even an explicit evaluation of the right-hand side of (28) in a specific problem is usually easier than trying to obtain the likelihood ratio by other methods. Furthermore, the formula has a great intuitive appeal, since $\hat{X}(t)$ is, by definition, the best estimate of the signal $X(t)$, and the first integral in the exponent is actually the so-called matched filter, a familiar object in communication engineering. ▲

The innovation process $V(t)$, or equivalently, the innovation martingale $U(t)$, is according to its definition completely determined by the observation process.

Previous discussion also suggests that the converse statement may be true as well. That is, the observation process can be thought of as being built up from the infinitely dense stream of random innovations and hence completely determined by its innovation process. In other words, it seems that the observation process must be *causally equivalent* to its innovation process in the sense that since for any $t \geq 0$, $V(t)$ can be obtained from $Y(t')$, $0 \leq t' \leq t$, so should $Y(t)$ be obtainable from $V(t')$, $0 \leq t' \leq t$. Thus, one could visualize a causal filter that, when $Y(t)$, $t \geq 0$, is applied to its input, will produce the innovation process $V(t), t \geq 0$, on its output. Then there should be another causal filter that can reconstruct the process $Y(t)$, $t \geq 0$, when $V(t), t \geq 0$, is applied to its input, as in Figure 4.5.2. Under rather moderate conditions, this is indeed true, but the proof of this statement involves too many fine points to be reproduced here.

However, there is an important special case where the heuristic argument for the causal equivalence can be given quite easily, namely when the signal process $X(t)$, $t \geq 0$, is a Gaussian process. Since the causal equivalence for such a case can be utilized to obtain many useful results in filtering and prediction, we feel it is worthwhile to at least sketch the chain of reasoning behind our next theorem.

Thus, let us assume from now on that in model (1)

$$Y(t) = X(t) + b(t)\dot{W}_0(t), \qquad t \geq 0 \tag{29}$$

in addition to assumptions (8), (9), and (10), the signal $X(t)$, $t \geq 0$, is also a *zero mean Gaussian process.*†

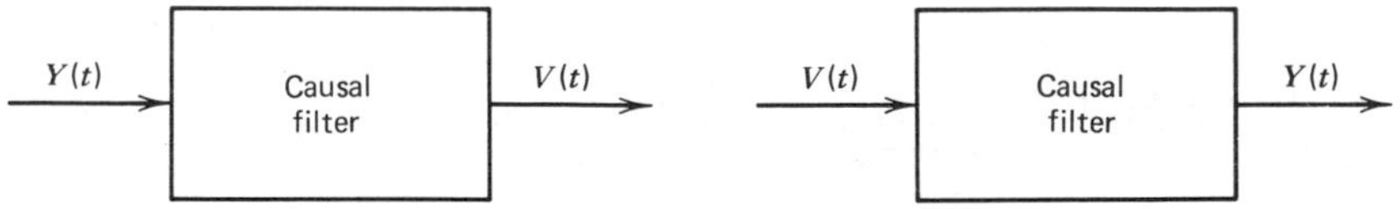

Figure 4.5.2

† It will turn out that we will be dealing exclusively with linear operations, and thus there is no loss of generality in assuming that the process $X(t)$ has a zero mean function.

To begin, let us first rewrite equation (29) in integrated form

$$Z(t) = \int_0^t X(\tau)\, d\tau + \int_0^t b(\tau)\, dW_0(\tau), \qquad t \geq 0.$$

Since both integrals on the right-hand side are now zero mean Gaussian processes, it follows that the integrated observation process is also a zero mean Gaussian process. But then the conditional expectation

$$\hat{X}(t) = \mathrm{E}[X(t) \mid Z(t'), 0 \leq t' \leq t] \tag{30}$$

is a linear functional of the past $Z(t')$, $0 \leq t' \leq t$, and as such could be approximated arbitrarily closely (in MS sense) by a linear combination of increments $dZ(t)$.

Let us again, as many times before, consider a partition $t_0 = 0 < t_1 < \cdots < t_n < t_{n+1} = t$ of a time interval $[0, t]$ and denote

$$dt_k = t_k - t_{k-1}, \; dZ(t_k) = Z(t_k) - Z(t_{k-1}), \; k = 1, \ldots, n.$$

The conditional expectation (30) can be approximated by the sum

$$\hat{X}(t_k) \doteq \sum_{j=0}^{k-1} h(t_k, t_j)\, dZ(t_j)$$

where $h(t_k, t_j)$, $j = 0, \ldots, k$ are the coefficients of the above linear combination. Then the expression for the increment of the innovation martingale

$$dU(t) = dZ(t) - \hat{X}(t)\, dt$$

can be approximated for $t = t_k$, $k = 0, \ldots, n-1$, by

$$dU(t_0) \doteq dZ(t_0)$$

$$dU(t_k) \doteq dZ(t_k) - \sum_{j=0}^{k-1} h(t_k, t_j)\, dZ(t_j)\, dt_k, \qquad k = 1, \ldots, n-1.$$

Dividing each equation by dt_k, this becomes

$$\frac{dV(t_0)}{dt_0} \doteq \frac{dZ(t_0)}{dt_0}, \tag{31a}$$

$$\frac{dV(t_k)}{dt_k} = \frac{dZ(t_k)}{dt_k} - \sum_{j=0}^{k-1} h(t_k, t_j) \frac{dZ(t_j)}{dt_j}\, dt_j, \qquad k = 1, \ldots, n-1. \tag{31b}$$

Now (31) can be looked upon as being a system of n linear equations in n unknowns

$$\frac{dZ(t_0)}{dt_0}, \ldots, \frac{dZ(t_{n-1})}{dt_{n-1}}.$$

A closer look at these equations reveals that the system is of a very special kind, the matrix of the coefficients is lower triangular with ones on the main diagonal,

$$\begin{bmatrix} 1 & 0 & \cdots & 0 & 0 \\ -h(t_1, t_0)\,dt_0 & 1 & \cdots & 0 & 0 \\ -h(t_2, t_0)\,dt_0 & -h(t_2, t_1)\,dt_1 & & & \\ \cdots & \cdots & \cdots & 1 & 0 \\ -h(t_{n-1}, t_0)\,dt_0 & -h(t_{n-1}, t_1)\,dt_1 & \cdots & -h(t_{n-1}, t_{n-2})\,dt_{n-2} & 1 \end{bmatrix}.$$

It is well known (and easily verified) that such a matrix always has an inverse that is again a lower triangular matrix with 1's on the main diagonal. But this means that the system (31) must have a unique solution, which is of the form

$$\frac{dZ(t_0)}{dt_0} \doteq \frac{dU(t_0)}{dt_0}, \tag{32a}$$

$$\frac{dZ(t_k)}{dt_k} \doteq \frac{dU(t_k)}{dt_k} + \sum_{j=0}^{k-1} g(t_k, t_j)\frac{dU(t_j)}{dt_j}\,dt_j, \qquad k = 1, \ldots, n-1. \tag{32b}$$

If we now compare equations (31) and (32) with $k = n - 1$, we have a pair of approximate equations

$$\frac{dU(t_n)}{dt_n} \doteq \frac{dZ(t_n)}{dt_n} - \sum_{j=0}^{n-1} h(t_n, t_j)\frac{dZ(t_j)}{dt_j}\,dt_j,$$

$$\frac{dZ(t_n)}{dt_n} \doteq \frac{dU(t_n)}{dt_n} + \sum_{j=0}^{n-1} g(t_n, t_j)\frac{dU(t_j)}{dt_j}\,dt_j. \tag{33}$$

If we now let $n \to \infty$ by making the partition $t_0 = 0 < t_1 < \cdots < t_n < t_{n+1} = t$ finer and finer, we would then expect the sums in (33) to become integrals and the ratios

$$\frac{dV(t_n)}{dt_n} \quad \text{and} \quad \frac{dZ(t_n)}{dt_n}$$

to become the formal derivatives $\dot{U}(t) = V(t)$, $\dot{Z}(t) = Y(t)$, respectively.

In other words, we expect to end up with the pair of relations

$$V(t) = Y(t) - \int_0^t h(t, \tau)Y(\tau)\,d\tau, \tag{34a}$$

$$Y(t) = V(t) + \int_0^t g(t, \tau)V(\tau)\,d\tau, \tag{34b}$$

valid for all $t \geq 0$.

Of course, the above is merely a plausible argument and far from any rigorous proof.

Nevertheless, note that the functions h and g in (34) can be thought of as the

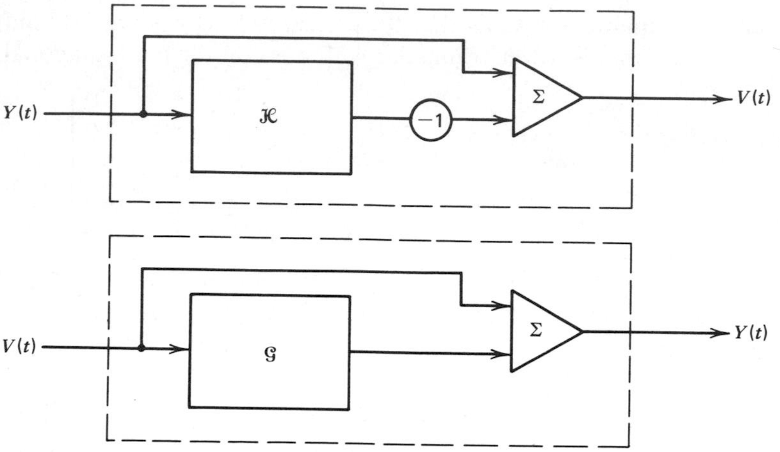

Figure 4.5.3

impulse-response functions of causal linear systems $\mathcal{H}$ and $\mathcal{G}$, respectively, and so equations (34) can be represented by block diagrams, as in Figure 4.5.3. The first diagram represents a *causal linear* filter that converts the observation process Y into white Gaussian noise (the innovation process). For this reason, such a filter is called a *whitening filter*.

The second diagram shows that it is possible to restore the observation process by passing the innovation process through another *causal linear* filter.

In other words, the portion of the innovation process $V(t')$, $0 \le t' \le t$, indeed contains all the information contained in the portion of the observation process $Y(t)$, $0 \le t' \le t$, and vice versa. We refer to such a situation by saying that the processes $Y(t)$ and $V(t)$ are *causally linearly equivalent*.

Remark: The block diagrams show that the causal filter $\mathcal{H}$ extracts from the past observations the predictable part, which is then subtracted from the current value $Y(t)$ to generate the instantaneous innovation $V(t)$.

On the other hand, the causal filter $\mathcal{G}$ recombines the past innovations, and the result is added to the current innovation to recreate the observation $Y(t)$.

These operators can be written symbolically as

$$V = (\mathcal{I} - \mathcal{H})Y, \qquad Y = (\mathcal{I} + \mathcal{G})V$$

where the symbol $\mathcal{I}$ represents the identity operator. But then the operator $(\mathcal{I} + \mathcal{G})$ should be the inverse of the operator $(\mathcal{I} - \mathcal{H})$, and from the analogy with geometric series, we might be tempted to write

$$(\mathcal{I} + \mathcal{G}) = (\mathcal{I} - \mathcal{H})^{-1} = \mathcal{I} + \mathcal{H} + \mathcal{H}^2 + \mathcal{H}^3 + \cdots, \tag{35}$$

with powers interpreted as repeated applications of the operator $\mathcal{H}$. In terms of block diagrams, this would look like Figure 4.5.4. It can be shown that for causal linear operators

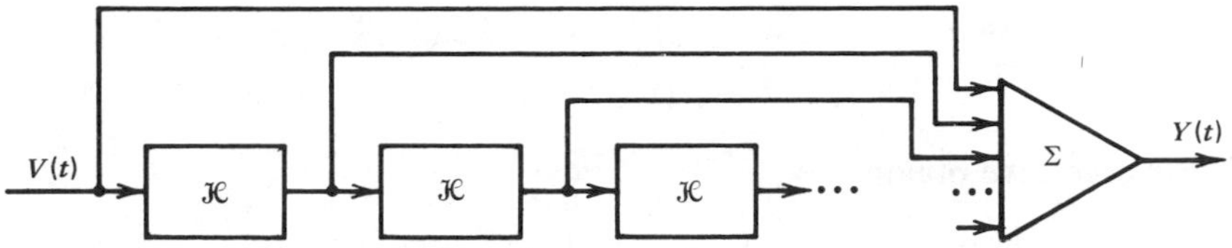

Figure 4.5.4

of the above kind, called the Volterra operators, the symbolic equation (35) is indeed meaningful. (The symbolic infinite series is known as Neumann series.) Я

Example 4.5.2

As an example of a simple but admittedly somewhat artificial pair of causally linearly equivalent transformations, suppose that the filter ℋ has the impulse-response function

$$h(t, \tau) = \begin{cases} \sigma(1 + \tau)^{\sigma - 1}(1 + t)^{-\sigma} & \text{if} \quad 0 \leq \tau < t, \\ 0 & \text{otherwise,} \end{cases} \tag{36}$$

where $\sigma > 0$ is a constant. Clearly, ℋ is a causal filter turning every finite-energy input into a finite-energy output, since for any $t_0 > 0$

$$\int_0^{t_0} \int_0^{t_0} h^2(t, \tau)\, d\tau\, dt < \infty.$$

Then, given

$$v(t) = y(t) - \int_0^t h(t, \tau)y(\tau)\, d\tau, \qquad t \geq 0, \tag{37}$$

with $y(t)$ a square-integrable function such that $y(0) = 0$, we wish to find the causally inverse relation

$$y(t) = v(t) + \int_0^t g(t, \tau)v(\tau)\, d\tau, \qquad t \geq 0,$$

that is, to determine the causal impulse-response function g.

With h as in (36), the relation (37) can be written as

$$(1 + t)^\sigma v(t) = (1 + t)^\sigma y(t) - \int_0^t \sigma(1 + \tau)^{\sigma - 1} y(\tau)\, d\tau. \tag{38}$$

Since $y(0) = 0$, integration by parts yields,

$$\int_0^t \sigma(1 + \tau)^{\sigma - 1} y(\tau)\, d\tau = (1 + t)^\sigma y(t) - \int_0^t (1 + \tau)^\sigma \dot{y}(\tau)\, d\tau,$$

and thus (38) becomes

$$(1 + t)^\sigma v(t) = \int_0^t (1 + \tau)^\sigma \dot{y}(\tau)\, d\tau.$$

Taking derivatives, we obtain

$$\sigma(1 + t)^{\sigma - 1} v(t) + (1 + t)^\sigma \dot{v}(t) = (1 + t)^\sigma \dot{y}(t),$$

whence upon dividing by $(1 + t)^\sigma$ and integrating, we have

$$y(t) = v(t) + \int_0^t \frac{\sigma}{1 + \tau} v(\tau)\, d\tau, \qquad t \geq 0.$$

This gives the desired causal inverse filter $\mathcal{G}$ with impulse-response function

$$g(t, \tau) = \begin{cases} \dfrac{\sigma}{1 + \tau} & \text{if } 0 \leq \tau < t, \\ 0 & \text{otherwise,} \end{cases}$$

which is clearly causal and again square-integrable:

$$\int_0^{t_0} \int_0^{t_0} g^2(t, \tau)\, d\tau\, dt < \infty \qquad \text{for all } t_0 > 0.$$

It should be noted, however, that such a simple pair of linear causal inverses is an exception, although general methods for finding an inverse are available. Nevertheless, if one of the filters, say $\mathcal{H}$, is available as a piece of hardware, it is a relatively simple matter to create the causal inverse transformation. One can simply put the filter $\mathcal{H}$ into a positive feedback loop with unit gain in the forward link as shown in Figure 4.5.5. Then symbolically, $V + \mathcal{H}Y = Y$, that is, $V = (\mathcal{I} - \mathcal{H})Y$, from which

$$Y = (\mathcal{I} - \mathcal{H})^{-1} V = (\mathcal{I} + \mathcal{G})V,$$

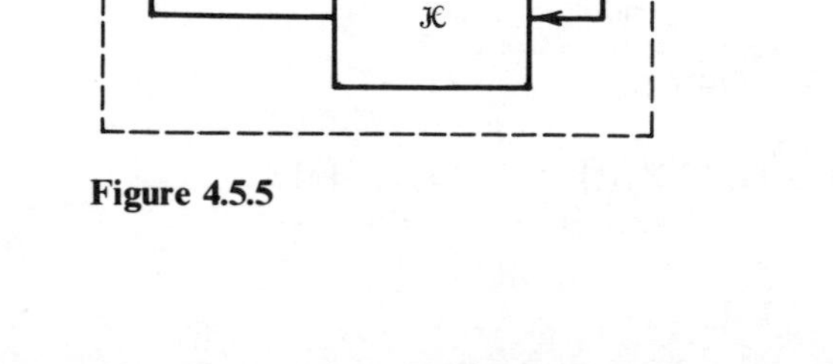

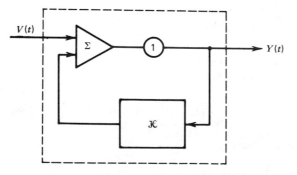

Figure 4.5.5

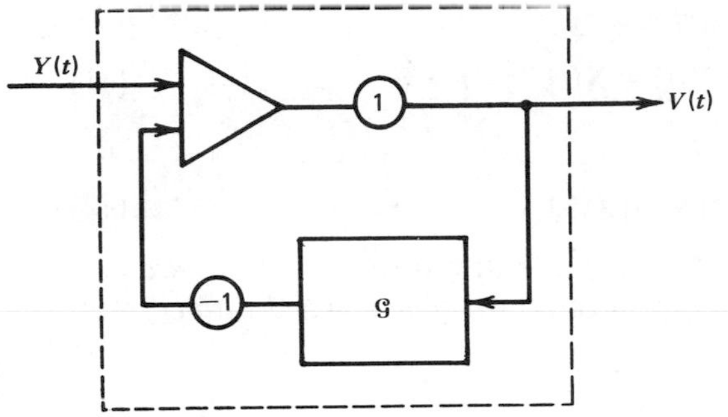

Figure 4.5.6

so that the transformation in Figure 4.5.5 is equivalent to that depicted in Figure 4.5.6.

Similarly, if the filter $\mathcal{G}$ is available, then using a negative feedback as in Figure 4.5.6, we find the transformation to be equivalent to that in Figure 4.5.5. ▲

Causal linear equivalence, together with the content of Theorem 4.5.1 specialized to a Gaussian signal, constitutes what is called the linear innovation theorem. Although we have so far considered only scalar-valued processes the entire previous discussion extends easily to vector-valued processes. Since these are what is needed in most applications, we will state the linear innovation theorem in this more general version.

THEOREM 4.5.2. *Let*

$$\mathbf{Y}(t) = \mathbf{X}(t) + \mathbf{b}(t)\dot{\mathbf{W}}_0(t), \qquad t \geq 0, \tag{39}$$

where

$$\left.\begin{array}{ll} \text{(i)} & \mathbf{X}(t),\ t \geq 0, \textit{ is a vector-valued Gaussian process with zero mean function and with correlation function } \mathbf{R}_\mathbf{X}(t_1, t_2) \textit{ such that each entry of the matrix } \mathbf{R}_\mathbf{X}(t, t) \textit{ is integrable over any interval } [0, t],\ t > 0. \\ \text{(ii)} & \dot{\mathbf{W}}_0(t),\ t \geq 0, \textit{ is vector-valued white Gaussian noise, and } \mathbf{b}(t),\ t \geq 0 \textit{ is a square matrix nonsingular for each } t > 0 \textit{ and such that each of its entries is square-integrable over any interval } [0, t],\ t > 0. \\ \text{(iii)} & \textit{For any } 0 \leq t_1 < t_2 \quad E[\mathbf{X}(t_1)\dot{\mathbf{W}}'_0(t_2)] = \mathbf{0}. \end{array}\right\} \tag{40}$$

Then the innovation process defined by

$$\mathbf{V}(t) = \mathbf{Y}(t) - \hat{\mathbf{X}}(t), \qquad t \geq 0, \tag{41}$$

where

$$\hat{\mathbf{X}}(t) = \mathrm{E}[\mathbf{X}(t) \mid \mathbf{Y}(t'), 0 \leq t' \leq t], \tag{42}$$

is equivalent to the process $\mathbf{b}(t)\dot{\mathbf{W}}_0(t)$, $t \geq 0$; *that is, it is a generalized vector-valued Gaussian process with zero mean function and with correlation function*

$$\mathbf{R}_{\mathbf{V}}(t_1, t_2) = \mathbf{b}(t_1)\mathbf{b}'(t_2)\,\delta(t_1 - t_2), \qquad t_1 \geq 0, \qquad t_2 \geq 0.$$

Also for any $0 \leq t_1 < t_2$

$$\mathrm{E}[\hat{\mathbf{X}}(t_1)\mathbf{V}'(t_2)] = \mathbf{0}.$$

Furthermore, the generalized processes $\mathbf{Y}(t)$ *and* $\mathbf{V}(t)$ *are causally linearly equivalent, which means that there exist two causal and square-integrable matrix-valued impulse-response functions* $\mathbf{h}$ *and* $\mathbf{g}$ *such that for all* $t \geq 0$,

$$\mathbf{V}(t) = \mathbf{Y}(t) - \int_0^t \mathbf{h}(t, \tau)\mathbf{Y}(\tau)\,d\tau, \tag{43a}$$

and

$$\mathbf{Y}(t) = \mathbf{V}(t) + \int_0^t \mathbf{g}(t, \tau)\mathbf{V}(\tau)\,d\tau. \tag{43b}$$

The reader may notice that assumptions (40) are merely restatements of the assumptions of Theorem 4.5.1 for the vector-valued case. Since the signal process $\mathbf{X}(t)$ is assumed zero mean, Gaussian assumption (40, iii) is equivalent to the assumption that the future increments $d\mathbf{W}_0(t)$ are independent of the past and present of the signal. Notice also that the definition of the innovation process (41), together with equations (43 a, b) imply that for any $t > 0$ the conditional expectation (42) can be expressed either as

$$\hat{\mathbf{X}}(t) = \int_0^t \mathbf{h}(t, \tau)\mathbf{Y}(\tau)\,d\tau, \tag{44}$$

or as

$$\hat{\mathbf{X}}(t) = \int_0^t \mathbf{g}(t, \tau)\mathbf{V}(\tau)\,d\tau. \tag{45}$$

This, of course, is a consequence of the causal linear equivalence of $\mathbf{Y}(t)$ and $\mathbf{V}(t)$, since conditioning upon $\mathbf{Y}(t')$, $0 \leq t' \leq t$, must then be the same as conditioning upon $\mathbf{V}(t')$, $0 \leq t' \leq t$, that is, we must also have

$$\hat{\mathbf{X}}(t) = \mathrm{E}[\mathbf{X}(t) \mid \mathbf{V}(t'), 0 \leq t' \leq t].$$

Remark. It would be more accurate to replace the upper limit t in the integrals (43a,b), (44), and (45) by $t-$, for as the approximating sums (33) indicate, the "last" forward increments $dZ(t_n)$ and $dU(t_n)$ were not included in the sums. We can, however, achieve the same goal by interpreting the term "causal filter $\mathcal{F}$" to mean that its impulse-response function $f(t, \tau) = 0$ for $\tau \geq t$. ∎

Since, by definition, $\hat{\mathbf{X}}(t)$, $t \geq 0$, is the best (mean square error) estimate of the signal $\mathbf{X}(t)$ based on the observation $\mathbf{Y}(t')$, $0 \leq t' \leq t$, either of equations (44) and (45) provides an immediate solution to the signal estimation problem. Of course, the snag here is that one must first obtain an expression for at least one of the filters $\mathcal{H}$ or $\mathcal{G}$ from a specification of the signal. Generally, this may be quite difficult, but there is one quite important class of problems where a natural specification of the signal allows one to solve the estimation problem in a manner especially suitable for actual implementation. This is the case of the famous Kalman-Bucy recursive filter. Since the purpose of this book is not to study the estimation problem in detail but rather to develop the appropriate probabilistic tools, we choose a considerably simplified version of such a problem merely to illustrate the use of innovation processes in this case.

KALMAN-BUCY RECURSIVE FILTER EQUATIONS. Suppose that the signal process $\mathbf{X}(t)$, $t \geq 0$, is defined as a state of a dynamic system that evolves according to the state equation

$$\dot{\mathbf{X}}(t) = \mathbf{a}(t)\mathbf{X}(t) + \mathbf{b}(t)\dot{\mathbf{W}}_0^{(s)}(t), \qquad t \geq 0, \tag{46}$$

where $\mathbf{a}(t)$ and $\mathbf{b}(t)$ are square matrices, and $\dot{\mathbf{W}}_0^{(s)}(t)$, $t \geq 0$, is vector-valued white Gaussian noise, referred to as the system noise. The initial state $\mathbf{X}(0)$ is assumed to be a Gaussian random vector with $\mathrm{E}[\mathbf{X}(0)] = 0$ and with covariance matrix

$$\mathrm{E}[\mathbf{X}(0)\mathbf{X}'(0)] = \boldsymbol{\sigma}_0^2.$$

These assumptions imply that $\mathbf{X}(t)$, $t \geq 0$, is a vector-valued Gauss-Markov process, which may seem to be a rather restrictive assumption. However, the majority of physical systems can, at least approximately, be described by a system of linear differential equations, which can always be represented in form (46) by a suitable choice of the state variables. Thus, our signal represents the state of quite a general linear dynamic system disturbed by white Gaussian noise, a model that has been successfully applied for well over a decade to a variety of practical problems.

However, the signal $\mathbf{X}(t)$, $t \geq 0$, cannot be observed directly; instead, we again have an observation process

$$\mathbf{Y}(t) = \mathbf{X}(t) + \dot{\mathbf{W}}_0^{(o)}(t), \qquad t \geq 0 \tag{47}$$

masking the signal by additive white Gaussian noise, referred to as the observation noise. We will assume here that the observation noise and the signal are mutually independent:

$$E[\mathbf{X}(t_1)\dot{\mathbf{W}}_0^{(0)\prime}(t_2)] = \mathbf{0} \qquad \text{for all } t_1 \geq 0, t_2 \geq 0, \tag{48}$$

and that the two noises are mutually independent as well:

$$E[\dot{\mathbf{W}}_0^{(s)}(t_1)\dot{\mathbf{W}}_0^{(o)\prime}(t_2)] = \mathbf{0} \qquad \text{for all } t_1 \geq 0, t_2 \geq 0. \tag{49}$$

Our goal is to obtain computationally feasible equations for the best estimate

$$\hat{\mathbf{X}}(t) = E[\mathbf{X}(t) \mid \mathbf{Y}(t'), 0 \leq t' \leq t], \qquad t \geq 0,$$

of the signal based on the past and present sample path of the observation process.

Let

$$\mathbf{V}(t) = \mathbf{Y}(t) - \hat{\mathbf{X}}(t), \qquad t \geq 0,$$

denote the innovation process of the observation process $\mathbf{Y}(t)$, and let $0 \leq t_1 < t_2$. In view of the causal linear equivalence of $\mathbf{V}(t)$ and $\mathbf{Y}(t)$, we then have

$$\begin{aligned} E[\hat{\mathbf{X}}(t_2)\mathbf{V}'(t_1)] &= E[E[\hat{\mathbf{X}}(t_2)\mathbf{V}'(t_1) \mid \mathbf{V}(t'), 0 \leq t' \leq t_2]] \\ &= E[E[\mathbf{X}(t_2) \mid \mathbf{V}(t'), 0 \leq t' \leq t_2]\mathbf{V}'(t_1)] \\ &= E[\mathbf{X}(t_2)\mathbf{V}'(t_1)]. \end{aligned} \tag{50}$$

Consequently, if we write, as in (45),

$$\hat{\mathbf{X}}(t_2) = \int_0^{t_2} \mathbf{g}(t_2, \tau)\mathbf{V}(\tau)\, d\tau, \tag{51}$$

then multiply both sides on the right by $\mathbf{V}'(t_1)$ and take the expectation, we get by (50)

$$E[\mathbf{X}(t_2)\mathbf{V}'(t_1)] = \int_0^{t_2} \mathbf{g}(t_2, \tau)E[\mathbf{V}(\tau)\mathbf{V}'(t_1)]\, d\tau = \mathbf{g}(t_2, t_1), \tag{52}$$

since the innovation process is, in view of (47), just white Gaussian noise, that is,

$$E[\mathbf{V}(\tau)\mathbf{V}'(t_1)] = \delta(\tau - t_1)\mathbf{I}$$

with $\mathbf{I}$ the identity matrix. Upon substitution from (52) into (51), we thus have the expression

$$\hat{\mathbf{X}}(t) = \int_0^{t} E[\mathbf{X}(t)\mathbf{V}'(\tau)]\mathbf{V}(\tau)\, d\tau, \qquad t \geq 0. \tag{53}$$

Taking the time derivative and using state equation (46), we obtain

$$\dot{\hat{\mathbf{X}}}(t) = \mathrm{E}[\mathbf{X}(t)\mathbf{V}'(t)]\mathbf{V}(t) + \mathbf{a}(t)\int_0^t \mathrm{E}[\mathbf{X}(t)\mathbf{V}'(\tau)]\mathbf{V}(\tau)\,d\tau$$

$$+ \mathbf{b}(t)\int_0^t \mathrm{E}[\dot{\mathbf{W}}_0^{(s)}(t)\mathbf{V}'(\tau)]\mathbf{V}(\tau)\,d\tau. \tag{54}$$

However, by (48), $\mathrm{E}[\dot{\mathbf{W}}_0^{(s)}(t)\mathbf{V}'(\tau)] = 0$, while by (53), the second term on the right-hand side equals $\mathbf{a}(t)\hat{\mathbf{X}}(t)$. Hence, if we denote the expectation in the first term on the right-hand side

$$\mathrm{E}[\mathbf{X}(t)\mathbf{V}'(t)] = \mathbf{k}(t) \tag{55}$$

equation (54) becomes

$$\dot{\hat{\mathbf{X}}}(t) = \mathbf{k}(t)\mathbf{V}(t) + \mathbf{a}(t)\hat{\mathbf{X}}(t), \qquad t \geq 0. \tag{56}$$

If we further substitute for the innovation process

$$\mathbf{V}(t) = \mathbf{Y}(t) - \hat{\mathbf{X}}(t), \tag{57}$$

we obtain, after minor rearrangement, the differential equation

$$\dot{\hat{\mathbf{X}}}(t) = (\mathbf{a}(t) - \mathbf{k}(t))\hat{\mathbf{X}}(t) + \mathbf{k}(t)\mathbf{Y}(t), \qquad t \geq 0. \tag{58}$$

Since the assumption $\mathrm{E}[\mathbf{X}(0)] = \mathbf{0}$ immediately yields the initial condition

$$\hat{\mathbf{X}}(0) = \mathbf{0},$$

we thus have a first order linear differential equation (or more precisely, a system of such equations, since $\hat{\mathbf{x}}(t)$ is a vector) for the estimate $\hat{\mathbf{X}}(t)$, $t \geq 0$.

This is the *main Kalman-Bucy equation* and from it we can clearly see why the filter is called recursive. There is no need to actually store past observations; to update our estimate, that is, to calculate the infinitesimal increment $d\hat{\mathbf{X}}(t)$, we only need the current estimate $\hat{\mathbf{X}}(t)$ and the current observation $\mathbf{Y}(t)$. Thus the recursive form is extremely well suited for real-time computation, which is one of the main reasons for the popularity of this filter.

It remains to determine the matrix function $\mathbf{k}(t)$ known as the *Kalman gain*, but as will be shown, this function can be computed entirely from the problem data, that is, $\mathbf{a}(t)$, $\mathbf{b}(t)$, $t \geq 0$, and $\boldsymbol{\sigma}_0^2$ in our simplified version of the problem, and is therefore not dependent in any way on the sample path of the observation process. Now the Kalman gain was defined by (55) to be

$$\mathbf{k}(t) = \mathrm{E}[\mathbf{X}(t)\mathbf{V}'(t)], \qquad t \geq 0.$$

Since by (57) and (47)

$$\mathbf{V}(t) = \mathbf{Y}(t) - \hat{\mathbf{X}}(t) = \mathbf{X}(t) - \hat{\mathbf{X}}(t) + \dot{\mathbf{W}}_0^{(o)}(t),$$

we obtain, upon substitution,

$$\mathbf{k}(t) = \mathrm{E}[\mathbf{X}(t)(\mathbf{X}(t) - \hat{\mathbf{X}}(t))'], \tag{59}$$

since by (48) $\mathrm{E}[\mathbf{X}(t)\dot{\mathbf{W}}_0^{(o)\prime}(t)] = 0$. But recall that $\hat{\mathbf{X}}(t)$ is the best mean square estimate of $\mathbf{X}(t)$, and in Exercise 1.38 you were asked to show that the expectation in (59) is then equal to the mean square error matrix

$$\boldsymbol{\varepsilon}^2(t) = \mathrm{E}[(\mathbf{X}(t) - \hat{\mathbf{X}}(t))(\mathbf{X}(t) - \hat{\mathbf{X}}(t))']. \tag{60}$$

Thus, in our simplified version† of the problem, the Kalman gain

$$\mathbf{k}(t) = \boldsymbol{\varepsilon}^2(t), \tag{61}$$

the mean square estimation error matrix. In Exercise 1.38 you were also asked to show that a mean square error matrix, such as the one in (60), can also be expressed as

$$\boldsymbol{\varepsilon}^2(t) = \mathrm{E}[\mathbf{X}(t)\mathbf{X}'(t)] - \mathrm{E}[\hat{\mathbf{X}}(t)\hat{\mathbf{X}}'(t)], \tag{62}$$

that is, as a difference of variance functions

$$\boldsymbol{\sigma}_{\mathbf{X}}^2(t) = \mathrm{E}[\mathbf{X}(t)\mathbf{X}'(t)], \qquad t \geq 0, \tag{63}$$

and

$$\boldsymbol{\sigma}_{\hat{\mathbf{X}}}^2(t) = \mathrm{E}[\hat{\mathbf{X}}(t)\hat{\mathbf{X}}'(t)], \qquad t \geq 0, \tag{64}$$

of the Gaussian processes $\mathbf{X}(t)$ and $\hat{\mathbf{X}}(t)$, respectively. (Recall that $\mathbf{X}(t)$ and hence also $\hat{\mathbf{X}}(t)$ have zero mean functions.) However, the process $\mathbf{X}(t)$, $t \geq 0$, is defined as the state of a dynamic system with white Gaussian noise input with the state equation (46). In Section 4.4, we have shown that its variance function then satisfies the differential equation

$$\dot{\boldsymbol{\sigma}}_{\mathbf{X}}^2(t) = \mathbf{a}(t)\boldsymbol{\sigma}_{\mathbf{X}}^2(t) + \boldsymbol{\sigma}_{\mathbf{X}}^2(t)\mathbf{a}'(t) + \mathbf{b}(t)\mathbf{b}'(t). \tag{65}$$

But the zero mean Gaussian process $\hat{\mathbf{X}}(t)$, $t \geq 0$, can also be regarded as the state of a dynamic system with state equation (56), that is,

$$\dot{\hat{\mathbf{X}}}(t) = \mathbf{a}(t)\hat{\mathbf{X}}(t) + \mathbf{k}(t)\mathbf{V}(t), \qquad t \geq 0,$$

with white Gaussian noise input $\mathbf{V}(t)$, $t \geq 0$. It follows that its variance function will also satisfy the differential equation (4.4.33) of Section 4.4, with the Kalman gain $\mathbf{k}(t)$ in place of the matrix $\mathbf{b}(t)$, that is,

$$\dot{\boldsymbol{\sigma}}_{\hat{\mathbf{X}}}^2(t) = \mathbf{a}(t)\boldsymbol{\sigma}_{\hat{\mathbf{X}}}^2(t) + \boldsymbol{\sigma}_{\hat{\mathbf{X}}}^2(t)\mathbf{a}'(t) + \mathbf{k}(t)\mathbf{k}'(t). \tag{66}$$

† In a more general case, the mean square error matrix, although not equal to the Kalman gain, is easily determined from it.

Subtracting (66) from (65) and using the fact that by (61) through (64)

$$\mathbf{k}(t) = \boldsymbol{\sigma}_{\mathbf{X}}^2(t) - \boldsymbol{\sigma}_{\hat{\mathbf{X}}}^2(t),$$

we obtain a differential equation for the Kalman gain matrix:

$$\dot{\mathbf{k}}(t) = \mathbf{a}(t)\mathbf{k}(t) + \mathbf{k}(t)\mathbf{a}'(t) + \mathbf{b}(t)\mathbf{b}'(t) - \mathbf{k}(t)\mathbf{k}'(t), \qquad t \geq 0. \tag{67}$$

The initial condition is now $\mathbf{k}(0) = \boldsymbol{\varepsilon}^2(0) = \boldsymbol{\sigma}_0^2$. This (system of) first order differential equation(s) is the *auxiliary Kalman-Bucy equation*, which as we have already said, allows us to determine the matrix function $\mathbf{k}(t)$, $t \geq 0$, from the problem data. Unlike the main equation (58), this onc is nonlinear because of the quadratic term $\mathbf{k}(t)\mathbf{k}'(t)$ on the right-hand side, and it is a differential equation known as a Riccati equation. Naturally, it will, in general, be more difficult to solve than a linear equation; for practical applications of the Kalman-Bucy filter, however, one usually does not need a general formula for the matrix function $\mathbf{k}(t)$ but rather its actual numerical values for some numerically given problem data. Efficient numerical algorithms for solving Riccati equations do exist, and since the function $\mathbf{k}(t)$ need not be computed in real time, the nonlinearity of the auxiliary equation is not a serious problem. After all, it is easily seen that both the main and auxiliary Kalman-Bucy equations can be readily implemented by an analog scheme as shown in Figure 4.5.7 or its corresponding digital counterpart.

Example 4.5.3

The state-space representation of almost any real world dynamic system requires at least two, and usually considerably more, components for its state vector. This is one of the reasons why the Kalman-Bucy equations are hardly ever used to calculate the impulse-response function h of the optimal filter (see equation (44)).

$$\hat{\mathbf{X}}(t) = \int_0^t \mathbf{h}(t, \tau)\mathbf{Y}(\tau)\, d\tau, \qquad t \geq 0. \tag{68}$$

Instead, as mentioned before, the equations are typically used to calculate the estimate recursively on a computer with numerical problem data. This, of course, is an advantage rather than a drawback as far as practical applications are concerned.

However, if we wish to present an example without engaging in tedious calculations, we are forced to choose an extremely simple case, which is, of necessity, rather artificial. We will therefore consider a scalar-valued case with both $a(t)$ and $b(t)$ constants equal to 1, specifically the state equation

$$\dot{X}(t) = X(t) + \dot{W}_0^{(s)}(t), \qquad t \geq 0,$$

with the initial variance $\sigma_0^2 = 1$. The observation process

$$Y(t) = X(t) + \dot{W}_0^{(o)}(t), \qquad t \geq 0$$

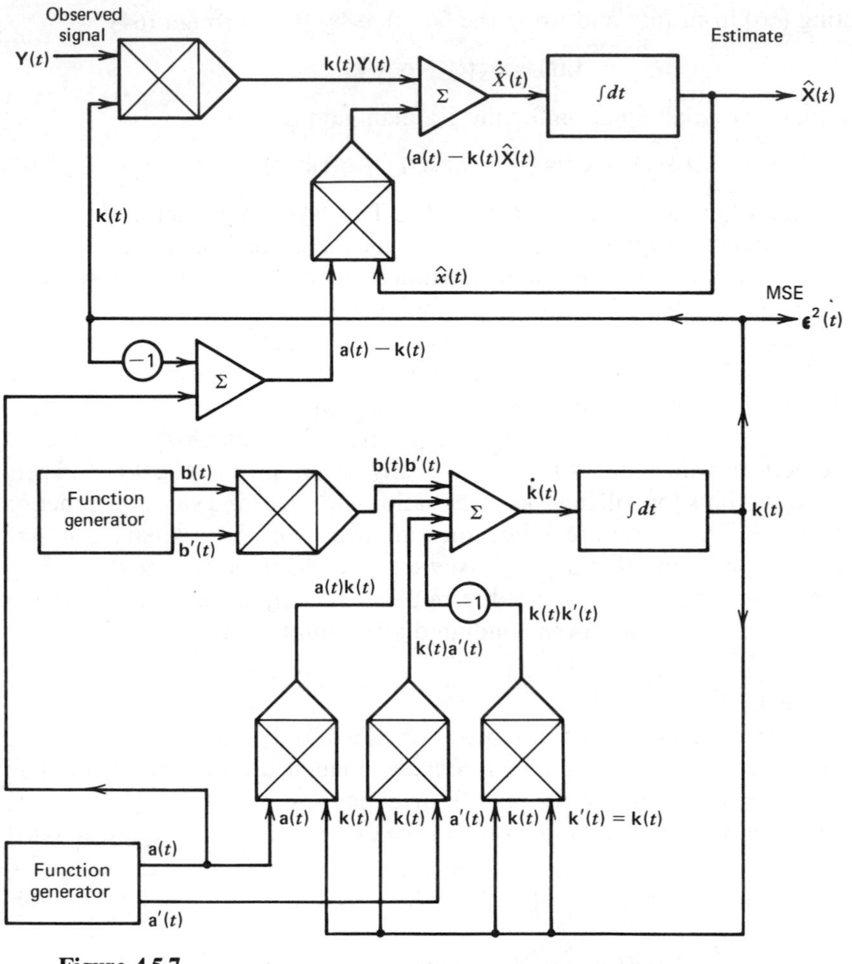

Figure 4.5.7

is also scalar-valued with all the remaining assumptions (E[$X(0)$] = 0, independence of noise, etc.) in force.

Let us begin with the auxiliary Kalman-Bucy equation (67). In our case the Kalman gain function $k(t)$, $t \geq 0$, is scalar-valued and since $a(t) = b(t) = 1$, equation (67) takes the form

$$\dot{k}(t) = 2k(t) + 1 - k^2(t), \qquad t \geq 0, \tag{69}$$

with the initial condition $k(0) = \sigma_0^2 = 1$. Now (69) is a Riccati differential equation with constant coefficients, and such an equation can be transformed into a homogeneous second order linear differential equation with constant coefficients by the substitution

$$k(t) = \frac{\dot{z}(t)}{z(t)}.$$

Applying this to equation (69), we obtain

$$\ddot{z}(t) - 2\dot{z}(t) - z(t) = 0,$$

which has the general solution

$$z(t) = C_1 e^{(1+\sqrt{2})t} + C_2 e^{(1-\sqrt{2})t}.$$

Differentiating and dividing by $z(t)$, we get back, after factoring e^t,

$$k(t) = 1 + \sqrt{2}\frac{C_1 e^{t\sqrt{2}} - C_2 e^{-t\sqrt{2}}}{C_1 e^{t\sqrt{2}} + C_2 e^{-t\sqrt{2}}}. \tag{70}$$

Since $k(0) = 1$, we must have $C_1 = C_2 = C$, and so (70) can be written as

$$k(t) = 1 + C\sqrt{2} \tanh t\sqrt{2}. \tag{71}$$

To determine the remaining constant C, we differentiate (71) to get

$$\dot{k}(t) = C\frac{2}{\cosh^2 t\sqrt{2}}, \tag{72}$$

and note that since $k(0) = 1$, we must have from (69)

$$\dot{k}(0) = 2 + 1 - 1 = 2.$$

It follows by setting $t = 0$ in (72) that $C = 1$, and so the desired solution of the auxiliary Kalman-Bucy equation (69) is

$$k(t) = 1 + \sqrt{2} \tanh t\sqrt{2}, \qquad t \geq 0. \tag{73}$$

Now the main Kalman-Bucy equation (58) for the recursive estimate $\hat{X}(t)$ can be written down, and with $a(t) = 1$ and $k(t)$ as in (73), it is

$$\begin{aligned}\dot{\hat{X}}(t) = {} & (-\sqrt{2} \tanh t\sqrt{2})\hat{X}(t) \\ & + (1 + \sqrt{2} \tanh t\sqrt{2})Y(t), \qquad t \geq 0\end{aligned} \tag{74}$$

with the initial condition $\hat{X}(0) = 0$. This equation would normally be used to continuously update the estimate by a computer, but in our simple case we can actually write down an explicit solution, for (74) is just a single first order linear differential equation that can always be solved, for instance by Lagrange's method of variation of parameters. We will not bother with details but just present the solution:

$$\hat{X}(t) = \frac{1}{\cosh t\sqrt{2}} \int_0^t (\cosh \tau\sqrt{2} + \sqrt{2}\sinh \tau\sqrt{2}) Y(\tau)\, d\tau \tag{75}$$

which can easily be verified by substitution into (74). Using the definition of hyperbolic sine and cosine, (75) can be written in a perhaps more appealing form:

$$\hat{X}(t) = \int_0^t \left(\frac{(1+\sqrt{2})e^{-(t-\tau)\sqrt{2}}}{1+e^{-2t}} + \frac{(1-\sqrt{2})e^{-(t+\tau)\sqrt{2}}}{1+e^{-2t}} \right) Y(\tau)\, d\tau, \qquad t \geq 0.$$

Comparison with (68) shows that the expression in brackets is the causal impulse-response function $h(t, \tau)$, which was thus explicitly found as promised. Note that for large t, we have approximately

$$h(t, \tau) \doteq (1 + \sqrt{2})e^{-(t-\tau)\sqrt{2}}, \qquad 0 < \tau < t,$$

a familiar time-invariant filter. Recall also that the Kalman gain $k(t)$ is equal to the mean square estimation error, and thus from (73) we see that as $t \to \infty$, it increases monotonically to its asymptotic value $1 + \sqrt{2}$. Since the estimate $\hat{X}(t)$ is, for each t, a Gaussian random variable with mean zero and variance $k(t)$, it follows, for instance, that the probability that at any fixed t the estimation error exceeds $3\sqrt{1+\sqrt{2}}$ is less than .0014. ▲

EXERCISES

Exercise 4.5

1. Suppose that the process $X(t)$ in equation (1) is just a standard Gaussian random variable $X(t) = X$ for all $t \geq 0$ independent of the noise $N(t) = \dot{W}_0(t)$. Argue that in this case

$$\hat{X}(t) = \frac{1}{1+t} \int_0^{t-} Y(\tau)\, d\tau$$

and use this to show directly that the innovation process $V(t)$ is indeed the standard white Gaussian noise. Can we conclude that $V(t) = N(t)$? (*Hint:* Express $V(t)$ in terms of $\dot{W}_0(t)$ and calculate the mean and correlation functions.)

2. Suppose that the signal $X(t)$ in (H_1) of Example 4.5.1 is a known deterministic waveform $X(t) = x(t)$, $0 \leq t \leq 1$. Show that Kailath's general likelihood formula (28) leads to a matched filter where the decision that the signal is present is made whenever the integral $\int_0^1 x(t)Y(t)\, dt$ exceeds some threshold c.

3. Show by direct substitution from (34a) to (34b) (or vice versa) that the filters $\mathcal{H}$ and $\mathcal{G}$ with impulse-response functions

$$h(t, \tau) = \alpha e^{t-\tau}, \qquad g(t, \tau) = \alpha e^{(1+\alpha)(t-\tau)},$$

$0 \leq \tau < t$, $\alpha > 0$ a constant, form a causally linearly equivalent pair.

4. Repeat Exercise 4.5.3 if the impulse-response functions are

$$h(t, \tau) = \frac{1}{\sqrt{\alpha}} \sinh \frac{t-\tau}{\sqrt{\alpha}},$$

$$g(t, \tau) = \frac{1}{\alpha}(t - \tau), \qquad 0 \leq \tau < t, \qquad \alpha > 0.$$

5. If $Y(t) = X(t) + \dot{W}_0(t)$, $t \geq 0$, where $X(t)$ is a zero mean Gaussian process with correlation function $R_X(t_1, t_2)$ independent of the noise $\dot{W}_0(t)$, show that for all $0 \leq \tau < t$,

$$h(t, \tau) + \int_0^t h(t, \tau')R_X(\tau', \tau)\, d\tau' = R_X(t, \tau),$$

where $h(t, \tau)$ is the impulse response function of the causal linear filter $\hat{X} = \mathcal{H}Y$. (*Hint:* Multiply the equation $\hat{X} = \mathcal{H}Y$ by $Y(\tau)$, take expectations, and use the orthogonality principle.)

6. With $Y(t)$ as in Exercise 4.5.1, verify that the impulse-response function $h(t, \tau)$ of the causal linear filter $\hat{X}(t) = -(1 + t)^{-1} \int_0^{t-} Y(\tau)\, d\tau$ indeed satisfies the equation of Exercise 4.5.5.

7. In the Kalman-Bucy filter equation, let both $X(t)$ and $Y(t)$ be scalar valued, and let $a(t) = b(t) = 0$, $\sigma_0^2 = 1$. Show that this is exactly the same situation as in Exercise 4.5.1, and hence verify that the MS estimate $\hat{X}(t) = (1 + t)^{-1} \int_0^{t-} Y(\tau)\, d\tau$ satisfies the main Kalman-Bucy equation (58). Calculate the MS error $\varepsilon^2(t)$.

8. In equations (46) and (47), let both $X(t)$ and $Y(t)$ be scalar valued; let $a(t) = a < 0$, $b(t) = b > 0$ be constants for all $t \geq 0$; and let $X(0) = 0$. Show that the Kalman-Bucy filter equation yields in this case the MMSE estimate $\hat{X}(t)$ for the Ornstein-Uhlenbeck process with additive independent white Gaussian observation noise. Calculate the mean square error $\varepsilon^2(t)$, $t \geq 0$.

9. Suppose that the observation process $\mathbf{Y}(t)$ is related to the state process $\mathbf{X}(t)$ by

$$\mathbf{Y}(t) = \mathbf{c}(t)\mathbf{X}(t) + \mathbf{d}(t)\dot{\mathbf{W}}_0^{(p)}(t), \qquad t \geq 0,$$

where $\mathbf{c}(t)$ and $\mathbf{d}(t)$ are not necessarily square matrix functions. Show that, in this case, the main Kalman-Bucy equation becomes

$$\dot{\hat{\mathbf{X}}}(t) = (\mathbf{a}(t) - \mathbf{k}(t)\mathbf{c}(t))\hat{\mathbf{X}}(t) + \mathbf{k}(t)\mathbf{Y}(t)$$

where the Kalman gain is defined by $\mathbf{k}(t) = \boldsymbol{\varepsilon}^2(t)\mathbf{c}'(t)(\mathbf{d}(t)\,\mathbf{d}'(t))^{-1}$ and the mean square error matrix function satisfies the auxiliary equation

$$\dot{\boldsymbol{\varepsilon}}^2(t) = \mathbf{a}(t)\boldsymbol{\varepsilon}^2(t) + \boldsymbol{\varepsilon}^2(t)\mathbf{a}(t) + \mathbf{b}(t)\mathbf{b}'(t) - \mathbf{k}(t)\,\mathbf{d}(t)\,\mathbf{d}'(t)\,\mathbf{k}'(t),$$

$$\boldsymbol{\varepsilon}^2(0) = \boldsymbol{\sigma}_0^2.$$

(*Hint:* Retrace the derivation in the text, making modifications as required.)

10. Suppose the observation process is $Y(t) = X(t) + \sigma \dot{W}_0(t)$, $t \geq 0$, where $X(t)$ is a random sine wave $X(t) = A \cos(\omega t + \Theta)$ (A Rayleigh, Θ uniform on $(0, 2\pi)$, A and Θ independent) independent of the noise $\dot{W}_0(t)$. Show how the Kalman-Bucy filter equation can be used to obtain the MMSE estimate $\hat{X}(t) = \mathrm{E}[X(t) \mid Y(t'), 0 \leq t' \leq t]$. (*Hint:* Recall that $X(t) = Z_1 \cos \omega t - Z_2 \sin \omega t$ with Z_1, Z_2 i.i.d. Gaussian, and use this to choose the state vector in the state-space representation.)

CHAPTER FIVE

SPECTRAL REPRESENTATION AND DECOMPOSITION OF STATIONARY PROCESSES

5.1. THE SPECTRAL REPRESENTATION

The reader is undoubtedly aware of the paramount importance of sinusoidal waveforms in both analysis and synthesis of electrical (and other, e.g., acoustical) systems. The reason for this is not only that sinusoidal waveforms are naturally associated with resonance phenomena but, more importantly, also that all waveforms encountered in practice can be represented as a (possibly infinite) linear combination of sinusoidal waveforms. The advantage of such a representation is further enhanced by the fact that the response of a linear system to a sinusoidal waveform is relatively easy to establish, and the superposition principle for linear systems then guarantees that the response to an arbitrary waveform will be a linear combination of its sinusoidal components.

The representations of a waveform in terms of sinusoidal components is generally referred to as harmonic or Fourier analysis. Although we assume that the reader is familiar with at least the fundamentals of this subject, we present here a few basic facts, mainly for the sake of introducing our notation.

Consider a sinusoidal waveform as a function of time t

$$x(t) = a \cos(\omega_0 t + \theta), \tag{1}$$

where $a > 0$ represents the amplitude; $0 \leq \theta < 2\pi$, the phase (in radians); and $\omega_0 = 2\pi\nu_0$, the radial frequency, with ν_0 the frequency in hertz and $t_0 = 1/\nu_0$ the period. For the purpose of harmonic analysis, it is convenient to use the formula for the cosine of the sum of two angles to rewrite (1) as

$$x(t) = a \cos \theta \cos \omega_0 t - a \sin \theta \sin \omega_0 t,$$

this being known as the *quadrature decomposition* of the waveform $x(t)$ into a cosine component $x_1(t) = a \cos \theta \cos \omega_0 t$ and a sine component $x_2(t) = a \sin \theta \sin \omega_0 t$. Even more convenient is to employ the complex exponential

$$e^{\imath\alpha} = \cos \alpha + \imath \sin \alpha$$

to write the sinusoidal waveform (1)

$$x(t) = z(\nu_0)e^{\iota\omega_0 t} + z(-\nu_0)e^{-\iota\omega_0 t}, \tag{2}$$

where the complex numbers

$$z(\nu_0) = \frac{a}{2}e^{\iota\theta}$$

and

$$z(-\nu_0) = \frac{a}{2}e^{-\iota\theta}, \tag{3}$$

are called *complex amplitudes* of the *phasors*

$$z(\nu_0)e^{\iota\omega_0 t} \qquad \text{and} \qquad z(-\nu_0)e^{-\iota\omega_0 t}$$

respectively. As long as the waveform $x(t)$ is real, we of course have

$$z(v_0) = z^*(-v_0), \qquad \text{the complex conjugate.}$$

The argument ν in the expression $z(\pm\nu) = (a/2)e^{\pm\iota\theta}$ merely serves to remind us that the complex amplitude is to be associated with $e^{\iota\omega t}$, $\omega = 2\pi\nu$ to form the phasor.

The geometric interpretation of (2) in the form of the two phasors in the complex plane rotating with angular frequencies $\pm\omega_0$ is shown in Figure 5.1.1. Note that the cosine component

$$x_1(t) = \mathcal{R}e\{z(\nu_0)e^{\iota\omega_0 t}\},$$

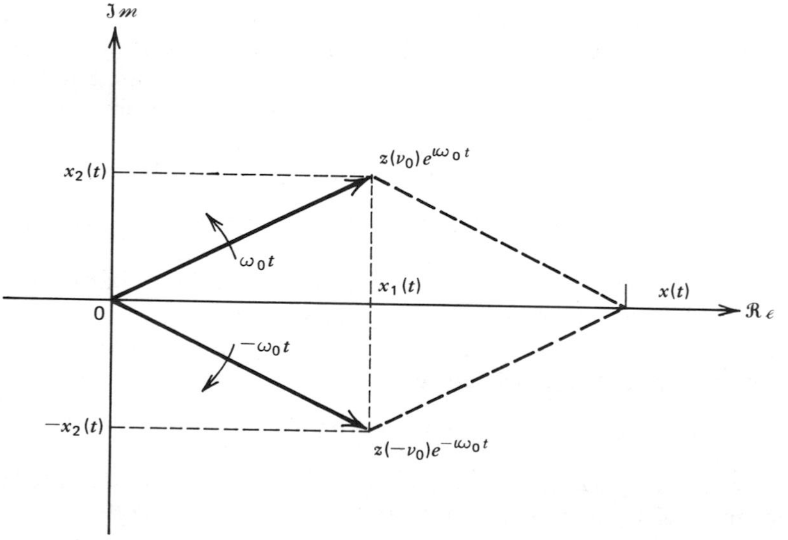

Figure 5.1.1

while the sine component

$$x_2(t) = \mathscr{I}m\{z(-\nu_0)e^{-\iota\omega_0 t}\} = \mathscr{R}e\{-\iota z(\nu_0)e^{\iota\omega_0 t}\},$$

with $\mathscr{R}e$ and $\mathscr{I}m$ indicating the real and imaginary parts respectively. Thus, if the sinusoidal waveform is given in quadrature form

$$x(t) = C_1 \cos \omega_0 t + C_2 \sin \omega_0 t,$$

we can immediately rewrite it in complex form (2) by setting

$$z(\nu_0) = C_1 + \iota C_2,$$

and

$$z(-\nu_0) = z^*(\nu_0) = C_1 - \iota C_2.$$

It is seen that a sinusoidal waveform

$$x(t) = a \cos(\omega_0 t + \theta), \qquad -\infty < t < \infty,$$

is completely specified by a pair of complex amplitudes (3), $z(\nu_0)$ and $z(-\nu_0)$ associated with the frequencies ν_0 and $-\nu_0$, respectively. The pair $(\nu_0, z(\nu_0))$ and $(-\nu_0, z(-\nu_0))$ is then referred to as the *complex spectrum* of the waveform (1).

Denoting the squared moduli of these complex numbers by

$$s(\nu_0) = |z(\nu_0)|^2 \qquad \text{and} \qquad s(-\nu_0) = |z(-\nu_0)|^2,$$

we then call the pair

$$\{(\nu_0, s(\nu_0)), (-\nu_0, s(-\nu_0))\}$$

the *power spectrum* of the waveform (1). Since $|e^{\pm\iota\omega_0 t}| = 1$, the square root $\sqrt{s(\pm\nu_0)}$ is just the magnitude of the rotating phasors $z(\pm\nu_0)e^{\pm\iota\omega_0 t}$ in Figure 5.1.1.

From (3) we have immediately

$$s(\nu_0) = s(-\nu_0) = \left(\frac{a}{2}\right)^2,$$

while from (1) the energy contained in the waveform over the period t_0 equals

$$\int_{-t_0/2}^{t_0/2} x^2(t)\, dt = a^2 \int_{-t_0/2}^{t_0/2} \cos^2(\omega_0 t + \theta)\, dt = 2t_0 \left(\frac{a}{2}\right)^2.$$

Thus $s(\nu_0)$ and $s(-\nu_0)$ each equal half the average power of the waveform (1)

$$\frac{1}{2}\frac{1}{t_0}\int_{-t_0/2}^{t_0/2} x^2(t)\, dt,$$

and thus can be interpreted as the average powers associated with the two frequencies ν_0 and $-\nu_0$.

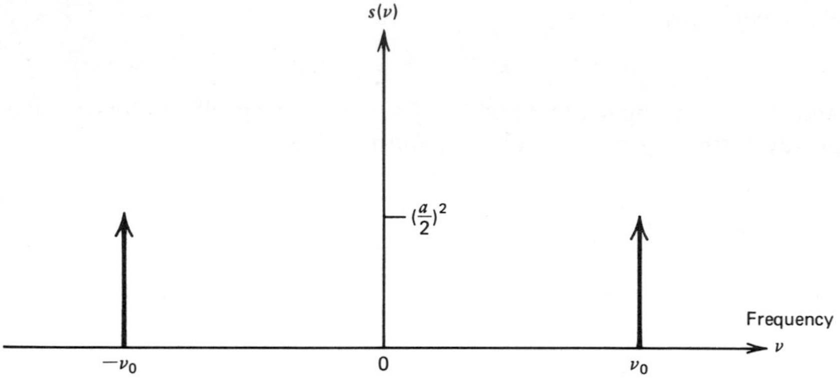

Figure 5.1.2

This power spectrum can be graphically displayed as in Figure 5.1.2, the two arrows being called power spectral lines. Note that for a direct current waveform of the same rms (root mean square) amplitude $x(t) = a/\sqrt{2}$, the two spectral lines merge at the origin $\nu = 0$, yielding the total power $s(0) = a^2/2$.

Turning now to periodic but not necessarily sinusoidal waveforms $x(t)$ with period t_0, under quite general assumptions such a waveform can be represented as a Fourier series, which in the complex form becomes

$$x(t) = \sum_{n=-\infty}^{\infty} z(n\nu_0)e^{\imath n\omega_0 t}, \qquad -\infty < t < \infty, \tag{4}$$

where $\omega_0 = 2\pi\nu_0$, $\nu_0 = 1/t_0$ is the basic frequency, and

$$z(n\nu_0) = \frac{1}{t_0}\int_{-t_0/2}^{t_0/2} x(t)e^{-\imath n\omega_0 t}\,dt, \qquad n = 0, \pm 1, \ldots. \tag{5}$$

The complex spectrum is now a doubly infinite sequence

$$\{(n\nu_0, z(n\nu_0)), \qquad n = 0, \pm 1, \pm 2, \ldots\},$$

where for real waveforms we have

$$z(-n\nu_0) = z^*(n\nu_0).$$

The corresponding power spectrum is also a doubly infinite sequence

$$\{(n\nu_0, z(n\nu_0)), \qquad n = 0, \pm 1, \pm 2, \ldots\},$$

with $s(n\nu_0) = |z(n\nu_0)|^2$ as before. It should be noted that the power spectral lines are $\nu_0 = 1/t_0$ apart, and so they become denser as the period t_0 increases; that is, the basic frequency of the periodic waveform decreases.

The average power of the waveform is now an infinite series

$$\frac{1}{t_0}\int_{-t_0/2}^{t_0/2} x^2(t)\,dt = \sum_{n=-\infty}^{+\infty} s(n\nu_0),$$

which follows from Parseval's identity for Fourier series

$$\frac{1}{t_0}\int_{-t_0/2}^{t_0/2} x(t)x(t+\tau)\,dt = \sum_{n=-\infty}^{\infty} |z(n\nu_0)|^2 e^{\imath n\omega_0\tau} \tag{6}$$

by setting $\tau = 0$.

For nonperiodic waveform $x(t)$ such that

$$\int_{-\infty}^{\infty} |x(t)|\,dt < \infty \tag{7}$$

a complex amplitude spectrum can be obtained from (5) for any fixed ν by letting $t_0 \to \infty$ and $n \to \infty$ such that

$$\frac{n}{t_0} = n\nu_0 = \nu.$$

We obtain

$$z(\nu) = \int_{-\infty}^{\infty} x(t)e^{-\imath\omega t}\,dt, \qquad -\infty < \nu < \infty \tag{8}$$

where $\omega = 2\pi\nu$. However, since $z(\nu)$ in (8) was obtained as the limit of $z(n/t_0)t_0$, its physical dimension is no longer that of the amplitude but rather amplitude $\times$ time or amplitude $\div$ frequency. That means, $z(\nu)$ is now a *density* of a complex amplitude distributed along the frequency axis.

Applying the same limiting process to the Fourier series (4)

$$\sum_{n=-\infty}^{\infty} z\left(\frac{n}{t_0}\right)t_0\, e^{\imath 2\pi(n/t_0)t}\,\frac{1}{t_0},$$

we see that with $n/t_0 = \nu$, $1/t_0 = d\nu \to 0$ the series turns into an integral,

$$x(t) = \int_{-\infty}^{\infty} z(\nu)e^{\imath\omega t}\,d\nu, \qquad -\infty < t < \infty \tag{9}$$

where again $\omega = 2\pi\nu$.

The integral transform (8) converting a time function $x(t)$ into a frequency function $z(\nu)$ is the familiar Fourier transform, and equation (9) is the inverse Fourier transform. The functions $x(t)$ and $z(\nu)$ are said to form a Fourier transform pair. We will occasionally indicate this relationship by writing

$$z(\nu) = \mathcal{F}\{x(t)\}, \qquad x(t) = \mathcal{F}^{-1}\{z(\nu)\}.$$

For real waveforms, we still have $z(-\nu) = z^*(\nu)$. Then Parseval's identity (with $\tau = 0$) takes the form

$$\int_{-\infty}^{\infty} x^2(t)\, dt = \int_{-\infty}^{\infty} |z(\nu)|^2\, d\nu.$$

But the left-hand side of this identity now represents the total energy of the waveform $x(t)$ rather than its average power

$$\lim_{t_0 \to \infty} \frac{1}{t_0} \int_{-t_0/2}^{t_0/2} x^2(t)\, dt,$$

which is now zero, at least for bounded waveforms satisfying (7). Thus, for nonperiodic waveforms of finite total energy, the frequency function

$$|z(\nu)|^2, \qquad -\infty < \nu < \infty,$$

represents a density of *energy*, rather than power, as distributed along the frequency axis.

However, the power spectrum can be defined even for nonperiodic waveforms $x(t)$ with infinite total energy as long as their average power

$$\lim_{t_0 \to \infty} \frac{1}{t_0} \int_{-t_0/2}^{t_0/2} x^2(t)\, dt = \bar{R}(0)$$

is finite. This can again be done for each fixed frequency ν by defining

$$s(\nu) = \lim_{\substack{t_0 \to \infty \\ n \to \infty}} t_0 \left| z\left(\frac{n}{t_0}\right) \right|^2, \qquad \frac{n}{t_0} = \nu, \tag{10}$$

with $z(n/t_0) = z(n\nu_0)$ defined by (5). For such a passage to the limit the right-hand side of Parseval's identity (6)

$$\sum_{n=-\infty}^{\infty} t_0 \left| z\left(\frac{n}{t_0}\right) \right|^2 e^{\imath 2\pi (n/t_0)\tau} \frac{1}{t_0},$$

approaches the integral

$$\int_{-\infty}^{\infty} s(\nu) e^{\imath \omega \tau}\, d\nu, \qquad \omega = 2\pi\nu, \tag{11}$$

while the left-hand side becomes a time-correlation function

$$\bar{R}(\tau) = \lim_{t_0 \to \infty} \frac{1}{t_0} \int_{-t_0/2}^{t_0/2} x(t) x(t + \tau)\, dt. \tag{12}$$

Note that by (10), $s(\nu)$ is again a density (having the dimensions power $\div$ frequency), but this time describing the distribution of the average power $\bar{R}(0)$

along the frequency domain. Further, since (11) and (12) are equal, this implies that the time-correlation function

$$\bar{R}(\tau) = \int_{-\infty}^{\infty} s(\nu) e^{i\omega\tau} \, d\nu, \qquad -\infty < \tau < \infty \tag{13}$$

is an inverse Fourier transform of the power spectral density $s(\nu)$. Consequently, if

$$\int_{-\infty}^{\infty} |\bar{R}(\tau)| \, d\tau < \infty$$

we can express the power spectral density as a Fourier transform

$$s(\nu) = \int_{-\infty}^{\infty} \bar{R}(\tau) e^{-i\omega\tau} \, d\tau, \qquad -\infty < \nu < \infty. \tag{14}$$

The Fourier transform pair (13) and (14) is known as the *Wiener-Khinchine relations.* Notice that for a periodic waveform with period t_0 the time-correlation function is also periodic with the same period t_0, so that

$$\bar{R}(\tau) = \frac{1}{t_0} \int_{-t_0/2}^{t_0/2} x(t) x(t + \tau) \, dt, \qquad -\infty < \tau < \infty.$$

Hence, in this case, we also have from (6)

$$\bar{R}(\tau) = \sum_{n=-\infty}^{\infty} s(n\nu_0) e^{in\omega_0\tau}, \qquad -\infty < \tau < \infty$$

and

$$s(n\nu_0) = \frac{1}{t_0} \int_{-t_0/2}^{t_0/2} \bar{R}(\tau) e^{-in\omega_0\tau} \, d\tau, \qquad n = 0, \pm 1, \ldots$$

Thus the Fourier relations between the power spectrum and the time-correlation function hold for any waveform with finite average power.

Our main interest, however, is not in deterministic waveforms but in stationary random processes. Since a stochastic process

$$X(t), \qquad -\infty < t < \infty, \tag{15}$$

is, by definition just an ensemble of sample paths

$$x(t) \qquad \text{on the time domain } -\infty < t < \infty, \tag{16}$$

together with a probability law describing their distribution, it may seem that in order to obtain a complex amplitude density

$$Z(\nu), \qquad -\infty < \nu < \infty, \tag{17}$$

for a stochastic process (15) we could define the former as a collection of sample paths

$$z(\nu) \qquad \text{on the frequency domain } -\infty < \nu < \infty,$$

which are simply Fourier transforms $z(\nu) = \mathcal{F}\{x(t)\}$ of the sample paths (16). Not even mentioning the obvious difficulty in determining the probability law of (17), this approach is bound to fail, since, except in rather special cases, we can hardly expect the sample paths (16) of a stationary process to satisfy the integrability condition

$$\int_{-\infty}^{\infty} |x(t)| \, dt < \infty.$$

On the other hand, if $X(t)$, $-\infty < t < \infty$, is a finite-variance stationary and ergodic process, then almost all of its sample paths $X(t) = x(t)$ will have finite and constant average power, since by the ergodicity property

$$\bar{R}(0) = \lim_{t_0 \to \infty} \frac{1}{t_0} \int_{-t_0/2}^{t_0/2} x^2(t) \, dt = R_X(0).$$

In fact, we even have for an ergodic stationary process

$$\bar{R}(\tau) = R_X(\tau) \qquad \text{for all } -\infty < \tau < \infty,$$

and this suggests that we might introduce the concept of power spectrum for a stationary process $X(t)$, $-\infty < t < +\infty$, by means of a Fourier transform of its correlation function $R_X(\tau)$. We could, in essence, follow the previous line of reasoning, but we prefer to be a bit more rigorous this time. Our starting point will be a theorem, credited to Bochner, the proof of which is sketched in the appendix to this section.

THEOREM 5.1.1 *Bochner's Theorem*
Let $\mathrm{R}(\tau)$, $-\infty < \tau < \infty$, *be a continuous function. Then this function is nonnegative definite if and only if there exists a distribution function*

$$S(\nu), \qquad -\infty < \nu < \infty,$$

such that for all $-\infty < \tau < \infty$

$$\frac{\mathrm{R}(\tau)}{\mathrm{R}(0)} = \int_{-\infty}^{\infty} e^{\iota\omega\tau} \, dS(\nu), \qquad \omega = 2\pi\nu. \tag{18}$$

This representation is unique.

The differential $dS(\nu)$ is the abbreviation for $[dS(\nu)/d\nu] \, d\nu$, which we shall use in this section for reasons that will become apparent. Note $S(\nu)$ is a distribution function, and its derivative may not exist, so the integral may involve Dirac delta functions. If $S(\nu)$ were the distribution function of a random variable, the integral is simply its characteristic function.

Recall that the term "distribution function" means that the function $S(\nu)$ must be:

1. nondecreasing, that is, $\nu_1 \le \nu_2 \Rightarrow S(\nu_1) \le S(\nu_2)$.
2. $\lim_{\nu \to -\infty} S(\nu) = 0$, $\lim_{\nu \to \infty} S(\nu) = 1$.
3. continuous from the right, that is, $\lim_{\nu \to \nu_0 +} S(\nu) = S(\nu_0)$.

From now on, let us consider a finite-variance wide-sense stationary process

$$X(t), \qquad -\infty < t < \infty,$$

and let us assume that:

(i) the mean function $\mu_X(t) = 0$,

(ii) the process is MS continuous so that its correlation function $R_X(\tau)$ is a continuous function of τ.

Let us denote for short

$$R_X(0) = \sigma^2$$

the variance of the process. Then, since any correlation function must be nonnegative definite, it follows from Bochner's theorem that there is a distribution function

$$S(\nu), \qquad -\infty < \nu < \infty,$$

such that

$$R_X(\tau) = \int_{-\infty}^{\infty} e^{i\omega\tau}\sigma^2 \, dS(\nu), \qquad -\infty < \tau < \infty. \tag{19}$$

Now what is the physical significance of the distribution function $S(\nu)$ with respect to the process $X(t)$?

In anticipation of what is to come, let us refer to the variable ν as *frequency* and the real line $-\infty < \nu < \infty$ on which $S(\nu)$ is defined as the *frequency domain.*

Now take two fixed frequencies

$$\nu_1 < \nu_2,$$

and consider an ideal bandpass filter, that is, a linear filter with the property that if a harmonic waveform of frequency ν_0,

$$x(t) = e^{i\omega_0 t}, \qquad \omega_0 = 2\pi\nu_0, \tag{20}$$

is applied to its input, it passes through undisturbed only if its frequency ν_0 lies within the frequency band $(\nu_1, \nu_2]$; otherwise, its passage is completely blocked. That is, we want the output of our bandpass filter to be

$$y(t) = \begin{cases} x(t) & \text{if} \quad \nu_1 < \nu_0 \leq \nu_2, \\ 0 & \text{otherwise.} \end{cases} \tag{21}$$

Let $h(\tau)$ be the impulse-response function of such a filter, that is,

$$y(t) = \int_{-\infty}^{\infty} h(t-\tau)x(\tau)\,d\tau. \tag{22}$$

(Note that we do not require the filter to be causal; after all, we seek an "ideal" bandpass filter.) If we choose

$$h(\tau) = \int_{\nu_1}^{\nu_2} e^{\iota\omega\tau}\,d\nu, \qquad -\infty < \tau < \infty,$$

which, of course, is a filter with bandpass transfer function

$$H(\tau) = \mathcal{F}\{h(\tau)\} = \begin{cases} 1 & \text{if} \quad \nu_1 < \nu \leq \nu_2 \\ 0 & \text{otherwise,} \end{cases}$$

then it is easy to verify that with input $x(t)$ defined by (20), the output (22) will indeed satisfy (21).

Let us now take our process $X(t)$ and let it pass through this filter. Since the filter is clearly linear and time-invariant, the output process

$$Y(t) = \int_{-\infty}^{\infty} h(t-\tau)X(\tau)\,d\tau, \qquad -\infty < t < \infty \tag{23}$$

will be a wide-sense stationary process (in general, complex valued) with zero mean and with constant variance

$$\mathrm{E}[|Y(t)|^2] = \mathrm{E}[|Y(0)|^2].$$

Now from (23)

$$\mathrm{E}[|Y(0)|^2] = \mathrm{E}\left[\left(\int_{-\infty}^{\infty} h(-\tau_1)X(\tau_1)\,d\tau_1\right)\left(\int_{-\infty}^{\infty} h(-\tau_2)X(\tau_2)\,d\tau_2\right)^*\right]$$

$$= \int_{-\infty}^{\infty}\int_{-\infty}^{\infty} h(-\tau_1)h^*(-\tau_2)R_X(\tau_1-\tau_2)\,d\tau_1\,d\tau_2,$$

so that

$$\mathrm{E}[|Y(0)|^2] = \int_{-\infty}^{\infty}\int_{-\infty}^{\infty}\int_{-\infty}^{\infty} h(-\tau_1)h^*(-\tau_2)e^{\iota\omega(\tau_1-\tau_2)}\,d\tau_1\,d\tau_2\,\sigma^2\,dS(\nu)$$

by substituting for the correlation R_X from (19). However, the triple integral can be written as

$$\int_{-\infty}^{\infty}\left(\int_{-\infty}^{\infty} h(-\tau_1)e^{\iota\omega\tau_1}\,d\tau_1\right)\left(\int_{-\infty}^{\infty} h(-\tau_2)e^{\iota\omega\tau_2}\,d\tau_2\right)^* \sigma^2\,dS(\nu) \tag{24}$$

and since h is the impulse-response function in the bandpass filter (22)

$$\int_{-\infty}^{\infty} h(-\tau)e^{\iota\omega\tau}\,d\tau = y(0) \qquad \text{for} \quad x(\tau) = e^{\iota\omega\tau}.$$

But then by (21) with $t = 0$

$$y(0) = \begin{cases} e^{\iota\omega 0} = 1 & \text{if} \quad \nu_1 < \nu \le \nu_2, \\ 0 & \text{otherwise,} \end{cases}$$

so that (24) becomes simply

$$\int_{\nu_1}^{\nu_2} \sigma^2\,dS(\nu) = \sigma^2(S(\nu_2) - S(\nu_1))$$

that is, the variance of the process $Y(t)$ is

$$\mathrm{E}[\,|\,Y(t)|^2] = \sigma^2(S(\nu_2) - S(\nu_1)). \tag{25}$$

Now if, for instance, the original process $X(t)$ represents a voltage process, then $Y(t)$ is also a voltage process and $|\,Y(t)|^2$ is then the power dissipated in a 1-ohm resistor. Since $Y(t)$ was obtained by passing the process through a bandpass filter, we can interpret equation (25) by saying that $\sigma^2(S(\nu_2) - S(\nu_1))$ is the expected power of the wide-sense stationary process $X(t)$, which is contained in the frequency band $(\nu_1, \nu_2]$.

In summary, the *function* $\sigma^2 S(\nu)$ *represents the distribution of the expected power of the process* $X(t)$ *over the frequency domain.* For this reason the function $S_X(\nu) = \sigma^2 S(\nu)$ is called the *power spectral distribution function* of the process $X(t)$.

Example 5.1.1.
Comparing equation (19)

$$R_X(\tau) = \int_{-\infty}^{\infty} e^{\iota\omega\tau}\sigma^2\,dS(\nu), \qquad -\infty < \nu < \infty, \qquad \omega = 2\pi\nu,$$

with the definition of a characteristic function $\psi(\zeta)$ of a random variable with distribution function $F(x)$, and recalling that the density $f(x) = dF(x)/dx$, we can write

$$\psi(\zeta) = \int_{-\infty}^{\infty} e^{\iota\zeta x}\,dF(x), \qquad -\infty < \zeta < \infty,$$

and we see that apart from the multiplicative constant σ^2, the relation between a power spectral distribution function $S_X(\nu)$ and the correlation function $R_X(\tau)$ is exactly the same as that between the distribution function $F(x)$, with $x = \nu$, and the characteristic function $\psi(\zeta)$, with $\zeta = 2\pi\tau$.

As a simple case, consider a process with power spectral distribution function

$$S_X(\nu) = \begin{cases} \sigma^2 & \text{if} \quad \nu \geq 0, \\ 0 & \text{if} \quad \nu < 0. \end{cases}$$

Dividing by the constant $\sigma^2 > 0$, this corresponds to the distribution function of a random variable that is identically equal to zero. The characteristic function of such a random variable equals

$$\psi(\zeta) = e^{\iota\zeta 0} = 1, \qquad -\infty < \xi < \infty,$$

and hence the correlation function of the wide-sense stationary process $X(t)$, $-\infty < t < \infty$, having the above power spectral distribution will be

$$R_X(\tau) = \sigma^2 \qquad \text{for all } -\infty < \tau < \infty.$$

But a zero mean process $X(t)$ with such a correlation function is just

$$X(t) = Z, \qquad -\infty < t < \infty,$$

where Z is a random variable with zero mean and variance σ^2. Thus the ensemble of this process will consist of sample paths $x(t) = z$, which are constants for all t. From the definition of its power spectral distribution function, we see that the expected power in any frequency interval $(\nu_1, \nu_2]$ not containing the frequency $\nu = 0$ is zero. In other words, all the expected power σ^2 is concentrated at the zero frequency. Such a process is appropriately called a DC process.

As a more interesting case, consider a Cauchy random variable with distribution function

$$F(x) = \frac{1}{2} + \frac{1}{\pi} \arctan \frac{x}{a}, \qquad -\infty < x < \infty, \qquad a > 0,$$

and characteristic function

$$\psi(\zeta) = e^{-a|\zeta|}, \qquad -\infty < \zeta < \infty.$$

(see Appendix Two).

Setting $a = \alpha/2\pi$, we see that a wide-sense stationary process $X(t)$, $-\infty < t < \infty$, with zero mean and the exponential correlation function

$$R_X(\tau) = \sigma^2 e^{-\alpha|\tau|}, \qquad -\infty < \tau < \infty,$$

has the power spectral distribution function

$$S_X(\nu) = \sigma^2\left(\frac{1}{2} + \frac{1}{\pi} \arctan \frac{2\pi\nu}{\alpha}\right), \qquad -\infty < \nu < \infty.$$

It follows that the expected power in a small interval $(\nu, \nu + d\nu]$ in the frequency domain is approximately

$$dS_X(\nu) = \frac{2\alpha\sigma^2}{\sigma^2 + (2\pi\nu)^2}\, d\nu,$$

decreasing monotonically to zero as the interval $(\nu, \nu + d\nu]$ moves away from 0 towards higher frequencies. ▲

Since, by Bochner's theorem, the correspondence between the correlation function $R_X(\tau)$ and the power spectral distribution function $S_X(\nu)$ is one-to-one, both functions provide the same information about the process $X(t)$. In particular, if the wide-sense stationary process $X(t)$ is *Gaussian* MS continuous with zero mean, then its probability law must be *completely specified* by its power spectral distribution function $S_X(\nu)$.

More generally, any property of a wide-sense stationary MS continuous process that is expressible in terms of its correlation function R_X must also be expressible in terms of its power spectral distribution function S_X, and vice versa.

For example, we know from Section 4.1 that a wide-sense stationary process $X(t)$ is MS differentiable if and only if its correlation function has a second derivative at $\tau = 0$. Differentiating the expression

$$R_X(\tau) = \int_{-\infty}^{\infty} e^{\imath\omega\tau}\, dS_X(\nu)$$

twice with respect to τ, we get

$$\frac{d^2}{d\tau^2} R_X(\tau) = -\int_{-\infty}^{\infty} \omega^2 e^{\imath\omega\tau}\, dS_X(\nu).$$

Setting $\tau = 0$, this becomes

$$\left.\frac{d^2}{d\tau^2} R_X(\tau)\right|_{\tau=0} = -\int_{-\infty}^{\infty} \omega^2\, dS_X(\nu) = -4\pi \int_{-\infty}^{\infty} \nu^2\, dS_X(\nu)$$

and thus the process $X(t)$ is MS differentiable if and only if the last integral exists, that is, if and only if

$$\int_{-\infty}^{\infty} \nu^2\, dS_X(\nu) < \infty.$$

Similar conditions can be obtained for sample continuity of the process $X(t)$ and other properties.

However, we are more interested in finding out whether a Fourier-type representation of some kind can be obtained for the process $X(t)$ itself. To this end, let us now consider two *disjoint* intervals

$$I = (\nu_1, \nu_2], \qquad J = (\nu_3, \nu_4]$$

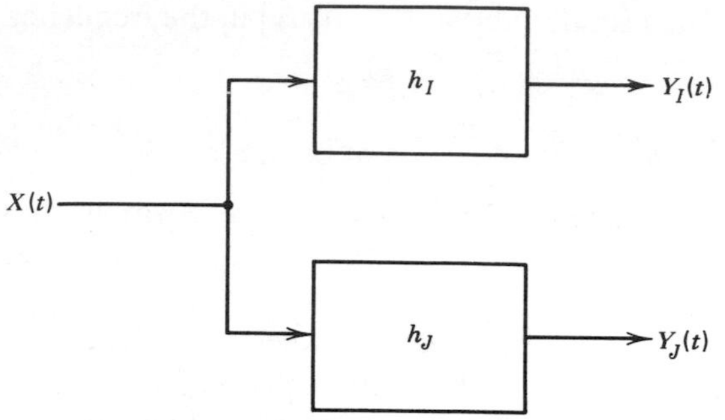

Figure 5.1.3

in the frequency domain, and let

$$h_I(\tau) \qquad \text{and} \qquad h_J(\tau)$$

be the impulse-response functions of ideal linear bandpass filters corresponding to the intervals I and J, respectively. Let

$$\begin{aligned} Y_I(t) &= \int_{-\infty}^{\infty} h_I(t-\tau)X(\tau)\,d\tau, \\ Y_J(t) &= \int_{-\infty}^{\infty} h_J(t-\tau)X(\tau)\,d\tau, \end{aligned} \tag{26}$$

be two wide-sense stationary processes obtained by passing the process $X(t)$ through the two filters, as pictured in Figure 5.1.3. Then these two processes will be uncorrelated. To see this, note that for any t_1, t_2, we have, by using (26) and (19),

$$\begin{aligned} \mathrm{E}[Y_I(t_1)Y_J^*(t_2)] = \int_{-\infty}^{\infty} \left[\int_{-\infty}^{\infty} h_I(t_1-\tau_1)e^{i\omega\tau_1}\,d\tau_1\right] \\ \times \left[\int_{-\infty}^{\infty} h_J(t_2-\tau_2)e^{i\omega\tau_2}\,d\tau_2\right]^* dS_X(\nu) = 0, \end{aligned}$$

since the integrals in brackets can both be nonzero only if $\nu = \omega/2\pi$ is in both I and J, which is impossible, since I and J are disjoint.

This all suggests that we should be able to decompose a wide-sense stationary zero mean process $X(t)$ into mutually uncorrelated processes by passing it through an array of bandpass filters.

To carry on this program suppose that the frequency domain has been partitioned into an infinite number of disjoint intervals of length $d\nu$,

$$I_n = (\nu_n, \nu_{n+1}], \qquad n = \ldots, -1, 0, 1, \ldots$$

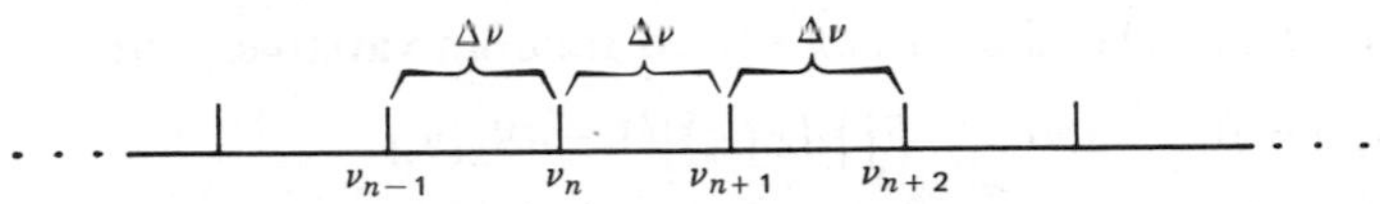

Figure 5.1.4

as pictured in Figure 5.1.4, and imagine that a wide-sense stationary, MS continuous zero mean process $X(t)$, $-\infty < t < \infty$, has been passed through a corresponding infinite array of ideal bandpass filters, as in Figure 5.1.5. We thus obtain a doubly infinite sequence

$$Y_n(t) = \int_{-\infty}^{\infty} h_n(t-\tau)X(\tau)\,d\tau, \qquad n = \ldots, -1, 0, 1, \ldots$$

of wide-sense stationary, MS continuous, zero mean processes, which will all be mutually uncorrelated.

Suppose further that the bandwidth dv of all these filters is very small, so that virtually only a single frequency v_n can pass through the nth filter. Then almost all time variation left in the process $Y_n(t)$ would have to be due to this single frequency, that is, we would have approximately for all $-\infty < t < \infty$

$$Y_n(t) \doteq dZ(v_n)e^{\iota\omega_n t}, \qquad \omega_n = 2\pi v_n,$$

where $dZ(v_n)$ would be a random complex amplitude. From our previous discussion, it follows that

$$\ldots dZ(v_n),\ dZ(v_{n+1}), \ldots$$

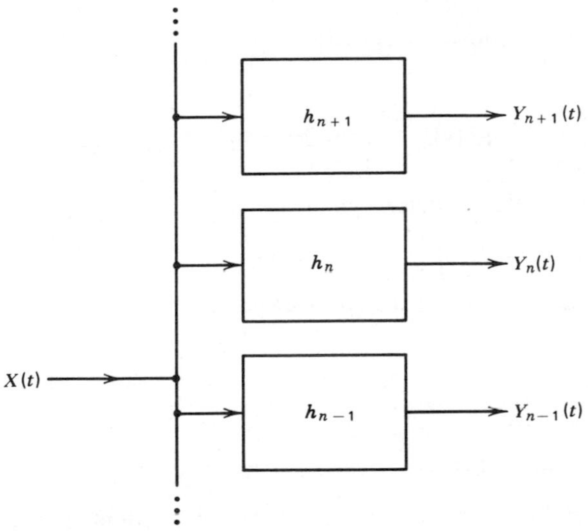

Figure 5.1.5

should be a sequence of uncorrelated complex valued random variables with

$$\mathrm{E}[dZ(\nu_n)] = 0 \qquad \text{and} \qquad \mathrm{E}[|dZ(\nu_n)|^2] = dS_X(\nu_n).$$

This suggests that we should be able to write approximately

$$X(t) \doteq \sum_{n=-\infty}^{\infty} e^{\imath\omega_n t}\, dZ(\nu_n),$$

or, as the bandwidth $d\nu \to 0$,

$$X(t) = \int_{-\infty}^{\infty} e^{\imath\omega t}\, dZ(\nu), \qquad \omega = 2\pi\nu, \tag{27}$$

where $Z(\nu)$ is a complex-valued process on the frequency domain $-\infty < \nu < \infty$, and the integral in (27) is defined as the MS limit

$$\int_{-\infty}^{\infty} e^{\imath\omega t}\, dZ(\nu) = \underset{d\nu \to 0}{\mathrm{l.i.m.}} \sum_{n=-\infty}^{\infty} e^{\imath\omega_n t}\, dZ(\nu_n), \tag{28}$$

$$dZ(\nu_n) = Z(\nu_n + d\nu) - Z(\nu_n)$$

This conjecture is indeed true and we now state it as a theorem.

THEOREM 5.1.2. *Spectral Representation Theorem*
Let $\mathrm{X(t)}$, $-\infty < \mathrm{t} < \infty$, *be a wide-sense stationary, MS continuous, zero mean process with power spectral distribution functions*

$$S_X(\nu), \qquad -\infty < \nu < \infty.$$

Then there exists a complex-valued finite-variance process $\mathrm{Z}(\nu)$ *on the frequency domain* $-\infty < \nu < \infty$, *called the* spectral process *of the process* $\mathrm{X(t)}$, *such that for all* $-\infty < \mathrm{t} < \infty$

$$X(t) = \int_{-\infty}^{\infty} e^{\imath\omega t}\, dZ(\nu), \qquad \omega = 2\pi\nu. \tag{29}$$

The spectral process has uncorrelated increments and

$$\mathrm{E}[dZ(\nu)] = 0, \qquad \mathrm{E}[|dZ(\nu)|^2] = dS_X(\nu), \qquad -\infty < \nu < \infty. \tag{30a,b}$$

If the process $\mathrm{X(t)}$ *is Gaussian, then so is the spectral process* $\mathrm{Z}(\nu)$.

Example 5.1.2

Let us consider the random sine wave

$$X(t) = A\cos(\omega_0 t + \Theta), \qquad -\infty < t < \infty,$$

with A a Rayleigh random variable with parameter $\sigma > 0$ independent of the random phase Θ uniformly distributed over $[0, 2\pi)$. We know that $X(t)$,

$-\infty < t < \infty$, is a zero mean stationary Gaussian process with correlation function

$$R_X(\tau) = \sigma^2 \cos \omega_0 \tau, \qquad -\infty < \tau < \infty.$$

Finding the power spectral distribution function $S_X(\nu)$ for this process is easy. Writing the correlation function in the form

$$R_X(\tau) = \frac{\sigma^2}{2} e^{-\iota\omega_0\tau} + \frac{\sigma^2}{2} e^{\iota\omega_0\tau},$$

where $\omega_0 = 2\pi\nu_0$, we see from (19) that we must have

$$S_X(\nu) = \begin{cases} \sigma^2 & \text{if} \quad \nu \geq \nu_0, \\ \dfrac{\sigma^2}{2} & \text{if} \quad -\nu_0 \leq \nu < \nu_0, \\ 0 & \text{if} \quad \nu \leq \nu_0, \end{cases}$$

that is, that the entire expected power σ^2 of this process is equally divided between the frequencies $-\nu_0$ and ν_0. In other words,

$$dS_X(\nu) = \begin{cases} \dfrac{\sigma^2}{2} & \text{if} \quad \nu = \pm\nu_0, \\ 0 & \text{otherwise.} \end{cases}$$

The complex-valued spectral process $Z(\nu)$, $-\infty < \nu < \infty$, will in this case have all sample paths constant except for two jumps $dZ(-\nu_0)$ and $dZ(\nu_0)$ at the frequencies $\pm\nu_0$. It follows that the spectral representation (29) is just

$$X(t) = e^{-\iota\omega_0 t}\, dZ(-\nu_0) + e^{\iota\omega_0 t}\, dZ(\nu_0), \qquad -\infty < t < \infty.$$

From the phasor diagram of this expression in Figure 5.1.6, we see that we must have $dZ(-\nu_0) = dZ^*(\nu_0)$. By comparing it with the expression

$$X(t) = A \cos(\omega_0 t + \Theta)$$

we conclude further that the complex-valued random variable $dZ(\nu_0)$ has magnitude $\frac{1}{2}A$ and argument Θ, $dZ(\nu_0) = \frac{1}{2}Ae^{\iota\Theta}$. Consequently, its real and imaginary parts are i.i.d. Gaussian random variables with mean zero and variance $\frac{1}{2}\sigma^2$. Thus, a typical sample path of, say, the real part of the spectral process $Z(\nu)$, $-\infty < \nu < \infty$, will look like the one shown in Figure 5.1.7, where the heights of the two jumps are the same, having a Gaussian distribution with zero mean and variance $\sigma^2/2$. A sample path of the imaginary part will also have two jumps at ν_0 and $-\nu_0$ of the same magnitude, except that the jump at ν_0 will go in the opposite direction of that at $-\nu_0$. ▲

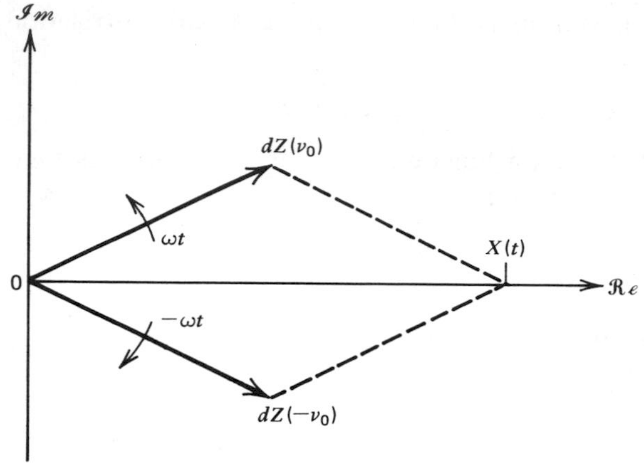

Figure 5.1.6

Remark: So far, we have been thinking of a stochastic process as a random quantity evolving with time t. In the case of the spectral process, however, the variable ν represents a frequency rather than time. Hence, we have to think of a spectral process as a (complex-valued) random function, not evolving with time, but rather chosen as a whole from the ensemble of the process. In fact, a spectral process is the random counterpart of a cumulative frequency spectrum of an ordinary deterministic signal.

Notice also that only the increments $dZ(\nu)$ appear in the representation (29), so that the spectral process is determined only up to an arbitrary additive random variable. Я

We will not give a rigorous proof of the above theorem. Instead, let us merely point out that $X(t)$, being defined by means of a linear operation (cf. 28) of a Gaussian process, must also be Gaussian. To verify that its power spectral distribution $S_X(\nu)$ indeed satisfies (30b), we note that if R_X is the correlation function of

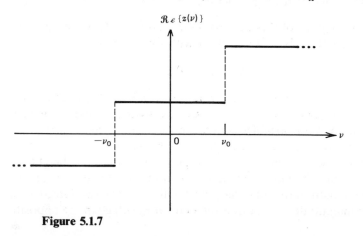

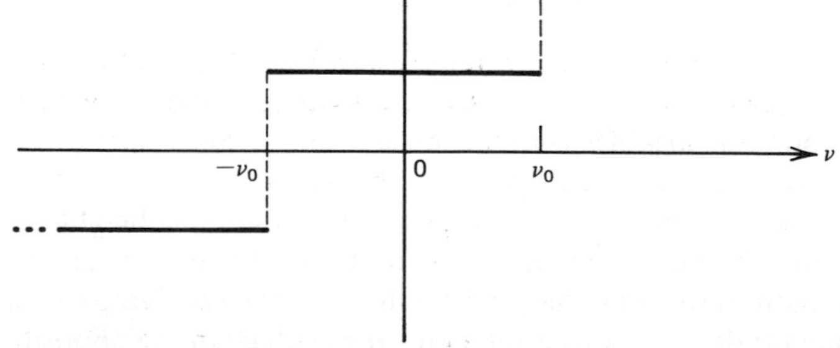

Figure 5.1.7

the process defined by (29), then by (28) for small $d\nu$

$$\begin{aligned}R_X(t_1, t_2) &= \mathrm{E}[X(t_1)X^*(t_2)]\\ &\doteq \mathrm{E}\left[\left(\sum_n e^{\iota\omega_n t_1}\, dZ(\nu_n)\right)\left(\sum_m e^{\iota\omega_m t_2}\, dZ(\nu_m)\right)^*\right]\\ &= \sum_n \sum_m e^{\iota(\omega_n t_1 - \omega_m t_2)}\mathrm{E}[dZ(\nu_n)\, dZ^*(\nu_m)]\\ &= \sum_n e^{\iota\omega_n(t_1 - t_2)}\, dS_X(\nu_n)\end{aligned}$$

since $Z(\nu)$ has uncorrelated increments so that

$$\mathrm{E}[dZ(\nu_n)\, dZ^*(\nu_m)] = \begin{cases}\mathrm{E}[\,|dZ(\nu_n)|^2] & \text{for } m = n,\\ 0 & \text{for } m \neq n.\end{cases}$$

Thus, as $d\nu \to 0$ the sum on the right-hand side approximates an integral, that is, we have

$$R_X(t_1, t_2) = \int_{-\infty}^{\infty} e^{\iota\omega(t_1 - t_2)}\, dS_X(\nu).$$

We see that X is a stationary process whose power spectral distribution function is $S_X(\nu)$ since, by Bochner's theorem, there is a one-to-one correspondence between correlation functions and power spectral functions.

The physical significance of the spectral representation theorem should be evident from the heuristic discussion we used to "derive" the theorem. Every zero mean wide-sense stationary MS continuous process can be regarded as a superposition of elementary harmonic oscillations $dZ(\nu)e^{\iota\omega t}$, with frequencies spread over the entire frequency domain. The elementary oscillations with frequencies in an infinitesimal interval ν to $\nu + d\nu$ have a random complex amplitude $dZ(\nu)$, the amplitudes of elementary oscillations with different frequencies being uncorrelated. These complex amplitudes have zero mean, and their expected energies are determined by the power spectral distribution function $\mathrm{E}[\,|dZ(\nu)|^2] = dS_X(\nu)$. Writing the complex amplitude $dZ(\nu)$ in polar form,

$$dZ(\nu) = |dZ(\nu)|\, e^{\iota\, d\Theta(\nu)}$$

it follows that the random phases $d\Theta(\nu)$ have no influence whatsoever on the correlation function of the process $X(t)$. Moreover, if we change each phase $d\Theta(\nu)$ by an arbitrary amount $d\gamma(\nu)$, then the new spectral process $\tilde{Z}(\nu)$ will still have uncorrelated increments $d\tilde{Z}(\nu) = e^{\iota\gamma(\nu)}\, dZ(\nu)$ satisfying

$$\mathrm{E}[d\tilde{Z}(\nu)] = e^{\iota\gamma(\nu)}\mathrm{E}[dZ(\nu)] = 0,$$

and

$$\mathrm{E}[\,|d\tilde{Z}(\nu)|^2] = |e^{\iota\gamma(\nu)}|^2\mathrm{E}[\,|dZ(\nu)|^2] = dS_X(\nu),$$

since $|e^{\imath\gamma(\nu)}| = 1$. Thus, *any changes of phases of the elementary oscillations are completely obliterated in the superposition.* This property, which stands in a sharp contrast with the spectral (i.e., Fourier) representation of deterministic waveforms, is often expressed by saying that the spectral process $Z(\nu)$ is *completely incoherent.*

The spectral representation theorem holds true even if the process $X(t)$ is complex valued.

However, for a *real-valued* process $X(t)$, more can be said about the power spectral distribution function $S_X(\nu)$ and the spectral process $Z(\nu)$.

If $X(t)$ is zero mean and real valued, its correlation function $R_X(\tau) = \mathrm{E}[X(t)X(t+\tau)]$ must be real. But then, since

$$R_X(\tau) = \int_{-\infty}^{\infty} e^{\imath\omega\tau}\, dS_X(\nu) = \int_{-\infty}^{\infty} \cos\omega\tau\, dS_X(\nu) + \imath \int_{-\infty}^{\infty} \sin\omega\tau\, dS_X(\nu)$$

the last integral must be zero. This implies that the power spectral distribution function $S_X(\nu)$ of a real-valued process $X(t)$ must be a symmetric distribution function, that is, for all $-\infty < \nu < \infty$

$$\frac{1}{\sigma^2} S_X(-\nu) = 1 - \frac{1}{\sigma^2} S_X(\nu).$$

Note that this implies that the correlation function

$$R_X(\tau) = \int_{-\infty}^{\infty} \cos\omega\tau\, dS_X(\nu) = 2\int_{0}^{\infty} \cos\omega\tau\, dS_X(\nu) \tag{31}$$

so that $R_X(-\tau) = R_X(\tau)$ as it should for a real-valued process.

The spectral process $Z(\nu)$ is still a complex-valued process. However, if we write it as

$$Z(\nu) = Z_1(\nu) + \imath Z_2(\nu)$$

with $Z_1(\nu)$ and $Z_2(\nu)$ being its real and imaginary parts, respectively, we get from (29)

$$X(t) = \int_{-\infty}^{\infty} \cos\omega t\, dZ_1(\nu) - \int_{-\infty}^{\infty} \sin\omega t\, dZ_2(\nu)$$
$$+ \imath\left[\int_{-\infty}^{\infty} \sin\omega t\, dZ_1(\nu) + \int_{-\infty}^{\infty} \cos\omega t\, dZ_2(\nu)\right].$$

Since the process $X(t)$ is real, the expression in brackets must be zero for all t. This implies that

$$Z_1(\nu) = Z_1(-\nu) \qquad \text{and} \qquad Z_2(\nu) = -Z_2(-\nu), \tag{32}$$

or, equivalently,

$$Z(-\nu) = Z^*(\nu) \qquad \text{for all frequencies } \nu. \tag{33}$$

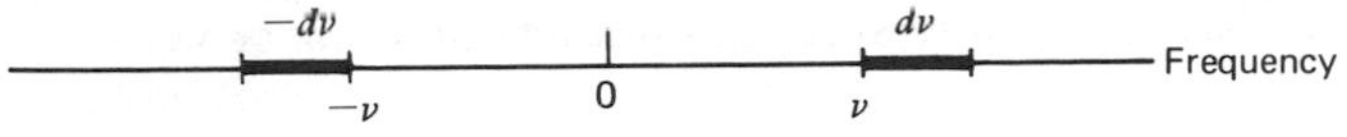

Figure 5.1.8

Consider the next two small intervals $d\nu$ and $-d\nu$ in the frequency domain that are symmetrically displaced about $\nu = 0$, as in Figure 5.1.8. Since these intervals are disjoint, $dZ(\nu)$ and $dZ(-\nu)$ are uncorrelated zero mean random variables,

$$E[dZ(\nu)\, dZ^*(-\nu)] = 0.$$

But by (33) also

$$\begin{aligned} E[dZ(\nu)\, dZ^*(-\nu)] &= E[(dZ(\nu))^2] \\ &= E[(dZ_1(\nu))^2] + 2\iota E[dZ_1(\nu)\, dZ_2(\nu)] - E[(dZ_2(\nu))^2], \end{aligned} \tag{34}$$

and hence

$$E[(dZ_1(\nu))^2] = E[(dZ_2(\nu))^2], \tag{35}$$

and

$$E[dZ_1(\nu)\, dZ_2(\nu)] = 0, \tag{36}$$

since both the real and the imaginary parts of (34) must be zero. Thus by (36) the two processes $Z_1(\nu)$ and $Z_2(\nu)$ are uncorrelated (they have, of course, both zero mean, since $E[dZ(\nu)] = 0$) and since

$$dS_X(\nu) = E[|dZ(\nu)|^2] = E[(dZ_1(\nu))^2] + E[(dZ_2(\nu))^2]$$

(35) implies that

$$E[(dZ_1(\nu))^2] = E[(dZ_2(\nu))^2] = \frac{1}{2}\, dS_X(\nu). \tag{37}$$

In particular, if the real-valued process is also Gaussian, we have the following corollary to the spectral representation theorem:

THEOREM 5.1.3. *Let* $X(t)$, $-\infty < t < \infty$, *be a real-valued stationary, MS continuous Gaussian process with zero mean and with power spectral distribution function* $S_X(\nu)$, $-\infty < \nu < \infty$. *Then the real and imaginary parts* $Z_1(\nu)$ *and* $Z_2(\nu)$ *of the spectral process*

$$Z(\nu) = Z_1(\nu) + \iota Z_2(\nu), \qquad -\infty < \nu < \infty,$$

are mutually independent, identically distributed Gaussian processes satisfying (32) *and* (37).

They both have zero means, and when restricted to $0 \le \nu < \infty$, *they also have independent increments. The spectral representation can be written as*

$$X(t) = 2\int_0^\infty \cos \omega t \, dZ_1(\nu) - 2\int_0^\infty \sin \omega t \, dZ_2(\nu), \qquad -\infty < t < \infty,$$

where $\omega = 2\pi\nu$.

Example 5.1.3

Let us consider the familiar zero mean stationary Ornstein-Uhlenbeck process, $X(t)$, $-\infty < t < \infty$. Since its correlation function is exponential

$$R_X(\tau) = \sigma^2 e^{-\alpha|\tau|}, \qquad -\infty < \tau < \infty,$$

we have from Example 5.1.1 that the power spectral distribution function $S_X(\nu)$ has increments

$$dS_X(\nu) = \frac{2\sigma^2\alpha}{\sigma^2 + (2\pi\nu)^2}\, d\nu.$$

It follows that the real and imaginary parts $Z_1(\nu)$ and $Z_2(\nu)$ of the spectral process, restricted to $0 \le \nu < \infty$, will be mutually independent zero mean Gaussian processes with independent increments such that

$$\mathrm{E}[(dZ_1(\nu))^2] = \mathrm{E}[(dZ_2(\nu))^2] = \frac{\sigma^2\alpha}{\sigma^2 + (2\pi\nu)^2}\, d\nu.$$

But then $Z_1(\nu)$ and $Z_2(\nu)$, $0 \le \nu < \infty$, must in fact be inhomogeneous Wiener processes with intensity

$$b(\nu) = \sqrt{\frac{\sigma^2\alpha}{\sigma^2 + (2\pi\nu)^2}}, \qquad 0 \le \nu < \infty.$$

Thus the spectral representation of the Ornstein-Uhlenbeck process can be written as a sum of two Wiener integrals,

$$X(t) = 2\int_0^\infty \cos \omega t \sqrt{\frac{\sigma^2\alpha}{\alpha^2 + \omega^2}}\, dW_{01}(\nu)$$

$$- 2\int_0^\infty \sin \omega t \sqrt{\frac{\sigma^2\alpha}{\alpha^2 + \omega^2}}\, dW_{02}(\nu), \qquad -\infty < t < \infty,$$

where $W_{01}(\nu)$ and $W_{02}(\nu)$, $0 \le \nu < \infty$, are two independent standard Wiener processes and $\omega = 2\pi\nu$.

Since, for fixed ν, the increments $dW_{01}(\nu)$ and $dW_{02}(\nu)$ are i.i.d. Gaussian random variables with zero mean and variance $d\nu$, the process

$$Y_\nu(t) = dW_{01}(\nu) \cos \omega t - dW_{02}(\nu) \sin \omega t, \qquad -\infty < t < \infty,$$

is the familiar random sine wave,

$$Y_\nu(t) = dA(\nu) \cos(\omega t + d\Theta(\nu)),$$

where $dA(\nu)$ is a Rayleigh random variable with the parameter $2\,d\nu$, and $d\Theta(\nu)$ is uniformly distributed over $[0, 2\pi)$ and independent of $dA(\nu)$. Thinking of the Wiener integral in terms of approximating an infinite series, we see that the process $X(t)$ can indeed be regarded as a superposition of independent random sine waves

$$X(t) \doteq 2 \sum_\nu \sqrt{\frac{\sigma^2\alpha}{\alpha^2 + \omega^2}}\, Y_\nu(t)\, d\nu,$$

where the Rayleigh amplitudes are weighted by the factor

$$2\sqrt{\frac{\sigma^2\alpha}{\alpha^2 + \omega^2}}.$$

Thus the contributions of higher frequencies tend monotonically to zero with rate inversely proportional to the frequency. ▲

Thus for real-valued Gaussian processes, a complex-valued spectral process can be avoided. However, just as in the case of Fourier analysis of deterministic signals, the spectral representation with a complex-valued spectral process is more elegant and easier to handle than its real form given above.

For a real-valued stationary Gaussian process $X(t)$, $-\infty < t < \infty$, the incoherence of its spectral process $Z(\nu)$ can be demonstrated even more clearly. If

$$X(t) = \int_{-\infty}^{\infty} e^{i\omega t}\, dZ(\nu), \qquad -\infty < t < \infty$$

is a spectral representation of such a process, then

$$X(t) = \int_{-\infty}^{\infty} e^{i(\omega t + \gamma(\nu))}\, dZ(\nu), \qquad -\infty < t < \infty, \tag{38}$$

is an equivalent spectral representation provided only that the phase function $\gamma(\nu)$ is odd, $\gamma(-\nu) = -\gamma(\nu)$, $-\infty < \nu < \infty$, to keep the process real valued.

If we write (38) as

$$X(t) = \int_{-\infty}^{\infty} e^{i\omega t}\, d\tilde{Z}(\nu),$$

where

$$d\tilde{Z}(\nu) = e^{i\gamma(\nu)}\, dZ(\nu), \tag{39}$$

then the real and imaginary parts of the increments

$$dZ(\nu) = dZ_1(\nu) + \imath\, dZ_2(\nu),$$
$$d\tilde{Z}(\nu) = d\tilde{Z}_1(\nu) + \imath\, d\tilde{Z}_2(\nu),$$

are related by

$$\begin{aligned} d\tilde{Z}_1(\nu) &= \cos \gamma(\nu)\, dZ_1(\nu) - \sin \gamma(\nu)\, dZ_2(\nu), \\ d\tilde{Z}_2(\nu) &= \sin \gamma(\nu)\, dZ_1(\nu) + \cos \gamma(\nu)\, dZ_2(\nu). \end{aligned} \tag{40}$$

But the vector

$$d\mathbf{Z}(\nu) = \begin{pmatrix} dZ_1(\nu) \\ dZ_2(\nu) \end{pmatrix}$$

viewed as a point in the complex plane has a circularly symmetric Gaussian distribution, and the transformation (40) is just a rotation of the coordinate system by the angle $-\gamma(\nu)$. Thus the vector

$$d\tilde{\mathbf{Z}}(\nu) = \begin{pmatrix} d\tilde{Z}_1(\nu) \\ d\tilde{Z}_2(\nu) \end{pmatrix}$$

has the same circularly symmetric Gaussian distribution. Since independence of increments is not affected by the modification (39) it follows that the spectral processes $Z(\nu)$ and $\tilde{Z}(\nu)$ have identical (Gaussian) probability laws.

Intuitively, the reason is easy to see: a real-valued stationary Gaussian process $X(t)$ can be regarded as a superposition of independent elementary random sine waves

$$Y_\nu(t) = dA(\nu) \cos(\omega t + d\Theta(\nu))$$

with Rayleigh amplitudes and uniform phases, the probability law of which is not changed by adding an arbitrary constant $\gamma(\nu)$ to the phase $d\Theta(\nu)$.

APPENDIX

SKETCH OF THE PROOF OF BOCHNER'S THEOREM. Let $R(\tau)$, $-\infty < \tau < \infty$, be the correlation function of a zero mean wide-sense stationary process. For any $-\infty < \omega < \infty$ and a positive integer $m = 1, 2, \ldots$, define a function

$$\alpha(t) = \frac{1}{\sqrt{R(0)}} e^{-\imath\omega t - (t/2m)^2}, \qquad -\infty < t < \infty.$$

Since a correlation function is nonnegative definite, we must have for any finite set of numbers

$$-\infty < t_0 < t_1 < \cdots < t_n < \infty$$

$$\sum_{j=0}^{n-1} \sum_{k=0}^{n-1} R(t_j - t_k)\alpha(t_j)\, \Delta t_j \alpha^*(t_k)\, \Delta t_k \geq 0,$$

where

$$\Delta t_j = t_{j+1} - t_j, \qquad j = 0, \ldots, n-1.$$

This also implies that

$$\int_{-\infty}^{\infty} \int_{-\infty}^{\infty} R(t' - t'')\alpha(t')\alpha^*(t'')\, dt'\, dt'' \geq 0 \tag{A1}$$

where the existence of this double integral follows from the continuity of $R(\tau)$ and the fact the absolute value of the integrand is bounded by an integrable function

$$e^{-(t'/2m)^2} e^{-(t''/2m)^2}.$$

Making the substitution $\tau = t' - t''$, $\phi = t' + t''$ into the integral (A1) and noticing that

$$\alpha(t')\alpha^*(t'') = \frac{1}{R(0)}\, e^{-\iota\omega\tau} e^{-\tau^2/2m^2} e^{-\phi^2/2m^2}$$

the integral becomes a product of two integrals

$$2\pi g_m(\omega) \int_{-\infty}^{\infty} e^{-\phi^2/2m^2}\, d\phi,$$

where we have denoted

$$g_m(\omega) = \frac{1}{2\pi} \int_{-\infty}^{\infty} e^{-\iota\omega\tau} \frac{R(\tau)}{R(0)} e^{-\tau^2/2m^2}\, d\tau. \tag{A2}$$

Since clearly $\int_{-\infty}^{\infty} e^{-\phi^2/2m^2}\, d\phi > 0$, it follows from (A1) that the function

$$g_m(\omega) \geq 0 \qquad \text{for all } -\infty < \omega < \infty \tag{A3}$$

Furthermore, upon multiplying both sides of (A2) by $e^{\iota\omega\zeta}$, $-\infty < \zeta < \infty$, and integrating $d\omega$, we get

$$\begin{aligned} \int_{-\infty}^{\infty} e^{\iota\omega\zeta} g_m(\omega)\, d\omega &= \int_{-\infty}^{\infty} \int_{-\infty}^{\infty} e^{-\iota\omega(\tau-\zeta)} \frac{R(\tau)}{R(0)} e^{-\tau^2/2m^2}\, d\tau\, d\zeta \\ &= \int_{-\infty}^{\infty} \delta(\tau - \zeta) \frac{R(\tau)}{R(0)} e^{-\tau^2/2m^2}\, d\tau \\ &= \frac{R(\zeta)}{R(0)} e^{-\zeta^2/2m^2} \end{aligned} \tag{A4}$$

since

$$\int_{-\infty}^{\infty} e^{-\imath\omega t}\, d\omega = 2\pi\, \delta(t).$$

Setting $\zeta = 0$, this gives

$$\int_{-\infty}^{\infty} g_m(\omega)\, d\omega = 1,$$

which, together with (A3), implies that the function $g_m(\omega)$ can be the probability density of some continuous random variable Ω_m. But that means that the right-hand side of (A4), that is, the function

$$\psi_m(\zeta) = \frac{R(\zeta)}{R(0)} e^{-\zeta^2/2m^2}, \qquad -\infty < \zeta < \infty,$$

is, in fact, the characteristic function of this random variable Ω_m, since by (A4)

$$\psi_m(\zeta) = \mathrm{E}[e^{\imath\zeta\Omega_m}] = \frac{1}{2\pi}\int_{-\infty}^{\infty} e^{\imath\zeta\omega} g_m(\omega)\, d\omega.$$

Now let $m \to \infty$. Then from (A4)

$$\lim_{m\to\infty} \psi_m(\zeta) = \frac{R(\zeta)}{R(0)} \qquad \text{for all } -\infty < \zeta < \infty,$$

which is a continuous function of ζ, since the correlation function R was assumed continuous at zero and hence everywhere. However, by Lévy's continuity theorem, if a sequence of characteristic functions converges to a continuous function, then the *limit must itself be the characteristic function* of some random variable Ω. Denoting its distribution function by $F_\Omega(\omega)$, this means that we can write

$$\frac{R(\zeta)}{R(0)} = \int_{-\infty}^{\infty} e^{\imath\zeta\omega}\, dF_\Omega(\omega), \qquad -\infty < \zeta < \infty,$$

which, upon renaming ζ by τ and calling $S(v) = F_\Omega(2\pi v)$, $-\infty < v < \infty$, becomes representation (18) of Bochner's theorem.

Still to be proved is the "if" part, that is, to show that if $S(v)$ is any distribution function, then

$$R(\tau) = R(0) \int_{-\infty}^{\infty} e^{\imath\omega\tau}\, dS(v) \tag{A5}$$

is a correlation function. However, this is quite easy; clearly from (A5)

$$|R(\tau)| \le R(0) \qquad \text{and} \qquad R(-\tau) = R^*(\tau)$$

so that we only have to show that $R(\tau)$ is nonnegative definite. Taking an arbitrary finite collection of complex numbers $\gamma_1, \ldots, \gamma_n$ and $-\infty < t_1 < \cdots < t_n < \infty$, we have from (A5)

$$\sum_{j=1}^{n}\sum_{k=1}^{n} R(t_j - t_k)\gamma_j\gamma_k^* = R(0)\int_{-\infty}^{\infty}\sum_{j=1}^{n}\sum_{k=1}^{n} e^{\iota\omega(t_j-t_k)}\,\gamma_j\gamma_k^*\,dS(\nu)$$

$$= R(0)\int_{-\infty}^{\infty}\left(\sum_{j=1}^{n}\gamma_j e^{\iota\omega t_j}\right)\left(\sum_{k=1}^{n}\gamma_k e^{\iota\omega t_k}\right)^* dS(\nu)$$

$$= R(0)\int_{-\infty}^{\infty}\left|\sum_{j=1}^{n}\gamma_j e^{\iota\omega t_j}\right|^2 dS(\nu),$$

which is necessarily nonnegative.

Thus Bochner's theorem is proved.

Notice that the theorem says, in fact, that any correlation function of a wide-sense stationary process with zero mean and unit variance can serve as the characteristic function of some probability distribution, and vice versa. In particular, it follows that any characteristic function must be nonnegative definite.

EXERCISES

Exercise 5.1

1. Let $X(t)$, $-\infty < t < \infty$, be a zero mean wide-sense stationary process with correlation function $R_X(\tau)$ defined below. For each case, find and sketch the power spectral distribution function $S_X(\nu)$, $-\infty < \nu < \infty$. α and σ are positive constants.
   (a) $R_X(\tau) = \sigma^2 e^{-\alpha\tau^2}$,
   (b) $R_X(\tau) = \sigma^2(\cos \alpha\tau)^3$,
   (c) $R_X(\tau) = \dfrac{\sigma^2}{1+\alpha\tau^2}\cdot$
   (*Hint:* Identify each $R_X(\zeta/2\pi)/\sigma^2$ as a characteristic function.)
2. A zero mean wide-sense stationary process $X(t)$, $-\infty < t < \infty$, has the power spectral distribution function as defined below. In each case find the correlation function $R_X(\tau)$, $-\infty < \tau < \infty$.

   (a) $S_X(\nu) = \begin{cases} 0 & \text{if} \quad \nu < -1, \\ \nu^3 + 1 & \text{if} \quad -1 \le \nu < 1, \\ 2 & \text{if} \quad \nu \ge 1. \end{cases}$

   (b) $S_X(\nu) = \begin{cases} \frac{1}{2}(1-\nu)e^{\nu} & \text{if} \quad \nu < 0, \\ 1 - \frac{1}{2}(1+\nu)e^{-\nu} & \text{if} \quad \nu \ge 0. \end{cases}$

   (c) $S_X(\nu)$ as in Figure 5.1.E.1.

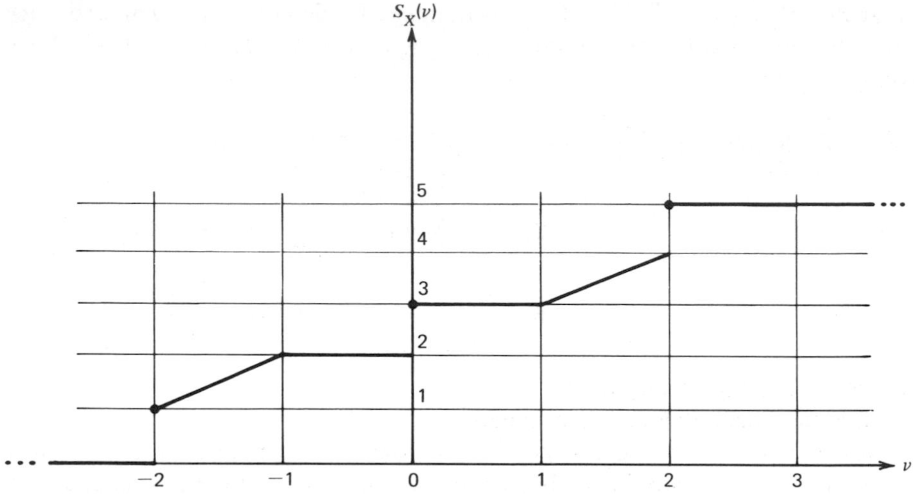

Figure 5.1.E.1

3. Show that for the nth MS derivative of a wide-sense stationary process $X(t)$, $-\infty < t < \infty$, to exist, it is necessary and sufficient that the integral

$$\int_{-\infty}^{\infty} \omega^{2n} \, dS_X(\nu)$$

be finite. What is the relationship between the power spectral distribution function of the nth MS derivative $X^{(n)}(t) = d^nX(t)/dt^n$ and $S_X(\nu)$?

4. Describe typical sample paths of the spectral process $Z(\nu) = Z_1(\nu) + \imath Z_2(\nu)$, $-\infty < \nu < \infty$, if $X(t)$, $-\infty < t < \infty$, is a zero mean stationary Gaussian process with correlation function $R_X(\tau) = (\cos \tau)^3$, $-\infty < \tau < \infty$.

5. Suppose $Y(t)$, $-\infty < t < \infty$, is purely imaginary, that is, $Y(t) = \imath X(t)$, $-\infty < t < \infty$, where $X(t)$ is a stationary zero mean MS continuous Gaussian process. What can be said about the power spectral distribution function $S_Y(\nu)$ and the corresponding spectral process $Z(\nu)$?

6. A source of incoherent monochromatic light can be imagined as consisting of a large number n of oscillators, each producing a sine wave $\cos(\omega t + \Theta_k)$, $k = 1, \ldots, n$, with independent phases Θ_k all uniformly distributed over $[0, 2\pi)$. Show that the superposition $X(t) = \sum_{k=1}^{n} \cos(\omega t + \Theta_k)$ is approximately a stationary Gaussian process of the form $X(t) = Z_1(\nu)\cos \omega t - Z_2(\nu)\sin \omega t$, $\omega = 2\pi\nu$, where $Z_1(\nu)$ and $Z_2(\nu)$ are i.i.d. Gaussian random variables. (*Hint:* Write $\cos(\omega t + \Theta_k)$ in the quadrature decomposition, show that the two components are uncorrelated, and apply the central limit theorem.)

5.2. SPECTRAL TYPES AND LINEAR TRANSFORMATIONS

The power spectral distribution function $S_X(\nu)$ of a wide-sense stationary MS-continuous process $X(t)$ is a distribution function of general type, that is it is a nondecreasing function such that

$$\lim_{\nu \to -\infty} S_X(\nu) = 0, \qquad \lim_{\nu \to +\infty} S_X(\nu) = \sigma^2$$

and it is continuous from the right. In other words, although its meaning is different, the function $(1/\sigma^2)S_X(\nu)$ could serve as the distribution function of a random variable. It could, therefore, be a distribution function of either a discrete or continuous or, in general, mixed-type random variable.

DISCRETE-TYPE POWER SPECTRAL DISTRIBUTIONS. If $S_X(\nu)$ is a *discrete type*, it must then be a step function, with at most a countable number of jumps of height $s_X(\nu_n)$ at discrete points ν_n in the frequency domain. It is then completely characterized by the doubly infinite sequence

$$\{(\nu_n, s_X(\nu_n)), \qquad n = 0, \pm 1, \pm 2, \ldots\}, \tag{1}$$

where the frequencies ν_n are labeled such that

$$\cdots < \nu_{-2} < \nu_{-1} < \nu_0 = 0 < \nu_1 < \nu_2 < \cdots \dagger$$

$$s_X(\nu) = S_X(\nu) - \lim_{\nu' \to \nu -} S_X(\nu') = \begin{cases} s_X(\nu_n) & \text{if} \quad \nu = \nu_n, \, n = 0, \pm 1, \pm 2, \ldots \\ 0 & \text{otherwise.} \end{cases} \tag{2}$$

The sequence (1) is then called a *discrete* (or line) *power spectrum* (the values $s_X(\nu_n)$ are referred to as *spectral lines*), and in analogy with the probability mass distribution of a discrete random variable, the function $s_X(\nu)$ defined by (2) is called the *power spectral mass function.*

Note that $s_X(\nu) \geq 0$ for all $-\infty < \nu < \infty$ and that

$$\sum_{n=-\infty}^{\infty} s_X(\nu_n) = \sigma^2.$$

For a real process $X(t)$, it must also be symmetric about the zero frequency

$$s_X(\nu) = s_X(-\nu), \qquad -\infty < \nu < \infty.$$

† If $S_X(\nu)$ has no jump at $\nu = 0$, we set $s_X(0) = 0$.

Using the Dirac delta function, we could also write symbolically

$$dS_X(\nu) = \sum_{n=-\infty}^{\infty} s_X(\nu_n)\, \delta(\nu - \nu_n)\, d\nu. \tag{3}$$

The integral representation of the correlation function

$$R_X(\tau) = \int_{-\infty}^{\infty} e^{\iota\omega\tau}\, dS_X(\nu),$$

then takes the form of an infinite series

$$R_X(\tau) = \sum_{n=-\infty}^{\infty} e^{\iota\omega_n\tau} s_X(\nu_n), \qquad \omega_n = 2\pi\nu_n, \tag{4}$$

or, for a real process $X(t)$, from (5.1.31),

$$R_X(\tau) = s_X(0) + 2\sum_{n=1}^{\infty} s(\nu_n) \cos \omega_n \tau. \tag{5}$$

The discrete nature of the power spectral distribution function is reflected in the structure of the spectral process $Z(\nu)$. From (3) we have in this case

$$\mathrm{E}[\,|dZ(\nu)|^2] = \sum_{n=-\infty}^{\infty} s_X(\nu_n)\, \delta(\nu - \nu_n)\, d\nu,$$

which is nonzero only if $\nu = \nu_n$ for some $n = 0, \pm 1, \pm 2, \ldots$. Thus the spectral process $Z(\nu)$, $-\infty < \nu < \infty$, can be defined in terms of a sequence

$$Z(n), \qquad n = 0, \pm 1, \pm 2, \ldots, \tag{6}$$

of zero mean uncorrelated complex-valued random variables such that $\mathrm{E}[\,|Z(n)|^2] = s_X(\nu_n)$.

The spectral representation can be written as a doubly infinite series

$$X(t) = \sum_{n=-\infty}^{\infty} e^{\iota\omega_n t} Z(n), \qquad -\infty < t < \infty, \qquad \text{where } \omega_n = 2\pi\nu_n. \tag{7}$$

If the process $X(t)$ is also Gaussian and real valued, the sequence (6) must further be such that

$$Z(n) = \begin{cases} Z_1(n) + \iota Z_2(n) & \text{if} \quad n = 0, 1, \ldots, \\ Z_1(-n) - \iota Z_2(-n) & \text{if} \quad n = 0, -1, \ldots, \end{cases}$$

where $Z_1(n)$, $n = 0, 1, \ldots$, and $Z_2(n)$, $n = 0, 1, \ldots$, are mutually independent sequences of independent real-valued Gaussian random variables such that

$$\mathrm{E}[Z_1^2(n)] = \mathrm{E}[Z_2^2(n)] = \frac{1}{2} s_X(\nu_n), \qquad n = 1, 2, \ldots,$$

$$\mathrm{E}[Z_1^2(0)] = s_X(0), \qquad \mathrm{E}[Z_2^2(0)] = 0.$$

The spectral representation (7) then becomes

$$X(t) = 2\sum_{n=0}^{\infty} Z_1(n)\cos \omega_n t - 2\sum_{n=0}^{\infty} Z_2(n)\sin \omega_n t, \qquad -\infty < t < \infty, \tag{8}$$

$\omega_n = 2\pi\nu_n$, which is an analog of the quadrature decomposition of a deterministic sinusoidal waveform into cosine components

$$X_1(t) = 2\sum_{n=0}^{\infty} Z_1(n)\cos \omega_n t,$$

and sine components

$$X_2(t) = 2\sum_{n=0}^{\infty} Z_2(n)\sin \omega_n t.$$

Note that $X_1(t)$ and $X_2(t)$ are mutually independent. In fact, we can actually write

$$X(t) = \sqrt{s_X(0)}\,Z_0 + 2\sum_{n=1}^{\infty} \sqrt{s_X(\nu_n)}\,A_n \cos(\omega_n t + \Theta_n), \tag{9}$$

where A_1, A_2, ... are independent standard† Rayleigh random variables, and Θ_1, Θ_2, ... are independent uniform $[0, 2\pi)$ random variables, also independent of the sequence A_1, A_2, ... and of the standard Gaussian random variable Z_0.

It can be seen that every stationary Gaussian process with a discrete-type spectrum can be written as the superposition of a DC component and independent random sine waves with frequencies ν_1, ν_2, ...

Notice that either (8) or (9) resembles a Fourier series expansion of a periodic function except that the frequencies ν_n, $n = 1, 2, \ldots$, need not be integral multiples of any basic frequency. For this reason, processes with discrete-type spectra are sometimes called *almost periodic.*

Example 5.2.1

A wide-sense stationary process $X(t)$, $-\infty < t < \infty$, with zero mean and constant correlation function

$$R_X(\tau) = \sigma^2, \qquad -\infty < \tau < \infty,$$

has, as we have seen in Example 5.1.1, the power spectral distribution function

$$S_X(\nu) = \begin{cases} \sigma^2 & \text{if} \quad \nu \geq 0, \\ 0 & \text{if} \quad \nu < 0. \end{cases}$$

Clearly, this is a simple discrete-type spectrum with a single spectral line $s_X(0) = \sigma^2$ at the zero frequency $\nu_0 = 0$ as in Figure 5.2.1.

† That is, such that $E[A_n^2] = 1$, $n = 1, 2, \ldots$.

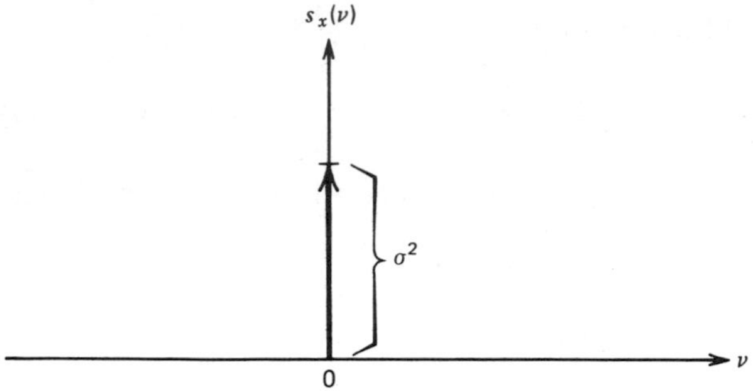

Figure 5.2.1

The random sine wave of Example 5.1.2 also has a discrete-type spectrum, this time with two spectral lines $(-\nu_0, \frac{1}{2}\sigma^2)$, $(\nu_0, \frac{1}{2}\sigma^2)$ as shown in Figure 5.2.2. Expression (5) yields its correlation function

$$R_X(\tau) = 2\frac{\sigma^2}{2} \cos \omega_0 \tau, \qquad -\infty < \tau < \infty,$$

while from (8) the spectral representation is

$$X(t) = Z_1(\nu_0)\cos \omega_0 t + Z_2(\nu_0)\sin \omega_0 t, \qquad -\infty < t < \infty.$$

Since the random sine wave is a Gaussian process, $Z_1(\nu_0)$ and $Z_2(\nu_0)$ are i.i.d. Gaussian random variables with zero mean and variance $\sigma^2/2$. Thus the above representation is just the familiar quadrature decomposition of the random sine wave. ▲

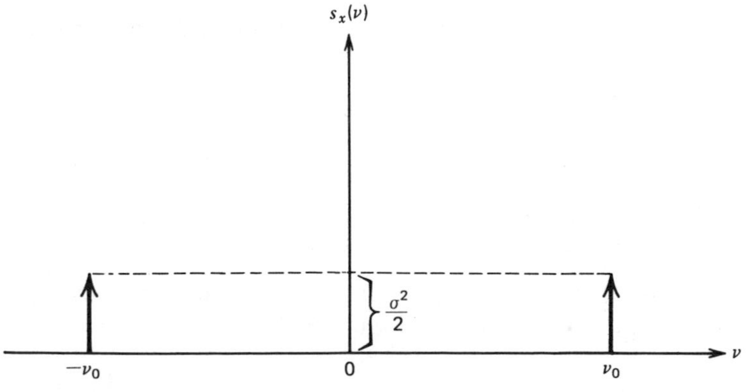

Figure 5.2.2

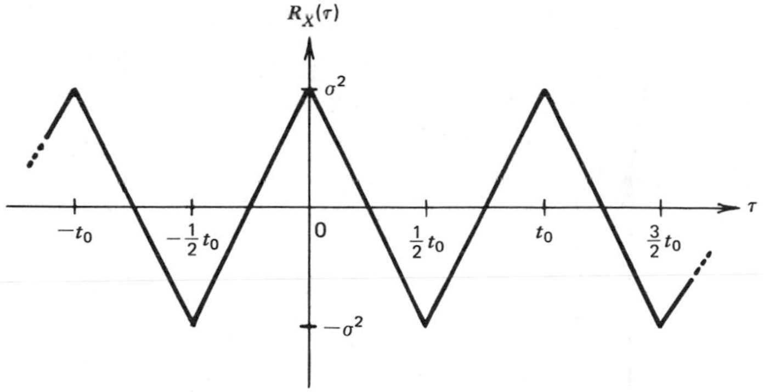

Figure 5.2.3

Example 5.2.2

For a case that is a little more complicated, consider the random square wave of Exercise 1.22. It is easily seen that this process $X(t)$, $-\infty < t < \infty$, is indeed a wide-sense stationary zero mean process with correlation function $R_X(\tau)$, $-\infty < \tau < \infty$, as pictured in Figure 5.2.3. Since this is a periodic function with period t_0, it can be expanded into a Fourier series

$$R_X(\tau) = \sum_{n=-\infty}^{\infty} c_n e^{\imath 2\pi n\tau}, \tag{10}$$

where

$$c_n = \frac{1}{t_0}\int_0^{t_0} R_X(\tau)e^{-\imath 2\pi n\tau/t_0}\,d\tau, \qquad n = 0, \pm 1, \ldots,$$

according to a well-known formula for the Fourier coefficients c_n. Comparing (10) with expression (4) it is seen that the spectrum of this process is discrete with power spectral mass function

$$s_X(\nu) = \begin{cases} c_n & \text{if} \quad \nu = \dfrac{n}{t_0}, \qquad n = 0, \pm 1, \ldots \\ 0 & \text{otherwise.} \end{cases}$$

Evaluating the integral

$$\begin{aligned} c_n &= \frac{1}{t_0}\int_0^{t_0} R_X(\tau)e^{-\imath 2\pi n\tau/t_0}\,d\tau \\ &= \frac{\sigma^2}{t_0}\int_0^{t_0/2}\left(1 - \frac{4\tau}{t_0}\right)e^{-\imath 2\pi n\tau/t_0}\,d\tau + \frac{\sigma^2}{t_0}\int_{t_0/2}^{t_0}\left(\frac{4\tau}{t_0} - 3\right)e^{-\imath 2\pi n\tau/t_0}\,d\tau, \end{aligned}$$

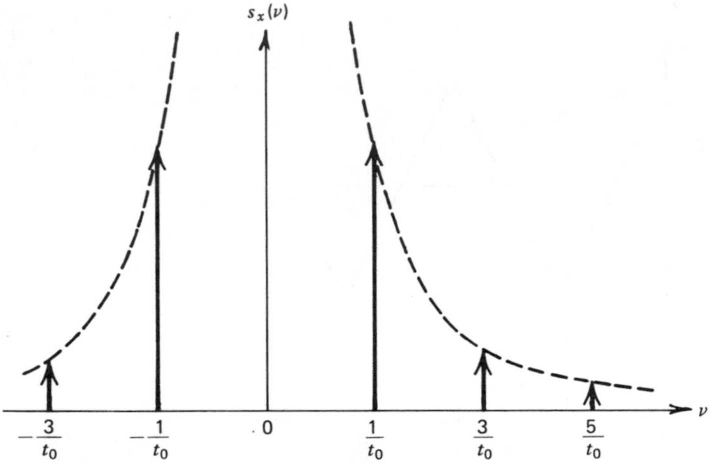

Figure 5.2.4

we find that $c_n = 0$ for n even, and

$$c_n = \left(\frac{2\sigma}{\pi n}\right)^2 \qquad \text{for } n \text{ odd.}$$

Thus the power spectral mass function is

$$s_X(\nu) = \begin{cases} \left(\dfrac{2\sigma}{\pi t_0 \nu}\right)^2 & \text{if} \quad \nu = \pm\dfrac{1}{t_0}, \pm\dfrac{3}{t_0}, \pm\dfrac{5}{t_0}, \ldots \\ 0 & \text{otherwise,} \end{cases}$$

that is, the power spectrum has an infinite number of spectral lines located at the odd multiples of the basic frequency $1/t_0$ as in Figure 5.2.4. Note that the envelope, shown as a dashed line in Figure 5.2.4, decreases as $1/\nu^2$. From the formula

$$1 + \frac{1}{3^2} + \frac{1}{5^2} + \frac{1}{7^2} + \cdots = \frac{\pi^2}{8},$$

(see Appendix Three), it can be verified that indeed

$$\sum_{n=-\infty}^{\infty} s_X(\nu_n) = \sigma^2.$$

From the spectral representation (8), it follows that the random square wave process can be expressed as

$$X(t) = 2\sum_{n=0}^{\infty} (Z_1(n)\cos(2n+1)\omega_0 t + Z_2(n)\sin(2n+1)\omega_0 t),$$

$-\infty < t < \infty$, where $\omega_0 = 2\pi/t_0$ and $Z_1(n)$, $Z_2(n)$, $n = 0, 1, \ldots$, are mutually uncorrelated zero mean random variables such that

$$E[Z_1^2(n)] = E[Z_2^2(n)] = \frac{2\sigma^2}{\pi^2(2n+1)^2}.$$

However, since the random square wave is not a Gaussian process, these random variables are neither Gaussian nor independent.

This example should convince the reader of the fact that whenever the correlation function of a wide-sense stationary process is periodic with period t_0, the resulting spectrum is necessarily discrete with spectral lines only at integral multiples of the basic frequency $1/t_0$.

However, a wide-sense stationary process, even a Gaussian one, can have a discrete-type spectrum without having a periodic correlation function or, for that matter, periodic sample paths. This is easily seen by taking the sum of two independent random sine waves with frequencies, say, $\sqrt{2}$ and $\sqrt{3}$. ▲

Example 5.2.3

If the process $X(t)$, $-\infty < t < \infty$, is not Gaussian, the random variables $Z(n)$, $n = 0, \pm 1, \pm 2, \ldots$, of its spectral process, although uncorrelated, may be highly dependent. To illustrate such a case, let

$$X(t) = \cos(\Omega t + \Theta), \qquad -\infty < t < \infty,$$

where both the frequency Ω and the phase Θ are independent random variables. The phase Θ is uniformly distributed over $[0, 2\pi)$, while the frequency has a geometric distribution

$$P(\Omega = n) = pq^{n-1}, \qquad n = 1, 2, \ldots,$$

where $0 < p < 1$.

Conditioning upon the random frequency $\Omega = n$, we find that

$$E[X(t) \mid \Omega = n] = \int_0^{2\pi} \cos(nt + \theta)\frac{d\theta}{2\pi} = 0,$$

and using the formula for the product of cosines,

$$\begin{aligned} E[X(t_1)X(t_2) \mid \Omega = n] &= \int_0^{2\pi} \cos(nt_1 + \theta)\cos(nt_2 + \theta)\frac{d\theta}{2\pi} \\ &= \frac{1}{2}\int_0^{2\pi} \cos n(t_1 - t_2)\frac{d\theta}{2\pi} + \frac{1}{2}\int_0^{2\pi} \cos(nt_1 + nt_2 + 2\theta)\frac{d\theta}{2\pi} \\ &= \frac{1}{2}\cos n(t_1 - t_2). \end{aligned}$$

Taking expectations, it follows that

$$\mu_X(t) = \mathrm{E}[X(t)] = 0, \qquad -\infty < t < \infty,$$

and

$$\mathrm{E}[X(t_1)X(t_2)] = \sum_{n=1}^{\infty} \mathrm{E}[X(t)\,|\,\Omega = n]pq^{n-1}$$

$$= \frac{1}{2}\sum_{n=1}^{\infty} pq^{n-1} \cos n(t_1 - t_2).$$

Thus $X(t)$, $-\infty < t < \infty$, is a zero mean wide-sense stationary process with correlation function

$$R_X(\tau) = \sum_{n=1}^{\infty} \frac{1}{2} pq^{n-1} \cos n\tau, \qquad -\infty < \tau < \infty.$$

Remark: The infinite series above happens to be a Fourier series for the periodic function

$$\frac{p}{2}\,\frac{\cos \tau - q}{q^2 - 2q \cos \tau + 1}.$$

Hence the correlation function can be expressed in closed form and plotted. The interested reader may do so. Я

Again, comparing the expression for the correlation function with (5) we see that the process has a discrete power spectrum with power spectral mass function

$$s_X(\nu) = \begin{cases} \dfrac{1}{4} pq^{n-1} & \text{if} \quad \nu = \dfrac{n}{2\pi}, \qquad n = \pm 1, \pm 2, \ldots, \\ 0 & \text{otherwise.} \end{cases}$$

To get the spectral representation

$$X(t) = 2\sum_{n=1}^{\infty} (Z_1(n)\cos nt + Z_2(n)\sin nt),$$

where clearly $Z_1(0) = Z_2(0) = 0$, since the process has no DC component, just write

$$X(t) = \cos(\Omega t + \Theta) = \cos \Theta \cos \Omega t - \sin \Theta \sin \Omega t, \qquad -\infty < t < \infty,$$

and set, for $n = 1, 2, \ldots$,

$$Z_1(n) = \begin{cases} \cos \Theta & \text{if} \quad \Omega = n, \\ 0 & \text{if} \quad \Omega \neq n, \end{cases}$$

$$Z_2(n) = \begin{cases} -\sin \Theta & \text{if} \quad \Omega = n, \\ 0 & \text{if} \quad \Omega \neq n. \end{cases}$$

Hence, each sample path of the random sequence

$$\begin{pmatrix} Z_1(n) \\ Z_2(n) \end{pmatrix}, \qquad n = 1, 2, \ldots,$$

consists of all zero vectors except one, which is then distributed as

$$\begin{pmatrix} \cos \Theta \\ -\sin \Theta \end{pmatrix}, \qquad \Theta \text{ uniform on } [0, 2\pi).$$

Thus both sequences $Z_1(n)$, $n = 1, 2, \ldots$, and $Z_2(n)$, $n = 1, 2, \ldots$, not only consist of dependent random variables but are also mutually dependent, since if $Z_1(n) \neq 0$, then $Z_1^2(n) + Z_2^2(n) = 1$. ▲

CONTINUOUS-TYPE POWER SPECTRAL DISTRIBUTIONS. If the power spectral distribution function $S_X(\nu)$ is of a *continuous type*, then it can be written as an integral

$$S_X(\nu) = \int_{-\infty}^{\nu} s_X(\nu')\, d\nu'.$$

The function

$$s_X(\nu) = \frac{dS_X(\nu)}{d\nu}, \qquad -\infty < \nu < \infty,$$

then has the dimension power ÷ frequency and is therefore called the *power spectral density* of the process $X(t)$. Note that this time

$$s_X(\nu) \geq 0 \qquad \text{and} \qquad \int_{-\infty}^{\infty} s_X(\nu)\, d\nu = \sigma^2 = R_X(0),$$

and that for a real process $X(t)$ it is symmetric about the zero frequency

$$s_X(-\nu) = s_X(\nu).$$

Upon writing

$$dS_X(\nu) = s_X(\nu)\, d\nu, \tag{11}$$

the correlation function is now related to the power spectral density by the formula

$$R_X(\tau) = \int_{-\infty}^{\infty} e^{\iota\omega\tau} s_X(\nu)\, d\nu, \qquad \omega = 2\pi\nu, \qquad -\infty < \tau < \infty \tag{12}$$

But that means that the correlation function is just the inverse Fourier transform of the power spectral density

$$R_X = \mathfrak{F}^{-1}\{s_X\},$$

and consequently, s_X is now expressible as a Fourier transform of the correlation function

$$s_X = \mathcal{F}\{R_X\},$$

or

$$s_X(\nu) = \int_{-\infty}^{\infty} e^{-\iota\omega\tau} R_X(\tau)\, d\tau, \qquad \omega = 2\pi\nu, \qquad -\infty < \nu < \infty. \tag{13}$$

Substituting $\nu = 0$ into the above expression, we get

$$s_X(0) = \int_{-\infty}^{\infty} R_X(\tau)\, d\tau = \sigma_X^2 \tau_{cor}$$

where

$$\tau_{cor} = \frac{1}{\sigma_X^2} \int_{-\infty}^{\infty} R_X(\tau)\, d\tau$$

is sometimes called the correlation time. If the power spectral density $s_X(\nu)$ is continuous at $\nu = 0$, this quantity must be finite. For a stationary Gaussian process $X(t)$, the condition

$$\int_{-\infty}^{\infty} |R_X(\tau)|\, d\tau < \infty$$

guarantees that the process $X(t)$ is ergodic (see Chapter One). Thus, for a *stationary Gaussian process, continuity of the power spectral density at* $\nu = 0$ *is a sufficient* (not necessary) *condition for the process to be ergodic.* Since for an ergodic process the time-correlation function $\bar{R}_X(\tau) = R_X(\tau)$, the power spectral density is also the Fourier transform of the time-correlation function and can therefore be estimated directly from a sample path $x(t)$ by the formula (5.1.10), that is,

$$s(\nu) = \lim_{T\to\infty} \int_{-T/2}^{T/2} x(t) e^{-\iota\omega t}\, dt, \qquad \omega = 2\pi\nu.$$

For a real process $X(t)$ the two relations (12) and (13) can also be written as what is known as the Fourier-cosine transform pair

$$\begin{aligned} R_X(\tau) &= 2 \int_0^{\infty} s_X(\nu) \cos \omega\tau\, d\nu, \\ s_X(\nu) &= 2 \int_0^{\infty} R_X(\tau) \cos \omega\tau\, d\tau, \end{aligned} \tag{14}$$

which is easily established from the symmetry of both R_X and s_X.

Example 5.2.4
As we have already seen in Example 5.1.1, a zero mean wide-sense stationary process $X(t)$, $-\infty < t < \infty$, with exponential correlation function

$$R_X(\tau) = \sigma^2 e^{-\alpha|\tau|}, \qquad -\infty < \tau < \infty, \qquad \alpha > 0,$$

has a continuous power spectral distribution function

$$S_X(\nu) = \sigma^2\left(\frac{1}{2} + \frac{1}{\pi}\arctan\frac{2\pi\nu}{\alpha}\right), \qquad -\infty < \nu < \infty.$$

Consequently, its spectrum is of the continuous type with power spectral density,

$$s_X(\nu) = \frac{dS_X(\nu)}{d\nu} = \frac{2\alpha\sigma^2}{\alpha^2 + (2\pi\nu)^2}, \qquad -\infty < \nu < \infty.$$

Indeed, the functions $R_X(\tau)$ and $s_X(\nu)$, pictured in Figure 5.2.5, form a Fourier transform pair, as is easily verified.

Notice that as α gets larger, the power spectral density becomes flatter, so that more power is transferred into a higher frequency range. As α increases the correlation time

$$\tau_{cor} = \frac{s_X(0)}{\sigma^2} = \frac{2}{\alpha}$$

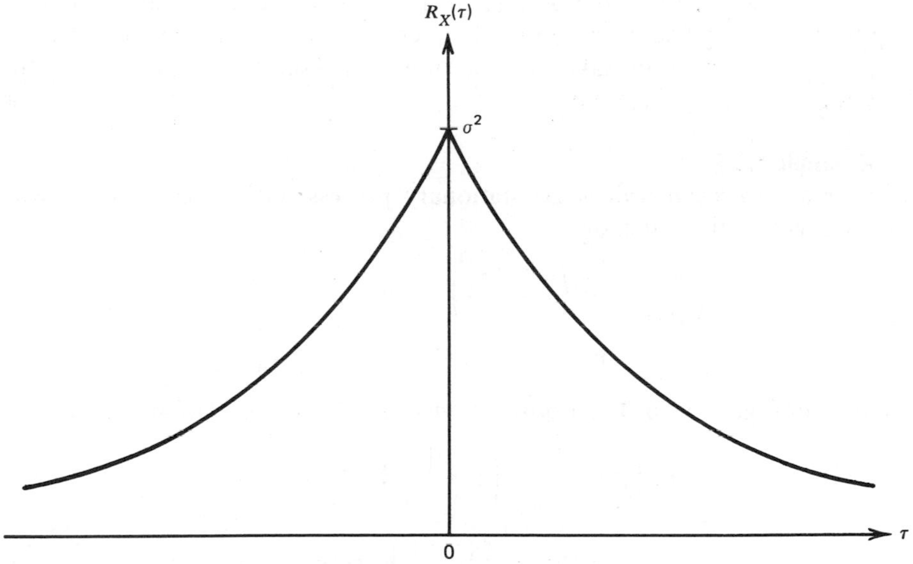

Figure 5.2.5a

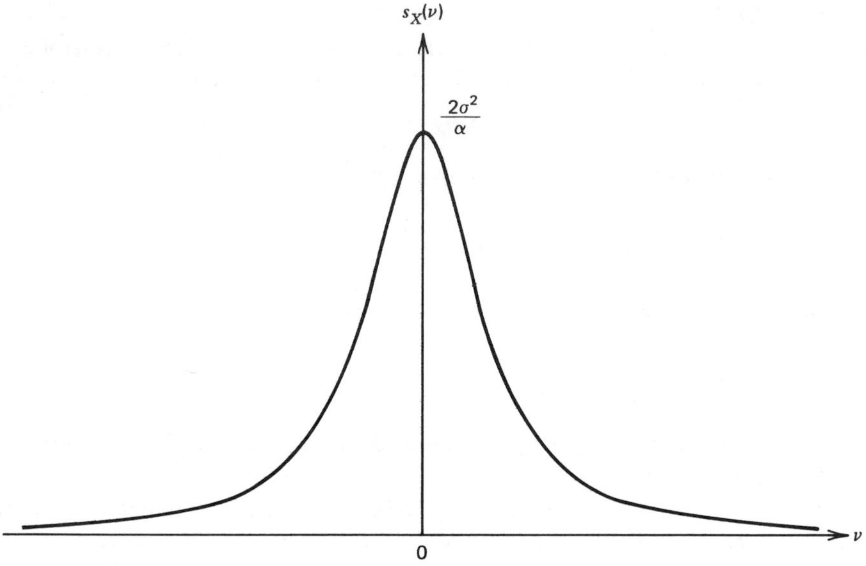

Figure 5.2.5b

decreases, and so the values of the process $X(t)$ become less and less correlated over a fixed time increment. This suggests that we can expect more rapid changes in a typical sample path of the process $X(t)$, the rapid changes being reflected in the power of higher frequencies in the frequency domain. This general feature of the spectral representation makes the interpretation of the power spectrum particularly attractive in applications. ▲

Example 5.2.5

Consider a zero mean wide-sense stationary process $X(t)$, $-\infty < t < \infty$, with triangular correlation function

$$R_X(\tau) = \begin{cases} \sigma^2\left(1 - \dfrac{|\tau|}{\tau_0}\right) & \text{if} \quad |\tau| < \tau_0, \\ 0 & \text{if} \quad \tau \geq \tau_0, \end{cases}$$

as shown in Figure 5.2.6. The Fourier transform of this correlation function,

$$\begin{aligned} \mathcal{F}\{R_X\} &= \int_{-\tau_0}^{\tau_0} \sigma^2\left(1 - \frac{|\tau|}{\tau_0}\right)e^{-\iota\omega\tau}\, d\tau \\ &= 2\sigma^2 \int_0^{\tau_0}\left(1 - \frac{\tau}{\tau_0}\right)\cos(2\pi\nu\tau)\, d\tau, \end{aligned}$$

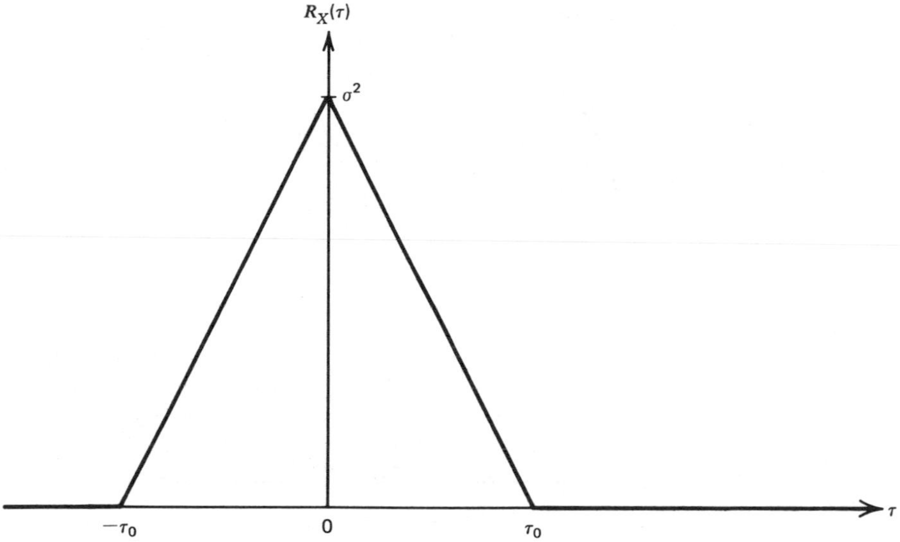

Figure 5.2.6

clearly exists, and hence the process has a continuous-type spectrum. Integrating, we find the power spectral density

$$s_X(\nu) = \sigma^2 \tau_0 \left(\frac{\sin \frac{\omega \tau_0}{2}}{\frac{\omega \tau_0}{2}} \right)^2, \qquad -\infty < \nu < \infty, \qquad \omega = 2\pi\nu.$$

Its graph is shown in Figure 5.2.7. Now, a Fourier transform is a symmetric operation in the sense that if a function $f(\tau)$, $-\infty < \tau < \infty$, has as its Fourier transform $g = \mathcal{F}\{f\}$ a function $g(\nu)$, $-\infty < \nu < \infty$, then the function $g(\tau)$, $-\infty < \tau < \infty$, will have as its Fourier transform $\mathcal{F}\{g\}$ the function $2\pi f(-2\pi\nu)$, $-\infty < \nu < \infty$. It follows that a zero mean wide-sense stationary process $Y(t)$, $-\infty < t < \infty$, with the triangular power spectral density (Fig. 5.2.8)

$$s_Y(\nu) = \begin{cases} s_Y(0)\left(1 - \dfrac{|\nu|}{\nu_0}\right) & \text{if} \quad |\nu| < \nu_0, \\ 0 & \text{if} \quad |\nu| \geq \nu_0, \end{cases}$$

will have as its correlation function

$$R_Y(\tau) = s_Y(0)\nu_0 \left(\frac{\sin 2\pi^2 \nu_0 \tau}{2\pi^2 \nu_0 \tau} \right)^2, \qquad -\infty < \tau < \infty,$$

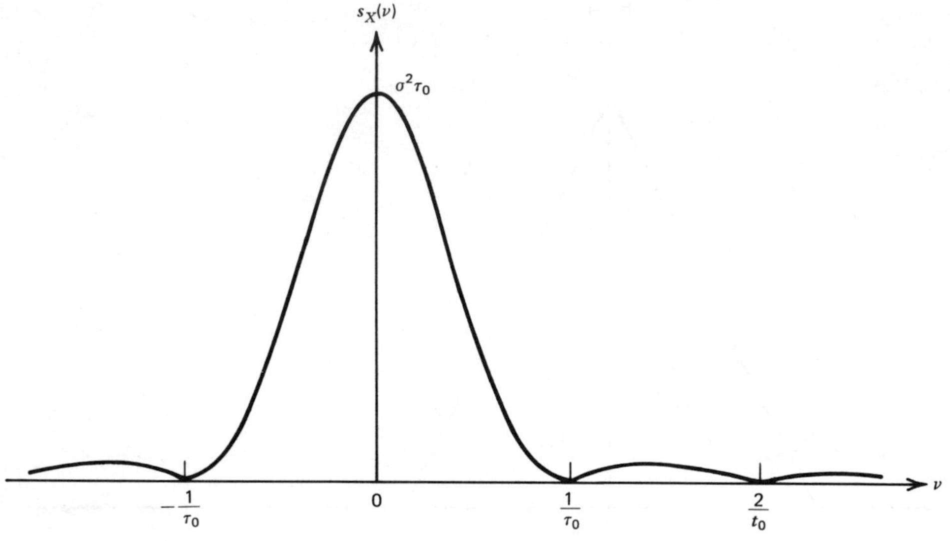

Figure 5.2.7

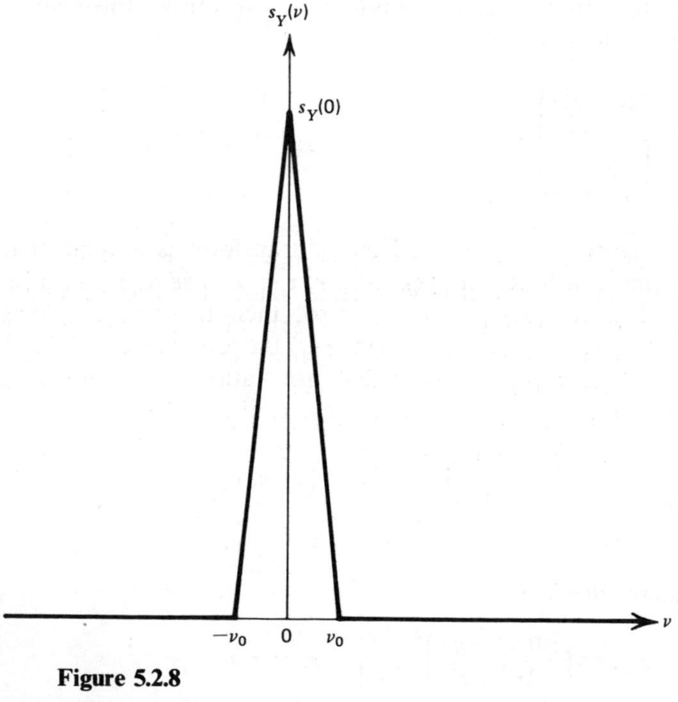

Figure 5.2.8

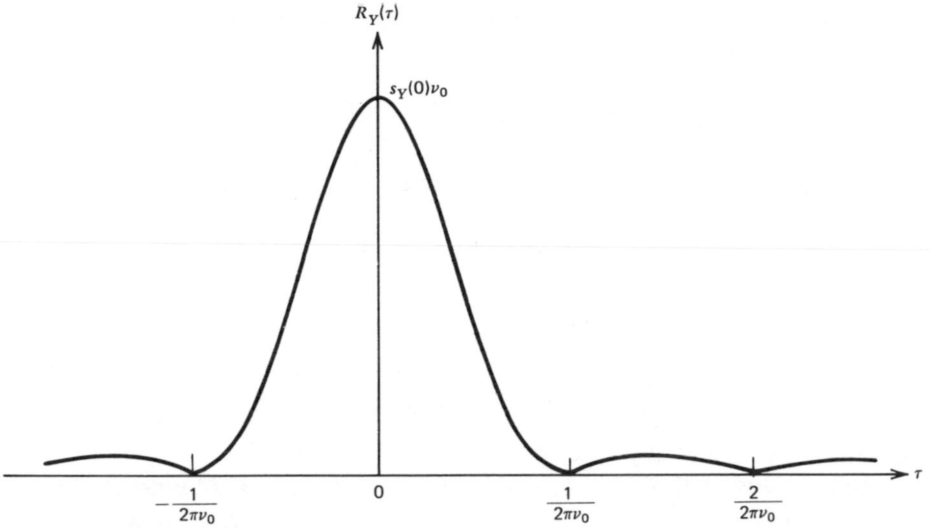

Figure 5.2.9

given in Figure 5.2.9. A zero mean wide-sense stationary process whose power spectrum is zero outside some frequency interval $(-\nu_0, \nu_0)$ is called *band-limited*, while a zero mean wide-sense stationary process with correlation function $R(\tau) = 0$ for $|\tau| \geq \tau_0 > 0$, is called *time-limited*.

The above example suggests another general feature of the spectral representation: *it is impossible for a zero mean wide-sense stationary process to be both time-limited and band-limited.*

Heuristically, the reason is easy to see, especially if the process $X(t)$ is Gaussian. If $R_X(\tau) = 0$ for $|\tau| \geq \tau_0$, then for such a τ, $X(t)$ and $X(t + \tau)$ are independent Gaussian random variables, and thus their values may differ by an arbitrarily large amount. To accommodate for such a change in value of the process considered to be a superposition of harmonic oscillations, arbitrarily large frequency components must be present in this superposition. Conversely, if all frequencies higher than some ν_0 are absent, that is, if the process is band-limited, then its values cannot be changing faster than the highest nonzero frequency component, and so the dependence between $X(t)$ and $X(t + \tau)$ cannot completely vanish over any finite time difference τ. ▲

Example 5.2.6

Consider the frequency function

$$g(\nu) = \begin{cases} e^{i\omega t} & \text{if} \quad |\nu| < \nu_0, \\ 0 & \text{if} \quad |\nu| \geq \nu_0, \end{cases}$$

where $\omega = 2\pi\nu$; $\nu_0 = 1/t_0 > 0$, a constant; and $-\infty < t < \infty$ fixed. Expanding this function into its Fourier series we have

$$g(\nu) = \sum_{n=-\infty}^{\infty} c_n e^{\iota\omega n t_0}, \qquad |\nu| < \nu_0.$$

Substituting into this series for the coefficients

$$c_n = \frac{1}{2\nu_0}\int_{-\nu_0}^{\nu_0} e^{\iota\omega t} e^{-\iota\omega n t_0}\, d\nu = \frac{\sin \omega_0(t - nt_0)}{\omega_0(t - nt_0)},$$

where $\omega_0 = 2\pi\nu_0$, we obtain the identity

$$e^{\iota\omega t} = \sum_{n=-\infty}^{\infty} e^{\iota\omega n t_0} \frac{\sin \omega_0(t - nt_0)}{\omega_0(t - nt_0)}, \tag{15}$$

valid for $|\nu| < \nu_0$ and any $-\infty < t < \infty$.

Suppose now that $X(t)$, $-\infty < t < \infty$, is a zero mean wide-sense stationary process band-limited to the frequency interval $(-\nu_0, \nu_0)$. Then its power spectral density $s_X(\nu) = 0$ for $|\nu| \geq \nu_0$, and hence its spectral representation can be written as

$$X(t) = \int_{-\nu_0}^{\nu_0} e^{\iota\omega t}\, dZ_X(\nu), \qquad -\infty < t < \infty.$$

Using the identity (15), this becomes, after interchanging integration with the infinite summation

$$X(t) = \sum_{n=-\infty}^{\infty} \frac{\sin \omega_0(t - nt_0)}{\omega_0(t - nt_0)} \int_{-\nu_0}^{\nu_0} e^{\iota\omega n t_0}\, dZ_X(\nu).$$

However, letting $t = nt_0$ in the spectral representation for $X(t)$

$$\int_{-\nu_0}^{\nu_0} e^{\iota\omega n t_0}\, dZ_X(\nu) = X(nt_0),$$

and hence

$$X(t) = \sum_{n=-\infty}^{\infty} X(nt_0) \frac{\sin \omega_0(t - nt_0)}{\omega_0(t - nt_0)}, \qquad -\infty < t < \infty,$$

a result known as the *sampling theorem.*

The significance of this theorem is twofold. First, it shows that a band-limited process is actually determined by its samples $X(nt_0)$, $n = \ldots, -1, 0, 1, \ldots$, taken at the rate $1/t_0 = 2\nu_0$, that is, twice the limiting frequency. Thus, a sample path of the process $X(t)$, $-\infty < t < \infty$, can be reconstructed from these samples, although an infinite number of samples is needed for such a reconstruction to have zero mean square error.

Second, the sampling theorem allows one to extend the entire theory of spectral representation to zero mean wide-sense stationary *discrete-time* processes. For with each such discrete-time process

$$X(n), \qquad n = \ldots, -1, 0, 1, \ldots,$$

we can associate a continuous time process $Y(t)$, $-\infty < t < \infty$, by defining

$$Y(t) = \sum_{n=-\infty}^{\infty} X(n) \frac{\sin 2\pi(t-n)}{2\pi(t-n)}.$$

Clearly, $Y(t)$ will be a zero mean wide-sense stationary process such that

$$Y(t) = X(n) \qquad \text{whenever } t = n, \qquad n = \ldots, -1, 0, 1, \ldots,$$

and it will also be band-limited to the frequency interval $(-\frac{1}{2}, \frac{1}{2})$. If $Z_Y(\nu)$ is the spectral process of $Y(t)$, we immediately have the spectral representation for the discrete-time process $X(n)$,

$$X(n) = Y(n) = \int_{-1/2}^{1/2} e^{\iota\omega n} \, dZ_Y(\nu), \qquad n = \ldots, -1, 0, 1, \ldots.$$

We can also define the power spectral density $s_X(\nu)$ (or, more generally, power spectral distribution function) of the discrete-time process $X(n)$ as the power spectral density $s_Y(\nu)$ of the continuous-time process $Y(t)$. Note that the power spectral density of a discrete-time process is therefore always restricted to the frequency interval $(-\frac{1}{2}, \frac{1}{2})$. Since the correlation function

$$R_X(n) = R_Y(n) = \int_{-1/2}^{1/2} e^{\iota\omega n} s_Y(\nu) \, d\nu,$$

we have

$$R_X(n) = \int_{-1/2}^{1/2} e^{\iota\omega n} s_X(\nu) \, d\nu, \qquad n = 0, \pm 1, \ldots,$$

and conversely,

$$s_X(\nu) = \sum_{n=-\infty}^{\infty} R_X(n) e^{-\iota\omega n}, \qquad -\frac{1}{2} < \nu < \frac{1}{2},$$

that is, the values of the correlation function $R_X(n)$ are just the coefficients of the Fourier series for the power spectral density $s_X(\nu)$. ▲

Example 5.2.7

Let $X(t)$, $-\infty < t < \infty$, be a zero mean wide-sense stationary process with correlation function

$$R_X(\tau) = \sigma^2 e^{-\alpha|\tau|} \cos \omega_0 \tau, \qquad -\infty < \tau < \infty,$$

where σ^2, α, and ω_0 are positive constants. Processes with such correlation functions are typically associated with a second order linear differential system, such as a resonant circuit, for instance, disturbed by a random noise.

To obtain the power spectral density of this process, we use the Fourier cosine transform

$$s_X(\nu) = 2\int_0^\infty R_X(\tau)\cos \omega\tau \, d\tau, \qquad \omega = 2\pi\nu.$$

Using the formula for the product of two cosines we have

$$\int_0^\infty e^{-\alpha\tau} \cos \omega_0 \tau \cos \omega\tau \, d\tau = \frac{1}{2}\int_0^\infty e^{-\alpha\tau} \cos(\omega + \omega_0)\tau \, d\tau$$
$$+ \frac{1}{2}\int_0^\infty e^{-\alpha\tau} \cos(\omega - \omega_0)\tau \, d\tau,$$

and since the definite integral

$$\int_0^\infty e^{-\alpha\tau} \cos \beta\tau \, d\tau = \frac{\alpha}{\alpha^2 + \beta^2},$$

the power spectral density equals

$$s_X(\nu) = \frac{\sigma^2\alpha}{\alpha^2 + 4\pi^2(\nu + \nu_0)^2} + \frac{\sigma^2\alpha}{\alpha^2 + 4\pi^2(\nu - \nu_0)^2}, \qquad -\infty < \nu < \infty,$$

where $\nu_0 = \omega_0/2\pi$. This function is a sum

$$s_X(\nu) = \frac{1}{2}g(\nu + \nu_0) + \frac{1}{2}g(\nu - \nu_0),$$

where

$$g(\nu) = \frac{2\alpha\sigma^2}{\alpha^2 + (2\pi\nu)^2}$$

is identical with the power spectral density of Example 5.2.4. Thus the graph of $s_X(\nu)$, $-\infty < \nu < \infty$, is obtained by shifting the graph of $g(\nu)$ to $+\nu_0$ and $-\nu_0$ and averaging the two curves, as shown in Figure 5.2.10. Those who are familiar with the properties of Fourier transforms should recognize this as a consequence of the general property that multiplication of a time function by $e^{i\omega_0 t}$ (modulation) results in a shift of its Fourier transform along the frequency axis. For comparison, the correlation function $R_X(\tau)$, $-\infty < \tau < \infty$, is sketched in Figure 5.2.11. ▲

Let us now look at the spectral representation. For the spectral process $Z(\nu)$, we now have from (11)

$$\mathrm{E}[|dZ(\nu)|^2] = s_X(\nu)\, d\nu. \tag{16}$$

so that $Z(\nu)$ is MS continuous, since by (13)

$$\mathrm{E}[|dZ(\nu)|^2] \to 0 \qquad \text{as } d\nu \to 0.$$

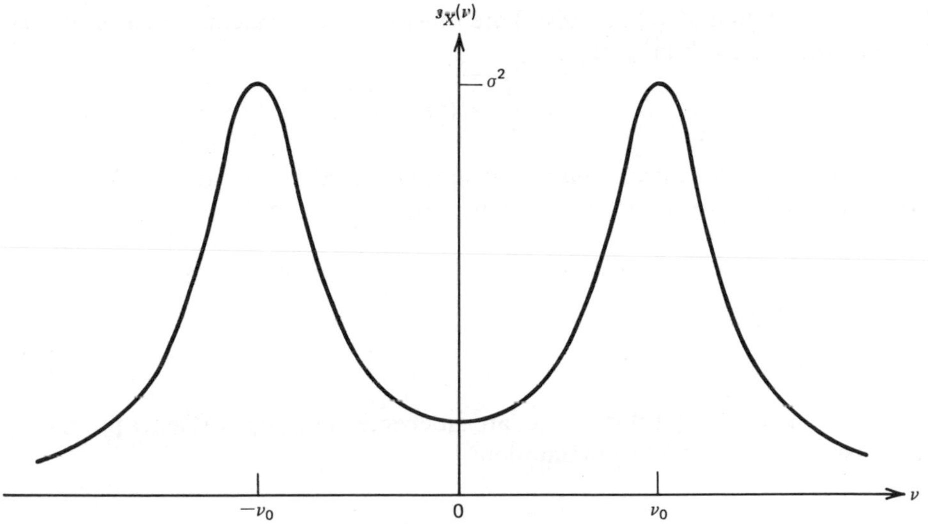

Figure 5.2.10

In particular, if $X(t)$ is real and Gaussian, then the real and imaginary parts $Z_1(\nu)$ and $Z_2(\nu)$ of the spectral process will be mutually independent, zero mean Gaussian processes with independent increments, which in the present case will also be MS continuous and such that

$$E[(dZ_1(\nu))^2] = E[(dZ_2(\nu))^2] = \frac{1}{2} s_X(\nu)\, d\nu.$$

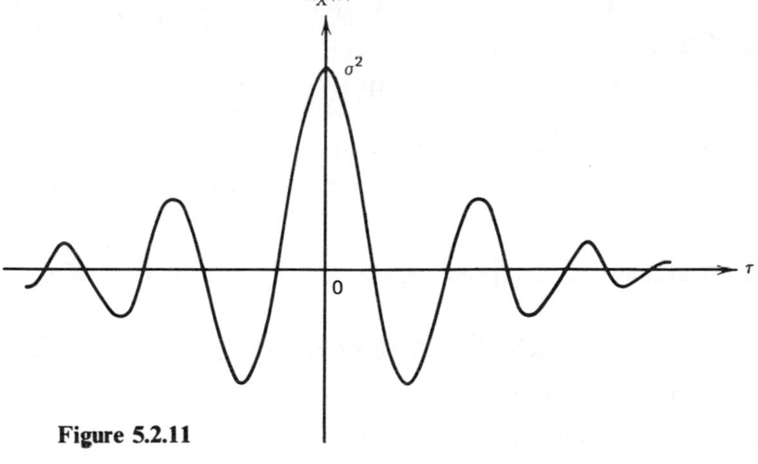

Figure 5.2.11

But this means that $Z_1(\nu)$ and $Z_2(\nu)$ are mutually independent, inhomogeneous Wiener processes with intensity

$$\sqrt{\frac{1}{2}s_X(\nu)}.$$

Hence, if we introduce a complex-valued standard Wiener process $W_c(\nu)$ as a process defined on the frequency domain $-\infty < \nu < \infty$ by†

$$W_c(\nu) = \begin{cases} \dfrac{1}{\sqrt{2}} W_{01}(\nu) + \dfrac{\imath}{\sqrt{2}} W_{02}(\nu) & \text{if} \quad \nu \geq 0, \\ \dfrac{1}{\sqrt{2}} W_{01}(-\nu) - \dfrac{\imath}{\sqrt{2}} W_{02}(-\nu) & \text{if} \quad \nu \leq 0, \end{cases} \tag{17}$$

where $W_{01}(\nu)$ and $W_{02}(\nu)$, $0 \leq \nu < \infty$, are independent standard Wiener processes, we can write the spectral representation

$$X(t) = \int_{-\infty}^{\infty} e^{\imath\omega t}\, dZ(\nu),$$

for a real-valued process as a Wiener integral

$$X(t) = \int_{-\infty}^{\infty} \sqrt{s_X(\nu)}\, e^{\imath\omega t}\, dW_c(\nu), \qquad -\infty < t < \infty, \tag{18}$$

with $\omega = 2\pi\nu$. Sometimes we may prefer to use the concept of white noise and write this as

$$X(t) = \int_{-\infty}^{\infty} \sqrt{s_X(\nu)}\, e^{\imath\omega t} \dot{W}_c(\nu)\, d\nu, \tag{19}$$

where the *complex-valued white Gaussian noise* $\dot{W}_c(\nu)$ on the frequency domain $-\infty < \nu < \infty$ is defined as a formal derivative

$$\dot{W}_c(\nu) = \frac{dW_c(\nu)}{d\nu} = \begin{cases} \dfrac{1}{\sqrt{2}} \dot{W}_{01}(\nu) + \dfrac{\imath}{\sqrt{2}} \dot{W}_{02}(\nu) & \text{if} \quad \nu \geq 0, \\ \dfrac{1}{\sqrt{2}} \dot{W}_{01}(\nu) - \dfrac{\imath}{\sqrt{2}} \dot{W}_{02}(\nu) & \text{if} \quad \nu \leq 0, \end{cases}$$

with $\dot{W}_{01}(\nu)$ and $\dot{W}_{02}(\nu)$ independent white Gaussian noises.

If we wish to avoid the complex-valued Wiener process $W_c(\nu)$, we can rewrite representation (18) in the form

$$X(t) = \int_0^{\infty} \sqrt{2s_X(\nu)} \cos \omega t \, dW_{01}(\nu) - \int_0^{\infty} \sqrt{2s_X(\nu)} \sin \omega t \, dW_{02}(\nu),$$

where, as usual, $\omega = 2\pi\nu$.

† The factor $1/\sqrt{2}$ guarantees that $\mathrm{E}[|dW_c(\nu)|^2] = d\nu$ so that $W_c(\nu)$ merits the term "standard."

As in the case of a discrete-type spectrum, this can be considered as the quadrature decomposition of the process $X(t)$ into its cosine component

$$X_1(t) = \int_0^\infty \sqrt{2s_X(\nu)} \cos \omega t \, dW_{01}(\nu),$$

and an independent sine component

$$X_2(t) = \int_0^\infty \sqrt{2s_X(\nu)} \sin \omega t \, dW_{02}(\nu).$$

Similarly, the symbolic integral (19) can be written

$$X(t) = \int_0^\infty \sqrt{2s_X(\nu)}[\dot{W}_{01}(\nu) \cos \omega t - \dot{W}_{02}(\nu) \sin \omega t] \, d\nu,$$

Regarding the symbolic process in brackets as a random sine wave with correlation function

$$\delta(\nu) \cos \omega\tau$$

we see again that $X(t)$ is a "superposition" of independent random sine waves, this time involving a continuum of frequencies.

Example 5.2.8
Suppose that $X(t)$, $-\infty < t < \infty$, is a zero mean stationary Gaussian process with power spectral density

$$s_X(\nu) = \frac{1}{2} s_0(\nu + \nu_0) + \frac{1}{2} s_0(\nu - \nu_0), \qquad -\infty < \nu < \infty, \tag{20}$$

where $\nu_0 > 0$ is a fixed frequency and $s_0(\nu)$ is a nonnegative, even, and integrable function, that is,

$$s_0(-\nu) = s_0(\nu) \quad \text{for all } \nu, \quad \text{and} \quad \int_{-\infty}^{\infty} s_0(\nu) \, d\nu < \infty.$$

Clearly, $s_X(\nu)$ is then also an even function, so that the process $X(t)$ is real valued. We have seen a process with this type of power spectral density in Example 5.2.1.

Suppose, further, that the function $s_0(\nu)$ is such that

$$s_0(\nu) = 0 \quad \text{for} \quad |\nu| > \Delta\nu_0, \tag{21}$$

where $\Delta\nu_0 > 0$ is much smaller than the frequency ν_0. Then, as seen in Figure 5.2.12, the power spectral density $s_X(\nu)$ will be confined to two narrow frequency bands of width $2\,\Delta\nu_0$ centered at the "carrier frequency" $\pm\nu_0$.

In such a case the process $X(t)$ is called a *narrowband process*. Narrowband processes are important in applications, especially in comunication engineering, due to the fact that if a random signal is passed through several tuned circuits, its frequency components other than those near the resonant frequency ν_0 will be

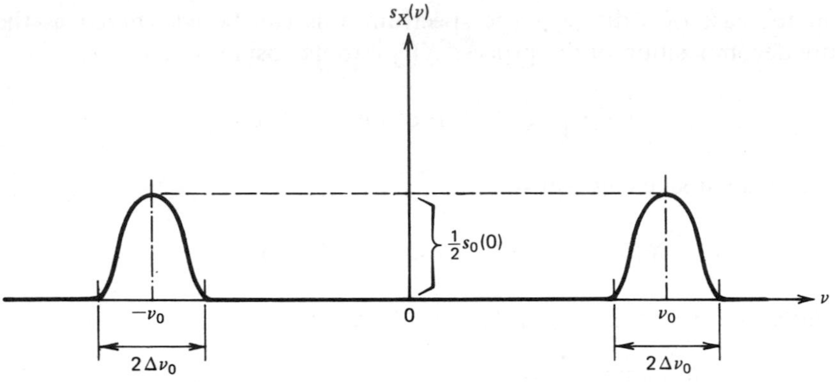

Figure 5.2.12

virtually eliminated. Hence, it is worth while investigating such processes in some detail.

In view of (20) and (21) the spectral representation (19) of a narrowband Gaussian process is

$$X(t) = \int_{\nu_0-\Delta\nu_0}^{\nu_0+\Delta\nu_0} \sqrt{\frac{1}{2}s_0(\nu-\nu_0)}\, e^{i\omega t}\dot{W}_c(\nu)\, d\nu + \int_{-\nu_0-\Delta\nu_0}^{-\nu_0+\Delta\nu_0} \sqrt{\frac{1}{2}s_0(\nu+\nu_0)}\, e^{i\omega t}\dot{W}_c(\nu)\, d\nu. \tag{22}$$

Substituting $\nu' = \nu - \nu_0$, the first integral above becomes

$$e^{i\omega_0 t}\int_{-\Delta\nu_0}^{\Delta\nu_0} \sqrt{\frac{1}{2}s_0(\nu')}\, e^{i\omega' t}\dot{W}_c(\nu'+\nu_0)\, d\nu',$$

where $\omega_0 = 2\pi\nu_0$. However, since $\Delta\nu_0 \ll \nu_0$, over the integration region $(-\Delta\nu_0, \Delta\nu_0)$ the complex white noise $\dot{W}_c$ in the integrand can also be written as

$$\dot{W}_c(\nu'+\nu_0) = \frac{1}{\sqrt{2}}\dot{W}_{01}(\nu') + \frac{i}{\sqrt{2}}\dot{W}_{02}(\nu'),$$

where $\dot{W}_{01}(\nu')$ and $\dot{W}_{02}(\nu')$ are independent white Gaussian noises defined on $(-\Delta\nu_0, \Delta\nu_0)$. Thus, dropping the prime and denoting by $Y(t)$, $-\infty < t < \infty$, a complex-valued process

$$Y(t) = \int_{-\Delta\nu_0}^{\Delta\nu_0} \frac{1}{2}\sqrt{s_0(\nu)}\, e^{i\omega t}[\dot{W}_{01}(\nu) + i\dot{W}_{02}(\nu)]\, d\nu \tag{23}$$

the first integral in the spectral representation (22) equals $e^{i\omega_0 t}Y(t)$. Similarly, making the substitution $\nu' = -(\nu+\nu_0)$ into the second integral in (22) and using the fact that $s_0(-\nu) = s_0(\nu)$ and $W_c(-\nu) = W_c^*(\nu)$, we find that the second integral is just the complex conjugate of the first, which after all could be seen immediately from the fact that $X(t)$ is real. It follows that we can write (22) as

$$X(t) = Y(t)e^{\iota\omega_0 t} + Y^*(t)e^{-\iota\omega_0 t}, \qquad -\infty < t < \infty, \tag{24}$$

or, upon denoting by $Y_1(t)$ and $Y_2(t)$ the real and imaginary parts

$$Y(t) = Y_1(t) + \iota Y_2(t),$$

also as

$$X(t) = 2Y_1(t)\cos\omega_0 t - 2Y_2(t)\sin\omega_0 t, \qquad -\infty < t < \infty. \tag{25}$$

Now from (23) we see that the process $Y(t)$ and hence also its real and imaginary parts $Y_1(t)$ and $Y_2(t)$ contain only frequencies in a narrow frequency band $(-\Delta\nu_0, \Delta\nu_0)$ around the zero frequency. It follows that their sample paths can be expected to vary with time very slowly relative to the period $t_0 = 1/\nu_0$ of the carrier frequency ν_0. We can therefore represent a narrowband Gaussian process $X(t)$ by a phasor diagram (as in Fig. 5.2.13), where $e^{\pm\iota\omega_0 t}$ are deterministic phasors

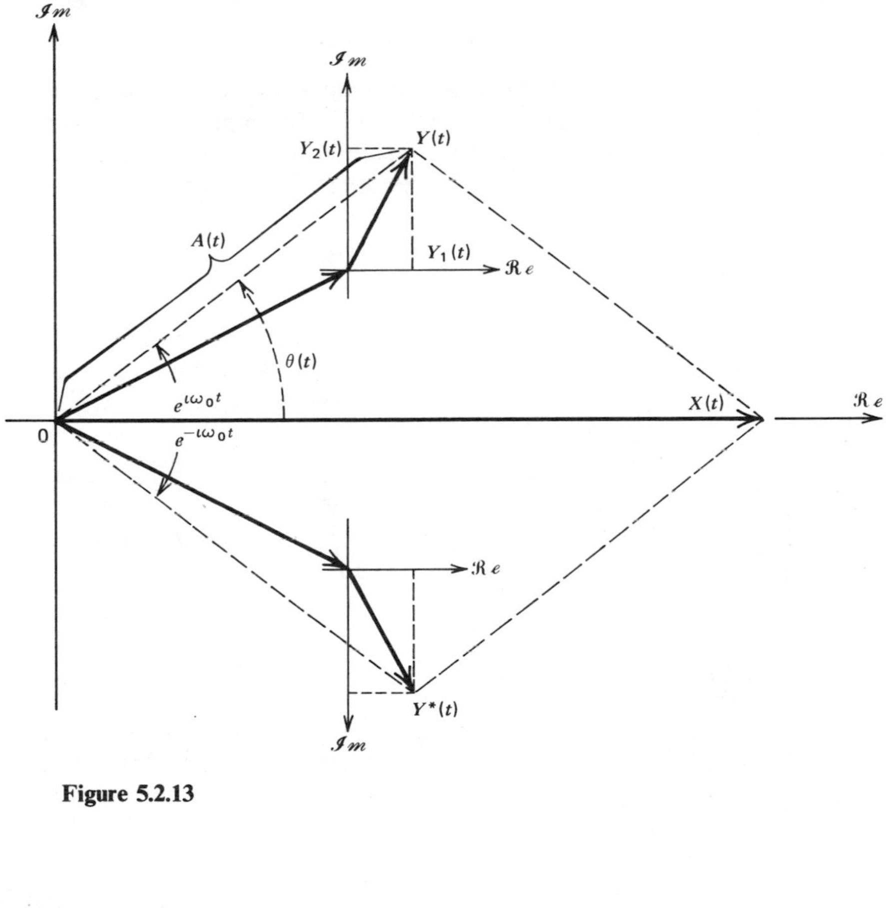

Figure 5.2.13

rotating with angular frequency ω_0, while $Y(t)$ is a slowly varying phasor modulating the amplitude and phase of $e^{\pm \imath\omega_0 t}$. In analogy with deterministic sinusoidal waveforms, equation (25) is then called the *quadrature decomposition of the narrowband process* $X(t)$, the slowly varying processes $X_1(t) = 2Y_1(t)$ and $X_2(t) = 2Y_2(t)$ being the amplitudes of the *cosine* and *sine components*, respectively. The analogy can be carried further, if we define a process

$$A(t) = \frac{1}{2}\sqrt{Y_1^2(t) + Y_2^2(t)}, \qquad -\infty < t < \infty$$

and a process $\Theta(t)$, $-\infty < t < \infty$, such that

$$\frac{Y_2(t)}{Y_1(t)} = \tan \Theta(t),$$

or equivalently, $Y(t) = \frac{1}{2}A(t)e^{\imath\Theta(t)}$, we have from (24)

$$X(t) = \frac{1}{2}A(t)e^{\imath(\omega_0 t + \Theta(t))} + \frac{1}{2}A(t)e^{-\imath(\omega_0 t + \Theta(t))}$$
$$= A(t)\cos(\omega_0 t + \Theta(t)), \qquad -\infty < t < \infty.$$

Clearly both $A(t)$ and $\Theta(t)$ will also be slowly varying relative to t_0, and hence we can expect a typical sample path of a narrowband Gaussian process $X(t)$ to be a sine wave with frequency ν_0 with slowly varying amplitude and phase as pictured in Figure 5.2.14. For this reason, the processes $A(t)$ and $\Theta(t)$ are called *the en-*

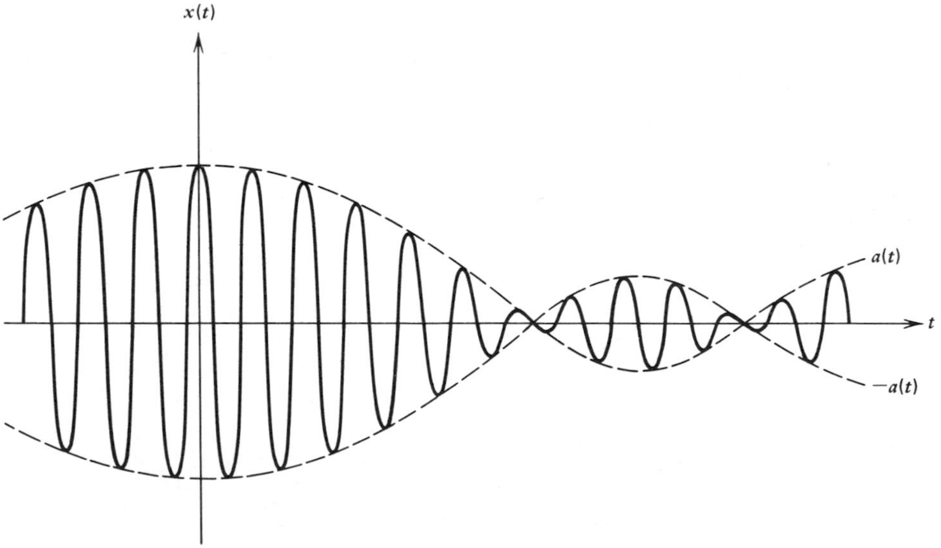

Figure 5.2.14

velope and *phase* of a narrowband Gaussian process $X(t)$. Notice that the amplitudes of the cosine and sine components of the quadrature decomposition (25)

$$X(t) = X_1(t)\cos \omega_0 t - X_2(t)\sin \omega_0 t$$

can also be expressed as

$$X_1(t) = A(t)\cos \Theta(t), \qquad X_2(t) = A(t)\sin \Theta(t).$$

From (23) we can also write

$$X_1(t) = \int_{-\Delta\nu_0}^{\Delta\nu_0} \sqrt{s_0(\nu)}(\dot{W}_{01}(\nu)\cos \omega t - \dot{W}_{02}(\nu)\sin \omega t)\, d\nu,$$

$$X_2(t) = \int_{-\Delta\nu_0}^{\Delta\nu_0} \sqrt{s_0(\nu)}(\dot{W}_{01}(\nu)\sin \omega t + \dot{W}_{02}(\nu)\cos \omega t)\, d\nu, \tag{26}$$

from which we see that both are zero mean Gaussian processes.

Using these expressions, we can calculate the correlation function

$$R_{X_1}(t_1, t_2) = \int_{-\Delta\nu_0}^{\Delta\nu_0}\int_{-\Delta\nu_0}^{\Delta\nu_0} \sqrt{s_0(\nu_1)}\sqrt{s_0(\nu_2)}\,\mathrm{E}[(\dot{W}_{01}(\nu_1)\cos \omega_1 t_1$$

$$- \dot{W}_{02}(\nu_1)\sin \omega_1 t_1)(\dot{W}_{01}(\nu_2)\cos \omega_2 t_2 - \dot{W}_{02}(\nu_2)\sin \omega_2 t_2)]\, d\nu_1\, d\nu_2$$

$$= \int_{-\Delta\nu_0}^{\Delta\nu_0} s_0(\nu)\cos \omega(t_1 - t_2)\, d\nu,$$

since

$$\mathrm{E}[\dot{W}_{01}(\nu_1)\dot{W}_{01}(\nu_2)] = \mathrm{E}[\dot{W}_{02}(\nu_1)\dot{W}_{02}(\nu_2)] = \delta(\nu_1 - \nu_2)$$

while

$$\mathrm{E}[\dot{W}_{01}(\nu_1)\dot{W}_{02}(\nu_2)] = 0.$$

Thus using (21)

$$R_{X_1}(\tau) = \int_{-\infty}^{\infty} s_0(\nu)\cos \omega\tau\, d\nu, \qquad -\infty < \tau < \infty, \tag{27}$$

and similarly

$$R_{X_2}(\tau) = \int_{-\infty}^{\infty} s_0(\nu)\cos \omega\tau\, d\nu, \qquad -\infty < \tau < \infty, \tag{28}$$

with $\omega = 2\pi\nu$ as usual. From (26) we can also get the cross-correlation function

$$R_{X_1, X_2}(\tau) = \int_{-\infty}^{\infty} s_0(\nu)\sin \omega\tau\, d\nu, \qquad -\infty < \tau < \infty.$$

Thus both the cosine and the sine components are identically distributed, zero mean stationary Gaussian processes with spectral density

$$s_{X_1}(v) = s_{X_2}(v) = \frac{1}{2} s_0(v), \qquad -\infty < v < \infty,$$

as can be seen by comparing (27) and (28) with (14). Finally since $R_{X_1, X_2}(0) = 0$, $X_1(t)$ and $X_2(t)$ are, for each t, independent Gaussian random variables with zero mean and with variance $\sigma^2 = \int_{-\infty}^{\infty} s_0(v)\, dv$. It follows that, for each t, the envelope $A(t)$ and the phase $\Theta(t)$ will be independent random variables, $A(t)$ Rayleigh with the parameter σ, and $\Theta(t)$ uniform over $[0, 2\pi)$. Unfortunately, the second and higher order distributions of the envelope and phase processes are quite complicated, and thus it is generally easier to work with the cosine and sine component processes. ▲

It may be of interest, and in fact is quite useful, to ask what might be the spectral representation of the white Gaussian noise itself. Of course, white Gaussian noise is a generalized process and Bochner's theorem (at least in the form stated) does not apply. On the other hand, white Gaussian noise

$$\dot{W}_0(t), \qquad -\infty < t < \infty$$

is used as if it were a stationary, zero mean Gaussian process with correlation function

$$R(\tau) = \delta(\tau),$$

and the Dirac delta function $\delta(\tau)$, $-\infty < \tau < \infty$, forms a Fourier transform pair with the frequency function $s(v) = 1$ for all $-\infty < v < \infty$. That is, we have

$$\delta(\tau) = \int_{-\infty}^{\infty} e^{\iota\omega\tau} 1\, dv \qquad \text{and} \qquad 1 = \int_{-\infty}^{\infty} e^{-\iota\omega\tau}\, \delta(\tau)\, d\tau.$$

This suggests that we may regard white Gaussian noise

$$b\dot{W}_0(t), \qquad -\infty < t < \infty,$$

with intensity $b > 0$, that is, with the correlation function

$$b^2\, \delta(\tau), \qquad -\infty < \tau < \infty,$$

as if it were a stationary Gaussian process with continuous type spectrum whose power spectral density

$$s(v) = b^2, \qquad -\infty < v < \infty, \tag{29}$$

is a constant over the entire frequency domain.

Recall that the intensity b^2 has the physical dimension of energy, while the

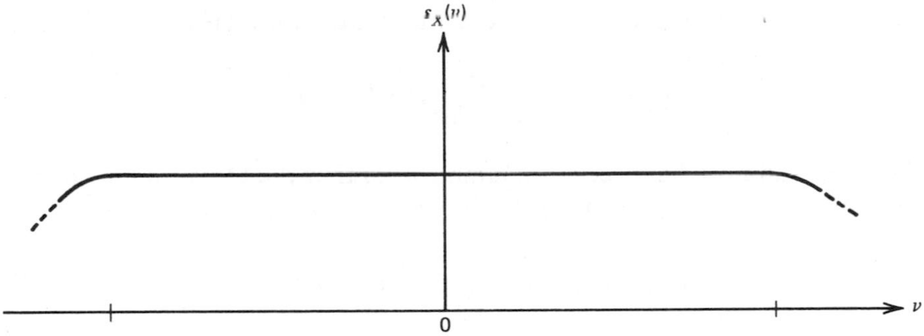

Figure 5.2.15

power spectral density has the dimension power ÷ frequency = power × time = energy, so that we do have agreement.

With the power spectral density (29) a constant, the spectral representations (18) and (19) become formally

$$\dot{W}_0(t) = \int_{-\infty}^{\infty} e^{\iota\omega t}\, dW_c(\nu), \qquad \dot{W}_0(t) = \int_{-\infty}^{\infty} e^{\iota\omega t} \dot{W}_c(\nu)\, d\nu. \tag{30}$$

Thus the spectral process of white Gaussian noise is the complex Wiener process $W_c(\nu)$ defined in (17). These formal relations explain why white Gaussian noise $\dot{W}_0(t)$ is called white. From (30), we see that it is a superposition of all harmonic frequencies, each frequency contributing an equal share, that is, each frequency having the same power. As white light is a uniform mixture of all frequencies (in the visible range at least), so is white noise. Of course, if the energy of each elementary oscillation with frequency in an interval $d\nu$ is to be a constant $d\nu$, then the total energy produced by the superposition must be infinite. Thus, we see again that, strictly speaking, white Gaussian noise cannot exist. However, if the power spectral density of a physical process is constant over a sufficiently large frequency range, then white Gaussian noise may be a good approximation (see Fig. 5.2.15). The benefit of such an idealization is that the subsequent mathematical operations are considerably simplified.

Example 5.2.9

Let us look again at a stationary zero mean Ornstein-Uhlenbeck process $X(t)$, $-\infty < t < \infty$. Since its correlation function

$$R_X(\tau) = \sigma^2 e^{-\alpha|\tau|}, \qquad -\infty < \tau < \infty,$$

is exponential, we know from Example 5.2.4 that it has a continuous spectrum, with power spectral density

$$s_X(\nu) = \frac{2\alpha\sigma^2}{\alpha^2 + \omega^2}, \qquad \omega = 2\pi\nu, \qquad -\infty < \nu < \infty.$$

Hence, its spectral representation can be written in the form (19)

$$X(t) = \int_{-\infty}^{\infty} \sqrt{\frac{2\alpha}{\alpha^2 + \omega^2}}\, e^{\imath\omega t} \dot{W}_c(v)\, dv. \tag{31}$$

On the other hand, we have seen in Example 4.3.4 that the Ornstein-Uhlenbeck process $X(t)$, $-\infty < t < \infty$, can also be expressed as an integral

$$X(t) = \sqrt{2\alpha\sigma^2} \int_{-\infty}^{t} e^{-\alpha(t-\tau)} \dot{W}_0(\tau)\, d\tau. \tag{32}$$

It should, therefore, be possible to obtain the spectral representation (31) from this expression. To this end, let us substitute for the white Gaussian noise in integral (32) its spectral representation (30)

$$\dot{W}_0(\tau) = \int_{-\infty}^{\infty} e^{\imath\omega\tau} \dot{W}_c(v)\, dv.$$

We get the double integral

$$\int_{-\infty}^{t} \int_{-\infty}^{\infty} e^{-\alpha(t-\tau)+\imath\omega\tau} \dot{W}_c(v)\, dv\, d\tau$$

$$= \int_{-\infty}^{\infty} \left(\int_{-\infty}^{t} e^{-\alpha(t-\tau)+\imath\omega\tau}\, d\tau \right) \dot{W}_c(v)\, dv$$

$$= \int_{-\infty}^{\infty} \left(\int_{0}^{\infty} e^{-(\alpha+\imath\omega)\tau'}\, d\tau' \right) e^{\imath\omega t} \dot{W}_c(v)\, dv,$$

where we have changed the order of integration and made the substitution $t - \tau = \tau'$ into the inner integral. Since $\alpha > 0$ the inner integral equals

$$\int_{0}^{\infty} e^{-(\alpha+\imath\omega)\tau'}\, d\tau' = \frac{1}{\alpha + \imath\omega},$$

so that

$$X(t) = \sqrt{2\alpha\sigma^2} \int_{-\infty}^{\infty} \frac{1}{\alpha + \imath\omega} e^{\imath\omega t} \dot{W}_c(v)\, dv. \tag{33}$$

This does have a general form of spectral representation, but it seems to be different from the representation (31). Actually, however, it is equivalent. If we transform the complex-valued function $1/(\alpha + \imath\omega)$ into its polar form

$$\frac{1}{\alpha + \imath\omega} = \frac{1}{\sqrt{\alpha^2 + \omega^2}} e^{\imath\gamma(\omega)},$$

where $\tan \gamma(\omega) = -\omega/\alpha$, we can write (33) as

$$X(t) = \sigma \int_{-\infty}^{\infty} \sqrt{\frac{2\alpha}{\alpha^2 + \omega^2}}\, e^{\imath(\omega t + \gamma(\omega))} \dot{W}_c(v)\, dv. \tag{34}$$

Now, we have shown earlier that for $X(t)$ real valued and Gaussian, the spectral representation

$$X(t) = \int_{-\infty}^{\infty} e^{\imath(\omega t + \gamma(\omega))}\, dZ(v) \qquad \text{with } \gamma(-\omega) = -\gamma(\omega)$$

as in the present case is equivalent to

$$X(t) = \int_{-\infty}^{\infty} e^{\imath \omega t}\, dZ(v).$$

It follows that (34), and hence also (33) is indeed equivalent to the spectral representation (31). ▲

MIXED-TYPE POWER SPECTRAL DISTRIBUTION. In general, the power spectral distribution function $s_X(v)$ can be of a *mixed type*, that is, neither a step function nor a continuous function, which can be expressed as an integral of a power spectral density. Although, strictly speaking, there is a third possibility,† most power spectral distribution functions can be written as a sum,

$$S_X(v) = S_X^{(d)}(v) + S_X^{(c)}(v), \qquad -\infty < v < \infty$$

of a power spectral distribution function $S_X^{(d)}(v)$ of the discrete type and a power spectral distribution function $S_X^{(c)}(v)$ of the continuous type.

Thus the power spectrum can be expressed as a corresponding mixture of discrete spectral lines

$$\{(s_X(v_n), v_n)): n = 0, \pm 1, \pm 2, \ldots\}$$

corresponding to the discrete component $S_X^{(d)}(v)$, and the power spectral density

$$s_X(v) = \frac{d}{dv} S_X^{(c)}(v).$$

Note that we must now have

$$\sum_{n=-\infty}^{\infty} s_X(v_n) + \int_{-\infty}^{\infty} s_X(v)\, dv = R_X(0).$$

† The so-called singular continuous distribution.

The correlation function $R_X(\tau)$ is then given by

$$R_X(\tau) = \sum_{n=-\infty}^{\infty} e^{\imath\omega_n\tau} s_X(\nu_n) + \int_{-\infty}^{\infty} e^{\imath\omega\tau} s_X(\nu)\, d\nu,$$

and the spectral representation of the process $X(t)$ is obtained by the corresponding sum of the spectral sequence $Z(n)$ and the MS-continuous spectral process $Z(\nu)$.

In particular, if $X(t)$ is real and Gaussian

$$X(t) = \sum_{n=-\infty}^{\infty} e^{\imath\omega_n t} Z(n) + \int_{-\infty}^{\infty} e^{\imath\omega t} s_X(\nu)\, dW_c(\nu),$$

where the spectral sequence $Z(n)$ and the complex Wiener process $W_c(\nu)$ are mutually independent.

Example 5.2.10

Consider a random PPM signal $X(t)$, $-\infty < t < \infty$, introduced in Exercise 1.21. Since the process $X(t)$ is neither wide-sense stationary nor does it have a zero mean function, we have to modify it first before we can talk about its spectrum. Such a modification can be accomplished by first assuming that the frequency markers, that is, the unmodulated periodic pulse train, is positioned at random with respect to the origin of the time domain, as in Figure 5.2.16. Here the marker $\times$, originally located at $t = 0$, is now displaced by a random variable Θ, which is

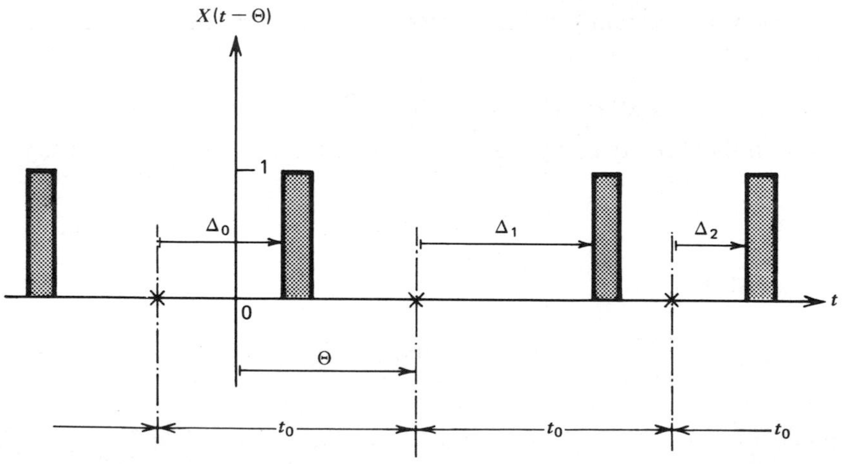

Figure 5.2.16

uniformly distributed over the interval $[0, t_0)$ and independent of the sequence Δn, $n = \ldots, -1, 0, 1, \ldots$. The new process $X(t - \Theta)$, $-\infty < t < \infty$, has mean function

$$\mathrm{E}[X(t-\Theta)] = \int_0^{t_0} \mathrm{E}[X(t-\Theta) \mid \Theta = \theta] \frac{d\theta}{t_0} = \frac{1}{t_0} \int_0^{t_0} \mu_X(t-\theta)\, d\theta,$$

where $\mu_X(t)$ is the mean function of the original process $X(t)$. However, $\mu_X(t)$ was seen to be a periodic function with period t_0 and hence the integral

$$\int_0^{t_0} \mu_X(t-\theta)\, d\theta$$

is a constant independent of t, namely the pulse width b. Thus

$$\mu = \mathrm{E}[X(t-\Theta)] = \frac{b}{t_0} \qquad \text{for all } -\infty < t < \infty.$$

Similarly, the correlation function is

$$\mathrm{E}[X(t_1-\Theta)X(t_2-\Theta)] = \int_0^{t_0} \mathrm{E}[X(t_1-\Theta)X(t_2-\Theta) \mid \Theta = \theta] \frac{d\theta}{t_0}$$

$$= \frac{1}{t_0} \int_0^{t_0} R_X(t_1-\theta, t_2-\theta)\, d\theta = \frac{1}{t_0} \int_{-t_1}^{t_0-t_1} R_X(-\theta, t_2-t_1-\theta)\, d\theta,$$

which depends only on the difference $t_2 - t_1$, since the integrand is again a periodic function of θ with period t_0.

Thus, if we define

$$Y(t) = X(t-\Theta) - \mathrm{E}[X(t-\Theta)], \qquad -\infty < t < \infty,$$

we have obtained a zero mean wide-sense stationary process with correlation function

$$R_Y(\tau) = \frac{1}{t_0} \int_0^{t_0} R_X(-\theta, \tau-\theta)\, d\theta = \mu^2.$$

Now if $|\tau| > t_0$, then the arguments of the correlation function $R_X(t_1, t_2)$ differ by at least t_0, in which case $X(t_1)$ and $X(t_2)$ are independent and hence

$$R_X(t_1, t_2) = \mu_X(t_1)\mu_X(t_2).$$

Thus, if we let

$$R_1(\tau) = \frac{1}{t_0} \int_0^{t_0} \mu_X(-\theta)\mu_X(\tau-\theta)\, d\theta - \mu^2, \qquad -\infty < \tau < \infty,$$

the correlation function $R_Y(\tau)$ can be written as a sum

$$R_Y(\tau) = R_1(\tau) + R_2(\tau), \qquad -\infty < \tau < \infty,$$

where

$$R_2(\tau) = R_Y(\tau) - R_1(\tau).$$

However, for $|\tau| > t_0$, $R_Y(\tau) = R_1(\tau)$, and hence $R_2(\tau) = 0$ for $|\tau| > t_0$. Furthermore, $R_2(\tau)$ is clearly a continuous and even function, and being actually the integral of the covariance $\text{Cov}[X(-\Theta), X(\tau - \Theta)]$, it is positive definite. Hence, its Fourier transform

$$s(\nu) = \int_{-t_0}^{t_0} e^{-\iota\omega\tau} R_2(\tau)\, d\tau, \qquad -\infty < \nu < \infty$$

is a nonnegative even frequency function that may therefore represent a power spectral density. On the other hand, on account of the periodicity of the mean function $\mu_X(t)$, we have for all $n = 0, \pm 1, \pm 2, \ldots,$

$$R_1(\tau + nt_0) = \frac{1}{t_0}\int_0^{t_0} \mu_X(-\theta)\mu_X(\tau + nt_0 - \theta)\, d\theta - \mu^2$$

$$= \frac{1}{t_0}\int_0^{t_0} \mu_X(-\theta)\mu_X(\tau - \theta)\, d\theta - \mu^2 = R_1(\tau),$$

so that $R_1(\tau)$ is a periodic function with period t_0. It may be expanded into a Fourier series

$$R_1(\tau) = \sum_{n=-\infty}^{\infty} s(\nu_n) e^{\iota\omega_n\tau},$$

where $\omega_n = 2\pi\nu_n$ and $\nu_n = n/t_0$ are integral multiples of the basic frequency $1/t_0$ of the unmodulated pulse train.

Consequently the correlation function $R_Y(\tau)$ can be written as a sum of two terms

$$R_Y(\tau) = \sum_{n=-\infty}^{\infty} s(\nu_n) e^{\iota\omega_n\tau} + \int_{-\infty}^{\infty} s(\nu) e^{\iota\omega\tau}\, d\nu.$$

It follows that the random PPM signal process $Y(t)$, $-\infty < t < \infty$ has a mixed-type spectrum with continuous power spectral density component $s_X(\nu)$, the Fourier transform of $R_2(\tau)$, and a discrete component with spectral lines $s_X(\nu_n)$ at integral multiples of the basic frequency $1/t_0$, as in Figure 5.2.17. The explicit expressions for both $s_X(\nu)$ and $s_X(\nu_n)$ can be obtained, but the computation, although in principle quite simple, is rather lengthy and is therefore omitted. ▲

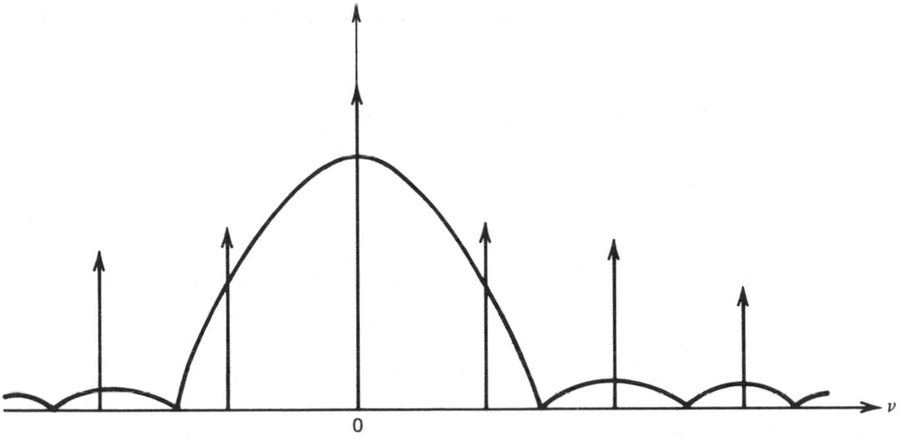

Figure 5.2.17

EFFECT OF LINEAR TRANSFORMATIONS ON THE SPECTRAL DISTRIBUTION. Let $X(t)$, $-\infty < t < \infty$, be a wide-sense stationary MS continuous process with power spectral distribution function $S_X(\nu)$, $-\infty < \nu < \infty$. Consider now a linear, time-invariant (not necessarily causal) filter with impulse-response function $h(t)$.

The output of this filter will be given by

$$Y(t) = \int_{-\infty}^{\infty} h(t - \tau)X(\tau)\, d\tau, \qquad -\infty < t < \infty \tag{35}$$

and if $h(t)$ is square-integrable, it will again be a wide-sense stationary MS-continuous process.

Let us now replace the process $X(\tau)$ in (35) by its spectral representation

$$X(\tau) = \int_{-\infty}^{\infty} e^{\iota\omega\tau}\, dZ_X(\nu), \qquad -\infty < \tau < \infty,$$

where $Z_X(\nu)$ is the spectral process of $X(t)$. We get

$$\begin{aligned} Y(t) &= \int_{-\infty}^{\infty} h(t - \tau) \int_{-\infty}^{\infty} e^{\iota\omega\tau}\, dZ_X(\nu)\, d\nu \\ &= \int_{-\infty}^{\infty} e^{\iota\omega t}\left[\int_{-\infty}^{\infty} h(t - \tau)e^{-\iota\omega(t-\tau)}\, d\tau\right] dZ_X(\nu), \end{aligned} \tag{36}$$

since $e^{\imath\omega t}e^{-\imath\omega(t-\tau)} = e^{\imath\omega\tau}$. The integral in brackets equals (by the substitution $t - \tau = t'$)

$$\int_{-\infty}^{\infty} h(t-\tau)e^{-\imath\omega(t-\tau)}\,d\tau = \int_{-\infty}^{\infty} h(t')e^{-\imath\omega t'}\,dt' = H(\nu) \tag{37}$$

which is recognized as the Fourier transform

$$H(\nu) = \mathcal{F}\{h(t)\} \tag{38}$$

of the impulse-response function $h(t)$. The complex-valued function $H(\nu)$, defined by (37), is known as the *transfer function*† of the time-invariant filter, and from (37)

$$H(\nu)e^{\imath\omega t} = \int_{-\infty}^{\infty} h(t-\tau)e^{\imath\omega\tau}\,d\tau, \qquad \omega = 2\pi\nu$$

it is clear that $H(\nu)$ is the complex amplitude of the harmonic output of the filter if the harmonic waveform $x(t) = e^{\imath\omega t}$ is applied to its input.

If the filter is causal, that is, $h(t) = 0$ for $t < 0$, then the transfer function equals

$$H(\nu) = \int_{0}^{\infty} h(t)e^{-\imath\omega t}\,dt$$

or, by calling $\imath\omega = \lambda$,

$$H\left(\frac{\lambda}{2\pi\imath}\right) = \int_{0}^{\infty} h(t)e^{-\lambda t}\,dt,$$

is the Laplace transform of $h(t)$. We assume that the reader is well acquainted with the notion of a transfer function, so the above facts are intended merely as reminders.

Returning now to equation (36), we can write it as

$$Y(t) = \int_{-\infty}^{\infty} e^{\imath\omega t}H(\nu)\,dZ_X(\nu).$$

Comparing this with the spectral representation for the output process

$$Y(t) = \int_{-\infty}^{\infty} e^{\imath\omega t}\,dZ_Y(\nu)$$

we conclude that the relation between the output and input spectral processes is simply

$$dZ_Y(\nu) = H(\nu)\,dZ_X(\nu), \qquad -\infty < \nu < \infty. \tag{39}$$

† The term "transfer function" is sometimes used to mean a Laplace rather than Fourier transform of a causal impulse-response function $h(t)$. However, in such a case, the latter is easily obtained from the former by replacing its argument with $\imath 2\pi\nu$.

This is exactly the same relationship as that between the Fourier transforms of output and input deterministic waveforms. In view of the interpretation of a spectral process, it is also easy to understand. If the input process $X(t)$ is regarded as a superposition of elementary harmonic oscillations with complex amplitudes $dZ_X(\nu)$, then the linearity of the filter implies that the output can be obtained by first passing each of these elementary harmonic oscillations through the filter separately and then forming their superposition. Since passage of a harmonic waveform through a linear filter does not change its frequency, but changes only its complex amplitude by multiplying it by the filter's response $H(\nu)$ to that frequency, relation (39) is intuitively obvious.

From (39), we readily obtain the relationship between the power spectral distribution functions $S_X(\nu)$ and $S_Y(\nu)$ of the input and output processes. Since

$$dS_X(\nu) = \mathrm{E}[|dZ_X(\nu)|^2] \qquad \text{and} \qquad dS_Y(\nu) = \mathrm{E}[|dZ_Y(\nu)|^2],$$

we have immediately

$$dS_Y(\nu) = |H(\nu)|^2\, dS_X(\nu), \qquad -\infty < \nu < \infty, \tag{40}$$

which is considerably simpler than the corresponding relation between the correlation functions

$$R_Y(t_2 - t_1) = \int_{-\infty}^{\infty}\int_{-\infty}^{\infty} h(t_1 - \tau_1)R_X(\tau_2 - \tau_1)h(t_2 - \tau_2)\, d\tau_1\, d\tau_2 . \tag{41}$$

Example 5.2.11

Suppose that the linear transformation averages the input process over the past time interval of length t_0; in symbols,

$$Y(t) = \frac{1}{t_0}\int_{t-t_0}^{t} X(t')\, dt', \qquad -\infty < t < \infty.$$

Such a linear time-invariant transformation is known as a moving average. Replacing the input process in the integral above by its spectral representation and interchanging the order of integration, we get

$$\begin{aligned} Y(t) &= \frac{1}{t_0}\int_{t-t_0}^{t}\int_{-\infty}^{\infty} e^{\imath\omega t'}\, dZ_X(\nu)\, dt' \\ &= \int_{-\infty}^{\infty}\left[\frac{1}{t_0}\int_{t-t_0}^{t} e^{\imath\omega t'}\, dt'\right] dZ_X(\nu). \end{aligned}$$

The integral in brackets equals

$$\frac{1 - e^{-\imath\omega t_0}}{\imath\omega t_0}\, e^{\imath\omega t},$$

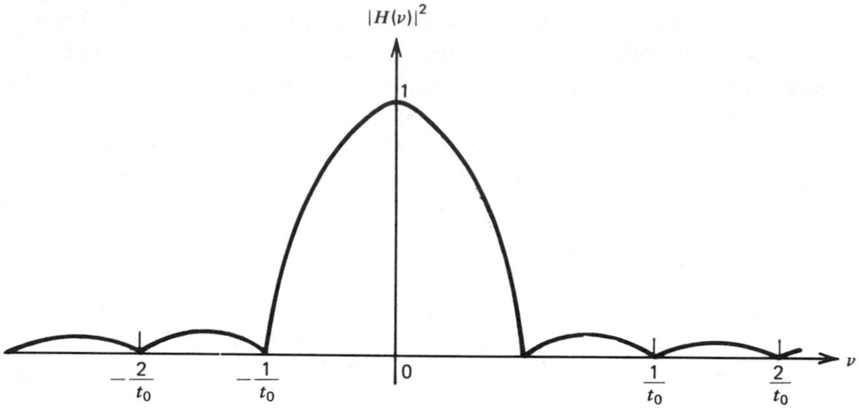

Figure 5.2.18

and hence

$$Y(t) = \int_{-\infty}^{\infty} e^{\iota\omega t} H(\nu)\, dZ_X(\nu),$$

where

$$H(\nu) = \frac{1 - e^{-\iota\omega t_0}}{\iota\omega t_0}, \qquad \omega = 2\pi\nu,$$

is identified as the transfer function of the moving average transformation. Expanding the complex-valued function into its real and imaginary parts

$$H(\nu) = \frac{\sin \omega t_0}{\omega t_0} - \iota \frac{1 - \cos \omega t_0}{\omega t_0}$$

we find that its square modulus equals

$$|H(\nu)|^2 = 2\frac{1 - \cos \omega t_0}{(\omega t_0)^2} = \left(\frac{\sin \frac{1}{2}\omega t_0}{\frac{1}{2}\omega t_0}\right)^2.$$

From Figure 5.2.18 and relation (40), we conclude that the moving average transformation filters out high frequencies of the input process. ▲

Example 5.2.12

Consider the simple passive *RC* filter shown in Figure 5.2.19. The transfer function of any passive lumped circuit containing resistances, capacitances, and induc-

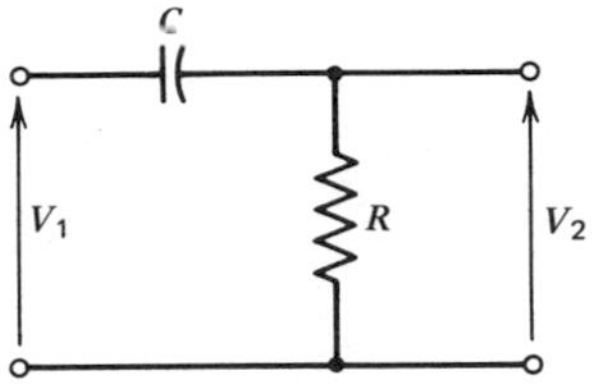

Figure 5.2.19

tances is easily found by replacing each capacitance C and each inductance L by their complex impedance $1/\imath\omega C$ and $\imath\omega L$, respectively, and applying Kirchhoff's laws. For our filter we have immediately

$$V_2 = \frac{R}{R + \dfrac{1}{\imath\omega C}} V_1,$$

with V_1 and V_2 the input and output voltage phasors. Hence the transfer function is

$$H(\nu) = \frac{R}{R + \dfrac{1}{\imath\omega C}} = \frac{\imath\omega}{\alpha + \imath\omega}, \qquad \omega = 2\pi\nu,$$

where $\alpha = 1/RC$. The squared magnitude of this complex-valued function equals

$$|H(\nu)|^2 = \left(\frac{\omega^2}{\alpha^2 + \omega^2}\right)^2 + \left(\frac{\alpha\omega}{\alpha^2 + \omega^2}\right)^2 = \frac{\omega^2}{\alpha^2 + \omega^2}.$$

Thus, if the input voltage is a wide-sense stationary zero mean process $X(t)$, $-\infty < t < \infty$, with power spectral distribution function $S_X(\nu)$, the power spectral distribution function $S_Y(\nu)$ of the output voltage $Y(t)$, $-\infty < t < \infty$, will satisfy

$$dS_Y(\nu) = \frac{\omega^2}{\alpha^2 + \omega^2} dS_X(\nu).$$

In particular, if the input process has exponential correlation function $R_X(\tau) = \sigma^2 e^{-\alpha|\tau|}$, that is, a power spectral density

$$s_X(\nu) = \frac{dS_X(\nu)}{d\nu} = \frac{2\alpha\sigma^2}{\alpha^2 + \omega^2},$$

then the output process will also have a continuous spectrum with power spectral density

$$s_Y(\nu) = \frac{2\alpha\sigma^2\omega^2}{(\alpha^2 + \omega^2)^2}, \qquad -\infty < \nu < \infty.$$

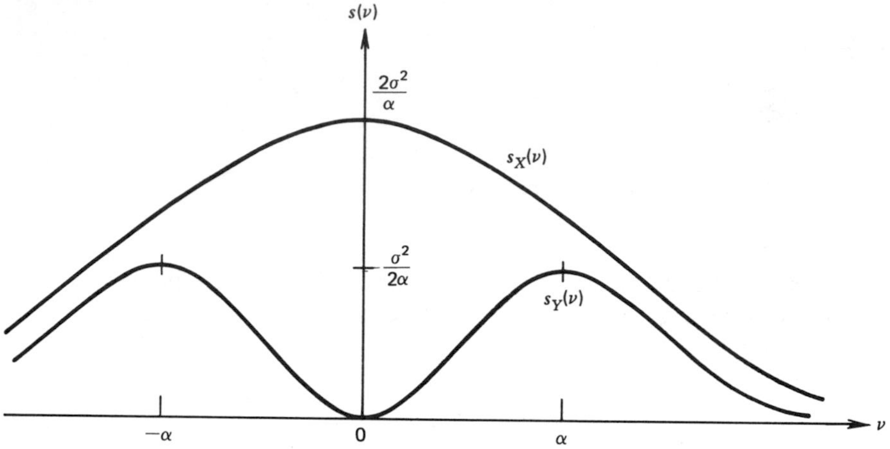

Figure 5.2.20

From Figure 5.2.20, it can be seen that, as expected, the filter tends to attenuate low frequencies. The output correlation function can be found as the inverse Fourier transform $R_Y = \mathcal{F}^{-1}\{s_Y\}$,

$$R_Y(\tau) = \frac{\sigma^2}{2}(1 - \alpha|\tau|)e^{-\alpha|\tau|}, \qquad -\infty < \tau < \infty. \qquad \blacktriangle$$

If the input process $X(t)$, $-\infty < t < \infty$, is a (zero mean) Gaussian process, then the output process $Y(t)$, $-\infty < t < \infty$, will be also Gaussian, as a result of the linearity of the transformation. Its entire probability law is then completely determined by the power spectral distribution function $S_Y(\nu)$, $-\infty < \nu < \infty$. Writing the complex-valued transfer function $H(\nu)$ in polar form

$$H(\nu) = |H(\nu)|e^{i\theta(\nu)},$$

where $\theta(\nu)$ is the argument of the complex number $H(\nu)$ (shown in Fig. 5.2.21), that is, the phase transfer function, we see from (40) that this function has no influence whatsoever on the probability law of the output process $Y(t)$.

In other words, if a stationary Gaussian process is applied to the inputs of two time-invariant linear transformations or filters with the same amplitude transfer functions $|H(\nu)|$ but with different phase transfer functions, the resulting two output processes will have identical Gaussian probability laws. Of course, each individual input sample path will not be transformed into two identical output sample paths by the two filters, since the phase transfer function does influence a deterministic input waveform. However, the ensembles of all sample values and probability laws of the two outputs will be identical, or roughly speaking, the

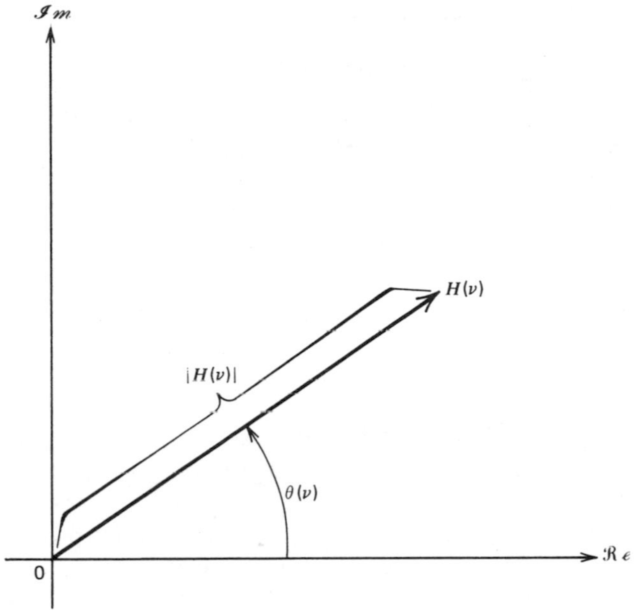

Figure 5.2.21

relative frequency of occurrence of any event defined by the output sample paths will be the same for the two filters.

Example 5.2.13

A simple example of this phenomenon is obtained by taking the linear transformation to be a constant delay,

$$Y(t) = X(t - t_0), \qquad -\infty < t < \infty,$$

where $t_0 > 0$. If the input process is strictly stationary, it is obvious that probability laws of the input and output processes will be identical. Indeed, if we write

$$X(t - t_0) = \int_{-\infty}^{\infty} e^{\imath\omega(t - t_0)}\, dZ_X(\nu),$$

that is,

$$Y(t) = \int_{-\infty}^{\infty} e^{\imath\omega t} e^{-\imath\omega t_0}\, dZ_X(\nu),$$

we see that the transfer function of this simple delay equals

$$H(\nu) = e^{-\imath\omega t_0}.$$

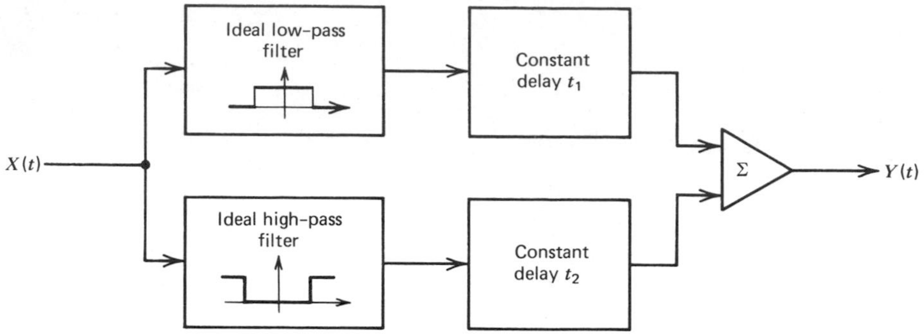

Figure 5.2.22

The amplitude transfer function is $|H(\nu)| = 1$ for all $-\infty < \nu < \infty$, while the phase transfer function is a linear function of the frequency $\theta(\nu) = -2\pi\nu t_0$, $-\infty < \nu < \infty$.

For a stationary Gaussian input, the probability law of the output process would not be affected even if the delay were dependent upon the frequency of the input, as, for example, in the arrangement in Figure 5.2.22. ▲

Because of (40) the real-valued function

$$|H(\nu)|^2 = H(\nu)H^*(\nu)$$

is sometimes called the *power transfer function* of the filter.

In terms of either the power spectral mass function or the power spectral density, relation (40) becomes

$$s_Y(\nu) = |H(\nu)|^2 s_X(\nu), \qquad -\infty < \nu < \infty.\dagger \tag{42}$$

Note that if the input process is white Gaussian noise $X(t) = \dot{W}_0(t)$, then

$$s_X(\nu) = 1 \qquad \text{for all } -\infty < \nu < \infty$$

and hence the output power spectral density is just a constant multiple of the power transfer function

$$s_Y(\nu) = |H(\nu)|^2.$$

Conversely, if we wish to design a *whitening filter*, that is, a filter that would turn a stationary MS continuous Gaussian process with power spectral density $s_X(\nu)$

† It should be mentioned that (42) can be obtained immediately by taking the Fourier transform of the double convolution (41)

$$R_Y = h \circledast R_X \circledast h$$

since the Fourier transform of a convolution is a product of Fourier transforms.

into white Gaussian noise, then the transfer function $H(\nu)$ of such a filter must satisfy the equation

$$1 = |H(\nu)|^2 s_X(\nu) \qquad \text{for all } -\infty < \nu < \infty.$$

Note however, that this equation determines only the amplitude transfer function $|H(\nu)|$ of the filter; the phase transfer function can be chosen arbitrarily.

Example 5.2.14
Suppose that white Gaussian noise $\dot{W}_0(t)$, $-\infty < t < \infty$, is applied to the input of a time-invariant linear system with transfer function

$$H(\nu) = \frac{\alpha}{\alpha + \imath\omega}, \qquad -\infty < \nu < \infty,$$

where $\omega = 2\pi\nu$ and $\alpha > 0$, a constant. For instance, a simple RC filter (Fig. 5.2.23) with $\alpha = 1/RC$ has such a transfer function. The output process $Y(t)$, $-\infty < t < \infty$, is then a zero mean stationary Gaussian process with power spectral density

$$s_Y(\nu) = |H(\nu)|^2 = \frac{\alpha^2}{\alpha^2 + \omega^2}.$$

It follows that the output process is a stationary Ornstein-Uhlenbeck process with parameters $\sigma^2 = \alpha/2$ and α. Since the impulse-response function $h = \mathfrak{F}^{-1}\{H\}$ is, in this case,

$$h(\tau) = \begin{cases} \alpha e^{-\alpha\tau} & \text{if} \quad \tau \geq 0, \\ 0 & \text{if} \quad \tau < 0, \end{cases}$$

we have actually rediscovered the already familiar integral representation

$$Y(t) = \sigma\sqrt{2\alpha} \int_{-\infty}^{t} e^{-\alpha(t-\tau)} \dot{W}_0(\tau)\, d\tau, \qquad -\infty < t < \infty$$

of a stationary Ornstein-Uhlenbeck process.

If we wanted to find a whitening filter for this Ornstein-Uhlenbeck process, that is, a filter with transfer function $H(\nu)$ such that

$$1 = |H(\nu)|^2 \frac{\alpha^2}{\alpha^2 + \omega^2}$$

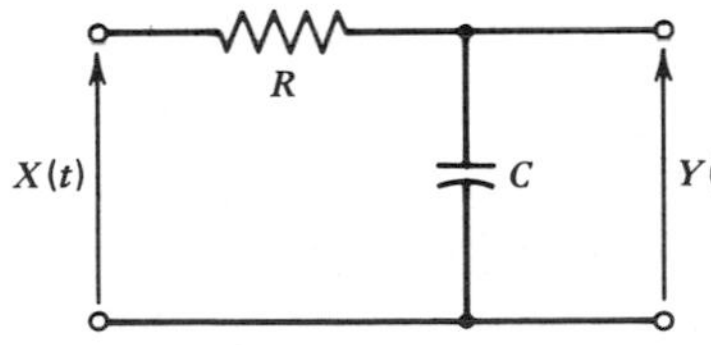

Figure 5.2.23

the obvious choice would be, for instance,

$$H(\nu) = \frac{\alpha \pm \imath\omega}{\alpha}.$$

From the formal identity

$$\int_{-\infty}^{\infty} (\alpha\, \delta(t) \pm \dot{\delta}(t))e^{-\imath\omega t}\, dt = \alpha \pm \imath\omega$$

we see that the corresponding impulse response function of such a whitening filter would be

$$h(t) = \frac{1}{\alpha}(\alpha\, \delta(t) \pm \dot{\delta}(t)).$$

Thus, if we write

$$\dot{W}_0(t) = \int_{-\infty}^{\infty} h(t-\tau)Y(\tau)\, d\tau = \frac{1}{\alpha}(\alpha Y(t) \pm \dot{Y}(t))$$

then by choosing the minus sign of the imaginary part of $H(\nu)$ we recover the Langevin equation

$$\dot{Y}(t) + \alpha Y(t) = \sigma\sqrt{2\alpha^2}\, \dot{W}_0(t),$$

for the Ornstein-Uhlenbeck process $Y(t)$. ▲

Example 5.2.15

Consider a two-port network, as in Figure 5.2.24, where the input voltage $V_1(t)$, $-\infty < t < \infty$, is a zero mean wide-sense stationary process with power spectral density

$$s_1(\nu) = \frac{1}{1+\omega^2}, \qquad -\infty < \nu < \infty.$$

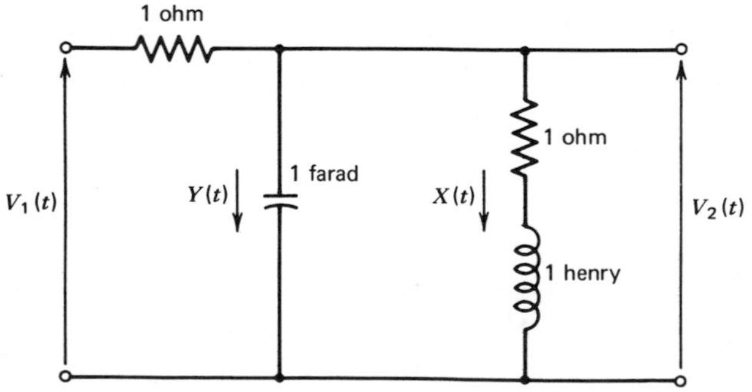

Figure 5.2.24

The transfer function $H(\nu)$ of this network is easily obtained from Kirchhoff's laws,

$$H(\nu) = \frac{Z_{RLC}}{1 + Z_{RLC}},$$

where

$$\frac{1}{Z_{RLC}} = \imath\omega + \frac{1}{1 + \imath\omega}$$

is the complex admittance of the resonant circuit. After a little algebra

$$H(\nu) = \frac{\imath\omega - \omega^2}{2\imath\omega + 1 - 2\omega^2},$$

and hence the power transfer function is

$$|H(\nu)|^2 = \frac{\omega^2(1 + \omega^2)}{1 + 4\omega^4}.$$

It follows that the power spectral density $s_2(\nu)$ of the output voltage process $V_2(t)$, $-\infty < t < \infty$, equals

$$s_2(\nu) = |H(\nu)|^2 s_1(\nu) = \frac{\omega^2}{1 + 4\omega^4}, \qquad -\infty < \nu < \infty.$$

It is seen that obtaining the output power spectral density of a lumped parameter network is no more difficult than the ordinary steady-state analysis with deterministic sinusoidal signals. Calculating the output correlation function is somewhat more difficult, as it requires computation of the inverse Fourier transform $\mathfrak{F}^{-1}\{s_2\}$, but it is usually still preferable to calculate it from the power spectral density than directly from formula (41). The reason for this is, of course, that it is generally much easier to get the transfer function of a given network than to get its impulse-response function.

Also, a considerable amount of information about the output process can be obtained directly from the power spectral density, for instance, the variance

$$\sigma^2 = R_2(0) = \int_{-\infty}^{\infty} s_2(\nu)\, d\nu,$$

often referred to as the output noise power. ▲

POWER CROSS-SPECTRAL DISTRIBUTION FUNCTION. Now consider two jointly wide-sense stationary MS continuous zero mean processes (possibly complex valued)

$$X(t), \qquad Y(t), \qquad -\infty < \tau < \infty.$$

and assume that their cross-correlation function

$$R_{X,Y}(t_2 - t_1) = \mathrm{E}[X(t_1)Y^*(t_2)]$$

is continuous at zero. Then Bochner's theorem still applies, so that there exists a unique distribution function $S_{X,Y}(v)$ on the frequency domain such that

$$R_{X,Y}(\tau) = \int_{-\infty}^{\infty} e^{\iota\omega\tau}\, dS_{X,Y}(v), \qquad \omega = 2\pi v, \qquad -\infty < \tau < \infty.$$

This function $S_{X,Y}$ is called the *power cross-spectral distribution function.*

However, unlike the ordinary power spectral distribution function, this function need no longer be real and positive, even if both $X(t)$ and $Y(t)$ are real valued. This is due to the fact that the cross-correlation function need not be symmetric. This is best seen in the case when the cross-correlation function $R_{X,Y}(\tau)$ has a Fourier transform, in which case the *power cross-spectral density*

$$s_{X,Y}(v) = \frac{d}{dv} S_{X,Y}(v)$$

exists and forms a Fourier transform pair with $R_{X,Y}(v)$:

$$s_{X,Y}(v) = \int_{-\infty}^{\infty} e^{-\iota\omega\tau} R_{X,Y}(\tau)\, d\tau, \tag{43a}$$

$$R_{X,Y}(\tau) = \int_{-\infty}^{\infty} e^{\iota\omega\tau} s_{X,Y}(v)\, dv. \tag{43b}$$

Even when the cross-correlation function $R_{X,Y}(\tau)$ is real, the lack of symmetry $R_{X,Y}(\tau) \neq R_{X,Y}(-\tau)$ implies that its Fourier transform $s_{X,Y}(v)$ is generally a complex-valued function. However, it can be shown that if $s_X(v)$ and $s_Y(v)$ are power spectral densities of $X(t)$ and $Y(t)$, respectively, then for all v

$$|s_{X,Y}(v)|^2 \leq s_X(v) s_Y(v), \tag{44}$$

an inequality reminiscent of the Schwarz inequality. Furthermore, since the cross-correlation function satisfies

$$R^*_{X,Y}(\tau) = R_{Y,X}(-\tau)$$

it is easy to see by taking the complex conjugate of (43a) that for all v

$$s_{Y,X}(v) = s^*_{X,Y}(v). \tag{45}$$

Hence, in the case under consideration (continuous spectra) the correlation matrix

$$\mathbf{R}_{\binom{X}{Y}}(\tau) = \begin{bmatrix} R_X(\tau) & R_{X,Y}(\tau) \\ R_{Y,X}(\tau) & R_Y(\tau) \end{bmatrix}$$

of the vector-valued process

$$\begin{pmatrix} X(t) \\ Y(t) \end{pmatrix}$$

has its Fourier counterpart in the *power spectral density matrix*

$$\mathbf{s}_{\binom{X}{Y}}(\nu) = \begin{bmatrix} s_X(\nu) & s_{X,Y}(\nu) \\ s_{Y,X}(\nu) & s_Y(\nu) \end{bmatrix}.$$

Both these matrices provide equivalent information about the process

$$\begin{pmatrix} X(t) \\ Y(t) \end{pmatrix}.$$

In particular, they specify completely its probability law if the process is Gaussian. Also, (44) implies that the power spectral density matrix is nonnegative definite, while (45) implies that its transpose equals its complex conjugate. (A matrix with this latter property is called Hermitian.)

We will not go into any more details, but it should be noted that the above discussion can easily be generalized to vector-valued processes with any number of components and that the spectral representation theorem can likewise be extended to vector-valued processes.

Example 5.2.16

In the two-port network of Example 5.2.15, let $X(t)$ and $Y(t)$, $-\infty < t < \infty$, be the currents through the inductive and capacitive branches, respectively, of the resonant circuit. If $Z_2(\nu)$, $-\infty < \nu < \infty$, is the spectral process of the output voltage $V_2(t)$, $-\infty < t < \infty$, the relations following from Ohm's law and (39)

$$dZ_X(\nu) = \frac{1}{1 + \imath\omega} \, dZ_2(\nu),$$
$$dZ_Y(\nu) = \imath\omega \, dZ_2(\nu),$$

allow us to write the spectral representations

$$X(t) = \int_{-\infty}^{\infty} e^{\imath\omega t} \frac{1}{1 + \imath\omega} \, dZ_2(\nu),$$

$$Y(t) = \int_{-\infty}^{\infty} e^{\imath\omega t} \imath\omega \, dZ_2(\nu).$$

It follows that the spectral densities are

$$s_X(\nu) = \left| \frac{1}{1 + \imath\omega} \right|^2 s_2(\nu) = \frac{s_2(\nu)}{1 + \omega^2},$$
$$s_Y(\nu) = |\imath\omega|^2 s_2(\nu) = \omega^2 s_2(\nu).$$

To obtain the power cross-spectral density, let us first write an expression for the cross-correlation function

$$R_{X,Y}(\tau) = \mathrm{E}[X(t)Y^*(t+\tau)] = \mathrm{E}\left[\left(\int_{-\infty}^{\infty} e^{\imath\omega t}\frac{1}{1+\imath\omega}\,dZ_2(\nu)\right)\right.$$
$$\left.\times\left(\int_{-\infty}^{\infty} e^{\imath\omega(t-\tau)}\imath\omega\,dZ_2(\nu)\right)^*\right].$$

Writing the product of integrals as a double integral

$$\int_{-\infty}^{\infty}\int_{-\infty}^{\infty} e^{\imath(\omega_1-\omega_2)t}e^{-\imath\omega_2\tau}\frac{-\imath\omega_2}{1+\imath\omega_1}\,dZ_2(\nu_1)\,dZ_2^*(\nu_2),$$

where $\omega_1 = 2\pi\nu_1$ and $\omega_2 = 2\pi\nu_2$, interchanging integration with the expectation operator, and using the fact that

$$\mathrm{E}[dZ_2(\nu_1)\,dZ_2^*(\nu_2)] = \delta(\nu_1-\nu_2)s_2(\nu_1)\,d\nu_1,$$

we obtain

$$R_{X,Y}(\tau) = \int_{-\infty}^{\infty} e^{-\imath\omega_1\tau}\frac{-\imath\omega_1}{1+\imath\omega_1}s_2(\nu_1)\,d\nu_1.$$

Making the trivial substitution $\nu = -\nu_1$ yields

$$R_{X,Y}(\tau) = \int_{-\infty}^{\infty} e^{\imath\omega\tau}\frac{\imath\omega}{1-\imath\omega}s_2(\nu)\,d\nu,$$

whence, by comparison with (43), we conclude that the power cross-spectral density equals

$$s_{X,Y}(\nu) = \frac{\imath\omega}{1-\imath\omega}s_2(\nu) = \frac{|\omega|s_2(\nu)}{\sqrt{1+\omega^2}}e^{-\imath/\omega}, \qquad -\infty<\nu<\infty.$$

Thus the power spectral density matrix of the vector-valued process

$$\begin{pmatrix} X(t) \\ Y(t) \end{pmatrix}, \qquad -\infty<t<\infty,$$

equals

$$\mathbf{s}_{\binom{X}{Y}}(\nu) = \begin{bmatrix} \dfrac{1}{1+\omega^2} & \dfrac{|\omega|}{\sqrt{1+\omega^2}}e^{-\imath/\omega} \\ \dfrac{|\omega|}{\sqrt{1+\omega^2}}e^{\imath/\omega} & \omega^2 \end{bmatrix} s_2(\nu).$$

Note that in this particular case

$$|s_{X,Y}(\nu)|^2 = s_X(\nu)s_Y(\nu),$$

which results from the fact that the currents are strongly correlated, both being deterministic functionals of the output voltage process $V_2(t)$, $-\infty < t < \infty$. ▲

As another example, let us examine the power cross-spectral density $s_{X,Y}(\nu)$ between the input $X(t)$ and the output $Y(t)$ of a linear time-invariant filter with impulse response function $h(t)$. Assuming that the input process $X(t)$ is a wide-sense stationary zero mean process with correlation function $R_X(\tau)$, we know from Section 4.2 that the cross-correlation function $R_{X,Y}(\tau)$ between the input and the output is given by

$$R_{X,Y}(\tau) = \int_{-\infty}^{\infty} h(\tau - \tau_1) R_X(\tau_1)\, d\tau_1.$$

(The filter is not assumed causal.) Since this is just the convolution integral

$$R_{X,Y} = h \circledast R_X,$$

if the input process $X(t)$ has power spectral density $s_X(\nu)$, the convolution theorem for Fourier transforms yields immediately

$$s_{X,Y}(\nu) = H(\nu) s_X(\nu), \qquad -\infty < \nu < \infty \tag{46}$$

where $H = \mathscr{F}^{-1}\{h\}$ is the transfer function of the filter. Since a transfer function $H(\nu)$ is generally a complex-valued function, we see again from (46) that a cross-spectral density need not be real. In Section 4.2 we also saw that the relation between the correlation function $R_Y(\tau)$ of the output and the cross-correlation function $R_{Y,X}(\tau)$ is a convolution

$$R_Y = R_{Y,X} \circledast h.$$

Hence we also have the relation for the output power spectral density

$$s_Y(\nu) = s_{Y,X}(\nu) H(\nu), \qquad -\infty < \nu < \infty.$$

Substituting from (46) and using (45), this gives

$$s_Y(\nu) = H^*(\nu) s_X(\nu) H(\nu),$$

or, (for a scalar-valued process),

$$s_Y(\nu) = |H(\nu)|^2 s_X(\nu), \qquad -\infty < \nu < \infty$$

a relation we have obtained before.

EXERCISES

Exercise 5.2

1. Let $X(t)$, $-\infty < t < \infty$, be a zero mean stationary Gaussian process with a discrete power spectrum $\{(\nu_n, s_X(\nu_n))\}$, where $\nu_n = n$, $n = 0, \pm 1, \pm 2, \ldots$. The power spectral mass distribution is

$$s_X(\nu_n) = a(1 + |\nu_n|)\gamma^{|\nu_n|},$$

where $a > 0$ and $0 < \gamma < 1$ are constants. Find the correlation function $R_X(\tau)$, $-\infty < \tau < \infty$.

2. Which frequencies contribute the largest power to the process of Exercise 5.2.1 if $\gamma = 3/5$? If $\gamma = 4/5$? What fraction of the total power do these frequencies contribute?
3. Suppose a zero mean wide-sense stationary process $X(t)$, $-\infty < t < \infty$, with correlation function $R_X(\tau)$ has a discrete power spectrum

$$\{(\nu_n, s_X(\nu_n))\}.$$

Describe how the correlation function changes in any of the following three cases:

(a) The power spectrum is subjected to a "red shift" that shifts each frequency ν_n to $\nu_n' = \gamma\nu_n$, $0 < \gamma < 1$, without changing the power $s_X(\nu_n)$.

(b) Each spectral line splits into two spectral lines displaced symmetrically about the original frequency ν_n, with each carrying half of the original power $s_X(\nu_n)$.

(c) Each spectral line with nonzero frequency ν_n produces an infinite number of harmonic frequencies $\nu_n, 2\nu_n, 3\nu_n, \ldots$, with the original power $s = s_X(\nu_n)$ being distributed into the harmonics according to the geometric series $\frac{1}{2}s, \frac{1}{4}s, \frac{1}{8}s, \ldots$.

4. For each of the correlation functions below, show that the process has continuous power spectrum and find the power spectral density.

(a) $R_X(\tau) = \sigma^2(1 + |\tau| + \frac{1}{3}\tau^2)e^{-|\tau|}$

(b) $R_X(\tau) = \sigma^2 \dfrac{\sin 2\tau}{\tau}$

(c) $R_X(\tau) = \sigma^2 e^{-|\tau|}(1 - \frac{1}{2}e^{-|\tau|})$

(d) $R_X(\tau) = \frac{1}{4}e^{-|\tau|}(\cos \tau + \sin|\tau|)$

5. Let $X(t)$, $-\infty < t < \infty$, be a wide-sense stationary complex-valued process defined by

$$X(t) = Ae^{i\Omega t}.$$

Here A and Ω are independent continuous random variables, A uniformly distributed over $(-\sqrt{3}, \sqrt{3})$ and Ω with some arbitrary probability density $f_\Omega(\omega)$. Show that the process $X(t)$ has a continuous spectrum, and find its power spectral density and its correlation function.

6. Explain why a frequency function such as $(1 + 2\nu^2)^{-1/2}$ cannot be a power spectral density of a genuine (as opposed to a symbolic) wide-sense stationary process. Show, on the other hand, that the time function $(1 + 2\tau^2)^{-1/2}$ is a legitimate correlation function. (*Hint:* Use Exercise 1.39.) Find conditions a correlation function must satisfy to be able to serve as a power spectral density (with ν replacing τ), and vice versa.

7. Suppose that the power spectral density of a zero mean stationary Gaussian process is defined as an infinite series

$$s_X(\nu) = \sum_{n=1}^{\infty} \frac{\gamma^n}{n^2 + (2\pi\nu)^2}, \qquad -\infty < \nu < \infty,$$

where $0 < \gamma < 1$ is a constant. Find the correlation function $R_X(\tau)$, $-\infty < \tau < \infty$, and show that the process $X(t)$, $-\infty < t < \infty$, can be regarded as an infinite linear combination of independent Ornstein-Uhlenbeck processes.

8. Let $X(n)$, $n = \ldots, -1, 0, 1, \ldots$, be a discrete-time zero mean wide-sense stationary process. Find the power spectral density $s_X(\nu)$, $-\frac{1}{2} < \nu < \frac{1}{2}$, for each of the following cases:
   (a) $X(n)$ is a sequence of i.i.d. random variables with variance σ^2.
   (b) $X(n)$ is a discrete-time Gauss-Markov process
   (c) $X(n)$ has the correlation function defined by the formula (2.1.52)
9. Let $X(t)$, $-\infty < t < \infty$, be a stationary zero mean Gaussian process with correlation function as in Exercise 5.2.4. For each of the four cases (a) through (d), write the spectral representation of the process $X(t)$ as a Wiener integral (18) or as a white noise integral (19). Rewrite it into the quadrature decomposition.
10. In practice we often speak of a narrowband process $X(t)$ even if the frequency function $s_0(\nu)$ in (21) is not quite zero for $|\nu| > \Delta\nu_0$ but only "small." For such a function, an *effective bandwidth* ν_{eff} is defined by the formula

$$\nu_{\text{eff}} = \frac{1}{2} \frac{\int_{-\infty}^{\infty} s_0(\nu)\, d\nu}{\max\limits_{-\infty<\nu<\infty} s_0(\nu)}.$$

Show that if $s_0(\nu)$ has a maximum at $\nu = 0$, then $\nu_{\text{eff}} = 1/2\tau_{\text{corr}}$, where $\tau_{\text{corr}} = [1/R_X(0)] \int_{-\infty}^{\infty} R_X(\tau)\, d\tau$ is the correlation time. If $s_0(\nu)$ is the power spectral density of a zero mean stationary Ornstein-Uhlenbeck process, what fraction of its power is actually outside the frequency interval $(-\nu_{\text{eff}}, \nu_{\text{eff}})$?
11. Suppose a zero mean stationary Gaussian process $X(t)$, $-\infty < t < \infty$, is defined by

$$X(t) = 6 \int_{t-1}^{t} (1 - t + \tau)\dot{W}_0(\tau)\, d\tau.$$

Find the spectral representation of this process both directly by substituting the spectral representation (30) of $\dot{W}_0(\tau)$ and by first computing the correlation function $R_X(\tau)$.
12. Suppose that a sequence of trigger pulses one second apart is applied to a circuit that responds to each trigger by producing a rectangular impulse of a random width. The

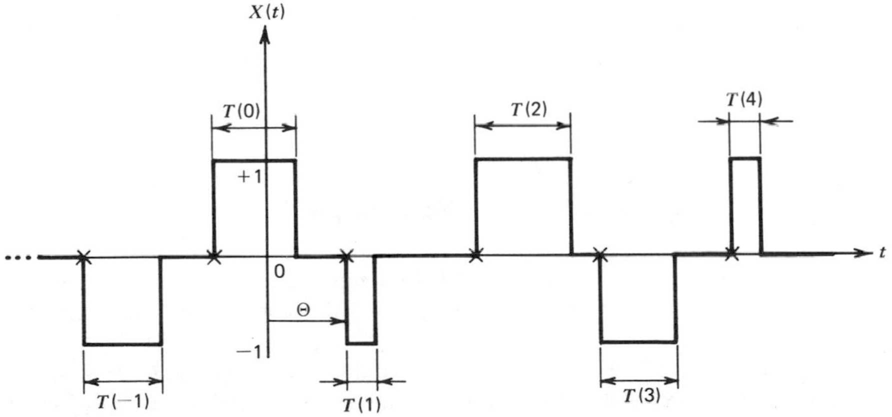

Figure 5.2.E.1

amplitudes of these impulses alternate between $+1$ and -1, and their widths $T(n)$, $n = \ldots, -1, 0, 1, \ldots$, are independent random variables uniformly distributed over $(0, 1)$. The trigger pulses are positioned at random with respect to time $t = 0$, just as in Example 5.2.10. The polarity of the rectangular pulse between the two trigger pulses astride the time $t = 0$ is either $+1$ or -1 with equal probability. Show that the resulting waveform $X(t)$, $-\infty < t < \infty$, as depicted in Figure 5.2.E.1, is a zero mean wide-sense stationary process with a mixed-type power spectrum, and find its power spectral distribution and density. Draw a sketch of the spectrum.

13. Suppose a zero mean wide-sense stationary process $X(t)$, $-\infty < t < \infty$, is passed through a time-invariant linear filter with the transfer function:

(a) $H(\nu) = \dfrac{\alpha}{2\omega_0 - \imath\omega^2}$ (second order Butterworth)

(b) $H(\nu) = \begin{cases} 1 & \text{if} \quad |\omega| \le \omega_0, \\ 0 & \text{if} \quad |\omega| > \omega_0, \end{cases}$ (ideal low-pass)

(c) $H(\nu) = \alpha e^{-(\omega^2 + \imath\omega\omega_0)}$ (Gaussian filter)

(d) $H(\nu) = \alpha\omega(\omega_0 - \imath\omega^2)^{-3/2}$

Here, $\omega = 2\pi\nu$, while ω_0 and α are positive constants. Assuming that the correlation function $R_X(\tau)$ of the input process is as in Exercise 5.2.4, find and sketch the power spectral density $s_Y(\nu)$ of the output process $Y(t)$, $-\infty < t < \infty$, for each of the 16 cases.

14. The thermal noise generated by a resistance of R ohms can be modeled as a voltage source $X(t)$, $-\infty < t < \infty$, with internal resistance R. The process $X(t)$, $-\infty < t < \infty$, is a stationary white Gaussian noise $X(t) = b\dot{W}_0(t)$ with intensity $b = \sqrt{2kTR}$, where T is the absolute temperature and $k = 1.37 \times 10^{-23}$ joule/degree Kelvin is the Boltzman constant. Suppose that the noise resistor is part of the circuit shown in Figure 5.2.E.2. Calculate the power spectral density $s_Y(\nu)$ of the voltage $Y(t)$ and the output noise power $R_Y(0)$ if the temperature of the resistor is 30 degrees Celsius.

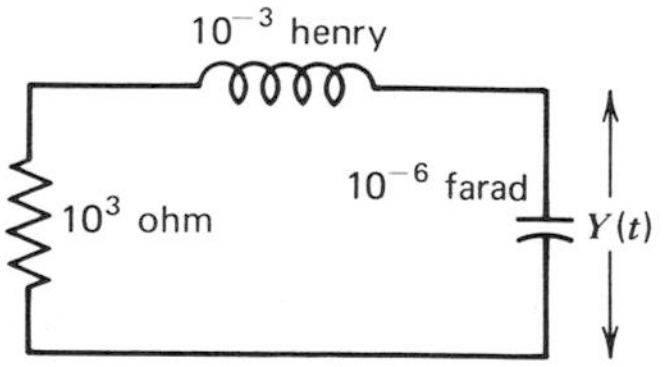

Figure 5.2.E.2

15. Find the power transfer function of a whitening filter for a zero mean stationary Gaussian process $X(t)$, $-\infty < t < \infty$, with correlation function $R_X(\tau) = (1 + |\tau|)e^{-|\tau|}$.

16. If $X(t)$, $-\infty < t < \infty$, is an MS differentiable zero mean wide-sense stationary process with spectral density $s_X(\nu)$, what is the cross-spectral density $s_{X,\dot{X}}(\nu)$ between

the process and its MS derivative? Use this result to derive the formula $s_{\dot{X}}(\nu) = \omega^2 s_X(\nu)$. (*Hint:* Pass the process through an ideal differentiator $H(\nu) = \imath\omega$.)

17. Let $X(t)$, $-\infty < t < \infty$, be a zero mean stationary Gaussian process with power spectral density $s_X(\nu)$. Define a zero mean wide-sense stationary process $Y(t) = X^2(t) - \sigma_X^2(t)$, $-\infty < t < \infty$, and show that its power spectral density equals

$$s_Y(\nu) = 2 \int_{-\infty}^{\infty} s_X(\nu - \nu') s_X(\nu')\, d\nu'.$$

(*Hint:* Show that $R_Y(\tau) = 2R_X^2(\tau)$, e.g., by differentiating the characteristic function of the process $X(t)$.)

18. Find the spectral density of the process $Y(t) = X(t)\dot{X}(t)$ if $X(t)$, $-\infty < t < \infty$, is a zero mean Gaussian process with correlation function $R_X(\tau) = \sigma^2 e^{-\alpha\tau^2}$, $\alpha > 0$. (*Hint:* Write $Y(t) = (1/2)(d/dt)X^2(t)$ and use Exercises 5.2.16 and 5.2.17.)

19. If $X(t)$, $-\infty < t < \infty$, is a zero mean wide-sense stationary process with power spectral density $s_X(\nu)$ and if $Y(t) = X(t + \Delta)$, $-\infty < t < \infty$, $\Delta > 0$, a constant, find the power spectral density matrix of the vector-valued process $(X(t), Y(t))'$.

5.3. STATIONARY GAUSSIAN PROCESSES

In this section we shall study continuous time processes, which are strictly stationary and Gaussian. We will also assume, as we have almost exclusively in this chapter, that the processes involved are sample continuous. (This, however, does not apply to generalized processes, where the concept of sample continuity is meaningless, anyway.)

As mentioned before, when dealing with stationary processes it is convenient to take for their time domain the entire real line $\mathscr{T} = (-\infty, +\infty)$. Thus, we will consider a sample-continuous stationary Gaussian process

$$X(t), \qquad -\infty < t < +\infty$$

with mean function

$$\mu_X(t) = \mu \tag{1}$$

and correlation function

$$R_X(\tau) = \mathrm{E}[X(t)X(t + \tau)], \qquad -\infty < \tau < \infty.$$

Since we will be dealing exclusively with linear operations, there is no loss of generality in assuming that the mean (1) is zero

$$\mu_X(t) = 0, \qquad -\infty < t < \infty$$

and we will do so from now on.

Let us begin by considering two examples. First, let us take a random sine wave, that is, a process defined by

$$X_s(t) = A \sin(\omega t + \Theta), \qquad -\infty < t < \infty \tag{2}$$

where $\omega = 2\pi\nu$ is a (nonrandom) radial frequency, the amplitude A is a Rayleigh random variable, and the phase Θ is a random variable uniformly distributed over $(0, 2\pi)$ and independent of the amplitude A. We have encountered this process several times before, and we know that that is indeed a zero mean Gaussian process. Its stationarity can be seen from the correlation function. Obviously, it is sample continuous.

Second, consider a stationary Gauss-Markov process

$$X_r(t), \qquad -\infty < t < \infty \tag{3}$$

with zero mean function and correlation function

$$R(\tau) = e^{-|\tau|}, \qquad -\infty < \tau < \infty$$

in other words, a stationary Ornstein-Uhlenbeck process.

In Section 4.1 we have seen that it is a sample-continuous process, and in Section 4.4 we concluded that such a process would result as a steady-state variation of voltage on the capacitor of an RC circuit with white noise Gaussian input. Putting it another way, this process would result if the capacitor was connected to the noise resistor in the infinitely remote past so that the system would be in steady state at any finite time $-\infty < t < \infty$.

If we look closely at these two processes, we note that they are of a fundamentally different kind. The sample paths of the process (2) are all sinusoids with a particular sample path uniquely determined by the values of the two random variables A and Θ. Thus, if we observe this process for an arbitrarily short interval of time, we can determine the sample path and hence predict the exact value $x(t)$ of the process for any future time t. (In fact, it is enough to observe only two values $X(t_1) = x(t_1)$ and $X(t_2) = x(t_2)$ provided the observation times t_1 and t_2 do not differ by a multiple of the period $1/\nu$.)

On the other hand, such errorless prediction into the future of the process (3) is impossible even if we were allowed to observe it for an infinitely long time, say during $(-\infty, t_0]$. This is because this process is constantly being innovated by the stream of white noise impulses, which, by their very nature, are unpredictable. Thus, if we tried to predict the value of $X(t)$ at some future time $t > t_0$, the exact prediction would not only be impossible, but our forecast would be less and less accurate the farther into the future we tried to predict.

We will now proceed to show that any stationary Gaussian process can, in fact, be decomposed into two processes of the above type, that is, a process $X_s(t)$, exactly predictable into the indefinite future, and a process $X_r(t)$, for which this is not possible and which can actually be thought of as a result of white Gaussian noise passed through a linear filter. As the subscripts suggest, the two processes will be called singular and regular components.

To begin, let t be a fixed time instant and let

$$t_0 < t$$

be an earlier time instant. Consider the conditional expectation of $X(t)$ given the past $X(t')$, $-\infty < t' \leq t_0$, up to the time t_0. To simplify our notation, let us denote this conditional expectation by

$$\hat{X}(t \mid t_0) = \mathrm{E}[X(t) \mid X(t'), -\infty < t' \leq t_0].$$

We know, that $\hat{X}(t \mid t_0)$ is the best MS prediction of $X(t)$ based on the past $X(t')$, $-\infty < t' \leq t_0$. Further, let

$$\tilde{X}(t \mid t_0) = X(t) - \hat{X}(t \mid t_0)$$

be the error of this prediction. Recall that, by the orthogonality principle (see Chapter One)

$$\mathrm{E}[\tilde{X}(t \mid t_0)Y] = \mathrm{E}[X(t)Y] \tag{4}$$

for any functional $Y = g(X(t'), -\infty < t' \leq t_0)$ of the conditioning past.

Note also, that because the process $X(t)$ is Gaussian and stationary, both $\hat{X}(t \mid t_0)$ and $\tilde{X}(t \mid t_0)$ will be Gaussian and will *depend only on the* time difference $t - t_0$. Let

$$\varepsilon^2(t - t_0) = \mathrm{E}[(\tilde{X}^2(t|t_0))] \tag{5}$$

be the mean square prediction error. Clearly

$$0 \leq \varepsilon^2(t - t_0) \leq \sigma^2 = R_X(0) \tag{6}$$

where the upper bound results from the fact that by predicting $X(t)$ as $\mu_X(t) = 0$, that is, disregarding any past information, the mean square error is the variance $\mathrm{Var}[X(t)] = R_X(0)$.

Suppose now that $t_0 \to -\infty$, while t is kept fixed. Then the mean square error $\varepsilon^2(t - t_0)$ cannot decrease with t_0, since, as $t_0 \to -\infty$, we are trying to predict farther and farther into the future or, if you prefer, using past observation farther and farther away from the present. Consequently by (6) the limit

$$\lim_{t_0 \to -\infty} \varepsilon^2(t - t_0) = \varepsilon^2(-\infty), \tag{7}$$

must exist and satisfy the inequality

$$0 \leq \varepsilon^2(-\infty) \leq \sigma^2.$$

But this implies that as $t_0 \to -\infty$ the random variable $\hat{X}(t \mid t_0)$ must converge in mean square to a random variable. To see this, take $t_1 < t_0$ and compute

$$\begin{aligned} \mathrm{E}[\tilde{X}(t|t_0)\tilde{X}(t|t_1)] &= \mathrm{E}[\tilde{X}(t|t_0)(\tilde{X}(t|t_0) + \hat{X}(t|t_1) - \hat{X}(t|t_0))] \\ &= \mathrm{E}[(\tilde{X}^2(t|t_0))] + \mathrm{E}[\tilde{X}(t|t_0)Y], \end{aligned} \tag{8}$$

where

$$Y = \hat{X}(t \mid t_1) - \hat{X}(t \mid t_0).$$

But Y is a linear functional of the past $x(t')$, $-\infty < t' \leq t_0$, and so, by the orthogonality principle (4), the last expectation in (8) is zero. Thus by (5)

$$\mathrm{E}[\tilde{X}(t \mid t_0)\tilde{X}(t \mid t_1)] = \varepsilon^2(t - t_0)$$

which by (7) converges as $t_0 \to -\infty$ and consequently also as $t_1 \to -\infty$. Therefore by the Loève criterion the MS limit

$$\underset{t_0 \to -\infty}{\text{l.i.m.}}\ \tilde{X}(t \mid t_0) = X(t) - \underset{t_0 \to -\infty}{\text{l.i.m.}}\ X(t \mid t_0),$$

exists for every t, since t was fixed but otherwise arbitrary. Hence, we can define two processes

$$X_s(t) \qquad \text{and} \qquad X_r(t), \qquad -\infty < t < \infty,$$

by

$$X_s(t) = \underset{t_0 \to -\infty}{\text{l.i.m.}}\ \hat{X}(t \mid t_0), \tag{9}$$

and

$$X_r(t) = X(t) - X_s(t) = \underset{t_0 \to -\infty}{\text{l.i.m.}}\ \tilde{X}(t \mid t_0). \tag{10}$$

Now both these two processes, being defined by means of conditional expectations and passages to the limit of a Gaussian process, are themselves Gaussian.

Furthermore, as already mentioned, $\hat{X}(t \mid t_0)$ depends only on the difference $t - t_0$, and so if we look upon

$$Z(\tau) = \hat{X}(t + \tau \mid t_0 + \tau)$$

as a stochastic process with time variable τ, $-\infty < \tau < +\infty$, this process will be stationary (and Gaussian), that is, its probability law will be invariant with respect to a time shift. But by (9) for any τ

$$\underset{t_0 \to -\infty}{\text{l.i.m.}}\ \hat{X}(t + \tau \mid t_0 + \tau) = \underset{t_0 \to -\infty}{\text{l.i.m.}}\ \hat{X}(t + \tau \mid t_0) = X_s(t + \tau)$$

that is, the probability law of the process $X_s(t)$ will also be invariant with respect to a time shift. This, together with definition (10) of $X_r(t)$, implies that both these processes are stationary.

Finally, taking $t_1 > t_0$ and $t_2 > t_0$, we have by the orthogonality principle

$$\mathrm{E}[\tilde{X}(t_1 \mid t_0)\hat{X}(t_2 \mid t_0)] = 0, \tag{11}$$

since $\hat{X}(t_2 \mid t_0)$ is a linear functional of the past $X(t')$, $-\infty < t' \leq t_0$. Hence letting $t_0 \to -\infty$, we also have by (10) and (11)

$$\mathrm{E}[X_r(t_1)X_s(t_2)] = 0 \qquad \text{for any } -\infty < t_1, t_2 < +\infty \tag{12}$$

and so, X_r and X_s being both zero mean and Gaussian, (12) implies that they are mutually independent.

To summarize, we present the following theorem.

THEOREM 5.3.1. (*Hanner Decomposition*)
Let X(t), $-\infty < t < +\infty$, *be a sample-continuous zero mean Gaussian stationary process. Then*

$$X(t) = X_s(t) + X_r(t), \qquad -\infty < t < \infty, \tag{13}$$

where $X_s(t)$ *and* $X_r(t)$ *are mutually independent, sample-continuous, zero mean Gaussian stationary processes, called the* singular *and* regular *part of the process* X(t). *Further, for any* $-\infty < t < +\infty$ *and* $-\infty < t_0 < t$

$$\mathrm{E}[X_s(t) \mid X(t'), -\infty < t' \le t_0] = X_s(t), \tag{14}$$

and

$$\operatorname*{l.i.m.}_{t_0 \to -\infty} \mathrm{E}[X_r(t) \mid X(t'), -\infty < t' \le t_0] = 0. \tag{15}$$

Equations (14) and (15) are easily proved by taking the conditional expectation of (9) and (10).

Let us now discuss the meaning of this decomposition. From (14) we see that the singular part $X_s(t)$ can be predicted with zero error from any portion of the past $X(t')$, $-\infty < t' < t_0$, no matter how remote. In fact, from (9) one can even say that $X_s(t)$ can be predicted with zero error from the infinitely remote past of the process $X(t)$. In other words, the singular part $X_s(t)$ is completely determined by the infinitely remote past. For this reason, $X_s(t)$ is sometimes also called a purely deterministic process. However, it is still a random (i.e., nondeterministic) process, and the term "purely deterministic nondeterministic process" sounds contradictory. We therefore prefer and shall use the term "singular."

Of course, in decomposition (13) either the singular or the regular component may be identically zero. For further discussion we will employ the following terminology for the process $X(t)$:

If $X(t) = X_s(t)$, i.e., if $X_r(t) = 0$ for all t, we will call $X(t)$ a *singular process*.

If $X(t) = X_r(t)$, that is, if $X_s(t) = 0$ for all t, we will call $X(t)$ a *purely regular process*.

We have added the adjective "purely" since the term "regular process" is often used to describe a process that is not singular, that is, such that its regular part is not identically zero. We prefer to call this kind of process simply a *nonsingular process*.†

Example 5.3.1

As a simple example, consider a stationary Gaussian process

$$X(t) = A \sin(\omega t + \Theta) + Y(t), \qquad -\infty < t < \infty, \tag{16}$$

which is the sum of a random sine wave (2) and a stationary Ornstein-Uhlenbeck process (3). More precisely, A is a Rayleigh random variable, Θ is a random variable uniformly distributed over $(0, 2\pi)$, and $Y(t)$, $-\infty < t < \infty$, is a zero mean stationary Gaussian process with correlation function

$$R_Y(\tau) = e^{-|\tau|}, \qquad -\infty < \tau < \infty.$$

Furthermore, the random variables A, Θ, and the process $Y(t)$, $-\infty < t < \infty$, are all mutually independent.

From what has been said in the introduction, we expect that the random sine wave will be the singular part and the Ornstein-Uhlenbeck process the regular part of the process $X(t)$. To see that this is indeed so, we first need to compute the conditional expectation

$$\hat{X}(t \mid t_0) = \mathrm{E}[X(t) \mid X(t'), -\infty < t' \le t_0] \tag{17}$$

for $-\infty < t_0 < t$. Now, obtaining an explicit expression for a conditional expectation given an infinite portion $X(t')$, $-\infty < t' \le t_0$, is usually a very difficult problem. However, in the present case, it turns out that

$$\mathrm{E}[X(t) \mid X(t'), -\infty < t' \le t_0] = \mathrm{E}[X(t) \mid A, \Theta, Y(t_0)], \tag{18}$$

the latter conditional expectation being easy to evaluate.

To understand why these two conditional expectations should be equal, one may argue as follows: the conditional expectation is determined by the amount of information that the conditioning provides about the distribution of $X(t)$. Knowing the past $X(t')$, $-\infty < t' \le t_0$, certainly cannot provide more information than if we knew the values of the random variables A and Θ and the past $Y(t')$, $-\infty < t' \le t_0$. However, the Ornstein-Uhlenbeck process $Y(t)$ is Markov, and hence knowing the entire past $Y(t')$, $-\infty < t' \le t_0$, tells us no more about the distribution of $Y(t)$ than just knowing the present value $Y(t_0)$. Since A, Θ, and the process $Y(t)$ are independent, the information about the distribution of $X(t)$ provided by the knowledge of the past $X(t')$, $-\infty < t' \le t_0$, can be no more than that provided by the knowledge of just A, Θ, and $Y(t_0)$. It remains to show that

† The alternate terminology also used in the literature is: deterministic for our singular, purely indeterministic for our purely regular, and indeterministic for our nonsingular.

the values of these three random variables can be determined from the past $X(t')$, $-\infty < t' \le t_0$. If this is so, then knowing $X(t')$, $-\infty < t' \le t_0$, provides the same amount of information about the distribution of $X(t)$ as knowing A, Θ, and $Y(t_0)$, and hence (18) must be true. How, then, can we determine the values of A, Θ, and $Y(t_0)$ from the portion $x(t')$, $\infty < t' \le t_0$, of a sample path of the process $X(t)$? After all, the sample path of the random sine wave will hardly be recognizable in the sample path $x(t')$, $-\infty < t' \le t_0$, being almost completely obliterated by the highly irregular sample path of the Ornstein-Uhlenbeck process. Nevertheless, there is a way to accomplish this. Take a fixed time instant $t_1 \le t_0$, and consider the average

$$\frac{1}{n}\sum_{k=0}^{n} x(t_1 - k\lambda),$$

where $\lambda = 2\pi/\omega$ is the wavelength of the random sine wave. The average is the sum of two averages

$$\frac{1}{n}\sum_{k=0}^{n} a \sin(\omega(t_1 - k\lambda) + \theta) + \frac{1}{n}\sum_{k=0}^{n} y(t_1 - k\lambda).$$

Since λ is the wavelength, $\sin(\omega(t_1 - k\lambda) + \theta) = \sin(\omega t_1 + \theta)$ for all $k = 0, 1, \ldots$, and hence the first average above is just $a \sin(\omega t_1 + \theta)$. On the other hand, the second average is the average of an ergodic stationary random sequence

$$Y(k) = Y(t_1 - k\lambda), \qquad k = 0, 1, \ldots,$$

and hence, as $n \to \infty$, it will converge to the mean of this sequence, that is, to zero. Thus we will have, as $n \to \infty$,

$$\frac{1}{n}\sum_{k=0}^{n} x(t_1 - \lambda k) \to a \sin(\omega t_1 + \theta),$$

and choosing a different time instant t_1, we can actually reconstruct the entire sample path $a \sin(\omega t' + \theta)$, $-\infty < t' \le t_0$, and hence determine the values a and θ of the random variables A and Θ. (Actually, we need only to choose two values of t_1 for this purpose.) Once a and θ are known, the value $y(t_0)$ of $Y(t_0)$ is found as

$$y(t_0) = x(t_0) - a \sin(\omega t_0 + \theta).$$

We have thus established that the equality (18) is true, and we can now calculate the conditional expectation (17) as

$$\begin{aligned}\hat{X}(t \mid t_0) &= \mathrm{E}[X(t) \mid A, \Theta, Y(t_0)] \\ &= A \sin(\omega t + \Theta) + \mathrm{E}[Y(t) \mid Y(t_0)] \\ &= A \sin(\omega t + \Theta) + Y(t_0)e^{-(t - t_0)}.\end{aligned}$$

Now, letting $t_0 \to -\infty$ and noticing that

$$\begin{aligned}\lim_{t_0 \to -\infty} \mathrm{E}[(Y(t_0)e^{-(t-t_0)})^2] &= \lim_{t_0 \to -\infty} e^{-2(t-t_0)}\mathrm{E}[Y^2(t_0)] \\ &= e^{-2t} \lim_{t_0 \to -\infty} e^{2t_0} = 0,\end{aligned}$$

that is, that

$$\underset{t_0 \to -\infty}{\text{l.i.m.}}\ Y(t_0)e^{-(t-t_0)} = 0,$$

we conclude that

$$\underset{t_0 \to -\infty}{\text{l.i.m.}}\ \hat{X}(t \mid t_0) = A \sin(\omega t + \Theta).$$

Thus, according to definition (9) the singular part $X_s(t)$ of the process (16) is indeed the random sine wave

$$X_s(t) = A \sin(\omega t + \Theta), \qquad -\infty < t < \infty,$$

and by (10) the regular part is the Ornstein-Uhlenbeck process

$$X_r(t) = Y(t), \qquad -\infty < t < \infty.$$

The reader may perhaps feel that we have spent a long time arriving at a result that was obvious from the very beginning. In a sense, this is true, but the main reason for doing so was to illustrate that definitions (9) and (10) really work.

The reader should also note that the procedure of extracting the singular component used here would apply even if the process $Y(t)$ in (16) were some other stationary ergodic process and if the random sine wave were replaced by a linear combination of several such sine waves or even by an infinite series

$$\sum_{k=0}^{\infty} (A_k \sin(\omega_k t + \Theta_k) + B_k \cos(\omega_k t + \Theta_k)). \qquad \blacktriangle$$

Let us now turn our attention to the regular part $X_r(t)$ of a nonsingular process $X(t)$. Consider a time instant t and a small time increment $dt > 0$. As usual, we denote by

$$dX_r(t) = X_r(t + dt) - X_r(t) \tag{19}$$

the forward increment of the process $X_r(t)$. Next call

$$dU(t) = dX_r(t) - \mathrm{E}[dX_r(t) \mid X(t'), -\infty < t' \leq t], \tag{20}$$

or using (19) and employing our previous notation,

$$dU(t) = X_r(t + dt) - \hat{X}_r(t + dt \mid t) = \tilde{X}_r(t + dt \mid t). \tag{21}$$

Thus $dU(t)$ is the error in predicting $X_r(t + dt)$ from the past $X_r(t')$, $-\infty < t' \le t$.

Since the process $X_r(t)$ is Gaussian, $dU(t)$ is a Gaussian random variable, and clearly

$$E[dU(t)] = 0. \tag{22}$$

Consider now two time instances $t_1 < t_2$ that are at least dt apart, that is,

$$t_1 + dt \le t_2. \tag{23}$$

Then by (20), $dU(t_1)$ is a linear functional of the past $X_r(t')$, $-\infty < t' \le t_1 + dt$ and hence by (23) also a linear functional of the past $X_r(t')$, $-\infty < t' \le t_2$. But then by the orthogonality principle

$$E[\tilde{X}_r(t_2 + dt \,|\, t_2)\, dU(t_1)] = 0$$

that is, by (21)

$$E[dU(t_2)\, dU(t_1)] = 0. \tag{24}$$

Since both $dU(t_1)$ and $dU(t_2)$ are zero mean Gaussian random variables, (24) implies that they must be independent.

Next the variance

$$E[(dU(t))^2] = E[(\tilde{X}_r^2(t + dt \,|\, t))],$$

must be, for fixed $dt > 0$, independent of the time t, since $X_r(t)$ is a stationary process. It cannot be zero, since it would then have to be zero for all t, and, being the mean square error of the prediction of $X_r(t + dt)$, this would mean that $X_r(t)$ is a singular process.

Thus if $d_\tau U(t) = U(t + \tau) - U(t)$, then the variance

$$E[(d_\tau U(t))^2] = v(\tau), \tag{25}$$

a positive function of τ for all $\tau > 0$, and, since $X(t)$ is an MS continuous process

$$\lim_{\tau \to 0} v(\tau) = v(0) = 0. \tag{26}$$

But for any $\tau_1 > 0$ and $\tau_2 > 0$, we have

$$d_{\tau_1 + \tau_2} U(t) = d_{\tau_1} U(t) + d_{\tau_2} U(t + \tau_1)$$

and since the increments are independent and zero mean with variance (25), this implies that for all $\tau_1 > 0$, $\tau_2 > 0$

$$v(\tau_1 + \tau_2) = v(\tau_1) + v(\tau_2). \tag{27}$$

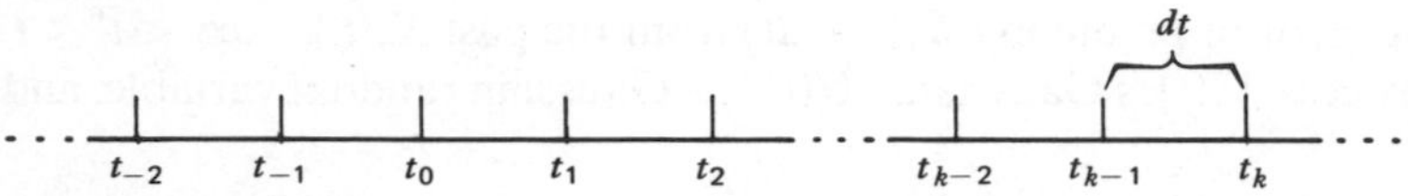

Figure 5.3.1

However,† the only function v satisfying (26) and (27) is a linear function $v(\tau) = b^2\tau$, $\tau > 0$, with $b^2 > 0$ a constant. Thus returning to our differential notation $dU(t) = U(t + dt) - U(t)$, we must have

$$\mathrm{E}[(dU(t))^2] = b^2\, dt. \tag{28}$$

Imagine now that the time domain $(-\infty, +\infty)$ has been partioned by a doubly infinite sequence of time instances $-\infty < \cdots t_{-2} < t_{-1} < t_0 < t_1 < t_2 < +\infty$, each t_k being dt apart from its neighbors, as in Figure 5.3.1.

Then the sequence of increments

$$dW(t_k) = \frac{1}{b^2}\, dU(t_k), \qquad k = \ldots, -1, 0, 1, \ldots, \tag{29}$$

will be a sequence of independent Gaussian random variables with (by (22) and (28))

$$\mathrm{E}[dW(t_k)] = 0 \qquad \text{and} \qquad \mathrm{E}[(dW(t_k))^2] = dt.$$

But this shows‡ that these increments should be increments of a standard Wiener process

$$W_0(t), \qquad -\infty < t < \infty.$$

Further, from (20) and (29)

$$dX_r(t_k) = b^2\, dW_0(t_k) + \mathrm{E}[dX_r(t_k)\,|\,X(t'), -\infty < t' \le t_k] \tag{30}$$

and since $X_r(t)$ is a zero mean Gaussian process, the conditional expectation in (30) is a homogeneous linear functional of the past and can therefore be approximated by a linear combination of the past increments $dX_r(t_{k-1}), dX_r(t_{k-2}), \ldots$.

Thus, for small dt, we can write

$$dX_r(t_k) \doteq \text{linear combination of } dW_0(t_k), dX_r(t_{k-1}), dX_r(t_{k-2}), \ldots \tag{31}$$

and similarly

$$dX_r(t_{k-1}) \doteq \text{linear combination of } dW_0(t_{k-1}), dX_r(t_{k-2}), dX_r(t_{k-3}), \ldots \tag{32}$$

† See Appendix Three.

‡ Recall Theorem 4.3.1.

However, the linear combination in (32) can be substituted for $dX_r(t_{k-1})$ into the linear combination in (31), thus obtaining

$$dX_r(t_k) \doteq \text{linear combination of } dW_0(t_k), dW_0(t_{k-1}), dX_r(t_{k-2}), \ldots \quad (33)$$

But (31) holds for any k, in particular

$$dX_r(t_{k-2}) \doteq \text{linear combination of } dW_0(t_{k-2}), dX_r(t_{k-3}), \ldots$$

so that by again substituting for $dX_r(t_{k-2})$ into the linear combination in (33), $dX_r(t_{k-2})$ is replaced by $dW_0(t_{k-2})$. Clearly this procedure can be continued indefinitely, which suggests that each increment $dX_r(t_k)$ can be approximated by a linear combination of the increments $dW(t_k), dW_0(t_{k-1}), dW_0(t_{k-2}), \ldots$.

Now, if all the increments

$$\ldots, dX_r(t_{k-2}), dX_r(t_{k-1}), dX(t_k),$$

can be approximated by linear combinations of the increments

$$\ldots, dW_0(t_{k-2}), dW_0(t_{k-1}), dW_0(t_k),$$

then we should be able to approximate the random variable $X_r(t_k)$ itself by a linear combination of increments

$$\ldots, dW_0(t_{k-2}), dW_0(t_{k-1}), dW_0(t_k). \quad (34)$$

The fact that $X_r(t)$ is the regular part guarantees, loosely speaking, that in the infinite linear combination, no residual of the process $X_r(t)$ will be left, for if it were, it would have to be infinitely remote in the past from the time t_k, and yet, being included in the linear combination for $X_r(t_k)$, it would thus provide some information about the value of $X_r(t_k)$. But for a regular part, no prediction of $X_r(t_k)$ could be made from the infinitely remote past of $X(t)$ and, a fortiori, of $X_r(t)$.

Now let the time increment dt shrink to zero while keeping $t_k = t$ fixed, and the approximation of $X_r(t_k) = X_r(t)$ by the linear combination of the increments (34) should become more and more accurate, which suggests that we should be able to write $X_r(t)$ as an integral

$$X_r(t) = \int_{-\infty}^{t} h(t, \tau)\, dW_0(\tau), \quad (35)$$

that is, as a Wiener integral. However, we know that $X_r(t)$ is a stationary process, and since we expect (35) to hold for any t, the function h in the integrand may possibly depend only on the time difference $t - \tau$; that is, it must be of the form

$$h(t - \tau).$$

It is hoped that the above heuristic discussion gives enough credibility to the theorem that we are about to present. We would only like to mention in passing that a rigorous proof of the forthcoming theorem can be based on the geometric interpretation of Gaussian conditional expectations as projections into a subspace of an appropriate linear space. The representation (35) is then obtained as an expansion of $X_r(t)$ with respect to the orthonormal basis formed by the increments $dW_0(t_k)$.

THEOREM 5.3.2. *Let* X(t), $-\infty < t < \infty$, *be a stationary, zero mean sample-continuous Gaussian process. Then the process is purely regular* $X(t) = X_r(t)$, $-\infty < t < \infty$, *if and only if it can be represented as a Wiener integral*

$$X_r(t) = \int_{-\infty}^{t} h(t-\tau)\, dW_0(\tau), \qquad -\infty < t < \infty, \tag{36}$$

where h *is a square-integrable function*

$$\int_0^{\infty} h^2(\tau)\, d\tau < \infty.$$

Example 5.3.2

Consider again a stationary Ornstein-Uhlenbeck process $X(t)$, $-\infty < t < \infty$, with the correlation function

$$R_X(\tau) = \sigma^2 e^{-\alpha|\tau|}.$$

In Example 5.3.1 we have seen that this is a purely regular process, and indeed we recall from Example 4.3.4 that it can be expressed as the integral

$$X(t) = \sigma\sqrt{2\alpha} \int_{-\infty}^{t} e^{-\alpha|t-\tau|}\, dW_0(\tau),$$

with

$$h(\tau) = \sigma\sqrt{2\alpha}\, e^{-\alpha\tau}, \qquad \tau \geq 0$$

clearly being a square-integrable function, since $\alpha > 0$.

As another simple example, consider the square-integrable function

$$h(\tau) = \begin{cases} 1 & \text{if} \quad 0 \leq \tau \leq \Delta, \\ 0 & \text{otherwise}, \end{cases}$$

where $\Delta > 0$ is a constant. Then the stationary process

$$X(t) = \int_{-\infty}^{t} h(t-\tau)\, dW_0(t) = W_0(t) - W_0(t-\Delta),$$

$-\infty < t < \infty$, is obviously purely regular, since the increment $W_0(t) - W_0(t-\Delta)$ of the standard Wiener process is independent of its entire past up to time $t - \Delta$. ▲

Since (36) is a Wiener integral, let us use white Gaussian noise $\dot{W}_0(t)$ and also write

$$X_r(t) = \int_{-\infty}^{t} h(t - \tau)\dot{W}_0(\tau)\, d\tau. \tag{37}$$

But since the increments $dW_0(t)$ of the Wiener process were defined by

$$dW_0(t) = \frac{1}{b^2}(dX_r(t) - \mathrm{E}[dX_r(t) \mid X_r(t'), -\infty < t' \leq t]) \tag{38}$$

with the right-hand side being a linear functional of the past $X_r(t')$, $-\infty < t' \leq t + dt$, it is at least plausible that the white Gaussian noise $\dot{W}_0(t)$, regarded as a formal derivative of the Wiener process $W_0(t)$, should be a linear functional of the past $X_r(t')$, $-\infty < t' \leq t$. More specifically, we expect to be able to write an inverse relation to (37), namely

$$\dot{W}_0(t) = \int_{-\infty}^{t} g(t - \tau)X_r(\tau)\, d\tau, \qquad -\infty < t < \infty \tag{39}$$

where g is again a square-integrable function

$$\int_0^{\infty} g^2(\tau)\, d\tau < \infty.$$

Remark: Since the linear transformation (39) results in white Gaussian noise, which is a generalized process, the impulse-response function g will typically involve the Dirac delta function and/or its derivatives. Я

Now, (37) and (39) mean that the process $X_r(t)$ and white Gaussian noise $\dot{W}_0(t)$ can be obtained from each other by passage through a causal linear time-invariant filter with impulse-response functions

$$h(\tau), \qquad \tau \geq 0, \qquad h(\tau) = 0 \qquad \text{for } \tau < 0,$$

and

$$g(\tau), \qquad \tau \geq 0, \qquad g(\tau) = 0 \qquad \text{for } \tau < 0,$$

respectively, as pictured in Figure 5.3.2. We refer to this situation by saying that the process $X_r(t)$ and the white Gaussian noise $\dot{W}_0(t)$ are *causally linearly equivalent*.

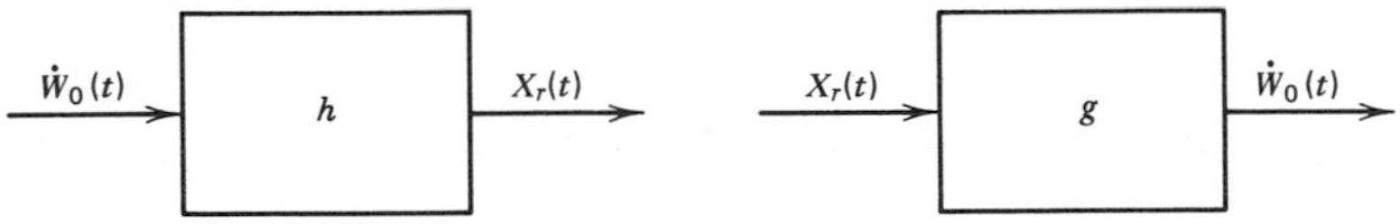

Figure 5.3.2

However, the reader should be cautioned against jumping to the conclusion that whenever a process $X_r(t)$, $-\infty < t < \infty$, can be expressed as an integral

$$X_r(t) = \int_{-\infty}^{t} h(t - \tau)\dot{W}_0(\tau)\, d\tau \tag{40}$$

with a square-integrable $h(\tau)$, then the causal inverse filter g, giving

$$\dot{W}_0(t) = \int_{-\infty}^{t} g(t - \tau)X_r(\tau)\, d\tau \tag{41}$$

can simply be obtained by inverting the filter h, that is, by solving equation (40) for $\dot{W}_0(t)$. This is not true, for clearly not every causal linear time-invariant filter has a causal inverse. Instead, the causal linear equivalence of the process $X_r(t)$ and white Gaussian noise $\dot{W}_0(t)$ means that *there always exists* such a pair of causal filters for which equations (40) and (41) form mutually inverse relations.

In other words, the square-integrable function h whose existence is assured by Theorem 5.3.2 is not unique. However, in the multitude of such functions representing a given process $X_r(t)$ as a Wiener integral (36), *there is always one that does have a causal inverse.*

If such a function is used in the representation (36) or (40), this representation is referred to as the *canonical representation* (or filter), and the inverse representation (41) is then called a *whitening filter.* The following example illustrates this point.

Example 5.3.3

Consider a zero mean stationary Gaussian process defined as a causal linear time-invariant transformation

$$X(t) = \int_{-\infty}^{t} e^{-\alpha(t-\tau)}(\cos \omega_0(t - \tau) + \gamma \sin \omega_0(t - \tau))\dot{W}_0(\tau)\, d\tau, \tag{42}$$

$-\infty < t < \infty$, where $\alpha > 0$, $\omega_0 > 0$, and γ are constants. Such processes are quite common in connection with resonance phenomena since the impulse-response function of the type

$$h(\tau) = e^{-\alpha\tau}(\cos \omega_0 \tau + \gamma \sin \omega_0 \tau), \qquad \tau \geq 0 \tag{43}$$

is typical of responses exhibiting exponentially decaying oscillations. Since

$$h^2(\tau) \leq e^{-2\alpha\tau}(1 + \gamma)^2$$

this function is trivially square-integrable, and hence by Theorem 5.3.2, $X(t)$, $-\infty < t < \infty$, is a purely regular process.

Let us now try to find the inverse transformation

$$\dot{W}_0(t) = \int_{-\infty}^{t} g(t - \tau)X(\tau)\, d\tau, \qquad -\infty < t < \infty,$$

by, at least formally, solving the integral equation

$$x(t) = \int_{-\infty}^{t} e^{-\alpha(t-\tau)}(\cos \omega_0(t-\tau) + \gamma \sin \omega_0(t-\tau))u(\tau)\,d\tau$$

for the unknown function $u(t)$. Taking the time derivative, we get by Leibnitz's rule

$$\dot{x}(t) = u(t) - \alpha x(t) + \int_{-\infty}^{t} e^{-\alpha(t-\tau)}(-\omega_0 \sin \omega_0(t-\tau) + \gamma\omega_0 \cos \omega_0(t-\tau))u(\tau)\,d\tau, \tag{44}$$

and differentiating this one more time

$$\ddot{x}(t) = \dot{u}(t) - \alpha\dot{x}(t) + \gamma\omega_0 u(t) - \alpha \int_{-\infty}^{t} e^{-\alpha(t-\tau)}(-\omega_0 \sin \omega_0(t-\tau) + \gamma\omega_0 \cos \omega_0(t-\tau))u(t)\,d\tau - \omega_0^2 x(t). \tag{45}$$

From (44), the integral in (46) equals $\dot{x}(t) - u(t) + \alpha x(t)$, and upon substitution in (46), we have, after a slight rearrangement,

$$\ddot{x}(t) + 2\alpha\dot{x}(t) + (\alpha^2 + \omega_0^2)x(t) = (\alpha + \gamma\omega_0)u(t) + \dot{u}(t).$$

Temporarily calling the left-hand side of this equation

$$y(t) = \ddot{x}(t) + 2\alpha\dot{x}(t) + (\alpha^2 + \omega^2)x(t) \tag{46}$$

we have for the unknown function $u(t)$ a first order linear differential equation

$$\dot{u}(t) + (\alpha + \gamma\omega_0)u(t) = y(t). \tag{47}$$

Now if $\alpha + \gamma\omega_0 > 0$, then

$$u(t) = \int_{-\infty}^{t} e^{-(\alpha+\gamma\omega_0)(t-\tau)}y(\tau)\,d\tau, \tag{48}$$

which clearly expresses $u(t)$ as a causal linear time-invariant transformation of the function $y(t)$ and hence, because of (46), also of $x(t)$. Actually, if we substitute from (46) into (48) and integrate by parts, we can write $u(t)$ as

$$u(t) = \dot{x}(t) + (\alpha^2 + \omega_0^2 + 3\alpha + \gamma\omega_0)x(t) + (\alpha + \gamma\omega_0)(3\alpha + \gamma\omega_0)\int_{-\infty}^{t} e^{-(\alpha+\gamma\omega_0)(t-\tau)}x(\tau)\,d\tau,$$

or, employing the Dirac delta function, in the form

$$u(t) = \int_{-\infty}^{t} g(t-\tau)x(\tau)\,d\tau,$$

where

$$g(\tau) = \dot{\delta}(\tau) + (\alpha^2 + \omega_0^2 + 3\alpha + \gamma\omega_0)\,\delta(\tau) + (\alpha + \gamma\omega_0)(3\alpha + \gamma\omega_0)e^{-(\alpha+\gamma\omega_0)\tau}, \qquad \tau \geq 0. \tag{49}$$

Thus, for the case $\alpha + \gamma\omega_0 > 0$, the impulse-response functions $h(\tau)$ and $g(\tau)$, given by (43) and (49), are a causally invertible pair, and hence the representation (42) is canonical and $g(\tau)$ is the impulse response of the corresponding whitening filter.

Suppose now that the constant γ is such that $\alpha + \gamma\omega_0 < 0$. Then in order to obtain a meaningful solution of equation (47), we would have to reverse the time flow and write instead of (48)

$$u(t) = \int_t^{\infty} e^{(\alpha+\gamma\omega_0)(t-\tau)}y(-\tau)\,d\tau.$$

But this would lead to a filter that would yield a present value of the output only if fed the entire future of the input rather than the past, an anticausal filter if you wish. Thus, for $\alpha + \gamma\omega_0 < 0$ the impulse-response function (43) does not have a causal inverse, and the representation (42), although still meaningful, is no longer canonical.

However, the process $X(t)$, $-\infty < t < \infty$, is a zero mean Gaussian process, and thus its entire probability law is completely specified by its correlation function. The latter can be calculated from (42), for instance, by using the formula

$$R_X(t_1, t_2) = \int_{-\infty}^{\min(t_1, t_2)} h(t_1 - \tau)h(t_2 - \tau)\,d\tau$$

with $h(\tau)$ as in (43). After a simple but somewhat tedious calculation, we find that it equals

$$R_X(\tau) = \frac{1}{4}e^{-\alpha|\tau|}(A\cos\omega_0\tau - B\sin\omega_0|\tau|), \qquad -\infty < \tau < \infty,$$

where

$$A = \frac{1}{\alpha}\left[1 + \frac{(\alpha + \gamma\omega_0)^2}{\alpha^2 + \omega_0^2}\right], \qquad B = \frac{1}{\omega_0}\left[1 - \frac{(\alpha + \gamma\omega_0)^2}{\alpha^2 + \omega_0^2}\right].$$

But this shows that the correlation function depends only on the square of the quantity $\alpha + \gamma\omega_0$, which means that if $\alpha + \gamma\omega_0 < 0$, then by changing the constant γ to

$$\gamma' = -\left(\frac{2\alpha}{\omega_0} + \gamma\right),$$

we will have

$$\alpha + \gamma'\omega_0 = -(\alpha + \gamma\omega_0) > 0,$$

while the correlation function $R_X(\tau)$ and hence the process $X(t)$, $-\infty < t < \infty$, remain the same. Thus, if $\alpha + \gamma\omega_0 < 0$, the canonical representation of the process $X(t)$ is

$$X(t) = \int_{-\infty}^{t} e^{-\alpha(t-\tau)}\left(\cos\,\omega_0(t-\tau) - \left(\frac{2\alpha}{\omega_0} + \gamma\right)\sin\,\omega_0(t-\tau)\right)\dot{W}_0(\tau)\,d\tau,$$

$-\infty < t < \infty$, with the corresponding whitening filter given by the impulse-response function

$$g(\tau) = \dot{\delta}(\tau) + (\alpha^2 + \omega_0^2 + \alpha - \gamma\omega_0)\,\delta(\tau)$$
$$- (\alpha + \gamma\omega_0)(\alpha - \gamma\omega_0)e^{(\alpha+\gamma\omega_0)\tau}, \qquad \tau \geq 0. \qquad \blacktriangle$$

There is yet another aspect of the canonical representation, which can be deduced from equation (38). First, since we know from Theorem 5.3.1 that the regular and singular parts of the process $X(t) = X_s(t) + X_r(t)$ are independent, knowing the singular part $X_s(t)$ can be of no use in predicting the regular part $X_r(t)$. Hence, the conditional expectation in (38) can be replaced by a conditional expectation given $X(t')$, $-\infty < t' \leq t$,

$$\mathrm{E}[dX_r(t)\,|\,X_r(t'), -\infty < t' \leq t] = \mathrm{E}[dX_r(t)\,|\,X(t'), -\infty < t' \leq t].$$

Next, substituting for $dX_r(t) = dX(t) - dX_s(t)$ and using the fact that by (14)

$$\mathrm{E}[dX_s(t)\,|\,X(t'), -\infty < t' \leq t] = dX_s(t)$$

we can rewrite (38) in the form

$$dW_0(t) = \frac{1}{b^2}\,(dX(t) - \mathrm{E}[dX(t)\,|\,X(t'), -\infty < t' \leq t]).$$

This implies that the increment $dW_0(t)$ can be interpreted as the new information, or *random innovation*, entering into the process $X(t)$ during the time interval $(t, t + dt)$. Thus the process $X(t)$ can be thought of as being generated by a random element in the infinitely remote past projected deterministically to the present time t (the singular part $X_s(t)$) and by a continuous stationary stream of random innovations $\dot{W}_0(t')$, $-\infty < t' \leq t$, combined in a linear fashion (the regular part $X_r(t)$). For this reason, white Gaussian noise $\dot{W}_0(t)$ in equation (40) is frequently referred to as the *innovation process* of a (nonsingular) process $X(t)$, as shown in the block diagram of Figure 5.3.3.

The whitening filter is a causal time-invariant linear filter from equation (41) and is shown in Figure 5.3.4.

The concept of the innovation process or, more specifically, the fact that the regular part $X_r(t)$ can be transformed into white Gaussian noise $\dot{W}_0(t)$ in a causal manner without any loss of information is of great theoretical and practical importance.

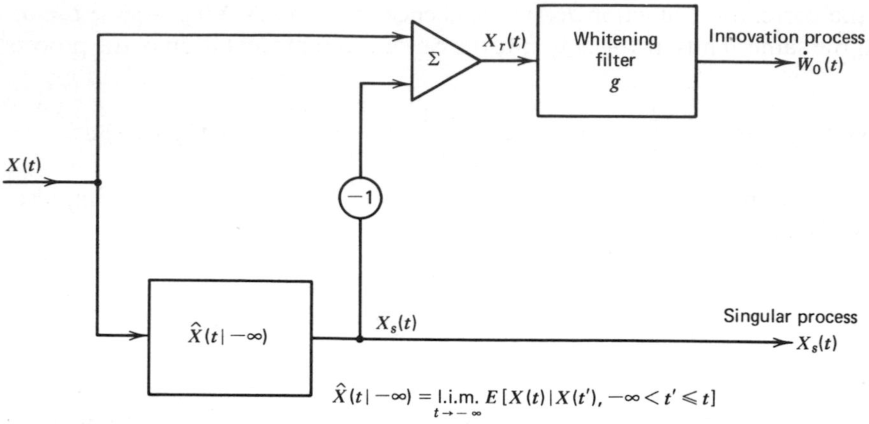

Figure 5.3.3

As an example suppose we are allowed to observe a stationary Gaussian process $X(t)$, $-\infty < t < \infty$, and we wish to optimally predict its value $X(t + \Delta t)$, $\Delta t > 0$ time units ahead from the observations $X(t')$, $-\infty < t' \leq t$, up to the present time t. More specifically, we wish to design a causal filter that would accept the process $X(t)$ as its input and produce at the output a process $\hat{X}(t + \Delta t\,|\,t)$ such that the mean square error

$$\varepsilon^2(t + \Delta t\,|\,t) = \mathrm{E}[(X(t + \Delta t) - \hat{X}(t + \Delta t\,|\,t))^2]$$

is minimized for all $-\infty < t < \infty$. Now we know that such a best predictor is provided by the conditional expectation

$$\hat{X}(t + \Delta t\,|\,t) = \mathrm{E}[X(t + \Delta t)\,|\,X(t), -\infty < t' \leq t] \tag{50}$$

and since the process $X(t)$ is stationary and Gaussian, it is natural to expect that the best predictor can be realized by a time-invariant causal linear filter, that is, that

$$\hat{X}(t + \Delta t\,|\,t) = \int_{-\infty}^{t} p_{\Delta t}(t - \tau)X(\tau)\,d\tau, \qquad -\infty < t < \infty. \tag{51}$$

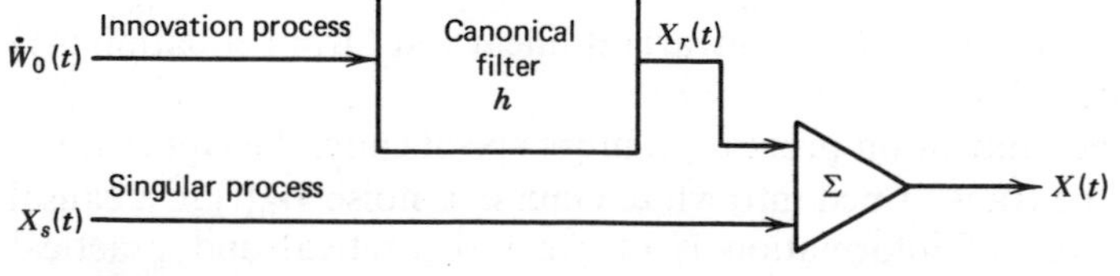

Figure 5.3.4

The problem is to determine the impulse-response function $p_{\Delta t}$. Now, by the orthogonality principle, we must have

$$\mathrm{E}[X(t+\Delta t)X(t')] = \mathrm{E}[\hat{X}(t+\Delta t \mid t)X(t')] \tag{52}$$

for all $-\infty < t' \le t$, which, upon substituting (51) into the right-hand side of (52) and interchanging the integration and expectation operators, yields the equation

$$R_X(t+\Delta t - t') = \int_{-\infty}^{t} p_{\Delta t}(t-\tau)R_X(\tau - t')\,d\tau, \qquad -\infty < t' \le t,$$

where $R_X(\tau)$ is the correlation function of the process $X(t)$. Minor modification (substituting $t-\tau = u$ and then renaming $t - t' = \tau$) gives a more pleasing form

$$R_X(\tau + \Delta t) = \int_{0}^{\infty} p_{\Delta t}(u)R_X(\tau - u)\,du, \qquad 0 \le \tau < \infty, \tag{53}$$

but we still have an integral equation (a form of the so called Wiener-Hopf equation) for the unknown function $p_{\Delta t}$, which is by no means easy to solve.

Let us now try to solve the problem using the innovation process $\dot{W}_0(t)$. Still assuming that $X(t)$ is a purely regular sample-continuous zero mean process, we have from (36) with $X_r(t) = X(t)$,

$$\begin{aligned} X(t+\Delta t) &= \int_{-\infty}^{t+\Delta t} h(t+\Delta t - \tau)\dot{W}_0(\tau)\,d\tau \\ &= \int_{-\infty}^{t} h(t+\Delta t-\tau)\dot{W}_0(\tau)\,d\tau + \int_{t}^{t+\Delta t} h(t+\Delta t-\tau)\dot{W}_0(\tau)\,d\tau \end{aligned} \tag{54}$$

Since the innovation process $\dot{W}_0(t)$ and the process $X(t)$ are causally linearly equivalent, the past $X(t')$, $-\infty < t' \le t$, contains the same information as the past $\dot{W}_0(t')$, $-\infty < t' \le t$. Thus the conditional expectation (50) can also be written as

$$\hat{X}(t+\Delta t \mid t) = \mathrm{E}[X(t+\Delta t) \mid \dot{W}_0(t'), -\infty < t' \le t].$$

Taking the indicated conditional expectation on both sides of (54), we obtain immediately

$$\hat{X}(t+\Delta t \mid t) = \int_{-\infty}^{t} h(t+\Delta t - \tau)\dot{W}_0(\tau)\,d\tau, \qquad -\infty < t < \infty, \tag{55}$$

since for $\tau > t$ the fact that, having a delta function as its correlation function, the white Gaussian noise behaves as if $\dot{W}_0(\tau)$ were independent of $\dot{W}_0(t')$, $-\infty < t' \le t$, implies

$$\mathrm{E}[\dot{W}_0(\tau) \mid \dot{W}_0(t'), -\infty < t' \le t] = \mathrm{E}[\dot{W}_0(\tau)] = 0.$$

Note that (55) is a result that intuitively makes a lot of sense. Since knowing the past $X(t')$, $-\infty < t' \le t$, is equivalent to knowing the past $\dot{W}_0(t')$, $-\infty < t' \le t$, the first integral on the right-hand side of (54) is known and represents the response of the filter h at the future time $t + \Delta t$ to the stream of white noise impulses having arrived up to the present time t. On the other hand, the second integral is the response to impulses that are yet to arrive, and since those are independent of the past and present the best we can do is to assume that they will average to zero.

Now equation (55), together with expression (41) for the whitening filter

$$\dot{W}_0(t) = \int_{-\infty}^{t} g(t - \tau)X(\tau)\, d\tau,$$

provides a complete solution to the prediction problem.

The best predictor is simply a whitening filter g followed by a time-invariant causal linear filter with impulse-response function

$$k_{\Delta t}(\tau) = \begin{cases} h(\tau + \Delta t) & \text{if} \quad \tau \geq 0 \\ 0 & \text{if} \quad \tau < 0. \end{cases}$$

These are shown in Figure 5.3.5. The optimal predictor is then implemented as in Figure 5.3.6, that is, its overall impulse-response function $p_{\Delta t}$ is the convolution

$$p_{\Delta t}(\tau) = \int_{-\infty}^{\infty} g(\tau - \tau')k_{\Delta t}(\tau')\, d\tau'.$$

Of course, to construct the optimal predictor we need to know the impulse-response functions h and g. In some cases, particularly if the process $X(t)$ is defined by means of a differential equation driven by white Gaussian noise, these functions are relatively easy to obtain.

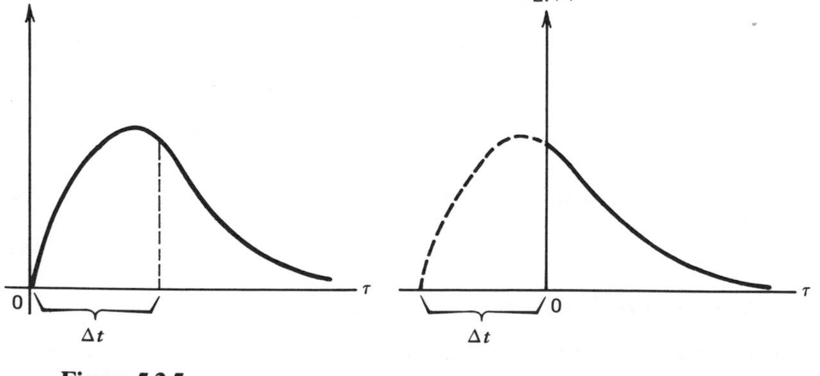

Figure 5.3.5

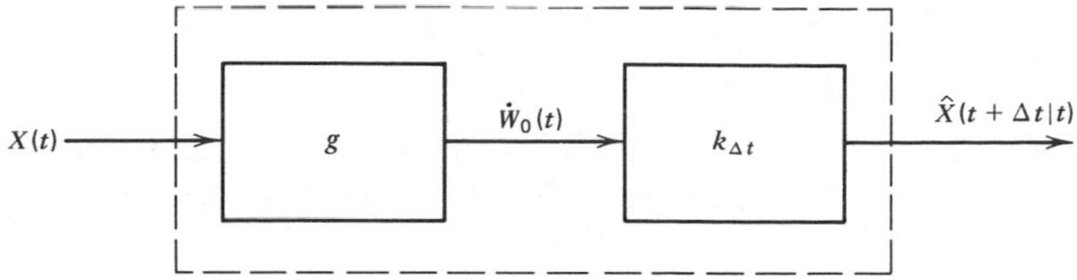

Figure 5.3.6

If the process $X(t)$ is defined by its correlation function $R_X(\tau)$, then by (36) we must have

$$R_X(\tau) = \int_0^\infty h(u)h(u+\tau)\,du, \qquad 0 \le \tau < \infty, \tag{56}$$

which is again an integral equation for the unknown function h. However, unlike equation (53), this one is considerably easier to solve, for instance, by means of Laplace or Fourier transforms. The impulse-response function g of the whitening filter, being the inverse of a canonical filter h, is obtained from the latter.

Example 5.3.4

Let us again take for the process $X(t)$, $-\infty < t < \infty$, the zero mean stationary Ornstein-Uhlenbeck process with correlation function

$$R_X(\tau) = \sigma^2 e^{-\alpha|\tau|}, \qquad -\infty < \tau < \infty.$$

Of course, in this case we already know that the best predictor should be

$$\hat{X}(t+\Delta t\,|\,t) = e^{-\alpha\,\Delta t} X(t), \tag{57}$$

since the Gauss-Markov character of an Ornstein-Uhlenbeck process allows us to directly evaluate the conditional expectation on the right-hand side of equation (50). Nevertheless, it may be instructive to see how the method developed in this section works in this simple case.

First, we need to represent the process as a Wiener integral. We have done this several times, and we know that

$$X(t) = \sigma\sqrt{2\alpha}\int_{-\infty}^{t} e^{-\alpha(t-\tau)}\dot{W}_0(\tau)\,d\tau. \tag{58}$$

We also know that the Ornstein-Uhlenbeck process satisfies a first order linear differential equation with white Gaussian noise as a forcing function. Indeed, if we differentiate both sides of (58), we get by Leibnitz's rule,

$$\dot{X}(t) = \sigma\sqrt{2\alpha}\,\dot{W}_0(t) - \alpha\sigma\sqrt{2\alpha}\int_{-\infty}^{t} e^{-\alpha(t-\tau)}\dot{W}_0(\tau)\,d\tau,$$

or, in other words,

$$\dot{X}(t) + \alpha X(t) = \sigma\sqrt{2\alpha}\,\dot{W}_0(t).$$

Thus, white Gaussian noise is obtained as a linear transformation of the process $X(t)$, so that representation (58) is canonical, and the causally invertible pair of impulse response functions h and g are, in this case,

$$h(\tau) = \begin{cases} \sigma\sqrt{2\alpha}\, e^{-\alpha\tau} & \text{if} \quad \tau > 0, \\ 0 & \text{if} \quad \tau \le 0, \end{cases}$$

$$g(\tau) = \frac{1}{\sigma\sqrt{2\alpha}}(\dot{\delta}(\tau) + \alpha\,\delta(\tau)), \qquad -\infty < \tau < \infty. \tag{59}$$

This means that the best predictor is obtained by first transforming $X(t)$ into its innovation process

$$\dot{W}_0(t) = \frac{1}{\sigma\sqrt{2\alpha}}(\dot{X}(t) + \alpha X(t))$$

and then passing this innovation process through a filter with impulse-response function

$$k_{\Delta t}(\tau) = \begin{cases} \sigma\sqrt{2\alpha}\, e^{-\alpha(\tau + \Delta t)} & \text{if} \quad \tau > 0, \\ 0 & \text{if} \quad \tau \le 0. \end{cases} \tag{60}$$

The result is

$$\hat{X}(T + \Delta t \mid t) = \int_{-\infty}^{t} k_{\Delta t}(t - \tau)\dot{W}_0(\tau)\, d\tau \quad (61)$$

$$= \int_{-\infty}^{t} \sigma\sqrt{2\alpha}\, e^{-\alpha(t-\tau+\Delta t)} \frac{1}{\sigma\sqrt{2\alpha}}(\dot{X}(\tau) + \alpha X(\tau))\, d\tau$$

$$= e^{-\alpha\,\Delta t} \int_{-\infty}^{t} e^{-\alpha(t-\tau)}(\dot{X}(\tau) + \alpha X(\tau))\, d\tau.$$

However, integrating by parts, we find

$$\int_{-\infty}^{t} e^{-\alpha(t-\tau)}\dot{X}(\tau)\, d\tau = e^{-\alpha(t-\tau)}X(\tau)\Big|_{\tau=-\infty}^{\tau=t}$$

$$- \int_{-\infty}^{t} \alpha e^{-\alpha(t-\tau)}X(\tau)\, d\tau = X(t) - \int_{-\infty}^{t} e^{-\alpha(t-\tau)}\alpha X(\tau)\, d\tau$$

so that the integral on the right-hand side of (61) equals $X(t)$, and we get equation (57).

The same result can be obtained by first evaluating the overall impulse-response function of the predictor,

$$p_{\Delta t}(\tau) = \int_{-\infty}^{\infty} g(\tau - \tau')k_{\Delta t}(\tau')\,d\tau'$$

which is a good exercise in manipulating the delta function. Substitution from (59) and (60) yields

$$p_{\Delta t}(\tau) = \int_{0+}^{\infty} (\dot{\delta}(\tau - \tau') + \alpha\,\delta(\tau - \tau'))e^{-\alpha(\tau' + \Delta t)}\,d\tau'$$

$$= \sigma\sqrt{2\alpha}\,e^{-\alpha\,\Delta t}\left[\int_{0+}^{\infty} \dot{\delta}(\tau - \tau')h(\tau')\,d\tau' + \alpha\int_{0+}^{\infty} \delta(\tau - \tau')h(\tau')\,d\tau'\right]. \quad (62)$$

For $\tau < 0$, both integrals in brackets are zero, and so $p_{\Delta t}(\tau) = 0$, as it should. For $\tau > 0$, $h(\tau) = \sigma\sqrt{2\alpha}\,e^{-\alpha\tau}$ is a continuous function and hence

$$\alpha\int_{0+}^{\infty} \delta(\tau - \tau')e^{-\alpha\tau'}\,d\tau' = \alpha e^{-\alpha\tau}$$

and

$$\int_{0+}^{\infty} \dot{\delta}(\tau - \tau')e^{-\alpha\tau'}\,d\tau' = \frac{d}{d\tau}e^{-\alpha\tau} = -\alpha e^{-\alpha\tau}$$

so that the expression in brackets, and hence also $p_{\Delta t}(\tau)$ is again zero. This leaves the case $\tau = 0$. Then

$$\int_{0+}^{\infty} \delta(0 - \tau')h(\tau')\,d\tau' = \lim_{\tau' \to 0-} h(\tau') = 0$$

while

$$\int_{0+}^{\infty} \dot{\delta}(0 - \tau')h(\tau')\,d\tau' = \frac{d}{d\tau'}h(\tau')\bigg|_{\tau = 0} = \delta(0)\frac{1}{\sigma\sqrt{2\alpha}}$$

since $h(\tau) = 0$ for $\tau < 0$ and has an upward jump of height $1/\sigma\sqrt{2\alpha}$ at $\tau = 0$. Thus, for $\tau = 0$, the expression in brackets in (62) equals $1/\sigma\sqrt{2\alpha}\,\delta(0)$, and so we can write for all $-\infty < \tau < \infty$

$$p_{\Delta t}(\tau) = e^{-\alpha\,\Delta t}\,\delta(\tau).$$

Then the best predictor is given by

$$\hat{X}(t + \Delta t\,|\,t) = \int_{-\infty}^{\infty} p_{\Delta t}(t - \tau)X(\tau)\,d\tau$$

$$= e^{-\alpha\,\Delta t}\int_{-\infty}^{\infty} \delta(t - \tau)X(\tau)\,d\tau = e^{-\alpha\,\Delta t}X(t)$$

as before. ▲

SPECTRA OF SINGULAR AND REGULAR PROCESSES. In the preceding discussion, we have found that any MS continuous zero mean Gaussian stationary process

$$X(t), \qquad -\infty < t < \infty$$

can be decomposed into its singular and regular parts $X_s(t)$ and $X_r(t)$. We have also seen that if the process $X(t)$ is regular, then its regular part $X_r(t)$ is causally linearly equivalent to its innovation process, which is white Gaussian noise,

$$X_r(t) = \int_{-\infty}^{t} h(t-\tau)\dot{W}_0(\tau)\,d\tau.$$

Since the probability law of a zero mean Gaussian process is completely determined by its correlation function, which for MS-continuous stationary processes is, in turn, uniquely determined by the power spectral distribution function $S_X(\nu)$, it is natural to ask whether and how the above decomposition is reflected in the function $S_X(\nu)$.

To get a feeling for such relationships, suppose that the power spectral distribution function is discrete with only a finite number of spectral lines at frequencies ν_n, $n = 0, 1, \ldots, m$, and suppose that the process $X(t)$ is real. Then, according to (5.2.8) we can write its spectral representation as

$$X(t) = 2\left[\sum_{n=0}^{m} Z_1(\nu_n)\cos\omega_n t - \sum_{n=0}^{m} Z_2(\nu_n)\sin\omega_n t\right] \tag{63}$$

for all $-\infty < t < \infty$. But from this we can conclude immediately that $X(t)$ must be a singular process, since each of its sample paths is forever determined by the values $Z_1(\nu_n) = z_1(\nu_n)$ and $Z_2(\nu_n) = z_2(\nu_n)$ of the $2m + 1$ random amplitudes.

In other words, we can predict the sample path of such a process indefinitely into the future with zero error once we have determined the values of the random amplitudes and the frequencies of the harmonic waveforms in (63). To do this, we only need to observe the sample path for an arbitrarily small period of time; in fact, we only need to observe it at a finite number of time instances within that period of time.

Using a little imagination, it is not hard to see that this can be done, even if the process has an infinite number of discrete spectral lines. Although it is true that in such a case the right-hand side of (63) is an infinite series, this series is convergent in MS† and so by observing the value of a sample path $x(t)$ at an increasing number of time instances all within a fixed period of time, we should be able to determine the random amplitudes and frequencies ν_n with increasing

† Recall that $\sum_{n=-\infty}^{\infty} \mathrm{E}[(dZ(\nu_n))^2] = R_X(0)$.

accuracy. Thus, an errorless prediction indefinitely into the future should again be possible.

We summarize this research as a theorem.

THEOREM 5.3.3. *If the process* X(t), $-\infty < t < \infty$, *is almost periodic, that is, if its power spectral distribution function is discrete, then the process is singular.*

In view of the above discussion, it is tempting to conjecture that processes with continuous power spectral distribution function should be purely regular. Surprisingly, this is not true. For example, a process with correlation function

$$R_X(\tau) = e^{-\tau^2/2}, \qquad -\infty < \tau < \infty \tag{64}$$

which is a process having power spectral density

$$s_X(v) = \int_{-\infty}^{\infty} e^{-\iota\omega\tau} e^{-\tau^2/2}\, d\tau = \sqrt{2\pi}\, e^{-2\pi^2 v^2}, \qquad -\infty < v < \infty \tag{65}$$

is singular. It is not an easy matter to give a heuristic explanation of such a strange phenomenon. Perhaps, in the above example, one can argue that since the correlation function has all derivatives at $\tau = 0$ the process $X(t)$ is infinitely MS differentiable, and hence by observing its sample path during a small period of time, one could determine all its derivatives at some time instant t. Then the Taylor expansion of $x(t)$ at the point t would provide the errorless prediction into the future. Unfortunately, this is not a very convincing argument, especially in view of the fact that a process may be singular without being MS differentiable.

Since processes with discrete power spectra are singular, just as some processes with continuous power spectra may be, some further condition is clearly required to make the process nonsingular. The following theorem, which we present without proof, provides such a condition.

THEOREM 5.3.4. Let X(t), $-\infty < t < \infty$, *be an MS continuous zero mean stationary Gaussian process; let*

$$S_X(v) = S_X^{(d)}(v) + S_X^{(c)}(v), \qquad -\infty < v < \infty$$

be its power spectral distribution function; and let

$$s_X(v) = \frac{d}{dv} S_X^{(c)}(v), \qquad -\infty < v < \infty$$

be the power spectral density of the continuous component.

Then the process is nonsingular if and only if

$$\left| \int_{-\infty}^{\infty} \frac{\ln s_x(\nu)}{1+\nu^2}\, d\nu \right| < \infty. \tag{66}$$

If this condition is satisfied, then $S_X^{(d)}(\nu)$ *and* $S_X^{(c)}(\nu)$ *are power spectral distribution functions of the singular and regular parts* $X_s(t)$ *and* $X_r(t)$ *respectively, of the process* $X(t)$.

Condition (66) is known as the *Paley-Wiener criterion*, and the integrand appearing there, that is, the frequency function

$$c_X(\nu) = \frac{\ln s_X(\nu)}{1+\nu^2}, \qquad -\infty < \nu < \infty$$

is often called the *cepstrum*† of the process $X(t)$.

Note that the theorem embraces all possible cases. If the power spectral distribution function $S_X(\nu)$ is discrete, then $S_X^{(c)}(\nu) = 0$ implies that the power spectral density $s_X(\nu)$ is also identically zero, the cepstrum $c_X(\nu) = -\infty$, and hence the process is singular.

If the power spectral distribution function $S_X(\nu)$ is of the continuous type, then the process is either purely regular or singular, depending on whether or not the Paley-Wiener criterion is satisfied. If the power spectral distribution function $S_X(\nu)$ is of mixed type and the Paley-Wiener criterion is satisfied, then the process is nonsingular with an almost periodic singular component.

Example 5.3.5

Consider a zero mean stationary Gaussian process $X(t)$, $-\infty < t < \infty$, with correlation function

$$R_X(\tau) = e^{-\alpha|\tau|} \cos \beta\tau, \qquad -\infty < \tau < \infty,$$

where $\alpha > 0$ and $\beta > 0$ are constants. This process has a continuous spectrum with the power spectral density

$$s_X(\nu) = \int_{-\infty}^{\infty} R_X(\tau) e^{-\iota\omega\tau}\, d\tau = 2\int_0^{\infty} e^{-\alpha\tau} \cos \beta\tau \cos \omega\tau\, d\tau$$

$$= 2\alpha\left(\frac{1}{\alpha^2 + (\omega+\beta)^2} + \frac{1}{\alpha^2 + (\omega-\beta)^2}\right), \qquad \omega = 2\pi\nu.$$

† Sometimes defined with an absolute value of the logarithm.

The cepstrum is a rather complicated function

$$c_X(v) = \frac{1}{1+v^2} \ln 2\alpha\left(\frac{1}{\alpha^2 + (2\pi v + \beta)^2} + \frac{1}{\alpha^2 + (2\pi v - \beta)^2}\right) \tag{67}$$

$-\infty < \beta < \infty$. However, the spectral density is everywhere positive and bounded, and hence the convergence of the integral

$$\int_{-\infty}^{\infty} c_X(v)\, dv$$

depends only on the behavior of $c_X(v)$ for large values of v. From (67) it can be seen that for large v the function $c_X(v)$ behaves like

$$\frac{\ln \dfrac{1}{v^2}}{v^2} = -\frac{2 \ln v}{v^2}$$

which is integrable over any interval (v_0, ∞), $v_0 > 0$. It follows that the integral of the cepstrum is finite, and hence the process $X(t)$ is purely regular.

On the other hand, consider a zero mean stationary Gaussian process with correlation function (64)

$$R_X(\tau) = e^{-\tau^2/2}, \qquad -\infty < \tau < \infty.$$

It also has a continuous spectrum, with the power spectral density found in (65) to be

$$s_X(v) = \sqrt{2\pi}\, e^{-2\pi^2 v^2}, \qquad -\infty < v < \infty.$$

The cepstrum in this case equals

$$c_X(v) = \frac{\ln(\sqrt{2\pi}\, e^{-2\pi^2 v^2})}{1+v^2} = \frac{\ln \sqrt{2\pi}}{1+v^2} - \frac{2\pi^2 v^2}{1+v^2},$$

and, as $v \to \pm\infty$, $c_X(v) \to -2\pi^2$. Thus

$$\int_{-\infty}^{\infty} c_X(\omega)\, d\omega = -\infty,$$

and the process is singular. ▲

Suppose now that the process $X(t)$ is purely regular. In that case, not only must its power spectral density $s_X(v)$ satisfy the Paley-Wiener criterion but the process must also be causally linearly equivalent to white Gaussian noise.

$$X(t) = \int_{-\infty}^{t} h(t-\tau)\dot{W}_0(\tau)\, d\tau.$$

But then we must have

$$s_X(\nu) = |H(\nu)|^2,$$

where $H(\nu)$ is the transfer function of a causal and causally invertible time-invariant linear filter. Thus the Paley-Wiener condition is also a necessary and sufficient condition for the existence of such a filter.

Equation (68) is just the Fourier transform of the integral equation (56), which we encountered in the prediction problem. If we write the complex-valued transfer function $H(\nu)$ as

$$H(\nu) = |H(\nu)|e^{\iota\theta(\nu)}$$

it is seen that equation (68) allows us only to determine the modulus, or gain-transfer function $|H(\nu)|$, namely

$$|H(\nu)| = \sqrt{s_X(\nu)},$$

while the phase-transfer function $\theta(\nu)$ remains undetermined, since $|e^{\iota\theta(\nu)}| = 1$ for any $\theta(\nu)$. The phase-transfer function must then be determined from the requirement that the filter be canonical, that is, causal and causally invertible. Although the existence of such a filter can be checked from the Paley-Wiener condition, the actual determination of the phase-transfer function employs techniques and results from the theory of functions of complex variables, which are beyond the scope of this text.

EXERCISES

Exercise 5.3

In this exercise $X(t)$, $-\infty < t < \infty$, is always a zero mean stationary Gaussian process.

1. Suppose that the correlation function $R_X(\tau)$ of the process $X(t)$ is such that $R_X(\tau) = 0$ if $|\tau| > \tau_0$, where τ_0 is positive constant. Show that $X(t)$ is a purely regular process.
2. The process $X(t)$ is defined as the Wiener integral

$$X(t) = \int_{-\infty}^{\infty} e^{-|t-\tau|}\, dW_0(\tau).$$

What kind of process (singular, nonsingular, purely regular) do you think this is? Calculate the correlation function $R_X(\tau)$, and check your intuition by applying Theorem 5.3.4.

3. The purely regular process $X(t)$ is defined by

$$X(t) = \int_{-\infty}^{t} (t-\tau)e^{-(t-\tau)}\dot{W}_0(\tau)\, d\tau.$$

Check if this is a canonical representation, and if so, find the whitening filter $g(\tau)$, $\tau \geq 0$.

4. The purely regular process $X(t)$ is defined by

$$X(t) = \int_{-\infty}^{t} h(t-\tau)\dot{W}_0(\tau)\, d\tau,$$

where

$$h(\tau) = e^{-\alpha\tau}(1 - e^{-\beta\tau}) \qquad \tau \geq 0,$$

with α and β positive constants. Check if $h(\tau)$ is a canonical filter (modify if necessary), and find the whitening filter $g(\tau)$.

5. Suppose the process $X(t)$ satisfies the differential equation

$$\ddot{X}(t) + 5\alpha\dot{X}(t) + 4\alpha^2 X(t) = \sigma\dot{W}_0(t) \quad -\infty < t < \infty,$$

where α and σ are positive constants. Find the canonical representation

$$X(t) = \int_{-\infty}^{t} h(t-\tau)\dot{W}_0(\tau)\, d\tau.$$

6. For the process $X(t)$ defined in Exercises 5.3.3 through 5.3.5, find the best predictor $\hat{X}(t + \Delta t \mid t)$, $-\infty < t < \infty$, $\Delta t > 0$.

7. If $X(t) = \int_{-\infty}^{t} h(t-\tau)\dot{W}_0(\tau)\, d\tau$ is the canonical representation of $X(t)$, show that the mean square prediction error is

$$\varepsilon^2(t + \Delta t \mid t) = \mathrm{E}[(X(t + \Delta t) - \hat{X}(t + \Delta t \mid t))^2] = \int_0^{\Delta t} |h(\tau)|^2\, d\tau.$$

8. Suppose the process $X(t)$ has correlation function $R_X(\tau) = \sigma^2(1 + |\tau|)e^{-|\tau|}$. Solve the integral equation (56) to find the canonical filter $h(\tau)$ for the process. (*Hint:* Compare (56) with Exercise 1.11.)

9. For the process of Exercise 5.3.8, find the whitening filter $g(\tau)$ and use it to obtain the best predictor $\hat{X}(t + 1 \mid t)$. Also evaluate the mean square error $\varepsilon^2(t + 1 \mid t)$.

10. Show that if $X(t)$ is a purely regular process, then it can always be represented in an "anticausal" form

$$X(t) = \int_{t}^{\infty} h'(t-\tau)\dot{W}_0(\tau)\, d\tau$$

where $h'(t)$, $t \leq 0$, is a square-integrable function.

11. Show that if the power spectral density $s_X(\nu)$ is such that $s_X(\nu) = 0$ if $|\nu| > \nu_0$ for ν_0 a positive constant (band-limited process), then the process $X(t)$ is singular. Show that this remains true as long as $s_X(\nu) = 0$ on some frequency interval, no matter how small.

12. Suppose the power spectral density of the process $X(t)$ equals

$$s_X(\nu) = \frac{1}{1 + \nu^2} e^{-1/|\nu|}.$$

Show that this process is singular and yet does not have even the first MS derivative.

13. Use Theorem 5.5.4 to classify the process $X(t)$ as singular, purely regular, or nonsingular if the correlation function equals:

(a) $R_X(\tau) = (1 + e^{-|\tau|})\cos \tau$,

(b) $R_X(\tau) = \dfrac{\sin \tau}{\tau}$,

(c) $R_X(\tau) = (1 + 2|\tau| + \frac{4}{3}\tau^2)e^{-2|\tau|}$,

(d) $R_X(\tau) = (1 + 2|\tau|)e^{-2|\tau|}$,

(e) $R_X(\tau) = \dfrac{\cos \tau}{2 - \cos \tau}$. (*Hint:* Expand into a geometric series.)

CHAPTER SIX

NONLINEAR SYSTEMS AND STOCHASTIC DIFFERENTIALS

6.1. INTRODUCTION

In Section 4.3, we argued that under certain technical assumptions a finite-variance sample-continuous process $X(t)$, $t \geq 0$, can be decomposed into the sum of two processes, a predictable process with "smooth" sample paths and an unpredictable noise process—a sample-continuous martingale. We have also seen that such a decomposition can, in most cases, be expressed in the form

$$X(t) = X(0) + \int_0^t A(\tau)\, d\tau + \int_0^t B(\tau)\, dW_0(\tau), \qquad t \geq 0, \tag{1}$$

or, using forward increments $dX(t) = X(t + dt) - X(t)$, in an intuitively more appealing form

$$dX(t) = A(t)\, dt + B(t)\, dW_0(t). \tag{2}$$

Recall that in both these representations, $A(t)$ and $B(t)$, were predictable processes, that is, functionals determined by the past portion $X(t')$, $0 \leq t' \leq t$, of the process $X(t)$; in symbols

$$A(t) = a_t(X(t'), 0 \leq t' \leq t), \tag{3a}$$

$$B(t) = b_t(X(t'), 0 \leq t' \leq t). \tag{3b}$$

The process $W_0(t)$ was a standard Wiener process, whose value $W_0(t)$ was also determined by the past $X(t')$, $0 \leq t' \leq t$, but whose forward increment $dW_0(t)$ was independent of this past and thus also independent of the corresponding past of $A(t)$ and $B(t)$.

Our objective in this section is to investigate the case in which the finite-variance sample-continuous process $X(t)$ represents the *state* of a dynamic system operating in continuous time and disturbed by a random noise and can thus be assumed to be a Markov process.

We assume that the reader is familiar with the basic concept of the state-space representation of a dynamic system. Nevertheless, let us briefly review the central idea.

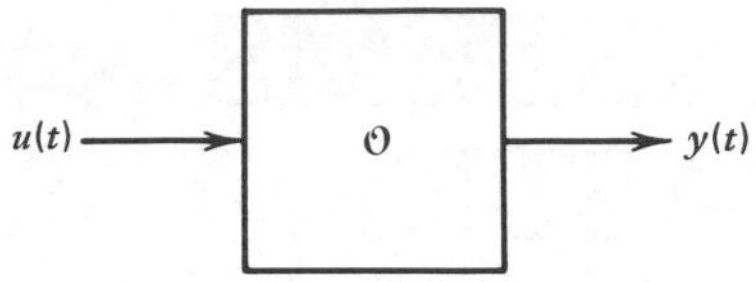

Figure 6.1.1

A dynamic system can generally be defined as an operator $\mathcal{O}$ that transforms an input waveform (function of time) $u(t)$ into an output waveform $y(t)$ (see Fig. 6.1.1), the input and output both being defined on the same time domain, which we take here as $t \geq 0$. The essence of the state-space representation of the system $\mathcal{O}$ is a specification of a *state variable* (function of time) $x(t)$ with the following properties:

1. The output $y(t)$ at time t depends only on the present state $x(t)$.
2. The future state $x(t'')$, $t'' > t$, depends only on the present state $x(t)$ and present and future input $u(t'')$, $t'' \geq t$, but neither on the past state $x(t')$, $t' < t$, nor on the past input $u(t')$, $t' < t$.

In other words, the state $x(t)$ summarizes the entire past history of the system up to time t, and the future evolution of the system can be determined from $x(t)$ and the input $u(t'')$, $t'' \geq t$.

In this chapter we are going to consider only dynamic systems for which the transformation $\mathcal{O}$ is smooth. More precisely, we wish to study differential systems, that is, those systems in which the future state is related to the present state by means of differential equations.

Such a differential system can be generally described by a pair of equations

$$\dot{x}(t) = f(t, x(t), u(t)), \tag{4a}$$

$$y(t) = g(t, x(t)), \qquad t \geq 0, \tag{4b}$$

where the function f is such that for a given input $u(t)$, $t \geq 0$, and given initial state $x(0)$ the state $x(t)$ is uniquely determined for all future times $x(t)$, $t > 0$. The reader should be reminded that in the above equation the state $x(t)$ is typically a *vector-valued* function of time, (the output and input often are as well) so that (4a) may, in fact, be a system of first-order differential equations for the state components.

However, in the following discussion, we regard the state $x(t)$ as a scalar. The reason for doing so is that our purpose here is to explain basic ideas rather than give a detailed treatment of the subject, and working with vector-valued state variables would necessarily tend to obscure these ideas and burden the notation. Besides, in some of the following topics, the extension to a vector-valued case is quite straightforward, although in some others, it is not. On occasion, we will indicate the possibility of such an extension.

We now wish to study stochastic differential systems, that is, differential systems whose state $x(t)$ is a stochastic process rather than a deterministic function of time. The randomness of the state variable may arise from a noise component of the input $u(t)$, a noise disturbance influencing the state equation (4a), and/or a random initial state $x(0)$. At any rate, if the system transformation is sufficiently smooth, it is natural to expect that the state $X(t)$ will be a finite-variance sample-continuous process and, more specifically, that it will be decomposable in the form (1) or (2). However, since $X(t)$ is to represent a state of the system now, the future evolution of the system may depend on its past history only through its present state $X(t)$. This means that the predictable processes $A(t)$ and $B(t)$ can no longer depend on the entire past $X(t')$, $0 \le t' \le t$, but only on the current value $X(t)$ of the state and possibly the time t. In other words, the relations (3) become

$$A(t) = a(X(t), t),$$
$$B(t) = b(X(t), t),$$

where $a(x, t)$ and $b(x, t)$ are now deterministic functions of two variables, the time t and the state x. Thus we have the decomposition

$$X(t) = X(0) + \int_0^t a(X(\tau), \tau)\, d\tau + \int_0^t b(X(\tau), \tau)\, dW_0(\tau), \tag{5}$$

or, in forward differentials,

$$dX(t) = a(X(t), t)\, dt + b(X(t), t)\, dW_0(t). \tag{6}$$

In particular, employing the concept of white Gaussian noise $\dot{W}_0(t) = dW_0(t)/dt$, we can formally write equations (6) and (4b) as

$$\dot{X}(t) = a(X(t), t) + b(X(t), t)\dot{W}_0(t), \tag{7a}$$
$$Y(t) = g(X(t), t), \qquad t \ge 0. \tag{7b}$$

This pair of equations now has the form of the canonical state-space representation of a differential system driven (or disturbed, if you prefer) by white Gaussian noise as an input. It may seem that in the process of getting (7a,b) from (4a,b) the input $u(t)$ has somehow been lost. What really happened is that the smooth component of the input has been incorporated into the functions a and b, while its noise component contributed to the white Gaussian noise $\dot{W}_0(t)$. Of course, for certain kinds of application, particularly in control theory, we would like to see the input $u(t)$ appear explicitly on the right-hand side of (7a). Also, we would like to have an explicit relation between the system function f in equation (4a) and the functions a, b in equation (7a). Later, we will briefly discuss this matter. For the time being, however, our main interest is in the stochastic systems described by (7a,b). Even more specifically, since equation (7b) is just a memoryless transformation, our main topic is the study of the state process $X(t)$ obeying equation (7a).

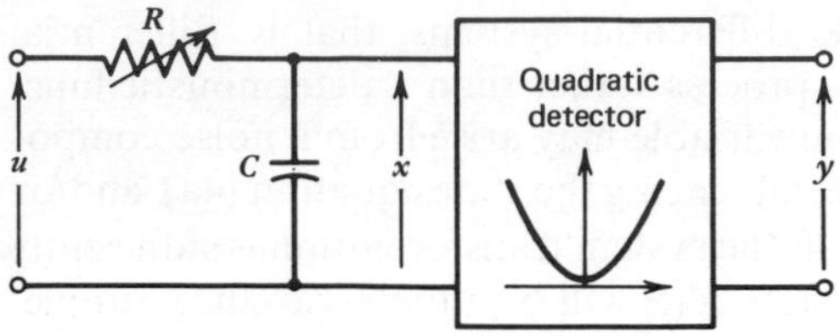

Figure 6.1.2

Example 6.1.1

Consider a simple *RC* circuit followed by a quadratic detector, as shown in Figure 6.1.2. Suppose further that the resistance R varies with time $R = r(t)$, while the capacitance changes with the voltage x across the capacitor $C = c(x)$. The detector is assumed to have an infinite input impedance. If $u(t)$ is the input voltage, then by Kirchhoff's law

$$u(t) = r(t)c(x)\dot{x}(t) + x(t)$$

since $c(x)\dot{x}(t)$ is the current through the resistor. The output voltage $y(t) = x^2(t)$, and so by designating the voltage $x(t)$ across the capacitor as the state of the system, we get the canonical state space representation

$$\dot{x}(t) = a(x(t), t) + b(x(t), t)u(t),$$
$$y(t) = x^2(t), \quad t \geq 0,$$

where

$$a(x, t) = -\frac{x(t)}{r(t)c(x(t))},$$

and

$$b(x, t) = -\frac{1}{r(t)c(x(t))}.$$

If the input is white Gaussian noise $\dot{W}_0(t)$, this is an example of a simple nonlinear stochastic system (7a,b). ▲

EXERCISES

Exercise 6.1

1. Consider an ideal current source, a resistance R, a capacitance C, and two semiconductor diodes, all connected in parallel as in Figure 6.1.E.1. Each diode is assumed to have infinite resistance in the reverse direction, while in the forward direction its volt-ampere characteristic is

$$i = i_s(e^{q_0 v/kT} - 1).$$

Here, v is the voltage across the diode; i, the current through the diode; and i_s, q_0, T, and k are parameters (saturation current, electron charge, absolute temperature, and Boltzman constant, respectively). The current source represents a thermal noise current

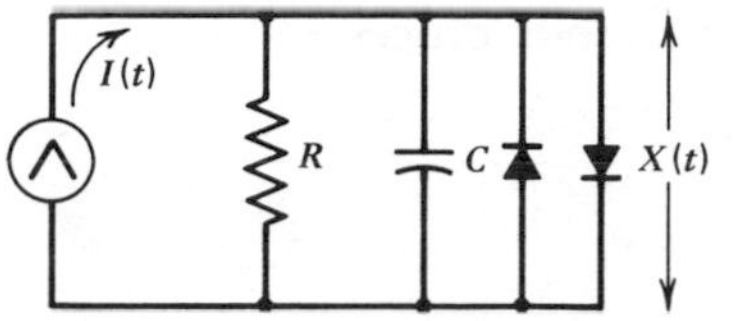

Figure 6.1.E.1

$I(t) = \sqrt{2kT}\,\dot{W}_0(t)$ of the resistor. Write the differential equation for the voltage across the diodes in the form of equation (7a).

2. Repeat Exercise 6.1.1 if one of the diodes is removed from the circuit.
3. A particle of mass m confined to move without friction along the straight line L in Figure 6.1.E.2 is subjected to two forces: (i) a central attractive force directed to the origin O and inversely proportional to the distance of the particle from the origin; and (ii) a random noise force (white Gaussian noise) in the direction perpendicular to the attractive force (i). Write the canonical state-space equations for the process $\mathbf{X}(t)$ representing the x-components of the position and velocity of the particle.

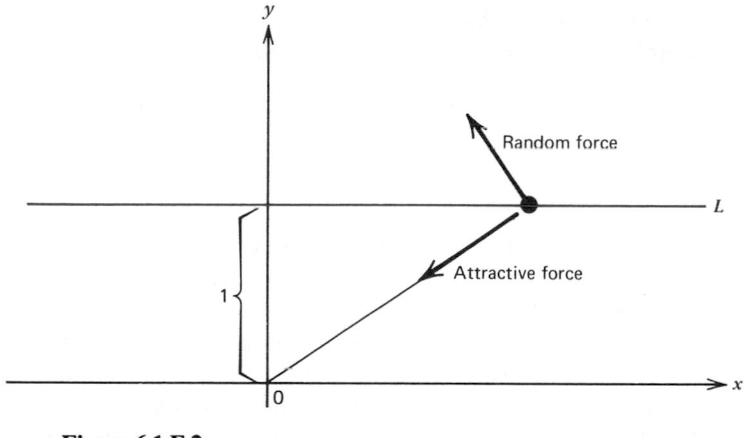

Figure 6.1.E.2

4. Consider a control system designed to keep an error variable x at state $x = 0$. If at some time t the error $x(t) \neq 0$, the system applies a correction $dx(t)$ proportional to the error $x(t)$. However, the correction is also subjected to an additive random disturbance increasing linearly with the magnitude of the error. Assuming that the disturbance can be modeled as white Gaussian noise, obtain the differential equation (7a) for the error process $X(t)$.

6.2. KOLMOGOROV DIFFERENTIAL EQUATIONS

Although the mathematically rigorous form of equation (6.1.7a) is the integral equation (6.1.5), we will, at least for the time being, consider the differential form (6.1.6), since it provides the most intuitive insight.

Thus, we have a finite-variance sample-continuous process

$$X(t), \qquad t \geq 0,$$

with (possibly random) initial value

$$X(0),$$

and such that its forward differential satisfies

$$dX(t) = a(X(t), t)\,dt + b(X(t), t)\,dW_0(t), \tag{1}$$

where $W_0(t)$, $t \geq 0$, is the standard Wiener process.

Now the increment $dW_0(t)$ of the Wiener process is independent of the past $X(t')$, $0 \leq t' \leq t$, as well as of its own past $W_0(t')$, $0 \leq t' \leq t$. Since $dX(t) = X(t + dt) - X(t)$ is a forward increment and as a result of equation (1), we see that the conditional distribution of $X(t + dt)$, given the past and present $X(t')$, $0 \leq t' \leq t$, can only depend on the present $X(t)$. But this is just the Markov property, so we conclude that the process $X(t)$ should be a *Markov process.*

Next, the probability law of a Markov process $X(t)$ with a continuous time domain $t \geq 0$ is completely specified by its *initial density*

$$f_0(x)$$

and by the family of *transition probability densities*

$$f_{t_2|t_1}(x_2 \,|\, x_1)$$

defined for all $0 \leq t_1 \leq t_2$ (see Chapter One). Its initial density $f_0(x)$ is just the density of the initial value $X(0)$ and is assumed to be given. Thus the transition probability densities should be determined by equation (1), that is, by the two functions $a(t, x)$ and $b(t, x)$. Since the process $X(t)$ is assumed to have continuous sample functions, it is not unreasonable to expect that under appropriate smoothness conditions imposed upon the functions $a(t, x)$ and $b(t, x)$, the transition probability densities will be continuous functions and involve no delta functions. Recall that the transition probability densities of a Markov process must satisfy the *Chapman-Kolmogorov equation*

$$f_{t_2|t_1}(x_2 \,|\, x_1) = \int_{-\infty}^{\infty} f_{t_2|t}(x_2 \,|\, x) f_{t|t_1}(x \,|\, x_1)\,dx \tag{2}$$

for all $t_1 < t < t_2$. Now fix x_2 and t_2, and for $t < t_2$, let us define

$$q(x, t) = f_{t_2|t}(x_2 \,|\, x). \tag{3}$$

The right-hand side of the Chapman-Kolmogorov equation (2) can be written as a conditional expectation

$$q(x_1, t_1) = \mathrm{E}[q(X(t), t) \,|\, X(t_1) = x_1]. \tag{4}$$

Set $t = t_1 + dt$ so that

$$X(t) = X(t_1) + X(t_1 + dt) - X(t_1) = X(t_1) + dX(t_1)$$

and expand the function $q(x, t)$ into its Taylor series in two variables

$$\begin{aligned} q(X(t_1) + dX(t_1), t_1 + dt_1) = q(X(t_1), t_1) &+ \frac{\partial}{\partial t_1} q(X(t_1), t_1)\, dt_1 \\ &+ \frac{\partial}{\partial x} q(X(t_1), t_1)\, dX(t_1) \\ &+ \frac{1}{2}\frac{\partial^2}{\partial x^2} q(X(t_1), t_1)(dX(t_1))^2 + \cdots . \end{aligned}$$

Substituting into the conditional expectation in (4), we get

$$\begin{aligned} q(x_1, t_1) = q(x_1, t_1) &+ \frac{\partial}{\partial t_1} q(x_1, t_1)\, dt_1 \\ &+ \frac{\partial}{\partial x_1} q(x_1, t_1) \mathrm{E}[dX(t_1) \,|\, X(t_1) = x_1] \\ &+ \frac{1}{2}\frac{\partial^2}{\partial x_1^2} q(x_1, t_1) \mathrm{E}[(dX(t_1))^2 \,|\, X(t_1) = x_1] + \cdots . \end{aligned} \tag{5}$$

From the fact that the increment of the standard Wiener process $dW_0(t)$ is independent of $X(t')$, $0 \le t' \le t$, we have

$$\mathrm{E}[dW_0(t) \,|\, X(t'), 0 \le t' \le t] = \mathrm{E}[dW_0(t)] = 0,$$

and

$$\mathrm{E}[(dW_0(t))^2 \,|\, X(t'), 0 \le t' \le t] = \mathrm{E}[(dW_0(t))^2] = dt,$$

so that from (1)

$$\mathrm{E}[dX(t_1) \,|\, X(t_1) = x_1] \doteq a(x_1, t_1)\, dt_1 \tag{6}$$

and using the fact that for small dt, $(dt)^2$ is negligible with respect to dt

$$\mathrm{E}[(dX(t_1))^2 \,|\, X(t_1) = x_1] \doteq b^2(x_1, t_1)\, dt_1. \tag{7}$$

Hence, substituting into (5) and noticing that the differentials of all the higher terms of the Taylor expansion would be of order at least $(dt)^2$, we obtain the partial differential equation

$$-\frac{\partial}{\partial t_1} q(x_1, t_1) = a(x_1, t_1)\frac{\partial}{\partial x_1} q(x_1, t_1) + \frac{1}{2} b^2(x_1, t_1) \frac{\partial^2}{\partial x_1^2} q(x_1, t_1). \tag{8}$$

Replacing the function q by the transition probability density (3), the equation becomes

$$-\frac{\partial}{\partial t_1} f_{t_2|t_1}(x_2|x_1) = a(x_1, t_1)\frac{\partial}{\partial x_1} f_{t_2|t_1}(x_2|x_1) + \frac{1}{2}b^2(x_1, t_1)\frac{\partial^2}{\partial x_1^2} f_{t_2|t_1}(x_2|x_1), \quad (9)$$

where $0 \le t_1 \le t_2$.

This equation is known as *Kolmogorov's backward equation* for the transition probability density. It is called a backward equation because the time t_2 and state $X(t_2) = x_2$ are fixed and all derivatives are taken with respect to an earlier time $t_1 < t_2$ and state $x_1 = X(t_1)$. In other words, the solution of this equation would have to proceed from a fixed terminal time t_2 and state x_2 backward in time.

A dual equation, that is, one where t_1 and $X(t_1) = x_1$ are fixed and the differentiation is with respect to the later time $t_2 > t_1$ and state $X(t_2) = x_2$, can also be derived. It has the form

$$\frac{\partial}{\partial t_2} f_{t_2|t_1}(x_2|x_1) = -\frac{\partial}{\partial x_2}[a(x_2, t_2) f_{t_2|t_1}(x_2|x_1)] + \frac{1}{2}\frac{\partial^2}{\partial x_2^2}[b^2(x_2, t_2) f_{t_2|t_1}(x_2|x_1)], \quad (10)$$

$0 \le t_1 \le t_2$, and is called the *Kolmogorov forward equation.*

If $t_1 = 0$ or if $t = t_2 - t_1$, and t_1 and x_1, being fixed, are suppressed, that is, if we denote

$$f(x, t) = f_{t_1 + t|t_1}(x|x_1),$$

this equation can be written in a (seemingly) simpler form:

$$\frac{\partial}{\partial t} f(x, t) = -\frac{\partial}{\partial x}[a(x, t) f(x, t)] + \frac{1}{2}\frac{\partial^2}{\partial x^2}[b^2(x, t) f(x, t)], \qquad t \ge 0. \quad (11)$$

In this form the equation is also known, especially in applications, as the *Fokker-Planck equation.*

Note that if $a(x, t) = \mu$ and $b(x, t) = \sigma > 0$, the forward equation (10) becomes

$$\frac{\partial f_{t_2|t_1}(x_2|x_1)}{\partial t_2} = -\mu \frac{\partial f_{t_2|t_1}(x_2|x_1)}{\partial x_2} + \frac{1}{2}\sigma^2 \frac{\partial^2}{\partial x_2^2} f_{t_2|t_1}(x_2|x_1), 0 \le t_1 < t_2.$$

The reader is invited to verify by substitution that the transition probability density

$$f_{t_2|t_1}(x_2|x_1) = \frac{1}{\sqrt{2\pi\sigma^2(t_2 - t_1)}} e^{-(x_2 - x_1 - \mu(t_2 - t_1))^2/2\sigma^2(t_2 - t_1)}$$

is a solution of this equation. But this is just the transition probability density of a Wiener process with drift parameter μ and intensity parameter σ. Since we have defined such a process by

$$X(t) = \sigma W_0(t) + \mu t$$

that is, by taking forward differentials

$$dX(t) = \mu \, dt + \sigma \, dW_0(t), \tag{12}$$

so that indeed $a(x, t) = \mu$ and $b(x, t) = \sigma$, the result is in agreement with our earlier derivation of the transition probability density for this Wiener process. In particular, if $\mu = 0$ and $\sigma = 1$, the forward equation with $x_1 = 0$, $t_1 = 0$ becomes the differential equation one arrives at in deriving the standard Wiener process from a symmetric random walk as was done in Volume I of this book.

Although Kolmogorov's forward equation may seem to be more natural inasmuch as it describes the evolution of the transition probability density from a fixed time t_1 and state x_1 into the future, it is actually the backward equation that is more general. For if we retrace our derivation of equation (8), we see that we have only used the fact that the function $q(x, t)$ satisfied relation (4) and not that it was actually defined by (3). In other words, if $q(x, t)$ is a function such that for $0 \leq t_1 \leq t \leq t_2$

$$q(x_1, t_1) = \mathrm{E}[q(X(t), t) \mid X(t_1) = x_1] \tag{13}$$

then this function ought to satisfy the differential equation

$$-\frac{\partial q(x, t)}{\partial t} = a(x, t)\frac{\partial q(x, t)}{\partial x} + \frac{1}{2} b^2(x, t)\frac{\partial^2 q(x, t)}{\partial x^2} \tag{14}$$

for $0 \leq t \leq t_2$, provided, of course, that this function is differentiable as required to justify the Taylor expansion used in the derivation.

Remark: In fact, we can even admit functions $q(x, t)$ such that for $0 \leq t_1 \leq t \leq t_2$

$$q(X(t_1), t_1) = \mathrm{E}[q(X(t), t) \mid X(t'), 0 \leq t' \leq t_1], \tag{15}$$

as long as $X(t)$ is a Markov process, and equation (14) will be satisfied. It may be worth pointing out that (15) implies that the process

$$q(X(t), t), \qquad t \geq 0$$

is a martingale. We will make use of this generalization later. Я

As a simple example, if we define, for x_2, t_2 fixed,

$$q(x, t) = P(X(t_2) \leq x_2 \mid X(t) = x), \qquad 0 \leq t \leq t_2,$$

then it is easy to verify, using the Markov property of the process $X(t)$, that (13) holds, and so (14) becomes a differential equation for the transition distribution function of the process. Of course, the same equation can be obtained by integrating the backward equation (9) with respect to x_2. However, the present derivation would be valid even when the process may not have an x_1-differentiable transition probability density, in which case equation (9) would not apply.

In fact, Kolmogorov's forward equation itself can be, at least formally, derived from this generalized backward equation (14). A sketch of the derivation is given in the appendix to this section.

The backward and forward equations (9) and (10) are frequently called *propagation equations* (for the transition probability density), since they describe the temporal and spatial evolution of the conditional density of the process.

If they are to be used to actually determine the unknown transition probability density, it is necessary to specify the initial and boundary condition that the density is required to satisfy. In the case of the Fokker-Planck equation, for example, this means that we must prescribe the function $f(x, t)$ for $t = 0$ and all x (initial condition), as well as for all x on or near the boundary of the state space and all $t \geq 0$ (boundary conditions). The initial condition is easy to formulate, since the sample paths are assumed continuous

$$\lim_{t \to 0} f(x, t) = \delta(x_0),$$

where $x_0 = X(0)$ is the initial state.

If the state space of the process is the entire real line, that is, if its "boundaries" are at $\pm\infty$, and if the functions $a(x, t)$ and $b(x, t)$ satisfy certain regularity conditions, then the boundary conditions are supplied by the requirement that $f(x, t)$ as a function of $x \in (-\infty, \infty)$ must be a probability density. However, in the general case, for instance, if the state space has finite boundaries, the boundary conditions may be quite complicated.

The same is true for the backward equation (9) or its generalized version (14). At any rate, obtaining an explicit solution of these equations is usually difficult and is beyond the scope of this book.

However, the importance of these equations, at least for the purpose of further discussion, is that under appropriate conditions they do determine the transition probabilities and hence the probability law of the Markov process $X(t)$. Thus, many important properties of this process can often be derived from these equations without actually obtaining an explicit solution.

Example 6.2.1

Consider the equation

$$\dot{X}(t) = -\alpha X(t) + \sigma \dot{W}_0(t), \qquad t \geq 0, \tag{16}$$

where $\alpha > 0$ and $\sigma > 0$ are constants. This equation is clearly a special case of equation (1) with $a(x, t) = -\alpha x$ and $b(x, t) = \sigma$, and so the process $X(t)$, $t \geq 0$, should be a finite-variance sample-continuous Markov process. Thus, its transition probabilities should satisfy the backward equation

$$-\frac{\partial}{\partial t_1} f_{t_2|t_1}(x_2|x_1) = -\alpha x_1 \frac{\partial}{\partial x_1} f_{t_2|t_1}(x_2|x_1) + \frac{1}{2}\sigma^2 \frac{\partial^2}{\partial x_1^2} f_{t_2|t_1}(x_2|x_1), \qquad 0 \le t_1 \le t_2, \tag{17}$$

and also the forward equation

$$\frac{\partial}{\partial t_2} f_{t_2|t_1}(x_2|x_1) = \frac{\partial}{\partial x_2}[\alpha x_2 f_{t_2|t_1}(x_2|x_1)] + \frac{1}{2}\frac{\partial^2}{\partial x_2^2}[\sigma^2 f_{t_2|t_1}(x_2|x_1)], \qquad 0 \le t_1 \le t_2. \tag{18}$$

However, equation (16) is the Langevin equation encountered in Section 4.4, where we have established that its solution $X(t)$, $t \ge 0$, is a (nonstationary) Ornstein-Uhlenbeck process with parameters α and $\sigma^2/2\alpha$. We know that this process is a zero mean Gauss-Markov process with correlation function

$$R_X(t_1, t_2) = \frac{\sigma^2}{2\alpha}(e^{-\alpha|t_1-t_2|} - e^{-\alpha(t_1+t_2)})$$

and hence that the conditional density of $X(t_2)$, given $X(t_1) = x_1$, is Gaussian with mean $x_1 e^{-\alpha(t_2-t_1)}$ and variance $(\sigma^2/2\alpha)(1 - e^{-2\alpha(t_2-t_1)})$. In other words, the transition probability density is

$$f_{t_2|t_1}(x_2|x_1) = \frac{1}{\sqrt{\frac{\pi\sigma^2}{\alpha}(1 - e^{-2\alpha(t_2-t_1)})}} \exp\left\{-\frac{\alpha(x_2 - x_1 e^{-\alpha(t_2-t_1)})^2}{\sigma^2(1 - e^{-2\alpha(t_2-t_1)})}\right\}$$

$$0 \le t_1 \le t_2, \quad -\infty < x_1 < \infty, \quad -\infty < x_2 < \infty.$$

It can now be verified by straightforward but rather lengthy differentiation that this function indeed satisfies both the backward and the forward equations (17) and (18). Note that the initial conditions

$$\lim_{t_2 \to t_1} f_{t_2|t_1}(x_2|x_1) = \lim_{t_1 \to t_2} f_{t_2|t_1}(x_2|x_1) = \delta(x_1 - x_2)$$

are satisfied as well, as are the boundary conditions $f_{t_2|t_1}(x_2|x_1) \to 0$ as either $x_2 \to \pm\infty$ or $x_1 \to \pm\infty$. ▲

Example 6.2.2

Consider the Fokker-Planck equation

$$\frac{\partial}{\partial t} f(x, t) = \frac{1}{2}\frac{\partial^2}{\partial x^2} f(x, t), \qquad t \ge 0, \tag{19}$$

with the initial condition

$$\lim_{t\to 0} f(x, t) = \delta(x), \qquad -\infty < x < \infty, \tag{20}$$

and with boundary conditions

$$\lim_{x\to\pm\infty} f(x, t) = 0, \qquad t \geq 0, \tag{21}$$

and

$$\lim_{x\to\alpha} f(x, t) = 0, \qquad t \geq 0, \tag{22}$$

where $\alpha > 0$ is a given constant. The Fokker-Planck equation (19) is seen to be the same as the one for the standard Wiener process and indeed the function

$$f_0(x, t) = \frac{1}{\sqrt{2\pi t}} e^{-x^2/2t}, \qquad t \geq 0, \qquad -\infty < x < \infty, \tag{23}$$

is easily seen to satisfy both equation (19) and the conditions (20) and (21). But the boundary condition (22) is clearly violated, and we must seek another solution.

To this end, note that if we define

$$f_\mu(x, t) = \frac{1}{\sqrt{2\pi t}} e^{-(x-\mu)^2/2t}, \qquad t \geq 0, \qquad -\infty < x < \infty, \tag{24}$$

a Gaussian density with mean μ and variance t, then this function will also satisfy equation (19) for any value $-\infty < \mu < \infty$, since

$$\frac{\partial}{\partial t} f_\mu(x, t) = \left(\frac{(x-\mu)^2}{2t^2} - \frac{1}{2t}\right) f_\mu(x, t),$$

and

$$\frac{\partial^2}{\partial x^2} f_\mu(x, t) = \left(\frac{(x-\mu)^2}{t^2} - \frac{1}{t}\right) f_\mu(x, t).$$

Notice also that if c_1 and c_2 are any two constants, then the linear combination of (23) and (24)

$$c_1 f(x, t) + c_2 f_\mu(x, t),$$

will satisfy the Fokker-Planck equation as well. Furthermore, boundary condition (21) will be satisfied too, and if $c_1 = 1$ and if we choose $\mu = 2\alpha$ and $c_2 = -1$, then we will also satisfy boundary condition (22), since

$$f(\alpha, t) = f_{2\alpha}(\alpha, t) \qquad \text{for all } t \geq 0.$$

Hence the function

$$f(x, t) = \frac{1}{\sqrt{2\pi t}} (e^{-x^2/2t} - e^{-(x-2\alpha)^2/2t}), \qquad t \geq 0 \tag{25}$$

will satisfy both (21) and (22), and as long as $x \leq \alpha$, it also satisfies initial condition (20). However, for $x > \alpha$ the function (25) would be negative and would also violate initial condition (20) at $x = 2\alpha$. Thus, for $x > \alpha$, we just set

$$f(x, t) = 0, \qquad t \geq 0,$$

and so the function defined for all $t \geq 0$ by

$$f(x, t) = \begin{cases} \dfrac{1}{\sqrt{2\pi t}}\left(e^{-x^2/2t} - e^{-(x-2\alpha)^2/2t}\right) & \text{if} \quad x < \alpha, \\ 0 & \text{if} \quad x \geq \alpha, \end{cases} \tag{26}$$

will satisfy the Fokker-Planck equation (19) as well as conditions (20), (21), and (22). Also, (26) is clearly nonnegative for all $t \geq 0$ and $-\infty < x < \infty$, but it still cannot be the probability density $f_t(x)$ of the process $X(t)$, $t \geq 0$, since for $t > 0$

$$\int_{-\infty}^{\infty} f(x, t)\, dx = \Phi\left(\frac{\alpha}{\sqrt{t}}\right) - \Phi\left(\frac{-\alpha}{\sqrt{t}}\right) < 1,$$

with $\Phi(z)$ denoting, as usual, the standard Gaussian distribution function. This means that we must have for $t > 0$

$$P(X(t) = \alpha) = 1 - \int_{-\infty}^{\infty} f(x, t)\, dx = 2\Phi\left(\frac{-\alpha}{\sqrt{t}}\right) \tag{27}$$

and so the distribution of $X(t)$, $t > 0$, will be of mixed type with density

$$f_t(x) = f(x, t) + 2\Phi\left(\frac{-\alpha}{\sqrt{t}}\right)\delta(x - \alpha), \qquad t \geq 0, \qquad -\infty < x < \infty,$$

where $f(x, t)$ is defined by (26). In view of the above calculation, we can now describe the nature of the sample paths of the process $X(t)$, $t \geq 0$. Each sample path begins at time $t = 0$ at the state $x = 0$ and is continuous like the sample paths of a standard Wiener process until it reaches the state $x = \alpha$. Once that happens the sample path remains there forever (see Fig. 6.2.1). No sample path can ever get beyond this level α, and since from (26)

$$\lim_{t \to \infty} f_t(x) = \delta(x - \alpha),$$

every sample will eventually reach and stay at this level with probability 1. For this reason the state $x = \alpha$ is referred to as an absorbing boundary. Notice also that $2\Phi(-\alpha/\sqrt{t})$ as a function of $t \geq 0$ is the distribution function of an inverse Gaussian distribution (see Appendix Two). This is not surprising, since if we call T the random time when the process $X(t)$ reaches the boundary α, then by (27)

$$P(T \leq t) = P(X(t) = \alpha) = 2\Phi\left(\frac{-\alpha}{\sqrt{t}}\right).$$

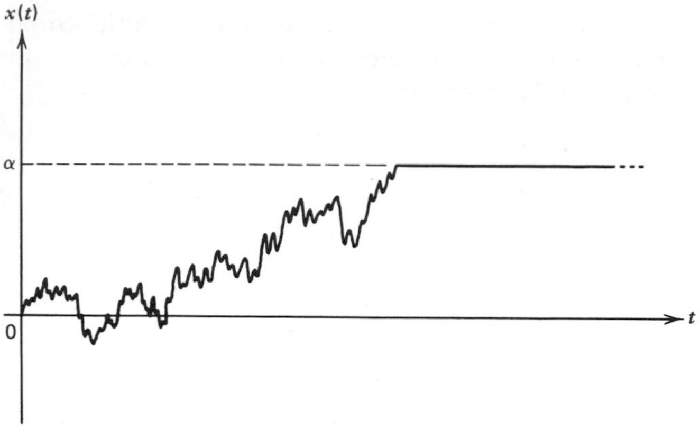

Figure 6.2.1

Indeed, this is the most common way to define the inverse Gaussian distribution—the distribution of the time a Wiener process reaches a given level. Thus the procedure above provides yet another way to derive the distribution of first-crossing times. The advantage of the present method is that it can be applied to a more general class of Markov processes provided that we can solve the corresponding Fokker-Planck equation. ▲

APPENDIX

SKETCH OF THE DERIVATION OF KOLMOGOROV'S FORWARD EQUATION. Define for $0 \le t \le t_m$, t_m fixed,

$$q(x, t) = \mathrm{E}[u(X(t_2)) \mid X(t) = x], \tag{A1}$$

where $u(x)$ is a bounded continuous function. Since $X(t)$ is a Markov process, we have for $0 \le t_1 \le t \le t_m$,

$$q(x_1, t_1) = \mathrm{E}[\mathrm{E}[u(X(t_m)) \mid X(t)] \mid X(t_1) = x_1] = \mathrm{E}[q(X(t), t) \mid X(t_1) = x_1]$$

so that (15) is satisfied. Assuming that the transition probability densities exist, we can write this as

$$q(x_1, t_1) = \int_{-\infty}^{\infty} q(x, t) f_{t|t_1}(x \mid x_1).$$

Differentiating with respect to t, we get

$$0 = \int_{-\infty}^{\infty} \frac{\partial q(x, t)}{\partial t} f_{t|t_1}(x \mid x_1)\, dx + \int_{-\infty}^{\infty} q(x, t) \frac{\partial}{\partial t} f_{t|t_1}(x \mid x_1)\, dx. \tag{A2}$$

Substituting equation (14) for $\partial q(x, t)/\partial t$ in the first integral, this integral becomes

$$-\int_{-\infty}^{\infty} \frac{\partial q(x, t)}{\partial x} a(x, t) f_{t|t_1}(x|x_1)\, dx - \frac{1}{2}\int_{-\infty}^{\infty} \frac{\partial^2 q(x, t)}{\partial x^2} b^2(x, t) f_{t|t_1}(x|x_1)\, dx. \tag{A3}$$

Applying integration by parts, the first of the two integrals in (A3) turns into

$$-q(x, t)a(x, t)f_{t|t_1}(x|x_1)\Big|_{x=-\infty}^{x=\infty} + \int_{-\infty}^{\infty} q(x, t)\frac{\partial}{\partial x}[a(x, t)f_{t|t_1}(x|x_1)]\, dx, \tag{A4}$$

while the second integral, after being integrated by parts twice, becomes

$$\begin{aligned} -\frac{1}{2}\frac{\partial q(x, t)}{\partial x} b^2(x, t) f_{t|t_1}(x|x_1)\Big|_{x=-\infty}^{x=\infty} \\ + \frac{1}{2}\frac{\partial^2 q(x, t)}{\partial x^2}\frac{\partial}{\partial x}[b^2(x, t)f_{t|t_1}(x|x_1)]\Big|_{x=-\infty}^{x=\infty} \\ - \frac{1}{2}\int_{-\infty}^{\infty} q(x, t)\frac{\partial^2}{\partial x^2}[b^2(x, t)f_{t|t_1}(x|x_1)]\, dx. \end{aligned} \tag{A5}$$

Since $f_{t|t_1}(x|x_1)$ is a probability density it, together with its derivative, should go to zero as $x \to \pm\infty$. Thus, we expect the first term in (A4) and the first two terms in (A5) to be zero, and so we obtain, upon substitution back into (A2), the equation

$$\begin{aligned} 0 = \int_{-\infty}^{\infty} q(x, t)\bigg\{\frac{\partial}{\partial x}[a(x, t)f_{t|t_1}(x|x_1)] - \frac{1}{2}\frac{\partial^2}{\partial x^2}[b^2(x, t)f_{t|t_1}(x|x_1)] \\ + \frac{\partial}{\partial t} f_{t|t_1}(x|x_1)\bigg\}\, dx. \end{aligned} \tag{A6}$$

However, the function $u(x)$ used in (A1) to define $q(x, t)$ was bounded and continuous but otherwise arbitrary, and so if the integral in (A6) is to be zero for various functions $q(x, t)$, the expression in braces must be identically zero. But

$$\frac{\partial}{\partial x}[a(x, t)f_{t|t_1}(x|x_1)] - \frac{1}{2}\frac{\partial^2}{\partial x^2}[b^2(x, t)f_{t|t_1}(x|x_1)] + \frac{\partial}{\partial t} f_{t|t_1}(x|x_1) = 0,$$

is nothing but Kolmogorov's forward equation (10) with $x = x_2$ and $t = t_2$.

EXERCISES

Exercise 6.2

1. The phase error $X(t)$ of the phase-lock loop circuit satisfies the differential equation

$$\dot{X}(t) + \alpha \sin X(t) = \sigma \dot{W}_0(t), \qquad t \geq 0,$$

where α and σ are positive constants and $X(0) = 0$. Write the Fokker-Planck equation (11) for the first order density $f_t(x)$ of the process $X(t)$.

2. Write the backward and forward Kolmogorov equations for the transition probability densities of the process described in Exercise 6.1.1.
3. Verify by substitution that the transition probability density $f_{t_2|t_1}(x_2 \mid x_1)$ of the Wiener process $X(t) = \mu t + \sigma W_0(t)$ satisfies both the forward and backward Kolmogorov equations.
4. Verify by substitution that the first order density $f_t(x)$ of the Ornstein-Uhlenbeck process of Example 6.2.1 satisfies the appropriate Fokker-Planck equation.
5. Consider the Fokker-Planck equation

$$\frac{\partial}{\partial t} f(t, x) = -\frac{\partial}{\partial x}(af(t, x)) + \frac{1}{2}\frac{\partial^2}{\partial x^2}(b^2 f(t, x)),$$

$t \geq 0$, with a and b as constants. The initial and boundary conditions are as in Example 6.2.2. Employ the method used in that example to find the solution of this equation. [*Hint:* Try again a linear combination $c_1 f_{\mu_1}(x, t) + c_2 f_{\mu_2}(x, t)$, where $f_\mu(x, t)$ is the Gaussian density with mean μ and variance $\sigma^2 = b^2$.)

6. Let $X(t)$, $t \geq 0$, be the process of Exercise 6.2.5. Describe the nature of the sample paths in both cases $a > 0$ and $a < 0$. In the former case, derive the density of the time T when the process reaches the level $x = \alpha$.
7. Let $\mathbf{X}(t) = \sigma(W_{01}(t), W_{02}(t), W_{03}(t))'$, $t \geq 0$, be a vector-valued process whose components are independent standard Wiener processes. Show that the transition probability densities $f_{t_2|t_1}(\mathbf{x}_2 \mid \mathbf{x}_1)$ satisfy the differential equation

$$\frac{\partial}{\partial t_2} f_{t_2|t_1}(\mathbf{x}_2 \mid \mathbf{x}_1) = \frac{\sigma^2}{2}\left[\frac{\partial^2}{\partial x_{21}^2} f_{t_2|t_1}(\mathbf{x}_2 \mid \mathbf{x}_1) + \frac{\partial^2}{\partial x_{22}^2} f_{t_2|t_1}(\mathbf{x}_2 \mid \mathbf{x}_1) + \frac{\partial^2}{\partial x_{23}^2} f_{t_2|t_1}(\mathbf{x}_2 \mid \mathbf{x}_1)\right],$$

$0 \leq t_1 < t_2$, where $\mathbf{x}_2 = (x_{21}, x_{22}, x_{23})'$. (*Hint:* Write the forward equation for each component, multiply by the densities of the other two components, and add.)

8. In Exercise 6.2.7, set $t_1 = 0$, $\mathbf{x}_1 = 0$, $t_2 = t > 0$, and $\mathbf{x}_2 = \mathbf{x} = (x_1, x_2, x_3)'$ and make the substitution $r = \sqrt{x_1^2 + x_2^2 + x_3^2}$ into the differential equation. Use this to show that the first order density $f_t(r)$ of the process $R(t) = \sigma\sqrt{W_{01}^2(t) + W_{02}^2(t) + W_{03}^2(t)}$ satisfies the equation

$$\frac{\partial f_t(r)}{\partial t} = -\sigma^2 \frac{\partial}{\partial r}\left[\frac{1}{r} f_t(r)\right] + \frac{\sigma^2}{2}\frac{\partial^2 f_t(r)}{\partial r^2}.$$

Recall that $f_t(r)$ should be the Maxwell density and verify the equation. (*Hint:* Note that

$$\frac{\partial^2 f}{\partial x_k^2} = \frac{\partial^2 f}{\partial r^2}\left(\frac{\partial r}{\partial x_k}\right)^2 + \frac{\partial f}{\partial r}\frac{\partial^2 r}{\partial x_k^2},$$

$k = 1, 2, 3$ and $f_{R(t)} = 4\pi r^2 f_{\mathbf{x}(t)}$.)

9. The Ornstein-Uhlenbeck process can be defined by means of the time transformation $X(t) = \sigma e^{-\alpha t} W_0(e^{2\alpha t} - 1)$, $t \geq 0$. Use this to derive the Kolmogorov backward and

forward equations for the Ornstein-Uhlenbeck process from those of the standard Wiener process.

10. Derive the Fokker-Planck equation for the process $X(t)$, $t \geq 0$, defined by $X(t) = (1 - \cos t)W_0(t + \sin t)$. Identify the functions $a(x, t)$ and $b(x, t)$. (*Hint:* Start with the backward equation.)

6.3. THE STRUCTURE OF THE PROCESS $X(t)$

Let us again examine the differential relation

$$dX(t) = a(x, t)\, dt + b(x, t)\, dW_0(t) \tag{1}$$

for a finite-variance sample-continuous Markov process $X(t)$, $t \geq 0$, where $x = X(t)$ and $dX(t) = X(t + dt) - x$. Now if the functions $a(x, t)$ and $b(x, t) > 0$ do not change rapidly with small changes of x and t during each small interval $(t_k, t_k + \Delta t)$ in the time domain, then during these small intervals they can be regarded as constants

$$\mu_k = a(x_k, t_k), \qquad \sigma_k = b(x_k, t_k), \tag{2}$$

where

$$x_k = X(t_k),$$

and $0 = t_0 < t_1 < t_2 < \cdots$ is a partition of the time domain $t \geq 0$ into small intervals of length Δt. Then during each of these small intervals, equation (1) becomes approximately

$$dX(t) = \mu_k\, dt + \sigma_k\, dW_0(t),$$

with $t_k \leq t < t_{k+1}$ and $0 < dt$ much smaller than Δt. But this means (see equation (6.2.12)) that during each of these small intervals $(t_k, t_k + \Delta t)$ the sample paths $x(t)$ of the Markov process $X(t)$ will be almost like those of the Wiener process starting at the point x_k at the time t_k and having drift and intensity parameters μ_k and σ_k, respectively.

Since the sample paths of the process $X(t)$ are continuous, they can thus be imagined as being patched together from infinitesimal pieces of paths of Wiener processes, with drift and intensity parameters changing from piece to piece (Fig. 6.3.1).

The process $X(t)$ will no longer have independent increments, since the parameters μ_{k+1} and σ_{k+1} that determine the (Gaussian) distribution of the $(k + 1)$st piece of the Wiener process depend on the value $x_k = X(t_k)$, that is, on the preceding kth piece.

Surprisingly, perhaps, the process $X(t)$ need not even be a Gaussian process. The reason is the same as above and is easily verified by looking at the first two

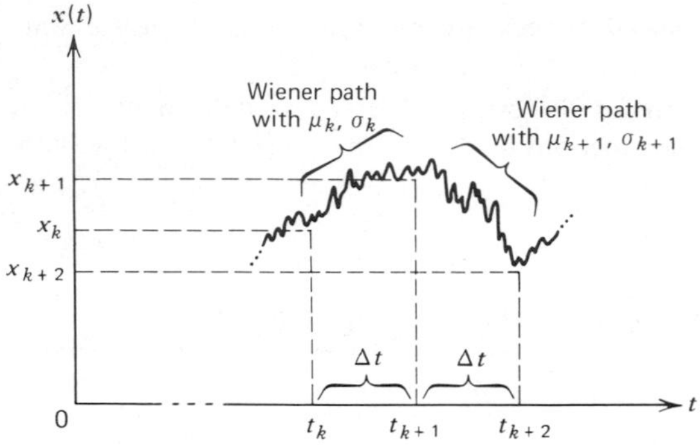

Figure 6.3.1

Wienerian pieces of the process $X(t)$ with initial value $X(0) = 0$. The density of $X(t_1)$ at time $t_1 = t_0 + \Delta t = \Delta t$ will be Gaussian,

$$f_{t_1}(x_1) = \frac{1}{\sqrt{2\pi\sigma_0^2 \Delta t}} e^{-(x_1 - \mu_0 \Delta t)^2/2\sigma_0^2 \Delta t},$$

as will the conditional density of $X(t_2)$ given $X(t_1) = x_1$

$$f_{t_2|t_1}(x_2 | x_1) = \frac{1}{\sqrt{2\pi\sigma_1^2 \Delta t}} e^{-(x_2 - x_1 - \mu_1 \Delta t)^2/2\sigma_1^2 \Delta t}$$

since $X(t_2) = W_0(\Delta t)$, which is the Wiener process with parameters μ_1, σ_1 starting at t_1, with $W(t_1) = x_1$. Therefore, the density of $X(t_2)$ must be given by

$$f_{t_2}(x_2) = \int_{-\infty}^{\infty} f_{t_2|t_1}(x_2 | x_1) f_{t_1}(x_1)\, dx_1,$$

which is not necessarily Gaussian inasmuch as μ_1 and σ_1 are functions of x_1.

On the other hand, the process $X(t)$, $t \geq 0$, constructed by patching together the Wienerian pieces, will clearly be Markov, since given a value $X(t) = x$ of the process at some present time $t \geq 0$, the probability law of the entire future of the process is completely determined by this value and hence independent of the past $X(t')$, $0 \leq t' < t$.

Furthermore, since for sufficiently small Δt, the conditional probability law of the forward increment $dX(t_k) = X(t_k + dt) - X(t_k)$, given $X(t_k) = x_k$, is approximately identical to the conditional probability law of the corresponding kth Wienerian piece, we have

$$\mathrm{E}[dX(t_k) | X(t_k) = x_k] \doteq \mu_k\, dt, \tag{3}$$

and

$$\mathrm{E}[(dX(t_k))^2 | X(t_k) = x_k] \doteq \sigma_k^2\, dt. \tag{4}$$

In addition, for any $\varepsilon > 0$ the Chebyshev inequality implies that

$$P(|dX(t_k)| \geq \varepsilon | X(t_k) = x_k) \leq \frac{\sigma_k^2 \, dt}{\varepsilon^2}. \tag{5}$$

Now μ_k and σ_k above were defined by (2), and hence as $\Delta t \to 0$ the assumption that $a(x, t)$ and $b(x, t)$ do not change rapidly implies that (5), (3), and (4) can be written in a more rigorous fashion. In particular, we should have

$$\lim_{dt \to 0} \frac{1}{dt} P(|dX(t)| \geq \varepsilon | X(t) = x) = 0, \tag{6a}$$

$$\lim_{dt \to 0} \frac{1}{dt} \mathrm{E}[dX(t) | X(t) = x] = a(x, t), \tag{6b}$$

$$\lim_{dt \to 0} \frac{1}{dt} \mathrm{E}[(dX(t))^2 | X(t) = x] = b^2(x, t), \tag{6c}$$

for all $\varepsilon > 0$, $0 \leq t$, and for all x in the state space of the process. Of course (6b,c) is just a rigorous way of writing (6.2.6) and (6.2.7) while (6a) allows us to neglect higher-order terms in the Taylor expansion, which we used to derive the Kolmogorov backward equation. This means that we have essentially come back to where we started and shown, albeit somewhat informally, *that a finite-variance sample-continuous process that satisfies differential equation (6.2.1) is identical with a Markov process that satisfies conditions (6).*

The latter processes are called *diffusion processes*, the functions $a(x, t)$ and $b^2(x, t)$ defined by the left-hand sides of (6b,c) being known as the *instantaneous mean and variance* of the diffusion process. Alternative terms, also used frequently, are *coefficient of drift* and *coefficient of diffusion*,† respectively.

The reason for this terminology is that processes of this kind were first encountered in physics in studying diffusion phenomena. It may be instructive to look at a simple model for diffusion to illustrate the connection.

Consider a thin column of partially ionized gas, for instance a fluorescent tube, and assume for simplicity that we wish to study the motion of electrons along the axis of this column. If there is an external electrostatic field, possibly due to a voltage applied to the electrodes at the ends of the column, the electrons will be attracted toward the anode and thus produce a macroscopic current in the direction of the axis. However, the electrons will also be constantly colliding with neutral molecules of the gas, and since the latter undergo thermal motion, the electrons will be bouncing back and forth in random fashion. Further, if the neutral molecules are far heavier than the electrons the direction and momentum of an electron after collision could be assumed to be independent of its own

† The function $b(x, t)$ is also called the intensity function of the process.

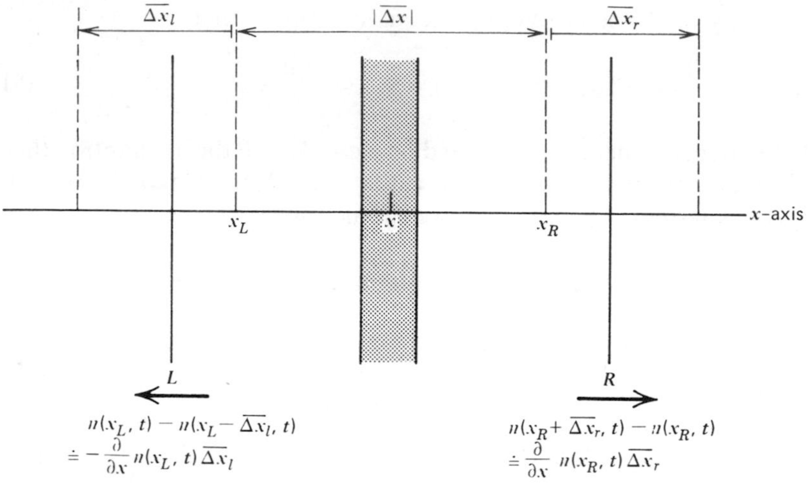

Figure 6.3.2

direction and momentum before the collision. In other words, the electron's motion will be Markovian.

Now imagine a disk of infinitesimal thickness located at the point x of the axis, and let $n(x, t)$ be the number of electrons inside the disk at some time t. Also, let $\overline{\Delta x_r}$ be the average distance traveled to the right (i.e., in the positive direction of the axis) by those electrons, and let $\overline{\Delta x_l}$ be the average distance traveled to the left, both during a time interval Δt.

Consider next two cross sections R and L to the right and left of the interval (x_L, x_R) (see Fig. 6.3.2). Then the average number of electrons crossing the two cross sections outwardly during Δt will be approximately

$$\frac{\partial n(x_R, t)}{\partial x} \overline{\Delta x_r} \qquad \text{and} \qquad -\frac{\partial n(x_L, t)}{\partial x} \overline{\Delta x_l} \tag{7}$$

and so the average net flow of electrons out of the interval (x_L, x_R) during Δt will be the sum of these two quantities.

Now let

$$\overline{\Delta x} = \overline{\Delta x_r} - \overline{\Delta x_l} \tag{8}$$

be the average net distance traveled in the positive direction during t, and let

$$\overline{|\Delta x|} = \overline{\Delta x_r} + \overline{\Delta x_l} \tag{9}$$

be the average magnitude of the displacement during the same time interval Δt.

From (8) and (9)

$$\overline{\Delta x_r} = \frac{1}{2}\overline{\Delta x} + \frac{1}{2}\overline{|\Delta x|}, \qquad \overline{\Delta x_l} = -\frac{1}{2}\overline{\Delta x} + \frac{1}{2}\overline{|\Delta x|}.$$

Substituting these into (7) and adding, we calculate the net average outflow of electrons from (x_L, x_R) to be

$$\frac{1}{2}\left[\frac{\partial n(x_R, t)}{\partial x} + \frac{\partial n(x_L, t)}{\partial x}\right]\overline{\Delta x} + \frac{1}{2}\left[\frac{\partial n(x_R, t)}{\partial x} - \frac{\partial n(x_L, t)}{\partial x}\right]\overline{|\Delta x|} \tag{10}$$

But continuity of motion of electrons requires that for small Δt, both $\overline{\Delta x}$ and $\overline{|\Delta x|}$ also be small, and thus we can write (10) as

$$\frac{\partial n(x, t)}{\partial x}\overline{\Delta x} + \frac{1}{2}\frac{\partial^2 n(x, t)}{\partial x^2}\overline{(\Delta x)^2}. \tag{11}$$

which is now the net outflow of electrons from the disk at the point x during the time interval $(t, t + \Delta t)$.

Since electrons can neither be created nor destroyed (we are neglecting effects like ionization and recombination), this quantity must be equal to the net decrease of their number during the time interval $(t, t + \Delta t)$, which is given by the time derivative

$$-\frac{\partial n(x, t)}{\partial t}\overline{\Delta t}. \tag{12}$$

Hence equating (12) and (13) and dividing by $\overline{\Delta t}$, we get

$$\frac{-\partial n(x, t)}{\partial t} = \frac{\partial n(x, t)}{\partial x}\frac{\overline{\Delta x}}{\Delta t} + \frac{1}{2}\frac{\partial^2 n(x, t)}{\partial x^2}\frac{\overline{(\Delta x)^2}}{\Delta t}. \tag{13}$$

To obtain a differential equation governing the diffusion of charge distribution

$$q(x, t) = n(x, t)\epsilon_0,$$

where ϵ_0 is the charge of an electron, we let $\Delta t \to 0$ and call

$$\lim_{\Delta t \to 0} \frac{\overline{\Delta x}}{\Delta t} = a(x, t), \qquad \lim_{\Delta t \to 0} \frac{\overline{(\Delta x)^2}}{\Delta t} = b^2(x, t). \tag{14}$$

Then, equation (13) becomes

$$-\frac{\partial q(x, t)}{\partial t} = a(x, t)\frac{\partial q(x, t)}{\partial x} + \frac{1}{2}b^2(x, t)\frac{\partial^2 q(x, t)}{\partial x^2} \tag{15}$$

which is the (backward) equation for diffusion of charge. We see that it is formally identical with the backward Kolmogorov equation (6.2.9). Note that the left-hand side of (15) represents a current, while $a(x, t)$ has the dimensions of velocity—the so-called drift velocity. The diffusion coefficient $b^2(x, t)$ can be expressed in terms of the thermal energy of the gas and the average time between collisions, and it is then the task of the physicist to establish the exact relationships between the drift and diffusion coefficient and measurable quantities such as the external field force, the temperature of the gas, and so forth.

It is now an easy matter to establish a link between this physical model for diffusion and the concept of a diffusion stochastic process as defined earlier. If we single out one particular electron and follow its x-coordinate as a function of time, we are defining a stochastic process $X(t)$, $t \geq 0$. From the above discussion, this process should be Markov, and since the motion of electrons is continuous, it should have continuous sample paths. This requires that for small time increments dt, large displacements $dX(t)$ must be improbable, which leads to condition (6a). Since we have assumed that there is no interaction between electrons themselves, the averages $\overline{\Delta x}$ and $\overline{(\Delta x)^2}$ can be replaced by their expectations (conditional upon the present position), and hence (14) leads to conditions (6b,c). Finally, dividing both sides of equation (13) by the total number of electrons N, we can interpret the fraction

$$\frac{n(x, t)}{N}$$

as the probability density of $X(t + \Delta t)$, given $X(t) = x$, and we end up with Kolmogorov's backward equation for the transition probability density of the diffusion process.

EXERCISES

Exercise 6.3

1. Verify by direct computation that the Ornstein-Uhlenbeck process $X(t)$, $t \geq 0$, of Example 6.2.1 satisfies the conditions (6) and hence is a diffusion process.
2. Let $X(t)$, $t \geq 0$, be a process defined by $X(t) = W_{01}^2(t) + W_{02}^2(t)$, where $W_{01}(t)$ and $W_{02}(t)$ are independent standard Wiener processes. Show that this process satisfies the conditions (6b,c) and determine the coefficients of drift and diffusion. (*Hint:* Write

$$dX(t) = (dW_{01}(t))^2 + (dW_{02}(t))^2 + 2W_{01}\, dW_{01}(t) + 2W_{02}(t)\, dW_{02}(t)$$

and condition first on $(W_{01}(t'), W_{02}(t'))$, $0 \leq t' \leq t$.)
3. Let $X(t)$, $t \geq 0$, be a Markov process with transition probability density

$$f_{t_2|t_1}(x_2 | x_1) = \frac{1}{\Gamma(t_2 - t_1)}(x_2 - x_1)^{t_2 - t_1 - 1} e^{-(x_2 - x_1)},$$

$0 < x_1 < x_2$, $0 \leq t_1 < t_2$, and such that $X(0) = 0$. Show that $X(t)$ is not a diffusion process. Which of the conditions (6a,b,c) is violated? (*Hint:* The process has independent increments with $dX(t)$ having the gamma density with parameters dt and 1.)

6.4. HOMOGENEOUS DIFFUSION PROCESSES

To gain a deeper insight into the properties of diffusion processes, we will now consider a special class of these processes, namely those for which the infinitesimal mean (6.3.6b) and infinitesimal variance (6.3.6c) do not depend on time. In other words, we wish to study finite-variance sample-continuous processes

$$X(t), \qquad t \geq 0, \tag{1}$$

whose forward infinitesimal increments satisfy the equation

$$dX(t) = a(X(t))\,dt + b(X(t))\,dW_0(t), \tag{2}$$

with $a(x)$ and $b(x)$ being defined on the state space of the process. We already know that the process (1) will be a Markov process. To keep things simple, we will assume that the transition probability densities are differentiable as needed.

Since for any fixed time t_1 the conditional distribution, given $X(t_1) = x_1$, of the right-hand side (2) can depend only on the time increment dt and not on t_1, it is clear that for any $t_2 > t_1$ the conditional distribution of $X(t_2)$, given $X(t_1) = x_1$, can only depend on the two time instances t_1 and t_2 via the difference

$$t = t_2 - t_1. \tag{3}$$

This means in particular that the transition probability densities

$$f_{t_2|t_1}(x_2 | x_1)$$

will also depend on t_1 and t_2 only through the difference (3), which is to say that the Markov process will have stationary transition probabilities. Taking advantage of this, we simplify our notation slightly and write

$$f_{t_2|t_1}(x_2 | x_1) = f_t(x_2 | x_1)$$

where $t = t_2 - t_1 > 0$. Since with this notation

$$\frac{\partial f_{t_2|t_1}(x_2 | x_1)}{\partial t_2} = \frac{\partial f_t(x_2 | x_1)}{\partial t},$$

and

$$\frac{\partial f_{t_2|t_1}(x_2 | x_1)}{\partial t_1} = -\frac{\partial f_t(x_2 | x_1)}{\partial t},$$

the forward and backward equations take the form

$$\frac{\partial f_t(x_2 | x_1)}{\partial t} = -\frac{\partial}{\partial x_2}[a(x_2) f_t(x_2 | x_1)] + \frac{1}{2}\frac{\partial^2}{\partial x_2^2}[b^2(x_2) f_t(x_2 | x_1)],$$

and

$$\frac{\partial f_t(x_2 \mid x_1)}{\partial t} = a(x_1)\frac{\partial f_t(x_2 \mid x_1)}{\partial x_1} + \frac{1}{2}b^2(x_1)\frac{\partial^2 f_t(x_2 \mid x_1)}{\partial x_1^2} \tag{4}$$

respectively.

If we now think of $X(t)$ as being the position of a diffusing particle at time t, then we know from our previous discussion that the infinitesimal mean $a(x)$ represents an external force pulling the particle in the positive direction whenever the particle reaches the state x. Let us see if we can transform the process in such a way that this force, that is, the infinitesimal mean $a(x)$, becomes identically zero.

Let us try the memoryless transformation

$$Y(t) = s(X(t)), \qquad t \geq 0, \tag{5}$$

where $s(x)$ is a monotonically increasing continuous function defined on the state space of the process. It is easy to see that the process (5) will again be a Markov process with continuous sample paths. Remember that we are using an abbreviated notation for the transition densities: $f_t(x_2 \mid x_1)$ represents the transition density for the process $X(t)$. To avoid confusion between the transition densities for $X(t)$ and $Y(t) = s(X(t))$, we denote them by $f_{X,t}(x_2 \mid x_1)$ and $f_{Y,t}(y_2 \mid y_1)$, respectively. We know that they must be related by the equation

$$f_{X,t}(x_2 \mid x_1) = \frac{ds(x_2)}{dx_2}\, f_{Y,t}(s(x_2) \mid s(x_1)). \tag{6}$$

(see Chapter One). If we now substitute from (6) into the backward equation (4), we obtain, abbreviating $f_Y = f_{Y,t}(s(x_2) \mid s(x_1))$,

$$\begin{aligned}\frac{ds(x_2)}{dx_2}\frac{\partial f_Y}{\partial t} &= a(x_1)\frac{ds(x_2)}{dx_2}\frac{\partial f_Y}{\partial y_1}\frac{ds(x_1)}{dx_1}\\ &\quad + \frac{1}{2}b^2(x_1)\frac{ds(x_2)}{dx_2}\frac{\partial^2 f_Y}{\partial y_1^2}\left(\frac{ds(x_1)}{dx_1}\right)^2\\ &\quad + \frac{1}{2}b^2(x_1)\frac{ds(x_2)}{dx_2}\frac{\partial f_Y}{\partial y_1}\frac{d^2 s(x_1)}{dx_1^2}.\end{aligned}$$

The derivative $ds(x_2)/dx_2$ cancels on both sides, and if we assume that the function s is such that

$$\frac{1}{2}b^2(x_1)\frac{d^2 s(x_1)}{dx_1^2} + a(x_1)\frac{ds(x_1)}{dx_1} = 0 \tag{7}$$

we have the backward equation for the process $Y(t)$, namely

$$\frac{\partial f_{Y,t}(y_2 \mid y_1)}{\partial t} = \frac{1}{2}b_Y^2(y_1)\frac{\partial^2 f_{Y,t}(y_2 \mid y_1)}{\partial y_1}, \tag{8}$$

with a new infinitesimal variance given by

$$b_Y^2(y) = b^2(x)\left(\frac{ds(x)}{dx}\right)^2\Bigg|_{x=s^{-1}(y)}, \tag{9}$$

where $b_X^2 = b^2$ is the original infinitesimal variance of the process $X(t)$. Thus, we see that the memoryless transformation

$$Y(t) = s(X(t))$$

will indeed result in a process with zero infinitesimal mean, provided we can find a monotonically increasing continuous function $s(x)$ satisfying the differential equation

$$\frac{1}{2}b^2(x)\frac{d^2s(x)}{dx^2} + a(x)\frac{ds(x)}{dx} = 0. \tag{10}$$

But this is an ordinary differential equation, and if we assume that $a(x)$ and $b^2(x)$ are continuous functions and $b^2(x) > 0$ for all x, then (10) is easily solved, yielding

$$\frac{ds(x)}{dx} = C_1 e^{-2\int \frac{a(x)}{b^2(x)}dx}, \tag{11}$$

and

$$s(x) = C_1 \int e^{-2\int \frac{a(x)}{b^2(x)}dx}\, dx + C_2, \tag{12}$$

where $C_1 > 0$ and C_2 are arbitrary constants. From (11), we see that

$$\frac{ds(x)}{dx} > 0,$$

so that $s(x)$ is continuous and monotonically increasing as required.

The equation

$$y = s(x)$$

represents a scale transformation of the state space of the process $X(t)$ and is therefore called a *scale function* of this process.† It is determined by formula (12) up to the constants $C_1 > 0$ and C_2, which may be chosen for convenience. For instance, if the initial state of the process $X(t)$ is

$$X(0) = x_0$$

† It has no relation to a spectral density, also denoted by s in Chapter Five. We are just running out of letters.

we may wish to choose these constants such that

$$s(x_0) = 0 \qquad \text{and} \qquad \frac{ds(x_0)}{dx} = 1. \tag{13a,b}$$

What, then, is the significance of this scale function?

To answer this question, let us first assume that the process $X(t)$ is *regular*, in the sense that starting from some point x_0 in the state space, there is a positive probability that the process will pass through any other point x at some future time. (This assumption is quite natural, since we can always define the state space such that the above property is satisfied by simply removing "inaccessible" points. Since the process $X(t)$ has continuous sample paths, the resulting state space will always be an interval.)

Now let $s(x)$ be a scale function of the process $X(t)$, and let

$$Y(t) = s(X(t)), \qquad t \geq 0.$$

As we have seen, the process $Y(t)$ will be Markov, with transition probability densities satisfying the backward equation (8). The corresponding relation for forward differentials is then

$$dY(t) = b_Y(Y(t))\, dW_0(t). \tag{14}$$

Furthermore, since the scale function is continuous, the regularity of the process $X(t)$ implies that the transformed process $Y(t)$ will also be regular.

Suppose that at some time t_0 the process is in state $y = Y(t_0)$, and consider an interval $I_Y(\eta_l, \eta_r)$ with endpoints η_l and η_r such that, $\eta_l < y < \eta_r$. We assume that this interval lies entirely in the state space of the process $Y(t)$. Since the process $Y(t)$ is regular, it must eventually exit from this interval. This means, if T_Y is the random time it takes the process to exit from the interval $I_Y(\eta_l, \eta_r)$,

$$T_Y = \min\{t - t_0\colon Y(t) = \eta_r \quad \text{or} \quad Y(t) = \eta_l\}, \tag{15}$$

then (see Fig. 6.4.1).

$$P(T_Y < \infty) = 1. \tag{16}$$

We will refer to T_Y as the exit time from the interval $I_Y(\eta_l, \eta_r)$.† Since the process $X(t)$, and hence also $Y(t)$, is Markov and homogeneous it is clear that the distribution of the exit time T_Y will not depend on time t_0. That is, the distribution of the exit time will be the same upon each visit to the state y. Thus, we can simply take $t_0 = 0$, that is, assume that the process starts at the state y. Furthermore, $Y(t)$ has continuous sample paths, and hence at the time T_Y, either

$$Y(T_Y) = \eta_r \qquad \text{with some probability } P_r(y),$$

† Remember that T_Y depends not only on the interval $I_Y(\eta_l, \eta_r)$ but also on the state $y \in I_Y(\eta_l, \eta_r)$.

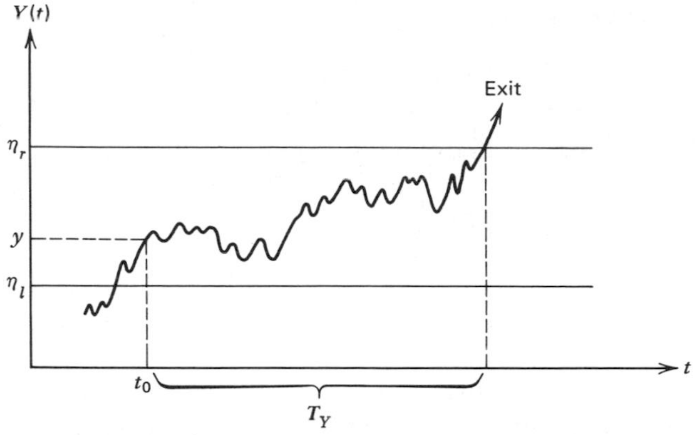

Figure 6.4.1

or

$$Y(T_Y) = \eta_l \qquad \text{with some probability } P_l(y),$$

where, because of (16),

$$P_r(y) + P_l(y) = 1.$$

This implies that the expectation

$$E[Y(T_Y)] = \eta_r P_r(y) + \eta_l P_l(y). \tag{17}$$

On the other hand, (14) implies that

$$E[dY(t) | Y(t)] = b_Y(Y(t))E[dW_0(t) | Y(t)] = 0$$

so that the process $Y(t)$ is a martingale. Since clearly T_Y is a stopping time for the process $Y(t)$ and

$$E[Y^2(T_Y)] \leq \eta_r^2 + \eta_l^2 < \infty,$$

the conditions of the optional stopping theorem (see Chapter One) are satisfied, and hence

$$E[Y(T_Y)] = E[Y(0)] = y. \tag{18}$$

Combining (18) with (17), we obtain the exit probabilities

$$P_r(y) = \frac{y - \eta_l}{\eta_r - \eta_l}, \qquad P_l(y) = \frac{\eta_r - y}{\eta_r - \eta_l}, \tag{19}$$

or if we denote the distances of the point y from the endpoints of the interval $I_Y(\eta_l, \eta_r)$ by

$$\Delta_l = y - \eta_l, \qquad \Delta_r = \eta_r - y$$

the exit probabilities can be written as

$$P_r(y) = \frac{\Delta_l}{\Delta_r + \Delta_l}, \qquad P_l(y) = \frac{\Delta_r}{\Delta_r + \Delta_l}.$$

In particular, if $\Delta_r = \Delta_l$, that is, if y is the midpoint of the interval $I_y(\eta_l, \eta_r)$, then

$$P_r(y) = P_l(y) = \frac{1}{2}.$$

But these are exactly the same exit probabilities obtained in Exercise 1.52 for the standard Wiener process. We refer to this feature by saying that the process $Y(t)$ is in *natural scale.*

Determining the exit probabilities for the original process $X(t)$ is now easy. Take an interval $I_X(\xi_l, \xi_r)$ with endpoints ξ_l, ξ_r in the state space of $X(t)$, and let

$$X(0) = x,$$

be a point in the interior of this interval. We wish to compute the probabilities $P_r(x)$ and $P_l(x)$ that the process $X(t)$ exits from the interval $I_X(\xi_l, \xi_r)$ via the right versus the left endpoint, respectively. Since the scale function s is monotonically increasing and continuous, it is apparent (see Fig. 6.4.2) that $X(t)$ exits through ξ_r

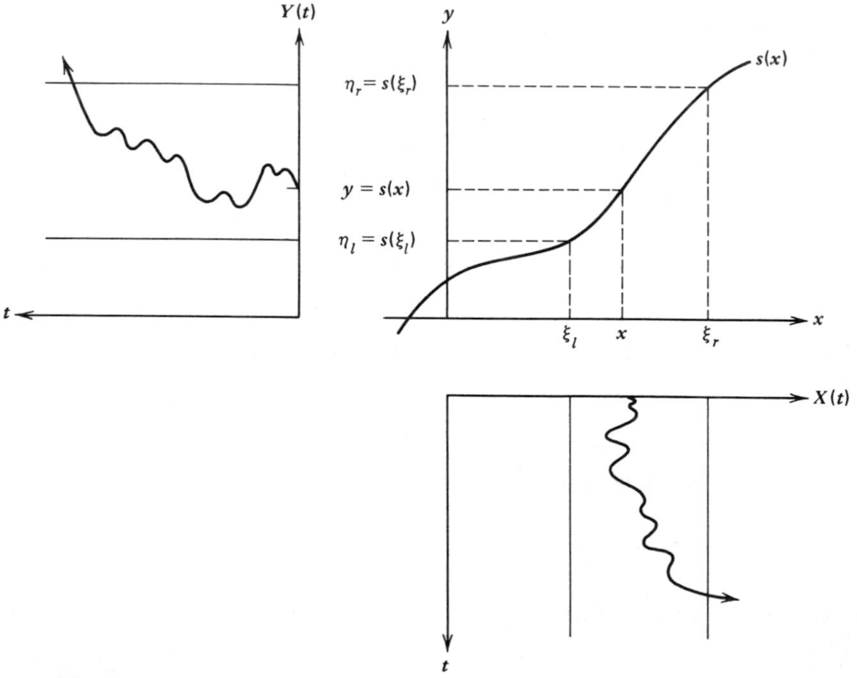

Figure 6.4.2

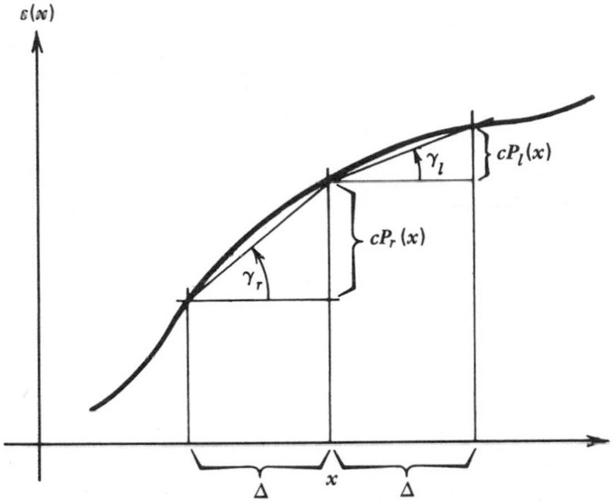

Figure 6.4.3

if and only if the process $Y(t) = s(X(t))$ exits through $s(\xi_r)$; the same is true for the other endpoint. Hence from (19), we have immediately

$$P_r(x) = \frac{s(x) - s(\xi_l)}{s(\xi_r) - s(\xi_l)}, \qquad P_l(x) = \frac{s(\xi_r) - s(x)}{s(\xi_r) - s(\xi_l)}. \tag{20}$$

Note that these expressions do not depend on the choice of the constants $C_1 > 0$ and C_2 in the formula (12) for $s(x)$.

The exit probabilities (20) can be read directly from the graph of the scale function $s(x)$. For instance, if we take an interval of length 2Δ with center at x (see Fig. 6.4.3), then it is easily seen from (20) that the exit probabilities $P_r(x)$ and $P_l(x)$ are proportional to the tangent of the angles γ_r and γ_l, respectively. Thus, if $\gamma_r > \gamma_l$, that is, if the scale function is concave in the interval $(x - \Delta, x + \Delta)$, then there is a greater tendency for the paths of the process to move in the positive direction, while if $s(x)$ is convex in the neighborhood of x, there is a greater tendency to move in the negative direction.

We can also conclude from (20) that the process $X(t)$ is in natural scale if and only if its scale function is linear:

$$s(x) = C_1 x + C_2, \qquad C_1 > 0.$$

Example 6.4.1
Consider the equation

$$\dot{X}(t) = -\tanh X(t) + \dot{W}_0(t), \tag{21}$$

which may arise, for instance, if a nonlinear memoryless element with input-output characteristic $y = \tanh x$ (e.g., a saturating linear amplifier) is inserted into

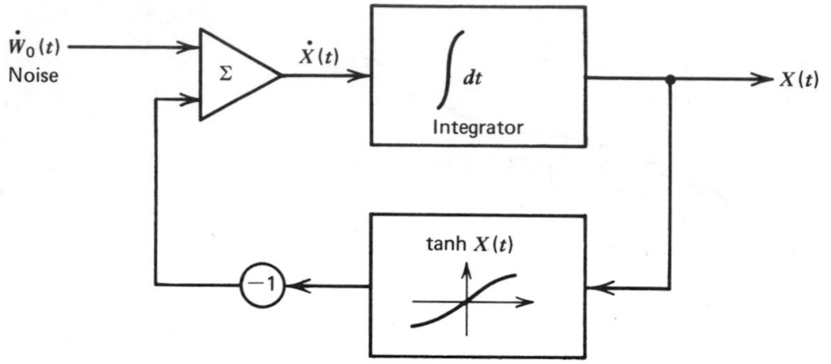

Figure 6.4.4

a negative feedback link in the circuit shown in Figure 6.4.4. Equation (21) has the general form

$$dX(t) = a(X(t))\,dt + b(X(t))\,dW_0(t),$$

with $a(x) = -\tanh x$ and $b(x) = 1$, and thus $X(t)$, $t \geq 0$, is a homogeneous diffusion process. To find its scale function $s(x)$, we first use (11), that is,

$$\frac{ds(x)}{dx} = C_1 e^{2\int \tanh x\,dx}.$$

Now, as is easily verified,

$$\int \tanh x\,dx = \ln \cosh x,$$

and hence

$$\frac{ds(x)}{dx} = C_1 e^{2 \ln \cosh x} = C_1(\cosh x)^2.$$

Suppose that the initial state $X(0) = 0$. Then, to satisfy (13b), we choose $C_1 = 1$, and by (12)

$$s(x) = \int (\cosh x)^2\,dx + C_2. \tag{22}$$

By the substitution $\cosh x = \frac{1}{2}(e^x + e^{-x})$, the indefinite integral in (22) equals $\frac{1}{2}x + \frac{1}{4}\sinh 2x$, and to satisfy (13a), we set $C_2 = 0$. Thus the scale function of the process $X(t)$, $t \geq 0$, is found to be

$$s(x) = \frac{1}{2}\left(x + \frac{1}{2}\sinh 2x\right), \qquad -\infty < x < \infty$$

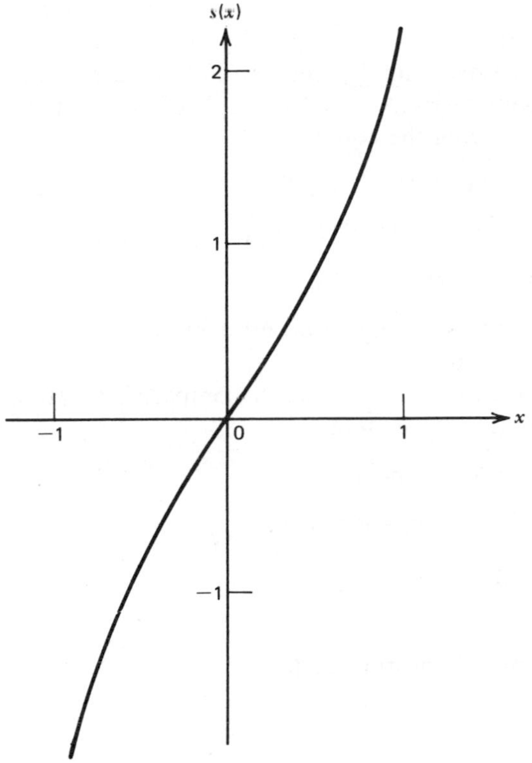

Figure 6.4.5

and is sketched in Figure 6.4.5. This function is concave for $x < 0$ and convex for $x > 0$, and hence, as expected, there is a tendency for the process $X(t)$ always to move toward zero, the tendency being greater the farther away from zero the process gets. The exit probabilities are easily computed from (20). In particular, for large x, we have approximately

$$s(x) \doteq \frac{1}{8} e^{2x},$$

and so the ratio of exit probabilities from an interval $(x - \frac{1}{2}, x + \frac{1}{2})$ becomes approximately

$$\frac{P_r(x)}{P_l(x)} \doteq \frac{1 - e^{-1}}{e - 1} \doteq .3678. \qquad \blacktriangle$$

Example 6.4.2

The scale transformation $Y(t) = s(X(t)), t \geq 0$, may result in a homogeneous diffusion process whose state space may be restricted by finite boundaries even if the original process had no such finite boundaries. A simple example of such a case is provided by a process $X(t), t \geq 0$, satisfying the equation

$$dX(t) = \mu\, dt + \sigma\, dW_0(t), \qquad t \geq 0, \tag{23}$$

where $\mu > 0$ and $\sigma > 0$ are constants, and $X(0) = 0$. Integrating, we have

$$X(t) = \mu t + \sigma W_0(t), \qquad t \geq 0,$$

so that $X(t), t \geq 0$, is just a Wiener process with a constant drift, that is, with a linearly increasing mean function $\mu t, t \geq 0$.

Let us now calculate the scale function for this process. Comparing equation (23) above with the general equation (2), we see that

$$a(x) = \mu \qquad \text{and} \qquad b(x) = \sigma.$$

The differential equation (7) then has constant coefficients

$$\frac{1}{2}\sigma^2 \frac{d^2 s(x)}{dx^2} + \mu \frac{ds(x)}{dx} = 0,$$

and it is then easily verified either by substitution or from formula (12) that a solution satisfying $s(0) = 0$ and $ds(x)/dx\,|_{x=0} = 1$ is

$$s(x) = \frac{\sigma^2}{2\mu}(1 - e^{-2\mu x/\sigma^2}), \qquad -\infty < x < \infty.$$

From the sketch of this function (Fig. 6.4.6), we can see that this scale function maps the state space $(-\infty, \infty)$ of the process $X(t)$ into an interval $(-\infty, \sigma^2/2\mu)$. In other words, the transformed process

$$Y(t) = s(X(t)), \qquad t \geq 0, \tag{24}$$

will have a boundary at level $\sigma^2/2\mu$ beyond which no sample path $y(t), t \geq 0$, can ever go. The reason for this is that the sample paths $x(t), t \geq 0$, of the original process all drift to $+\infty$, while the transformed process (24) must be in natural scale, that is, its sample paths must not show any drift whatsoever. This is accomplished by progressively compressing the state space more and more for larger values of x. However, since every sample path of the process $X(t)$ eventually drifts away to $+\infty$, the compression must, so to speak, pull $+\infty$ down to a finite number, since otherwise the sample paths of $Y(t)$ would also drift to $+\infty$. Naturally, no sample path $x(t)$ can go "beyond" plus infinity, and so the transformed $+\infty$ appears as a finite boundary.

The positive drift of the process $X(t), t \geq 0$, is clearly demonstrated by com-

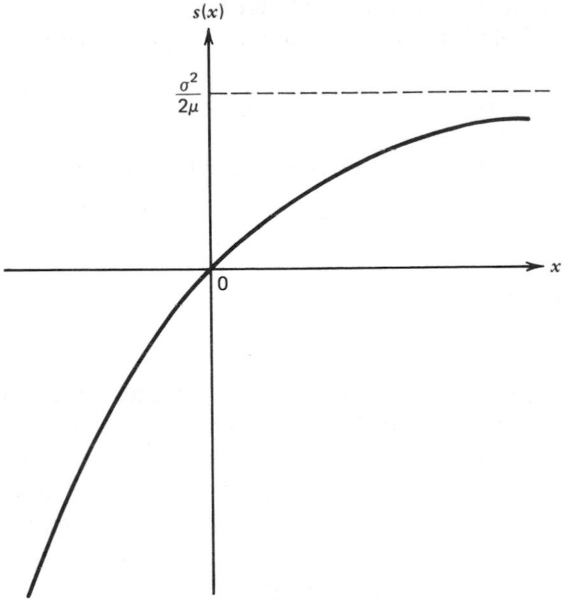

Figure 6.4.6

puting the exit probabilities $P_r(x)$ and $P_l(x)$ from formula (20). In particular, taking $X(0) = x = 0$, we have for $\xi_l < 0 < \xi_r$

$$P_l(0) = \frac{s(\xi_r)}{s(\xi_r) - s(\xi_l)} = \frac{1 - e^{-(2\mu/\sigma^2)\xi_r}}{e^{-(2\mu/\sigma^2)\xi_l} - e^{-(2\mu/\sigma^2)\xi_r}}$$

and

$$P_r(0) = \frac{e^{-(2\mu/\sigma^2)\xi_l} - 1}{e^{-(2\mu/\sigma^2)\xi_l} - e^{-(2\mu/\sigma^2)\xi_r}}.$$

Letting $\xi_l \to -\infty$, we find that

$$P_r(0) \to 1 \qquad \text{for any } \xi_r > 0,$$

while letting $\xi_r \to +\infty$, we have

$$P_l(0) \to e^{(2\mu/\sigma^2)\xi_l} < 1 \qquad \text{since } \xi_l < 0.$$

Thus, every level $\xi_r > 0$ on the positive side of the initial state $X(0) = 0$ is eventually reached with probability 1, while the probability of the process $X(t)$ ever reaching a negative level $\xi_l < 0$ equals only

$$e^{(2\mu/\sigma^2)\xi_l}.$$

This is clearly true for any initial state $X(0) = x$, since levels on the positive side of x are reached and crossed with probability 1, while for any level $\xi_l < x$, there is a positive probability

$$1 - e^{(2\mu/\sigma^2)(\xi_l - x)}$$

that ξ_l is never reached. Since the process $X(t)$ is Markov, once $X(t) = x$ for some x, this state can be regarded as a new initial state, and so every sample path $x(t)$, $t \geq 0$, drifts to $+\infty$. ▲

Now let us study further the transformed process

$$Y(t) = s(X(t)), \qquad t \geq 0.$$

We already know that this process is in natural scale so that its exit probabilities $P_l(y)$ and $P_r(y)$ are the same as if it were a standard Wiener process. That is, if the process starts at the midpoint of some interval $I_r(\eta_l, \eta_r)$, then it has equal probability to exit through the left or right endpoint. However, the process may still differ from the standard Wiener process in the time it takes to exit from this interval.

As before, denote $I_Y(\eta_l, \eta_r)$, an interval with endpoints $\eta_l < \eta_r$, and let T_Y be the time it takes the process $Y(t)$ to exit from this interval provided it started at the point y, $\eta_l < y < \eta_r$. We wish to compute the expected value of this exit time T_Y,

$$\mathrm{E}[T_Y \mid Y(0) = y].$$

To this end, we introduce the function $q(y_1, t_1)$, defined for $t_1 \geq 0$ and $\eta_l < y_1 < \eta_r$ to be the conditional expectation

$$q(y_1, t_1) = \mathrm{E}[T_Y \mid T_Y > t_1,\ Y(t_1) = y_1]. \tag{25}$$

Using the smoothing property of conditional expectations and the Markov property of the process $Y(t)$, we have for any $0 \leq t_1 < t$

$$\begin{aligned} q(Y(t_1), t_1) &= \mathrm{E}[\mathrm{E}[T_Y \mid T_Y > t,\ Y(t),\ Y(t_1)] \mid T_Y > t_1,\ Y(t_1)] \\ &= \mathrm{E}[\mathrm{E}[T_Y \mid T_Y > t,\ Y(t)] \mid T_Y > t_1,\ Y(t_1)] \\ &= \mathrm{E}[q(Y(t), t) \mid T_Y > t_1,\ Y(t_1)]. \end{aligned}$$

Taking conditional expectations $\mathrm{E}[\quad \mid Y(t_1) = y_1]$ on both sides of this equation leaves

$$q(y_1, t_1) = \mathrm{E}[q(Y(t), t) \mid Y(t_1) = y_1], \qquad 0 \leq t_1 \leq t,$$

which is condition (6.2.13).

Consequently, the function $q(y_1, t_1)$ must satisfy the backward equation

$$-\frac{\partial q(y_1, t_1)}{\partial t_1} = \frac{1}{2} b_Y^2(y_1) \frac{\partial^2 q(y_1, t_1)}{\partial y_1^2}, \tag{26}$$

for $t_1 \geq 0$ and $\eta_l < y_1 < \eta_r$. In particular, the equation must hold true for $t_1 = 0$, in which case, as seen from (25),

$$q(y, 0) = \mathrm{E}[T_Y \mid Y(0) = y]$$

is the desired expected exit time.

To obtain the partial derivative with respect to time, we note that since $Y(t)$ is a homogeneous Markov process,

$$\begin{aligned} q(y, t_1) &= \mathrm{E}[T_Y \mid T_Y > t_1,\ Y(t_1) = y] \\ &= \mathrm{E}[T_Y + t_1 \mid Y(0) = y] = t_1 + q(y, 0), \end{aligned}$$

and thus

$$\left.\frac{\partial q(y, t_1)}{\partial t_1}\right|_{t_1 = 0} = \lim_{t_1 \to 0} \frac{1}{t_1}(q(y, t_1) - q(y, 0)) = 1.$$

Writing $q(y)$ as an abbreviation for $q(y, 0)$, equation (26), with $t_1 = 0$ and $y_1 = y$, turns into an ordinary differential equation

$$-1 = \frac{1}{2} b_Y^2(y) \frac{d^2 q(y)}{dy^2}, \tag{27}$$

where $\eta_l < y < \eta_r$. In order to solve this equation we also need to know the behavior of the function $q(y)$ as y approaches either of the two endpoints of the interval $I_Y(\eta_l, \eta_r)$. Since we have defined the function $q(y)$ as

$$q(y) = \mathrm{E}[T_Y \mid Y(0) = y]$$

we have

$$q(\eta_l) = q(\eta_r) = 0,$$

because $T_Y = 0$ for $y = \eta_l$ or $y = \eta_r$ by the definition (15) of the exit time.

Obtaining the solution of our differential equation (27) is now an easy matter. First, for convenience, rewrite this equation as

$$\frac{d^2 q(y)}{dy^2} = -2 v_Y(y),$$

where we have denoted

$$v_Y(y) = \frac{1}{b_Y^2(y)}. \tag{28}$$

Integrating once yields

$$\frac{dq(y)}{dy} = -2 \int_{\eta_l}^{y} v_Y(z)\, dz + C,$$

where C is as yet an unspecified integration constant. Integrating once more, we get

$$q(y) = -2 \int_{\eta_l}^{y} \left(\int_{\eta_l}^{u} v_Y(z)\, dz \right) du + C(y - \eta_l). \tag{29}$$

Note that we have already satisfied the condition $q(\eta_l) = 0$, and it remains to determine the constant C in order that we also have $q(\eta_r) = 0$.

To this end, let us first reverse the order of integration in the iterated integral in (29) to get

$$\int_{\eta_l}^{y} \left(\int_{\eta_l}^{u} v_Y(z)\, dz \right) du = \int_{\eta_l}^{y} \left(\int_{z}^{y} v_Y(z)\, du \right) dz = \int_{\eta_l}^{y} (y - z) v_Y(z)\, dz. \tag{30}$$

Substituting into (29) and setting $y = \eta_2$, we find that

$$0 = q(\eta_r) = -2 \int_{\eta_l}^{\eta_r} (\eta_r - z) v_Y(z)\, dz + C(\eta_r - \eta_l),$$

whence

$$C = \frac{2}{\eta_r - \eta_l} \int_{\eta_l}^{\eta_r} (\eta_r - z) v_Y(z)\, dz.$$

Thus (29) becomes (using (30))

$$\begin{aligned} q(y) &= -2 \int_{\eta_l}^{y} (y - z) v_Y(z)\, dz + 2 \int_{\eta_l}^{\eta_r} \frac{(\eta_r - z)(y - \eta_l)}{(\eta_r - \eta_l)} v_Y(z)\, dz \\ &= +2 \int_{\eta_l}^{y} \left(\frac{(\eta_r - z)(y - \eta_l)}{\eta_r - \eta_l} - (y - z) \right) v_Y(z)\, dz \\ &\quad + 2 \int_{y}^{\eta_r} \frac{(\eta_r - z)(y - \eta_l)}{\eta_r - \eta_l} v_Y(z)\, dz. \end{aligned} \tag{31}$$

Using the easily verified algebraic identity

$$(\eta_r - z)(y - \eta_l) - (y - z)(\eta_r - \eta_l) = (\eta_r - y)(z - \eta_l),$$

we can rewrite the first integral on the right-hand side of (31) as

$$\int_{\eta_l}^{y} \left(\frac{(\eta_r - z)(y - \eta_l)}{\eta_r - \eta_l} - (y - z) \right) v_Y(z)\, dz = \int_{\eta_l}^{y} \frac{(\eta_r - y)(z - \eta_l)}{\eta_r - \eta_l} v_Y(z)\, dz,$$

and we end up with the final expression for the expected exit time

$$q(y) = \mathrm{E}[T_Y \mid Y(0) = y]$$

from the interval $I_Y(\eta_l, \eta_r)$, namely

$$E[T_Y \mid Y(0) = y] = 2 \int_{\eta_l}^{y} \frac{(\eta_r - y)(z - \eta_l)}{(\eta_r - \eta_l)} v_Y(z)\, dz + 2 \int_{y}^{\eta_r} \frac{(y - \eta_l)(\eta_r - z)}{\eta_r - \eta_l} v_Y(z)\, dz. \tag{32}$$

Note that if $Y(t) = W_0(t)$ is a standard Wiener process starting at state y, then from (28) $v_Y(y) = 1$, and we obtain from (32) the expected exit time

$$E[T_{W_0} \mid W_0(0) = y] = (\eta_r - y)(y - \eta_l). \tag{33}$$

As with the exit probabilities, the formula for the expected exit time for a homogeneous diffusion process $X(t)$, $t \geq 0$, not necessarily in natural scale is now easy to obtain. If T_X denotes the time it takes the process $X(t)$ to exit from an interval $I_X(\xi_l, \xi_r)$, provided that $X(0) = x$ is an interior point, $\xi_l < x < \xi_r$ of this interval, then the process $Y(t) = s(X(t))$ will exit at the same time from the interval $I_Y(\eta_l, \eta_r)$ with $\eta_l = s(\xi_l)$ and $\eta_r = s(\xi_r)$. Since $X(0) = x$ implies $Y(0) = s(x)$, we have from (32)

$$E[T_x \mid X(0) = x] = 2 \int_{s(\xi_l)}^{s(x)} \frac{(s(\xi_r) - s(x))(z - s(\xi_l))}{s(\xi_r) - s(\xi_l)} v_Y(z)\, dz + 2 \int_{s(x)}^{s(\xi_r)} \frac{(s(x) - s(\xi_l))(s(\xi_r) - z)}{s(\xi_r) - s(\xi_l)} v_Y(z)\, dz.$$

Making the substitution $z = s(u)$ into both integrals, this formula becomes

$$E[T_X \mid X(0) = x] = 2 \int_{\xi_l}^{x} \frac{(s(\xi_r) - s(x))(s(u) - s(\xi_l))}{s(\xi_r) - s(\xi_l)} v_Y(s(u)) s'(u)\, du + 2 \int_{x}^{\xi_r} \frac{(s(x) - s(\xi_l))(s(\xi_r) - s(u))}{s(\xi_r) - s(\xi_l)} v_Y(s(u)) s'(u)\, du, \tag{34}$$

where $s'(u) = ds(u)/du$.

The function appearing in the integrands

$$v(x) = v_Y(s(x)) \frac{ds(x)}{dx}, \tag{35}$$

and defined on the state space of the process $X(t)$ is called a *speed function*† of this process. In view of relations (9) and (28), it can also be calculated directly from the infinitesimal variance and the scale function by

$$v(x) = \frac{1}{b^2(x) \dfrac{ds(x)}{dx}}. \tag{36}$$

† More accurately, a density function of the speed distribution.

From expression (34), we can see that the speed function $v(x)$, together with the scale function, tells us how long a typical sample path of the process $X(t)$ is expected to remain in the vicinity of a state x. In fact, if we evaluate the expected exit time $E[T_X \mid X(0) = x]$ from a small interval $I_x(x - \Delta, x + \Delta)$ with $\Delta > 0$ small, we obtain approximately the value

$$\Delta^2 v(x) \frac{ds(x)}{dx}. \tag{37}$$

In other words, for small Δ the expected time spent in the vicinity of x before exit is approximately $v(x)[ds(x)/dx]$ times the expected time spent in the same interval by a standard Wiener process. The speed function v, together with the scale function s, completely characterizes the homogeneous diffusion process $X(t)$, $t \geq 0$. For one thing, given the scale and speed functions, we can calculate the infinitesimal mean $a(x)$ and the infinitesimal variance $b^2(x)$ from the pair of equations (10) and (36). In fact, it turns out that scale and speed functions provide a more convenient characterization of a homogeneous diffusion than do the infinitesimal mean and variance, especially if the state space has finite boundaries. However, rather than going into more details in this respect, we prefer to discuss the meaning of the speed function v in terms of the behavior of the process.

Example 6.4.3

Consider a process $X(t)$, $t \geq 0$, satisfying the equation

$$\dot{X}(t) = a(X(t)) + \dot{W}_0(t), \tag{38}$$

where

$$a(x) = \begin{cases} -1 & \text{if} \quad x > 0, \\ 0 & \text{if} \quad x = 0, \\ +1 & \text{if} \quad x < 0. \end{cases}$$

Note that $-a(x)$ is the input-output characteristic of an ideal hard limiter, and so the process $X(t)$ may appear in a simple control system, as depicted in Figure 6.4.7. If the noise term $\dot{W}_0(t)$ were absent from equation (38), the system would, from any initial state $X(0)$, linearly approach the state $x = 0$ and stay in that state forever. Thus, with the initial state $X(0)$ being the only random element in the equation

$$\dot{X}(t) = a(X(t)), \qquad t \geq 0,$$

the sample paths $x(t)$, $t \geq 0$, would all look like those shown in Figure 6.4.8. Clearly, the white Gaussian noise $\dot{W}_0(t)$ in equation (38) will prevent the system from staying at state 0, but we would still expect to see at least a tendency for the process $X(t)$ to move toward $x = 0$.

To examine this, let us calculate the scale and speed functions of this process.

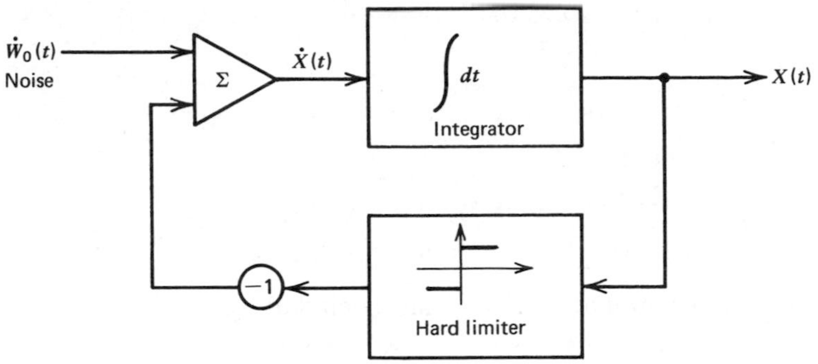

Figure 6.4.7

For the scale function $s(x)$, we use formula (12). The indefinite integral

$$\int \frac{a(x)}{b^2(x)}\,dx = \int a(x)\,dx = -|x|, \qquad -\infty < x < \infty,$$

since $b(x) = 1$, and the derivative of $-|x|$ is -1 if $x > 0$ and $+1$ if $x < 0$, that is, the function $a(x)$.† Consequently, by (11)

$$\frac{ds(x)}{dx} = C_1 e^{-2\int \frac{a(x)}{b^2(x)}dx} = e^{2|x|}, \qquad -\infty < x < \infty,$$

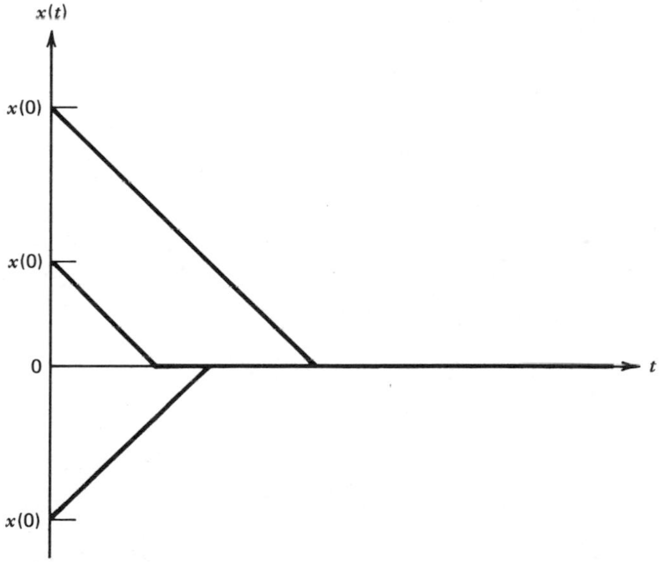

Figure 6.4.8

† Recall that the value $a(x)$ at $x = 0$ has no effect on the integral.

where we have chosen $C_1 = 1$ to assure that

$$\left.\frac{ds(x)}{dx}\right|_{x=0} = 1.$$

Integrating once more we obtain

$$s(x) = \begin{cases} \frac{1}{2}e^{2x} + C_2 & \text{if} \quad x > 0, \\ -\frac{1}{2}e^{-2x} + C_2 & \text{if} \quad x < 0. \end{cases}$$

To make the scale function continuous at $x = 0$ and such that $s(0) = 0$, we have to choose $C_2 = -\frac{1}{2}$ if $x \geq 0$ and $C_2 = \frac{1}{2}$ if $x < 0$, thus obtaining

$$s(x) = \begin{cases} \frac{1}{2}(e^{2x} - 1) & \text{if} \quad x \geq 0, \\ \frac{1}{2}(1 - e^{-2x}) & \text{if} \quad x < 0. \end{cases} \tag{39}$$

This scale function is sketched in Figure 6.4.9 and is a continuously differentiable function over $-\infty < x < \infty$, although its second derivative at $x = 0$ does not exist. Note that $s(x)$ is an odd function, that is,

$$s(x) = -s(-x) \qquad \text{for all } x$$

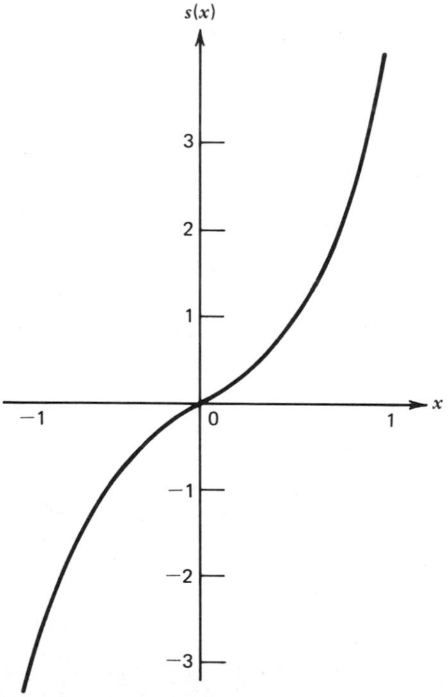

Figure 6.4.9

so that for any interval $I_x(\xi_l, \xi_r)$ with midpoint $x = 0$, that is, with $-\xi_l = \xi_r > 0$, we will have exit probabilities

$$P_r(0) = P(0) = \frac{1}{2}.$$

In fact, the scale function is quite similar to that of Example 6.4.1, concave for $x > 0$ and convex for $x < 0$, so that there is indeed a tendency for the sample paths to always move toward zero.

It may be interesting to look at the process

$$Y(t) = s(X(t)), \qquad t \geq 0, \tag{40}$$

transformed to its natural scale. From formula (9) for the new infinitesimal variance

$$b_Y^2(y) = b^2(x)\left(\frac{ds(x)}{dx}\right)^2\bigg|_{x=s^{-1}(y)}$$

we have

$$b_Y^2(y) = (1 + 2|y|)^2, \qquad -\infty < y < \infty,$$

since $b(x) = 1$, and from (39)

$$y = s^{-1}(x) = \begin{cases} \ \ \frac{1}{2}\ln(1 + 2y) & \text{if} \quad y > 0, \\ -\frac{1}{2}\ln(1 - 2y) & \text{if} \quad y < 0. \end{cases}$$

Thus the process (40) satisfies the equation

$$dY(t) = (1 + 2|Y(t)|)\, dW_0(t), \qquad t \geq 0,$$

and has the speed function

$$v_Y(y) = \frac{1}{b_Y^2(y)} = (1 + 2|y|)^{-2}, \qquad -\infty < y < \infty.$$

The speed function $v(x)$ of the original process can now be computed from either (35) or (36):

$$v(x) = e^{-2|x|}, \qquad -\infty < x < \infty$$

(see Fig. 6.4.10). Note that $v(x)$, $-\infty < x < \infty$, is actually a probability density, since

$$\int_{-\infty}^{\infty} v(x)\, dx = 1.$$

Finally, let us consider an interval $I_X(-\xi, \xi)$, $\xi > 0$, with midpoint $x = 0$, and let us calculate the expected exit time

$$\mathrm{E}[T_X | X(0) = 0].$$

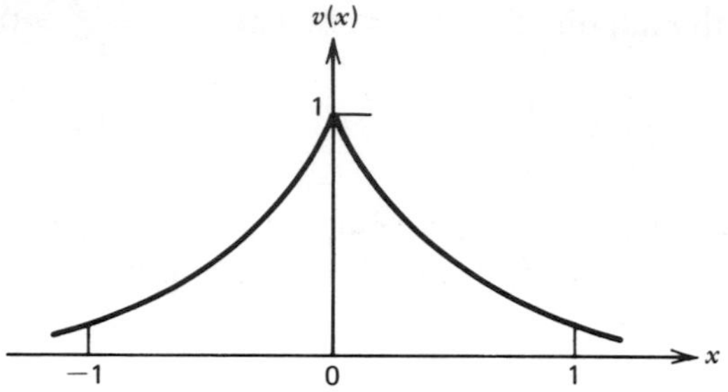

Figure 6.4.10

We could use formula (34), but it is somewhat simpler to first calculate the expected exit time $E[T_Y \mid Y(0) = 0]$ for the process $Y(t)$, $t \geq 0$, from the interval $I_Y(-\eta, \eta)$, $\eta > 0$, which for

$$\eta = s(\xi) = \frac{1}{2}(e^{2\xi} - 1) \tag{41}$$

must be the same as $E[T_X \mid X(0) = 0]$. We use formula (32). Since $\eta_r = \eta, \eta_l = -\eta$, $v_Y(-z) = v_Y(z)$ and $y = 0$, we have

$$\begin{aligned}
E[T_Y \mid Y(0) = 0] &= 2\int_{-\eta}^{0} \frac{(\eta - 0)(z - (-\eta))}{\eta - (-\eta)} v_Y(z)\, dz \\
&\quad + 2\int_{0}^{\eta} \frac{(0 - (-\eta))(\eta - z)}{\eta - (-\eta)} v_Y(z)\, dz \\
&= 2\int_0^{\eta} (\eta - z) v_Y(z)\, dz = 2\int_0^{\eta} \frac{\eta - z}{(1 + 2z)^2}\, dz.
\end{aligned}$$

This integral is easily evaluated; for example, by the substitution $u = (1 + 2z)^2$. We obtain

$$E[T_Y \mid Y(0) = 0] = \eta + \frac{1}{2}\ln(1 + 2\eta), \qquad \eta > 0. \tag{42}$$

Since $Y(t)$ is in natural scale, we can compare this with the expected exit time from the same interval $I_Y(-\eta, \eta)$ for the standard Wiener process (formula (33))

$$E[T_{W_0} \mid W_0(0) = 0] = \eta^2, \qquad \eta > 0.$$

Figure 6.4.11 illustrates the difference; clearly our process $Y(t)$ spends more time not far away from zero but less time near zero than does the standard Wiener

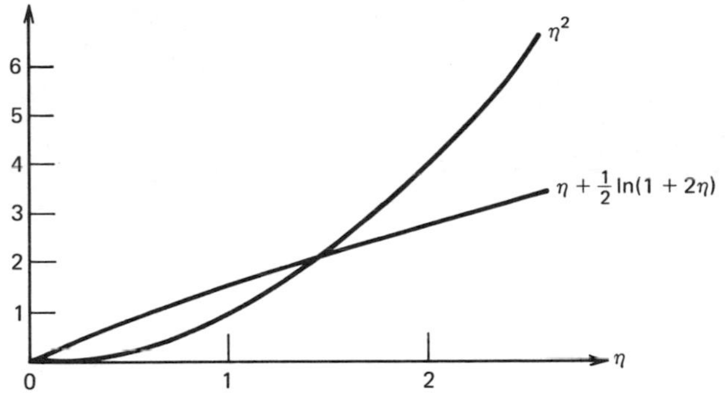

Figure 6.4.11

process. Calculation of the expected exit time for the original process $X(t)$ is now easy; just substitute into (42) for η from (41) to obtain

$$E[T_X | X(0) = 0] = \frac{1}{2}e^{2\xi} + \xi - \frac{1}{2}, \qquad \xi > 0. \qquad \blacktriangle$$

Example 6.4.4

Suppose we are given a homogeneous diffusion process $X(t)$, $t \geq 0$, with state space $\mathcal{S} = (-\infty, \infty)$ in terms of the scale and speed functions; in particular,

$$s(x) = \frac{x}{1 + |x|} \qquad \text{and} \qquad v(x) = (1 + |x|)^{\gamma},$$

$-\infty < x < \infty$, where the exponent γ is left unspecified for the time being. These functions are plotted in Figures 6.4.12 through 6.4.14.

Let us first calculate the infinitesimal mean and variance of the process $X(t)$. From (36)

$$b^2(x) = \left(v(x)\frac{ds(x)}{dx}\right)^{-1},$$

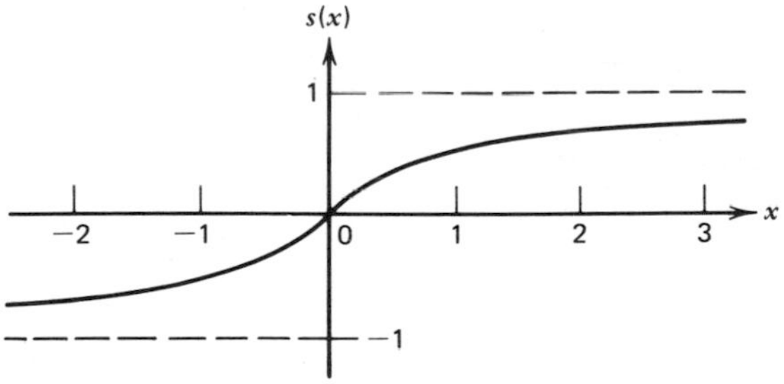

Figure 6.4.12

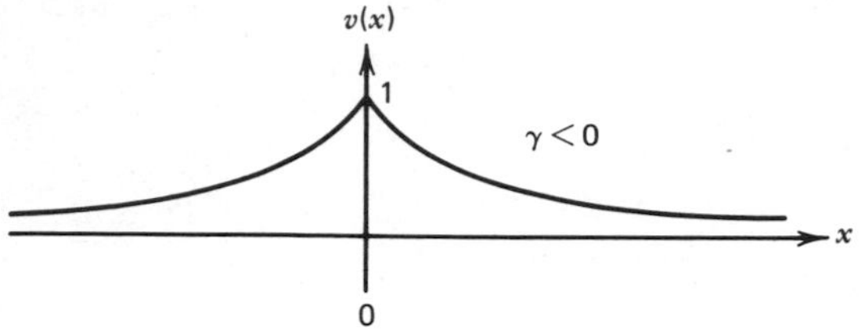

Figure 6.4.13

and since

$$\frac{ds(x)}{dx} = (1 + |x|)^{-2},$$

the infinitesimal variance equals

$$b^2(x) = (1 + |x|)^{2-\gamma}.$$

For the infinitesimal mean we have, from equation (10),

$$a(x) = -\frac{1}{2} b^2(x) \frac{\dfrac{d^2 s(x)}{dx^2}}{\dfrac{ds(x)}{dx}},$$

and since

$$\frac{d^2 s(x)}{dx^2} = \begin{cases} -2(1 + x)^3 & \text{if} \quad x > 0, \\ \ \ 2(1 - x)^3 & \text{if} \quad x < 0, \end{cases}$$

we get

$$a(x) = 2 \text{ sign } x(1 + |x|)^{1-\gamma},$$

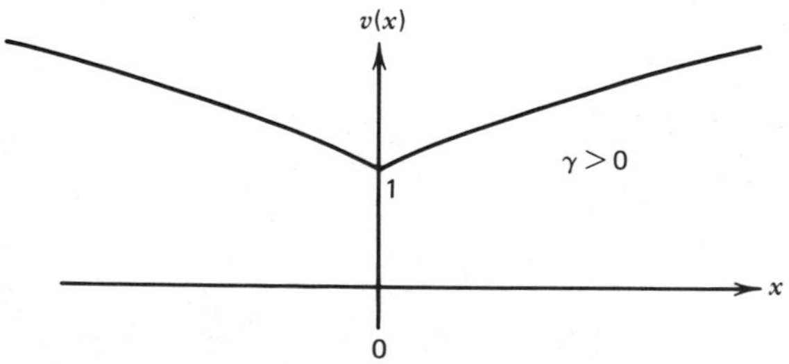

Figure 6.4.14

where

$$\text{sign } x = \begin{cases} 1 & \text{if} \quad x > 0, \\ 0 & \text{if} \quad x = 0, \\ -1 & \text{if} \quad x < 0. \end{cases}$$

In other words the differential equation for the process $X(t)$, $t \geq 0$, is

$$\dot{X}(t) = (\text{sign } X(t))(1 + |X(t)|)^{1-\gamma} + (1 + |X(t)|)^{1-\gamma/2}\dot{W}_0(t).$$

Next, let us draw some conclusions about the process $X(t)$ from its scale and speed functions. Since both these functions are symmetric about the origin $x = 0$ of the state space, the ensemble $\mathcal{E}_{X(t)}$ of all sample paths of the process should also be symmetric. That is, a portion of a typical sample path near a state $x \geq 0$ should be the mirror image of its portion near the state $-x$. Furthermore, since the scale function is concave for $x > 0$ and convex for $x < 0$, we expect an overall tendency to move away from the origin. What conclusion can we obtain from the speed function? By (37) the expected time a typical sample path spends in a small interval $(x - \Delta, x + \Delta)$ before exiting is approximately

$$v(x)\frac{ds(x)}{dx}\Delta^2 = (1 + |x|)^{\gamma-2}\,\Delta^2. \tag{43}$$

It would then seem that if $\gamma < 2$ the farther from the origin the state x is, the sooner will a typical sample path leave its vicinity, while if $\gamma > 2$, exactly the opposite will be true. Since a sample path is more likely to leave in the direction away from the origin than toward it, this seems to indicate that if $\gamma < 2$, a typical sample path will be racing faster and faster away from zero. Such a case suggests the possibility that a sample path may actually be blown out of the state space to $\pm\infty$. To check whether this could happen, let us take an interval $I_X(-\xi, \xi)$, $\xi > 0$, and calculate the expected exit time $\mathrm{E}[T_X | X(0) = 0]$ from the formula (34). We find that for $\gamma \neq 0$

$$\begin{aligned} \mathrm{E}[T_X | X(0) = 0] &= 2\int_0^\xi (s(\xi) - s(z))v(z)\,dz \\ &= 2(1+\xi)^{-1}\int_0^\xi (\xi - z)(1+z)^{\gamma-1}\,dz \\ &= 2\int_0^\xi (1+z)^{\gamma-1}\,dz - 2(1+\xi)^{-1}\int_0^\xi (1+z)^\gamma\,dz \\ &= \frac{2}{\gamma(\gamma+1)}(1+\xi)^\gamma + \frac{2}{\gamma+1}(1+\xi)^{-1} - \frac{2}{\gamma}. \end{aligned} \tag{44}$$

For $\gamma = 0$, we similarly obtain

$$\mathrm{E}[T_X | X(0) = 0] = 2\ln(1+\xi) - 2\xi(1+\xi)^{-1}.$$

Hence, the limit

$$\lim_{\xi \to \infty} \mathrm{E}[T_X \,|\, X(0) = 0], \tag{45}$$

is infinite as long as $\gamma \geq 0$, but for $\gamma < 0$ it becomes finite. It follows that if $\gamma < 0$ the expected exit time from an interval $I_X(-\xi, \xi)$, no matter how large, is bounded by a finite constant. Consequently, the anticipated phenomenon indeed occurs in this case: the process $X(t)$ is, with probability 1, blown out from the state space at some random time. This phenomenon is called explosion of the process, and it can be shown that the limit (45) actually provides a generally applicable test for such an explosion, discovered by W. Feller:

> *A homogeneous diffusion process explodes with probability one or zero depending on whether the limit (45) is finite or infinite.*

In Figure 6.4.15, we attempt to illustrate two exploding sample paths of such an exploding process. It should be remembered that an explosion is an entirely different phenomenon from a drift to $\pm\infty$. An exploding sample path completely disappears from the state space at some random time T_e (called the explosion time); it does not continue beyond T_e. On the other hand, a drifting sample path remains in the state space forever, although (in case it drifts to $+\infty$) for every level x, no matter how large, there is a random time after which the sample path stays forever above this level.

This is exactly what happens to our process if $\gamma \geq 0$; each sample path drifts to either $+\infty$ or $-\infty$, although the rate of this drift is generally not a constant (as

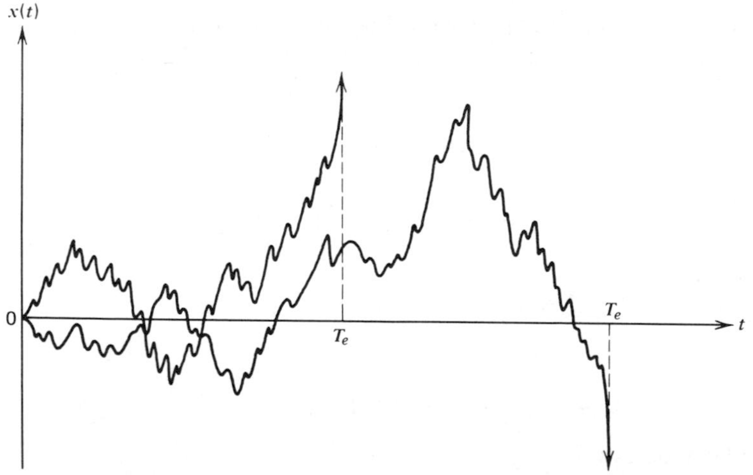

Figure 6.4.15

in Example 6.4.2) but changes from state to state. Is there anything special about the case $\gamma = 2$? There indeed is, for we can see from (43) that in that case the expected exit time from a small interval $(x - \Delta, x + \Delta)$ is approximately Δ^2 regardless of x. This suggests that a small increment $dX(t)$ of the process should be of order $\sqrt{dt}$, just like for a standard Wiener process. The same heuristic argument can be based on the expected exit time (44); since the expected exit time from $(-\xi, \xi)$ is of order ξ^γ, the average distance traveled by a sample path during dt should be of order $(dt)^{1/\gamma}$, that is, $\sqrt{dt}$ for $\gamma = 2$. We will elaborate on this kind of argument in the sequel. ▲

It is time to ask why the function v is called a speed function when, in fact, it is used to evaluate the expected exit time or, which is the same thing, the expected time the process spends in an interval before exiting.

Consider a homogeneous diffusion process $Y(t)$, $t \geq 0$, in natural scale with speed function $v(y)$ and a standard Wiener process $W_0(t)$, $t \geq 0$. Suppose that at some time t, both processes are at the state y, and consider an interval $I(y - \Delta, y + \Delta)$ where $\Delta > 0$ is small. Then, according to (32) and (33), the expected exit time will be approximately

$$\Delta^2 \, v(y)$$

for the process $Y(t)$ and

$$\Delta^2$$

for the standard Wiener process. Since both these processes are in natural scale, each will have equal probability $\frac{1}{2}$ to exit in either direction. Thus the only difference is the expected exit time. Now imagine that the sample path of each of the two processes is being recorded on a strip of paper driven by a clockwork mechanism. If we did not know that we were recording two different processes, we may have attributed the difference in expected exit time to the possibility that the clock driving the recorder for the process $Y(t)$ was running faster or slower than the one driving the recorder for the Wiener process. In other words, if we make the speed of the clock driving the latter recorder equal to $v(y)$ times the speed of the former clock, the two recordings will look like two sample paths of the same process.

This suggests that we may generate the process $Y(t)$ by accelerating or slowing the flow of time of the standard Wiener process. However, as the speed $v(y)$ is a function of the state, the change of the time flow will generally depend on the current state of the process and therefore will, in general, be random.

Thus, suppose that to a real time t there corresponds a random time $T(t)$ such that $Y(t) = y$ whenever $W_0(T(t)) = y$. To keep the two processes in phase, when the real time increases by the exit time $dt = \Delta^2 \, v(y)$ of the process $Y(t)$, the random time must increase by an increment dT equal to the exit time Δ^2 of the

Wiener process. That is, for each t such that $W_0(T(t)) = y$, the increments dt and dT must satisfy the equation $dt = v(y)\,dT$. Substituting for $y = W_0(T(t))$, we get

$$v(W_0(T(t))\,dT(t) = dt$$

and integrating this from 0 to t yields

$$\int_0^{T(t)} v(W_0(T(\tau)))\,d\tau = t. \tag{46}$$

This equation defines, for each $t \geq 0$, a random variable $T(t)$ or, in other words, a stochastic process

$$T(t), \qquad t \geq 0, \qquad \text{with } T(0) = 0.$$

Note that since $v(y) > 0$, it follows from (46) that this process will have continuous and increasing sample paths. Furthermore, from the way this process was constructed,

$$Y(t) = W_0(T(t)), \qquad t \geq 0, \tag{47}$$

so that our homogeneous diffusion process $Y(t)$ is now indeed obtained by randomly changing the time flow of the standard Wiener process.

Recalling that the process $Y(t)$ was obtained from an arbitrary homogeneous diffusion process

$$X(t), \qquad t \geq 0,$$

by the memoryless transformation

$$Y(t) = s(X(t)), \tag{48}$$

where the scale function s is continuous and increasing, we can combine (47) and (48) into a single representation

$$X(t) = s^{-1}(W_0(T(t))), \qquad t \geq 0, \tag{49}$$

of the homogeneous diffusion process $X(t)$. For obvious reasons, the process $T(t)$ is called an *intrinsic time* (*or clock*) of the process $X(t)$. As seen from (46), its probability law is completely determined by the speed function v, which further corroborates our earlier statement that the scale and speed functions completely characterize a homogeneous diffusion process.

Yet another way of looking at representation (49) is to observe that the sample paths of the process $X(t)$ can be obtained by continuous transformation, affecting both the state space and time axis, of the paths of the standard Wiener process. This gives further evidence of the predominant role the Wiener process plays in this chapter.

As a final topic concerning homogeneous diffusion processes we would like to ask whether such a process approaches a steady state (in some sense) as time increases to infinity. Since the processes studied in this section typically represent a state of a randomly disturbed dynamic system, this question is clearly important in ascertaining the stability of such a system.

Again, let $X(t)$, $t \geq 0$, be a homogeneous diffusion process, and assume that the transition probability densities

$$f_t(x_2 \mid x_1), \qquad t > 0, \tag{50}$$

exist. Since the process is homogeneous, (50) can also be regarded as the density $f_t(x)$ of $X(t)$, provided that the process started at the state x_1 at $t = 0$. It is then natural to say that the process *approaches a steady-state distribution* with probability density $\pi(x)$ if

$$\lim_{t \to \infty} f_t(x_2 \mid x_1) = \pi(x_2)$$

exists and is independent of the initial state x_1.

For example, such is the case for the Ornstein-Uhlenbeck process, where $f_t(x_2 \mid x_1)$ is a Gaussian density with mean $x_1 e^{-\alpha t}$ and variance $\sigma^2(1 - e^{-2\alpha t})$, $\alpha > 0$, which, as $t \to \infty$, converges to a Gaussian density with zero mean and variance σ^2 regardless of the initial state x_1.

On the other hand, if $X(t) = W_0(t)$ is the standard Wiener process, then no steady-state distribution exists, since in this case

$$f_t(x_2 \mid x_1) = \frac{1}{\sqrt{2\pi t}} e^{-(x_2 - x_1)^2/2t},$$

and although the limit of this expression exists and is independent of x_1, the limit $\pi(x_2) = 0$ for all x_2, and so it cannot be a density.

A closely related question is whether or not there exists an initial distribution, that is, a distribution of $X(0)$ with density $\pi(x)$ such that the process $X(t)$, $t \geq 0$, is strictly stationary. From the Chapman-Kolmogorov equation, it follows that such will be the case if and only if there is a probability density $\pi(x)$ satisfying the equation

$$\pi(x) = \int f_t(x \mid x_1)\pi(x_1)\, dx_1,$$

for all $t > 0$ and all x. Such a density, if it exists, is then called the *stationary density* of the process.

Again, the reader may verify that in the case of the Ornstein-Uhlenbeck process the Gaussian density with zero mean and variance σ^2 is its stationary density, while no such stationary density exists for the standard Wiener process.

To answer this question, let us first transform the process $X(t)$ to natural scale, that is, let us consider the process

$$Y(t) = s(X(t)), \qquad t \geq 0,$$

where $s(x)$ is the scale function of $X(t)$. We know that under sufficient differentiability conditions the transition probability density

$$f_{Y,t}(y \mid y_0)$$

of this process must satisfy the Kolmogorov forward equation

$$\frac{\partial f_{Y,t}(y \mid y_0)}{\partial t} = \frac{1}{2} \frac{\partial^2}{\partial y^2} [b_Y^2(y) f_{Y,t}(y \mid y_0)]. \tag{51}$$

Now if the transition probability density $f_{Y,t}(y \mid y_0)$ is to converge to a density $\pi_Y(y)$ independent of y_0, it is at least plausible to expect that the time derivative

$$\frac{\partial f_{Y,t}(y \mid y_0)}{\partial t} \to 0 \qquad \text{as } t \to \infty.$$

We would thus obtain an ordinary differential equation

$$\frac{d^2}{dy^2} (b_Y^2(y) \pi_Y(y)) = 0. \tag{52}$$

Similarly, assuming that there is a stationary density $\pi_Y(y)$ satisfying

$$\pi_Y(y) = \int f_{Y,t}(y \mid y_0) \pi_Y(y_0) \, dy_0$$

then by multiplying both sides of (51) by $\pi_Y(y_0)$ and integrating with respect to y_0, we would again end up with equation (52).

Now this equation has the general solution

$$b_Y^2(y) \pi_Y(y) = C_1 y + C_0$$

and since the left-hand side must be nonnegative, the constant C_1 must be zero.† Hence

$$\pi_Y(y) = \frac{C_0}{b_Y^2(y)} = C_0 v(y) \tag{53}$$

where

$$v_Y(y) = \frac{1}{b_Y^2(y)}$$

is the speed function. Thus we can conclude that if a homogeneous diffusion process in natural scale has a steady-state or stationary density, then this density

† Assuming that the state space $S_Y = (-\infty, \infty)$. A little more sophisticated argument would show that $C_1 = 0$ in general.

must be proportional to the speed function of this process. Consequently, a *necessary condition for the existence of such a density is that the integral*

$$\int v_Y(y)\, dy \tag{54}$$

over the state space of the process $Y(t)$ be finite.

Of course, we then have from (53)

$$\pi_Y(y) = \frac{v_Y(y)}{\int v_Y(y)\, dy}.$$

It is now easy to get back to the original process $X(t)$. We simply use the fact that $X(t) = s^{-1}(Y(t))$ to get

$$\pi_X(x) = \pi_Y(s(x)) \frac{ds(x)}{dx},$$

the steady-state or stationary density $\pi_X(x)$ of the process $X(t)$. Note that since, according to (35), the speed function

$$v(x) = v_Y(s(x)) \frac{ds(x)}{dx} \tag{55}$$

we can also obtain the distribution $\pi_X(x)$ directly from this speed function

$$\pi_X(x) = \frac{v(x)}{\int v(x)\, dx},$$

provided again that the speed function $v(x)$ is integrable. Thus the *steady-state distribution, if it exists, is proportional to the speed function*, which provides yet another interpretation of the latter function.

Remark: The integrability condition (54) alone is *not sufficient* for the existence of a steady-state (or stationary) distribution. After all, it is clear from (55) that the integral

$$\int v_Y(y)\, dy = \int v(x)\, dx,$$

and with the process of Example 6.4.4, we have

$$\int v(x)\, dx = \int_{-\infty}^{\infty} (1 + |x|)^{\gamma}\, dx < \infty$$

as long as $\gamma < -1$. However, we have seen that for such values of the exponent γ the process is certain to explode, and so no steady state of the process can possibly exist.

Nevertheless it can be shown that if a homogeneous diffusion process passes Feller's test for nonexplosion (cf. Example 6.4.4), then the integrability of its speed function over the state space is indeed sufficient for the steady-state distribution to exist. Я

Example 6.4.5

Consider the process of Example 6.4.1. The speed function is easily computed from (36)

$$v(x) = \left(b^2(x)\frac{ds(x)}{dx}\right)^{-1}, \tag{56}$$

and since $b^2(x) = 1$ and $ds(x)/dx = \cosh^2 x$, we have

$$v(x) = \frac{1}{\cosh^2 x}, \qquad -\infty < x < \infty.$$

Since $1/\cosh^2 x$ is the derivative of $\tanh x$, we see that

$$\int_{-\infty}^{\infty} v(x)\, dx = \tanh x \Big|_{-\infty}^{\infty} = 2,$$

so that the speed function is integrable. It follows that the steady-state or stationary density, if it exists, must be

$$\pi(x) = \frac{1}{2\cosh^2 x}, \qquad -\infty < x < \infty. \tag{57}$$

To confirm its existence, let us apply Feller's test. Taking an interval $I_X(-\xi, \xi)$, $\xi > 0$, we have from (34)

$$\mathrm{E}[T_X | X(0) = 0] = 2\int_0^{\xi} (s(\xi) - s(z))v(z)\, dz \tag{58}$$

since the scale function

$$s(x) = \frac{1}{2}\left(x + \frac{1}{2}\sinh 2x\right) \tag{59}$$

is odd: $s(-\xi) = -s(\xi)$. Substitution from (56) and (59) into (58) yields

$$\mathrm{E}[T_X | X(0) = 0] = \left(\xi + \frac{1}{2}\sinh 2\xi\right)\int_0^{\xi}\frac{dz}{\cosh^2 z} - \int_0^{\xi}\left(z + \frac{1}{2}\sinh 2z\right)\frac{dz}{\cosh^2 z}.$$

However, it is easily verified that

$$\int_0^{\infty}\frac{dz}{\cosh^2 z} \qquad \text{and} \qquad \int_0^{\infty}\left(z + \frac{1}{2}\sinh 2z\right)\frac{dz}{\cosh^2 z}$$

are both finite, and so

$$\lim_{\xi\to\infty} \mathrm{E}[T_x | X(0) = 0] = \infty.$$

Therefore the process cannot explode, and (57) is indeed its steady-state or stationary distribution. ▲

EXERCISES

Exercise 6.4

1. Calculate the scale and speed functions for the Ornstein-Uhlenbeck process $\dot{X}(t) = -\alpha X(t) + \sigma \dot{W}_0(t), t \geq 0$. Set $\alpha = 1, \sigma = 1$, and calculate the probability that the process exits from the interval $[-1, 2]$ through the right endpoint, given that $X(0) = 0$.
2. Consider the process $\dot{X}(t) = -\alpha/X(t) + \sigma \dot{W}_0(t)$, $t \geq 0$, $X(0) = 1$, where α and σ are positive constants. Calculate the scale function for this process and use it to obtain the left exit probability $P_l(1)$ from the interval $[1 - \Delta_l, 1 + \Delta_r]$, where $0 < \Delta_l < 1$. Show that $P_l(1) \to 1$ as $\Delta_r \to \infty$.
3. Repeat Exercise 6.4.2 with α as a negative constant.
4. Calculate the scale and speed function for the process

$$\dot{X}(t) = -X(t) + \sqrt{2(1 + X^2(t))}\,\dot{W}_0(t), \qquad t \geq 0, \qquad X(0) = 0.$$

Transform the process to the natural scale and find the infinitesimal variance $b_Y^2(y)$.
5. Calculate the infinitesimal variance $b_Y^2(y)$ of the transformed process (24) of Example 6.4.2. Use the result to obtain a solution of the differential equation

$$\dot{Y}(t) = \sigma^2(1 + \alpha Y(t))^2 \dot{W}_0(t), \qquad t \geq 0, \qquad Y(0) = 0,$$

where α is an arbitrary constant.
6. For the Wiener process with a constant drift, calculate the expected exit time from an interval $[-\alpha, \beta]$, $\alpha > 0$, $\beta > 0$, given that $X(0) = 0$.
7. Consider the process $X(t)$, $t \geq 0$, satisfying the equation $\dot{X}(t) = -a(X(t)) + \dot{W}_0(t)$, where the function $a(x)$ is the same as in Example 6.4.3. Calculate the scale and speed functions for this process. Transform to the natural scale and test for explosion.
8. Suppose a process $Y(t)$, $t \geq 0$, satisfies the equation $\dot{Y}(t) = \sigma(1 + Y^2(t))^{-1} \dot{W}_0(t)$, $Y(0) = 0$. For $\Delta > 0$, find the expected time it takes the process to exit from the interval $[-\Delta, \Delta]$.
9. Calculate the functions $a(x)$ and $b^2(x)$ for a homogeneous diffusion process with scale function $s(x) = \arctan x$ and speed function $v(x) = (1 + x^2)^\gamma$, where $\gamma \geq 0$. Test for explosion.
10. Repeat Exercise 6.4.9, this time with $s(x) = (\text{sign } x) \ln(1 + |x|)$ and $v(x) = (1 + |x|)^\gamma$.
11. Show that if the intrinsic time $T(t)$ of a *homogeneous* diffusion process is deterministic (i.e., not a random process), then it is a linear function $T(t) = ct$, $t \geq 0$, where c is a positive constant. (*Hint:* Recall condition (6.3.6) for a diffusion process.)
12. Verify that the steady-state density $\pi(x) = \lim_{t \to \infty} f_t(x|x_1)$ of the Ornstein-Uhlenbeck process

$$\dot{X}(t) = -\alpha X(t) + \sigma \dot{W}_0(t), \qquad t \geq 0, \qquad X(0) = 0,$$

satisfies the equation

$$\pi(x) = \int_{-\infty}^{\infty} f_t(x \mid x_1)\pi(x_1)\,dx_1, \qquad -\infty < x < \infty.$$

13. Find the steady-state density $\pi(x)$, $-\infty < x < \infty$, for the process of Example 6.4.3.
14. Check if the process $X(t)$, $t \geq 0$, defined in Exercise 6.4.4, has a steady-state distribution and, if so, find the density $\pi(x)$, $-\infty < x < \infty$.
15. Check if the error process $X(t)$, $t \geq 0$, of the control system described in Exercise 6.1.4 has a steady-state distribution and, if so, find the general form of the density $\pi(x)$, $-\infty < x < \infty$.

6.5. ITŌ STOCHASTIC CALCULUS

In the introduction to this chapter, we said that our problem was to find the time evolution of the state $x(t)$ of a dynamic system defined by a differential relation

$$\dot{x}(t) = g(t, x(t), u(t)), \qquad t \geq 0, \tag{1}$$

where $u(t)$ was the input to this system. Specifically, we have assumed that the right-hand side of (1) has the standard form of the state-space representation

$$\dot{x}(t) = a(x(t), t) + b(x(t), t)u(t),$$

and that the input $u(t)$ is a random disturbance, in particular white Gaussian noise

$$u(t) = \dot{W}_0(t).$$

We have seen that under these circumstances the state $X(t)$ was a sample-continuous Markov process whose transition probabilities satisfy Kolmogorov backward and forward equations. (Provided, of course, that $a(x, t)$ and $b(x, t)$ are sufficiently well behaved functions.) Since transition probabilities, together with the (distribution of the) initial state $X(0)$, determine the probability law of the process $X(t)$, $t \geq 0$, the forward or backward Kolmogorov equations furnish a complete solution of the problem posed above. Although we have derived these equations for the scalar case only, whereas the state-space representation of most dynamic systems typically requires that the state be a vector, this does not constitute any considerable difficulty, since both the backward and the forward equations easily generalize to a vector-valued process.

Evidently, the main obstacle is in actually solving these equations. Except for a few rather simple cases, finding an explicit solution of these equations is extremely difficult, especially in the vector-valued cases. Even an approximate numerical solution requires very extensive computational effort. It is therefore only natural to look for an alternate means of obtaining at least some information about the state process $X(t)$.

To this end, let us again examine the differential equation for the state of the system

$$\dot{x}(t) = a(x(t), t) + b(x(t), t)u(t), \qquad t \geq 0, \tag{2}$$

and let the initial state be $x(0) = x_0$. Assuming temporarily that the input $u(t)$, $t \geq 0$, is a deterministic function of time, we might be able to find the solution of this equation for any reasonably smooth input function $u(t)$. Since (2) is a first order ordinary differential equation, the reader will probably agree that finding its solution may be easier than solving the Kolmogorov equations, which are second order partial differential equations.

How then would a solution of (2) look? Since the state $x(t)$ at some time instant $t > 0$ must be determined by the initial state x_0 and by the past and present input $u(t')$, $0 \leq t' \leq t$, $x(t)$ will generally be a functional

$$x(t) = h(t, x_0; u(t'), 0 \leq t' \leq t),$$

depending on the present time t, initial state x_0, and the portion of the input $u(t')$, $0 \leq t' \leq t$.

Now, if we can actually determine the form of this functional h, it seems likely that by replacing the deterministic input $u(t')$, $0 \leq t' \leq t$ by the corresponding portion of white Gaussian noise $\dot{W}_0(t')$, $0 \leq t' \leq t$, we might obtain an explicit expression for the process $X(t)$, namely

$$X(t) = h(t, x_0; \dot{W}_0(t'), 0 \leq t' \leq t). \tag{3}$$

In fact, we have already successfully used this approach in Section 4.4, where we found a solution of the Langevin equation

$$\dot{X}(t) = -X(t) + \dot{W}_0(t), \qquad t \geq 0, \qquad x(0) = 0,$$

in a form of the Wiener integral

$$X(t) = \int_0^t e^{-(t-\tau)}\dot{W}_0(\tau)\,d\tau = \int_0^t e^{-(t-\tau)}\,dW_0(\tau),$$

from which we were able to determine the probability law of the process $X(t)$ (which turned out to be the Ornstein-Uhlenbeck process). Of course, in this particular instance, we made use of the fact that the functional h was linear, which we can hardly expect to always be the case. Nevertheless, expressing the process $X(t)$ in form (3) would certainly be helpful in deducing at least some of its properties.

Let us now try to use this idea on a simple example. One of the simplest differential equations, which the reader has probably seen on many occasions, is the equation

$$\dot{x}(t) = x(t)u(t), \qquad t \geq 0, \qquad x(0) = 1, \tag{4}$$

where $u(t)$, $t \geq 0$, is a given function whose integral $\int_0^t u(t)\, dt$ is assumed to exist.

Since the corresponding equation with white Gaussian noise as an input would have the form

$$\dot{X}(t) = X(t)\dot{W}_0(t), \qquad t \geq 0, \qquad X(0) = 1, \tag{5}$$

or, in terms of differentials,

$$dX(t) = X(t)\, dW_0(t),$$

we first rewrite equation (4) as

$$dx(t) = x(t)\, dw(t) \tag{6}$$

where $dw(t) = u(t)\, dt$, that is, $w(t) = \int_0^t u(\tau)\, d\tau$.

The usual method of solving (6) is to use the fact that

$$d \ln x(t) = \frac{1}{x(t)}\, dx(t), \tag{7}$$

and then substitute for $dx(t)$ from (6) to obtain an equality between differentials

$$d \ln x(t) = dw(t). \tag{8}$$

This implies (by integration)

$$\ln x(t) - \ln x(0) = w(t) - w(0)$$

and since $\ln x(0) = \ln 1 = 0$ and $w(0) = 0$, we get

$$\ln x(t) = w(t),$$

or

$$x(t) = e^{w(t)}, \qquad t \geq 0$$

as the unique solution.

Now carrying out the program outlined earlier, we replace the deterministic function $w(t)$ by the standard Wiener process $W_0(t) = \int_0^t \dot{W}_0(\tau)\, d\tau$ and conclude that the solution of (5) should be the process

$$X(t) = e^{W_0(t)}, \qquad t \geq 0. \tag{9}$$

Since (9) is just a memoryless transformation of the standard Wiener process $W_0(t)$, it is clearly a sample-continuous Markov process, and we can easily compute the probability density $f_t(x)$ of $X(t)$. Since the density of $W_0(t)$ is Gaussian with mean zero and variance t, we get

$$f_t(x) = \frac{1}{x\sqrt{2\pi t}}\, e^{-(\ln x)^2/2t}, \qquad x > 0. \tag{10}$$

But before we accept this result, let us check if the transition probabilities of the process (9) satisfy the Kolmogorov equations. Since the process $X(t)$ is homogeneous, the density $f_t(x)$ is equal to the transition probability density $f_t(x|x_0)$ with $x_0 = 1$, and hence it should satisfy the Fokker-Planck equation. From the differential form

$$dX(t) = X(t)\, dW_0(t)$$

we see that the infinitesimal mean and variance are $a(x) = 0$ and $b^2(x) = x^2$, respectively, so that the Fokker-Planck equation for this process is

$$\frac{\partial f_t(x)}{\partial t} = \frac{1}{2}\frac{\partial^2}{\partial x^2}[x^2 f_t(x)]. \tag{11}$$

But now we are in for a rather nasty surprise. If we take the density (10) and perform the indicated differentiation, we find that

$$\frac{\partial f_t(x)}{\partial t} = \left(\frac{(\ln x)^2}{4t^2} - \frac{1}{2t}\right) f_t(x),$$

while

$$\frac{1}{2}\frac{\partial^2}{\partial x^2}[x f_t(x)] = \left(\frac{(\ln x)^2}{2t^2} - \frac{\ln x}{2t} - \frac{1}{2t}\right) f_t(x).$$

No matter how hard we try, these two expressions are not equal, and so our density (10) does not satisfy the Fokker-Planck equation. Clearly, something went wrong.

The trouble originates from the fact that in solving equation (6), we have treated the function $w(t)$ as if it were a nice differentiable function and then we replaced it by the Wiener process, whose sample paths are nowhere differentiable. To pinpoint the exact place where the discrepancy has arisen, let us reexamine equation (7). If we interpret the differential $d \ln x(t)$ as a forward increment

$$d \ln x(t) = \ln(x(t) + dx(t)) - \ln x(t),$$

then the Taylor expansion of the logarithm gives

$$d \ln x(t) = \frac{1}{x(t)}\, dx(t) - \frac{1}{2x^2(t)}(dx(t))^2 + \frac{1}{3x^3(t)}(dx(t))^3 - \cdots.$$

Substituting $dx(t) = x(t)\, dw(t)$ then yields

$$d \ln x(t) = dw(t) - \frac{1}{2}(dw(t))^2 + \frac{1}{3}(dw(t))^3 - \cdots \tag{12}$$

Now if $w(t)$ is a differentiable function, then the second and higher powers of the increment $dw(t)$ are negligible as compared with the time increment dt, and so, for infinitesimal increments dt, only the first term remains on the right-hand side of (12). That is, we get equation (8).

However, if $w(t)$ is a typical sample path of a Wiener process, then *the square of its forward increment is no longer negligible with respect to dt*. In fact, the Lévy oscillation property $(dW_0(t))^2 \doteq dt$ states that this term is actually of order dt. This means that for infinitesimal increments dt, *both the first and the second term* on the right-hand side of (12) must be retained. Then from $(dW_0(t))^2 \doteq dt$, we get instead of (8) the equation

$$d \ln X(t) = dW_0(t) - \frac{1}{2}\,dt,$$

Integrating this from 0 to t, we now have

$$\ln X(t) = W_0(t) - \frac{1}{2}\,t.$$

Thus, this time we obtain the expression

$$X(t) = e^{W_0(t) - t/2}, \qquad t \geq 0. \tag{13}$$

Clearly, $X(t)$ is again a sample-continuous Markov process, and the initial condition $X(0) = 1$ is satisfied. Since (13) is still a memoryless transformation of the standard Wiener process, the probability density of $X(t)$ is easily found to be

$$f_t(x) = \frac{1}{x\sqrt{1\pi t}} \exp\left\{-\left(\ln x + \frac{t}{2}\right)^2 \Big/ 2t\right\}, \qquad x > 0.$$

If we now calculate the partial derivatives

$$\frac{\partial f_t(x)}{\partial t} = \left(\frac{(\ln x)^2}{2t^2} - \frac{1}{2t} - \frac{1}{8}\right) f_t(x),$$

and

$$\frac{1}{2}\frac{\partial^2}{\partial x^2}[x^2 f_t(x)] = \left(\frac{(\ln x)^2}{2t^2} - \frac{1}{2t} - \frac{1}{8}\right) f_t(x).$$

we find that the Fokker-Planck equation (11) is satisfied.

The moral of this story is that although the approach suggested at the beginning of this section is basically sound, a provision must be made for the peculiar character of the Wiener process. In solving the differential equation

$$dX(t) = a(X(t), t)\,dt + b(X(t), t)\,dW_0(t)$$

we cannot just use the familiar formulas for differentials of various functions; instead, these formulas must be modified to take into account Lévy's oscillation property of the Wiener process.

The preceding example also suggests how this modification can be accomplished.

Consider a function $g(x)$ of a real variable x, and assume that the function is three times differentiable. Then, according to Taylor's formula, we have for $\Delta x > 0$ and fixed x

$$g(x + \Delta x) - g(x) = g'(x)\,\Delta x + \frac{1}{2}g''(x)(\Delta x)^2 + \frac{1}{6}g'''(x_0)(\Delta x)^3, \tag{14}$$

where $x \le x_0 \le x + \Delta x$. Here, to avoid future confusion, we used primes to indicate derivatives with respect to x. Denote the increment

$$\Delta g(x) = g(x + \Delta x) - g(x)$$

and write (14) as

$$|\Delta g(x) - g'(x)\,\Delta x| \le (\Delta x)^2 K, \tag{15}$$

where K is constant such that

$$\left|\frac{1}{2}g''(x) + \frac{1}{6}g'''(x_0)\,\Delta x\right| \le K \qquad \text{for all } x \le x_0 \le x + \Delta x.$$

Suppose now that x is a function of another variable, say time $x = x(t)$, and suppose further that it satisfies the so-called Lipschitz condition:

$$|\Delta x(t)| \le C|\Delta t|, \tag{16}$$

where C is a constant and

$$\Delta x(t) = x(t + \Delta t) - x(t).$$

Then $(\Delta x)^2$ is of order $(\Delta t)^2$ and hence negligible with respect to Δt. (That is $(\Delta x)^2/\Delta t \to 0$ as $t \to 0$.) This makes the right-hand side of (15) arbitrarily small as $\Delta t \to 0$, and so if we replace the symbol Δ by the symbol d to indicate infinitesimal increments, we get for $\Delta t \to 0$

$$dg(x(t)) = g'(x)\,dx(t) \tag{17}$$

which, of course, is the expression for the differential of the function g at the point x. In particular, if $x(t)$ has a derivative at time t, we also have

$$dx(t) = \dot{x}(t)\,dt \tag{18}$$

and substituting (18) into (17), we end up with the familiar chain rule for differentials

$$dg(x(t)) = g'(x(t))\dot{x}(t)\,dt. \tag{19}$$

Now suppose that the function $x(t)$ does not satisfy the Lipschitz condition (16) but instead

$$|\Delta x(t)|^2 \le C|\Delta t|.$$

Then $(\Delta x)^2$ is no longer negligible with respect to Δt, and formula (17), although still true for $\Delta x \to 0$, is no longer true when $\Delta t \to 0$, and neither then is the chain rule (19). However, if instead of (15) we take

$$|\Delta g(x) - g'(x)\,\Delta x - \frac{1}{2}g''(x)(\Delta x)^2| \le (\Delta x)^3 K$$

where now

$$\left|\frac{1}{6}g'''(x_0)\right| \le K \qquad \text{for } x \le x_0 \le x + \Delta x,$$

then $(\Delta x)^3$ is of order $(\Delta t)^{3/2}$ and thus negligible with respect to Δt as $\Delta t \to 0$. Instead of (17), we now have for $\Delta t \to 0$ the new "differential"

$$dg(x(t)) = g'(x(t))\,dx(t) + \frac{1}{2}g''(x(t))(dx(t))^2. \tag{20}$$

Of course, if the function $x(t)$ does not have a derivative at t, then $dx(t)$ cannot be interpreted as a differential as defined by standard calculus. In fact, the expression (9) acquires meaning, although still symbolic, only if $dx(t)$ is replaced by a *stochastic differential*:

$$dX(t) = a(X(t), t)\,dt + b(X(t), t)\,dW_0(t).$$

Since

$$\begin{aligned}(dX(t))^2 = {} & a^2(X(t), t)(dt)^2 + 2a(X(t), t)\,dt\,dW_0(t) \\ & + b^2(X(t), t)(dW_0(t))^2,\end{aligned} \tag{21}$$

the Lévy oscillation property

$$dW_0(t) \doteq \sqrt{dt}$$

implies that the first two terms on the right-hand side of (21) are of order $(dt)^2$ and $(dt)^{3/2}$ and hence negligible with respect to dt as $dt \to 0$. Thus we can write

$$(dX(t))^2 = b^2(X(t), t)\,dt,$$

and if we substitute for $dX(t)$ and $(dX(t))^2$ into (20), we obtain

$$\begin{aligned}dg(X(t)) = {} & \left[g'(X(t))a(X(t), t) + \frac{1}{2}g''(X(t))b^2(X(t), t)\right] dt \\ & + g'(X(t))b(X(t), t)\,dW_0(t).\end{aligned} \tag{22}$$

This is the so-called *Itō differential formula* for the forward differential of the process

$$Y(t) = g(X(t)),$$

and it represents a stochastic version of the chain rule of ordinary calculus. We can even generalize it slightly by allowing the function g to depend also on the time variable t. That is, if we define a process

$$Y(t) = g(X(t), t)$$

where, as before,

$$dX(t) = a(X(t), t)\,dt + b(X(t), t)\,dW_0(t), \tag{23}$$

then the Itō differential formula becomes

$$dY(t) = \left[\dot{g}(X(t), t) + g'(X(t), t)a(X(t), t) + \frac{1}{2}g''(X(t), t)b^2(X(t), t)\right] dt$$
$$+ g'(X(t), t)b(X(t), t)\,dW_0(t). \tag{24}$$

Here again, the dot indicates time derivatives

$$\dot{g}(x, t) = \frac{\partial g(x, t)}{\partial t},$$

while primes indicate derivatives with respect to x,

$$g'(x, t) = \frac{\partial g(x, t)}{\partial x}, \qquad g''(x, t) = \frac{\partial^2 g(x, t)}{\partial x^2},$$

and it is assumed that these derivatives exist and are continuous.

The Itō differential formula now allows us, in some cases, to carry out our original idea of expressing the solution of an equation

$$\dot{X}(t) = a(X(t), t) + b(X(t), t)\dot{W}_0(t)$$

as a functional of white Gaussian noise or equivalently of the standard Wiener process.†

We illustrate its use for this purpose with a few simple examples.

Example 6.5.1

Consider a process $X(t)$, $t \geq 0$, satisfying the differential equation

$$dX(t) = \alpha(t)X(t) + \beta(t)X(t)\,dW_0(t) \tag{25}$$

with $X(0) = 1$, where $\alpha(t)$ and $\beta(t) \neq 0$ are given deterministic functions of time $t \geq 0$.

† This does not mean that we can always obtain an explicit expression for such a functional. The Itō formula is merely used to transform the differential $dX(t)$ into a form that can hopefully be recognized as the differential of a known function.

Suppose we tentatively define a new process $Y(t)$, $t \geq 0$, by means of a memoryless transformation

$$Y(t) = \ln X(t). \tag{26}$$

Applying Itō differential formula (22) with $g(x) = \ln x$ and $a(x, t) = x\alpha(t)$, $b(x, t) = x\beta(t)$, we find that

$$dY(t) = \left(\frac{1}{X(t)} X(t)\alpha(t) + \frac{1}{2} - \frac{1}{X^2(x)} (X(t)\beta(t))^2\right) dt$$
$$+ \frac{1}{X(t)} X(t)\beta(t)\, dW_0(t),$$

or upon cancellation,

$$dY(t) = \left(\alpha(t) - \frac{1}{2}\beta^2(t)\right) dt + \beta(t)\, dW_0(t).$$

But this shows that $Y(t)$, $t \geq 0$, ought to be an inhomogeneous Wiener process

$$Y(t) = \int_0^t \left(\alpha(\tau) - \frac{1}{2}\beta^2(\tau)\right) d\tau + \int_0^t \beta(\tau)\, dW_0(\tau), \qquad t \geq 0, \tag{27}$$

provided the deterministic functions $\alpha(t)$ and $\beta(t)$ are such that the above integrals exist. Assuming that to be the case, we have from (26)

$$X(t) = e^{Y(t)},$$

and hence

$$X(t) = \exp\left\{\int_0^t \left(\alpha(\tau) - \frac{1}{2}\beta^2(\tau)\right) d\tau + \int_0^t \beta(\tau)\, dW_0(\tau)\right\}, \qquad t \geq 0. \tag{28}$$

Note that $X(0) = e^0 = 1$ as required, and so a solution of our differential equation (25) appears to be the process (28), a memoryless transformation of the inhomogeneous Wiener process (27). For reasonably well behaved functions $\alpha(t)$ and $\beta(t)$, it is indeed a solution, in fact, the unique solution of (25). ▲

Example 6.5.2

Suppose we are given the differential equation

$$\dot{X}(t) = -\frac{1}{2}\left(X(t) + t + \frac{1}{2}\right) + \sqrt{X(t) + t}\, \dot{W}_0(t), \qquad t \geq 0, \tag{29}$$

with the initial condition $X(0) = 0$. Rewritten in differential form (23), we see that

$$a(x, t) = -\frac{1}{2}x - \frac{1}{2}t - \frac{1}{4} \qquad \text{and} \qquad b(x, t) = \sqrt{x + t},$$

and we are looking for a memoryless transformation

$$y = g(x, t)$$

which would change equation (29) into an equation for a familiar process. In view of Itō formula (24), we may try a function $g(x, t)$ that would make the new infinitesimal variance

$$g'(x, t)b(x, t)\Big|_{y=g(x,t)}$$

of the transformed process independent of x. Since $b(x, t) = \sqrt{x + t}$, we need a function such that

$$g'(x, t) = \frac{1}{\sqrt{x + t}}.$$

Clearly, such a function is

$$g(x, t) = 2\sqrt{x + t}.$$

Its time derivative and the second x derivative are, respectively,

$$\dot{g}(x, t) = \frac{1}{\sqrt{x + t}}, \qquad g''(x, t) = \frac{-3}{2(x + t)^{3/2}},$$

and hence the Itō formula (24) applied to the process

$$Y(t) = 2\sqrt{X(t) + t} \tag{30}$$

yields

$$dY(t) = \left[\frac{1}{\sqrt{X(t) + t}} - \frac{\frac{1}{2}X(t) + \frac{t}{2} + \frac{1}{4}}{\sqrt{X(t) + t}} - \frac{3(X(t) + t)}{4(X(t) + t)^{3/2}}\right] dt + dW_0(t).$$

However, the expression in brackets equals

$$-\frac{2X(t) + 2t}{\sqrt{X(t) + t}} = -2\sqrt{X(t) + t} = -Y(t)$$

and so the transformed process satisfies the equation

$$dY(t) = -Y(t) + dW_0(t), \qquad t \geq 0.$$

But this is a familiar Langevin equation, and since by (30), $X(0) = 0$ implies $Y(0) = 0$, we see that the process $Y(t)$, $t \geq 0$, should be an Ornstein-Uhlenbeck process

$$Y(t) = \int_0^t e^{-(t-\tau)}\, dW_0(\tau).$$

Since from (30)

$$X(t) = \frac{1}{2} Y^2(t) - t$$

we thus obtain for the process $X(t)$, $t \geq 0$, satisfying the differential equation (29), the expression

$$X(t) = \frac{1}{2}\left(\int_0^t e^{-(t-\tau)}\, dW_0(\tau)\right)^2 - t, \qquad t \geq 0. \qquad \blacktriangle$$

Remark: Two preceding examples may create a false impression that any stochastic differential equation can be easily "solved" by a clever use of the Itō rule. This is not so; the Itō rule merely serves as a stochastic analog of a substitution formula for ordinary first order differential equations and can do no more for stochastic differential equations than a substitution does for a deterministic differential equation (see also the preceding footnote). Я

The Itō differential rule is not restricted to processes defined by the differential equation

$$\dot{X}(t) = a(X(t), t) + b(X(t), t)\dot{W}_0(t).$$

In fact, such a differential equation, or its form in terms of forward differentials

$$dX(t) = a(X(t), t)\, dt + b(X(t), t)\, dW_0(t),$$

is but a symbolic, although intuitively appealing, way to write an integral equation

$$X(t) - X(0) = \int_0^t a(X(\tau), \tau)\, d\tau + \int_0^t b(X(\tau), \tau)\, dW_0(\tau).$$

Since, as mentioned earlier, the solution of this equation will generally be, for each t, a functional

$$X(t) = h(t, X(0); W_0(t'), 0 \leq t' \leq t),$$

so will be the two integrands $a(X(t), t)$ and $b(X(t), t)$. This means the process $X(t)$ will generally have the form

$$X(t) = X(0) + \int_0^t A(\tau)\, d\tau + \int_0^t B(\tau)\, dW_0(\tau), \qquad t \geq 0 \tag{31}$$

where $A(t)$ and $B(t)$ are functionals of the portion $W_0(t')$, $0 \leq t' \leq t$ of the Wiener process and possibly also of the time t and the initial value $X(0)$. Here we assume that the integrals

$$\int_0^t \mathrm{E}[|A(\tau)|]\, d\tau \qquad \text{and} \qquad \int_0^t \mathrm{E}[B^2(\tau)]\, d\tau$$

exist for all t and that $X(0)$, when random, is independent of the entire Wiener process $W_0(t)$.

Processes of this form are often called Itō processes. We have already met such a process in Section 4.3, where it was pointed out that the second integral on the right-hand side of (31)

$$\int_0^t B(\tau)\, dW_0(\tau)$$

is an integral of a special kind—the so-called Itō integral. Like the Wiener integral in Section 4.3, it can be defined, at least for sample-continuous $B(\tau)$, as an MS limit of approximating sums

$$\text{l.i.m.} \sum_{k=0}^{n} B(\tau_k)[W_0(\tau_{k+1}) - W_0(\tau_k)] \tag{32}$$

over a partition $0 = \tau_0 < \tau_1 < \cdots < \tau_n < \tau_{n+1} = t$.

Unlike the Wiener integral, however, the integrand $B(t)$ is now a stochastic process rather than just a deterministic function of time. But $B(\tau_k)$ is also a functional of the portion $W_0(t')$, $0 \le t' \le \tau_k$, of the Wiener process and hence independent of the forward increment $W_0(\tau_{k+1}) - W_0(\tau_k)$. It is exactly this fact that accounts for the existence and special properties of this integral.

We will not prove here the existence of the Itō integral but merely point out some of its properties. Some of these properties are shared with any other integral, for instance, linearity in integrand and domain

$$\int_0^t (c_1 B_1(\tau) + c_2 B_2(\tau))\, dW_0(\tau) = c_1 \int_0^t B_1(\tau)\, dW_0(\tau) + c_2 \int_0^t B_2(\tau)\, dW_0(\tau),$$

$$\int_0^{t_1} B(\tau)\, dW_0(\tau) + \int_{t_1}^{t_2} B(\tau)\, dW_0(\tau) = \int_0^{t_2} B(\tau)\, dW_0(\tau).$$

Also, viewed as a function of the upper limit, the Itō integral defines a stochastic process

$$\int_0^t B(\tau)\, dW_0(\tau) = U(t), \qquad t \ge 0,$$

which is a finite-variance sample-continuous martingale.† Furthermore, it is easily seen from (32) that

$$\mathrm{E}\left[\int_0^t B(\tau)\, dW_0(\tau)\right] = 0,$$

and

$$\mathrm{E}\left[\left(\int_0^t B(\tau)\, dW_0(\tau)\right)^2\right] = \int_0^t \mathrm{E}[B^2(\tau)]\, d\tau.$$

† Recall that in Section 4.3, we actually arrived at the Itō integral as a *representation* for finite-variance sample-continuous martingales.

On the other hand, the Itō integral has its peculiarities. Suppose, for instance, that

$$B(t) = W_0(t)$$

which is trivially a functional of $W_0(t')$, $0 \le t' \le t$. In analogy with ordinary integrals we might be tempted to conclude that

$$\int_0^t W_0(\tau)\, dW_0(\tau) = \frac{1}{2} W_0^2(\tau)\Big|_{\tau=0}^{\tau=t} = \frac{1}{2} W_0^2(t).$$

However, if we use the definition

$$\int_0^t W_0(\tau)\, dW_0(\tau) = \text{l.i.m.} \sum_{k=0}^{n} W_0(\tau_k)(W_0(\tau_{k+1}) - W_0(\tau_k)),$$

and the algebraic identity

$$W_0(\tau_k)(W_0(\tau_{k+1}) - W_0(\tau_k)) = \frac{1}{2}(W_0^2(\tau_{k+1}) - W_0^2(\tau_k)) - \frac{1}{2}(W_0(\tau_{k+1}) - W_0(\tau_k))^2,$$

we find that

$$\sum_{k=0}^{n} W_0(\tau_k)(W_0(\tau_{k+1}) - W_0(\tau_k)) = \frac{1}{2} W_0^2(t) - \frac{1}{2} \sum_{k=0}^{n} (W_0(\tau_{k+1}) - W_0(\tau_k))^2.$$

Hence

$$\int_0^t W_0(\tau)\, dW_0(\tau) = \frac{1}{2} W_0^2(t) - \frac{1}{2} t, \tag{33}$$

since the sum of squares of forward increments of the standard Wiener process satisfies

$$\sum_{k=0}^{n} (W_0(\tau_{k+1}) - W_0(\tau_k))^2 = t,$$

for any partition of $[0, t]$. But notice that if we rewrite (33) as

$$W_0^2(t) = \int_0^t 2W_0(\tau)\, dW_0(\tau) + t,$$

and take the formal differential

$$d(W_0^2(t)) = 2W_0(t)\, dW_0(t) + dt,$$

we recognize this as the Itō differential rule applied to the process

$$Y(t) = g(W_0(t)),$$

with

$$g(x) = x^2.$$

More generally now, if we finally define a stochastic differential $dX(t)$ as the symbolic notation for the Itō process (31)

$$dX(t) = A(t)\,dt + B(t)\,dW_0(t),$$

and if $Y(t)$, $t \geq 0$, is a new process defined by

$$Y(t) = g(X(t), t),$$

then $Y(t)$ is again an Itō process with stochastic differential obtained from the Itō differential rule

$$dY(t) = \left[\dot{g}(X(t), t) + g'(X(t), t)A(t) + \frac{1}{2}g''(X(t), t)B^2(t)\right] dt$$
$$+ g'(X(t), t)B(t)\,dW_0(t). \tag{34}$$

Note that (34), if rewritten back into an integral version, again has the general form (31).

In particular, if $A(t) = 0$, $B(t) = 1$, that is, $X(t) = W_0(t)$ and $g'(x) = h(x)$, then rewriting the Itō rule

$$dg(W_0(t)) = \frac{1}{2}h'(W_0(t))\,dt + h(W_0(t))\,dW_0(t),$$

in integral form

$$\int_0^t h(W_0(\tau))\,dW_0(\tau) = g(W_0(t)) - g(0) - \frac{1}{2}\int_0^t h'(W_0(\tau))\,d\tau, \tag{35}$$

allows us to express the Itō integral on the left by means of an ordinary integral.

Example 6.5.3

Formula (35) is a stochastic analog of a by-parts integration rule for ordinary integrals. We may use it, for instance, to calculate Itō integrals of powers of a Wiener process.

Specifically, taking in (35)

$$g(x) = \frac{1}{3}x^3,$$

so that

$$h(x) = g'(x) = x^2$$

and

$$h'(x) = 2x$$

we get (since $W_0^3(0) = 0$)

$$\int_0^t W_0^2(\tau)\,dW_0(\tau) = \frac{1}{3}W_0^3(t) - \int_0^t W_0(\tau)\,d\tau. \tag{36}$$

We can also calculate iterated integrals such as, for example,

$$\begin{aligned} I_3(t) &= \int_0^t \left(\int_0^{\tau_1} \left(\int_0^{\tau_2} dW_0(\tau_3) \right) dW_0(\tau_2) \right) dW_0(\tau_1) \\ &= \int_0^t \left(\int_0^{\tau_1} W_0(\tau_2)\, dW_0(\tau_2) \right) dW_0(\tau_1) \\ &= \int_0^t \frac{1}{2} (W_0^2(\tau_1) - \tau_1)\, dW_0(\tau_1) \end{aligned} \tag{37}$$

by (33). Now by (36)

$$\int_0^t \frac{1}{2} W_0^2(\tau_1)\, dW_0(\tau_1) = \frac{1}{6} W_0^3(t) - \frac{1}{2} \int_0^t W_0(\tau)\, d\tau, \tag{38}$$

while

$$\frac{1}{2} \int_0^t \tau_1\, dW_0(\tau_1) = \frac{1}{2} t W_0(t) - \frac{1}{2} \int_0^t W_0(\tau)\, d\tau, \tag{39}$$

since the latter is a Wiener integral and thus an ordinary by-parts integration formula applies. Substitution from (38) and (39) into (37) easily yields

$$I_3(t) = \frac{1}{6} W_0^3(t) - \frac{1}{2} t W_0(t), \qquad t \geq 0.$$

Similarly, a general formula for the nth iterated integral can be developed, which in turn can be used to establish a series expansion for functionals of a Wiener process. ▲

One of the main advantages of the Itō stochastic calculus approach is that it is easily extended to vector-valued Itō processes and consequently also to systems of differential equations

$$\dot{\mathbf{X}}(t) = \mathbf{a}(\mathbf{X}(t), t) + \mathbf{b}(\mathbf{X}(t), t)\dot{\mathbf{W}}_0(t)$$

where now, $\mathbf{x}(t)' = (x_1(t), \ldots, x_m(t))$ is a state vector and $\mathbf{a}(\mathbf{x}, t)$ and $\mathbf{b}(\mathbf{x}, t)$ are matrices of appropriate dimension. This is so because the Itō differential rule can be extended to functions $g(x_1, \ldots, x_m; t)$ of m state variables by simply expanding such a function into a Taylor series in all variables up to and including partial derivatives of the second order. The increments dx_j, $j = 1, \ldots, m$, appearing in such an expansion are then again replaced by the corresponding stochastic differentials. We will not discuss this topic any further but only present a simple example.

Example 6.5.4
Consider a pair of coupled equations

$$\begin{aligned} dX_1(t) &= X_2(t)\, dW_0(t), \\ dX_2(t) &= X_1(t)\, dW_0(t), \end{aligned} \tag{40}$$

$t \geq 0$, with initial condition $X_1(0) = X_2(0) = 1$. Note that the same Wiener process is involved in both equations. Similarity with the scalar equation $dX(t) = X(t)\, dW_0(t)$ suggests that we might try to solve the system by a logarithmic transformation. This time, however, we take a function $y = g(x_1, x_2)$ of two variables, namely

$$y = \ln(x_1 + x_2). \tag{41}$$

As before, we need to expand this function into a Taylor series up to and including the second order terms:

$$\begin{aligned} dy &\doteq \frac{\partial y}{\partial x_1} dx_1 + \frac{\partial y}{\partial x_2} dx_2 + \frac{1}{2}\frac{\partial^2 y}{\partial x_1^2}(dx_1)^2 + \frac{\partial^2 y}{\partial x_1\, \partial x_2} dx_1\, dx_2 + \frac{1}{2}\frac{\partial^2 y}{\partial x_2^2}(dx_2)^2 \\ &= \frac{1}{x_1 + x_1}(dx_1 + dx_2) + \frac{1}{2}\frac{-1}{(x_1 + x_2)^2}((dx_1)^2 + 2\, dx_1\, dx_2 + (dx_2)^2). \end{aligned}$$

Substitution from (40) for $dX_1(t)$, $dX_2(t)$ and

$$\begin{aligned} (dX_1(t))^2 &= X_2^2(t)(dW_0(t))^2 \doteq X_2^2(t)\, dt, \\ (dX_2(t))^2 &= X_1^2(t)(dW_0(t))^2 \doteq X_1^2(t)\, dt, \\ dX_1(t)\, dX_2(t) &= X_1(t)X_2(t)(dW_0(t))^2 \doteq X_1(t)X_2(t)\, dt \end{aligned}$$

results in massive cancellation, and we end up with the equation

$$dY(t) = -\frac{1}{2}\, dt + dW_0(t).$$

This is easily recognized as an equation for a Wiener process with a constant drift

$$Y(t) = W_0(t) - \frac{1}{2}t, \qquad t \geq 0,$$

and since by (41)

$$Y(t) = \ln(X_1(t) + X_2(t)) \tag{42}$$

we have

$$X_1(t) + X_2(t) = e^{W_0(t) - (1/2)t}. \tag{43}$$

To get $X_1(t)$ and $X_2(t)$ we need one more equation. Evidently, we should try something like

$$Z(t) = \ln(X_1(t) - X_2(t)), \tag{44}$$

which leads to the equation

$$dZ(t) = \frac{1}{2}\,dt + dW_0(t). \tag{45}$$

Thus

$$Z(t) = W_0(t) + \frac{1}{2}t, \qquad t \geq 0,$$

and hence

$$X_1(t) - X_2(t) = e^{W_0(t) + (1/2)t}. \tag{46}$$

From (43) and (46), we obtain easily

$$X_1(t) = e^{W_0(t)} \sinh \frac{t}{2},$$

$$X_2(t) = e^{W_0(t)} \cosh \frac{t}{2},$$

but, alas, this shows that the transformation (44) is not legitimate since $\sinh t/2 < \cosh t/2$ for all t, that is, $X_1(t) - X_2(t)$, would be always negative. Fortunately, we can easily fix that, merely replacing (44) by

$$Z(t) = \ln(X_2(t) - X_1(t)). \tag{47}$$

Instead of (45) we now have

$$dZ(t) = -\frac{1}{2}\,dt - dW_0(t),$$

whence

$$Z(t) = -W_0(t) - \frac{1}{2}t, \qquad t \geq 0,$$

and

$$-X_1(t) + X_2(t) = e^{-W_0(t) - (1/2)t}. \tag{48}$$

From (43) and (48) we now have

$$\begin{aligned} X_1(t) &= e^{-t/2} \sinh W_0(t), \\ X_2(t) &= e^{-t/2} \cosh W_0(t), \end{aligned} \tag{49}$$

which is now all right, since $X_2(t) > X_1(t) \geq 0$, as needed for both transformations (42) and (47). It only remains to satisfy the initial conditions: from (49), $X_2(0) = 1$, but $X_1(0) = 0$, so we need to add 1 to the first expression in (49). Thus

$$X_1(t) = 1 + e^{-(1/2)t} \sinh W_0(t),$$

$$X_2(t) = e^{-(1/2)t} \cosh W_0(t), \qquad t \geq 0,$$

is the desired solution of system (40). ▲

Before we close this section the reader should be warned that, out of necessity, we have carried the entire discussion on a heuristic level only. A number of important topics have been omitted. For instance, we have not even touched on the question of existence and uniqueness of a solution of the differential equations we discussed. Nor have we considered the obvious, important question of how one should properly interpret a stochastic differential equation in view of the fact that white Gaussian noise is an idealized process. Readers who may in their future work come across some problems involving stochastic differential equations are therefore advised to consult a text dealing with this topic in sufficient detail and rigor.

EXERCISES

Exercise 6.5

1. Consider the process defined by $X(t) = W_0^2(t) - t, t \geq 0$. Apply Itō differential rule (24) to show that

$$dX(t) = 2\sqrt{X(t) + t}\, dW_0(t) \qquad X(0) = 0.$$

Write the Fokker-Planck equation for this process, and verify that the first order density $f_t(x)$ satisfies this equation.

2. Express the process $X(t)$, $t \geq 0$, satisfying the equation

$$dX(t) = \frac{1}{2} \cos^2 X(t)(\sin 2X(t)\, dt + 2\, dW_0(t)), \qquad X(0) = 0,$$

as a memoryless transformation of the standard Wiener process. (*Hint:* Try $y = \tan x$.)

3. Show that the process $X(t)$, $t \geq 0$, satisfying the equation

$$\dot{X}(t) = 1 - 2X(t) + 2\sqrt{X(t)}\, \dot{W}_0(t),$$

is a memoryless transformation of an Ornstein-Uhlenbeck process. (*Hint:* Use the Itō rule.)

4. Use the Itō rule to establish the recurrence

$$(W_0(t))^n = n \int_0^t (W_0(\tau))^{n-1}\, dW_0(\tau) + \binom{n}{2} \int_0^t (W_0(\tau))^{n-2}\, d\tau,$$

$n = 2, 3, \ldots.$ (*Hint:* Rewrite in differential form.)

5. Use formula (35) to express the Itō integral $\int_0^t \cos W_0(\tau)\, dW_0(\tau)$ as an ordinary stochastic integral of the Wiener process.

CHAPTER SEVEN

POINT PROCESSES

7.1. BASIC CONCEPTS

In the preceding chapters, we have always defined a stochastic process as an ensemble of sample paths together with a probability law specifying the probabilities of events defined in terms of the sample paths. Typically, we have thought of the sample paths as being functions of time, such as one might obtain by recording the noise signal at a receiver, a position or velocity of a randomly moving object, or a temperature or atmospheric pressure on appropriate recording devices. Implicit in all these examples was the assumption that the sample paths were continuous functions of time, and this assumption was made almost exclusively throughout the whole of Chapters Four, Five, and Six. With discrete-time processes, we generally used either the case where such a process is obtained by periodic sampling of a continuous-time process with continuous paths or the case in which the random variables $X(n)$ represent a sequence of successive states of a system actually operating in discrete-time steps. A common feature of all these cases was that for each time instant t, discrete or continuous, a random quantity $X(t)$ could be observed, and our main objective was to study models of this quantity varying over time.

There are, however, many physical situations in which the main aspect of interest is the repeated occurrence of some specific observable phenomenon at randomly varying time instants. Specificly, such a phenomenon may be the emission of an alpha particle from a radioactive material, the arrival of an electron at the anode of a vacuum tube, the emission of a photon from an excited atom, an electrical discharge from a nerve cell, a demand for access to a particular address in a computer memory, the arrival of a phone call at an exchange, or other such phenomena.

The emphasis in this chapter, at least for the initial phase of analysis, is not on the phenomenon itself but only on the stream of random time instants at which the phenomenon occurs. Since a time instant can also be referred to as a point in the time domain $\mathcal{T}$, the occurrence of the phenomenon in question is commonly called a *point-event*, and the random stream of successive occurrences of such point-events is called a *point process*.

Although a point process is a stochastic process of a rather special kind, we would still like to follow the general idea and define it as an ensemble of sample paths together with a probability law. Now a sample path of a stochastic process was defined as a possible record or realization obtained when an experiment

Figure 7.1.1

producing the process is performed. Unlike the processes studied earlier in this book, the record of such a point process will simply consist of a list (or set) of time instants at which the point-events occurred. Graphically, it may look like Figure 7.1.1 where the $\times$'s signify times of occurrence of the point-events. Since the ensemble of a stochastic process is the collection of all possible realizations of the process, we can define the ensemble for a point process as follows:

Definition 7.1.1
The ensemble of a point process is a collection of discrete sets of points from the time domain $\mathcal{T}$. The elements of each of these sets are called *epochs*.

Thus an epoch is just a shorter name for the time instant at which a point-event of a particular realization of the point process has occurred.

Comment Our usage of the term "discrete" is as follows. A discrete subset of the time domain $\mathcal{T}$ is a finite or infinite set of points (epochs) $t \in \mathcal{T}$ such that:

1. it consists of distinct isolated points only,
2. no interval in $\mathcal{T}$ of finite length can contain an infinite number of points.

We are thus excluding not only the possibility that several point-events can occur simultaneously (at the same epoch) but also the possibility that the set of epochs may have a limit point; that is, that the elements get closer and closer to some time instant t^*. For example, if the epochs formed a sequence such as $1 - (1/n)$, $n = 1, 2, \ldots$ (see Fig. 7.1.2), they would then have $t^* = 1$ as a limit point. A case such as this is somewhat dramatically referred to as an *explosion* of the point process at the time t^* and is far from being merely a mathematical curiosity—just think of the point-events representing emissions of neutrons from a critical mass of plutonium 239. However, we are, a priori, excluding this type of point process from our consideration simply because we cannot cover every aspect in a text of limited size.

For the same reason, we also omit the possibility that a point process may be defined on domains other than the time domain; point-events may be occurring in space—epicentra of earthquakes provide a good example.

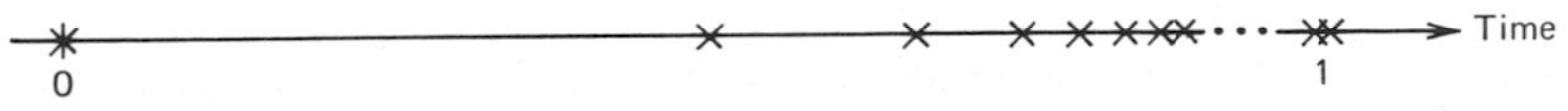

Figure 7.1.2

As a final restriction, we also exclude from further discussion point processes with discrete time domain. The reason for this is that a point process with time domain such as $\mathcal{T} = \{n: n = 0, 1, \ldots\}$ can be completely described by a stochastic sequence $X(n)$, $n = 0, 1, \ldots$, of Bernoulli random variables with $X(n) = 1$ being interpreted as the occurrence of a point-event at epoch $t = n$.

From now on, we will only consider point processes with continuous time domain of the form

$$\mathcal{T} = (-\infty, \infty) \qquad \text{or} \qquad \mathcal{T} = (0, \infty)$$

with ensemble as given by Definition 7.1.1.

Having defined the ensemble for a point process, our next task is to find a way to characterize its probability law. One of the simplest means to this end is to associate with each point process a stochastic process $N(t)$, $t \in \mathcal{T}$, of the ordinary kind (see Chapter One), which counts the number of point-events from some fixed time t_0 up to and including the current time instant t.

Such a process $N(t)$, $t \in \mathcal{T}$, can be imagined as being the time-continuous output of a counter whose input is triggered by each incoming point-event.

We will call this process

$$N(t), \qquad t \in \mathcal{T}$$

a *counting process* associated with a given point process. Formally, it can be defined by choosing a fixed time t_0 and defining the random variable $N(t)$ as follows:

For $t > t_0$

$$N(t) = \text{number of point-events with epochs } \tau \text{ such that } t_0 < \tau \leq t, \qquad \text{(1a)}$$

For $t \leq t_0$

$$-N(t) = \text{number of point-events with epochs } \tau \text{ such that } t < \tau \leq t_0. \qquad \text{(1b)}$$

Note that with this definition for any $t_1 < t_2$ in time domain $\mathcal{T}$, the forward increment $N(t_2) - N(t_1)$ equals the number of point-events with epochs τ such that $t_1 < \tau \leq t_2$, regardless of the choice of the time instant t_0 in (1). Unless there are some reasons to do otherwise, the obvious choice is $t_0 = 0$, and thus *from now on we will always assume that* $N(0) = 0$. We will do so even for point processes defined on the time domain $\mathcal{T} = [0, \infty)$, thus extending in this case the time domain of the associated counting process by attaching the point $t = 0$ to the interval $(0, \infty)$.

A typical sample path $n(t)$ of the counting process resulting from a particular realization of a point process is depicted in Figure 7.1.3.

It should be noted that the ensemble of sample paths of a counting process $N(t)$, $t \in \mathcal{T}$, will consist of step functions having unit jumps at epochs of the point

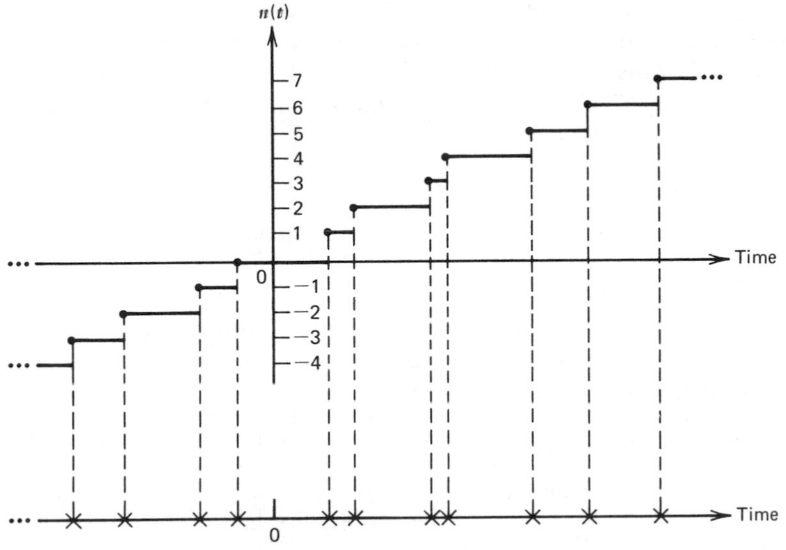

Figure 7.1.3

process and that because of the way $N(t)$ was defined in (1), the sample paths will all be continuous from the right. More importantly, however, there is a one-to-one correspondence between the ensemble of the point process and the ensemble of sample paths of the associated counting process. Consequently, the probability law of the point process can be given in terms of the probability law of the associated counting process, which in turn can be done by means of the family of joint distributions (see Chapter One).

Since $N(t_2) - N(t_1)$ is the number of point events in the interval $(t_1, t_2]$, the probability law of the counting process allows us to determine the probability distribution of the number of point-events in any† interval with endpoints $t_1 < t_2$ in the time domain. We say that the associated counting process describes the *counting properties* of the point process.

Although the association of the counting process $N(t)$, $t \in \mathcal{T}$, with a given point process is quite natural, it is not the only way in which the probability law of a point process could be specified. Since the epochs of any particular realizations of a point process form a discrete set of points in the time domain $\mathcal{T}$, they can be arranged in a sequence $\{t_n\}$ of time instants such that if t_n is an epoch of a point-event, then t_{n+1} is the epoch of the first point-event occurring subsequent to

† Note that, for instance, the number of point-events in an interval of the type $[t_1, t_2]$, that is, with epochs τ such that $t_1 \leq \tau \leq t_2$, will be given by $N(t_2) - N(t_1 -)$, where $N(t_1 -) = \lim N(t)$ as $t \to t_1$ from the left.

t_n. In other words, the epochs can be labeled in the natural order of occurrence of successive point-events. To do this, it is necessary to select an epoch to be labeled t_1, and this is most naturally done by choosing the epoch of the first point-event occurring after time $t = 0$. With this choice, each realization of a point process can be specified by a sequence of epochs

$$0 < t_1 < t_2 < \cdots \tag{2}$$

for the time domain $\mathscr{T} = (0, \infty)$, or by the sequence of epochs

$$\cdots < t_{-1} < t_0 \leq 0 < t_1 < t_2 < \cdots \tag{3}$$

for the time domain $\mathscr{T} = (-\infty, \infty)$, and there is again one-to-one correspondence between these sequences of epochs and the ensemble of a point process.

Hence, we can define a stochastic sequence

$$T(n), \qquad n = 1, 2, \ldots \qquad \text{or} \qquad n = \cdots, -1, 0, 1, \ldots \tag{4}$$

whose ensemble consists of sequences of epochs (2) or (3) and specify the probability law of the point process by means of the probability law of the stochastic sequence (4). We refer to this sequence (4) as the *random epoch sequence*† associated with the point process.

There is an obvious modification of this specification, namely to introduce the times between consecutive epochs rather than the epochs themselves. Thus, with the labeling of epochs as above, if we define

$$X(n) = T(n+1) - T(n), \qquad n = 1, 2, \ldots,$$
$$X(0) = T(1)$$

for the point process with time domain $\mathscr{T} = (0, \infty)$, or just

$$X(n) = T(n+1) - T(n), \qquad n = \ldots, -1, 0, 1, \ldots$$

for the time domain $\mathscr{T} = (-\infty, \infty)$, we could also specify the probability law of the point process by means of the law of the *waiting time sequence*‡

$$X(n), \qquad n = 0, 1, \ldots \qquad \text{or} \qquad n = \ldots, -1, 0, 1, \ldots, \tag{5}$$

of positive random variables (waiting times) $X(n)$.

Note that in the case of the time domain $\mathscr{T} = (-\infty, \infty)$, one cannot determine the exact location of the origin on the time axis from a sample path $x(n)$, $n = \ldots, -1, 0, 1, \ldots$, of the waiting time sequence (5). Therefore, in this case, it is also necessary to specify the time of occurrence of some particular point-event, for instance, the first one after the time $t = 0$. That means that the random epoch $T(1)$

† Also called the "arrival time sequence."

‡ Alternatively, the "interarrival time sequence."

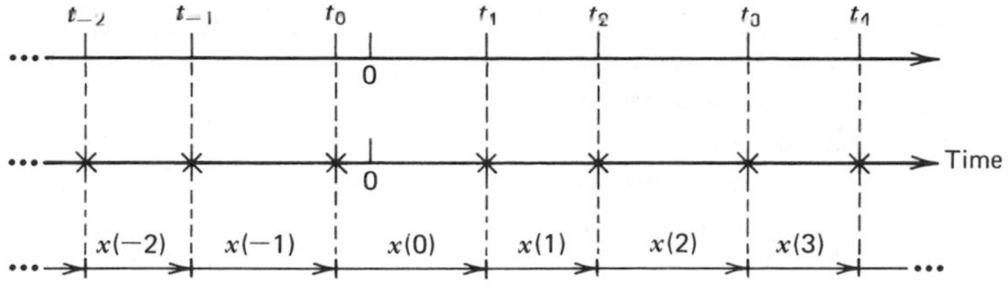

Figure 7.1.4

must be attached to the sequence (5) to obtain a complete specification of the probability law of the underlying point process.

The relationships between a typical realization of a point process and its associated random epoch sequence and waiting time sequence are shown in Figure 7.1.4.

Since both the waiting time sequence and random epoch sequence are defined in terms of time intervals, the properties of the point process that are directly expressible in terms of either of the two sequences are referred to as *interval properties*.

It should be kept in mind that the counting process, the random epoch sequence, and the waiting time sequence with $T(1)$ attached are all equivalent in the sense that each alone provides a complete specification of the probability law of the underlying point process and thus also of the other two processes. This does not mean, however, that it is an easy matter to pass from one of these specifications to another. In fact, many problems in the theory of point processes are actually concerned with expressing probabilities of some counting property in terms of interval properties and vice versa.

Remark: A special class of point processes is obtained if the ensemble is such that all sets of epochs are finite. This would mean that for each realization of the point process, only a finite number of point-events can occur. The probability law of such a *finite point process* can be specified much more simply by

1. giving the probability mass distribution $P(K = k)$, $k = 1, 2, \ldots$, of the total number K of point-events, and
2. giving the conditional density functions

$$f_{T(1), \ldots, T(k)|K}(t_1, \ldots, t_k | k)$$

of the random epochs $T(n)$ of these point-events.

If the ensemble of a point process is such that there is a positive probability of occurrence of both finite and infinite numbers of point-events, then we have another interesting class of point processes, namely *point processes with extinction*. That is, if a particular realization of such a point process contains only a finite number of epochs, we may say that the process has become extinct at the instant of the last epoch. By assumption the probability of extinction is positive, but in contrast with finite point processes, this probability must be strictly less than 1, since extinction may or may not occur. In a sense, extinction is the opposite of explosion, but interesting as both these cases may be, we will further restrict our subsequent discussion to point processes without extinction. Я

For a point process with time domain $\mathcal{T} = (-\infty, +\infty)$, we shall usually assume that the epochs extend both ways to plus and minus infinity. More precisely, we wish to assume that there is no first epoch or, in picturesque language, we say the point process was born in the infinitely remote past. Yet another way to say the same thing is that the process should not ever become extinct both in forward and in reversed time. We may perhaps call such a point process *everlasting*, thereby implying no explosions and no extinction in either direction of time.

With these additional restrictions, each realization of our point process or equivalently each set of epochs in the ensemble of the process must satisfy the following conditions:

(i) All epochs are distinct.
(ii) Any finite time interval can contain only a finite number of epochs.
(iii) Any infinite time interval always contains an infinite number of epochs.

Condition (i) precludes multiple occurrences of point-events at the same time instant, condition (ii) prevents explosions, and (iii) prevents extinction of the process (in either time direction, in the case, $\mathcal{T} = (-\infty, \infty)$).

It is worthwhile noticing how these conditions look in terms of the three processes associated with the point process.

For the counting process $N(t)$, $t \in \mathcal{T}$, the reader may easily verify that the corresponding conditions become

(i) $N(t) - N(t-) \leq 1$ for all $t \in \mathcal{T}$.
(ii) $N(t_2) - N(t_1) < \infty$ for any $t_1 < t_2$ from the time domain $\mathcal{T}$.
(iii) $N(t) \to \infty$ as $t \to \infty$, and for the time domain $\mathcal{T} = (-\infty, +\infty)$, also $N(t) \to -\infty$ as $t \to -\infty$.

For the random epoch sequence $T(n)$ or the waiting times sequence $X(n)$, condition (i) is simply

$$\cdots < T(-1) < T(0) \leq 0 < T(1) < T(2) < \cdots$$

or

$$X(n) > 0, \qquad n = \ldots, -1, 0, 1, \ldots$$

with the obvious modification if $\mathcal{T} = (0, \infty)$. Conditions (ii) and (iii) require that both $T(n)$ and $X(n)$ be infinite or double infinite sequences, such that

$$T(n) \to \pm\infty \qquad \text{as} \qquad n \to \pm\infty,$$

and for the waiting time sequence

$$\sum_{n=0}^{\infty} X(n) = \infty, \qquad \text{and also} \qquad \sum_{n=-1}^{-\infty} X(n) = \infty$$

in case $\mathcal{T} = (-\infty, \infty)$. Thus, if a probability law for a point process is to be defined in terms of the probability law of one of the associated processes, care must be taken to assure that the above conditions are met.

There is still one somewhat degenerate type of point process that should be eliminated from further discussion. An extreme case of this degeneracy would be a point process where the point-events can occur only at some fixed deterministic sequence of epochs $\ldots < t_{-1} < t_0 < t_1 < \ldots$, and at no other times. Such a point process is clearly equivalent to a point process with a discrete time domain consisting of the sequence of indices of the deterministic sequence of epochs. Since, as mentioned earlier, point processes with discrete time domain can be treated as ordinary discrete-time processes of a rather simple kind, we prefer to exclude this kind of degeneracy altogether.

Actually, we wish to completely eliminate even the possibility that there is some *fixed* time instant at which a point-event may occur with positive probability.

In terms of the associated counting process $N(t)$, $t \in \mathcal{T}$, this condition requires that

(iv) $$P(N(t) - N(t-) > 0) = 0 \qquad \text{for all } t \in \mathcal{T}.$$

We shall refer to this condition by saying that *point-events do not occur at predetermined times*. The reader may verify that if the associated counting process has a finite variance, condition (iv) is equivalent to MS continuity of $N(t)$, $t \in \mathcal{T}$ (see Chapter Four).

Equivalently, the random epochs $T(n)$ or the waiting times $X(n)$ must then have continuous distributions.

From now on, we shall assume that the point processes we will be dealing with satisfy conditions (i) through (iv), which we list again for future reference:

Assumptions for a Point Process

(*i*) *All epochs are distinct.*
(*ii*) *Any finite time interval contains only a finite number of epochs.*
(*iii*) *Any infinite time interval contains an infinite number of epochs.*
(*iv*) *Point-events do not occur at predetermined times.*

Unless stated otherwise, these four assumptions will be considered satisfied throughout this entire chapter and referred to as assumptions (i) through (iv).

EXERCISES

Exercise 7.1

1. Consider a clock that produces a regular stream of pulses (regarded as points in time) with frequency 1 hertz. The clock is turned on at time $t = 0$; but the first pulse appears only after a random delay T whose density is $f_T(t) = 2(1 - t)$, $0 \leq t < 1$. Describe the ensemble of the point process whose point-events correspond to the clock pulses. Show that this process satisfies all four assumptions (i) through (iv).
2. For the point process of Exercise 7.1.1, sketch a few sample paths of the associated counting process. Define the corresponding random epoch sequence and waiting time sequence and use either of these to specify the probability law of the point process.
3. Suppose that the sequence of clock pulses described in Exercise 7.1 is subjected to random jitter so that each pulse, starting with the second, is slightly advanced or retarded in time. Assume specifically that the displacement of each pulse from its original epoch is uniformly distributed over the interval $(-\frac{1}{4}, \frac{1}{4})$, independently for each pulse. Describe the ensemble of the resulting point process and specify its probability law.
4. You are asked to construct a simple model for occurrences of lightning discharges. Assume that these discharges can occur only during daytime (6 A.M. to 6 P.M.), independently from day to day, most likely around noontime. At most one (possibly none) main discharge can occur each day, and if it does, it may be followed by several secondary discharges. Describe the ensemble of the resulting point process and specify its probability law by making further assumptions of your own choice. Try to make a reasonable compromise between realism and mathematical tractability of your model.
5. Suppose that the probability law of a point process on the time domain $\mathscr{T} = (-\infty, \infty)$ is specified by the family of Kth order densities of its random epoch sequence. Obtain a general formula for the Kth order densities of the waiting time sequence. Also derive a formula for the second order probability mass distributions of the associated counting process.
6. Verify that the Poisson process $N(t)$, $t > 0$ (see Chapter One), may serve as a counting process of a point process satisfying assumptions (i) through (iv). Specify the probability laws of the corresponding random epoch sequence and the waiting time sequence.
7. Show that any point process satisfying assumptions (i) through (iv) can be made to explode by applying a deterministic time transformation $N'(t) = N(g(t))$ to its associated counting process. Specifically, consider a Poisson process (Exercise 7.1.6) and make it explode at the time $t = 1$. Do you think one can always turn an exploding point process into a nonexploding one by such a deterministic time transformation?
8. A certain substance has the property that if exposed to ultraviolet radiation, a random number of its atoms, geometrically distributed with parameter p, are brought to an excited state. Each of these atoms remains excited for a random length of time, independently of all other atoms, these times being exponentially distributed with

parameter α. When relapsing back into its normal state each atom emits a photon. Show that the instants of photon emissions form a finite point process and specify its probability law. Also find the density of the epoch of the last emission (extinction time of the process).

9. Suppose the epochs of a point process on $\mathcal{T} = (0, \infty)$ are defined as the time instants $t_1, t_2, t_3, \ldots$ at which a standard Wiener process $W_0(t)$, $t \geq 0$, first crosses the levels a, $2a$, $3a$, ..., where $a > 0$ is a constant. (Only the first crossing of each level counts.) Specify the probability law of this point process by means of the associated counting process, the random epoch sequence, and the waiting time sequence. Verify the truth of assumptions (i) through (iv).

10. Show that if a point process has an MS-continuous associated counting process, then condition (iv) is satisfied. (*Hint:* Use Chebyshev's inequality.)

7.2. STATIONARITY AND EVOLUTION WITHOUT AFTEREFFECTS

Since the counting process $N(t)$, associated with a point process on the time domain $\mathcal{T}$, is an ordinary continuous time stochastic process, that is, $N(t)$ is, for each t, a random variable, it is meaningful to define its mean function

$$\mu(t) = \mathrm{E}[N(t)], \qquad t \geq 0 \qquad \text{or} \qquad -\infty < t < \infty,$$

provided the expectation is finite. We then refer to this function as the *mean function of the point process* itself. Since we have agreed to choose $N(0) = 0$, we will always have

$$\mu(0) = 0,$$

and since for $t_1 < t_2$ the increment $N(t_2) - N(t_1)$ is, by definition, the number of point-events with epoch in the interval $(t_1, t_2]$ the difference

$$\mu(t_2) - \mu(t_1) = \mathrm{E}[N(t_2) - N(t_1)]$$

is the expected number of these point-events. It then follows that a mean function $\mu(t)$ of a point process must be a nondecreasing function of t. Furthermore, it must also be a continuous function, since if it were discontinuous at some fixed time instant t, the counting process $N(t)$ would have to have a jump at this fixed time instant, the possibility of which we have excluded by assumption (iv). Finally, assumption (iii) implies that

$$\mu(t) \to \infty \qquad \text{as} \qquad t \to \infty,$$

and, in the case of the time domain $\mathcal{T} = (-\infty, \infty)$, also $\mu(t) \to -\infty$ as $t \to -\infty$, since if the limit $\lim_{t \to \infty} \mu(t)$, which must exist for a nondecreasing function, were finite, the expectation and hence also the number of point-events with epochs in

$(0, \infty)$ would have to be finite. We thus summarize these findings with the following theorem.

THEOREM 7.2.1. *The mean function of a point process defined as the expectation*

$$\mu(t) = \mathrm{E}[N(t)], \qquad t \geq 0 \qquad \text{or} \qquad -\infty < t < \infty,$$

if it exists, is a continuous nondecreasing function such that

$$\mu(0) = 0, \qquad \lim_{t \to \infty} \mu(t) = \infty,$$

and in the case of the time domain $\mathcal{T} = (-\infty, \infty)$ *also* $\lim_{t \to -\infty} \mu(t) = -\infty$.

Remark: The fact that the mean function of a point process is nondecreasing does not eliminate the possibility that its graph has some flat parts, that is, that for some $t_1 < t_2$, we may have $\mu(t_1) = \mu(t_2)$. But in such a case, $\mathrm{E}[N(t_2) - N(t_1)] = 0$, which in turn implies that with probability 1 there would be no point-events with epochs in the interval $(t_1, t_2]$. This represents a rather minor nuisance that is easily eliminated by cutting out any such interval from the time domain $\mathcal{T}$ and glueing the remaining parts back together.

Thus from now on, we shall assume that this surgery, if needed, has been performed on the point process, and consequently we will take for granted that the mean function $\mu(t)$ is actually a *strictly increasing function of t.* Я

Now, except in rather pathological cases, a continuous function can be written as an indefinite integral; in our particular case, since $\mu(0) = 0$, as an integral

$$\mu(t) = \int_0^t \lambda(\tau)\, d\tau, \qquad t \geq 0 \qquad \text{or} \qquad -\infty < t < \infty,\dagger \tag{1}$$

where $\lambda(t)$ is another function of time. Since for any time instant t and an infinitesimal time increment $dt > 0$ we have approximately

$$d\mu(t) \doteq \mu(t + dt) - \mu(t) = \int_t^{t+dt} \lambda(\tau)\, d\tau \doteq \lambda(t)\, dt,$$

or by the definition of the mean function

$$\mathrm{E}[dN(t)] \doteq \lambda(t)\, dt, \tag{2}$$

the function $\lambda(t)$ may be interpreted as the expected number of point-events with epochs in an interval $(t, t + dt]$ divided by the time increment dt. Thus, it may be aptly called the expected rate of occurrence of point-events in the point process. However, it is customary to use a related term and call the function $\lambda(t)$ the

† For $t < 0$, we use the familiar convention of integral calculus $\int_0^t \lambda(\tau)\, d\tau = -\int_{-t}^0 \lambda(\tau)\, d\tau$.

intensity function of the point process. Since the mean function $\mu(t)$ in (1) is strictly increasing, the intensity function must be *positive*,

$$\lambda(t) > 0$$

for all $t \geq 0$ or $-\infty < t < \infty$, with the possible exception of a discrete set of time instances.

The intensity function of a point process admits yet another important interpretation. According to assumptions (i) and (ii), all epochs of point-events must be distinct and cannot form an infinitely dense cluster around any time instant t. This suggests that for a sufficiently small time interval $(t, t + dt]$ the probability $P(dN(t) > 1)$ of having more than one point-event occurring in such an interval should be negligible as compared to the probability $P(dN(t) \leq 1)$ of having zero or one point-event occurring there. In other words, we might expect something like

$$P(dN(t) \leq 1) \to 1 \qquad \text{as} \qquad dt \to 0. \tag{3}$$

This in turn indicates that as the time increment dt approaches zero, the expectation $\mathrm{E}[dN(t)]$ could be approximated with increasing accuracy by the probability $P(dN(t) = 1)$. But then from (2), we should have also

$$P(dN(t) = 1) \doteq \lambda(t)\, dt \tag{4}$$

so that the intensity $\lambda(t)$ can be looked upon as the probability rate of occurrence of point-events. Then, (3) together with (4) would also imply that

$$P(dN(t) = 0) \doteq 1 - \lambda(t)\, dt. \tag{5}$$

Of course, the idea of $P(dN(t) > 1)$ being negligible compared to $P(dN(t) \leq 1)$ would have to be made precise. One possibility might be to express it as

$$\lim_{dt \to 0} \frac{P(dN(t) > 1)}{dt} = 0. \tag{6}$$

Equations (4) and (5), together with (6) or some similar relation formalizing the negligibility of $P(dN(t) > 1)$, are generally referred to as the *infinitesimal conditions for the point process*. The reader should be cautioned, however, against jumping to the conclusion that these conditions follow automatically from our assumptions (i) through (iv); *this is not so*. Generally, some further assumptions must be made about the point process, but these are too technical to be discussed here.

There is, however, an important class of point processes where assumptions (i) through (iv) are indeed sufficient for the infinitesimal conditions to be true, namely the class of stationary point processes. Now, as in the case of an ordinary stochastic process, a point process is called stationary if its probability law is invariant with respect to a shift of time. Formally, this is most conveniently expressed in terms of the associated counting process $N(t)$.

Definition 7.2.1
A point process is said to be (strictly) *stationary* if for any collection of time instants $t_0 < t_1 < \cdots < t_k$ from the time domain $\mathcal{T}$, and any $k = 1, 2, \ldots$ the joint distribution of the increments

$$N(t_1 + \tau) - N(t_0 + \tau),\ N(t_2 + \tau) - N(t_1 + \tau),\ \ldots,\ N(t_k + \tau) - N(t_{k-1} + \tau)$$

does not depend on the time shift $\tau \geq 0$.

Example 7.2.1
In applications, one frequently encounters point processes generated as a stream of crossings of an ordinary stochastic process $Y(t)$, $t \in \mathcal{T}$, through some given level. More spececifically, if $Y(t)$, $t \in \mathcal{T}$, is a stochastic process with continuous sample paths, then each time instant at which a sample path $y(t)$ crosses the level zero, that is, changes its sign, could be considered an epoch of a point-event. (See Fig. 7.2.1.) However, a point process so obtained need not generally satisfy assumptions (i) through (iv), and even if it does, relating its probability law to that of the process $Y(t)$, $t \in \mathcal{T}$, is usually a difficult problem.

On the other hand, it is easy to see that *if the process $Y(t)$, $t \in \mathcal{T}$ is strictly stationary, so must be the resulting zero-crossing point process*. To see this, merely

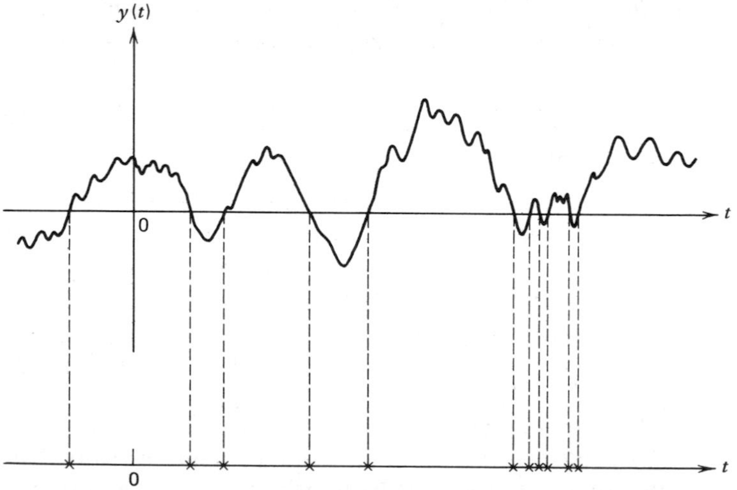

Figure 7.2.1

notice that any particular sequence of epochs of the point process is completely determined by the sample path $y(t)$, $t \in \mathcal{T}$, and if the sample path $y(t)$ is shifted by an amount $\tau > 0$ along the time domain, the sequence of epochs is shifted by the same amount with respect to the origin. If $Y(t)$, $t \in \mathcal{T}$, is strictly stationary, then the shifted process $Y(t + \tau)$, $t \in \mathcal{T}$, has by definition the same probability law as $Y(t)$, $t \in \mathcal{T}$, and thus the joint distribution of the increments in Definition 7.2.1 is also independent of the time shift τ. ▲

From Definition 7.2.1, it is clear that for any $t_1 < t_2$ the expectation $\mathrm{E}[N(t_2) - N(t_1)]$, if finite, can only depend on the difference $t_2 - t_1$. In view of (1), this implies that the intensity function of a stationary point process must be a positive constant

$$\lambda(t) = \lambda > 0 \qquad \text{for all } t,$$

or, equivalently, that the mean function must be a linear function of time,

$$\mu(t) = \lambda t.$$

The above-mentioned assertion concerning the infinitesimal conditions will now be stated as a theorem, in its original form due to Khintchine, Korolyuk, and Dobrushin.

THEOREM 7.2.2. *For a stationary point process satisfying assumptions* (*i*) *through* (*iv*) *and such that the expected number of point-events with epochs in any finite interval is finite, the infinitesimal conditions* (4), (5), *and* (6) *hold.*

Although the proof of this theorem is not particularly difficult, we will not present it here.† Instead, because of the obvious applied importance of point processes generated by a stream of crossings as described in Example 7.2.1, we shall derive a formula for the intensity of such a process.

Let $Y(t)$, $-\infty < t < \infty$, be a stationary zero mean Gaussian process with correlation function

$$R_Y(\tau), \qquad -\infty < \tau < \infty.$$

Assume further that the second derivative of the correlation function at $\tau = 0$

$$\left.\frac{d^2}{d\tau^2} R_Y(\tau)\right|_{\tau=0} = \ddot{R}(0)$$

† The interested reader may consult a paper by K. L. Chung in *American Mathematical Monthly*, Vol. 79, pp. 867–877, for an elegant proof and further generalizations.

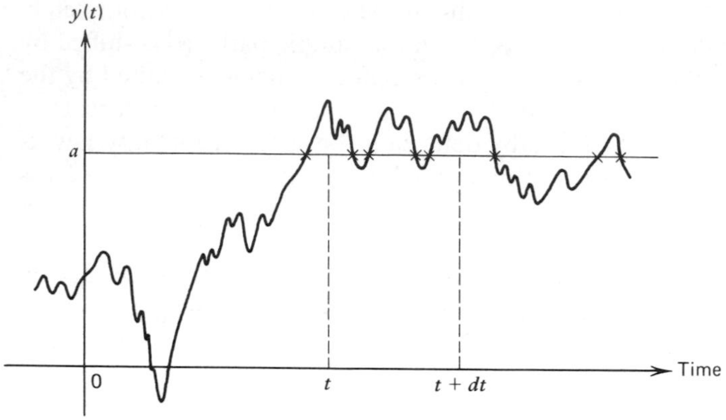

Figure 7.2.2

exists and is finite. As we have seen in Section 4.1 this guarantees that the process $Y(t)$ is MS differentiable and also sample continuous.

Now take a small time interval $(t, t + dt]$ and some constant a (possibly zero); we wish to find the expected number of crossings of level a by the process $Y(t)$ during the interval $(t, t + dt]$ (see Fig. 7.2.2).

Keeping in mind the stationarity of the point process generated by these crossings, let us denote the number of crossings in the interval $(t, t + dt]$ by

$$dN(t).$$

Now it can be shown that the assumed existence of the second derivative $R(0)$ or, equivalently, the MS differentiability of the Gaussian process $Y(t)$ entails the fact that the number of crossings $dN(t)$ must be finite. It then follows from Theorem 7.2.2 that for sufficiently small dt

$$\mathrm{E}[dN(t)] \doteq P(dN(t) = 1),$$

that is, that the probability of having more than one crossing in a small interval $(t, t + dt]$ is negligible. Also for small dt, we have approximately

$$Y(t + dt) \doteq Y(t) + \dot{Y}(t)\, dt,$$

and from Figure 7.2.3, it is then clear that $dN(t) = 1$, that is, that a crossing will occur, if and only if either

$$Y(t) < a \quad \text{and} \quad Y(t) + \dot{Y}(t)\, dt > a,$$

or

$$Y(t) > a \quad \text{and} \quad Y(t) + \dot{Y}(t)\, dt < a.$$

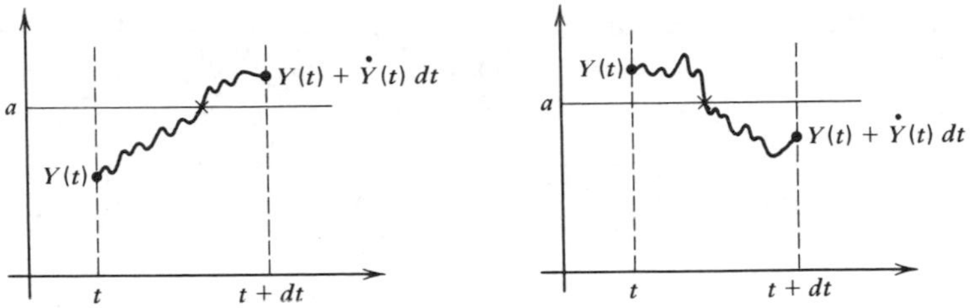

Figure 7.2.3

For these inequalities to be satisfied the vector $(Y(t), \dot{Y}(t))$ must lie in the shaded region shown in Figure 7.2.4, and hence the probability $P(dN(t) = 1)$ can be evaluated by integrating the joint density $f_{Y(t), \dot{Y}(t)}(y, \dot{y})$ over this region,

$$P(dN(t) = 1) = \int_0^{\infty} \left[\int_{a - \dot{y}\,dt}^{a} f_{Y(t), \dot{Y}(t)}(y, \dot{y})\, dy \right] d\dot{y} + \int_{-\infty}^{0} \left[\int_{a}^{a - \dot{y}\,dt} f_{Y(t), \dot{Y}(t)}(y, \dot{y})\, dy \right] d\dot{y}. \tag{7}$$

Since the stationary process $Y(t)$, $-\infty < t < \infty$, is a zero mean Gaussian process, it follows from Section 4.1 that $Y(t)$ and $\dot{Y}(t)$ are both Gaussian random variables with zero means, variances

$$\operatorname{Var}[Y(t)] = R_Y(0), \qquad \operatorname{Var}[\dot{Y}(t)] = R_{\dot{Y}}(0) = -\ddot{R}_Y(0),$$

and covariance

$$\operatorname{Cov}[Y(t), \dot{Y}(t)] = R_{Y, \dot{Y}}(0) = \dot{R}_Y(0).$$

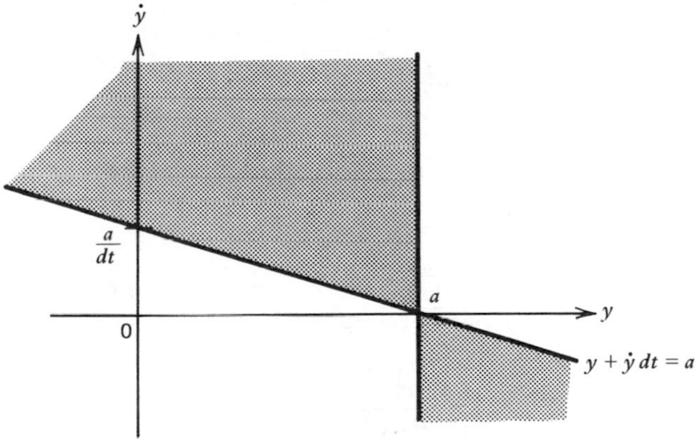

Figure 7.2.4

Furthermore, the existence of the second derivative $\ddot{R}_Y(0)$, together with the symmetry property of the correlation function $R_Y(\tau)$, implies that $\dot{R}_Y(0) = 0$, so that the random variables $Y(t)$ and $\dot{Y}(t)$ are independent. Hence the joint density

$$f_{Y(t),\,\dot{Y}(t)}(y, \dot{y}) = \frac{1}{\sigma_1}\phi\left(\frac{y}{\sigma_1}\right)\frac{1}{\sigma_2}\phi\left(\frac{\dot{y}}{\sigma_2}\right),$$

where

$$\phi(z) = \frac{1}{\sqrt{2\pi}}e^{-z^2/2}, \qquad -\infty < z < \infty,$$

is the standard Gaussian density and

$$\sigma_1 = \sqrt{R_Y(0)}, \qquad \sigma_2 = \sqrt{-\ddot{R}_Y(0)}.$$

We can now evaluate the probability $P(dN(t) = 1)$. Let us begin with the first integral on the right-hand side of (7). Upon substitution of $y = a - x\,dt$ for the integration variable y in the inner integral, we get

$$\int_0^\infty \frac{1}{\sigma_2}\phi\left(\frac{\dot{y}}{\sigma_2}\right)\left[\int_0^{\dot{y}} \frac{dt}{\sigma_1}\phi\left(\frac{a - x\,dt}{\sigma_1}\right)dx\right]d\dot{y}. \tag{8}$$

For small dt, however, the integrand inside the integral in brackets is approximately a constant independent of x, and thus (8) becomes

$$\frac{dt}{\sigma_1}\phi\left(\frac{a}{\sigma_1}\right)\int_0^\infty \frac{\dot{y}}{\sigma_2}\phi\left(\frac{\dot{y}}{\sigma_2}\right)d\dot{y} = \frac{dt}{\sigma_1}\phi\left(\frac{a}{\sigma_1}\right)\sqrt{\frac{\sigma_2}{2\pi}}$$

since for the standard Gaussian density $\int_0^\infty z\phi(z)\,dz = 1/\sqrt{2\pi}$. The second integral on the right-hand side of (7) can be evaluated by exactly the same method, and it turns out that its value is the same. Thus

$$P(dN(t) = 1) = 2\frac{dt}{\sigma_1}\phi\left(\frac{a}{\sigma_1}\right)\sqrt{\frac{\sigma_2}{2\pi}}. \tag{9}$$

As a result of the infinitesimal condition for the stationary point process

$$P(dN(t) = 1) \doteq \lambda\,dt = \mathrm{E}[dN(t)]$$

it follows by substituting back into (9) for σ_1, σ_2, and $\phi(z)$ that the intensity λ of the point process generated by level crossings equals

$$\lambda = \frac{1}{\pi}e^{-a^2/2R_Y(0)}\sqrt{\frac{-\ddot{R}_Y(0)}{R_Y(0)}} \tag{10}$$

This expression is known as *Rice's formula* and was derived by S. O. Rice as early as 1945. Note that the crossing intensity decreases with increasing level a but increases with increasing variance $R_{\dot{Y}}(0) = -\ddot{R}_Y(0)$ of the MS-derivative process $\dot{Y}(t)$, $-\infty < t < \infty$, which makes a lot of sense.

Actually, it is possible to obtain much more information about stationary point processes generated by level crossings, but the derivations are quite involved and lengthy and so will not be treated further.

EVOLUTION WITHOUT AFTEREFFECTS. Point processes are frequently used in applications to model random disturbances that are of impulsive character. A typical example is the so-called shot noise—a random current between the anode and cathode of a diode vacuum tube caused by random emissions of electrons from the incandescent cathode. As with ordinary stochastic processes, inherent in the idea of random noise is the concept of independence; that is, the past behavior of the noise process should in no way influence its future. In terms of a point process, this means that having observed the occurrences of point-events up to some present time t_p, we should have no information about the epochs of point-events yet to occur after this present time t_p.

It may seem that this goal might be accomplished by requiring that the waiting times $X(n)$ between consecutive epochs be independent random variables. But this, except in the rather special case of exponential waiting times, will not suffice. The reason for this is that knowledge of the time of the last (say nth) observed point-event at epoch t_n, prior to our present time instant t_p, would allow us to reevaluate the original distribution of the waiting time $X(n + 1)$ to the forthcoming point-event by conditioning upon the observation that $X(n + 1) > t_p - t_n$, that is, that we have already waited for $t_p - t_n$ time units for the subsequent point-events to occur. If the waiting times are independent, this conditional distribution would have to be the same as the unconditional distribution and, except for the exponential density, this condition cannot be met.

Example 7.2.2

Suppose that the waiting times $X(n)$, $n = \ldots, -1, 0, 1, \ldots$, are independent, identically distributed random variables with Rayleigh density

$$f_X(x) = \begin{cases} 2xe^{-x^2} & \text{if} \quad x > 0, \\ 0 & \text{if} \quad x \le 0, \end{cases}$$

and let $t_n \le t_p$ be the epoch of the last point-event that occurred before the present time t_p. Then the epoch of the next point-event to occur in the future is $T(n + 1) = t_n + X(n)$, and granted that $T(n) = t_n$ was the last observed epoch, we already know that $T(n+1) > t_p$ or, equivalently, that $X(n) > t_p - t_n$. Thus, for any $t > t_p$, we have

$$P(T(n+1) > t \mid T(n+1) > t_p,\ T(n) = t_n) = P(X(n) > t - t_n \mid X(n) > t_p - t_n)$$
$$= \frac{P(X(n) > t - t_n)}{P(X(n) > t_p - t_n)} = \frac{1 - F_X(t - t_n)}{1 - F_X(t_p - t_n)},$$

where $F_X(x) = 1 - e^{-x^2}$, $x > 0$, is the Rayleigh distribution function. Hence

$$P(t(n+1) > t \mid \text{last observed epoch at } t_n) = \frac{e^{-(t-t_n)^2}}{e^{-(t_p-t_n)^2}},$$

and differentiating with respect to t, we obtain the conditional density of the next epoch $T(n+1)$, given that the last one occurred at $t_n \le t_p$, namely

$$f(t) = \begin{cases} 2(t - t_n)\dfrac{e^{-(t-t_n)^2}}{e^{-(t_p-t_n)^2}} & \text{if } t > t_p, \\ 0 & \text{if } t \le t_p. \end{cases}$$

From this expression, or from the graph of this density in Figure 7.2.5, it is evident that the distribution of the next epoch does depend on the observed occurrence of the past point-events, in particular on the time t_n of the last observed epoch. Note that the density $f(t)$ has the shape of the Rayleigh density shifted to the point t_n, having the initial portion chopped off at the point t_p and subsequently normalized by the constant $e^{(t_p-t_n)^2}$. ▲

Since the waiting times do not work, let us try instead to use the counting properties of the point process. Suppose we require that the *numbers of point-events in any finite collection of nonoverlapping time intervals be independent random variables.*

Then, indeed, having observed the point process up to the present time t_p will be of no use in predicting future occurrences, since the number of point-events with epochs in any future time interval will be independent of our observation simply because any future time interval cannot overlap with any time interval or intervals lying entirely on the past side of the present t_p. To illustrate this more clearly, suppose, for example, that prior to the present time t_p, n point-events with epochs $T_1 = t_1, \ldots, T_n = t_n$ have occurred. Then the conditional probability of the next $(n+1)$st point-event occurring later than $t + \tau$, given the observations, is the same as the unconditional probability of the same event, since $T_{n+1} > t + \tau$ is the same as having no point-event in the time interval $(t, t+\tau]$, which by assumption is independent of whatever happened in past time intervals.

Point processes having the property that the numbers of point-events with epochs in nonoverlapping time intervals are independent random variables are

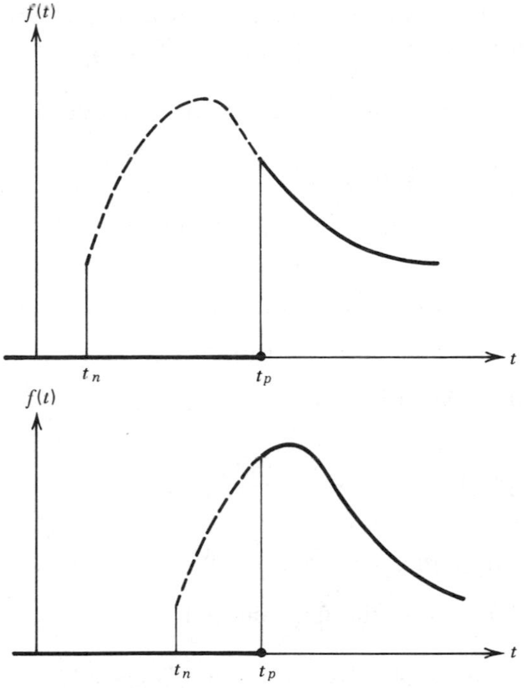

Figure 7.2.5

said to *evolve without aftereffects.* The formal definition is best expressed in terms of the associated counting process, namely that such a process has independent increments.

Definition 7.2.2
A point process is said to *evolve without aftereffects* if for any collection of time instants $t_0 < t_1 < \cdots < t_{k+1}$ from the time domain $\mathcal{T}$, and any $k = 1, 2, \ldots$, the increments

$$N(t_1) - N(t_0),\ N(t_2) - N(t_1),\ \ldots,\ N(t_{k+1}) - N(t_k)$$

of the associated counting process $N(t)$, $t \in \mathcal{T}$, are independent random variables.

If a point process evolves without aftereffects, then according to this definition the numbers of point-events in nonoverlapping time intervals are independent random variables. Since any time interval with endpoints $t_1 < t_2$ can be divided into an arbitrary number of nonoverlapping subintervals, with the number of point-events in the large interval being the sum of the numbers of point-events in its subintervals, it is clear that evolution without aftereffects will require that the probability law of the point process be of a very special kind. In fact, as we now show, the probability law will be of the Poisson type.

To begin with, let us consider a point process that evolves without aftereffects, and let

$$N(t), \qquad t \in \mathscr{T},$$

be its associated counting process. Next, let $t_0 \in \mathscr{T}$ be a fixed time instant, and let us define the function

$$g(t) = -\ln P(N(t) - N(t_0) = 0), \qquad t \geq t_0. \tag{11}$$

Since a counting process has right-continuous sample paths, we must have

$$\lim_{t \to t_0+} P(N(t) - N(t_0) = 0) = 1$$

that is,

$$g(t_0) = 0.$$

Since clearly†

$$P(N(t_1) - N(t_0) = 0) > P(N(t_2) - N(t_0) = 0)$$

whenever $t_0 < t_1 < t_2$, we see from (11) that the function $g(t)$, $t \geq t_0$ must be strictly increasing.

Finally, the function $g(t)$, $t \geq t_0$, must also be *continuous*. To see this, note that by (11)

$$g(t + \tau) - g(t) = -\ln \frac{P(N(t + \tau) - N(t_0) = 0)}{P(N(t) - N(t_0) = 0)}, \tag{12}$$

and

$$\begin{aligned} P(N(t + \tau) - N(t_0) = 0) &= P(N(t + \tau) - N(t) = 0, N(t) - N(t_0) = 0) \\ &= P(N(t + \tau) - N(t) = 0)P(N(t) - N(t_0) = 0), \end{aligned} \tag{13}$$

the last equality following from the assumption that the process evolves without aftereffects. Thus

$$g(t + \tau) - g(t) = -\ln P(N(t + \tau) - N(t) = 0),$$

and by letting $\tau \to 0$, the right-hand side of (13) approaches zero, since

$$\begin{aligned} P(N(t + \tau) - N(t) = 0) &= 1 - P(N(t + \tau) - N(t) > 0) \\ &\to 1 - P(N(t+) - N(t) > 0) = 1 \end{aligned}$$

by assumption (iv) that point-events do not occur at predetermined times.

† Recall that we have eliminated any time interval having a zero probability of point-events occurring in it.

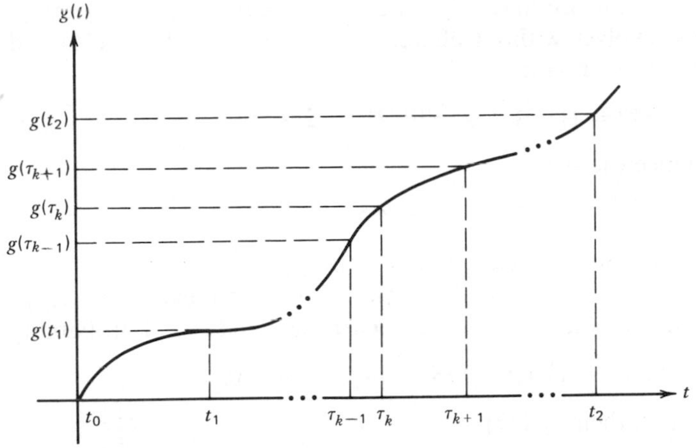

Figure 7.2.6

To summarize, the graph of the function $g(t)$ must look like Figure 7.2.6.

Next, choose two time instants $t_0 \leq t_1 \leq t_2$ and partition the interval $(t_1, t_2]$ into n subintervals

$$t_1 = \tau_0 < \tau_1 < \cdots < \tau_n = t_2, \tag{14}$$

such that for all $k = 1, \ldots, n$

$$g(\tau_k) - g(\tau_{k-1}) = \frac{1}{n}[g(t_2) - g(t_1)].\dagger \tag{15}$$

Since the process evolves without aftereffects, we have, as in (13),

$$P(N(\tau_k) - N(\tau_{k-1}) = 0) = \frac{P(N(\tau_k) - N(t_0) = 0)}{P(N(\tau_{k-1}) - N(t_0) = 0)}$$

and thus, from (12) and (15), we obtain

$$\begin{aligned} P(N(\tau_k) - N(\tau_{k-1}) = 0) &= e^{-[g(\tau_k) - g(\tau_{k-1})]} \\ &= e^{-[g(t_2) - g(t_1)]/n} \end{aligned} \tag{16}$$

for all $k = 1, \ldots, n$. Note that these probabilities do not depend on k but only on n.

Let us next define n random variables

$$Z_1^{(n)}, Z_2^{(n)}, \ldots, Z_n^{(n)}$$

by

$$Z_k^{(n)} = \begin{cases} 1 & \text{if } N(\tau_k) - N(\tau_{k-1}) \geq 1, \\ 0 & \text{if } N(\tau_k) - N(\tau_{k-1}) = 0. \end{cases} \tag{17}$$

† Note that the subintervals (14) need not have equal lengths. See Figure 7.2.6.

Note that $Z_k^{(n)}$, $k = 0, 1, \ldots, n$, are mutually independent Bernoulli random variables, since the process evolves without aftereffects, and as seen from (16) and (17), are identically distributed with

$$P(Z_k^{(n)} = 1) = 1 - e^{-[g(t_2) - g(t_1)]/n} = p_n \tag{18}$$

for all $k = 1, \ldots, n$. Hence the sum

$$S_n = Z_1^{(n)} + \cdots + Z_n^{(n)}$$

is a binomial random variable with parameters n and p_n.

Now let A_n be the event that in at least one of the subintervals $(\tau_{k-1}, \tau_k]$, $k = 1, \ldots, n$, more than one point-event occurs. Then the conditional probability

$$P(N(t_2) - N(t_1) = m \,|\, \overline{A_n}) = P(S_n = m), \qquad m = 0, 1, \ldots,$$

since if A_n does not occur, then by (17)

$$Z_k^{(n)} = N(\tau_k) - N(\tau_{k-1}) \qquad \text{for all } k = 1, \ldots, n.$$

Then, by the formula for total probability,

$$P(N(t_2) - N(t_1) = m) = P(N(t_2) - N(t_1) = m \,|\, A_n)P(A_n) + P(S_n = m)P(\overline{A_n}). \tag{19}$$

We now let n, the number of subintervals (14), go to infinity. Then

$$P(A_n) \to 0 \tag{20}$$

since if $P(A_n) \geq \varepsilon > 0$ for infinitely many n, there would be a positive probability of having point-events with epochs arbitrarily close together in time, so that there would be an infinite number of them in a finite time interval. But (20) and (19) imply that as $n \to \infty$

$$|P(N(t_2) - N(t_1) = m) - P(S_n = m)| \to 0 \tag{21}$$

for any $m = 0, 1, \ldots$, so that the distribution of $N(t_2) - N(t_1)$, that is, the distribution of the number of point-events with epochs in $(t_1, t_2]$, is the same as the limiting distribution of the binomial random variable S_n as $n \to \infty$.

From (18) we see that as $n \to \infty$

$$p_n \to 0,$$

and, by expanding the exponential function in (18) into its Taylor's series

$$np_n = [g(t_2) - g(t_1)] - \frac{1}{2!}\frac{[g(t_2) - g(t_1)]^2}{n} + \frac{1}{3!}\frac{[g(t_2) - g(t_1)]^3}{n^2} - \frac{1}{4!}\frac{[g(t_2) - g(t_1)]^4}{n^3} + \cdots,$$

so that

$$np_n \to g(t_2) - g(t_1)$$

as $n \to \infty$. As is easily seen (Chapter One), this means that the distribution of S_n converges to the Poisson distribution with parameter equal to

$$g(t_2) - g(t_1),$$

so that by (21), $N(t_2) - N(t_1)$ is a Poisson random variable with the same parameter.

Furthermore, since the parameter of a Poisson random variable is equal to its expected value, we also have

$$\mathrm{E}[N(t_2) - N(t_1)] = g(t_2) - g(t_1),$$

and consequently the function g is actually the mean function of the process. We now summarize this entire derivation as a theorem.

THEOREM 7.2.3. *If a point process [satisfying assumptions (i) through (iv)] evolves without aftereffects then, for any* $t_1 < t_2$ *from the time domain* $\mathcal{T}$, *the number of point-events with epochs in* $(t_1, t_2]$ *has the Poisson distribution*

$$P(N(t_2) - N(t_1) = n) = \frac{[\mu(t_2) - \mu(t_1)]^n}{n!} e^{-[\mu(t_2) - \mu(t_1)]}$$

$n = 0, 1, \ldots,$ *with* $\mu(t) = E[N(t)]$ *the mean function of the point process.*

Since the process is assumed to evolve without aftereffects, its entire probability law is completely specified by the distribution of $N(t_2) - N(t_1)$ for any $t_1 < t_2$, that is, in view of Theorem 7.2.3, by the function $\mu(t)$, $t \in \mathcal{T}$. Thus, for example, for any $t_1 < t_2 < t_3 < t_4$ the probability that there are exactly n_1 point-events with epochs in $(t_1, t_2]$ and exactly n_2 point-events with epochs in $(t_3, t_4]$ equals

$$P(N(t_2) - N(t_1) = n_1, N(t_4) - N(t_3) = n_2)$$
$$= e^{-[\mu(t_2) - \mu(t_1)]} \frac{[\mu(t_2) - \mu(t_1)]^{n_1}}{n_1!} e^{-[\mu(t_4) - \mu(t_3)]} \frac{[\mu(t_4) - \mu(t_3)]^{n_2}}{n_2!},$$

$n_1 = 0, 1, \ldots, n_2 = 0, 1, \ldots.$

The point process, whose probability law we just derived, is naturally called a *Poisson point process*. Its probability law is thus completely specified by its mean function $\mu(t)$, $t \in \mathcal{T}$, or assuming as we shall from now on that

$$\mu(t) = \int_0^t \lambda(\tau)\, d\tau, \qquad t \in \mathcal{T},$$

so it may also be specified by its intensity function $\lambda(t)$, $t \in \mathcal{T}$.

Furthermore, according to Theorem 7.2.3, *it is the only point process* [satisfying (i) through (iv)] *that evolves without aftereffects.*

Generally, the intensity function $\lambda(t)$ varies with time, in which case the point process (or, alternatively, the counting process) is called an *inhomogeneous Poisson process.*

If the intensity function is a constant

$$\lambda(t) = \lambda > 0, \qquad t \in \mathcal{T},$$

that is, if the point process is stationary, it is called a *homogeneous Poisson process* with intensity λ. Its associated counting process $N(t)$, $t \geq 0$, is then the already familiar Poisson process (see Chapter One).

Again, Theorem 7.2.3 implies that a homogeneous Poisson process is the only stationary point process that evolves without aftereffects. A homogeneous Poisson process with intensity $\lambda = 1$ will be referred to as the *standard Poisson process*, and its associated counting process will be denoted by $N_0(t)$, $t \in \mathcal{T}$.

The reader will soon recognize that the standard Poisson process $N_0(t)$, $t \in \mathcal{T}$, occupies a fundamental position among point processes that is similar in many respects to that of the standard Wiener process $W_0(t)$, $t \in \mathcal{T}$ among processes with continuous sample paths.

Example 7.2.3

If a point process with time domain $\mathcal{T} = (0, \infty)$ is a homogeneous Poisson process with intensity λ, then the sequence

$$X(n), \qquad n = 0, 1, \ldots,$$

of waiting times is a sequence of independent random variables each with exponential density

$$f_X(x) = \begin{cases} \lambda e^{-\lambda x} & \text{if } x > 0, \\ 0 & \text{if } x \leq 0. \end{cases}$$

Suppose that, as in Example 7.2.2, the process has been observed up to the present time t_p, the last observed epoch being $T(n) = t_n \leq t_p$. Then the conditional probability for $t > t_p$ is

$$P(T(n+1) > t \mid \text{last observed epoch at } t_n) = \frac{1 - F_X(t - t_n)}{1 - F_X(t_p - t_n)} = \frac{e^{-\lambda(t - t_n)}}{e^{-\lambda(t_p - t_n)}}$$

$$= e^{-\lambda(t - t_p)},$$

and hence the conditional density of the next epoch $T(n + 1)$, given the observation, equals

$$f(t) = \begin{cases} e^{-\lambda(t - t_p)} & \text{if } t > t_p, \\ 0 & \text{if } t \leq t_p. \end{cases}$$

This time, the density does not depend on t_n at all, and therefore observing the process up to the present time t_p provides no further information about the distribution of the next epoch to occur in the future. ▲

EXERCISES

Exercise 7.2

1. Calculate the mean function and the intensity function of the point process defined in Exercise 7.1.1.
2. Calculate the mean function and the intensity function of the point process defined in Exercise 7.1.3.
3. For a point process on $\mathscr{T} = (0, \infty)$, establish the formula for the intensity function $\lambda(t) = \sum_{n=1}^{\infty} f_{T(n)}(t)$, where $f_{T(n)}$ is the density of the nth random epoch. Generalize to a point process on $\mathscr{T} = (-\infty, \infty)$. (*Hint:* Exploit the equivalence $[T(n) \le t] \Leftrightarrow [N(t) \ge n]$ and Exercise 1.7.)
4. Calculate the mean function and the intensity function for the point process defined in Exercise 7.1.9. Leave the answer in the form of an infinite series.
5. Suppose that the waiting time sequence $X(n)$, $n = 0, 1, \ldots,$ of a point process on $\mathscr{T} = (0, \infty)$ is a sequence of independent random variables, each having the same density as the square of a Gaussian random variable with mean zero and variance σ^2. Calculate the intensity function for this point process. Leave the answer in the form of an infinite series.
6. Consider a point process on $\mathscr{T} = (-\infty, \infty)$ such that for any $0 < t < \infty$

$$P(N(t) = k) = e^{-\alpha t}(1 - e^{-\alpha t})^k,$$

$k = 0, 1, \ldots,$ where $\alpha > 0$ is a constant. Find the mean function and the intensity function for this process.

7. Consider a stationary point process on $\mathscr{T} = (-\infty, \infty)$ such that for all $-\infty < t_1 < t_2 < \infty$ $\quad P(N(t_2) - N(t_1) = k) = \binom{k+3}{k}\left(\frac{\alpha}{\alpha + t_2 - t_1}\right)^3\left(\frac{t_2 - t_1}{\alpha + t_2 - t_1}\right)^k,$ $k = 0, 1, \ldots,$ where $\alpha > 0$ is a constant. Find the mean function and intensity for this process. Prove directly from Definition 7.2.2 that the process does not evolve without aftereffects.
8. A point process is called Kth order stationary ($K \ge 1$, integer) if Definition 7.2.1 applies only to $k = 1, \ldots, K$. It is called wide-sense stationary if its intensity function is a constant. Prove the chain of implications: strictly stationary ⇒ $(K + 1)$st order stationary ⇒ Kth order stationary ⇒ wide-sense stationary. Prove that for a Poisson process, reversed implications are also true.
9. Suppose that an inhomogeneous Poisson process on $\mathscr{T} = (0, \infty)$ has a periodic intensity function $\lambda(t + k\Delta) = \lambda(t)$, $t > 0$, $k = 1, 2, \ldots,$ where $\Delta > 0$ is the period. Define a new point process with an associated counting process $N'(t) = N(t + \Theta)$, $t > 0$, where $N(t)$ is the original Poisson counting process and Θ is a random variable uniformly distributed on $(0, \Delta)$ and independent of $N(t)$, $t > 0$. Show that this new point process

is strictly stationary, and evaluate its intensity λ'. Is this still a Poisson process? (*Hint:* First calculate $E[N'(t) | \Theta]$.)

10. Consider a random sine wave $Y(t) = A \sin(\omega t + \Theta)$, $-\infty < t < \infty$, with Rayleigh amplitude A independent of the uniform $(0, 2\pi)$ phase Θ. Calculate the intensity of the point process generated by the crossings of level $\alpha > 0$. (*Hint:* There are either two or zero crossings during each period.) Recall that $Y(t)$ is a zero mean stationary Gaussian process, so you can check your result against Rice's formula.

11. Let $Y(t)$, $-\infty < t < \infty$, be a stationary zero mean Gaussian process with correlation function

$$R_Y(\tau) = \sigma^2 e^{-\beta\tau^2}, \qquad -\infty < \tau < \infty,$$

where σ and β are positive constants. Use Rice's formula to obtain the intensity of a point process generated by crossing the level $\alpha = 1$. Sketch the intensity as a function of the parameter β.

12. A point process is generated by the passages of the stationary process $Y(t)$, $-\infty < t < \infty$, through the level zero. Assuming that $Y(t)$ is a Gaussian process with zero mean and correlation function

$$R_Y(\tau) = \sigma^2 e^{-|\tau|/10}\left(\cos 0.7\tau + \frac{1}{7}\sin 0.7|\tau|\right),$$

where the time is measured in seconds, find the expected number of point-events occurring during a ten-minute interval.

13. A noise voltage $Y(t)$ in a digital circuit causes a malfunction (e.g., an indetermined logic state) each time it increases above 0.7 volt. Assuming that $Y(t)$, $-\infty < t < \infty$, is a zero mean stationary Gaussian process with correlation function

$$R_Y(\tau) = \sigma^2 e^{-10^3|\tau|}(1 + 10^3|\tau|)$$

find the maximum permissible noise power σ^2 for which the intensity of the point process of crossings is no larger than 10^{-6}/second.

14. Suppose a point process on $\mathscr{T} = (0, \infty)$ evolves without aftereffects. If the intensity function of the process is $\lambda(t) = (1 + t)^{-1}$, find the smallest time $t_0 > 0$ such that the probability of no point event occurring during the interval $(t_0, t_0 + 1]$ is at least .99.

15. Consider m adjacent intervals, each of the same length, on the time domain $\mathscr{T}$ of a stationary point process, and let Z_j be the number of point events occurring in the jth of these time intervals. Assuming that the point process evolves without aftereffects, show that the conditional probability mass distribution of the vector $Z' = (Z_1, \ldots, Z_m)$, given that $Z_1 + \cdots + Z_m = n$, is the Maxwell-Boltzmann statistic discussed in Exercise 1.8.

7.3. THE POISSON PROCESS

The Poisson process plays a prominent role among point processes not only because of its genesis as a point process evolving without aftereffects but also because it serves as a basic building block for more complicated point processes. It is therefore worthwhile to investigate this process in some detail.

Recall from the previous section that the probability law of an inhomogeneous Poisson process is completely characterized by the following two properties:

1 For any $t_1 < t_2$ from the time domain $\mathcal{T}$ the number of point-events with epochs in the interval $(t_1, t_2]$ has the Poisson distribution with parameter $g(t_2) - g(t_1)$, where $g(t)$, $t \in \mathcal{T}$, is an increasing continuous function.

2 The number of point-events with epochs in nonoverlapping time intervals are independent random variables.

Upon choosing the function $g(t)$ such that $g(0) = 0$, this function coincides with the mean function $\mu(t)$ of the process. In terms of the intensity function $\lambda(t)$, we can thus write the Poisson probability for the number of point-events in an interval $(t_1, t_2]$ as

$$P(N(t_2) - N(t_1) = n) = e^{-\int_{t_1}^{t_2} \lambda(\tau)\, d\tau} \frac{[\int_{t_1}^{t_2} \lambda(\tau)\, d\tau]^n}{n!}, \tag{1}$$

$n = 0, 1, \ldots, t_1 < t_2$.

We can also immediately write the characteristic function for the Poisson distribution (see Appendix Two)

$$\psi_{N(t_2)-N(t_1)}(\zeta) = e^{(e^{i\zeta} - 1)\int_{t_1}^{t_2} \lambda(\tau)\, d\tau}, \qquad -\infty < \zeta < \infty. \tag{2}$$

Of course, we already know that

$$\mathrm{E}[N(t_2) - N(t_1)] = \int_{t_1}^{t_2} \lambda(\tau)\, d\tau, \tag{3}$$

and since the expectation and the variance of a Poisson distribution are the same, we also have

$$\mathrm{Var}[N(t_2) - N(t_1)] = \int_{t_1}^{t_2} \lambda(\tau)\, d\tau. \tag{4}$$

Denoting, as usual,

$$dN(t) = N(t + dt) - N(t), \qquad dt > 0$$

an infinitesimal forward increment of the counting process, we may also rewrite (1) through (4) in the perhaps more suggestive differential notation:

$$P(dN(t) = n) = e^{-\lambda(t)\, dt} \frac{[\lambda(t)\, dt]^n}{n!}, \qquad n = 0, 1, \ldots, \tag{5}$$

$$\psi_{dN(t)}(\zeta) = e^{(e^{i\zeta} - 1)\lambda(t)\, dt}, \qquad -\infty < \zeta < \infty, \tag{6}$$

$$\mathrm{E}[dN(t)] = \lambda(t)\, dt, \tag{7}$$

$$\mathrm{Var}[dN(t)] = \lambda(t)\, dt. \tag{8}$$

Notice that because of the equality of mean and variance, we may also write, at least for $t \geq 0$,

$$\begin{aligned} d \operatorname{Var}[N(t)] &= \operatorname{Var}[N(t+dt)] - \operatorname{Var}[N(t)] \\ &= \mathrm{E}[N(t+dt)] - \mathrm{E}[N(t)] \\ &= \mathrm{E}[dN(t)] = \operatorname{Var}[dN(t)]. \end{aligned}$$

Thus the variance of an increment equals an increment of the variance, which after all is true for any process with independent increments. (Recall that the Wiener process has the same property.)

From (5) we also have

$$\begin{aligned} P(dN(t) > 1) &= 1 - P(dN(t) = 0) - P(dN(t) = 1) \\ &= 1 - e^{-\lambda(t)\,dt}(1 + \lambda(t)\,dt), \end{aligned}$$

whence, upon dividing both sides by dt and using the fact that for any constant a

$$\lim_{x \to 0} \frac{1 - e^{ax}}{x} = -a$$

we obtain

$$\lim_{dt \to 0} \frac{P(dN(t) > 1)}{dt} = 0. \tag{9}$$

Also from (5), by setting $n = 0$ and $n = 1$ and expanding the exponential function into a Taylor series

$$e^{-\lambda(t)\,dt} = 1 - \lambda(t)\,dt + \frac{[\lambda(t)\,dt]^2}{2!} - \frac{[\lambda(t)\,dt]^3}{3!} + \cdots,$$

we obtain

$$P(dN(t) = 0) \doteq 1 - \lambda(t)\,dt, \tag{10}$$

and

$$P(dN(t) = 1) \doteq \lambda(t)\,dt. \tag{11}$$

But (9), (10), and (11) are nothing but the infinitesimal conditions discussed in Section 7.2.

Thus a Poisson process with intensity function $\lambda(t)$ is another instance of a point process for which the infinitesimal conditions hold without any additional assumptions. These conditions are the ones usually assumed in deriving the Poisson distribution (with $\lambda(t) = \lambda$, a constant). In fact, the procedure used in the previous section to establish the fact that a point process evolving without after-effects must be a Poisson process is essentially the same argument. The crux of the matter is that if the time domain $\mathcal{T}$ is partitioned into intervals $(t, t + dt]$ of sufficiently small lengths, then the following two assumptions are used:

(a) The probability of having more than one point-event with epoch in such a small interval is negligible as compared to having either exactly one or no point-event there, and hence the occurrence of a point-event in such an interval can be regarded as a "success" in a Bernoulli trial with $p_t \doteq \lambda(t)\, dt$.

(b) Since the process evolves without aftereffects, these Bernoulli trials corresponding to distinct intervals are independent.

Several important results can be deduced from this picture. First, by a suitable time transformation we should be able to make all of these Bernoulli trials identically distributed and thus obtain any inhomogeneous Poisson process from a homogeneous one, in particular from the standard Poisson process. This is actually quite easy; let $\lambda(t)$, $t \in \mathcal{T}$, be an intensity function, and let

$$\mu(t) = \int_0^t \lambda(\tau)\, d\tau, \qquad t \in \mathcal{T} \tag{12}$$

be the corresponding mean function.

Now take a standard Poisson process with counting process

$$N_0(t), \qquad t \in \mathcal{T},$$

and define

$$N(t) = N_0(\mu(t)), \qquad t \in \mathcal{T}. \tag{13}$$

Then for any $t_1 < t_2$ from the time domain $\mathcal{T}$

$$\begin{aligned} \mathrm{E}[N(t_2) - N(t_1)] &= \mathrm{E}[N_0(\mu(t_2)) - N_0(\mu(t_1))] \\ &= \mu(t_2) - \mu(t_1) \\ &= \int_{t_1}^{t_2} \lambda(\tau)\, d\tau, \end{aligned} \tag{14}$$

and since the mean function is continuous and increasing, it is seen from Figure 7.3.1 that any collection of nonoverlapping intervals $(t_1, t_2]$, $(t_2, t_3]$, ... will correspond to nonoverlapping intervals $(t'_1, t'_2]$, $(t'_2, t'_3]$, ... Since the standard Poisson process evolves without aftereffects, the same will be true for the new point process, with the associated counting process $N(t)$ defined by (13). This, together with (14), implies that this process will be an inhomogeneous Poisson process with intensity function $\lambda(t)$. Thus, any *inhomogeneous Poisson process can be obtained as a time transformation* (13) *from a standard Poisson process.*

This result can be used to derive many properties of an inhomogeneous

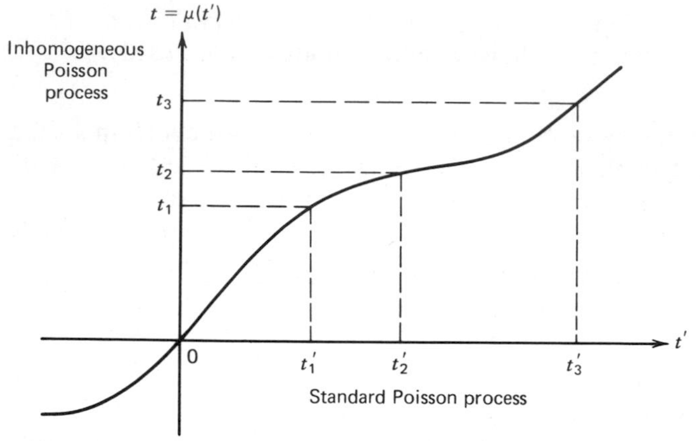

Figure 7.3.1

Poisson process from the corresponding properties of the standard Poisson process, as shown in the following example.

Example 7.3.1

Consider an inhomogeneous Poisson process with time domain $t > 0$ and intensity function $\lambda(t)$, and let $T(n)$, $n = 1, 2, \ldots$, be the nth random epoch. From the time transformation (13) it is clear that

$$T_0(n) = \mu(T(n)),$$

where $T_0(n)$ is the nth random epoch of the standard Poisson process on $\mathcal{T} = (0, \infty)$ and $\mu(t)$ is given by (12) (see Fig. 7.3.2). We know that $T_0(n)$ has the Erlang distribution (see Appendix Two) with density

$$f_{T_0(n)}(\tau) = \frac{1}{(n-1)!}\,\tau^{n-1}e^{-\tau}, \qquad \tau > 0.$$

Since $\tau = \mu(t)$ is increasing with $d\mu(t)/dt = \lambda(t) > 0$, we have immediately

$$f_{T(n)}(t) = \frac{\lambda(t)}{(n-1)!}\,(\mu(t))^{n-1}e^{-\mu(t)}, \qquad t > 0,$$

for the density of $T(n)$. ▲

Next, let $N(t)$, $t \in \mathcal{T}$, again be a counting process associated with an inhomogeneous Poisson process, and let $t_1 < t_2$ be two time instances in the time domain $\mathcal{T}$. Since the process evolves without aftereffects the increment $N(t_2) - N(t_1)$ must be independent of any increment $N(t') - N(t'')$ with $t'' < t' \le t_1$ and thus, in fact, of the entire past history $N(t')$, $t' \le t_1$ of the counting process $N(t)$.

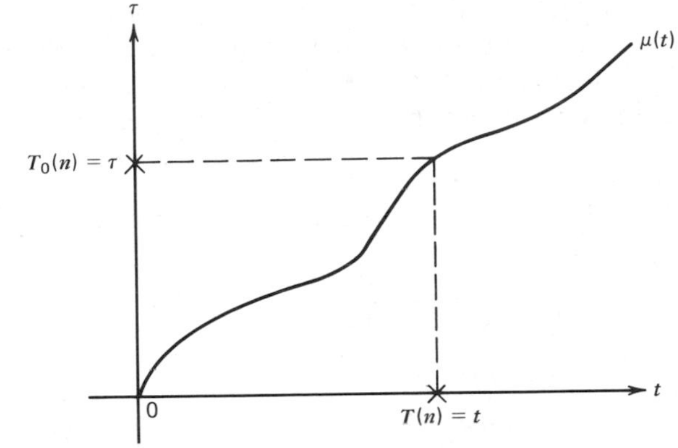

Figure 7.3.2

This implies that the conditional expectation

$$E[N(t_2) - N(t_1) | N(t'), t' \leq t_1] = E[N(t_2) - N(t_1)].$$

In other words, the forward increments are MS unpredictable from the past. Surprisingly, perhaps, the converse of this statement is also true: *if the forward increments of a point process are MS unpredictable from the past, then the process evolves without aftereffects*. The rigorous proof of this result is too technical to be reproduced here, but the reason behind it is simple. From assumption (a) above for a sufficiently small time interval $(t_1, t_2]$, the increment $N(t_2) - N(t_1)$ is "almost" a Bernoulli random variable, and so the expectations $E[N(t_2) - N(t_1) | N(t'), t' \leq t_1]$ and $E[N(t_2) - N(t_1)]$ should differ very little from the probabilities $P(N(t_2) - N(t_1) = 1 | N(t'), t' \leq t_1)$ and $P(N(t_2) - N(t_1) = 1)$, respectively. But if these two probabilities are equal, that is, if the conditional probability of the event "occurrence of a point-event in $(t_1, t_2]$", given the past Bernoulli trials in the small intervals prior to t_1, is the same as its unconditional probability, then the Bernoulli trials corresponding to the partition of the time domain must be independent. Thus, we get the property (b) above, or evolution without aftereffects.

The rigorous proof of this statement was given by S. Watanabe and F. Papangelou.

***THEOREM** 7.3.2. Let* N(t), t ∈ $\mathcal{T}$ *be a counting process associated with a point process.*† *Then the point process evolves without aftereffects (and hence is an inhomogeneous Poisson process) if and only if*

$$E[N(t_2) - N(t_1) | N(t'), t' \leq t_1] = E[N(t_2) - N(t_1)] < \infty, \quad (15)$$

for any $t_1 < t_2$ *from the time domain* $\mathcal{T}$.

† Assuming that we deal only with point processes satisfying assumptions (i) through (iv) of Section 7.1.

The significance of this theorem is greater than it may seem. We will show in Section 7.7 that it provides a basic stepping stone toward more general point processes. For the time being, we merely note that if $N(t)$, $t > 0$, is a counting process with mean function $\mu(t) = \mathrm{E}[N(t)]$, then condition (15) is equivalent to the statement that the process

$$X(t) = N(t) - \mu(t), \qquad t > 0,$$

is a martingale. This is so because

$$\mathrm{E}[N(t_2) - N(t_1)] = \mu(t_2) - \mu(t_1) \tag{16}$$

and by the properties of conditional expectation

$$\mathrm{E}[N(t_2) - N(t_1) \,|\, N(t'), t' \leq t_1] = \mathrm{E}[N(t_2) \,|\, N(t'), t' \leq t_1] - N(t_1) \tag{17}$$

for any $0 \leq t_1 \leq t_2$. From (15), (16), and (17), we obtain

$$\mathrm{E}[N(t_2) - \mu(t_2) \,|\, N(t'), t' \leq t_1] = N(t_1) - \mu(t_1),$$

or

$$\mathrm{E}[X(t_2) \,|\, X(t'), t' \leq t_1] = X(t_1), \qquad 0 \leq t_1 < t_2.$$

The theorem can thus be restated by saying that a point process with associated counting process $N(t)$, $t > 0$, is a Poisson process if and only if the process

$$X(t) = N(t) - \mathrm{E}[N(t)], \qquad t > 0$$

is a martingale. In particular, the standard Poisson process is the only point process with the property that if $N_0(t)$, $t > 0$, is its associated counting process, then

$$N_0(t) - t, \qquad t > 0$$

is a martingale. We restate this result in a slightly modified equivalent form as the following:

Corollary

A point process with an associated counting process $N(t)$, $t \in \mathcal{T}$, is a standard Poisson process if and only if for all $t \in \mathcal{T}$

$$\mathrm{E}[dN(t) \,|\, N(t'), t' \leq t] \doteq dt.$$

Comparing this result with Lévy-Doob's Theorem 4.3.1 further corroborates our earlier remark that the standard Poisson process plays the same role for point processes as does the standard Wiener process for processes with continuous sample paths.

As a last application of the basic picture of the Poisson process [assumptions

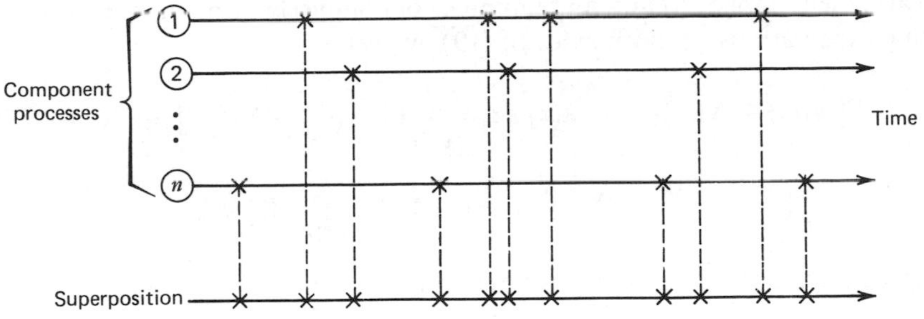

Figure 7.3.3

(a) and (b) above], we consider what may be called a central limit theorem for point processes.

Suppose we are given n mutually independent point processes with the same time domain $\mathcal{T}$. If

$$N_1(t), N_2(t), \ldots, N_n(t)$$

are the respective associated counting processes, then the sum

$$N(t) = N_1(t) + N_2(t) + \cdots + N_n(t), \qquad t \in \mathcal{T} \tag{18}$$

is again a counting process associated with some point process. We call the latter point process a *superposition* of the n independent point processes. The realization of such a superposition is obtained by combining all epochs of the n point processes into one stream of epochs, as shown in Figure 7.3.3. Superposition of point processes arises quite naturally in many applications; the component point processes may, for instance, represent demands for access to a computer center from each individual subscriber so that the superposition would then be the stream of incoming demands as viewed from the computer center.

From (18) it should be obvious that if each of the component processes evolves without aftereffects, the superposition process will also, as long as the component processes are mutually independent. To convince the reader, let $(t_1, t_2]$ and $(t_3, t_4]$ be two nonoverlapping time intervals. Then from (18) we have

$$\begin{aligned} N(t_2) - N(t_1) &= \sum_{k=1}^{n} [N_k(t_2) - N_k(t_1)] \\ N(t_4) - N(t_3) &= \sum_{k=1}^{n} [N_k(t_4) - N_k(t_3)]. \end{aligned} \tag{19}$$

Then the $2n$ random variables that appear in brackets on the right-hand sides of (19) are all mutually independent, and hence by the hereditary property of independence, so must be the two random variables on the left-hand side. Clearly, the same argument applies to any finite number of nonoverlapping time intervals.

Taking expectations on both sides of (19), we get

$$E[N(t_2) - N(t_1)] = \int_{t_1}^{t_2} \lambda(\tau)\, d\tau = \sum_{k=1}^{n} E[N_k(t_2) - N_k(t_1)]$$

$$= \sum_{k=1}^{n} \int_{t_1}^{t_2} \lambda_k(\tau)\, d\tau = \int_{t_1}^{t_2} \sum_{k=1}^{n} \lambda_k(\tau)\, d\tau$$

and hence we have proved the following theorem.

***THEOREM** 7.3.3. The superposition of* n *independent Poisson processes with intensity functions* $\lambda_1(t), \ldots, \lambda_n(t)$ *is a Poisson process with intensity function*

$$\lambda(t) = \lambda_1(t) + \lambda_2(t) + \cdots + \lambda_n(t).$$

Example 7.3.2

Consider a point process that is the superposition of two independent Poisson processes with intensity functions $\lambda_1(t)$ and $\lambda_2(t)$, respectively. Given a particular point-event of the superposition process, we may inquire about the probability that the point-event came from the first component process. Let τ be the epoch of the point-event in question, and let $p_1(\tau)$ and $p_2(\tau)$ be the probabilities that it came from the first or second component process, respectively. Consider an infinitesimal interval $(t, t + dt]$ containing the epoch τ. If the point-event came from the first component process, we must have for sufficiently small dt,

$$dN_1(t) = 1 \quad \text{and} \quad dN_2(t) = 0.$$

Since the component Poisson processes are independent, the probability of this occurring is a product,

$$\begin{aligned} P(dN_1(t) = 1, dN_2(t) = 0) &= P(dN_1(t) = 1)P(dN_2(t) = 0) \\ &\doteq \lambda_1(t)\, dt(1 - \lambda_2(t)\, dt) \\ &\doteq \lambda_1(t)\, dt, \end{aligned}$$

upon neglecting $(dt)^2$, as compared to dt, and assuming that $\lambda_1(t)$ is a continuous function, at least in a neighborhood of the epoch τ. Similarly,

$$P(dN_1(t) = 0, dN_2(t) = 1) \doteq \lambda_2(t)\, dt,$$

so that the ratio

$$\frac{p_1(\tau)}{p_2(\tau)} = \frac{\lambda_1(\tau)}{\lambda_2(\tau)}.$$

Since clearly $p_1(\tau) + p_2(\tau) = 1$, we obtain

$$p_1(\tau) = \frac{\lambda_1(\tau)}{\lambda(\tau)}, \qquad p_2(\tau) = \frac{\lambda_2(\tau)}{\lambda(\tau)},$$

where $\lambda(t) = \lambda_1(t) + \lambda_2(t)$ is the intensity function of the superposition Poisson process.

As an application of this result, consider a dynamic system with two possible states (a bistable multivibrator, for instance) labeled 1 and 2 and two inputs also labeled 1 and 2. Each time a trigger impulse arrives at one of these inputs the system transits to the state with the same label as the label of the triggered input. More specifically, the system operates according to the following table:

		PRESENT STATE	NEXT STATE
Input label of the present trigger impulse	1	1	1
		2	1
	2	1	2
		2	2

Suppose now that trigger impulses arriving at the two inputs are represented by two independent homogeneous Poisson processes with intensities λ_1 and λ_2, respectively. It then follows from the preceding discussion and the fact that a Poisson process evolves without aftereffects that the sequence of consecutive states of the system will follow a two-state Markov process with transition diagram as depicted in Figure 7.3.4.

The transition probabilities are

$$p_1 = \frac{\lambda_1}{\lambda}, \qquad p_2 = \frac{\lambda_2}{\lambda}, \qquad \lambda = \lambda_1 + \lambda_2,$$

and transitions are made upon each occurrence of a point-event of the superposition Poisson process with intensity λ. Since $p_1 + p_2 = 1$, it follows from the result

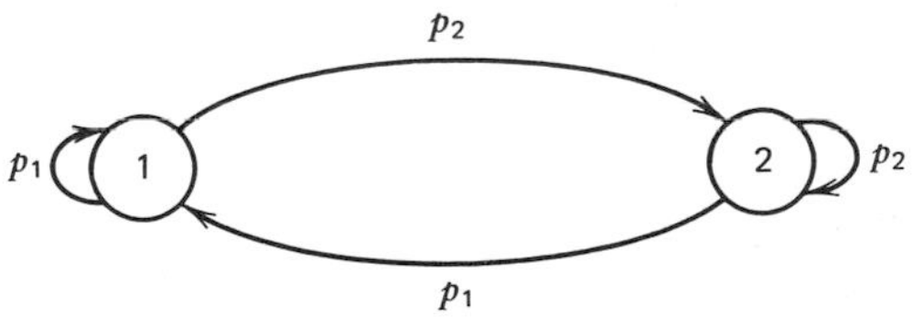

Figure 7.3.4

of Exercise 1.6 that at any time t the probability of the system being in state 1 is just p_1. ▲

Since, according to Theorem 7.3.3, the superposition of Poisson processes is again a Poisson process, a given Poisson process can in turn be regarded as being such a superposition. In other words, a Poisson process can be decomposed into independent Poisson processes. Example 7.3.2 indicates how such a decomposition can be accomplished. Suppose we wish to decompose a Poisson process with a continuous intensity function $\lambda(t)$ into n independent Poisson processes with continuous intensity functions $\lambda_1(t), \ldots, \lambda_n(t)$, where, of course, $\lambda_1(t) + \cdots + \lambda_n(t) = \lambda(t)$. Then we merely have to assign each point-event of our Poisson process to one of the n component processes at random and independently for each consecutive point-event. A point-event with epoch τ should be assigned to the kth component process ($k = 1, \ldots, n$), with probability

$$p_k(\tau) = \frac{\lambda_k(\tau)}{\lambda(\tau)}.$$

Example 7.3.3

Consider a particle performing a simple random walk on the set of all integers moving one step to the right with probability p $(0 < p < 1)$ or one step to the left with probability $q = 1 - p$, as in Figure 7.3.5. However, unlike the case of the Bernoulli random walk, the particle executes its motion only upon each occurrence of a point-event of a homogeneous Poisson process with intensity $\lambda > 0$. Assuming that at time $t = 0$ the particle is at the origin, we wish to find the probability distribution of its position $S(t)$ at some later time $t > 0$.

One way to do this would be to use the result that conditional upon the particle performing n steps by time t, the probability that it is at position k equals

$$\binom{n}{\frac{n+k}{2}} p^{(n+k)/2} q^{(n-k)/2}, \qquad k = -n, \ldots, n,$$

with the binomial coefficient being defined as zero if $n + k$ is odd. Since the number of steps n is the value of a Poisson process $n = N(t)$, we would then have the expression

$$P(S(t) = k) = \sum_{n=0}^{\infty} \binom{n}{\frac{n+k}{2}} p^{(n+k)/2} q^{(n-k)/2} e^{-\lambda t} \frac{(\lambda t)^n}{n!}$$

for the desired probability.

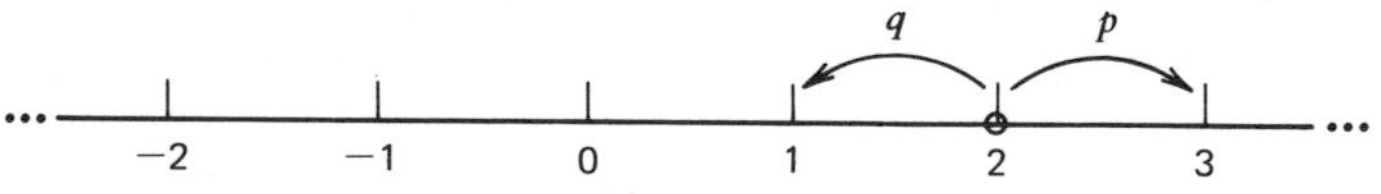

Figure 7.3.5

However, there is an alternate approach we might take. We can decompose the Poisson process into two independent homogeneous Poisson processes with intensities $\lambda_1 = p\lambda$ and $\lambda_2 = q\lambda$, respectively. In view of the mechanism of such a decomposition, as described above, we may now regard the particle as being pushed to the right upon each occurrence of a point-event of the first component process and to the left upon each occurrence of a point-event of the second component process. In fact, we can think of it as if the particle were bombarded by two independent Poisson streams of molecules moving in opposite directions. Denoting by $N_1(t)$ and $N_2(t)$ the associated counting processes, it is then clear that the position $S(t)$ of the particle at time $t > 0$ is simply

$$S(t) = N_1(t) - N_2(t). \tag{20}$$

Using the formula for the difference of two independent discrete random variables

$$P(S(t) = k) = \sum_{n=-\infty}^{\infty} P(N_1(t) = n + k)P(N_2(t) = n),$$

we obtain, upon substitution for the respective Poisson probabilities,

$$P(S(t) = k) = \sum_{n} e^{-p\lambda t} \frac{(p\lambda t)^{n+k}}{(n+k)!} e^{-q\lambda t} \frac{(q\lambda t)^n}{n!},$$

the summation going from 0 to ∞ if $k \geq 0$ and from k to ∞ if $k < 0$.

Taking first the case $k \geq 0$, this expression can be written as

$$P(S(t) = k) = e^{-\lambda t}(p\lambda t)^k \sum_{n=0}^{\infty} \frac{(\lambda t\sqrt{pq})^{2n}}{n!\,(n+k)!}.$$

The infinite series on the right cannot be summed to any elementary function. However, as the reader may find in any advanced calculus text, this is a Taylor series for the *modified Bessel function of order k*, $k = 0, 1, \ldots$, usually denoted by the symbol I_k,

$$I_k(x) = \sum_{n=0}^{\infty} \frac{1}{n!\,(n+k)!}\left(\frac{x}{2}\right)^{2n+k}, \qquad -\infty < x < \infty.$$

With $x = 2\lambda t\sqrt{pq}$, we thus have

$$P(S(t) = k) = e^{-\lambda t}\left(\frac{p}{q}\right)^{k/2} I_{|k|}(2\lambda t\sqrt{pq}),$$

$k = 0, \pm 1, \pm 2, \ldots$, since, as the reader is invited to verify, the same expression is also obtained for $k < 0$.

From (20) we can also easily obtain the characteristic function of $S(t)$

$$\begin{aligned}\psi_{S(t)}(\zeta) &= e^{(e^{i\zeta}-1)p\lambda t}e^{(e^{-i\zeta}-1)q\lambda t}\\ &= e^{\lambda t(pe^{i\zeta}+qe^{-i\zeta}-1)},\end{aligned}$$

and hence we can calculate moments of the random variable $S(t)$. Note that for $p = q = \frac{1}{2}$ the characteristic function becomes

$$\psi_{S(t)}(\zeta) = e^{\lambda t(\cos \zeta - 1)}. \qquad \blacktriangle$$

Let us now consider what happens if we have a superposition of component point processes which, although still mutually independent, can no longer be assumed to evolve without aftereffects. This clearly is quite an important question, since we generally know very little about the component processes except that they originated from independent sources. (The example of subscribers to a computer center mentioned above is a good illustration.) Thus, we may be quite willing to assume that the component point processes are mutually independent but rather reluctant to say that they should, in fact, be Poisson processes.

Clearly, the exact probability law of the superposition process will depend on the probability laws of the component processes, and if these are no longer Poisson processes, there is generally very little we can say about the law of the superposition process. Suppose, however, that we have a very large number n of component processes and, at the same time, that each component process is such that point-events occur only very rarely. Then it is plausible to argue that in the superposition process two or more consecutive point-events are very unlikely to come from the same component process. In fact, if n is sufficiently large and the contribution of each component process sufficiently rare, then it is likely that in a finite time interval $(t_1, t_2]$ each point-event of the superposition process came from a different component process. But if we again imagine that the time interval $(t_1, t_2]$ is partitioned into a large number of small subintervals, the occurrences of point-events of the superposition process in each of these subintervals can again be regarded as Bernoulli trials. Furthermore, since the component processes are by assumption independent, each contributing, at most, one point-event to the superposition, the Bernoulli trials will also be independent. Thus the superposition process will possess both the features (a) and (b) of the basic picture of a Poisson process, so that under the above assumptions the probability law of the superposition process will be approximately Poisson. We omit the exact statement of the asymptotic result hinted at by the above discussion but only roughly indicate its content.

If

$$N_n(t) = \sum_{k=1}^{n} N_{k,n}(t), \qquad t \in \mathcal{T}$$

is a counting process associated with a point process obtained as a superposition of n mutually independent point processes with counting processes $N_{k,n}(t)$, $k = 1, \ldots, n$, and if for any finite time interval $(t_1, t_2]$

$$\max_{k=1,\ldots,n} P(N_{k,n}(t_2) - N_{k,n}(t_1) > 0) \to 0, \tag{21}$$

and

$$\sum_{k=1}^{n} P(N_{k,n}(t_2) - N_{k,n}(t_1) > 1) \to 0, \tag{22}$$

as $n \to \infty$, then for large n the probability law of the superposition process is approximately that of an inhomogeneous Poisson process.

Condition (22) merely guarantees that the component processes do not contribute their point-events to the superposition in arbitrarily dense clusters, thus preventing an explosion of the superposition process.

However, condition (21) is crucial, since it assures that in each component process the point-events occur only very rarely, so that the contribution of each component process is negligible when compared to the total. If a collection of component processes satisfies this condition, the component processes are aptly called *uniformly sparse*.†

We can then summarize this result by saying that a *superposition of a large number of independent uniformly sparse point processes is approximately a Poisson process*.

The main significance of this result is that it further justifies the use of the Poisson process as an appropriate model for an impulsive random noise as well as for any other situation where the point process in question can be assumed to arise from the superposition of a large number of independent uniformly sparse sources. Again, we see the analogy between the roles of the Poisson and Wiener processes in modeling impulsive and continuous noises, respectively.

EXERCISES

Exercise 7.3

1. An inhomogeneous Poisson process on $\mathcal{T} = (0, \infty)$ has intensity function $\lambda(t) = 1 + \gamma \sin \pi t$, where $0 < \gamma < 1$ is a constant. Find the expectation and the variance of the difference between the number of point-events occurring during the intervals $(0, 1]$ and $(1, 2]$. Also find the conditional probability that no point-event occurs during $(0, 1]$, given that at least one does occur during the period $(0, 2]$.
2. Consider an inhomogeneous Poisson process with mean function $\mu(t) = 2\sqrt{1 + 10^3 t}$, $t > 0$. Compute the following probabilities:
   (a) $P(N(6) = 10)$
   (b) $P(N(6) = 10, N(20) = 15)$

† Readers who like exotic words may prefer the term "holospaudic," coined by K. L. Chung.

(c) $P(N(20) = 15 \mid N(6) = 10)$
(d) $P(N(6) = 10 \mid N(20) = 15)$
(e) $P(N(25) = 18 \mid N(12) - N(8) = 2)$

3. An inhomogeneous Poisson process on $\mathcal{T} = (0, \infty)$ has linearly increasing intensity $\lambda(t) = \alpha t$, where $\alpha > 0$. Calculate the probability mass distribution of $N(t)$, $t > 0$, and show that $P(N(t) < \infty) = 1$ for all $t > 0$, so that the process does not explode.
4. Consider a binary counter with two states, 0 and 1. Pulses are arriving at the input of the counter according to an inhomogeneous Poisson process with intensity function $\lambda(t) = \alpha(1 + \beta t)^{-1/2}$, $t \geq 0$, where α and β are positive constants. Calculate the probability that at time $t > 0$ the counter will be at the same state as at time $t = 0$.
5. Show that for a standard Poisson process the probability that an odd number of point-events occurs during a time interval of length $\tau > 0$ is $e^{-\tau} \sinh \tau$. Use the time transformation to generalize this result to an inhomogeneous Poisson process with (positive) intensity function $\lambda(t)$.
6. Suppose you have a random number generator that can produce a sequence of i.i.d. uniform (0, 1) random variables. You want to generate a stream of point-events of an inhomogeneous Poisson process with a prescribed intensity function $\lambda(t)$, $t > 0$. Show that the time transformation of the standard Poisson process provides a convenient algorithm for such a generation. (*Hint:* Recall that (-1) times the logarithm of a uniformly distributed random variable is exponential.)
7. Let $N(t)$, $t > 0$, be an inhomogeneous Poisson process with intensity function $\lambda(t)$. Find the first order characteristic function of the martingale $X(t) = N(t) - \mathrm{E}[N(t)]$, $t > 0$, and use it to calculate the first four moments $\mathrm{E}[X^m(t)]$, $m = 1, 2, 3, 4$.
8. Show that a homogeneous Poisson process is ergodic in the sense that $N(t)/t \to \lambda$ as $t \to \infty$ in MS (or even almost surely). (*Hint:* Use the fact that $N(t) - \lambda t$ is a martingale.)
9. A particle counter is recording two streams of particles, high-energy ones and low-energy ones. Assuming that the two streams are independent homogeneous Poisson processes with intensities λ_H and λ_L, respectively, find the conditional probability mass distribution of the number of high-energy particles registered during $(0, t]$, given that a total of n particles were recorded during this time interval. If a low-energy particle has energy ε_L and a high-energy one ε_H, find the expected energy of a particle just registered.
10. A capacitor is being charged by a homogeneous Poisson stream of unit charges with intensity λ_C. At random time instances it is being instantaneously discharged to zero, these time instances being epochs of an independent homogeneous Poisson process with intensity λ_D. Find the probability mass distribution of the charge on the capacitor just prior to its discharge.
11. Consider a random walk as in Example 7.3.3 assuming now that the particle moves one step to the right at each epoch of a Poisson process with intensity $\lambda_r(t) = 1 + \cos t$, $t > 0$, and to the left at each epoch of an independent Poisson process with intensity $\lambda_l(t) = 1 - \cos t$, $t > 0$. Obtain the distribution and the characteristic function of the particle position $S(t)$, $t > 0$.

12. Give a convincing argument to justify the assumption that the incoming calls to a telephone exchange, the stream of photons emitted by an incoherent light source, the stream of electrons tunneling through a junction of an Esaki diode, or the electromagnetic quanta bounced off a rough surface in one particular direction can be regarded as a Poisson point process.
13. An inhomogeneous Poisson process on $\mathscr{T} = (0, \infty)$ with intensity function $\lambda(t)$ is passed through a digital filter (a binary counter, for instance) that permits the passage of only every second of the original stream of point-events. Find an expression for the first order probability mass distributions of the resulting associated counting process. Is this new process still Poisson?
14. Consider an inhomogeneous Poisson process on $\mathscr{T} = (0, \infty)$ with intensity function $\lambda(t) = \lambda_0(1 + e^{-\alpha t})$, where λ_0 and α are positive constants. Show that this process can be regarded as a superposition of a homogeneous Poisson process and an inhomogeneous Poisson process that is finite, that is, that eventually becomes extinct. What is the probability mass distribution of the total number of point-events that can be attributed to this "transient" component process during $(0, \infty)$.
15. In a simplified version of an optical tracking device the image of the object to be tracked is projected to a focal plane. The azimuth tracking is accomplished by letting the light pass through a thin vertical slot in the center of the plane and counting the number of photons behind the slot during each millisecond. As soon as no photon count is recorded during a millisecond, a tracking mechanism is activated to readjust the azimuth. Suppose that if the image of the object is x millimeters off center of the focal plane, the intensity of the Poisson process of the photon count decreases as $(1 + x)^{-1}$. (See Fig. 7.3E.1.) Assume that at time $t = 0$ the image is at the center and then begins to drift to the left with a constant speed of 1 meter/second. Find the probability mass distribution of the time T_a (measured in multiples of a millisecond) when the readjustment mechanism is activated.

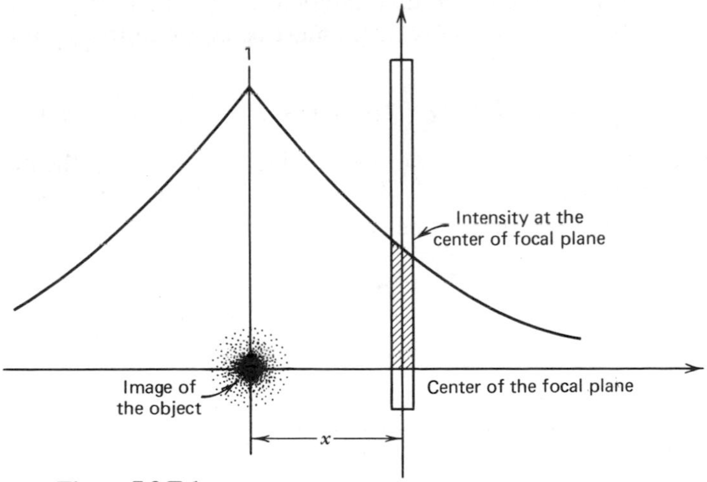

Figure 7.3.E.1

7.4. INTERVAL PROPERTIES OF A POISSON PROCESS

In the previous section we studied various aspects of a Poisson process and worked exclusively with the associated counting process $N(t)$. Now, we shall look at the interval properties of this process, in particular the random epoch sequence and the waiting time sequence.

We begin with an inhomogeneous Poisson process with time domain $\mathcal{T} = (0, +\infty)$ and with intensity function

$$\lambda(t), \qquad t > 0.$$

Let $N(t)$, $t > 0$, again be the associated counting process, and let

$$T(n), \qquad n = 1, 2, \dots \tag{1}$$

be the random epoch sequence. Recall from Section 7.1 that epoch $T(n)$ is the time (measured from $t = 0$) of occurrence of the nth point-event and that (1) is an increasing sequence of positive random variables

$$0 < T(1) < T(2) < \dots.$$

Let n be a fixed positive integer, and let us consider the conditional probability

$$P(T(n+1) > t_{n+1} \mid T(n) = t_n,\ T(n-1) = t_{n-1}, \dots, T(1) = t_1), \tag{2}$$

where $0 < t_1 < t_2 < \cdots < t_n < t_{n+1}$. Now if the nth epoch $T(n)$ occurred at the time t_n, while the $(n+1)$st epoch $T(n+1)$ occurs later than time t_{n+1}, then there can be no point-event occurring in the time interval $(t_n, t_{n+1}]$ (see Fig. 7.4.1). In other words, the conditional probability in (2) must be equal to the conditional probability

$$P(N(t_{n+1}) - N(t_n) = 0 \mid T(n) = t_n,\ T(n-1) = t_{n-1}, \dots, T(1) = t_1). \tag{3}$$

However, the condition $T(1) = t_1, \dots, T(n) = t_n$ uniquely determines the portion of the sample path $N(t) = n(t)$, $0 < t \le t_n$, of the counting process and vice versa (as in Fig. 7.4.2), and thus (3) can also be written as

$$P(N(t_{n+1}) - N(t_n) = 0 \mid N(t) = n(t),\ 0 < t \le t_n). \tag{4}$$

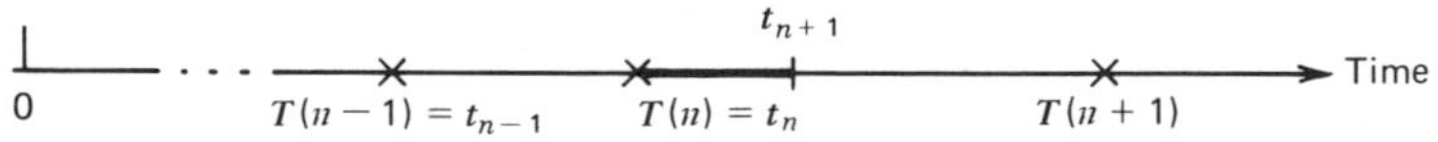

Figure 7.4.1

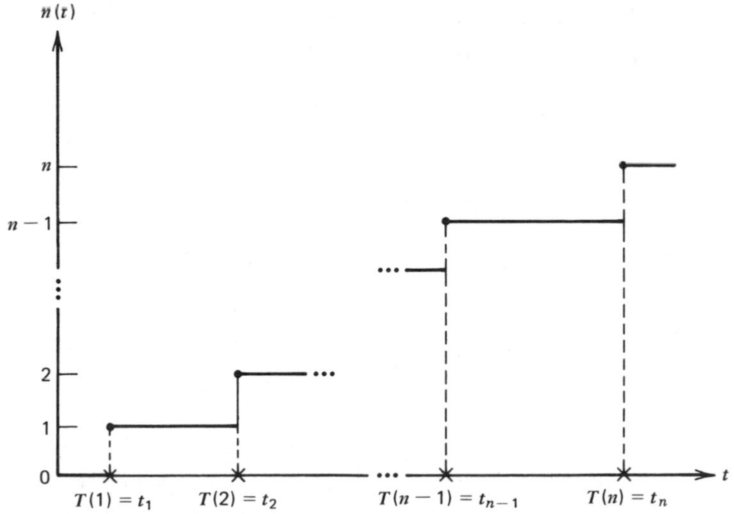

Figure 7.4.2

But since the Poisson process evolves without aftereffects, the increment $N(t_{n+1}) - N(t_n)$ is independent of the past $N(t)$, $0 < t \le t_n$,† so the conditional probability in (4) equals the unconditional probability

$$P(N(t_{n+1}) - N(t_n) = 0) = e^{-[\mu(t_{n+1}) - \mu(t_n)]}, \tag{5}$$

where

$$\mu(t) = \int_0^t \lambda(\tau)\, d\tau. \tag{6}$$

Thus we conclude from (2) through (5) that

$$P(T(n+1) > t_{n+1} \mid T(n) = t_n,\ T(n-1) = t_{n-1},\ \ldots,\ T(1) = t_1) = e^{-[\mu(t_{n+1}) - \mu(t_n)]},$$

or in terms of the conditional distribution function,

$$P(T(n+1) \le t_{n+1} \mid T(n) = t_n,\ T(n-1) = t_{n-1},\ \ldots,\ T(1) = t_1) = 1 - e^{-[\mu(t_{n+1}) - \mu(t_n)]},$$

$$0 < t_1 < \cdots < t_n < t_{n+1}. \tag{7}$$

† Actually, we are cheating a little bit here. True, for any *fixed* time t_n the forward increment $N(t_{n+1}) - N(t_n)$ is independent of $N(t)$, $0 < t \le t_n$, but in our case, t_n is not really fixed but rather it is the time of a random epoch $T(n) = t_n$. In fact, it is a stopping time for the process $N(t)$. Fortunately, it turns out that for a Poisson process the forward increments of the counting process are independent of its past even if the increment is counted from a stopping time of the process.

Since the right-hand side of this expression depends only on t_{n+1} and t_n and not on any of the previous epochs $t_1, \ldots, t_{n-1}$, we see that actually

$$P(T(n+1) \le t_{n+1} \,|\, T(n) = t_n, T(n-1) = t_{n-1}, \ldots, T(1) = t_1)$$
$$= P(T(n+1) \le t_{n+1} \,|\, T(n) = t_n).$$

But this is the Markov property; the random epoch sequence $T(n)$, $n = 1, 2, \ldots$, is a Markov sequence with transition probabilities given by (7). It remains to find the distribution of the first random epoch $T(1)$. This is quite easy:

$$P(T(1) > t_1) = P(N(t_1) = 0) = e^{-\mu(t_1)}, \qquad t_1 > 0, \tag{8}$$

since $\mu(0) = 0$ by (6). We have thus obtained a complete specification of the random epoch sequence for an inhomogeneous Poisson process.

THEOREM 7.4.1. *The random epoch sequence* T(n), n = 1, 2, ..., *of an inhomogeneous Poisson process with intensity function* λ(t) *defined on the time domain* $\mathcal{T} = (0, \infty)$ *is a Markov sequence with transition probability densities*

$$f_{T(n+1)|T(n)}(t_{n+1} \,|\, t_n) = \lambda(t_{n+1}) e^{-\int_{t_n}^{t_{n+1}} \lambda(\tau)\, d\tau}, \tag{9}$$

$0 < \mathrm{t_n} < \mathrm{t_{n+1}}$, n = 1, 2, ..., *and initial probability density*

$$f_{T(1)}(t_1) = \lambda(t_1) e^{-\int_0^{t_1} \lambda(\tau)\, d\tau}, \qquad 0 < t_1. \tag{10}$$

Expressions (9) and (10) for the densities are obtained by differentiating (7) and (8).

The joint density of $T(1), \ldots, T(n)$ can now be easily evaluated. Since $T(n)$, $n = 1, 2, \ldots$, is a Markov sequence

$$f_{T(1), \ldots, T(n)}(t_1, \ldots, t_n) = f_{T(n)|T(n-1)}(t_n \,|\, t_{n-1}) f_{T(n-1)|T(n-2)}(t_{n-1} \,|\, t_{n-2})$$
$$\cdots f_{T(2)|T(1)}(t_2 \,|\, t_1) f_{T(1)}(t_1).$$

Substituting from (9) and (10), we get at once

$$f_{T(1), \ldots, T(n)}(t_1, \ldots, t_n) = \lambda(t_1) \cdots \lambda(t_n) e^{-\int_0^{t_n} \lambda(\tau)\, d\tau}, \tag{11}$$

$0 < t_1 < \cdots < t_n$.

From (11) we can obtain the marginal density of $T(n)$,

$$f_{T(n)}(t) = \frac{\lambda(t)}{(n-1)!} \left[\int_0^t \lambda(\tau)\, d\tau \right]^{n-1} e^{-\int_0^t \lambda(\tau)\, d\tau}, \qquad t > 0 \tag{12}$$

which for a homogeneous Poisson process is seen to reduce to the already familiar Erlang density.

Example 7.4.1

Consider again the random walk of Example 7.3.3, this time with $p = q = \frac{1}{2}$. To compensate for this less general case, let us assume, on the other hand, that the Poisson process involved is inhomogeneous with a positive intensity function $\lambda(t)$, $t > 0$. The initial position of the particle is still at the origin. We wish to find the distribution of the time instant $T > 0$ at which the particle returns to its initial position for the first time. Clearly, T will also be the random time between any two consecutive visits of the particle to the origin, or more generally to any given position.

By a simple combinatorial argument it can be shown that the probability that exactly $2n$ steps are required for the particle to return to the origin equals

$$\frac{1}{2n}\binom{2(n-1)}{n-1}\frac{1}{2^{2(n-1)}}, \qquad n = 1, 2, \ldots.$$

Return to the origin in an odd number of steps is, of course, impossible. However, since the random walk is symmetric, a return to the origin is certain because the above probabilities add to 1.

Since the particle executes its moves at each occurrence of a point-event of the inhomogeneous Poisson process, the time at which it makes its $2n$th step is the random epoch $T(2n)$. Thus the density $f_T(t)$ of the time T of the first return to the origin will be given by

$$f_T(t) = \sum_{n=1}^{\infty} \frac{1}{2n}\binom{2(n-1)}{n-1}\frac{1}{2^{2(n-1)}} f_{T(2n)}(t), \qquad t > 0,$$

where, according to (12),

$$f_{T(2n)}(t) = \frac{\lambda(t)}{(2n-1)!}(\mu(t))^{2n-1}e^{-\mu(t)}, \qquad t > 0,$$

$$\mu(t) = \int_0^t \lambda(\tau)\,d\tau.$$

Substituting into the infinite series above, we obtain, after some cancellation,

$$f_T(t) = \lambda(t)e^{-\mu(t)} \sum_{n=1}^{\infty} \frac{1}{2n-1}\frac{1}{n!(n-1)!}\left(\frac{\mu(t)}{2}\right)^{2n-1}.$$

This is rewritten as

$$f_T(t) = \lambda(t)e^{-\mu(t)} \sum_{n=0}^{\infty} \frac{1}{2n+1}\frac{1}{(n+1)!n!}\left(\frac{\mu(t)}{2}\right)^{2n+1}$$

and notice that for any $x \geq 0$, the infinite series

$$\sum_{n=0}^{\infty} \frac{1}{2n+1} \frac{1}{(n+1)!\,n!} \left(\frac{x}{2}\right)^{2n+1} = \sum_{n=0}^{\infty} \frac{1}{(n+1)!\,n!} \int_0^x \left(\frac{y}{2}\right)^{2n} \frac{1}{2}\,dy$$
$$= \int_0^x \sum_{n=0}^{\infty} \frac{1}{(n+1)!\,n!} \left(\frac{y}{2}\right)^{2n+1} \frac{1}{y}\,dy$$
$$= \int_0^x \frac{I_1(y)}{y}\,dy,$$

where $I_1(y)$ is the modified Bessel function of order $k = 1$ introduced in Example 7.3.3.

It follows that the density of T equals

$$f_T(t) = e^{-\mu(t)}\lambda(t)\bar{I}_1(\mu(t)), \qquad t > 0,$$

where we have denoted

$$\bar{I}_1(x) = \int_0^x \frac{I_1(y)}{y}\,dy, \qquad x > 0.$$

In particular for a homogeneous Poisson process with intensity $\lambda > 0$, the density is

$$f_T(t) = e^{-\lambda t}\lambda\bar{I}_1(\lambda t), \qquad t > 0.$$

Recalling from Example 7.3.3 that the position $S(t)$ of the particle at time $t > 0$ can also be written as a difference of two independent Poisson counting processes, in the present case with the same intensity functions $\frac{1}{2}\lambda(t)$, it follows that the random variable T is also the time between two consecutive equalizations of the numbers of point-events of the two processes. ▲

For our next topic, let us fix some time instant $t > 0$ and find the distribution of the random epochs in the interval $(0, t]$ conditional upon the event that there are exactly n of them in this interval. In other words, we want the conditional density

$$f_{T(1),\ldots,T(n)|N(t)}(t_1, \ldots, t_n | n),$$

where $0 < t_1 < \cdots < t_n \leq t$ and $n = 1, 2, \ldots$. By Bayes' formula, this density equals

$$\frac{P(N(t) = n \,|\, T(1) = t_1, \ldots, T(n) = t_n) f_{T(1),\ldots,T(n)}(t_1, \ldots, t_n)}{P(N(t) = n)}. \tag{13}$$

The conditional probability in the numerator is easily seen to be the same as

$$P(N(t) - N(t_n) = 0 \,|\, T(1) = t_1, \ldots, T(n) = t_n),$$

which differs from (3) only by notation; we have written t instead of t_{n+1}. Consequently, using the same argument as before (evolution without aftereffects), we have

$$P(N(t) = n \mid T(1) = t_1, \ldots, T(n) = t_n) = P(N(t) - N(t_n) = 0) = e^{-\int_{t_n}^{t} \lambda(\tau)\, d\tau}. \tag{14}$$

If we now substitute from (14) and (11) into the numerator of (13) and (since $N(t)$ is a Poisson random variable)

$$P(N(t) = n) = \frac{[\int_0^t \lambda(\tau)\, d\tau]^n}{n!} e^{-\int_0^t \lambda(\tau)\, d\tau} \tag{15}$$

into the denominator, after some cancellation, we get the expression

$$f_{T(1), \ldots, T(n) \mid N(t)}(t_1, \ldots, t_n \mid n) = n! \frac{\lambda(t_1) \cdots \lambda(t_n)}{[\int_0^t \lambda(\tau)\, d\tau]^n}, \tag{16}$$

$0 < t_1 < \cdots < t_n < t$, $n = 1, 2, \ldots$. Now, however, the right-hand side of this expression is the density of the order statistics† of n independent random variables each distributed over the interval $(0, t]$ with the same density

$$f(\tau) = \frac{\lambda(\tau)}{\int_0^t \lambda(\tau)\, d\tau}, \qquad 0 < \tau \le t.$$

We have thus found yet another interesting feature of the Poisson process which we state slightly more generally as a theorem.

THEOREM 7.4.2. *For any fixed time interval* $(t_1, t_2]$, *the conditional distribution of the epochs, given that there are exactly* n *point-events in this interval, is the same as if the epoch times were selected independently each with the same density*

$$f(\tau) = \frac{\lambda(\tau)}{\int_{t_1}^{t_2} \lambda(\tau)\, d\tau}, \qquad t_1 < \tau \le t_2. \tag{17}$$

Note that for a homogeneous Poisson process the density in (17) is uniform over $(t_1, t_2]$.

Example 7.4.2

An optical communication system is used to transmit a string of binary digits "0" and "1". Transmission of each digit lasts for a unit of time and is accomplished by modulating the intensity of a light source by a sinusoidal signal of radial frequency ω_0 if "0" is being transmitted, and ω_1 if "1" is being transmitted. The receiver

† See Exercise 1.16.

contains a photodetector emitting a stream of electrons assumed to be point-events of a Poisson process with intensity function

$$\lambda(t) = 1 + \gamma \sin \omega t,$$

where $\omega = \omega_0$ or ω_1, depending on the binary digit being transmitted at time t, and $0 < \gamma < 1$ is a constant.

Now suppose that during a time interval, say $(0, 1]$, corresponding to a transmission of a single binary digit n point-events (i.e., electron emissions) at epochs $0 < t_1 < \cdots < t_n \leq 1$ have been recorded. This information is fed into a decoder that must determine which digit was being transmitted during this time interval. It is known from the classical theory of signal detection that an optimal decoder should compute the value of the likelihood ratio and compare it with a threshold value.

The likelihood ratio has the general form

$$L(\mathfrak{D}) = \frac{f_0(\mathfrak{D})}{f_1(\mathfrak{D})},$$

where the symbol $\mathfrak{D}$ stands for the observed data, and f_0 and f_1 are the densities of the data if the transmitted digit is "0" or "1" respectively. In the present case the observed data consist of the number $n = N(1)$ of point-events that occurred during $(0, 1]$ and their epochs $t_1 = T(1), \ldots, t_n = T(n)$. The density $f(t_1, \ldots, t_n; n)$, which is of mixed-type since the $T(k)$'s are continuous random variables while $N(1)$ is discrete, is obtained by multiplying (16) and (15). Thus

$$f(t_1, \ldots, t_n; n) = \lambda(t_1) \cdots \lambda(t_n) e^{-\int_0^1 \lambda(\tau)\, d\tau},$$

for $0 < t_1 < \cdots < t_n \leq 1$ provided $n > 0$.

If $N(1) = n = 0$, that is, if no emission has been observed, the density is $f(n) = \delta(n)P(N(1) = 0)$, where

$$P(N(1) = 0) = e^{-\int_0^1 \lambda(\tau)\, d\tau}.$$

The likelihood ratio is then given by

$$L(\mathfrak{D}) = \prod_{k=1}^{n} \frac{1 + \gamma \sin \omega_0 t_k}{1 + \gamma \sin \omega_1 t_k} e^{-\gamma \int_0^1 (\sin \omega_0 \tau - \sin \omega_1 \tau)\, d\tau},$$

where we define the product to be 1 if $n = 0$. The decoder then compares the quantity $L(\mathfrak{D})$ with a threshold value C, deciding that "0" was transmitted if $L(\mathfrak{D}) > C$ and that "1" was transmitted if $L(\mathfrak{D}) < C$. This, in turn, is equivalent to the decoding scheme:

$$\left.\begin{array}{l}\text{decide "0"}\\ \text{decide "1"}\end{array}\right\} \quad \text{if} \quad \sum_{k=1}^{n} \ln \frac{1 + \gamma \sin \omega_0 t_k}{1 + \gamma \sin \omega_1 t_k} \quad \begin{cases} > C' \\ < C', \end{cases}$$

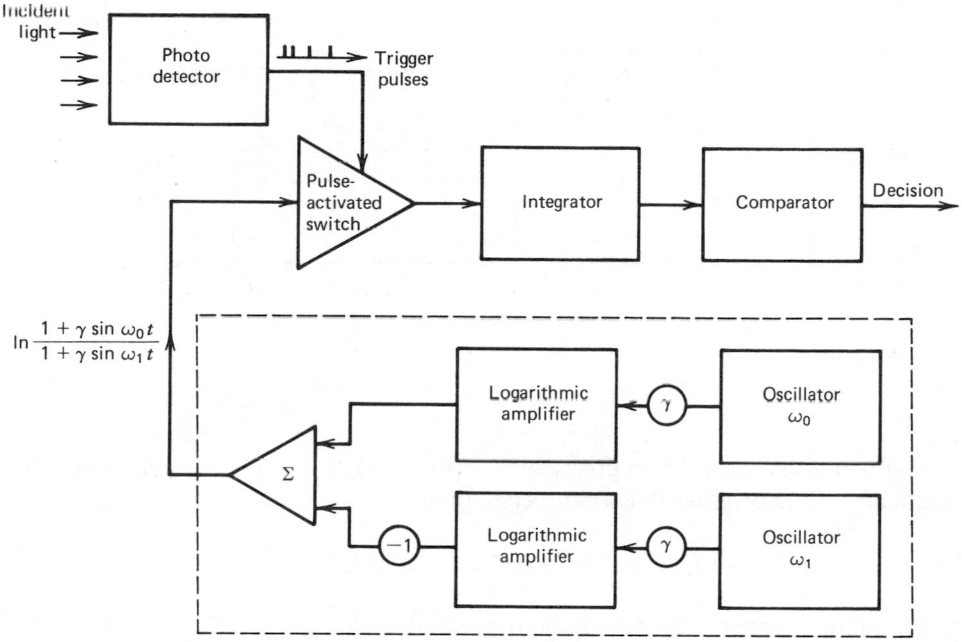

Figure 7.4.3

which is obtained by taking the logarithm of $L(\mathfrak{D})$ and incorporating the exponential term (which does not depend on the observed data) into the threshold. In this form, the decoding scheme is easily implemented as indicated in Figure 7.4.3.

Here, the circuit in the bottom half of the picture generates the function $\ln(1 + \gamma \sin \omega_0 t)/(1 + \gamma \sin \omega_1 t)$ whose values are successively sampled at time instants $t_1, \ldots, t_n$ realized as narrow pulses on the photodetector output. Thus the output of the integrator at the end of the time interval (0, 1] equals the sum

$$\sum_{k=1}^{n} \ln \frac{1 + \gamma \sin \omega_0 t_k}{1 + \gamma \sin \omega_1 t_k} \quad \text{if} \quad n > 0,$$

and is zero if $n = 0$. After a decision ("0" or "1") is made, the system is reset to its initial state and is ready for the next transmission interval. ▲

Example 7.4.3

Suppose that the point-events of an inhomogeneous Poisson process on $\mathcal{T} = (0, \infty)$ are subjected to random delays, the delays of each point-event being independent and identically distributed random variables, also independent of the

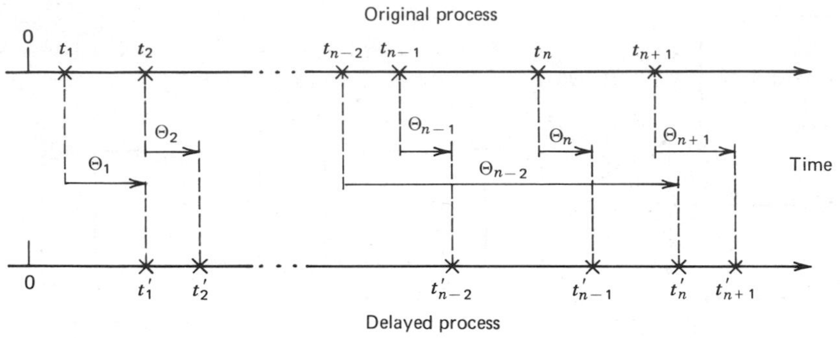

Figure 7.4.4

Poisson process itself. More precisely, if $T(n)$, $n = 1, 2, \ldots$, is the random epoch sequence of the original Poisson process, then

$$T'(n) = T(n) + \Theta(n), \qquad n = 1, 2, \ldots$$

is the random epoch sequence of the process after delay. The delays $\Theta(n)$, $n = 1, 2, \ldots$, are i.i.d. nonnegative random variables with distribution function F_Θ. Such a situation may be encountered, for instance, when incoming calls are placed into a buffer and released only after a random hold time. This operation can be visualized graphically, as shown in Figure 7.4.4. Notice that the point-events are emerging from the buffer in an order that is not necessarily the same as the one in which they entered. One may also think about each incoming point-event as an instant at which some operation is initiated in a computer. The associated delay then represents the time needed to complete this operation, so that the delayed point process represents the stream of point-events signifying the completion of each operation.

Let $N(t)$, $t > 0$, and $N'(t)$, $t > 0$, respectively, be the counting processes of the original and delayed point processes, and let $0 < t_1 < t_2$ be two fixed time instances. Suppose now that a point-event of the original Poisson process occurs at an epoch τ, $0 < \tau \le t_2$. Then this point-event will reappear with its epoch within the interval $(t_1, t_2]$ if and only if its delay Θ satisfies the inequality $t_1 < \tau + \Theta \le t_2$ (see Fig. 7.4.5). The probability of this occurring equals

$$P(t_1 < \tau + \Theta \le t_2) = F_\Theta(t_2 - \tau) - F_\Theta(t_1 - \tau).$$

Conditioning upon $N(t_2) = n$, it follows from Theorem 7.4.2 that each epoch τ of this kind has density (17), and hence the probability that any one of the n point-

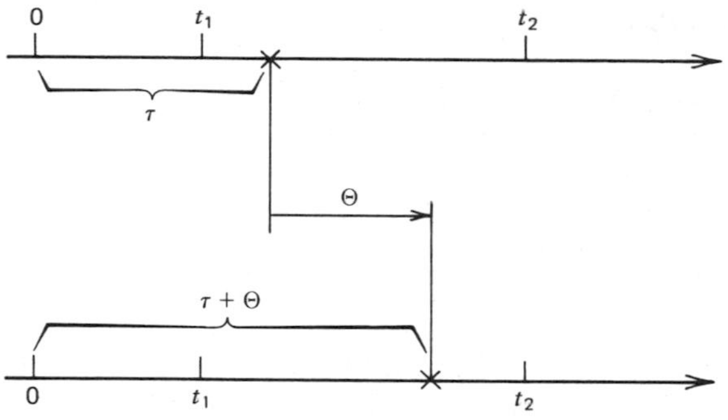

Figure 7.4.5

events of the original process will reappear in the interval $(t_1, t_2]$ is

$$p = \int_0^{t_2} P(t_1 < \tau + \Theta \leq t_2) f(\tau)\, d\tau$$

$$= \frac{\int_0^{t_2} [F_\Theta(t_2 - \tau) - F_\Theta(t_1 - \tau)]\lambda(\tau)\, d\tau}{\int_0^{t_2} \lambda(\tau)\, d\tau}, \tag{18}$$

where $\lambda(t)$ is the intensity of the original Poisson process. Furthermore, the epochs τ can be regarded as being selected independently, so the number of point events reappearing in $(t_1, t_2]$ is just like the number of successes in n independent Bernoulli trials with probability of success p. Thus (remember we have conditioned upon $N(t_2) = n$),

$$P(N'(t_2) - N'(t_1) = k \mid N(t_2) = n) = \binom{n}{k} p^k q^{n-k}, \qquad k = 0, \ldots, n.$$

where $q = 1 - p$. Obtaining the unconditional probability is now easy, since $N(t_2)$ is a Poisson random variable. We have

$$P(N'(t_2) - N'(t_1) = k) = \sum_{n=k}^{\infty} \binom{n}{k} p^k q^{n-k} P(N(t_2) = n),$$

$k = 0, 1, \ldots,$ where

$$P(N(t_2) = n) = \frac{[\int_0^{t_2} \lambda(\tau)\, d\tau]^n}{n!} e^{-\int_0^{t_2} \lambda(\tau)\, d\tau}$$

and p is given by (18). Abbreviating temporarily,

$$\mu(t_2) = \int_0^{t_2} \lambda(\tau)\, d\tau$$

we can evaluate the above infinite series as follows:

$$\sum_{n=k}^{\infty}\binom{n}{k}p^k q^{n-k}\frac{[\mu(t_2)]^n}{n!}e^{-\mu(t_2)}=\left(\frac{p}{q}\right)^k\frac{1}{k!}e^{-\mu(t_2)}\sum_{n=k}^{\infty}\frac{[q\mu(t_2)]^n}{(n-k)!}=\frac{[p\mu(t_2)]^k}{k!}e^{-p\mu(t_2)}$$

since

$$\sum_{n=k}^{\infty}\frac{[q\mu(t_2)]^n}{(n-k)!}=[q\mu(t_2)]^k\sum_{j=0}^{\infty}\frac{[q\mu(t_2)]^j}{j!}=[q\mu(t_2)]^k e^{q\mu(t_2)}.$$

It follows that the random variable $N'(t_2)-N'(t_1)$, that is, the number of point-events emerging in the interval $(t_1, t_2]$ also has a Poisson distribution with parameter

$$p\mu(t_2)=\int_0^{t_2}[F_\Theta(t_2-\tau)-F_\Theta(t_1-\tau)]\lambda(\tau)\,d\tau.$$

Since the delays Θ are nonnegative random variables $F_\Theta(\theta)=0$ for $\theta<0$, the integral is actually a difference:

$$\int_0^{t_2}F_\Theta(t_2-\tau)\lambda(\tau)\,d\tau-\int_0^{t_1}F_\Theta(t_1-\tau)\lambda(\tau)\,d\tau. \tag{19}$$

If the original Poisson is homogeneous, $\lambda(t)=\lambda>0$, (19) further simplifies into $\lambda\int_{t_1}^{t_2}F_\Theta(\tau)\,d\tau$. In this case the delayed point process is, in fact, again a Poisson process with intensity function

$$\lambda(t)=\lambda F_\Theta(t), \qquad t>0.$$ ▲

Let us now turn to the sequence of waiting times for an inhomogeneous Poisson process. Since we are still considering the process on the time domain $\mathcal{T}=(0,\infty)$, the relation between the random epoch sequence

$$T(n), \qquad n=1,2,\ldots$$

and the sequence of waiting times

$$X(n), \qquad n=0,1,\ldots \tag{20}$$

is very simple:

$$\begin{aligned}T(1)&=X(0)\\T(2)&=X(0)+X(1)\\T(3)&=X(0)+X(1)+X(2)\\&\cdots\\T(n)&=X(0)+\cdots+X(n-1)\\&\cdots\end{aligned}$$

Since the Jacobian of this transformation is unity, the joint density of the first n waiting times can be readily obtained from the joint density (11) of the first n random epochs. We get

$$f_{X(0), \ldots, X(n-1)}(x_0, \ldots, x_{n-1}) = \lambda(x_0)\lambda(x_0 + x_1)\lambda(x_0 + x_1 + x_2)$$
$$\cdots \lambda(x_0 + \cdots + x_{n-1})e^{-\int_0^{x_0 + \cdots + x_{n-1}} \lambda(\tau)\, d\tau}, \qquad x_k > 0, \qquad k = 0, \ldots, n-1. \tag{21}$$

Since this formula holds for any $n = 1, 2, \ldots$, the entire probability law of the waiting time sequence (20) has been found.

In the case of a homogeneous Poisson process,

$$\lambda(t) = \lambda, \qquad t > 0$$

expression (21) yields

$$f_{X(0), \ldots, X(n-1)}(x_0, \ldots, x_{n-1}) = \lambda^n e^{-(x_0 + \cdots + x_n)\lambda},$$

which is readily seen to be a product of n exponential densities

$$f_{X(k)}(x_k) = \lambda e^{-\lambda x_k}, \qquad x_k > 0, \qquad k = 0, \ldots, n-1.$$

This confirms an already familiar result:

THEOREM 7.4.3. The sequence of waiting times for a homogeneous Poisson process with time domain $\mathcal{T} = (0, \infty)$ is a sequence of independent, identically distributed exponential random variables with parameter λ equal to the intensity of the process.

It should be noted that for an *inhomogeneous* Poisson process the *waiting times are no longer independent*, since the joint density (21) cannot be factored into the product of its marginal densities unless $\lambda(t)$ is a constant. Thus, *if the waiting times of a Poisson process are independent, then the Poisson process must be homogeneous.*

Example 7.4.4

It follows from the above remark that if we constructed a point process on the time domain $\mathcal{T} = (0, 1)$ by choosing the waiting time sequence

$$X(n), \qquad n = 0, 1, \ldots$$

to be a sequence of independent, exponentially distributed random variables, the point process would *not* be a Poisson process and hence would not evolve without aftereffects unless all the exponential waiting times had the *same* parameter λ.

For instance, suppose that all these waiting times have an exponential distribution with $\lambda = 2$ except for $X(0)$, which as an exponential distribution with $\lambda = 1$. Let us now calculate the distribution of $N(t)$, the number of point-events in an interval $(0, t]$. To this end, we first compute the density $f_{T(n)}$ of the random epoch

$$T(n) = X(0) + \cdots + X(n-1), \qquad n = 1, 2, \ldots.$$

For $n = 1$, we have $T(1) = X(0)$, that is,

$$f_{T(1)}(t) = e^{-t}, \qquad t > 0.$$

For $n > 1$, we use the fact that $X(1) + \cdots + X(n-1)$, being a sum of $n-1$ i.i.d. exponential random variables, will have the Erlang density with parameters $n-1$ and 2. Hence the density of $T(n)$ is obtained from the convolution integral

$$f_{T(n)}(t) = \int_0^t \frac{2^{n-1}}{(n-2)!} \tau^{n-2} e^{-2\tau} e^{-(t-\tau)}\, d\tau, \qquad t > 0.$$

Factoring $2^{n-1}e^{-t}$ from the integral and using the identity

$$\int_0^t \frac{1}{(n-2)!} \tau^{n-2} e^{-\tau}\, d\tau = e^{-t} \sum_{k=n-1}^{\infty} \frac{t^k}{k!}$$

we have for $n > 1$,

$$f_{T(n)}(t) = 2^{n-1} e^{-2t} \sum_{k=n-1}^{\infty} \frac{t^k}{k!}, \qquad t > 0.$$

Since with $n = 1$ the right-hand side reduces to e^{-t}, the above formula holds for all $n = 1, 2, \ldots$.

To obtain the distribution of $N(t)$ we have immediately

$$P(N(t) = 0) = P(T(1) > t) = e^{-t},$$

while for $n > 0$

$$\begin{aligned} P(N(t) = n) &= P(T(n+1) > t) - P(T(n) > t) \\ &= \int_t^{\infty} f_{T(n+1)}(\tau)\, d\tau - \int_t^{\infty} f_{T(n)}(\tau)\, d\tau. \end{aligned}$$

Upon integrating the first of these two integrals by parts, we obtain

$$\begin{aligned} \int_t^{\infty} 2^n e^{-2\tau} \sum_{k=n}^{\infty} \frac{\tau^k}{k!}\, d\tau &= -2^{n-1} e^{-2\tau} \sum_{k=n}^{\infty} \frac{\tau^k}{k!} \Bigg|_{\tau=t}^{\tau=\infty} + \int_t^{\infty} 2^{n-1} e^{-2\tau} \sum_{k=n}^{\infty} \frac{\tau^{k-1}}{(k-1)!}\, d\tau \\ &= 2^{n-1} e^{-2t} \sum_{k=n}^{\infty} \frac{t^k}{k!} + \int_t^{\infty} f_{T(n)}(\tau)\, d\tau. \end{aligned}$$

Hence for any $n = 0, 1, \ldots$ and $t > 0$

$$P(N(t) = n) = 2^{n-1} e^{-2t} \sum_{k=n}^{\infty} \frac{t^k}{k!}.$$

Notice that

$$P(N(t) = n) = \frac{1}{2} f_{T(n+1)}(t)$$

just as in a Poisson process, but the distribution of $N(t)$ is no longer Poisson. ▲

Finally, let us consider a Poisson process on the time domain $\mathcal{T} = (-\infty, \infty)$. Take two finite collections of nonoverlapping time intervals such that the intervals in the first collection lie entirely to the right of the time instant $t = 0$, while the intervals of the second collection lie entirely to the left of the origin $t = 0$. Since we are dealing with a Poisson process the numbers of point-events with epochs in each of these intervals will be independent Poisson random variables. This means that we have, in fact, two Poisson processes, one defined on the time domain $\mathcal{T}_+ = (0, \infty)$ and the other one defined on the time domain $\mathcal{T}_- = (-\infty, 0)$, that is, running backward in time. Furthermore, these two Poisson processes will be *mutually independent*, since the two collections of time intervals are separated by the origin $t = 0$, and the evolution without aftereffects requires, roughly speaking, that whatever happens before time $t = 0$ can have no effect whatsoever on anything after this fixed time instant.

This idea provides us with a procedure for constructing a Poisson process on the entire time domain $\mathcal{T} = (-\infty, \infty)$ with a given intensity function

$$\lambda(t), \qquad -\infty < t < \infty.$$

We simply take two Poisson processes on the time domain $\mathcal{T} = (0, \infty)$ with intensity functions

$$\lambda_+(t) = \lambda(t), \qquad t > 0, \tag{22}$$

and

$$\lambda_-(t) = \lambda(-t), \qquad t > 0, \tag{23}$$

reverse the time flow of the second process, and glue them together at the point $t = 0$ in an independent fashion. Addition of the point $t = 0$ to the two time domains causes no difficulty, since by our basic assumption (iv), the probability of a point-event occurring at any predetermined time instant is zero.

Let us now look at the doubly infinite random epoch sequence

$$T(n), \qquad n = \ldots, -1, 0, 1, \ldots \tag{24}$$

of the Poisson process so constructed. Since in Section 7.1 we agreed upon indexing the random epochs such that

$$T(0) \le 0 < T(1)$$

the sequence

$$T(n), \qquad n = 1, 2, \ldots$$

will correspond to the random epoch sequence of the first Poisson process (with intensity $\lambda_+(t)$, $t > 0$) while the sequence

$$-T(n), \qquad n = 0, -1, -2, \ldots$$

will correspond to the second Poisson process (with intensity $\lambda_-(t)$, $t > 0$). These two sequences will be mutually independent with joint densities given by (11)

$$f_{T(1),\ldots,T(n)}(t_1, \ldots, t_n) = \lambda_+(t_1) \cdots \lambda_+(t_n) e^{-\int_0^{t_n} \lambda_+(\tau)\, d\tau},$$

$0 < t_1 < \cdots < t_n$, and

$$f_{-T(0),\ldots,-T(-m)}(t_0, \ldots, t_{-m}) = \lambda_-(t_0) \cdots \lambda_-(t_{-m}) e^{\int_0^{t_{-m}} \lambda_-(\tau)\, d\tau},$$

$$0 < t_0 < t_{-1} < \cdots < t_{-m}.$$

Reversing the time in the second sequence, using independence, and substituting for λ_+ and λ_- from (22) and (23), we get the joint probability density for the random epoch sequence (24)

$$f_{T(-m),\ldots,T(n)}(t_{-m}, \ldots, t_n) = \lambda(t_{-m}) \cdots \lambda(t_n) e^{-\int_{t_{-m}}^{t_n} \lambda(\tau)\, d\tau}, \tag{25}$$

$t_{-m} < \cdots < t_0 \leq 0 < t_1 < \cdots < t_n$ for any $n = 1, 2, \ldots$ and $m = 0, 1, \ldots$. Thus the probability law of the random epoch sequence $T(n)$, $n = \ldots, -1, 0, 1, \ldots$, is completely specified.

It remains to explore the sequence of waiting times

$$X(n), \qquad n = \ldots, -1, 0, 1, \ldots \tag{26}$$

of a Poisson process with time domain $\mathcal{T} = (-\infty, \infty)$. We could, of course, use the defining relation

$$X(n) = T(n+1) - T(n), \qquad n = \ldots, -1, 0, 1, \ldots,$$

and obtain the joint densities of the sequence (26) from the joint densities (25) of the random epoch sequence. However, it is more instructive to start afresh from the construction of a Poisson process with the time domain $\mathcal{T} = (-\infty, \infty)$ from the two independent Poisson processes with time domains $\mathcal{T}_+ = (0, \infty)$ and $\mathcal{T}_- = (-\infty, 0)$. Let us denote the sequences of the waiting times of these two processes by

$$X_+(n), \qquad n = 0, 1, \ldots,$$

and

$$X_-(n), \quad n = 0, -1, \ldots,$$

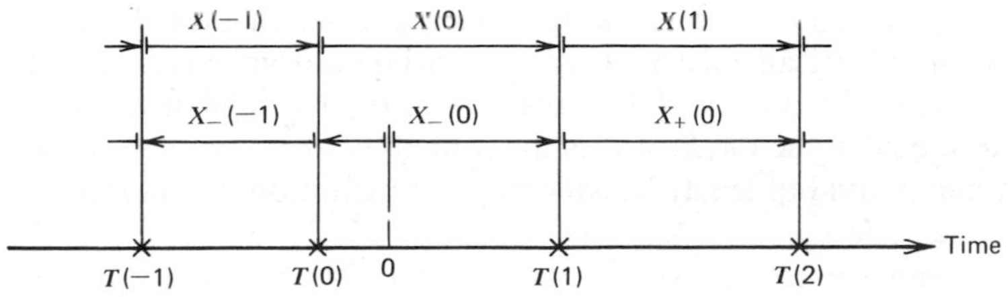

Figure 7.4.6

respectively. These two sequences will again be mutually independent with joint densities given by (21) with intensity functions $\lambda_+(t)$ and $\lambda_-(t)$, respectively. Further, as can be seen from Figure 7.4.6, we have

$$X(n) = X_+(n) \qquad \text{for } n = 1, 2, \ldots,$$

and

$$X(n) = X_-(n) \qquad \text{for } n = -1, -2, \ldots.$$

However, since the first waiting time of a point process with semi-infinite time domain is always measured from time $t = 0$, we have

$$X(0) = X_-(0) + X_+(0). \tag{27}$$

Thus the distribution of the waiting time between the two point-events embracing the time instant $t = 0$ will generally be different from that of the rest of the waiting times. This anomaly becomes especially conspicuous if we consider the sequence of waiting times for a *homogeneous* Poisson process on $\mathscr{T} = (-\infty, \infty)$. Then, according to Theorem 7.4.3, the two sequences $X_+(n)$, $n = 0, 1, \ldots$, and $X_-(n)$, $n = 0, -1, \ldots$, will both consist of independent, identically distributed exponential random variables with the same parameter λ. Since these two sequences are also mutually independent, the random variables of the doubly infinite waiting times sequence

$$X(n), \qquad n = \ldots, -1, 0, 1, \ldots$$

will all be independent. They will also be exponentially distributed with the same parameter λ, that is, *all of them but* $X(0)$. This particular waiting time, being a sum of two independent exponentially distributed random variables from (27), will have the Erlang distribution with density

$$f_{X(0)}(x) = \lambda^2 x e^{-\lambda x}, \qquad x > 0.$$

Thus, for a homogeneous Poisson process on $\mathscr{T} = (-\infty, \infty)$, the waiting times, although still independent, are *no longer identically distributed*, since waiting time $X(0)$ is different from the rest.

Yet, if we had decided to measure the waiting times forward and backward *from some particular fixed epoch*, the resulting sequence would indeed be a sequence of independent, identically distributed exponential random variables with no exception. This is so because if X is a waiting time measured from some particular epoch to the next, then $P(X > x)$ is the probability that no point-event occurs during a time interval of length x, which by the definition of a homogeneous Poisson process equals $e^{-\lambda x}$.

One heuristic explanation of this *waiting time paradox* can be given by imagining Nature creating the homogeneous Poisson process. She has a magic inexhaustible bag full of intervals of random lengths, the lengths being i.i.d. exponentially distributed. Starting in the infinitely remote past, she picks one interval after another and stacks them along the time axis. Since a homogeneous Poisson process is a stationary point process (recall Definition 7.2.1), Nature is completely oblivious about the location of the origin $t = 0$; she simply sees the time axis as a line stretching to infinity in both directions. From Nature's point of view, our choice of the origin $t = 0$ is actually random. It is then clear that we are more likely to find the randomly chosen time instant $t = 0$ in a longer interval of the stack than in a shorter one. Since we tied our labeling of the waiting time intervals to this particular time instant $t = 0$, it may no longer be so surprising that the waiting time $X(0)$ should be different from the rest.

Remark: It should be clear from the preceding discussion that the appearance of the waiting time paradox is not restricted to the homogeneous Poisson process but may be encountered in various forms when dealing with waiting times of any point process. It is therefore useful to introduce a few more random times related to the interval properties of a point process so that the waiting time paradox may be more conveniently described. Я

Consider an arbitrary point process with time domain $\mathcal{T}$, and let $T(n)$ be its random epoch sequence as defined in Section 7.1. Let $t \in \mathcal{T}$ be a fixed time instant, and let $K(t)$ be a random integer such that

$$T(K(t)) \le t < T(K(t) + 1).$$

In words, $K(t)$ is the index of the epoch of the last point-event occurring before or at the fixed time t. (Incidently, $K(t)$ is a stopping time for the random epoch sequence.)

Let us now define three random times as follows (see Fig. 7.4.7).

Spent waiting time at t:

$$Y_-(t) = t - T(K(t)).$$

Excess waiting time at t:

$$Y_+(t) = T(K(t) + 1) - t.$$

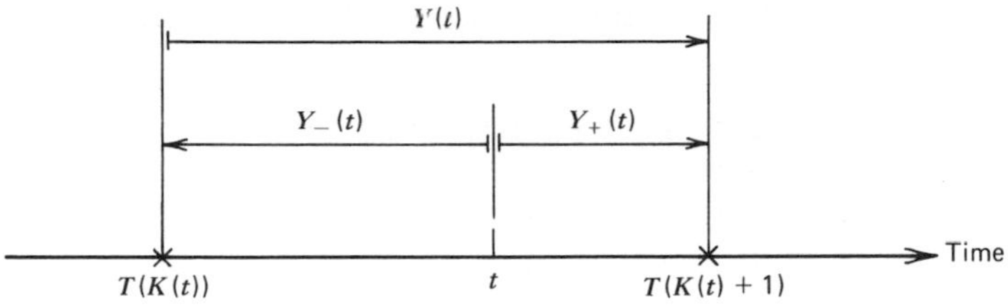

Figure 7.4.7

Current waiting time at t:

$$Y(t) = T(K(t) + 1) - T(K(t)) = Y_-(t) + Y_+(t).$$

Clearly, all three of these random variables are nonnegative and have continuous distributions. The distributions of these three random times are easily obtained from the distribution of the associated counting process $N(t)$, $t \in \mathcal{T}$; we just use the obvious relationship (see Fig. 7.4.7)

$$P(Y_+(t) > y_+, Y_-(t) \geq y_-) = P(N(t + y_+) - N(t - y_-) = 0),$$

from which the joint density of $Y_+(t)$ and $Y_-(t)$ can be obtained by differentiation.

In particular, for a homogeneous Poisson process on $\mathcal{T} = (-\infty, \infty)$ with intensity λ, we have the following result.

THEOREM 7.4.4. *For any* $t \in \mathcal{T}$, $Y_-(t)$ *and* $Y_+(t)$ *are independent, and both are exponentially distributed with parameter* λ, *while the current waiting time at* t *has the Erlang distribution with the density*

$$f_{Y(t)}(y) = \lambda^2 y e^{-\lambda y}, \qquad y > 0. \tag{28}$$

Example 7.4.5

It may be instructive to calculate the distributions of the spent, excess, and current waiting times at some $t > 0$ for a homogeneous Poisson process with time domain $\mathcal{T} = (0, \infty)$. Unlike the stationary Poisson process on $\mathcal{T} = (-\infty, +\infty)$, these distributions may now depend on the time t, since for instance, the spent waiting time cannot in this case exceed the value t.

Let us begin by calculating the joint density $f_{Y_-(t), Y_+(t)}$ of the spent and excess waiting times. The event $[Y_-(t) \geq y_-, Y_+(t) > y_+]$, where $0 < y_- < t$ and $0 < y_+$, occurs if and only if there is no point-event with epoch in the time interval $(t - y_-, t + y_+]$. Hence the probability

$$P(Y_-(t) \geq y_-, Y_+(t) > y_+) = e^{-\lambda(y_+ + y_-)},$$

for $0 < y_- < t$ and $0 < y_+ < \infty$, while

$$P(Y_-(t) \geq y_-, Y_+(t) > y_+) = 0 \qquad \text{for } y_- \geq t.$$

It is seen that for any values of y_- and y_+, this probability factors into the product of

$$P(Y_-(t) \geq y_-) = \lim_{y_+ \to 0} P(Y_-(t) \geq y_-, \; Y_+(t) > y_+)$$

$$= \begin{cases} e^{-\lambda y_-} & \text{if } 0 < y_- < t, \\ 0 & \text{if } y_- \geq t, \end{cases}$$

and

$$P(Y_+(t) > y_+) = \lim_{y_- \to 0} P(Y_-(t) \geq y_-, \; Y_+(t) > y_+) = e^{-\lambda y_+}.$$

It follows that the spent and excess waiting times are independent random variables, the excess waiting time $Y_+(t)$ having an exponential distribution with parameter λ, while the spent waiting time has the density function

$$f_{Y_-(t)}(y_-) = \begin{cases} \lambda e^{-\lambda y_-} + \delta(t - y_-)e^{-\lambda t} & \text{if } 0 < y_- \leq t, \\ 0 & \text{otherwise.} \end{cases}$$

Note that the latter density is of the mixed type with

$$P(Y_-(t) = t) = e^{-\lambda t}$$

corresponding to the probability that no point-event will occur in the interval $(0, t]$.

The density of the current waiting time

$$Y(t) = Y_-(t) + Y_+(t)$$

can be calculated as the convolution

$$f_{Y(t)}(y) = \int_0^{y+} \lambda e^{-\lambda(y-u)} f_{Y_-(t)}(u) \, du, \qquad y > 0.$$

Thus for $0 < y < t$, we have

$$f_{Y(t)}(y) = \int_0^y \lambda e^{-\lambda(y-u)} \lambda e^{-\lambda u} \, du = \lambda^2 y e^{-\lambda y}$$

while if $y \geq t$,

$$f_{Y(t)}(y) = \int_0^t \lambda e^{-\lambda(y-u)} \lambda e^{-\lambda u} \, du$$

$$+ \int_0^y \lambda e^{-\lambda(y-u)} e^{-\lambda t} \, \delta(t-u) \, du = \lambda(1 + \lambda t)e^{-\lambda y}.$$

Notice that as t becomes large, the distribution of the spent waiting time approaches the exponential, while the distribution of the current time approaches the Erlang distribution (28), thus showing the diminishing influence of the origin $t = 0$. ▲

EXERCISES

Exercise 7.4

1. It can be shown that the probability mass distribution of the time N that a symmetric Bernoulli random walk first returns to zero equals

$$P(N = 2n) = \frac{1}{2n-1}\binom{2n}{n}\frac{1}{2^{2n}}, \qquad n = 1, 2, \ldots.$$

In Example 7.4.1, we have used

$$\frac{1}{2n}\binom{2(n-1)}{n-1}\frac{1}{2^{2(n-1)}}.$$

Show that if we define $\binom{0}{0} = 1$ the two expressions are equal for all $n = 1, 2, \ldots$.

2. Suppose a Poisson process on $\mathcal{T} = (0, \infty)$ has intensity function $\lambda(t) = 2\alpha(1 + \alpha t)^{-1}$, where $\alpha > 0$ is a constant. Calculate the conditional expectations

$$\mathrm{E}[T(n+1) \mid T(1) = t_1, \ldots, T(n) = t_n]$$

and

$$\mathrm{E}[X(n) \mid X(0) = x_0, \ldots, X(n-1) = x_{n-1}].$$

3. Consider an RC circuit as shown in Figure 7.4.E.1. The switch is controlled by a homogeneous Poisson process, with $\lambda = 20$, instantaneously discharging the capacitor at each point event of the process. Between point events the switch is open. Find the expected voltage and the charge on the capacitor just prior to a discharge.

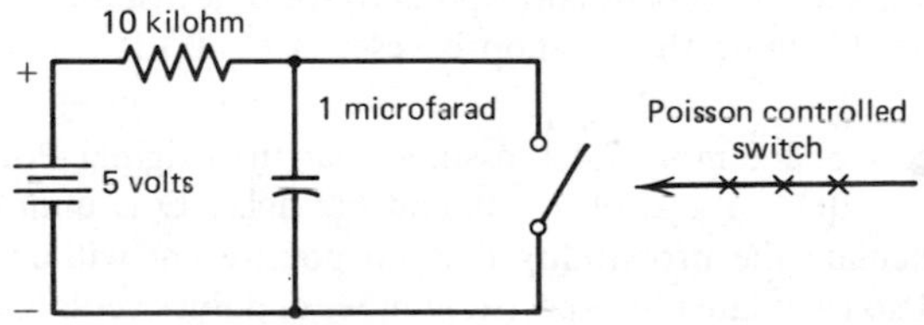

Figure 7.4.E.1

4. Consider again the symmetric ($p = q = \frac{1}{2}$) random walk driven by a homogeneous Poisson process as in Example 7.4.1. Find the density $f_r(t)$ of the time T_r when the random walk first reaches level r ($r = 1, 2, \ldots$). Show that this density has the reproductive property $f_{r_1}(t) \circledast f_{r_2}(t) = f_{r_1+r_2}(t)$, where $\circledast$ denotes the convolution. (*Hint:* Use the probability mass distribution

$$\frac{r}{n}\binom{n}{\frac{n-r}{2}}\frac{1}{2^n}$$

for the step N when the walk first reaches r.)

5. Use expressions (11) and (12), with n replaced by $n + 1$, to show that the conditional density of $T(1), \ldots, T(n)$, given $T(n + 1) = t$, is the same as the conditional density, given $N(t) = n$, in expression (16). Explain why this must be so.
6. For a homogeneous Poisson process find the conditional density for any k $(1 \le k \le n)$ of the first n waiting times, given $N(t) = n$.
7. For the Poisson process with intensity function $\lambda(t) = t$, $t \ge 0$, calculate the mean function $\mu_X(n) = \mathrm{E}[X(n)]$, $n = 0, 1, \ldots$, of the waiting time sequence. What happens as $n \to \infty$?
8. Let $X(n)$, $n = 0, 1, \ldots$, be the waiting time sequence of a homogeneous Poisson process on $\mathcal{T} = (0, \infty)$, and let $\Delta > 0$ be a constant. For $n = 0, 1, \ldots$ fixed, let $Y_\Delta(n)$ be the number of waiting times $X(0), \ldots, X(n)$ of duration exceeding Δ. Find the probability mass distribution of the random variable $Y_\Delta(n)$.
9. With the setting described in Exercise 7.4.8, now let $Z_\Delta(t)$, $t > 0$ be the number of waiting times exceeding Δ counted up to time t. In other words, $Z_\Delta(t) = Y_\Delta(N(t))$. Find the expectation and the variance of $Z_\Delta(t)$ for fixed $t > 0$. (*Hint:* Condition on $N(t) = n$.)
10. For a homogeneous Poisson process with intensity λ, find the conditional density of the shortest among the waiting times $X(n)$, $n = 0, 1, \ldots$, during an interval $(0, t]$ given that exactly n point events occur in this interval. (*Hint:* Recall Exercise 7.4.6.)
11. The so-called *RS* flip-flop has two states 1 (called "set") and 0 (called "reset") and two inputs labeled R and S. If an impulse arrives at the S input the flip-flop either changes its state to 1 (if it was at state 0) or remains in state 1 (if it was at state 1). Similarly, an impulse arriving at the R input causes the flip-flop either to change or to remain in the state 0. Suppose now that impulses are arriving at the two inputs according to two independent homogeneous Poisson processes with intensities λ_S and λ_R, respectively. Find the first order distributions of the waiting time sequence $X(n)$, $n = 0, 1, \ldots$, of the point process whose epochs correspond to the time instances when the flip-flop changes its state. (Assume the flip-flop is reset at $t = 0$.) Is this point process Poisson?
12. For the delayed point process of Example 7.4.3, assume that the original Poisson process has intensity function $\lambda(t) = 1 + \sin t$, $t > 0$, and the delay Θ is uniformly distributed over $(0, 2\pi)$. Calculate the probability that no point-event will emerge during the interval $(0, 2\pi]$. Also calculate the expected number of point-events emerging during an interval $(t, t + 2\pi]$ as a function of $t \ge 0$.
13. For an inhomogeneous Poisson process on $\mathcal{T} = (-\infty, \infty)$, find the general formula for the first order densities $f_{X(n)}(x)$ of the waiting time sequence $X(n)$, $n = \ldots, -1, 0, 1, \ldots$. Also obtain an expression for the densities of the spent, excess, and current waiting times at some arbitrary instant $t_0 \in \mathcal{T}$. (*Hint:* Translate t_0 to the origin by the time substitution $t' = t - t_0$ into the intensity function.)
14. In a particle accelerator with two colliding beams, an interaction occurs only if two particles, one from each beam, arrive at the cross section within a time interval whose length does not exceed some small positive constant δ. Assuming that the particles in each beam are arriving according to two independent homogeneous Poisson processes with the same intensity λ, find the expectation and the variance of the number of interactions per unit of time.

15. For simulation purposes, it is required to create a portable source of a homogeneous Poisson process. The idea is to record a sample of such a process as a stream of impulses on a magnetic tape running with a constant speed 10 centimeters/second. After 60 seconds of recording the tape is to be cut and its end glued to its beginning to create an endless loop. Each simulation will then start at some arbitrary place on the loop and will run for no more than 40 seconds. Explain why the stream of impulses so obtained will not quite correspond to that of a Poisson process. What is the probability that it will? Can you think of a way to remedy the situation?

7.5. MARKED POINT PROCESSES AND COMPOUND POISSON PROCESSES

In the preceding sections, we have regarded a point process as a random stream of point-events, where all the point-events were of the same kind. That is, the only characteristic of these point-events were their epochs—the times of their occurrence, which we assumed to be distinct.

However, there are many physical situations involving impulsive random phenomena in which it may not be adequate merely to describe the times at which these phenomena are occurring. It may be that the individual point-events have some additional numerical characteristic that should be incorporated into the model. For instance, if the point-events represent arrivals of particles at a counter or electrical impulses at a receiver, we may also wish to include in our model the feature that the particles may have different energies or that the pulses may vary in polarity or more generally in amplitude. Or we might wish to consider the possibility that several point-events may be occurring at the same epoch; such would be the case if we wanted to model emissions of secondary electrons in a photomultiplier.

All this can be accomplished by assuming that each point-event carries with it an additional random variable, appropriately called its *mark*. The ensemble of such a *marked point process* will then consist of discrete sets of epochs where now each epoch will be accompanied by a real number, its mark.†

If we think of point-events as a stream of random impulses with marks being their amplitudes, then a typical realization of this marked point process will look like the one depicted in Figure 7.5.1. Such a marked point process can be completely specified by the probability law of the sequence

$$(T(n),\ U(n)),\ n = \ldots,\ -1,\ 0,\ 1,\ \ldots,$$

where $T(n)$ is the nth random epoch (defined as in Section 7.1) and the random variable $U(n)$ is its mark.

An equivalent representation can be obtained by defining an ordinary continuous time stochastic process $X(t)$, $t \in \mathcal{T}$, such that for each $t_1 < t_2$ the increment

† More generally, a mark could be a vector or even a function.

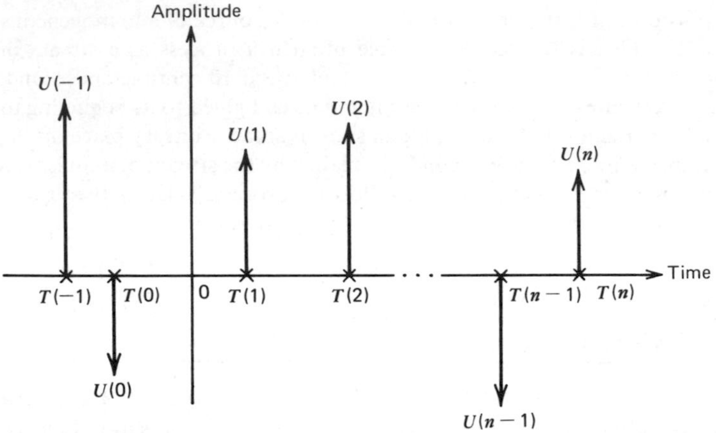

Figure 7.5.1

$X(t_2) - X(t_1)$ equals the sum of all marks corresponding to point-events with epochs in $(t_1, t_2]$.

In particular, if we are dealing with a marked point process with time domain $\mathcal{T} = (0, \infty)$,† we can define $X(t)$, $t > 0$, to be the sum of all marks corresponding to point-events with epochs in $(0, t]$. If we think of marks as amplitudes of a stream of random impulses (point-events) arriving at the input of an ideal integrator, then $X(t)$ would be the output of the integrator at time t, as in Figure 7.5.2. Note that if

$$N(t), \qquad t \geq 0$$

is the associated counting process of the original unmarked point process with the origin $t = 0$ included in the time domain by defining $N(0) = 0$, then the process $X(t)$, $t \geq 0$, can be expressed as a random sum

$$N_U(t) = \sum_{n=0}^{N(t)} U(n), \qquad t \geq 0, \tag{1}$$

provided we also define $U(0) = 0$. To see this, notice that if $N(t) = 0$, then there are no point-events with epochs in $(0, t]$, and so $N_U(t) = 0$, while if $N(t) = n > 0$, then there are exactly n point events with epochs $0 < T(1) < \cdots < T(n) \leq t$, and hence the sum of their marks is

$$N_U(t) = U(1) + \cdots + U(n).$$

The process $N_U(t)$, $t > 0$, defined by (1) has been called by D. Snyder a *mark accumulator process*.

† Only this type of time domain will be discussed in this section. The generalization to the time domain $\mathcal{T} = (-\infty, \infty)$ is quite straightforward.

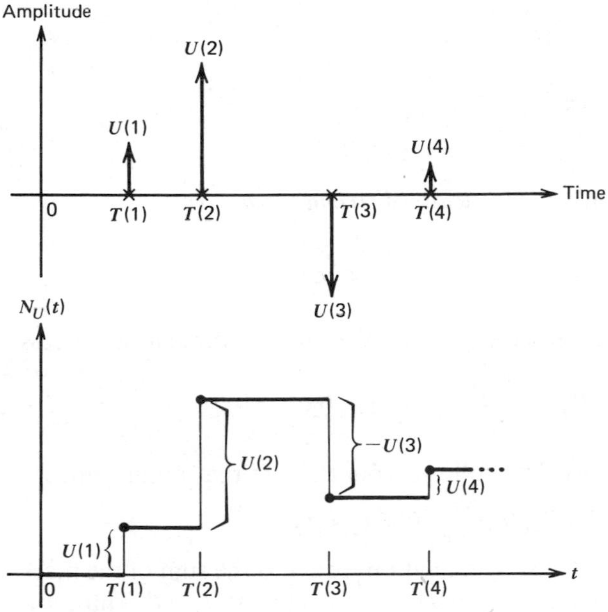

Figure 7.5.2

It should be clear from Figure 7.5.2 that the sample paths of this process are all jump functions (i.e., constant except for isolated jumps), with jumps occurring at the random epochs of the point process. The height of a jump at $T(n)$ is equal to the value of the corresponding mark $U(n)$. This implies that we can get the sequence

$$(T(n), U(n)), \qquad n = 1, 2, \ldots,$$

from the process

$$N_U(t), \qquad t \geq 0,$$

and vice versa. Consequently, the probability law of the mark accumulator process $N_U(t)$, $t \geq 0$, completely specifies the probability law of the marked point process on $\mathscr{T} = (0, \infty)$.

In general, the mark sequence $U(n)$, $n = 1, 2, \ldots$, and the random epoch sequence $T(n)$, $n = 1, 2, \ldots$, may be dependent, and thus the probability law of the marked point process may be extremely complicated. We therefore begin by making some simplifying assumptions:

(A) *The unmarked point process evolves without aftereffects.*

As we already know, this means that it must be an inhomogeneous Poisson process. We will denote the associated counting process by

$$N(t), \qquad t > 0, \tag{2}$$

and the intensity function

$$\lambda(t), \qquad t > 0.$$

(*B*) *The sequence of marks*

$$U(n), \qquad n = 1, 2, \ldots, \tag{3}$$

is a sequence of independent, identically distributed random variables with common density function

$$f_U(x), \qquad -\infty < x < \infty.$$

These two assumptions together imply that the mark accumulator process

$$N_U(t) = \sum_{n=0}^{N(t)} U(n), \qquad t \geq 0$$

will have *independent increments*. This follows from the fact that an increment

$$N_U(t_2) - N_U(t_1), \qquad 0 < t_1 < t_2,$$

of the mark accumulator process is, by definition, equal to the sum of the marks corresponding to the point-events with epochs in the interval $(t_1, t_2]$. Thus, if we consider nonoverlapping intervals on the time domain $\mathcal{T} = (0, \infty)$, the numbers of epochs in these intervals are independent by assumption (A), while the marks corresponding to these epochs are independent by assumption (B).

Finally, we shall also assume that

(*C*) *The mark sequence* (3) *and the unmarked point process* [*expressed, e.g., by its associated counting process* (2)] *are mutually independent.*

This means that the entire probability law of the marked point process is now completely specified by only two functions: the intensity function $\lambda(t)$, $t \geq 0$, and the density function of the marks $f_U(x)$, $-\infty < x < \infty$.

Definition 7.5.1
A marked point process satisfying assumptions (A), (B), and (C) is called a *compound Poisson process.*

Example 7.5.1
Consider an inhomogeneous Poisson process with intensity function $\lambda(t)$, $t > 0$, and let $N(t)$, $t > 0$, be its associated counting process. Given any m-tuple of nonnegative numbers $p_1, \ldots, p_m$ such that $p_1 + \cdots + p_m = 1$, we saw in section 7.3 that the Poisson process may be regarded as a superposition

$$N(t) = \sum_{k=1}^{m} N_k(t), \qquad t > 0,$$

of m independent Poisson processes with intensity functions $p_1 \lambda(t), \ldots, p_m \lambda(t)$, $t > 0$, respectively. Suppose now that we define a stochastic process $X(t), t > 0$, as a linear combination

$$X(t) = \sum_{k=1}^{m} \alpha_k N_k(t)$$

of the component Poisson counting processes $N_k(t)$ with $\alpha_1, \ldots, \alpha_m$ being given constants. It is then clear that the sample paths of the process $X(t)$ will all be jump functions, with jumps occurring at epochs of the point-events of the superposition Poisson process $N(t)$, $t > 0$. The heights of these jumps will be given by the constants $\alpha_1, \ldots, \alpha_m$; more specifically, each jump at the epoch contributed by an occurrence of the point-event of the kth component Poisson process will have height equal to α_k. Now we saw in Example 7.3.2 that the probability of a particular point-event with epoch τ coming from the kth component process equals $p_k \lambda(\tau)/\lambda(\tau) = p_k$.

This means that we can construct any sample path of the process $X(t), t > 0$, by taking the superposition Poisson process with intensity function $\lambda(t)$ and at each epoch of this Poisson process letting the sample path of $X(t)$ undergo a jump of height α_k selected at random with probability p_k, $k = 1, \ldots, m$, and independently for each such epoch.

But then the process $X(t), t > 0$, is identical with a mark accumulator process $N_U(t)$, $t > 0$, of a compound Poisson process with intensity function $\lambda(t)$ and with discrete mark distribution

$$P(U(n) = \alpha_k) = p_k, \qquad k = 1, \ldots, m.$$

It can be shown that the above discussion can be extended even to infinite linear combinations, and thus we have an important result, namely that a mark accumulator process $N_U(t)$, $t > 0$, of any compound Poisson process with a discrete mark distribution can be expressed as an infinite series

$$N_U(t) = \sum_{k=1}^{\infty} \alpha_k N_k(t), \qquad t > 0,$$

of independent Poisson processes and, conversely, that any such infinite series defines a compound Poisson mark accumulator process. ▲

We now proceed to derive the probability law of a compound Poisson process, or rather that of its associated mark accumulator process

$$N_U(t) = \sum_{n=0}^{N(t)} U(n), \qquad t \geq 0.$$

Clearly, since $N(0) = 0$, and by our convention $U(0) = 0$, we have

$$N_U(0) = 0.$$

Let us first derive the mean function and the correlation function of the mark accumulator process $N_U(t)$. Since by assumption (C) the mark sequence is independent of the counting process $N(t)$, by applying the formula for the mean of random sums we have

$$E[N_U(t)] = E[U]E[N(t)] = E[U] \int_0^t \lambda(\tau)\, d\tau, \tag{4}$$

where $E[U]$ is the common expectation of the mark variables $U(n)$.

Similarly, from the formula for the variance of random sums,

$$\begin{aligned} \text{Var}[N_U(t)] &= E[N(t)]\text{Var}[U] + \text{Var}[N(t)](E[U])^2 \\ &= E[U^2]E[N(t)] = E[U^2] \int_0^t \lambda(\tau)\, d\tau, \end{aligned} \tag{5}$$

since for a Poisson process $\text{Var}[N(t)] = E[N(t)]$. Finally, taking $0 < t_1 < t_2$ and writing

$$N_U(t_1)N_U(t_2) = N_U^2(t_1) + N_U(t_1)[N_U(t_2) - N_U(t_1)]$$

we have, upon taking expectations and using the independence of increments of the process, the correlation function

$$\begin{aligned} R_{N_U}(t_1, t_2) &= E[N_U(t_1)N_U(t_2)] \\ &= E[N_U^2(t_1)] + E[N_U(t_1)]E[N_U(t_2) - N_U(t_1)] \\ &= \text{Var}[N_U(t_1)] + E[N_U(t_1)]E[N_U(t_2)] \\ &= E[U^2] \int_0^{t_1} \lambda(\tau)\, d\tau + \left(E[U] \int_0^{t_1} \lambda(\tau)\, d\tau\right)^2, \qquad 0 < t_1 < t_2. \end{aligned} \tag{6}$$

Unless $E[U] = 0$ the covariance function has a somewhat more appealing form. Since

$$\text{Cov}[N_U(t_1), N_U(t_2)] = E[N_U(t_1)N_U(t_2)] - E[N_U(t_1)]E[N_U(t_2)]$$

we have from (6) and (5)

$$\text{Cov}[N_U(t_1), N_U(t_2)] = E[U^2] \int_0^{\min(t_1, t_2)} \lambda(\tau)\, d\tau. \tag{7}$$

Formulas (4), (5), and (7) are most conveniently expressed by using a differential formalism

$$E[dN_U(t)] = E[U]\lambda(t)\, dt, \tag{8}$$

$$\text{Var}[dN_U(t)] = E[U^2]\lambda(t)\, dt, \tag{9}$$

$$\text{Cov}[dN_U(t_1), dN_U(t_2)] = E[U^2]\, \delta(t_2 - t_1)\lambda(t_1)\, dt_1\, dt_2, \tag{10}$$

where δ is the Dirac delta function and $dN_U(t) = N(t + dt) - N(t)$ is an infinitesimal forward increment of the process $N_U(t)$.

Next, let us derive the entire probability law for the compound Poisson process. Since we have already established that the process $N_U(t)$ must have independent increments, its probability law will be completely specified by the distribution of its increments

$$N_U(t_2) - N_U(t_1), \qquad \text{for any } 0 \leq t_1 < t_2 .$$

Denote the distribution function of this increment by

$$F_{t_1, t_2}(u) = P(N_U(t_2) - N_U(t_1) \leq u), \qquad -\infty < u < \infty.$$

Conditioning on the number of point-events with epochs in $(t_1, t_2]$, we have

$$F_{t_1, t_2}(x) = \sum_{n=0}^{\infty} P(N_U(t_2) - N_U(t_1) \leq u \,|\, N(t_2) - N(t_1) = n) P(N(t_2) - N(t_1) = n). \tag{11}$$

However, given that there are exactly n point-events with epochs in $(t_1, t_2]$, the increment $N_U(t_2) - N_U(t_1)$ is the sum of exactly n marks corresponding to these epochs. Furthermore, by assumptions (B) and (C), the conditional distribution function

$$P(N_U(t_2) - N_U(t_1) \leq u \,|\, N(t_2) - N(t_1) = n) \tag{12}$$

will coincide with the distribution function of the sum of n independent, identically distributed random variables with common distribution function $F_U(u)$. Denoting the latter by $F_U^{(n)}(u)$, that is

$$F_U^{(n)}(u) = P(U(1) + \cdots + U(n) \leq u), \qquad -\infty < u < \infty \tag{13}$$

we see that (12) equals (13), for $n \geq 1$. The case $n = 0$ requires extra treatment; if there are no point-events with epochs in $(t_1, t_2]$, then clearly $N_U(t_2) - N_U(t_1) = 0$, so that (12) equals 0 for $u < 0$ and 1 for $u \geq 0$. Hence, if we define the degenerate distribution function†

$$F_U^{(0)}(u) = \begin{cases} 1 & \text{if } u \geq 0, \\ 0 & \text{if } u < 0, \end{cases} \tag{14}$$

we can write

$$P(N_U(t_2) - N_U(t_1) \leq u \,|\, N(t_2) - N(t_1) = n) = F_U^{(n)}(u) \tag{15}$$

for all $n = 0, 1, \ldots$. Finally, by assumption (A), $N(t), t \geq 0$, is the counting process of a Poisson process with intensity function $\lambda(t)$, $t \geq 0$ so that

$$P(N(t_2) - N(t_1) = n) = \frac{[\int_{t_1}^{t_2} \lambda(\tau)\, d\tau]^n}{n!} e^{-\int_{t_1}^{t_2} \lambda(\tau)\, d\tau} \tag{16}$$

† Note that this is actually the distribution function of $U(0) = 0$.

Substituting from (15) and (16) into (11), we obtain

$$F_{t_1, t_2}(u) = e^{-\int_{t_1}^{t_2} \lambda(\tau)\, d\tau} \sum_{n=0}^{\infty} F_U^{(n)}(u) \frac{[\int_{t_1}^{t_2} \lambda(\tau)\, d\tau]^n}{n!}, \tag{17}$$

$-\infty < u < \infty$, $0 \leq t_1 < t_2$, and the probability law of the compound Poisson process is completely determined.

Since the density function for the increment $N_U(t_2) - N_U(t_1)$ is given by the derivative of its distribution function and the same is true for the marks, note that (17) also gives the relationship between the density of the increment and the density of the marks. The same result could be arrived at by repeating the argument for densities rather than for distribution functions.

Example 7.5.2

Suppose that point-events of an inhomogeneous Poisson process with intensity function $\lambda(t)$, $t > 0$, represent arrivals of particles at a particle counter. Suppose further that each particle has a probability p $(0 < p < 1)$ of being actually registered by the counter, the registrations of distinct particles being assumed independent events.

The counting process $N_U(t)$, $t > 0$, of registered particles is then a compound Poisson process with a Bernoulli mark sequence $U(n)$, $n = 1, 2, \ldots$, where $U(n) = 1$ or $U(n) = 0$ correspond to registration or nonregistration of the nth particle respectively. Then the sum $U(1) + \cdots + U(n)$ is a binomial random variable with the probability mass distribution

$$P^{(n)}(U = u) = \binom{n}{u} p^u q^{n-u}, \qquad u = 0, \ldots, n, \tag{18}$$

where

$$p = 1 - q = P(U(n) = 1).$$

The probability mass distribution of the number $N_U(t_2) - N_U(t_1)$ of particles registered during the time interval $(t_1, t_2]$ is then according to (17),

$$P(N_U(t_2) - N_U(t_1) = u) = e^{-\int_{t_1}^{t_2} \lambda(\tau)\, d\tau} \sum_{n=u}^{\infty} \binom{n}{u} p^u q^{n-u} \frac{[\int_{t_1}^{t_2} \lambda(\tau)\, d\tau]^n}{n!},$$

$u = 0, 1, \ldots$, provided we interpret

$$\binom{0}{u} = 1 \quad \text{if} \quad u = 0 \quad \text{and} \quad \binom{0}{u} = 0 \quad \text{if} \quad u > 0,$$

so that (18) holds also for $n = 0$. (Recall that $U(0) = 0$.)

Since

$$\sum_{n=u}^{\infty} \binom{n}{u} p^u q^{n-u} \frac{[\int_{t_1}^{t_2} \lambda(\tau)\, d\tau]^n}{n!} = \frac{[p \int_{t_1}^{t_2} \lambda(\tau)\, d\tau]^u}{u!} \sum_{n=u}^{\infty} \frac{[q \int_{t_1}^{t_2} \lambda(\tau)\, d\tau]^{n-u}}{(n-u)!}$$
$$= \frac{[p \int_{t_1}^{t_2} \lambda(\tau)\, d\tau]^n}{u!} e^{q \int_{t_1}^{t_2} \lambda(\tau)\, d\tau}$$

we obtain

$$P(N_U(t_2) - N_U(t_1) = u) = \frac{[p \int_{t_1}^{t_2} \lambda(\tau)\, d\tau]^u}{u!} e^{-p \int_{t_1}^{t_2} \lambda(\tau)\, d\tau}$$

$u = 0, 1, \ldots$. Since the numbers of particles registered during nonoverlapping time intervals will clearly be independent random variables, it follows that the point process of particle registration is again an inhomogeneous Poisson process with intensity function $p\lambda(t)$, $t > 0$. ▲

Example 7.5.3

Consider a stream of particles impinging on a target according to a homogeneous Poisson process with intensity λ. Suppose that each particle carries a positive charge, or energy if you prefer, that we assume to be exponentially distributed† with parameter $\alpha > 0$. The total charge or energy delivered to the target up to time $t > 0$ can then be described as a compound Poisson counting process $N_U(t)$, $t > 0$, with exponentially distributed marks $U(n)$, $n = 1, 2, \ldots$.

The probability density $f_U^{(n)}(u)$ of the sum of n such marks is then

$$f_U^{(n)}(u) = \frac{\alpha^n}{(n-1)!} u^{n-1} e^{-\alpha u}, \qquad u > 0, \qquad n = 1, 2, \ldots.$$

Hence by (17), for the density of $N_U(t)$, we have the expression

$$f_{0,t}(u) = e^{-\lambda t}\left[\delta(u) + \sum_{n=1}^{\infty} \frac{\alpha^n}{(n-1)!} u^{n-1} e^{-\alpha u} \frac{(\lambda t)^n}{n!}\right], \qquad u \geq 0,$$

where the delta function is used to represent the density of the degenerate distribution (14). After factoring $\alpha\lambda t e^{-\alpha u}$ from the infinite series in brackets, we can write this as

$$\sum_{n=0}^{\infty} \frac{(\alpha\lambda u t)^n}{n!(n+1)!} = \frac{1}{\sqrt{\alpha\lambda u t}} \sum_{n=0}^{\infty} \frac{1}{n!(n+1)!}\left(\frac{2\sqrt{\alpha\lambda u t}}{2}\right)^{2n+1}$$

$$= \frac{1}{\sqrt{\alpha\lambda u t}} I_1(2\sqrt{\alpha\lambda u t}),$$

where

$$I_1(x) = \sum_{n=0}^{\infty} \frac{1}{n!(n+1)!}\left(\frac{x}{2}\right)^{2n+1}$$

† This may not be a very realistic assumption; a Maxwell distribution of energies might be perhaps more appropriate. However, we retain the exponential assumption to keep the calculation reasonably simple.

is the modified Bessel function of order $k = 1$ introduced in Example 7.3.3. Thus

$$f_{0,t}(u) = e^{-\lambda t}\,\delta(u) + \sqrt{\alpha\lambda\frac{t}{u}}\,e^{-(\alpha u + \lambda t)} I_1(2\sqrt{\alpha\lambda u t}), \qquad u \ge 0,$$

is the density of $N_U(t)$ for any fixed $t > 0$. Notice that this distribution is of a mixed type with

$$P(N_U(t) = 0) = e^{-\lambda t},$$

as will always be the case for a continuous mark distribution. ▲

The obvious disadvantage of formula (17) is that it requires knowledge of the distribution function $F_U^{(n)}(x)$ for all $n = 1, 2, \ldots$. As the reader is well aware, the explicit form of these distributions is in general very difficult to obtain.

This can be avoided by resorting to the characteristic function

$$\psi_{t_1, t_2}(\zeta) = \mathrm{E}[e^{\iota\zeta(N_U(t_2) - N_U(t_1))}], \qquad -\infty < \zeta < \infty, \tag{19}$$

$0 \le t_1 < t_2$, of the increment $N_U(t_2) - N_U(t_1)$, which provides an equivalent specification of the probability law. If we write the expectation (19) as

$$\mathrm{E}[e^{\iota\zeta(N_U(t_2) - N_U(t_1))}] = \sum_{n=0}^{\infty} \mathrm{E}[e^{\iota\zeta(N_U(t_2) - N_U(t_1))} \mid N(t_2) - N(t_1) = n]\, P(N(t_2) - N(t_1) = n), \tag{20}$$

and again use the fact that, conditioned upon $N(t_2) - N(t_1) = n$, the increment $N_U(t_2) - N_U(t_1)$ is the sum of n independent, identically distributed random variables, we see that the conditional expectation

$$\mathrm{E}[e^{\iota\zeta(N_U(t_2) - N_U(t_1))} \mid N(t_2) - N(t_1) = n] \tag{21}$$

is just the characteristic function of such a sum. Thus if

$$\psi_U(\zeta) = \int_{-\infty}^{\infty} e^{\iota\zeta x} f_U(x)\, dx \tag{22}$$

is the common characteristic function of the i.i.d. random variables $U(n)$, $n = 1, 2, \ldots$, it follows that the conditional expectation (21) is just the nth power of the function (22). This holds even for $n = 0$, since

$$(\psi_U(\zeta))^0 = 1, \qquad -\infty < \zeta < \infty,$$

is the characteristic function of the degenerate distribution (14). Substituting into (20) we get

$$\mathrm{E}[e^{\iota\zeta(N_U(t_2) - N_U(t_1))}] = \sum_{n=0}^{\infty} (\psi_U(\zeta))^n P(N(t_2) - N(t_1) = n),$$

and from this and (16), we obtain for the characteristic function

$$\psi_{t_1, t_2}(\zeta) = e^{(\psi_U(\zeta) - 1)\int_{t_1}^{t_2} \lambda(\tau)\, d\tau}, \qquad -\infty < \zeta < \infty, \qquad 0 \le t_1 < t_2. \tag{23}$$

Since the mark accumulator process $N(t)$, $t \ge 0$, has independent increments and $N_U(0) = 0$, we can write for any $0 \le t_1 < t_2$

$$\psi_{0, t_2}(\zeta) = \psi_{0, t_1}(\zeta)\psi_{t_1, t_2}(\zeta) \tag{24}$$

and thus the probability law of the process $N_U(t)$, $t \ge 0$, is completely specified by the characteristic function

$$\psi_t(\zeta) = \mathrm{E}[e^{i\zeta N_U(t)}] = \psi_{0,t}(\zeta) \tag{25}$$

for all $-\infty < u < \infty$ and $t > 0$. Setting $t_1 = 0$ and $t_2 = t$ in (23), we get for the characteristic function (25)

$$\psi_t(\zeta) = e^{(\psi_U(\zeta) - 1)\int_0^t \lambda(\tau)\, d\tau}, \qquad -\infty < \zeta < \infty, \qquad t \ge 0. \tag{26}$$

This formula provides a most compact and elegant characterization of the probability law of a compound Poisson process.

Notice that for a *homogeneous* compound Poisson process, (26), together with (24), implies that the characteristic function $\psi_t(\zeta)$ must satisfy the condition

$$\psi_{t_1 + t_2}(\zeta) = \psi_{t_1}(\zeta)\psi_{t_2}(\zeta) \qquad \text{for any } t_1 > 0,\ t_2 > 0.$$

Probability distributions whose characteristic functions satisfy this condition are called *infinitely divisible*, and a study of their properties is one of the most elegant and fruitful topics in probability theory.

Example 7.5.4

Suppose that each occurrence of a point-event in a standard Poisson process generates a burst of secondary point-events occurring simultaneously, that is, at the same epoch as the primary point-event. Suppose further that the number $U(n)$ of all point-events thus occurring at epoch $T(n)$ is a random variable with probability mass distribution

$$P(U = u) = \begin{cases} \left(\ln \dfrac{1}{p}\right)^{-1} \dfrac{q^u}{u} & \text{if } u = 1, 2, \ldots, \\ 0 & \text{otherwise} \end{cases} \tag{27}$$

where $p = 1 - q$ is a parameter, $0 < p < 1$.

Remark: This distribution is known as the logarithmic distribution for the obvious reason that its values at $u = 1, 2, \ldots$ are proportional to the terms in the Taylor expansion of the function $-\ln(1 - q)$. Notice that the probabilities $P(U = u)$ approach zero faster than those of a geometric distribution but considerably slower than those of a Poisson distribution. It makes its appearance, for instance, in some physical models of ionization and recombination of charged particles. Я

Assuming that the random variables $U(n)$, $n = 1, 2, \ldots$, are also independent, the total number $N_U(t)$ of point-events occurring in the time interval $(0, t]$ represents a mark accumulator process of a compound Poisson process.

To obtain the characteristic function (26), we first need to evaluate the characteristic function $\psi_U(\zeta)$ of the logarithmic distribution (27). This is quite easy; we have, using the expansion of the logarithm,

$$\begin{aligned}\psi_U(\zeta) = \mathrm{E}[e^{\iota\zeta U}] &= \left(\ln \frac{1}{p}\right)^{-1} \sum_{u=1}^{\infty} e^{\iota\zeta u}\frac{q^u}{u} \\ &= (\ln p)^{-1} \sum_{u=1}^{\infty} \frac{-(qe^{\iota\zeta})^u}{u} \\ &= (\ln p)^{-1} \ln(1 - qe^{\iota\zeta}).\end{aligned}$$

Substitution in (26) now yields the characteristic function of $N_U(t)$,

$$\psi_t(\zeta) = e^{[(\ln p)^{-1}\ln(1-qe^{\iota\zeta})-1]t}.$$

The exponent can be written as

$$[\ln(1 - qe^{\iota\zeta}) - \ln p]\frac{t}{\ln p} = \ln\left(\frac{p}{1 - qe^{\iota\zeta}}\right)^{t(\ln 1/p)^{-1}},$$

and hence

$$\psi_t(\zeta) = \left(\frac{p}{1 - qe^{\iota\zeta}}\right)^{t(\ln 1/p)^{-1}}, \qquad -\infty < \zeta < \infty$$

is the desired characteristic function. Note that $t(\ln 1/p)^{-1} > 0$, and if $t = k \ln 1/p$, with k a positive integer, this characteristic function is the same as the characteristic function of a negative binomial distribution with parameters k and p. For nonintegral values of the exponent $\alpha = t(\ln 1/p)^{-1}$, the probability mass distribution of $N_U(t)$, $t > 0$, equals

$$P(N_U(t) = n) = \frac{\Gamma(n + \alpha)}{n!\,\Gamma(\alpha)} p^\alpha q^n, \qquad n = 0, 1, \ldots$$

which is an extension of the negative binomial distribution to nonintegral values of the parameter $\alpha > 0$, just as the gamma distribution is an extension of the Erlang distribution. ▲

Remark: Above, we have defined a compound Poisson process only on the time domain $\mathcal{T} = (0, \infty)$, but it is easy to see that extending the definition to the time domain

$\mathcal{T} = (-\infty, \infty)$ should present no difficulty. We simply take an unmarked Poisson process on $\mathcal{T} = (-\infty, \infty)$ and a doubly infinite sequence

$$U(n), \qquad n = \ldots, -1, 0, 1, \ldots$$

of i.i.d. random variables (marks), independent of the unmarked Poisson process. The mark accumulator process

$$N_U(t), \qquad -\infty < t < \infty$$

can then be defined as a process with independent increments such that for any $-\infty < t_1 < t_2 < \infty$ the increment

$$N_U(t_2) - N_U(t_1) \tag{28}$$

equals the sum of marks corresponding to epochs in the interval $(t_1, t_2]$. The characteristic function of the increment (28) is still given by expression (23), this time valid for all $-\infty < t_1 < t_2 < \infty$.

If we further decide to anchor the mark accumulator process by setting

$$N_U(0) = 0$$

then the characteristic function of the increment (28) can be expressed solely in terms of the characteristic function

$$\psi_t(\zeta) = \mathrm{E}[e^{i\zeta N_U(t)}], \qquad -\infty < \zeta < \infty, \qquad -\infty < t < \infty$$

which is still given by the formula

$$\phi_t(\zeta) = e^{(\psi_U(\zeta) - 1)\int_0^t \lambda(\tau)\, d}$$

EXERCISES

Exercise 7.5

1. Show that the position $S(t)$, $t \geq 0$, of a particle performing a random walk defined in Example 7.3.3 can be regarded as a mark accumulator process with mark values ± 1. Use (26) to calculate the characteristic function of $S(t)$ and compare it with the one derived in Example 7.3.3.

2. Consider a compound Poisson process on $\mathcal{T} = (0, \infty)$, where the point-events are occurring with a constant intensity λ, and the marks have Poisson probability mass distribution $P(U = k) = e^{-\alpha}\alpha^k/k!$, $k = 0, 1, \ldots$, $\alpha > 0$. Find the expectation, the variance, and the characteristic function of the increment $N_U(t_2) - N_U(t_1)$, $0 \leq t_1 < t_2$.

3. Under the assumptions of Exercise 7.5.2, calculate the probability $P(N_U(t) = n)$ for $n = 0, 1, 2$ and $t > 0$. More generally, show that for any $n = 0, 1, \ldots$, the above probability mass distribution can be written as $P(N_U(t) = n) = e^{-c(t)}\alpha^n/n!\, m_n(t)$, where $m_n(t)$ is the nth moment function of the homogeneous Poisson process with intensity $\lambda e^{-\alpha}$.

4. Suppose that cosmic ray showers are occurring at a certain place according to a homogeneous Poisson process with intensity λ. Suppose further that the number of particles in each shower are independent random variables, each having a geometric distribution with parameter p. Argue that the total number of particles recorded

during a time interval $(0, t]$ is a compound Poisson process, and find its mean function, covariance function, and characteristic function.

5. Under the assumptions of Exercise 7.5.4, obtain an explicit formula for the probability mass distribution of the number of particles recorded up to time $t > 0$. (*Hint:* Recall that a sum of i.i.d. geometric random variables is negative binomial.)

6. Electrical impulses are arriving at a recording device according to an inhomogeneous Poisson process with intensity $\lambda(t) = \sin^2(t\pi)$, $t > 0$. Amplitudes of these impulses are independent random variables each having the Rayleigh density with parameter $\sigma = 1$ volt. However, an impulse is registered only if its amplitude is no less than 0.5 volt and no greater than 2 volts. Impulses with amplitudes outside this range have no effect on the recording device. Describe the probability law of the point process of recorded impulses. Find the expectation and the variance of the number of impulses recorded during a unit of time.

7. Under the assumptions of Exercise 7.5.6, suppose that each time a pulse with amplitude exceeding 2 volts arrives, there is a probability $p = \frac{1}{4}$ that some damage to the recorder may occur. The recorder is automatically inspected for damage every unit of time $t = 1, 2, \ldots$. Find the probability mass distribution of the time T when such a damage is first detected. What is the expected number of impulses recorded by this time?

8. Consider a compound Poisson process on $\mathcal{T} = (0, \infty)$ where the point process is homogeneous with intensity λ and its mark distribution is uniform on $\left(-\sqrt{\frac{3}{\lambda}}, \sqrt{\frac{3}{\lambda}}\right)$. For $0 \leq t_1 < t_2$, calculate the characteristic function $\psi_{t_1, t_2}(\zeta)$ of the increment $N_U(t_2) - N_U(t_1)$. Show that as $\lambda \to \infty$, the probability law of the process $N_U(t)$, $t \geq 0$, approaches that of the standard Wiener process $W_0(t)$, $t \geq 0$.

9. Consider a particle performing a symmetric random walk on the set of all integers starting from the origin. The time instances of its jumps are epochs of a homogeneous Poisson process with intensity λ, and each jump takes the particle k units to the right or to the left with probability $2^{-(k+1)}$ independently for each jump. Denoting by $S(t)$ its position at time $t > 0$, show that $S(t)$ can be written as an infinite series $S(t) = S_1(t) + 2S_2(t) + 3S_3(t) + \ldots$, where $S_1(t), S_2(t), S_3(t), \ldots$ are independent compound Poisson processes (simple symmetric walks) with marks $P(U = 1) = P(U = -1) = \frac{1}{2}$ and with intensities $\lambda/2^k$, $k = 1, 2, \ldots$. (*Hint:* Show that the characteristic function can be factored into a product.)

10. A stream of unit pulses is arriving at a counter according to a homogeneous Poisson process with intensity λ. The polarities of these pulses are i.i.d. symmetric Bernoulli random variables. Each positive pulse advances the counter by one, and each negative pulse resets the counter to zero. Show that the state of the counter at time t can be modeled as a mark accumulator process but not as a compound Poisson process. Show, nevertheless, that the probability law of this mark accumulator process can be specified quite easily, and do so.

11. Show that if $N_U(t)$, $t \in \mathcal{T}$, is a compound Poisson process, then for any $t_1 < t_2$ and $\varepsilon > 0$

$$P(N_U(t_2) - N_U(t_1) \geq \varepsilon) \leq \min_{x \geq 0} \exp\left\{-x\varepsilon + \mathrm{E}[e^{xU} - 1] \int_{t_1}^{t_2} \lambda(\tau)\, d\tau\right\}.$$

Evaluate the right-hand side for the process in Example 7.5.3

12. Find the mean function and the variance function of the compound Poisson process of Example 7.5.4.
13. For the process of charge buildup described in Example 7.5.3, let T_a be the time when the total charge delivered to the target first exceeds a level $a > 0$. Find an expression for the density $f_{T_a}(t)$. (*Hint:* Make use of the equivalence $[T_a \le t] \Leftrightarrow [N_U(t) \ge a]$.)
14. Consider a process $X(t)$, $t \ge 0$, which is an infinite linear combination

$$X(t) = \sum_{k=-\infty}^{\infty} kN_k(t)$$

of independent homogeneous Poisson processes with intensities $\lambda_k = 1/k^2$. Show that $X(t)$ is a compound Poisson process, and find the characteristic function $\psi_{t_1, t_2}(\zeta)$ of its increments. (*Hint:* Look at the Fourier series in Appendix Three.)

15. A volume of gas contains atoms of element A and atoms of element B. Energies of these atoms are assumed to be i.i.d. random variables having the gamma density

$$f(x) = \sqrt{\frac{x}{2\pi}} e^{-x/2}, \qquad x > 0,$$

in appropriately chosen units. Collisions between A atoms and B atoms are occurring according to a homogeneous Poisson process with intensity $\lambda = 5$ collisions per millisecond. Each time atoms of A and B collide, a molecule of substance C is produced, provided the total energy of the two colliding atoms is greater than 2.5 units. Find the smallest length of time during which at least 1000 molecules are produced with probability at least 0.99. (*Hint:* Use the Gaussian approximation.)

7.6. LINEAR TRANSFORMATIONS OF A POISSON PROCESS

In this section we will study the response of a linear dynamic system to a Poisson point process applied at its input. This kind of problem is clearly very important for a large variety of applications. One of the earliest models of this nature was a model for shot noise. Shot noise is a randomly fluctuating component of a current in a circuit containing a vacuum tube whose cause can be attributed to a random stream of electrons traveling from the cathode to the anode of the tube. As the simplest example of such noise, consider a diode operating with an ideal resistance load, as in Figure 7.6.1.

Assume that the bias voltage V is large enough so that each electron emitted

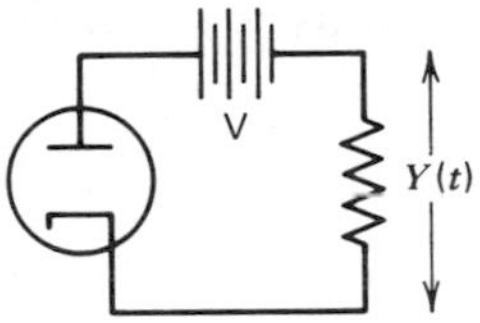

Figure 7.6.1

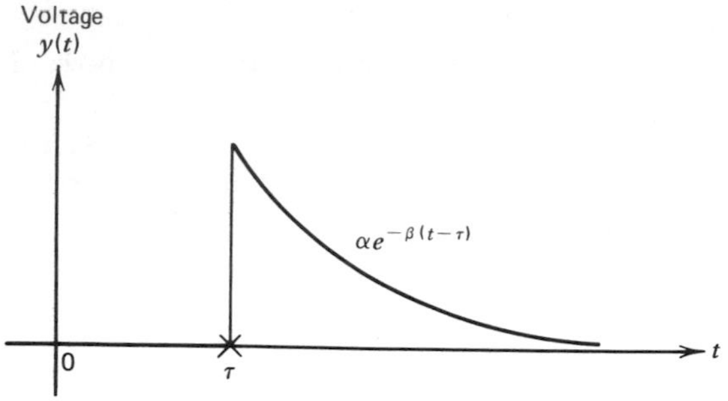

Figure 7.6.2

from the cathode is immediately pulled toward the anode. As a result, the interaction between electrons emitted from the cathode can be neglected. It can then be shown that the stream of electrons arriving at the anode can be regarded as a Poisson point process. Emission of an electron from the cathode at time τ produces a voltage impulse across the resistor, the shape of which is determined by the geometrical configuration of the diode, stray capacitances and inductances of the circuit, and a variety of other physical factors. Let us assume, just for the sake of argument, that this voltage impulse has a simple form of decaying exponential, as shown in Figure 7.6.2. If no nonlinearities are present in the circuit, the voltage $y(t)$ across the resistor at some time t will be a superposition of such pulses caused by electrons emitted at all times τ_n prior to the present time t. In other words, the resulting voltage waveform will be given by

$$y(t) = \sum_{\tau_n < t} \alpha e^{-\beta(t-\tau_n)},$$

which can be seen more clearly from Figure 7.6.3. Since the instances τ_n at which electrons are emitted are, by assumption, epochs of a Poisson process, the voltage waveform $Y(t)$ will be an ordinary stochastic process with a continuous time domain $\mathcal{C}$. Our goal in this section is to investigate the properties of stochastic processes generated in this fashion. But first we wish to put the problem into a more general framework.

Suppose that we have a linear dynamic system with input $x(t)$ and output $y(t)$ and that the system is causal and stable. Suppose further that the system can be described by its impulse response function

$$h(t, \tau),$$

so that the relation between the input and output waveforms is given by the integral transformation

$$y(t) = \int_{-\infty}^{t} h(t, \tau)x(\tau)\, d\tau. \tag{1}$$

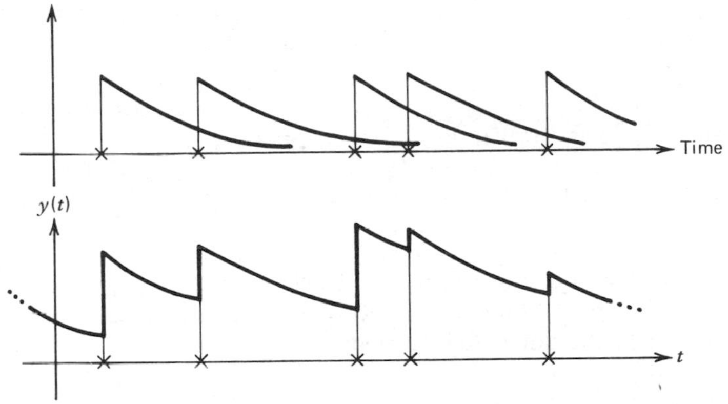

Figure 7.6.3

Let us now take for the input $x(t)$ a Poisson stream of impulses occurring at random epochs $T(n)$ and having random amplitudes $U(n)$, commencing at time $t = 0$. More precisely, let us assume that the input process is a compound Poisson process with time domain $\mathcal{T} = (0, \infty)$.

Recall that the impulse response function $h(t,\tau)$ physically represents the response $y(t)$ of the system caused by a unit impulse arriving at the input at time $\tau < t$. Therefore, the response at time t to a unit impulse (i.e., point-event) with epoch $T(n) < t$ will be $h(t, T(n))$, and if this impulse has amplitude $U(n)$, the response will be $h(t, T(n))U(n)$ by the linearity of the system. Suppose now that during the time interval $(0, t]$, we had $N(t) = m > 0$ impulses with epochs $0 < T(1) < \cdots < T(m) < t$. Then the linearity of the system implies that the output $Y(t)$ at time t will be a superposition of the m responses, that is,

$$Y(t) = \sum_{n=1}^{m} h(t, T(n))U(n)$$

Since by convention $U(0) = 0$, we can actually write the above sum by starting from $n = 0$ and thus include the case $N(t) = m = 0$. We then obtain a general expression for the output process

$$Y(t) = \sum_{n=0}^{N(t)} h(t, T(n))U(n), \qquad t \geq 0, \tag{2}$$

where $h(t, \tau)$ is the impulse-response function of the system and $T(n)$, $n = 1, 2, \ldots$, and $U(n)$, $n = 1, 2, \ldots$, are the random epoch sequence and mark sequence of a compound Poisson process. (Since $U(0) = 0$, there is no need to define $T(0)$, but we may as well also set $T(0) = 0$.)

Remark: It should be easy to verify that if the input process is a marked Poisson process with time domain $\mathcal{T} = (-\infty, \infty)$, the output process of the linear system is given by

$$Y(t) = \sum_{n=-\infty}^{N(t)} h(t, T(n))U(n), \qquad -\infty < t < \infty. \tag{3}$$

Of course, we then would no longer have $U(0) = 0$ and $T(0) = 0$. ∎

Remark: Expression (2) for the output process may also be derived directly from the integral transformation (1). If we think of the input process as a stream of impulses with epochs $T(n)$ and amplitudes $U(n)$, we can write it formally as

$$X(t) = \sum_{n=0}^{\infty} \delta(t - T(n))U(n), \qquad t \geq 0 \tag{4}$$

where δ is the Dirac delta function. Substitution into (1) immediately yields expression (2) as a result of the sifting property of the delta function. Of course, because of the presence of the delta function in (4), $X(t)$ is not really a stochastic process but rather a generalized stochastic process just like those we encountered in Chapter Four. In fact, if

$$T(n), \qquad n = \ldots, -1, 0, 1, \ldots$$

is a random epoch sequence of a homogeneous Poisson process with intensity λ defined on the time domain $\mathcal{T} = (-\infty, \infty)$, then the generalized process

$$X(t) = \sum_{n=-\infty}^{\infty} \delta(t - T(n)), \qquad -\infty < t < \infty \tag{5}$$

is frequently referred to as *white Poisson noise*. The analogy between white Gaussian noise and white Poisson noise actually goes much farther. Just as white Gaussian noise is a formal derivative of the Wiener process, so is white Poisson noise (5) a formal derivative of the homogeneous Poisson counting process

$$X(t) = \dot{N}(t), \qquad -\infty < t < \infty.$$

Similarly, it is seen that the generalized process (4) is a formal derivative of the mark accumulator process

$$X(t) = \dot{N}_U(t), \qquad t \in \mathcal{T}$$

of the corresponding compound Poisson process. This allows us to write, at least formally, the output process $Y(t)$ of the linear system in integral form

$$Y(t) = \int_{-\infty}^{t} h(t, \tau)\, dN_U(\tau), \qquad t \in \mathcal{T}. \tag{6}$$

Although in the present context the integral above is merely a symbolic representation of the random sum (2) or (3), it does behave as a genuine integral and allows us to perform many heuristic calculations. Moreover, the reader may have noticed that it is actually a

Poisson analog of the Wiener integral and can, in fact, also be defined as the MS limit of the approximating sums

$$\sum_{\tau_n<t} h(t, \tau_n)[N_U(\tau_{n+1}) - N_U(\tau_n)],$$

just like the Wiener integral.

This should be at least intuitively plausible, since for sufficiently small intervals $(\tau_n, \tau_{n+1}]$, loosely speaking, there is either no point-event with an epoch in this interval, in which case $N_U(\tau_{n+1}) - N_U(\tau_n) = 0$, or there is exactly one point-event with an epoch there, in which case the increment $N_U(\tau_{n+1}) - N_U(\tau_n)$ is equal to the value of the corresponding mark. Thus, for a sufficiently fine partition $0 < \cdots < \tau_{n-1} < \tau_n < \cdots < t$, we should have

$$\sum_{t_n<t} h(t, \tau_n)[N_U(\tau_{n+1}) - N_U(\tau_n)] \doteq \sum_{n \le N(t)} h(t, T(n))U(n)$$

as asserted. ∎

Let us now return to the expression for the output process

$$Y(t) = \sum_{n=0}^{N(t)} h(t, T(n))U(n), \qquad t \ge 0, \tag{7}$$

and let us try to obtain the probability law of this process. We begin again with the characteristic function

$$\psi_t(\zeta) = \mathrm{E}[e^{\iota\zeta Y(t)}].$$

Conditioning upon $N(t) = m$, we get from (7)

$$\psi_t(\zeta) = \sum_{m=0}^{\infty} \mathrm{E}[e^{\iota\zeta \sum_{n=0}^{m} h(t, T(n))U(n)} \mid N(t) = m]P(N(t) = m). \tag{8}$$

Now we know from Theorem 7.4.2 that for a Poisson process the conditional distribution of the epochs $T(1), \ldots, T(m)$, given that there are exactly $m > 0$ point-events occurring in the interval $(0, t]$, is the same as the distribution of m time instances chosen independently, each with density

$$f(\tau) = \frac{\lambda(\tau)}{\int_0^t \lambda(\tau)\,d\tau}, \qquad 0 < \tau \le t. \tag{9}$$

Furthermore, since the marks are also i.i.d. random variables and independent of the random epoch sequence, the conditional expectation

$$\mathrm{E}[e^{\iota\zeta \sum_{n=0}^{m} h(t, T(n))U(n)} \mid N(t) = m] = \mathrm{E}\left[\prod_{n=1}^{m} e^{\iota\zeta h(t, T(n))U(n)} \mid N(t) = m\right], \qquad m > 0, \tag{10}$$

equals the product of m expectations

$$\mathrm{E}[e^{\iota\zeta h(t, T)U}],$$

where T and U are independent random variables. The random variable T has density (9), while the random variable U has density function $f_U(u)$, the common density function of the mark sequence. Consequently, we find

$$\begin{aligned} \mathrm{E}[e^{\iota\zeta h(t,\,T)U}] &= \int_{-\infty}^{\infty}\int_{-\infty}^{\infty} e^{\iota\zeta h(t,\,\tau)u} f_U(u)\,du f(\tau)\,d\tau \\ &= \int_{-\infty}^{\infty} \mathrm{E}[e^{\iota\zeta h(t,\,\tau)U}] f(\tau)\,d\tau. \end{aligned} \tag{11}$$

But

$$\mathrm{E}[e^{\iota\zeta h(t,\,\tau)U}] = \psi_U(\zeta h(t, \tau)),$$

where ψ_U is the characteristic function of the mark distribution, and so from (11) and (9)

$$\mathrm{E}[e^{\iota\zeta h(t,\,T)U}] = \frac{\int_0^t \psi_U(\zeta h(t, \tau))\lambda(\tau)\,d\tau}{\int_0^t \lambda(\tau)\,d\tau}. \tag{12}$$

Thus the conditional expectation (10) is equal to the mth power of the right-hand side of (12). Substituting back into (8) with

$$P(N(t) = m) = e^{-\int_0^t \lambda(\tau)\,d\tau} \frac{[\int_0^t \lambda(\tau)\,d\tau]^m}{m!}, \qquad m = 0, 1, \ldots$$

we get

$$\psi_t(\zeta) = e^{-\int_0^t \lambda(\tau)\,d\tau} \sum_{m=0}^{\infty} \frac{1}{m!}\left(\int_0^t \psi_U(\zeta h(t, \tau))\lambda(\tau)\,d\tau\right)^m,$$

or recognizing the series as an expansion for the exponential function, we have

$$\psi_t(\zeta) = e^{\int_0^t [\psi_U(\zeta h(t,\,\tau)) - 1]\lambda(\tau)\,d\tau}, \qquad -\infty < \zeta < \infty, \qquad t \geq 0. \tag{13}$$

Thus, we have found the characteristic function of $Y(t)$ for any $t \geq 0$, and consequently, we have determined the first order distribution of the output process.

Notice the similarity between the characteristic function (13) and the characteristic function (7.5.26) of the mark accumulator process of a component Poisson process. In fact, if

$$h(t, \tau) = \begin{cases} 1 & \text{for} \quad t > \tau, \\ 0 & \text{for} \quad t \leq \tau, \end{cases} \tag{14}$$

then the two characteristic functions are identical. This is as it should be, for (14) is the impulse-response function of an ideal integrator, and as mentioned in the previous section, the mark accumulator process can be obtained by passing a marked point process through such an ideal integrator.

Example 7.6.1

Suppose that a homogeneous Poisson process with intensity λ represents a stream of trigger pulses arriving at the input of a linear system. The response of the system to each trigger impulse is a rectangular pulse of width $\Delta > 0$, unit amplitude, and random polarity $P(+) = P(-) = \frac{1}{2}$. The output process $Y(t)$, $t > 0$, is then a superposition of these rectangular pulses. A typical sample path $Y(t) = y(t), t > 0$, is depicted in Figure 7.6.4.

The output process $Y(t), t > 0$, can thus be considered the response of a linear time-invariant system with impulse-response function

$$h(t, \tau) = \begin{cases} 1 & \text{if} \quad 0 < t - \tau < \Delta, \\ 0 & \text{otherwise,} \end{cases}$$

to a marked Poisson process with marks $U(n)$, $n = 1, 2, \ldots$, taking on values $+1$ and -1, each with probability $\frac{1}{2}$.

The characteristic function of the mark distribution is then

$$\psi_U(\zeta) = \frac{1}{2} e^{i\zeta} + \frac{1}{2} e^{-i\zeta} = \cos \zeta, \qquad -\infty < \zeta < \infty,$$

and hence for $t > \Delta$, the integral in the exponent of (13) equals

$$\int_0^t [\psi_U(\zeta h(t, \tau)) - 1]\lambda \, d\tau = \int_{t-\Delta}^t (\cos \zeta - 1)\lambda \, d\tau = \lambda \, \Delta(\cos \zeta - 1).$$

Thus the characteristic function of $Y(t)$ for $t > \Delta$ is, from (13),

$$\psi_t(\zeta) = e^{\lambda \, \Delta(\cos \zeta - 1)}, \qquad -\infty < \zeta < \infty.$$

Note that (for $t > \Delta$) the characteristic function, and hence also the first order distributions of the process $Y(t)$, does not depend on t. A characteristic function of this form has already been encountered in Example 7.3.3, and in view of the

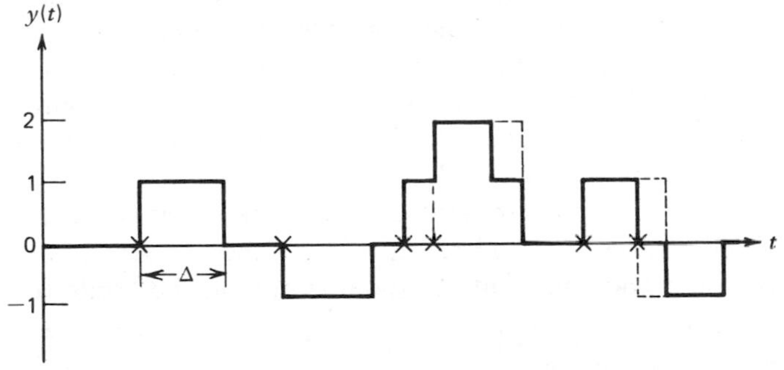

Figure 7.6.4

one-to-one correspondence between characteristic functions and probability mass distributions, it follows that for $t > \Delta$

$$P(Y(t) = k) = e^{-\lambda\Delta} I_{|k|}(\lambda\,\Delta), \qquad k = 0, \pm 1, \pm 2, \ldots,$$

where $I_k(x)$ is again the modified Bessel function of order k. Differentiating the characteristic function, we easily find that (not surprisingly)

$$\mathrm{E}[Y(t)] = 0,$$

while the variance

$$\mathrm{Var}[Y(t)] = -\frac{d^2}{d\zeta^2}\,\psi_t(\zeta)\bigg|_{\zeta=0} = \lambda\,\Delta. \qquad \blacktriangle$$

The characteristic function (8) has been derived for the case when the input compound Poisson process is defined on the time domain $\mathcal{T} = (0, \infty)$. However, the derivation is easily extended to the case of the time domain $\mathcal{T} = (-\infty, \infty)$ and to linear systems with not necessarily causal impulse-response function, that is, to output processes of the form

$$Y(t) = \sum_{n=-\infty}^{\infty} h(t, T(n))U(n), \qquad -\infty < t < \infty.$$

The characteristic function $\psi_t(\zeta) = \mathrm{E}[e^{i\zeta Y(t)}]$ is, in this case,

$$\psi_t(\zeta) = e^{\int_{-\infty}^{\infty} [\psi_U(\zeta h(t,\tau)) - 1]\lambda(\tau)\,d\tau}, \qquad -\infty < \zeta < \infty, \tag{15}$$

provided, of course, that the integral in the exponent converges.

Example 7.6.2

Consider a situation where unit electrical charges of random polarities are distributed at random along the entire real line $-\infty < t < \infty$ according to a homogeneous Poisson process with intensity λ. In this case, the random epochs $T(n)$, $n = \ldots, -1, 0, 1, \ldots$, actually represent locations of these charges (with respect to the origin $t = 0$) rather than time instances. Since in this text we have become accustomed to thinking of stochastic processes as evolving with time, we may imagine that the charges were deposited on the real line by a radioactive source travelling along the line with a constant speed of 1 meter/second and emitting the charges according to a Poisson process.

Suppose now that we wish to find for some fixed location t, say at the origin $t = 0$, the potential V of the electrostatic field resulting from these charges.

A unit charge located at τ will contribute to the potential at the origin the amount

$$v(\tau) = \frac{\pm c}{|\tau|}$$

where c is a proportionality constant ($c = 9 \times 10^9$ Newton $\times$ meter^2/coulomb^2), and thus by the superposition principle for electrostatic potential

$$V = c \sum_{n=-\infty}^{\infty} \frac{\pm 1}{|T(n)|}.$$

This expression, however, can also be written as

$$V = \sum_{n=-\infty}^{\infty} U(n)h(0, T(n))$$

where $U(n)$, $n = \ldots, -1, 0, 1, \ldots$ are symmetric Bernoulli random variables and $h(t, \tau) = c/|t - \tau|$, $-\infty < t < \infty$, $-\infty < \tau < \infty$, that is, as the output at $t = 0$ of a linear system with a noncausal but stable time-invariant impulse-response function driven by a stationary Poisson process. Since

$$\psi_U(\zeta) = \cos \zeta$$

the characteristic function of the random variable V is by (15)

$$\psi_V(\zeta) = \exp\left\{\lambda \int_{-\infty}^{\infty} \left(\cos \frac{c\zeta}{|\tau|} - 1\right) d\tau\right\}, \qquad -\infty < \zeta < \infty.$$

Now the integral in the exponent equals

$$2 \int_0^{\infty} \left(\cos \zeta \frac{c}{\tau} - 1\right) d\tau = -2c|\zeta| \int_0^{\infty} \frac{1 - \cos z}{z^2},$$

the last equality following from the substitution $\tau = |\zeta|c/z$. The last integral clearly converges, in fact,

$$\int_0^{\infty} \frac{1 - \cos z}{z^2} dz = \frac{\pi}{2},$$

and hence the characteristic function is

$$\psi_V(\zeta) = e^{-(c/\pi)\lambda|\zeta|}, \qquad -\infty < \zeta < \infty.$$

But this is the characteristic function of a Cauchy distribution, and so the probability density of the potential V equals

$$f_V(v) = \frac{1}{\pi} \frac{1}{1 + \left(\dfrac{\pi v}{c\lambda}\right)^2}, \qquad -\infty < v < \infty.$$

It should be clear from the assumed stationarity of the Poisson process that the density of the potential $V(t)$ at some arbitrary location t will be the same as at the origin $t = 0$; in fact, $V(t)$, $-\infty < t < \infty$, will be a stationary process. ▲

Of course, the characteristic function (13) specifies only the first order distributions of the output process $Y(t)$, $t \geq 0$. To specify the entire probability law of this process, we would need to know the joint characteristic functions

$$\psi_{t_1, \ldots, t_k}(\zeta_1, \ldots, \zeta_k) = \mathrm{E}[e^{\imath(\zeta_1 Y(t_1) + \cdots + \zeta_k Y(t_k))}] \tag{16}$$

for all $0 \leq t_1 < t_2 < \cdots < t_k$ and all $k = 1, 2, \ldots$. It is indeed possible to evaluate these characteristic functions by essentially the same method used to calculate the characteristic function (13). For example, for $k = 2$, we would first write the exponent as

$$\imath[\zeta_1 Y(t_1) + \zeta_2 Y(t_2)] = \imath\zeta_1 \sum_{n=0}^{N(t_1)} h(t_1, T(n))U(n) + \imath\zeta_2 \sum_{n=0}^{N(t_2)} h(t_2, T(n))U(n)$$

$$= \sum_{n=0}^{N(t_1)} \imath[\zeta_1 h(t_1, T(n)) + \zeta_2 h(t_2, T(n))]U(n) + \sum_{n=N(t_1)+1}^{N(t_2)} \imath\zeta_2 h(t_2, T(n))U(n).$$

Then we would use the independence of the two sums on the right-hand side of the above expression to decompose the expectation

$$\mathrm{E}[e^{\imath(\zeta_1 Y(t_1) + \zeta_2 Y(t_2))}]$$

into a product of two expectations, and apply procedures (8) through (13) to each of these expectations. The final result of this would be the joint characteristic function

$$\psi_{t_1, t_2}(\zeta_1, \zeta_2) = \exp\left\{\int_0^{t_1} [\psi_U(\zeta_1 h(t_1, \tau) + \zeta_2 h(t_2, \tau)) - 1]\, d\tau \right.$$

$$\left. + \int_{t_1}^{t_2} [\psi_U(\zeta_2 h(t_2, \tau)) - 1]\lambda(\tau)\, d\tau \right\}, \qquad 0 \leq t_1 \leq t_2. \tag{17}$$

Similar procedures can clearly be used for $k = 3, 4, \ldots$, but the final expression for the joint characteristic function would become progressively more unwieldy.

Although, as indicated above, we can calculate the entire family of joint characteristic functions (17) and so specify the probability law of the output process $Y(t)$, we are frequently satisfied with just the mean function $\mu_Y(t)$ and the correlation function $R_Y(t_1, t_2)$ of this process.

These can be derived in several ways; we can, of course, obtain these functions by differentiating the characteristic functions (13) and (17):

$$\mu_Y(t) = \frac{1}{\imath} \frac{d}{d\xi} \psi_t(\zeta) \Big|_{\zeta=0}, \qquad t \geq 0,$$

$$R_Y(t_1, t_2) = -\frac{\partial^2}{\partial\zeta_1\, \partial\zeta_2} \psi_{t_1, t_2}(\zeta_1, \zeta_2) \Big|_{\zeta_1=\zeta_2=0}, \qquad 0 \leq t_1 \leq t_2.$$

We can also derive them directly from expression (7) for the output process by essentially following the same line of reasoning as the one used to obtain the characteristic functions.

The simplest, albeit not quite rigorously justified derivation is to appeal directly to the integral representation

$$Y(t) = \int_{-\infty}^{t} h(t, \tau)\, dN_U(\tau), \qquad t \in \mathcal{T}. \tag{18}$$

Taking expectations on both sides of (18), interchanging it with the integral sign, and using formula (7.5.8), that is,

$$\mathrm{E}[dN_U(\tau)] = \mathrm{E}[U]\lambda(\tau)\, d\tau,$$

we have immediately

$$\mathrm{E}[Y(t)] = \mathrm{E}[U] \int_{-\infty}^{t} h(t, \tau)\lambda(\tau)\, d\tau.$$

Similarly from (7.5.10),

$$\mathrm{Cov}[dN_U(\tau_1)\, dN_U(\tau_2)] = \mathrm{E}[U^2]\, \delta(\tau_2 - \tau_1)\lambda(\tau_1)\, d\tau_1 d\tau_2$$

so we have for $t_1 \le t_2$

$$\mathrm{Cov}[Y(t_1), Y(t_2)] = \mathrm{E}[U^2] \int_{-\infty}^{t_1} h(t_1, \tau)h(t_2, \tau)\lambda(\tau)\, d\tau.$$

These results, originally derived for a stationary shot noise model by Campbell, can be summarized as a theorem.

THEOREM 7.6.1. *Let* Y(t), t ∈ $\mathcal{T}$, *be the output process of a stable and causal linear system with impulse-response function* h(t, τ), *whose input is a compound Poisson process. If the common distribution of the mark sequence is the same as that of a random variable* U *with finite variance, then the mean function of the output process is*

$$\mu_Y(t) = \mathrm{E}[U] \int_{-\infty}^{t} h(t, \tau)\lambda(\tau)\, d\tau,$$

and the correlation function of the output process equals

$$R_Y(t_1, t_2) = \mathrm{E}[U^2] \int_{-\infty}^{t_1} h(t_1, \tau)h(t_2, \tau)\lambda(\tau)\, d\tau + \mu_Y(t_1)\mu_Y(t_2),$$

$t_1 \le t_2$, $t_1 \in \mathcal{T}$, $t_2 \in \mathcal{T}$, *where* λ(t), t ∈ $\mathcal{T}$, *is the intensity function of the Poisson process.*

The significance of this theorem arises from the fact that under certain conditions, quite often satisfied in many applied problems, the *output process $Y(t)$ tends to be approximately Gaussian.* Heuristically, this can be attributed to the fact that $Y(t)$ is a superposition of a large number of random responses, and although these individual responses are, strictly speaking, not independent, their Poisson character is enough to compensate for this defect. Since the explicit expressions for joint characteristic functions of the output process $Y(t)$ are available, the exact conditions for the validity of this Gaussian approximation in each particular case can be obtained by investigating the limiting behavior of these characteristic functions.

Example 7.6.3

Consider the shot noise process as described at the beginning of this section, assuming for greater generality that the emission of an electron at some time τ results in a voltage impulse of some arbitrary shape, such as that given in Figure 7.6.5, where $g(t-\tau) = 0$ for $t \leq \tau$. Assume further that the impulse is bounded, that it has finite energy,

$$\int_0^\infty g^2(z)\, dz < \infty$$

and that the electron emissions follow a stationary Poisson process on $\mathcal{T} = (-\infty, \infty)$, that is, a steady-state condition.

Setting $U = 1$ and $h(t, \tau) = g(t - \tau)$, it then follows from Theorem 7.6.1 that the voltage process $Y(t)$, $-\infty < t < \infty$, has constant mean and variance

$$\mathrm{E}[Y(t)] = \lambda \int_0^\infty g(z)\, dz, \qquad \mathrm{Var}[Y(t)] = \lambda \int_0^\infty g^2(z)\, dz.$$

This is actually the original version of Campbell's result giving the average shot noise voltage (or current) and its mean square variation in terms of the response

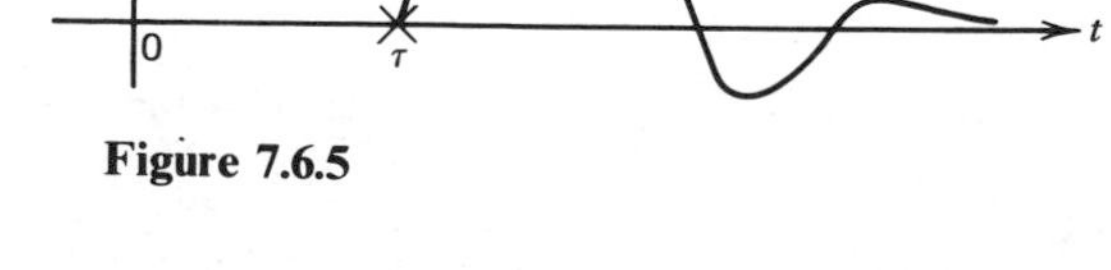

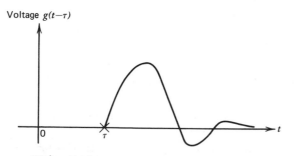

Figure 7.6.5

to a single electron emitted from the cathode. In particular, if the mean is interpreted as a dc signal and the standard deviation $\sqrt{\mathrm{Var}[Y(t)]}$ as a root mean square noise power, then the *signal-to-noise ratio* of the shot noise equals

$$\sqrt{\lambda}\,\frac{\int_0^\infty g(\tau)\,d\tau}{\sqrt{\int_0^\infty g^2(\tau)\,d\tau}},$$

and thus, for a given shape of the response pulse, increases as the square root of the intensity of the Poisson process. ▲

Example 7.6.4

From (13) the characteristic function $\psi_t(\zeta)$ of the shot noise process of Example 7.6.3 equals

$$\psi_t(\zeta) = e^{\lambda \int_0^\infty (e^{\imath\zeta g(\tau)} - 1)\,d\tau}$$

Represent the mean and the standard deviation of $Y(t)$ by

$$\mu = \lambda \int_0^\infty g(\tau)\,d\tau, \qquad \sigma = \sqrt{\lambda}\left(\int_0^\infty g^2(\tau)\,d\tau\right)^{1/2},$$

and define the standardized random variable

$$Z(t) = \frac{Y(t) - \mu}{\sigma}.$$

The characteristic function $\psi_{Z(t)}(\zeta)$ of $Z(t)$ is

$$\psi_{Z(t)}(\zeta) = e^{-\imath\zeta(\mu/\sigma)}\psi_t\left(\frac{\zeta}{\sigma}\right),$$

and its logarithm for ζ near zero

$$\ln \psi_{Z(t)}(\zeta) = -\imath\zeta\frac{\mu}{\sigma} + \lambda \int_0^\infty (e^{\imath(\zeta/\sigma)g(\tau)} - 1)\,d\tau.$$

From Taylor's formula for the complex exponential, it follows that

$$e^{\imath(\zeta/\sigma)g(\tau)} - 1 = \imath\frac{\zeta}{\sigma}g(\tau) - \frac{1}{2}\left(\frac{\zeta}{\sigma}g(\tau)\right)^2 + \frac{1}{3!}\left(\frac{\zeta}{\sigma}g(\tau)\right)^3 z,$$

where z is a complex number such that $|z| < 1$. Integrating with respect to τ from 0 to ∞, it follows that

$$\ln \psi_{Z(t)}(\zeta) = -\frac{1}{2}\zeta^2 + \frac{\lambda}{3!}\left(\frac{\zeta}{\sigma}\right)^3 \int_0^\infty g^3(\tau)z\,d\tau.$$

Hence the absolute value of the difference

$$\left|\ln \psi_{Z(t)}(\zeta) - \left(-\frac{1}{2}\zeta^2\right)\right| \le \frac{|\zeta|^3}{3!}\frac{\lambda}{\sigma^3}\int_0^\infty g^3(\tau)\,d\tau$$

which implies that if

$$\frac{\lambda}{\sigma^3}\int_0^\infty g^3(\tau)\,d\tau = \frac{1}{\sqrt{\lambda}}\frac{\int_0^\infty g^3(\tau)\,d\tau}{(\int_0^\infty g^2(\tau)\,d\tau)^{3/2}}$$

approaches zero the characteristic function $\Psi_{Z(t)}(\zeta)$ approaches $e^{-(1/2)\zeta^2}$, the characteristic function of the standard Gaussian distribution. Therefore, we can conclude that if the intensity λ is much larger than the quantity

$$\gamma = \frac{(\int_0^\infty g^3(\tau)\,d\tau)^2}{(\int_0^\infty g^2(\tau)\,d\tau)^3}$$

the first-order distributions of the shot noise process are approximately Gaussian with mean μ and variance σ^2 as given above.

It can be shown that the condition

$$\frac{\gamma}{\lambda} \to 0$$

actually guarantees that the same will also be true for higher-order distributions, that is, that the entire probability law of the shot noise process $Y(t)$, $-\infty < t < \infty$, approaches a Gaussian probability law. ▲

NONLINEAR RESPONSE TO MARK VARIABLES. So far, we have been thinking of the mark variable $U(n)$ as the amplitude of the impulse occurring at the epoch $T(n)$. Linearity of the system then required that the response to such an impulse must be

$$h(t, T(n))U(n).$$

However, when discussing the notion of a marked point process, we have mentioned that the marks may generally represent any additional numerical characteristic of point-events. For instance, if the point-events are interpreted as arrivals of particles at a counter, the marks may correspond to the energies of these particles. It is then possible that the response of the counter depends on the energy in some nonlinear fashion. For example, the counter may respond only if the energy of the incoming particle exceeds a certain threshold, or more generally, the entire shape of the response may depend on the energy of the particle.

To incorporate this feature into our model, we assume that the response $y(t)$

of the system at time t to a point-event with epoch $T(n) = \tau$ and mark $U(n) = u$ is a function of all three of these variables:

$$y(t) = h(t, \tau, u).$$

We still assume, however, that the system is *linear* in the sense that the response $y(t)$ to point-events with epochs $T(n) = \tau_n$ and $T(m) = \tau_m$ and corresponding marks $U(n) = u_n$ and $U(m) = u_m$, $n \neq m$, is a *superposition* of the individual responses:

$$y(t) = h(t, \tau_n, u_n) + h(t, \tau_m, u_m).$$

The stability condition required here then takes the form

$$\int_{-\infty}^{\infty} \mathrm{E}[|h(t, \tau, U)|^2]\, d\tau < \infty \quad \text{for all } t,$$

where, again, U is a random variable leaving the common distribution of the mark sequence.

We will also assume that

$$h(t, \tau, 0) = 0,$$

that is, that a mark with value zero elicits no response.

If the input to the system is a compound Poisson process with random epochs $T(n)$ and marks $U(n)$, then the superposition principle implies that the output process $Y(t)$ is given by the expression

$$Y(t) = \sum_{n}^{\infty} h(t, T(n), U(n)), \qquad t \in \mathcal{T}, \tag{19}$$

where $N(t)$ is again the associated counting process, and the lower limit of the sum depends on the time domain, that is, $n = 0$ for $\mathcal{T} = [0, \infty)$ and $n = -\infty$ for $\mathcal{T} = (-\infty, \infty)$.

We can now proceed to derive various properties of the output process. Since the derivation requires only minor modifications of what we have already done, we leave the details to the reader and only state the results.

THEOREM 7.6.2. *Let* $\mathrm{Y(t)}$, $\mathrm{t} \in \mathcal{T}$, *be the output process defined by* (19). *Then the mean function and the correlation function of this process are*

$$\mu_Y(t) = \int_{-\infty}^{\infty} \mathrm{E}[h(t, \tau, U)]\lambda(\tau)\, d\tau, \qquad t \in \mathcal{T}, \tag{20}$$

and

$$R_Y(t_1, t_2) = \int_{-\infty}^{\infty} \mathrm{E}[h(t_1, \tau, U)h(t_2, \tau, U)]\lambda(\tau)\, d\tau + \mu_Y(t_1)\mu_Y(t_2), \tag{21}$$

$t_1 \in \mathcal{T}$, $t_2 \in \mathcal{T}$, *where* $\lambda(t)$, $t \in \mathcal{T}$, *is the intensity function, and* U *is a random variable having the common distribution of the mark sequence.*

The first order characteristic function

$$\psi_t(\zeta) = E[e^{\iota\zeta Y(t)}], \qquad -\infty < \zeta < \infty, \qquad t \in \mathcal{T}$$

is, in this case,

$$\psi_t(\zeta) = \exp\left\{\int_{-\infty}^{\infty} E[e^{\iota\zeta h(t,\tau,U)} - 1]\,\lambda(\tau)\,d\tau\right\}. \tag{22}$$

The second or, more generally, kth order joint characteristic functions can also be obtained, and thus the entire probability law of the process (19) can actually be specified. The only thing that does not extend to this case without a major modification is the integral representation (6) of the output process.

Example 7.6.5

Suppose that each point-event of a homogeneous Poisson process on $\mathcal{T} = (0, \infty)$ produces a secondary point-event in such a manner that if the nth primary point-event occurs at an epoch $T(n)$, then the corresponding secondary point-event occurs at the time $T(n) + U(n)$, where $U(n)$ is the mark of the nth primary point-event. Thus we can regard the primary process as a compound Poisson process with intensity λ and mark distribution function $F_U(u)$ ($F_U(u) = 0$ for $u \leq 0$). Letting $N(t)$, $t > 0$, denote the counting process of the point process of primary and secondary point-events combined (see Fig. 7.6.6), we wish to determine the distribution of $N(t)$.

Although the problem may be handled by the same technique as was used in Example 7.4.3, the results of the present section provide an alternate and quicker method.

Recall that the counting process of an arbitrary point process can be constructed as a superposition of unit steps, each commencing at the epochs of the point-events. Similarly, our combined counting process $N(t)$, $t > 0$, can be obtained as a superposition of two-step functions, the first step commencing at each

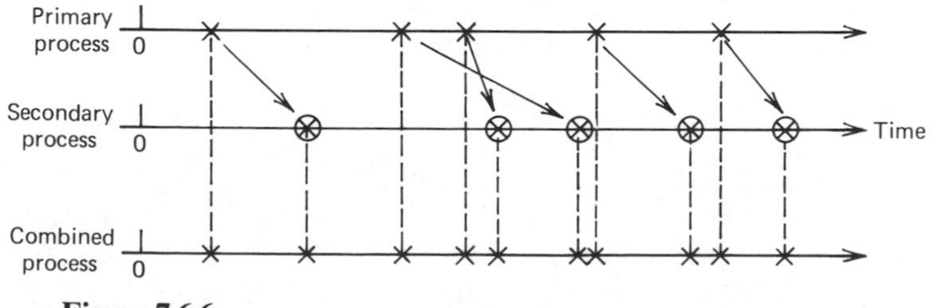

Figure 7.6.6

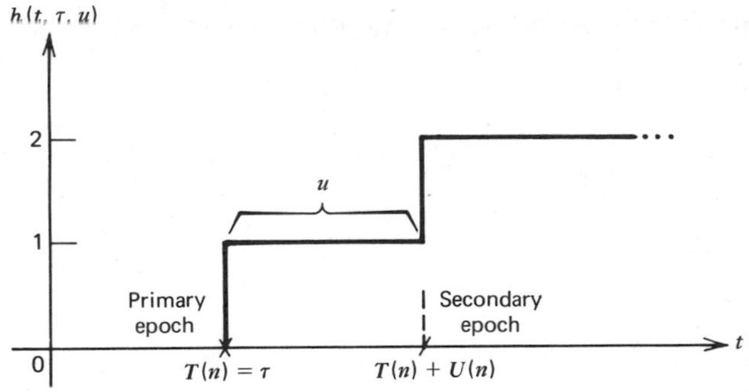

Figure 7.6.7

epoch of the primary process and the second step at the epoch of the corresponding secondary epoch, as depicted in Figure 7.6.7. In other words, if we define the function

$$h(t, \tau, u) = \begin{cases} 1 & \text{if} \quad 0 \le t - \tau < u, \\ 2 & \text{if} \quad 0 < u \le t - \tau, \\ 0 & \text{otherwise,} \end{cases}$$

we can write the counting process as in (19)

$$N(t) = \sum_{n=0}^{N(t)} h(t, T(n), U(n)), \qquad t > 0.$$

For $t \ge \tau$, we have

$$P(h(t, \tau, U) = 1) = P(U > t - \tau) = 1 - F_U(t - \tau),$$
$$P(h(t, \tau, U) = 2) = P(U \le t - \tau) = F_U(t - \tau),$$

so we can easily evaluate the mean and correlation functions of the process $N(t)$, $t > 0$, from (20) and (21).

We leave this to the reader and derive the characteristic function instead. Using the above probabilities

$$\mathrm{E}[e^{\iota\zeta h(t, \tau, U)} - 1] = e^{\iota\zeta}(1 - F_U(t - \tau)) + e^{\iota 2\zeta} F_U(t - \tau) - 1,$$

and hence the integral

$$\int_0^t \mathrm{E}[e^{\iota\zeta h(t, \tau, U)} - 1]\lambda \, d\tau = \lambda e^{\iota\zeta} \int_0^t (1 - F_U(u)) \, du + \lambda e^{\iota 2\zeta} \int_0^t F_U(u) \, du - \lambda t$$

$$= (e^{\iota\zeta} - 1)\lambda \int_0^t (1 - F_U(u)) \, du + (e^{\iota 2\zeta} - 1)\lambda \int_0^t F_U(u) \, du.$$

Thus we can write the characteristic function (22) as a product

$$\psi_t(\zeta) = \psi_1(\zeta)\psi_2(2\zeta), \qquad -\infty < \zeta < \infty,$$

where

$$\psi_1(\zeta) = e^{(e^{i\zeta} - 1)\lambda \int_0^t (1 - F_U(u))\, du},$$

and

$$\psi_2(\zeta) = e^{(e^{i\zeta} - 1)\lambda \int_0^t F_U(u)\, du},$$

are both characteristic functions of a Poisson distribution. Consequently $N(t)$ can, for each fixed $t > 0$, be written as a linear combination of two independent Poisson random variables

$$N(t) = N_1(t) + 2N_2(t) \tag{23}$$

with

$$\mathrm{E}[N_1(t)] = \lambda \int_0^t (1 - F_U(u))\, du,$$

and

$$\mathrm{E}[N_2(t)] = \lambda \int_0^t F_U(u)\, du.$$

Note that $N_1(t)$, $t > 0$, is *not* the counting process of the primary Poisson point process. Nor can we hastily conclude that $N(t)$ might be a compound Poisson process, since we do not know whether the processes $N_1(t)$ and $N_2(t)$, $t > 0$, evolve without aftereffects or whether they are mutually interdependent. Nevertheless, the result (23) is quite interesting and useful, for instance for obtaining the mean, variance, or even the entire distribution of $N(t)$.

Actually, a careful examination of equation (23) reveals that $N_1(t)$ is, in fact, the number of primary point-events that by time t have not yet produced secondary point-events. $N_2(t)$ is then the number of primary point-events that did produce secondary point-events by this time, and hence each of them contributed two point-events to the combined process. ▲

EXERCISES

Exercise 7.6

1. Find the mean function and the correlation function of the shot noise $Y(t)$, $-\infty < t < \infty$, generated by the diode circuit as pictured in Figure 7.6.3. Assume that the epochs of electron emissions form a stationary Poisson process on $\mathcal{T} = (-\infty, \infty)$. Under what conditions is $Y(t)$ approximately Gaussian?

2. Under certain assumptions, it can be shown that an idealized diode acts as a linear system with impulse-response function

$$h(t, \tau) = \begin{cases} \dfrac{c}{\Delta}(t - \tau) & \text{if } \tau \le t \le \tau + \Delta, \\ 0 & \text{otherwise}, \end{cases}$$

where c and Δ are positive constants. Assuming that the input to this system is a homogeneous Poisson process with intensity λ, find the mean and variance functions of the output process $Y(t)$, $-\infty < t < \infty$.

3. Repeat Exercise 7.6.2 under the assumptions that the input is an inhomogeneous Poisson process with intensity function $\lambda(t) = \lambda_0(1 + \gamma \cos \omega t)$, $-\infty < t < \infty$, with $\lambda_0 > 0$, $\omega > 0$, and $0 < \gamma < 1$ as given constants. (This may correspond to a sinusoidal bias voltage applied to the diode.) Assume $\omega = 2\pi\Delta^{-1}$.

4. Under the assumptions of Example 7.6.2 the electrostatic field strength at the origin is given by the formula

$$Y = c \sum_{n=-\infty}^{\infty} \frac{\text{sign } T(n)}{T^2(n)} U(n).$$

Show that the random variable Y has a symmetrized inverse Gaussian distribution.

5. Suppose that a stochastic process $X(t), t \geq 0, X(0) = 0$, with continuous sample paths is sampled at random time instances corresponding to epochs of an inhomogeneous Poisson process with intensity function $\lambda(t)$. The result is then a marked point process with marks $U(n) = X(T(n)), n = 0, 1, \ldots$. Show that its mark accumulator process can be written as the integral

$$\int_0^t X(\tau)\, dN(\tau), \qquad t \geq 0$$

and evaluate its first two moment functions in terms of the mean function $\mu_X(t)$ and correlation function of the process $X(t)$.

6. For the process $Y(t)$, $t > 0$, defined in Example 7.6.1, calculate the second order characteristic function $\psi_{t_1, t_2}(\zeta_1, \zeta_2)$. Use it to obtain the correlation function $R_Y(t_1, t_2)$, and compare your result with the one obtained directly from Theorem 7.6.1.

7. Arrivals of photons at a detector are assumed to form a compound Poisson process with intensity $\lambda > 0$ and with mark density function

$$f_U(u) = \begin{cases} \frac{1}{2}u^{-3/2} & \text{if} \quad u \geq 1, \\ 0 & \text{if} \quad u < 1. \end{cases}$$

The marks are supposed to represent the energies of the photons. The arrival of a photon at the detector at time $T(n) = \tau$ produces a voltage impulse of the shape shown in Figure 7.6.E.1. The constant c is such that the area of the impulse equals the mark value $U(n) = u$. Find the mean function and the correlation function of the process $Y(t)$, $t > 0$, representing the voltage at the output of the detector. (*Hint:* Use the substitution $v = u^{-1/2}$.)

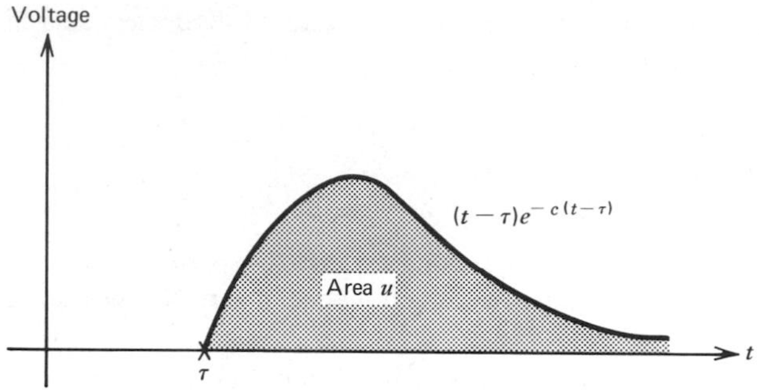

Figure 7.6.E.1

8. Find the mean function and the correlation function of the counting process $N(t)$ of Example 7.6.5.
9. Messages of random length are arriving at a central processing unit according to a stationary compound Poisson process with marks representing lengths of messages. These lengths are assumed exponentially distributed with parameter α. Call $Y(t)$ the number of messages processed at the time t. Assuming that the processor can handle a potentially unlimited number of messages, find the characteristic function $\psi_t(\zeta)$ of the process $Y(t)$, $-\infty < t < \infty$. What is the first order distribution of $Y(t)$? (*Hint:* Set $h(t, \tau, u) = 1$ if $0 < t - \tau < u$ and zero otherwise.)
10. If

$$Y(t) = \int_{-\infty}^{\infty} h(t, \tau)\, dN_U(t)$$

and $\varphi_t(\zeta) = \ln \psi_t(\zeta)$ is the logarithmic characteristic function of the random variable $Y(t)$, show that for all $n = 1, 2, \ldots$

$$(-\imath)^n \frac{d^n}{d\zeta^n} \varphi_t(\zeta)\bigg|_{\zeta=0} = \mathrm{E}[U^n] \int_{-\infty}^{\infty} h^n(t, \tau)\lambda(\tau)\, d\tau$$

provided $\mathrm{E}[U^n]$ and the integral exist.

11. Suppose a stationary Poisson process on $\mathcal{T} = (0, \infty)$ is used to manufacture a new point process by splitting each of the original point events into two new point-events displaced symmetrically by the random amount equal to the mark value of the original point events. Thus, each epoch $T(n)$ of the original process is replaced by two epochs $T(n) - U(n)$, $T(n) + U(n)$ in the new process. (See Fig. 7.6.E.2.) Follow the method of Example 7.6.5 to find the characteristic function of the new counting process $N(t)$, $t \geq 0$. Also evaluate the intensity function of this process.

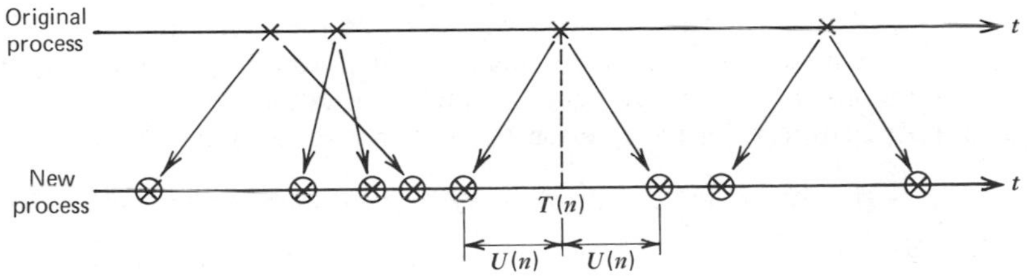

Figure 7.6.E.2

12. Suppose that at random time instances $T(n)$, bursts of radiation occur in space at random distances $U(n)$ from the origin of the coordinate system. Each such burst gives rise to an exponentially damped electromagnetic wave contributing to the electrostatic potential at the origin at time t the amount $h(t, T(n), U(n))$ where the function

$$h(t, \tau, u) = \begin{cases} \dfrac{a}{u} e^{-\beta(t-\tau-u/c)} \sin \omega\left(t - \tau - \dfrac{u}{c}\right) & \text{if} \quad t - \tau - \dfrac{u}{c} > 0, \\ 0 & \text{otherwise,} \end{cases}$$

with a, β, ω, and c being positive constants. Assuming that $T(n)$ and $U(n)$, $n = \ldots, -1, 0, 1, \ldots$, are, respectively, random epochs and marks of a stationary compound Poisson process with intensity λ and Erlang $r = 2$, $\alpha > 0$ mark distribution, find the expected potential at the origin of the coordinate system.

13. Show that the counting process $N(t)$ of an inhomogeneous Poisson process with intensity function $\lambda(t)$, $-\infty < t < \infty$, can be represented formally as an integral

$$N(t) = \int_0^{\infty} h(t, \tau)\dot{N}_0(\tau)\, d\tau,$$

where $\dot{N}_0(t)$ is the standard white Poisson noise. Compare this with the representation of an inhomogeneous Wiener process discussed in Section 4.4. How should the function $h(t, \tau)$ be chosen? (*Hint:* Recall the time transformation (7.3.13).)

14. Let $Y(t)$, $t > 0$, be the process obtained by passing a compound Poisson process through an RC filter with impulse-response function

$$h(t, \tau) = \begin{cases} 1 - e^{-\alpha(t-\tau)} & \text{if} \quad 0 < \tau < t, \\ 0 & \text{otherwise,} \end{cases}$$

where $\alpha > 0$. Assume the Poisson process is homogeneous with intensity λ and its marks are nonnegative random variables with mean μ_U and variance σ_U^2. Note that all sample paths of the process $Y(t)$ increase to infinity, so that we can define a random time T_a when the process $Y(t)$ first crosses a level $a > 0$. For large a and λ, use the Gaussian approximation to obtain an approximate expression for the density $f_{T_a}(t)$. (*Hint:* $T_a \leq t$ if and only if $Y(t) \geq a$.)

15. Consider a very long copper wire that originally was perfectly homogeneous with resistance $\rho = 10$ milliohms/centimeter. After a period of storage, tiny cracks appeared in some locations $T(n)$ along the wire. A crack at location $T(n) = \tau$ changes the resistance per centimeter $\rho(t)$ at location t, $-\infty < t < \infty$ according to the function

$$\rho(t) = 10^{-2}(U(n)(1 + U^2(n)(t - \tau)^2)^{-1} + 1) \text{ ohms/centimeter}$$

with $U(n)$ being a random variable uniformly distributed over the interval (0, 1). The effects of each such crack on the overall resistance are assumed additive. Treating $T(n)$ and $U(n)$ as epochs and marks of a compound Poisson process, calculate the mean and the variance of the new resistance per centimeter of the wire if, on the average, there are about ten cracks per meter.

7.7. PAST-DEPENDENT EVOLUTION OF A POINT PROCESS

Although an evolution without aftereffects is an adequate assumption for modeling an impulsive random noise, there are other random phenomena of an impulsive character where such an assumption would not be realistic. For instance, the point-events may represent instants at which errors are occurring during the execution of some lengthy calculation on a computer. It is then likely that an error occurring after a period of flawless calculation may be the cause of several further errors during the subsequent period. Another example is provided by considering atmospheric disturbance resulting from lightning discharges or the time stream of occurrences of earthquakes in some region. Since a lightning discharge is typically followed by several secondary discharges and an earthquake by a series of aftershocks, such a stream of random point-events does not evolve without aftereffects.

A past-dependent evolution of a point process is also obtained if we consider a Poisson process whose intensity function $\lambda(t)$ is the sample path of another random process. Such a case frequently arises in optical communication systems when the point-events represent emissions of photons from the optical transmitter and the rate of their emission is dependent upon the signal to be transmitted. Since the signal is assumed to carry some information, then unlike a noise process, its values usually exhibit a high degree of correlation over time. This correlation, or dependence, is then reflected in the past-dependent evolution of the resulting stream of photons.

Finally, if we consider the more general case of marked point processes, the past-dependent evolution of such a process may also be caused by the possible dependence between marks or, even more generally, by a dependence between the mark sequence and the sequence of random epochs of the process.

Clearly, there is a tremendous variety of ways in which past-dependent evolution may be introduced into the specification of a point process. Each usually requires its own specific treatment, and a detailed analysis or even a classification of various types of past-dependent point processes is beyond the scope of this book. However, we will discuss some rather general features of past-dependent point processes and briefly look into some special cases.

Our point of departure is Theorem 7.3.2, which says, in effect, that a point process with an associated counting process

$$N(t), \qquad t \in \mathscr{T}$$

evolves without aftereffects *if and only if*

$$\mathrm{E}[N(t_2) - N(t_1) \mid N(t), t \le t_1] = \mathrm{E}[N(t_2) - N(t_1)] \tag{1}$$

for all $t_1 < t_2$. Now the conditional expectation on the left-hand side of (1) being a functional of the past $N(t)$, $t \le t_1$, is, in general, a random variable or, when regarded as a function of the time variable t_1 with $t_2 = t_1 + \tau$ and τ fixed, is actually a stochastic process.

Let us assume for the sake of simplicity that the time domain of the point process is $\mathscr{T} = (0, \infty)$. Let us also assume that the stochastic process mentioned above is "sufficiently smooth" so that we can write it as an integral

$$\mathrm{E}[N(t_2) - N(t_1) \mid N(t), 0 < t \le t_1] = \int_{t_1}^{t_2} \Lambda(\tau)\, d\tau \tag{2}$$

of another stochastic process

$$\Lambda(t), \qquad t > 0.$$

Setting $t_1 = t$ and $t_2 = t + dt$, with $dt > 0$ an infinitesimal time increment, we can then rewrite (2) in the more suggestive differential notation

$$\mathrm{E}[dN(t) \mid N(t'), 0 < t' \le t] \doteq \Lambda(t)\, dt, \qquad t \ge 0, \tag{3}$$

where, as usual, $dN(t) = N(t + dt) - N(t)$ is an infinitesimal forward increment of the counting process.

Taking expectations on both sides of (3), we get

$$\mathrm{E}[dN(t)] = \mathrm{E}[\Lambda(t)]\, dt. \tag{4}$$

We can see that the mean function of the process $\Lambda(t)$, $t > 0$,

$$\lambda(t) = \mathrm{E}[\Lambda(t)], \qquad t > 0 \tag{5}$$

is actually the intensity function of the point process. We can thus restate Theorem 7.3.2 by saying that a point process with counting process

$$N(t), \qquad t > 0$$

evolves without aftereffects if and only if

$$\Lambda(t) = \lambda(t) \qquad \text{for all } t > 0,$$

that is, if and only if the process $\Lambda(t)$, $t > 0$, is, in fact, a deterministic function of time, the intensity function of the point process. For this reason, it is only natural to call the process $\Lambda(t)$, $t \geq 0$, defined by equation (3), the *intensity process* of the underlying general point process, that is, a point process that need not evolve without aftereffects.

From definition (3) and the fact that for a counting process we always have $dN(t) \geq 0$, it follows that

$$\Lambda(t) \geq 0 \qquad \text{for all } t > 0.$$

Because of our earlier assumption, the integral

$$\int_0^t \Lambda(\tau)\, d\tau$$

should also exist for all $t > 0$, and so we shall assume that the sample paths of an intensity process are at least piecewise continuous.

We want to emphasize that since the intensity process $\Lambda(t)$, $t > 0$, is for each time t defined by means of the conditional expectation

$$\mathrm{E}[dN(t) \mid N(t'),\, 0 < t' \leq t],$$

it is a functional of the past

$$\Lambda(t) = g_t(N(t'),\, 0 < t' \leq t). \tag{6}$$

However, the past $N(t')$, $0 < t' \leq t$, is, for each fixed $t > 0$, uniquely determined by the number $N(t) = n$ of point-events that occurred during the time interval $(0, t]$ and by their respective epochs

$$0 < T(1) = t_1 < T(2) = t_2 < \cdots < T(n) = t_n \leq t.$$

Thus the functional g_t in (6) can be defined by specifying for each integer $n = 0$, $1, \ldots$ a function of $n + 1$ real variables

$$g(t, n; t_1, \ldots, t_n), \tag{7}$$

where $n = 0, 1, \ldots$ is the number of point-events in $(0, t]$, $t > 0$ is the current time, and $0 < t_1 < \cdots < t_n \leq t$ are the times of the n random epochs. (Of course, for $n = 0$, g is just a function of the current time t.)

The intensity process can thus be written more explicitly as

$$\Lambda(t) = g(t, N(t); T(1), \ldots, T(N(t))), \qquad t \geq 0. \tag{8}$$

We make use of this explicit expression later in this section.

From definition (3) of the intensity process, it is clear that its probability law must be determined by the probability law of the original point process. But the converse of this statement is also true, that is, the probability law of the intensity process determines the probability law of the point process.

To see why this must be so, we again invoke the general idea of partitioning the time domain $\mathcal{T}$ of the point process into time intervals $(t_k, t_k + dt]$ so small that essentially one point-event, at most, can occur in each of these intervals. That is, we again approximate the occurrence or nonoccurrence of a point-event in such an interval by a Bernoulli trial. For a past-dependent point process, however, the occurrence or nonoccurrence of a point-event in a particular interval $(t_k, t_k + dt]$ is no longer independent of the occurrences of point-events in the preceding time intervals. It follows that our approximating Bernoulli trials are now *dependent*. To describe the evolution of the point process, we therefore need the conditional probability

$$P\,(\text{occurrence of a point-event in } (t_k, t_k + dt] \,|\, \text{evolution of the point process up to the time } t_k), \tag{9}$$

for each of these small time intervals. The occurrence of a point-event in $(t_k, t_k + dt]$ means that the counting process $N(t)$ will have a unit jump in that interval

$$dN(t_k) = 1,$$

and if no more than one point-event can occur in such a small interval (i.e., either $dN(t_k) = 1$ or $dN(t_k) = 0$), then the conditional probability (9) must be equal to the conditional expectation

$$\mathrm{E}[dN(t_k) \,|\, N(t'), 0 < t' \leq t_k].$$

Since this is approximately $\Lambda(t_k)\,dt$, we should have for infinitesimal $dt > 0$ the relation

$$P(dN(t) = 1 \,|\, N(t'), 0 < t' \leq t) \doteq \Lambda(t)\,dt, \tag{10}$$

and consequently also

$$P(dN(t) = 0 \,|\, N(t'), 0 < t' \leq t) \doteq 1 - \Lambda(t)\,dt, \tag{11}$$

for every $t > 0$. Thus the knowledge of the intensity $\Lambda(t)$ as a functional of the past $N(t')$, $0 < t' \leq t$, would allow us to determine the conditional probabilities (9) and hence the entire probability law of the point process.

Of course, the above reasoning is grossly heuristic. It can be shown, however, that under rather technical conditions (needed to guarantee the existence of an intensity process) differential relations (10) and (11) are indeed valid.

Example 7.7.1

Suppose that a point process on the time domain $\mathcal{T} = (0, \infty)$ evolves as a homogeneous Poisson process with intensity λ as long as the waiting times $X(1)$, $X(2)$, ... between consecutive random epochs are greater than some constant $\Delta > 0$. Upon the first occurrence of a point-event whose epoch $T(n)$ follows the

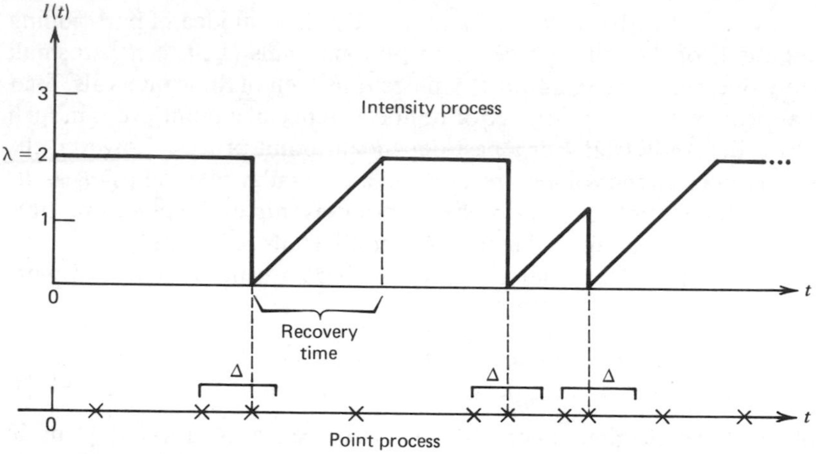

Figure 7.7.1

epoch $T(n-1)$ of the preceding point-event in a time no larger than Δ, that is, as soon as $X(n) = T(n) - T(n-1) \le \Delta$, the intensity of the process drops to zero. From there on, it begins to increase linearly with rate α until it either reaches its original value λ or until another two point-events occur with epochs not farther apart than Δ. That is, the intensity drops to zero at each epoch whose waiting time is no larger than Δ and then starts to increase again. (Such a point process may serve as a rough model for a physical process when rapid consecutive emissions of particles momentarily deplete the source, each depletion then being followed by a linear recovery.)

As the intensity of such a point process depends at each time t on the spacing of the preceding random epochs, the intensity is clearly a stochastic process, the intensity process $\Lambda(t)$, $t > 0$. An example of a particular realization of the point process, together with the corresponding sample path $l(t) = \Lambda(t)$ of the intensity process, is depicted in Figure 7.7.1 with $\lambda = 2$, $\Delta = 1$, and a recovery rate $\alpha = 1$.

In symbols, the intensity process $\Lambda(t)$, $t > 0$, defined as the conditional probability (10)

$$P(dN(t) = 1 \mid N(t'), 0 < t' \le t) \doteq \Lambda(t)\, dt,$$

can be defined as follows:

1. If $N(t) < 2$, then $\Lambda(t) = \lambda$, since if no more than one point-event has occurred during $(0, t]$ the process at t is still the original Poisson process.
2. If $N(t) = n \ge 2$, let m be the largest index $2 \le m \le n$ such that

$$X(m-1) = T(m) - T(m-1) \le \Delta.$$

If no such index exists, that is, if $X(2) > \Delta$, $X(3) > \Delta, \ldots, X(n-1) > \Delta$, then, as before, the process at t is still Poisson, and so $\Lambda(t) = \lambda$. If there is such an index, then

$$\Lambda(t) = \begin{cases} \alpha[t - T(m)] & \text{if} \quad t - T(m) \leq \dfrac{\lambda}{\alpha}, \\ \lambda & \text{if} \quad t - T(m) > \dfrac{\lambda}{\alpha}, \end{cases}$$

where λ/α is the recovery time, that is, the time needed for the intensity to reach the level λ from a previous drop to zero provided no further drop to zero occurred during this time.

Admittedly, the definition sounds rather complicated. But it shows that at any time $t > 0$ the value $l(t) = \Lambda(t)$ of the intensity process is indeed determined in a precise fashion by the epochs of the joint process during the interval $(0, t]$.

In fact, the above description actually defines the family of functions $g(t, n; t_1, \ldots, t_n)$. For instance with $\Delta = 1$, $\lambda = 2$, and $\alpha = 1$, we have

$$g(6, 5; 1.5, 2.1, 4, 4.5, 5.7) = 1.5$$

since, in this case, $n = 5 > 2$, $m = 4$, $t_m = 4.5$, and $t - t_m = 1.5 < \frac{2}{1}$, so that $l(t) = \alpha(t - t_m) = 1.5$. As the value of this functional can clearly be obtained for any values of its variables, the functional is indeed completely specified. Thus the probability law of the intensity process is also specified, although expressing it in some manageable form would not be an easy matter. ▲

We can actually go much farther in our investigation of a past-dependent point process. First, let us take

$$0 < t_0 \leq t \qquad \text{and} \qquad dt > 0$$

and consider the conditional probability

$$P(N(t + dt) - N(t_0) = 0 \mid N(t'), 0 < t' \leq t_0),$$

where $N(t)$, $t \geq 0$, is the counting process associated with our past-dependent point process on $\mathcal{T} = (0, \infty)$. Using the rules for conditional probabilities, we can write

$$\begin{aligned} P(N(t + dt) - N(t_0) = 0 \mid N(t'), 0 < t' \leq t_0) \\ &= P(dN(t) = 0, N(t) - N(t_0) = 0 \mid N(t'), 0 < t' \leq t_0) \\ &= P(dN(t) = 0 \mid N(t) - N(t_0) = 0; N(t'), 0 < t' \leq t_0) \\ &\quad \times P(N(t) - N(t_0) = 0 \mid N(t'), 0 < t' \leq t_0). \quad (12) \end{aligned}$$

But the conditional probability

$$P(dN(t) = 0 \mid N(t) - N(t_0) = 0;\ N(t'),\ 0 < t' \leq t_0) \tag{13}$$

is, in fact, conditioned upon the past extending up to time t, since $N(t) - N(t_0) = 0$ clearly implies that $N(t') = N(t_0)$ for all $t_0 < t' \leq t$. Thus according to (11), we can write it as

$$1 - \Lambda(t)\, dt,$$

where the intensity is given by the functional (8)

$$\Lambda(t) = g(t, N(t_0); T(1), \ldots, T(N(t_0))), \tag{14}$$

because $N(t) = N(t_0)$. Hence, substituting conditional probability (13) into (12), after minor rearrangement, we obtain

$$\begin{aligned} P(N(t + dt) - N(t_0) = 0 \mid N(t'),\ 0 < t' \leq t_0) \\ - P(N(t) - N(t_0) = 0 \mid N(t'),\ 0 < t' \leq t_0) \\ \doteq -\Lambda(t) P(N(t) - N(t_0) = 0 \mid N(t'),\ 0 < t' \leq t_0)\, dt. \end{aligned} \tag{15}$$

Temporarily denote

$$q(t) = P(N(t) - N(t_0) = 0 \mid N(t'),\ 0 < t' \leq t_0). \tag{16}$$

Dividing both sides of (15) by dt and letting $dt \to 0$, we get a first order linear differential equation

$$\frac{dq(t)}{dt} = -\Lambda(t) q(t), \qquad t \geq t_0 \tag{17}$$

where $t_0 \geq 0$ is fixed. Since clearly

$$q(t) = 1 \qquad \text{for} \quad t = t_0,$$

the solution of (17) is easily found to be

$$q(t) = e^{-\int_{t_0}^{t} \Lambda(\tau)\, d\tau}, \qquad t \geq t_0,$$

or, substituting back for $q(t)$ from (16) and for $\Lambda(t)$ from (14),

$$\begin{aligned} P(N(t) - N(t_0) = 0 \mid N(t'),\ 0 < t' \leq t_0) \\ = e^{-\int_{t_0}^{t} g(\tau, N(t_0); T(1), \ldots, T(N(t_0)))\, d\tau}, \qquad t \geq t_0. \end{aligned} \tag{18}$$

Suppose now that

$$X(n), \qquad n = 0, 1, \ldots$$

is the waiting time sequence of our past-dependent point process. Since, for a point process with time domain $\mathcal{T} = (0, \infty)$ we have, by definition,

$$X(0) + \cdots + X(n-1) = T(n), \qquad n = 0, 1, \ldots,$$

and since for any $x > 0$

$$X(n) > x \quad \text{if and only if} \quad N(x + T(n)) - N(T(n)) = 0$$

we have from (18) for $n = 1, 2, \ldots$

$$P(X(n) > x \mid X(0) = x_0, \ldots, X(n-1) = x_{n-1})$$

$$= e^{-\int_{t_n}^{x+t_n} g(\tau, n; t_1, \ldots, t_n)\, d\tau} = e^{-\int_0^x g(\tau + t_n, n; t_1, \ldots, t_n)\, d\tau} \quad x > 0, \quad \text{(19a)}$$

with $t_1 = x_0, \ldots, t_n = x_0 + \cdots + x_{n-1}$, and for $n = 0$,

$$P(X(0) > x) = e^{-\int_0^x g(\tau, 0)\, d\tau}, \qquad x > 0. \qquad \text{(19b)}$$

Thus, we see that knowledge of the family of functions g, as defined in (7), allows us to determine the distribution function of the first waiting time $X(0)$ and also the conditional distribution function of $X(n)$, $n = 1, 2, \ldots$, given the values of all previous waiting times $X(0) = x_0, \ldots, X(n-1) = x_{n-1}$. Consequently, the probability law of the waiting time sequence $X(n)$, $n = 0, 1, \ldots$, and hence also that of the entire point process, is determined. This shows in a little more detail that (under some technical conditions mentioned before) the probability law of a past-dependent point process and the probability law of its intensity process are equivalent in the sense that one determines the other.

Example 7.7.2

Consider a point process on $\mathcal{T} = (0, \infty)$ that begins as an inhomogeneous Poisson process with intensity decreasing in inverse proportion to the elapsed time $\lambda_0(t) = (1 + t)^{-1}$, $t > 0$. Upon the occurrence of the first point-event at epoch $T(1) = t_1$, a new inhomogeneous Poisson process is born, which then begins to contribute point-events independently of the original process. Its intensity $\lambda_1(t)$ is the same as the original process except that being born at the time t_1, the decrease begins at this time, that is,

$$\lambda_1(t) = (1 + t - t_1)^{-1}, \qquad t > t_1.$$

The next point-event with epoch $T(2) = t_2$, regardless of whether it came from the first or the second process currently in operation, again gives rise to a new inhomogeneous Poisson process with intensity

$$\lambda_2(t) = (1 + t - t_2)^{-1}, \qquad t > t_2,$$

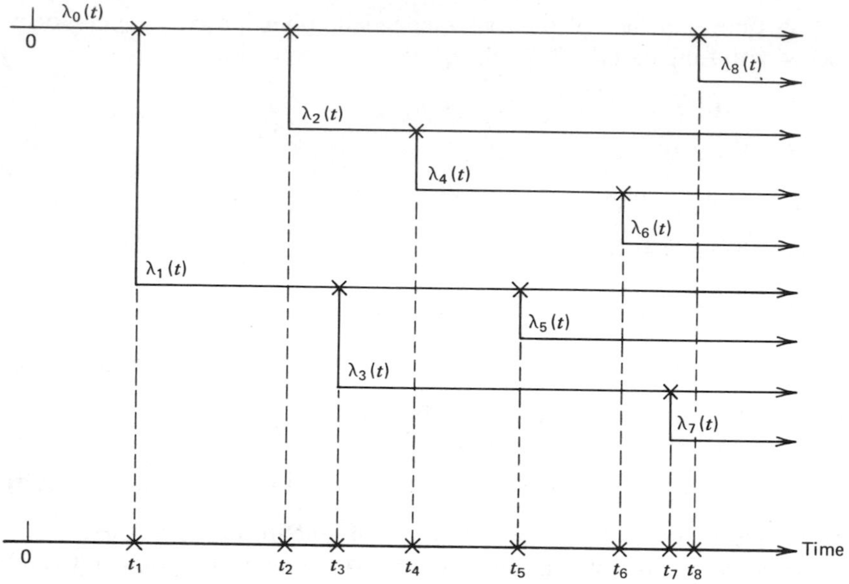

Figure 7.7.2

and independent of the first two processes. This procedure continues indefinitely, so that after the occurrence of the nth point-event, we have $n + 1$ independent inhomogeneous Poisson processes, each of which may be responsible for the subsequent point-event.

The reader may perhaps gain a better understanding of this mechanism from the scheme in Figure 7.7.2, which also suggests that this kind of situation may be suitable for modeling an avalanche-type phenomena. For instance, one may imagine a chunk of radioactive matter emitting neutrons with decaying intensity. Each emission in turn activates another piece of matter, which becomes radioactive and begins emitting neutrons in the same decaying rate.

Suppose that by some time $t > 0$ there were $N(t) = n$ occurrences of point-events of the resulting point process and that their respective epochs were $0 < T(1) = t_1 < T(2) = t_2 < \cdots < T(n) = t_n \leq t$. Then at the time t, one has $n + 1$ independent Poisson processes with intensities $\lambda_0(t) = (1 + t)^{-1}$, $\lambda_1(t) = (1 + t - t_1)^{-1}$, ..., $\lambda_n(t) = (1 + t - t_n)^{-1}$, contributing their point-events to the point process. It follows that the resulting intensity at time t will be the sum $\lambda_0(t) + \lambda_1(t) + \cdots + \lambda_n(t)$. Thus the conditional probability

$$P(dN(t) = 1 \mid N(t) = n, T(1) = t_1, \ldots, T(n) = t_n) \doteq \left(\sum_{k=0}^{n} \lambda_k(t) \right) dt,$$

where

$$\lambda_k(t) = (1 + t - t_k)^{-1}, \qquad k = 0, 1, \ldots, n \qquad \text{with } t_0 = 0.$$

It follows that the intensity process $\Lambda(t)$, $t > 0$, can be defined by

$$\Lambda(t) = \sum_{k=0}^{N(t)} (1 + t - T(k))^{-1}, \qquad t > 0,$$

provided we agree to call $T(0) = t_0 = 0$. But this means that the functional g_t defining the intensity process is

$$g(t, n; t_1, \ldots, t_n) = \sum_{k=0}^{n} (1 + t - t_k)^{-1},$$

$n = 0, 1, \ldots, 0 < t_1 < \cdots < t_n \leq t$, and hence by (19a,b), we have for the waiting times

$$P(X(0) > x) = e^{-\int_0^x (1+\tau)^{-1}\, d\tau} = e^{-\ln(1+x)} = (1 + x)^{-1},$$

$x > 0$, and for $n = 1, 2, \ldots,$

$$P(X(n) > x \mid X(0) = x_0, \ldots, X(n-1) = x_{n-1})$$

$$= e^{-\int_0^x \sum_{k=0}^{n} (1 + \tau + t_n - t_k)^{-1}\, d\tau} = e^{-\sum_{k=0}^{n} \ln(1 + x + t_n - t_k)}$$

$$= \prod_{k=0}^{n} (1 + x + t_n - t_k), \qquad x > 0,$$

where $t_n - t_k = x_k + x_{k+1} + \cdots + x_{n-1}$, $k = 0, \ldots, n - 1$, if we prefer to have the right-hand side expressed entirely in terms of the x's. Thus, we have obtained the family of distribution functions

$$F_{X(0)}(x_0) = 1 - (1 + x_0)^{-1}, \qquad x_0 > 0,$$

$$F_{X(n)|X(0), \ldots, X(n-1)}(x_n \mid x_0, \ldots, x_{n-1}) = 1 - \prod_{k=0}^{n} (1 + x_k + x_{k+1} + \cdots + x_n)^{-1},$$

$x_k > 0$; $k = 0, \ldots, n$; $n = 1, 2, \ldots,$ from which the conditional and hence also the joint densities of the waiting time sequence are easily obtained, although the resulting expression is somewhat complicated. We end this example by clarifying a point that may have bothered the reader from the very beginning. How do we know that this point process satisfies assumption (ii) of Section 7.1, that is, that it does not explode? After all, the possible physical situation of the avalanche phenomenon seems to suggest a possibility of such an explosion. Surprisingly perhaps, this will not happen, as can be seen from the following simple argument. From (4) and (5), we have

$$\mathrm{E}[dN(t)] \doteq \lambda(t)\, dt$$

where $\lambda(t) = \mathrm{E}[\Lambda(t)]$ and $\mathrm{E}[dN(t)] = d\mathrm{E}[N(t)] = d\mu(t)$, with $\mu(t) = \mathrm{E}[N(t)]$, $t > 0$, being the mean function of the point process. However

$$\Lambda(t) = \sum_{k=0}^{N(t)} (1 + t - T(k))^{-1} \leq 1 + N(t),$$

and so by taking expectations, $\lambda(t) \leq 1 + \mu(t)$. Thus

$$d\mu(t) \leq (1 + \mu(t))\, dt, \qquad t \geq 0,$$

with $\mu(0) = 0$. But this means that the nonnegative function $\mu(t)$, $t > 0$, is upper-bounded by the unique solution of the differential equation

$$\dot{y}(t) = 1 + y(t), \qquad t \geq 0, \qquad y(0) = 0,$$

in other words, that for every $t \geq 0$,

$$0 \leq \mu(t) \leq e^t - 1.$$

If the expectation $\mu(t) = \mathrm{E}[N(t)]$ of the number of point-events with epochs in an interval $(0, t]$ is finite, then this number, that is, the random variable $N(t)$, must itself be finite with probability 1, and so the nonexplosion assumption (ii) is indeed satisfied. ▲

The fact that the probability laws of a point process and its intensity process can be obtained from each other does not mean, however, that a particular sample path of a point process can be determined from the sample path of its intensity process. Clearly, given a sample path of a point process, for example, as the sample path of its associated counting process, we can construct the corresponding sample path of its intensity process by calculating the conditional expectations (3). For instance, the portion of the sample path $\Lambda(t) = l(t)$ between the nth and the $(n + 1)$st point-event is simply

$$l(t) = g(t, n; t_1, \ldots, t_n), \qquad t_n < t < t_{n+1},$$

where $t_1 = T(1)$, $t_2 = T(2)$, ... are the values of the random epoch sequence of the process.

But the converse is not true; even if we knew the sample path $l(t) = \Lambda(t)$, $t \geq 0$, of the intensity process, we would not be able to determine the instances $t_1 = T(1)$, $t_2 = T(2)$, ... of random epochs. This is immediately obvious in the particular case when $\Lambda(t) = \lambda$, $t > 0$, is a constant, that is, if the point process is a homogeneous Poisson process. Then knowing the value of λ amounts to knowing the probability law of the point process but not what the actual epochs of its point-events are going to be.

To get a better understanding of the relationship between a point process and its intensity process, let us look again at the definition of the latter, namely

$$\mathrm{E}[dN(t) \mid N(t'), 0 \leq t' \leq t] \doteq \Lambda(t)\, dt.$$

Subtracting $\Lambda(t)\,dt$ on both sides, we see that

$$\mathrm{E}[dN(t) - \Lambda(t)\,dt \mid N(t'), 0 \le t' \le t] = 0$$

or, which amounts to the same thing, that the process

$$U(t) = N(t) - \int_0^t \Lambda(\tau)\,d\tau, \qquad t > 0, \tag{20}$$

is a martingale.† Thus the equation

$$N(t) = U(t) + \int_0^t \Lambda(\tau)\,d\tau, \qquad t > 0,$$

represents the Doob-Meyer decomposition of the counting process $N(t)$ into a predictable process

$$\Theta(t) = \int_0^t \Lambda(\tau)\,d\tau, \qquad t \ge 0$$

and a martingale "noise" $U(t)$, $t > 0$. Thus the portion of the intensity process $\Lambda(t')$, $0 < t' \le t$, can be looked upon as containing all the information that can be extracted from the past $N(t')$, $0 < t' \le t$, and used to predict the future of the counting process. The martingale $U(t)$ can similarly be interpreted as the unpredictable or noisy component of the counting process.

Unfortunately, the additive decomposition

$$N(t) = U(t) + \Theta(t) \tag{21}$$

of the counting process is not well suited for point processes, since neither of the two components $U(t)$ and $\Theta(t)$ is a counting-type process and hence cannot be readily interpreted in terms of the random stream of point-events.

Nevertheless, it can be shown that the nature of the unpredictable noise contained in a past-dependent point process is of a Poisson type. One way to see this is to notice from (19) that, conditional upon the past epochs $T(1) = t_1, \ldots, T(n) = t_n$, the distribution of the next waiting time $X(n)$ is the same as it would be if the process were an inhomogeneous Poisson process with intensity function

$$\lambda(\tau) = g(\tau + t_n, n; t_1, \ldots, t_n).$$

Therefore, a past-dependent point process can be viewed as an inhomogeneous Poisson process whose intensity function at any time t depends on past random epochs of the process itself.

Even a better picture of a past-dependent point process may be obtained by considering a past-dependent transformation of time. This is perhaps best described by visualizing the time domain $\mathcal{T} = (0, \infty)$ as a highway with point-

† The symbol $U(t)$ used to denote process (20) has nothing to do with the mark sequence of preceding sections.

events as markers alongside and by imagining a traveler driving on the highway in the positive direction. The traveler carries a counter driven by a clock, which registers and counts the marks as the traveler passes by.

If the traveler moves with a constant speed $v = 1$, then obviously† her counter will register the counting process $N(t)$, $t > 0$, associated with the original point process.

Suppose now that we instruct the traveler to keep changing her speed in inverse proportion to the current value of the intensity process $\Lambda(t)$, $t > 0$, speeding up where the intensity is small and slowing down where it is large. That is, if $\Lambda(t) = l(t)$ at some point t on the highway, her instantaneous speed at this point should be‡

$$v(t) = \frac{1}{l(t)}.$$

Since $l(t)$ depends only on the positions of the markers in the portion $(0, t]$ of the highway the traveler has already covered, we are not asking anything impossible; she does not have to anticipate the future in any way. What kind of process will her counter register now?

Suppose that when her clock is showing time $\tilde{t}$, she has reached point t on the highway. Let us denote the number of markers recorded by her counter at that time by $M(\tilde{t})$. Clearly, $M(\tilde{t}) = N(t)$, while her speed at time $\tilde{t}$ is

$$v(\tilde{t}) = \frac{1}{l(t)}.$$

During the subsequent small time interval $(\tilde{t}, \tilde{t} + d\tilde{t}]$, she will have covered the additional distance

$$dt = v(\tilde{t})\, d\tilde{t} = \frac{d\tilde{t}}{l(t)} \tag{22}$$

and her counter will register the increment

$$dM(\tilde{t}) = M(\tilde{t} + d\tilde{t}) - M(\tilde{t}) = N(t + dt) - N(t) = dN(t).$$

Then the conditional expectation

$$\begin{aligned} \mathrm{E}[dM(\tilde{t}) \mid M(\tilde{t}'), 0 < \tilde{t}' \le \tilde{t}] &= \mathrm{E}[dN(t) \mid N(t'), 0 < t' \le t] \\ &= \Lambda(t)\, dt = d\tilde{t} \end{aligned} \tag{23}$$

since at the point t, $\Lambda(t) = l(t)$ and $dt = d\tilde{t}/l(t)$ by (22). But by the corollary to Theorem 7.3.2, (23) implies that the process $M(\tilde{t})$, $\tilde{t} > 0$, is a Poisson process with

† No "relativistic effects" are contemplated here.

‡ We assume here that $l(t) > 0$, that is, the portions of the highway where $l(t) = 0$ should be deleted.

intensity function $\lambda = 1$. That is, the traveler's counter will register the *standard Poisson process* $N_0(t)$.

This discussion shows how past-dependent point processes are constructed. Take a process $\Lambda(t)$, $t > 0$, such that for all $t \geq 0$

$$\Lambda(t) > 0 \qquad \text{and} \qquad \int_0^t \Lambda(\tau)\, d\tau < \infty, \tag{24}$$

and a standard Poisson process $N_0(t)$, $t > 0$, independent of the process (24). Define a random time transformation

$$\Theta(t) = \int_0^t \Lambda(\tau)\, d\tau, \qquad t > 0. \tag{25}$$

Then the process

$$N(t) = N_0(\Theta(t)), \qquad t > 0 \tag{26}$$

will be a past-dependent counting process having the process $\Lambda(t)$, $t > 0$, as its intensity process. To check this, note merely that (25) is a one-to-one transformation and so

$$\begin{aligned} \mathrm{E}[dN(t) \mid N(t'), \Theta(t'), 0 < t' \leq t] &= \mathrm{E}[dN_0(\Theta(t)) \mid N_0(t'), \Theta(t'), 0 < t' \leq t] \\ &= d\Theta(t) = \Lambda(t)\, dt, \end{aligned}$$

where the second equality follows from the independence of $N_0(t)$ and $\Lambda(t)$ and the properties of the standard Poisson process.

Notice also that the time transformation (25) is the same as the predictable component in the Doob-Meyer decomposition (21). However, representation (26) is more instructive, since it clearly shows the Poisson character of the "noisy" component of a point process.

Example 7.7.3

Consider again the point process of Example 7.7.2. Its intensity process was shown to be

$$\Lambda(t) = \sum_{k=0}^{N(t)} (1 + t - T(k))^{-1}, \qquad t > 0.$$

A typical sample path $l(t) = \Lambda(t)$, $t > 0$, is then a superposition of functions of the form $(1 + \tau)^{-1}$, $\tau > 0$, the first one commencing at time $t = 0$ and each subsequent one being added at times $t_k = T(k)$, $k = 1, 2, \ldots$, as in Figure 7.7.3. Consequently, the predictable process (25)

$$\Theta(t) = \int_0^t \Lambda(\tau)\, d\tau, \qquad t > 0,$$

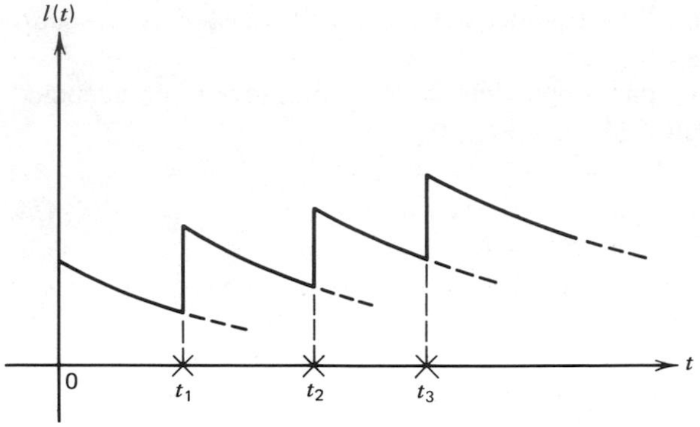

Figure 7.7.3

will have sample path $\theta(t) = \Theta(t)$, $t > 0$, consisting of the corresponding superposition of logarithmic functions

$$\int_0^\tau (1 + \tau')^{-1}\, d\tau' = \ln(1 + \tau), \qquad \tau > 0$$

commencing at times $0, t_1, t_2, \ldots$. The resulting increasing function of time is shown in Figure 7.7.4. Since the points depicted by $\otimes$ have coordinates $\tilde{t}_k = \theta(t_k)$, they represent epochs of a standard Poisson process obtained by the random time transformation

$$N(t) = N_0(\Theta(t)), \qquad t > 0.$$

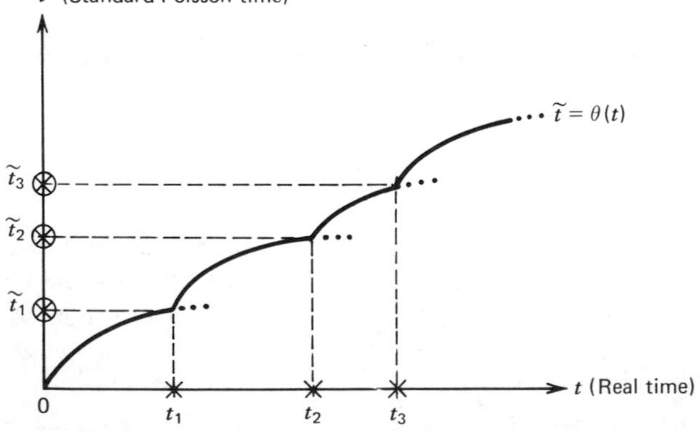

Figure 7.7.4

Notice that the smooth piece of the function $\theta(t)$ in an interval $[t_{n-1}, t_n]$ is given by the formula

$$\theta(t) = \sum_{k=0}^{n-1} \ln(1 + t - t_k), \qquad t_{n+1} \leq t \leq t_n,$$

$n = 1, 2, \ldots.$ This can be used, for instance, to simulate the point process of Example 7.7.2, using just a generator of a standard Poisson process, or even to derive further properties of the past-dependent point process from the well-known properties of the standard Poisson process. ▲

We will end this informal inquiry into the nature of point processes by once more emphasizing the fundamental role of the Poisson process and the Wiener process in modeling random noise. We hope to have convinced the reader that the common usage of Poisson and Wiener processes for this purpose is not just a matter of convenience but is actually a consequence of a more primitive concept of unpredictability, which we take to be the basic characteristic of truly random noise.

Remark: In this entire section we have been rather freely using differential relations such as

$$P(dN(t) = 1 \mid N(t'), 0 < t' \leq t) \doteq \Lambda(t)\, dt, \tag{27}$$

while mentioning, almost in passing, "certain technical conditions" that are needed to guarantee the existence of the intensity process $\Lambda(t)$ and hence also the validity of differential relations such as the one above. Although we are not going to spell out these "technical conditions," we feel that we owe the reader at least an example showing that an intensity process need not always exist.

Consider the very simple point process on $\mathcal{T} = (0, \infty)$ constructed as follows: the epoch $T(1)$ of the first point event is a random variable uniformly distributed over the interval $(0, 1)$, while all the subsequent point-events occur in a deterministic fashion, each exactly one time unit from the preceding one. In other words, the epochs $T(n)$ for $n = 2, 3, \ldots$ are just

$$T(n) = T(1) + (n - 1),$$

or equivalently, the waiting times

$$X(n) = 1 \qquad \text{for all } n = 1, 2, \ldots$$

as pictured in Figure 7.7.5. Let us now take $t_1 \geq 1$ and $t_2 > t_1$ and compute the conditional expectation

$$\mathrm{E}[N(t_2) - N(t_1) \mid N(t'), 0 < t' \leq t_1]$$

with $N(t)$, $t > 0$, being the associated counting process. Since the entire counting process is completely determined by the epoch $T(1) < 1$ of its first jump, which is included in the past, the conditional expectation as a function of t_2 is a step function with unit jumps exactly one

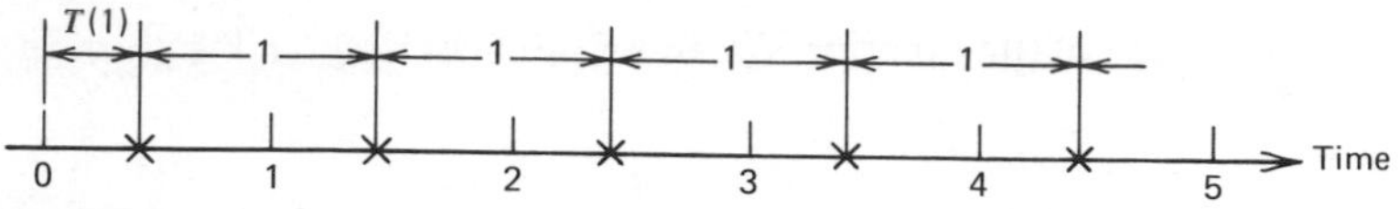

Figure 7.7.5

time unit apart. But then it cannot be written as the integral of an intensity process, that is, in the form

$$E[N(t_2) - N(t_1) | N(t'), 0 < t' \leq t_1] = \int_{t_1}^{t_2} \Lambda(\tau)\, d\tau \tag{28}$$

as we did in equation (2). The differential version (3)

$$E[dN(t) | N(t'), 0 < t' \leq t] \doteq \Lambda(t)\, dt$$

then does not make sense either, that is, the intensity process does not exist. True, we could employ the Dirac delta function and define

$$\Lambda(t) = \sum_{n=1}^{\infty} \delta(t - nT(1)), \qquad t > 0$$

in which case (28) would hold true. But this would make the intensity process an entirely new object, a generalized stochastic process, and we would face the necessity of developing either an appropriate theory or at least a set of rules for formal manipulation with such a process. This indeed can be done, but it would lead us too far and too deeply into the subject.

We therefore merely caution the reader that differential relations like (27), although often quite illuminating, must not be indiscrimantly used in any rigorous argument. Я

EXERCISES

Exercise 7.7

1. Consider the process of Example 7.7.1 with the following modifications:
   (a) Each time the intensity drops to zero, it begins to increase exponentially rather than linearly, that is, it increases as the function $\lambda(1 - e^{-\alpha(t-t_0)})$, where λ is the original Poisson intensity and t_0 is the time instant the intensity dropped to zero.
   (b) The parameter α above is inversely proportional to the average rate of occurrence of point-events, that is, if $N(t_0) = n$, then $\alpha = c(t_0/n)$, $c > 0$.

   Define the intensity process $\Lambda(t)$, $t > 0$, and sketch some of its sample paths. With $\Delta = 1$, $c = 1$ and $\lambda = 2$, evaluate the function $g(6, 5; 1.5, 2.1, 4, 4.5, 5.7)$.
2. Derive the formula for the second moment of the counting process

$$E[N^2(t)] = \int_0^t \lambda(\tau)\, d\tau + 2 \int_0^t E[\Lambda(\tau)N(\tau)]\, d\tau.$$

(*Hint:* Partition the interval $(0, t]$ into a large number of small subintervals, write

$$N^2(t) = \sum_k (dN(t_k))^2 + 2 \sum_k N(t_k)\, dN(t_k)$$

and compute the expectation in the last term by conditioning on $N(t')$, $0 < t' \leq t_k$.)

3. Use the time transformation (26) and conditioning on $\Theta(t)$ to show that

$$\psi_{N(t)}(\zeta) = \psi_{\Theta(t)}\left(\frac{e^{\iota\zeta} - 1}{\iota}\right).$$

4. Use the time transformation (26) and conditioning on $\Theta(t)$ to show that for all $n = 0$, $1, \ldots$

$$P(N(t) = n) = \frac{1}{n!}\, \mathrm{E}[\Theta^n(t) e^{-\Theta(t)}].$$

5. Suppose that the intensity process $\Lambda(t)$, $t > 0$, is such that the random variable $\Theta(t) = \int_0^t \Lambda(\tau)\, d\tau$ has the gamma density

$$f_{\Theta(t)}(\theta) = \frac{\alpha^t}{\Gamma(t)}\, \theta^{t-1} e^{-\alpha\theta}, \qquad \theta \geq 0,$$

with α a positive constant. Calculate the probability mass distribution $P(N(t) = n)$, $n = 0, 1, \ldots$. (*Hint:* Use Exercise 7.7.4.)

6. A monostable multivibrator has two states, stable 0 and unstable 1. When triggered by an input impulse, it changes its state from 0 to 1 and stays in the state 1 for a fixed period of time $\Delta > 0$ before returning back to the stable state 0. When it is in the unstable state 1, trigger pulses have no effect on the circuit. Suppose now that the trigger pulses are arriving according to an inhomogeneous Poisson process with intensity function $\lambda(t)$, $t > 0$. The output point process has epochs that are the time instances at which the circuit changes from state 0 to 1. (See Fig. 7.7.E.1.) Define the intensity process $\Lambda(t)$, $t > 0$, of the output point process in terms of the functions $g(t, n; t_1, \ldots, t_n)$. Sketch a typical sample path of the intensity process if $\lambda(t) = 1 + \cos(t\pi)$.

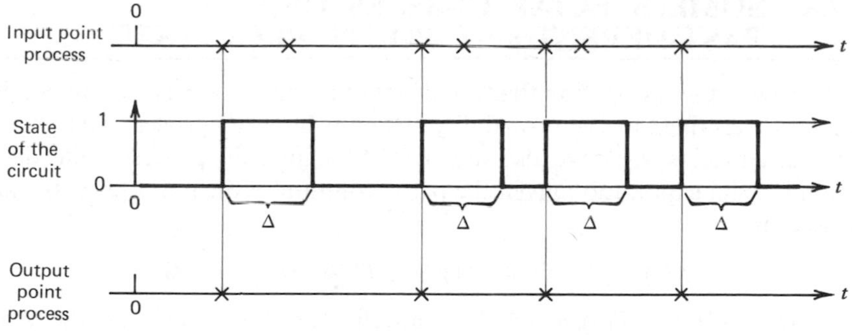

Figure 7.7.E.1

7. In Exercise 7.7.6, assume that the input Poisson process is homogeneous with intensity λ_i. Characterize the probability law of the output point process by specifying the waiting time sequence $X(n)$, $n = 0, 1, \ldots$.
8. Generalize Exercise 7.7.6 by assuming that the successive times the multivibrator stays in state 1 are independent random variables uniformly distributed on the interval (0.9, 1.1). Assuming that the input Poisson process is homogeneous with $\lambda = 2$, calculate the probability density of the second random epoch $T(2)$.
9. Use equations (19a and b) to derive a general formula for the joint density of the first n waiting times $X(0), \ldots, X(n-1)$ and the first n random epochs $T(1), \ldots, T(n)$.
10. Let $X(n)$, $n = 0, 1, \ldots$, be the waiting time sequence of a past-dependent point process. Suppose that its intensity process $\Lambda(t)$, $t > 0$, is defined by

$$\Lambda(t) = \begin{cases} 1 & \text{if} \quad N(t) = 0, \\ \dfrac{1}{n} \sum_{k=0}^{n-1} X(k) & \text{if} \quad N(t) = n > 0. \end{cases}$$

Calculate the joint densities of the waiting time sequence $X(n)$, $n = 0, 1, \ldots$, or the random epoch sequence $T(n)$, $n = 1, 2, \ldots$, whichever you prefer.
11. For $\Delta > 0$, a constant, suppose that the intensity process is a linear function $\Lambda(t) = 1 + K(t)$, $t > 0$, of the number $K(t)$ of point-events with epochs in the interval $(t - \Delta, t]$ or $(0, t]$ if $t < \Delta$. Calculate the conditional probability of no point-event occurring during the time interval $(t, t + s]$, given $N(t) = n$ and $T(1) = t_1, \ldots, T(n) = t_n$. How would you expect the point process to behave?
12. Suppose that the waiting time sequence $X(n)$, $n = 0, 1, \ldots$, of a point process is a Markov process such that $X(0)$ is uniformly distributed on (0, 1), while for each $n = 0, 1, \ldots$, $X(n+1)$, given $X(n) = x_n$, is uniformly distributed over the interval $(1 - x_n, 1)$. Argue that the intensity process depends only on the present time t and, at most, two of the most recent random epochs. Obtain an expression for the functions $g(t, n; t_1, \ldots, t_n)$.

7.8. SOME SPECIAL CLASSES OF PAST-DEPENDENT POINT PROCESSES

We saw in the last section that the probability law of a past-dependent point process is determined by the probability law of its intensity process $\Lambda(t)$, provided that the latter exists. We have also shown that the intensity process, being defined as a conditional expectation given the past evolution of the point process, can be expressed as

$$\Lambda(t) = g(t, N(t); T(1), \ldots, T(N(t))), \qquad t > 0, \tag{1}$$

where $N(t)$, $t \geq 0$, and $T(n)$, $n = 1, 2, \ldots$, are the counting process and the random epoch sequence, respectively, of the past-dependent point process. Therefore, as shown in the last section, the probability law of such a point process is determined by specifying the family of functions

$$g(t, 0), \qquad t > 0$$
$$g(t, n: t_1, \ldots, t_n), \qquad 0 < t_1 < \cdots < t_n < t, \qquad n = 1, 2, \ldots, \tag{2}$$

from which, for example, the probability law of the waiting time sequence $X(n)$, $n = 0, 1, \ldots$, may be obtained by using (7.7.19):

$$P(X(0) > x) = e^{-\int_0^x g(\tau, 0)\, d\tau}, \qquad x > 0. \tag{3a}$$

$$P(X(n) > x \mid X(0) = x_0, \ldots, X(n-1) = x_{n-1})$$
$$= e^{-\int_0^x g(\tau + t_n, n; t_1, \ldots, t_n)\, d\tau}, \quad x > 0, \tag{3b}$$

where

$$t_k = x_0 + \cdots + x_{k-1}, \qquad k = 1, \ldots, n.$$

and

$$x_k > 0 \qquad \text{for } k = 0, \ldots, n-1.$$

Since the left-hand sides of both (3a) and (3b) must be nonincreasing functions of x and must converge to zero as $x \to \infty$, the functions g in (2) must be nonnegative, integrable with respect to t, and such that both integrals

$$\int_0^x g(\tau, 0)\, d\tau \qquad \text{and} \qquad \int_0^x g(\tau + t_n, n; t_1, \ldots, t_n)\, d\tau$$

diverge to $+\infty$ as $x \to \infty$. Except for these requirements, however, the functions g may be quite arbitrary. Various special classes of past-dependent point processes can be obtained by imposing further conditions on these functions, for instance, allowing them to depend only on some restricted subsets of their arguments.

The most severe restriction would be to assume that they are all equal to a constant or, somewhat less severely, that they all depend only on their first argument t, that is, the present time. But in that case, from (1) we would have for the intensity process

$$\Lambda(t) = g \qquad \text{or} \qquad \Lambda(t) = g(t),$$

so that the intensity process would be a constant or a deterministic function of time. This, in turn, would imply that the point process would evolve without aftereffects and hence be a Poisson process. Therefore, in order to obtain a genuine past-dependent point process, the functions g in (2) must be allowed to depend on at least one of the arguments other than the first. A particularly simple choice of this kind leads to an important class of past-dependent point processes—the renewal processes.

RENEWAL POINT PROCESSES. Let

$$h(\tau), \qquad \tau > 0, \tag{4}$$

be a nonnegative function such that

$$\lim_{x \to \infty} \int_0^x h(\tau)\, d\tau = +\infty.$$

We define the functions g in (2) by

$$g(t, 0) = h(t),$$

and

$$g(t, n; t_1, \ldots, t_n) = h(t - t_n), \qquad n = 1, 2, \ldots.$$

Substituting into (3a,b), we have

$$P(X(0) > x) = e^{-\int_0^x h(\tau)\, d\tau}, \qquad x > 0,$$

and

$$P(X(n) > x \mid X(0) = x_0, \ldots, X(n-1) = x_{n-1}) = e^{-\int_0^x h(\tau)\, d\tau}, \qquad x > 0,$$

for all $x_k > 0$, $k = 0, 1, \ldots, n - 1$; $n = 1, 2, \ldots$. But this implies that the *waiting time sequence*

$$X(n), \qquad n = 0, 1, \ldots$$

of such a past-dependent point process *is a sequence of independent, identically distributed random variables* with common distribution function

$$F_X(x) = 1 - e^{-\int_0^x h(\tau)\, d\tau}, \qquad x > 0. \tag{5}$$

Since the only point process evolving without aftereffects and whose waiting times are independent and identically distributed is a homogeneous Poisson process, a point process with the waiting time sequence above is a genuine past-dependent point process unless the function $h(\tau)$ is a constant. (For $h(\tau) = \lambda > 0$, a constant, (5) is an exponential distribution function, and thus the point process is Poisson with intensity λ.)

Example 7.8.1

Suppose that the function $h(\tau)$, $\tau > 0$, is

$$h(\tau) = \begin{cases} \dfrac{1}{1 - \tau} & \text{if} \quad 0 < \tau < 1, \\ 0 & \text{if} \quad \tau \geq 1. \end{cases}$$

Then the integral

$$\int_0^x h(\tau)\,d\tau = \int_0^x \frac{1}{1-\tau}\,d\tau = -\ln(1-x)$$

if $0 < x < 1$, while if $x \geq 1$ the integral diverges to $-\infty$. Hence, the distribution function for the waiting times is, according to (5)

$$F_X(x) = 1 - e^{\ln(1-x)} = x \qquad \text{if} \quad 0 < x < 1,$$

and

$$F_X(x) = 1 - e^{-\infty} = 1 \qquad \text{for} \quad x \geq 1.$$

Of course, $F_X(x) = 0$ for $x \leq 0$, since the waiting times are continuous and non-negative random variables. This is the distribution function of a uniform random variable over (0, 1), and thus the waiting time sequence $X(n)$, $n = 0, 1, \ldots$, is, in this case, a sequence of i.i.d. uniformly distributed random variables. ▲

A point process, whose waiting time sequence is a sequence of i.i.d. random variables is called a *renewal point process*. The name comes from the probabilistic model of a situation, where some piece of equipment, such as a light bulb, is constantly being replaced as soon as it fails.

The waiting times are then the lifetimes of each consecutive bulb (the bulbs being of identical type), and the point-events correspond to successive replacements (renewals) of the bulb.

The distribution function (5) is easily seen to be the distribution function of a continuous random variable,† and its corresponding density is

$$f_X(x) = h(x)e^{-\int_0^x h(\tau)\,d\tau}, \qquad x > 0. \tag{6}$$

Comparing this with (5), we can write

$$h(x) = \frac{f_X(x)}{1 - F_X(x)} \tag{7}$$

and since for small dx we have approximately

$$f_X(x)\,dx \doteq P(x \leq X \leq x + dx)$$

we get from (7)

$$h(x)\,dx \doteq \frac{P(x \leq X \leq x + dx)}{P(X \geq x)} = P(x \leq X \leq x + dx \mid X \geq x).$$

† Renewal processes with discrete waiting time distributions are also important in applications. But this would require some generalization of our concept of a point process, so they will not be considered here.

This provides an interpretation for the function h in (4), in terms of the lifetimes $X(n)$:

> $h(x)\,dx$ is approximately the conditional probability that the piece of equipment currently in operation will fail during the forthcoming time unit of length dx, given that it has already functioned for x time units.

For this reason, the function h is called the *hazard function* of the renewal process.

Example 7.8.2

From the discussion in the introduction to this section, it follows that if the hazard function for a renewal process is a constant, $h(t) = \lambda > 0$, the renewal process is a homogeneous Poisson process with intensity λ. It may then be natural to ask what kind of renewal process has the hazard function

$$h(t) = \alpha t, \qquad t > 0,$$

where $\alpha > 0$ is a constant. In terms of lifetimes $X(n)$ of a component, this would correspond to the case where the chance of failure increases linearly with the age of the component. From equation (6), the lifetime density is

$$f_X(x) = \alpha x e^{-\alpha x^2/2}, \qquad x > 0,$$

which is recognized as a Rayleigh density with parameter $\sigma^2 = 1/\alpha$. Note that it is generally an easy matter to obtain a density $f_X(x)$ from the hazard function $h(t)$, and vice versa, since they are related by equations (6) and (7). ▲

Another important characteristic of a renewal point process is the *renewal function* $\mu(t)$. As the notation indicates, it is just another name for the *mean function* of the renewal point process, that is, for the expected number of point-events with epochs in $(0, t]$,

$$\mu(t) = \mathrm{E}[N(t)], \qquad t \geq 0, \tag{8}$$

with $\mu(0) = 0$, since by convention $N(0) = 0$.

Remark: Since we have earlier made the assumption that the point processes discussed in this text are such that their associated counting processes do have finite expectation, we can simply take the existence of (8) as assumed. However, as will be shown later, the expectation (8) is finite even if the renewal process is defined as having an i.i.d. sequence of waiting times with no further assumption on their distribution. Я

The renewal function $\mu(t)$ is tied to the distribution function $F_X(x)$ of the waiting times by the integral equation

$$\mu(t) = F_X(t) + \int_0^t \mu(t - x) f_X(x)\,dx, \qquad t \geq 0, \tag{9}$$

which is called the *renewal equation*. This equation is easy to derive. From definition (8) and the properties of conditional expectation, we have

$$\mu(t) = \mathrm{E}[\mathrm{E}[N(t) \mid X(0)]] = \int_0^\infty \mathrm{E}[N(t) \mid X(0) = x] f_X(x)\, dx. \tag{10}$$

Now if $X(0) = x > t$, then there is no point-event with epoch in $(0, t]$, that is, $N(t) = 0$, and so

$$\mathrm{E}[N(t) \mid X(0) = x] = 0 \qquad \text{for } x > t. \tag{11}$$

On the other hand, if $X(0) = x \le t$, then the number of point-events with epochs in $(0, t]$ is 1 plus the number of point events with epochs in $(x, t]$, so that

$$N(t) = 1 + N(t) - N(x).$$

However, the conditional distribution of $N(t) - N(x)$, given $X(0) = x$, must be the same as the unconditional distribution of $N(t - x)$, since the waiting times $X(n)$, $n = 0, 1, \ldots$, are i.i.d. random variables. Consequently

$$\mathrm{E}[N(t) \mid X(0) = x] = 1 + \mu(t - x) \qquad \text{for } x \le t, \tag{12}$$

and substituting from (11) and (12) into (10), we get

$$\mu(t) = \int_0^t (1 + \mu(t - x)) f_X(x)\, dx = F_X(t) + \int_0^t \mu(t - x) f_X(x)\, dx,$$

the renewal equation (9).

Example 7.8.3

Consider a renewal process with the hazard function of Example 7.8.1, that is, such that the i.i.d. waiting time random variables $X(n)$, $n = 0, 1, \ldots$, are uniformly distributed over the interval (0, 1). Since in this case we have for $t \ge 0$

$$F_X(t) = \begin{cases} t & \text{if} \quad 0 \le t < 1, \\ 1 & \text{if} \quad t \ge 1, \end{cases}$$

and the density is

$$f_X(t) = \begin{cases} 1 & \text{if} \quad 0 < t < 1, \\ 0 & \text{if} \quad t \ge 1, \end{cases}$$

the renewal equation (9) becomes

$$\mu(t) = t + \int_0^t \mu(t - x)\, dx \qquad \text{if} \quad 0 \le t \le 1,$$

and

$$\mu(t) = 1 + \int_0^1 \mu(t - x)\, dx \qquad \text{if} \quad t \ge 1.$$

Substituting $y = t - x$ into the integrals gives

$$\mu(t) = t + \int_0^t \mu(y)\, dy \qquad \text{if} \qquad 0 \le t \le 1$$

and

$$\mu(t) = 1 + \int_{t-1}^t \mu(y)\, dy \qquad \text{if} \qquad t \ge 1.$$

Differentiating both sides with respect to t, we get

$$\dot{\mu}(t) = 1 + \mu(t) \qquad \text{if} \qquad 0 \le t \le 1, \tag{13}$$

and

$$\dot{\mu}(t) = \mu(t) - \mu(t-1) \qquad \text{if} \qquad t \ge 1. \tag{14}$$

Since $\mu(0) = 0$, (13) is a simple linear differential equation the solution of which is easily found to be

$$\mu(t) = e^t - 1, \qquad 0 \le t \le 1.$$

Then for $1 \le t \le 2$, $\mu(t-1) = e^{t-1} - 1$, and so differential equation (14) becomes a first order linear differential equation

$$\dot{\mu}(t) - \mu(t) = 1 - e^{t-1}, \qquad 1 \le t \le 2,$$

with initial condition $\mu(1) = e - 1$. Since the general solution of a first order linear differential equation

$$\dot{y}(t) - y(t) = q(t), \qquad t_0 \le t$$

with the initial condition $y(t_0)$ has the form

$$y(t) = e^t[y(t_0)e^{-t_0} - \int_{t_0}^t e^{-\tau} q(\tau)\, d\tau], \tag{15}$$

we get

$$\mu(t) = e^t - 1 - (t-1)e^{t-1}, \qquad 1 \le t \le 2.$$

Now the renewal function $\mu(t)$ has been determined up to time $t = 2$, and thus in the subsequent unit time interval $2 \le t \le 3$, we again have, from (14), a first order linear differential equation

$$\dot{\mu}(t) - \mu(t) = e^{t-1} - 1 - (t-2)e^{t-2},$$

with initial condition $\mu(2) = e^2 - 1 - e$. Once more applying general formula (15), we obtain

$$\mu(t) = e^t - 1 - (t-1)e^{t-1} + \frac{1}{2}(t-2)^2 e^{t-2}, \qquad 2 \le t \le 3,$$

and we can proceed to the next interval $3 < t < 4$. Soon a pattern begins to emerge, and we see the general formula for the renewal function is

$$\mu(t) = \sum_{k=0}^{n} (-1)^k \frac{(t-k)^k}{k!} e^{t-k} - 1, \qquad n \le t \le n+1, \qquad n = 0, 1, \ldots,$$

the validity of which can be verified by showing that it indeed satisfies the renewal equation for all $t \geq 0$. Introducing the symbol

$$(t-a)_+ = \begin{cases} t-a & \text{if} \quad t \geq a, \\ 0 & \text{if} \quad t < a, \end{cases}$$

we can also write the expression for the renewal function as

$$\mu(t) = \sum_{k=0}^{\infty} (-1)^k \frac{(t-k)_+^k}{k!} e^{t-k} - 1, \qquad t \geq 0.$$

This case, however, is one of the rare occasions when an explicit solution of the renewal equation can be obtained. ▲

Example 7.8.4

The derivation of the renewal equation presented earlier, sometimes known as the renewal argument, can be used to derive equations for many other quantities associated with a renewal process. As an example, consider the so-called paralyzable counter, that is, a particle counter where each arrival of a particle locks the counter for some time $\Delta > 0$, during which no particle can be registered. In other words, a particle is registered by the counter only if at least Δ time units have elapsed since the arrival of the preceding particle regardless of whether or not the preceding particle was registered (see Fig. 7.8.1).

Assuming that the instants of particle arrivals can be represented as epochs of a renewal point process, the times between successive particle registrations will be independent, identically distributed positive random variables, and we would like to calculate their common distribution. Let $X(n)$, $n = 0, 1, \ldots$, be the waiting time sequence of the renewal point process of particle arrivals, and let $f_X(x)$ be the density of the i.i.d. random variables $X(n)$. Suppose that a particle arriving at epoch $T(n) = t$ found the counter open and was therefore registered. Let $Z > 0$ be the waiting time to the next registration, while $X = X(n) = T(n+1) - T(n)$ is the waiting time to the next particle arrival. Let $z \geq \Delta$ be fixed. If $X \geq \Delta$, then clearly $Z = X$, and hence

$$P(Z > z \mid X \geq \Delta) = P(X \geq z \mid X \geq \Delta).$$

On the other hand if $X = x < \Delta$, then the arriving particle renews the locked state of the counter, and hence

$$P(Z > z \mid X = x < \Delta) = P(Z > z - x).$$

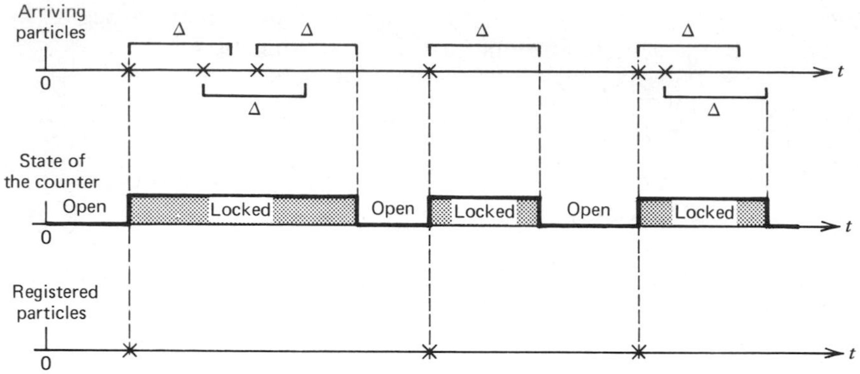

Figure 7.8.1

It follows that for $z \geq \Delta$

$$P(Z > z) = P(Z > z \,|\, X \geq \Delta)P(X \geq \Delta) + \int_0^\Delta P(Z > z \,|\, X = x < \Delta) f_X(x)\, dx$$

$$= P(X \geq z) + \int_0^\Delta P(Z > z - x) f_X(x)\, dx.$$

Since obviously $Z \geq \Delta$, we trivially have $P(Z > z) = 1$ if $z < \Delta$. Differentiating with respect to z, we obtain the integral equation for the density $f_Z(z)$, a renewal-type equation

$$f_Z(z) = f_X(z) + \int_0^\Delta f_Z(z - x) f_X(x)\, dx, \qquad z \geq \Delta,$$

with $f_Z(z) = 0$ if $z < \Delta$. ▲

Example 7.8.5

Renewal-type equations contain a convolution integral which suggests that solutions may be found by means of Laplace or Fourier transforms. For instance, the equation of Example 7.8.4 can be solved by a Fourier transform, in effect obtaining the characteristic function

$$\psi_Z(\zeta) = \mathrm{E}[e^{\imath\zeta Z}] = \int_\Delta^\infty e^{\imath\zeta z} f_Z(z)\, dz, \qquad -\infty < \zeta < \infty,$$

of the waiting times between successive particle registrations. Multiplying both sides of this renewal-type equation by $e^{\imath\zeta z}$ and then integrating with respect to z from Δ to ∞, we get

$$\psi_Z(\zeta) = \int_\Delta^\infty e^{\imath\zeta z} f_X(z)\, dz + \int_\Delta^\infty \int_0^\Delta f_Z(z - x) f_X(x)\, dx e^{\imath\zeta z}\, dz. \tag{16}$$

Reversing the order of integration in the double integral above, we can write it as

$$\int_0^{\Delta} \left[\int_{\Delta}^{\infty} f_Z(z-x) e^{\imath\zeta(z-x)} \, dz \right] f_X(x) e^{\imath\zeta x} \, dx.$$

Upon substituting $z - x = y$, the integral in brackets becomes

$$\int_{\Delta - x}^{\infty} f_Z(y) e^{\imath\zeta y} \, dy = \int_{\Delta}^{\infty} f_Z(y) e^{\imath\zeta y} \, dy = \psi_Z(\zeta)$$

since for $x > 0$, $\Delta - x < \Delta$, and $f_Z(y) = 0$ for $y < \Delta$. Thus the double integral becomes the product

$$\psi_Z(\zeta) \int_0^{\Delta} e^{\imath\zeta x} f_X(x) \, dx,$$

and the transformed equation (16) can be solved for $\psi_Z(\zeta)$, yielding the formula for the characteristic function,

$$\psi_Z(\zeta) = \frac{\int_{\Delta}^{\infty} e^{\imath\zeta x} f_X(x) \, dx}{1 - \int_0^{\Delta} e^{\imath\zeta x} f_X(x) \, dx}, \qquad -\infty < \zeta < \infty.$$

In particular, suppose that $f_X(x) = \lambda e^{-\lambda x}$, $x > 0$, that is, that the particles are arriving at the counter according to a homogeneous Poisson process with intensity $\lambda > 0$. In this case, the two integrals are easily evaluated, and we obtain, after some simplification,

$$\psi_Z(\zeta) = \left(1 - \frac{\imath\zeta}{\lambda} e^{(\lambda - \imath\zeta)\Delta}\right)^{-1}, \qquad -\infty < \zeta < \infty.$$

This characteristic function can actually be inverted and a (rather complicated) expression for the density $f_Z(z)$ written down, but we shall be satisfied with the mean of the waiting time distribution. We obtain

$$\mathrm{E}[Z] = \frac{1}{\imath} \frac{d}{d\zeta} \psi_Z(\zeta) \bigg|_{\zeta = 0} = \frac{1}{\lambda} e^{\lambda\Delta}.$$

Needless to say, the variance and higher moments can be calculated with no difficulty. ▲

The renewal equation can be shown to establish a one-to-one correspondence between the distribution function $F_X(x)$ and the renewal function $\mu(t)$. Consequently, the renewal function $\mu(t)$ uniquely determines the distribution function $F_X(x)$ and, hence, the entire probability law of the renewal point process. The converse is, of course, obvious since the renewal process has, by definition, independent waiting times with common distribution function $F_X(x)$, and therefore its entire probability law, a fortiori its mean function $\mu(t)$, must be completely specified by this distribution function.

Actually, we can write down an explicit formula for the renewal function. Since, for each $t > 0$, $N(t)$ is a discrete random variable with values 0, 1, ..., we know that its expectation equals

$$E[N(t)] = \sum_{n=1}^{\infty} P(N(t) \geq n). \tag{17}$$

However, $N(t) \geq n$ if and only if $T(n) \leq t$, and since $T(n) = X(0) + \cdots + X(n-1)$, the distribution functions of the continuous random variable $T(n)$,

$$F_{T(n)}(t) = P(T(n) \leq t), \qquad n = 1, 2, \ldots, \tag{18}$$

can, at least in principle, be calculated by repeated convolution of the density $f_X(x)$. Thus, we have from (17) and (18),

$$\mu(t) = \sum_{n=1}^{\infty} F_{T(n)}(t), \qquad t \geq 0. \tag{19}$$

Notice that (19) implies that the renewal function must be nondecreasing and continuous and that $\mu(0) = 0$ and $\mu(t) \to \infty$ as $t \to \infty$, but we already know this from Section 7.2.

Example 7.8.6

Consider an integrator with white Gaussian noise applied to its input at $t = 0$. The output of the integrator is fed into a circuit that generates an impulse each time the integrator output reaches level $\alpha > 0$ while at the same time resetting the integrator back to zero. The resulting stream of impulses forms a renewal point process with inverse Gaussian waiting time distribution

$$f_X(x) = \frac{\alpha}{\sqrt{2\pi x^3}} e^{-\alpha^2/2x}, \qquad x > 0,$$

which should be obvious from Figure 7.8.2 and the fact that the integrator output between consecutive resets is a standard Wiener process.

Now as can be verified from the characteristic function the sum of independent inverse Gaussian random variables with parameters α_1 and α_2 is again inverse Gaussian random with parameter $\alpha_1 + \alpha_2$. Consequently, the nth random epoch $T(n) = X(0) + \cdots + X(n-1)$ of our renewal process has density

$$f_{T(n)}(t) = \frac{\alpha n}{\sqrt{2\pi t^3}} e^{-(\alpha n)^2/2t}, \qquad t > 0.$$

The corresponding distribution function is easily found to be

$$F_{T(n)}(t) = \sqrt{\frac{2}{\pi}} \int_{t^{-1/2}}^{\infty} \alpha n e^{-(\alpha n \tau)^2/2}\, d\tau, \qquad t > 0,$$

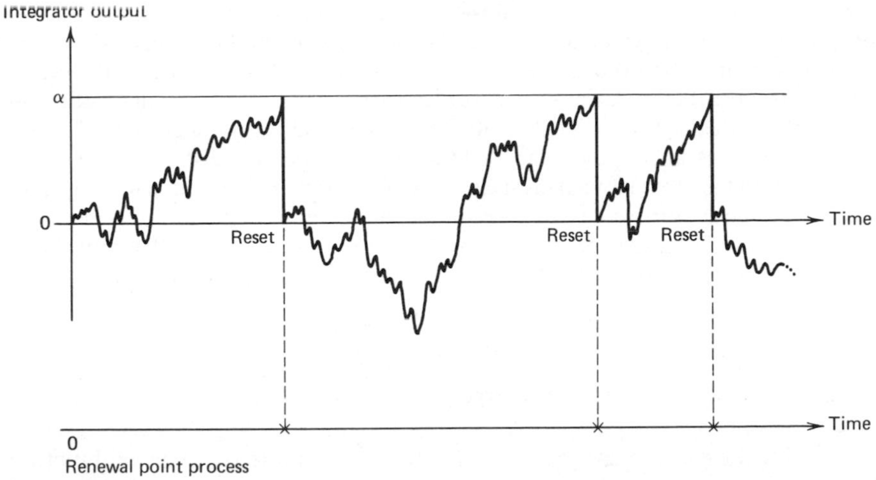

Figure 7.8.2

and hence we have, from (19), an expression for the renewal function

$$\mu(t) = \frac{2}{\pi} \sum_{n=1}^{\infty} \int_{t^{-1/2}}^{\infty} \alpha n e^{-(\alpha n \tau)^2/2}\, d\tau, \qquad t > 0.$$

Although this expression cannot be simplified, it can be used to estimate the asymptotic behavior of the renewal function for large t. Interchanging integration with summation, we have the convergent series

$$\sum_{n=1}^{\infty} \alpha n e^{-(\alpha n \tau)^2/2},$$

the sum of which may be roughly approximated by the integral

$$\int_0^{\infty} \alpha y e^{-(\alpha y \tau)^2/2}\, dy = \frac{1}{\alpha \tau^2}.$$

Since

$$\int_{t^{-1/2}}^{\infty} \frac{1}{\tau^2}\, d\tau = \sqrt{t},$$

we may expect that for large t the renewal function $\mu(t)$ increases at the rate $\sqrt{t}$.

Although our method can hardly be called precise (and much better ways can be designed), this asymptotic rate is basically correct. ▲

Remark: Equation (19) can be used to prove an earlier statement that the renewal function $\mu(t)$, $t \geq 0$, is always finite or, which is the same thing, that a renewal process always satisfies assumption (ii) of Section 7.1; it cannot explode. In Exercise 1.10 we saw that if $F_n(x)$ is the distribution function of n i.i.d. random variables with a common distribution function $F(x)$, then $F_n(x) \leq (F(x))^n$ for all x. Thus, if $t \geq 0$ in (19) is such that $F_X(t) < 1$, then the infinite series on the right-hand side of (19) is upperbounded by a convergent geometric series and hence is finite. If t is such that $F_X(t) = 1$, then none of the waiting times $X(n)$, $n = 0, 1, \ldots$, can ever have value exceeding t. But being i.i.d. positive random variables the sum of a sufficiently large number, say k, of them must with positive probability exceed t. Thus, $P(T(k) \leq t) = F_k(t) < 1$. Consequently, as above, $F_{2k}(t) \leq (F_k(t))^2$, $F_{3k}(t) \leq (F_k(t))^3$, and so forth, and since (see Exercise 1.10)

$$F_1(t) \geq F_2(t) \geq F_3(t) \geq \cdots$$

the infinite series in (19) must again converge. Я

The renewal equation can be used to actually calculate the renewal function $\mu(t)$ from the given distribution function $F(x)$. But this may be quite difficult and is, in fact, almost as difficult as calculating it directly from (19). However, the renewal equation is useful in finding the asymptotic behavior of the renewal function $\mu(t)$ for large t.

Denoting by

$$\mu_X = \int_0^\infty x f_X(x)\, dx$$

the common expectation of the identically distributed waiting times $X(n)$, $n = 0, 1, \ldots$, it can be shown, although we will not prove it here, that the following *renewal theorem* is true.

THEOREM 7.8.1. *As* $t \to \infty$,

$$\frac{N(t)}{t} \to \frac{1}{\mu_X} \qquad \textit{(almost surely)}, \tag{20a}$$

$$\frac{\mu(t)}{t} \to \frac{1}{\mu_X}, \tag{20b}$$

and for any $\tau > 0$ *the increment*

$$\mu(t+\tau) - \mu(t) \to \frac{\tau}{\mu_X}. \tag{20c}$$

The theorem remains true even if

$$\mu_X = \int_0^\infty x f_X(x)\, dx = \infty$$

in which case, all three limits above are zero. Note that the statements of this renewal theorem have a great intuitive appeal, for instance, (20c) says, in effect, that for large t, the mean rate of occurrence of point-events equals approximately the reciprocal of the mean waiting time between consecutive occurrences.

Example 7.8.7

By Theorem 7.8.1 the renewal function $\mu(t)$ of the renewal process of Example 7.8.3 is asymptotically linear,

$$\frac{\mu(t)}{t} \to 2t \qquad \text{and} \qquad \mu(t+\tau) - \mu(t) \to 2\tau \qquad \text{as } t \to \infty.$$

For the renewal process of registered particles by a paralyzable counter with Poisson input, we found in Example 7.8.5 that

$$\mu_Z = \frac{1}{\lambda} e^{\lambda \Delta}.$$

It follows that for large t, its renewal function satisfies

$$\mu(t+1) - \mu(t) \doteq \lambda e^{-\lambda \Delta}.$$

Since λ is the average number of particles arriving at the counter in a unit of time, this means that for large t (" in the steady state " if you prefer), on the average, only 100 $e^{-\lambda \Delta}$ percent of the arriving particles are registered. Also, we could use Theorem 7.8.1 to estimate the actual intensity λ of the Poisson process of particle arrivals by observing the number $N(t)$ of registrations during a sufficiently long period of time t. An estimate for the unknown intensity can then be calculated from the equation

$$\frac{N(t)}{t} \doteq \lambda e^{-\lambda \Delta}.$$

Finally, notice that for the renewal process of Example 7.8.6, we have

$$\frac{N(t)}{t} \to 0, \qquad \text{as } t \to \infty,$$

since the inverse Gaussian distribution has infinite expectation. Thus the point-events of this process will be occurring at a very low rate. This, of course, is due to the fact that about half of the time the Wiener process on the integrator output will make a long excursion below the zero level before it returns and reaches the positive level α. ▲

Remark: From the simple fact that for any $t > 0$ and $n = 1, 2, \ldots$

$$N(t) \geq n \qquad \text{if and only if} \qquad T(n) \leq t$$

we could actually obtain an expression for the distribution of $N(t)$. This is quite easy; for $n = 1, 2, ..$ and $t > 0$,

$$P(N(t) = n) = P(N(t) \geq n) - P(N(t) \geq n + 1)$$
$$= P(T(n) \leq t) - P(T(n + 1) \leq t) = F_{T(n)}(t) - F_{T(n+1)}(t), \qquad (21a)$$

while for $n = 0$,

$$P(N(t) = 0) = P(T(1) > t) = 1 - F_{T(1)}(t). \qquad (21b)$$

Since $T(n) = X(0) + \cdots + X(n - 1)$ with the waiting times i.i.d. with a common density $f_X(x)$, we could in principle obtain the density of $T(n)$ by repeated convolution and thus also evaluate the distribution of $N(t)$ from (21). Unfortunately, an explicit expression for the density of $T(n)$ for all $n = 1, 2, \ldots$ can be obtained only in a few cases, and thus the expression (21) is of little practical use.

However, if the common distribution $F_X(x)$ of the waiting time sequence possesses a finite variance, the central limit theorem implies that, for large n, the distribution of $T(n)$ will be approximately Gaussian. This suggests that for large $t > 0$, the distribution of $N(t)$ should also be approximately Gaussian. This is indeed the case, and although a formal proof is not particularly difficult, we merely present this useful result as the following theorem.

Я

THEOREM 7.8.2. *If* N(t), t ≥ 0, *is the counting process associated with a renewal point process whose waiting time distribution has mean* $\mu_X < \infty$ *and variance* $\sigma_X^2 < \infty$, *then for large* t *the distribution of* N(t) *is approximately Gaussian with mean* t/μ_X *and variance* $t\sigma_X^2/\mu_X^3$.

Example 7.8.8

In the renewal process of Example 7.8.3, the waiting times were assumed to be uniformly distributed over (0, 1). Thus $\mu_X = \frac{1}{2}$ and $\sigma_X^2 = \frac{1}{12}$, and so by Theorem 7.8.2 the number of point events $N(t)$ is, for large t, approximately Gaussian with mean $2t$ and variance $\frac{2}{3}t$.

For the renewal process of Example 7.8.6, Theorem 7.8.2 cannot be used, since the waiting time distribution has infinite variance. However, this is one of the rare occasions where the distribution of $N(t)$ can be found explicitly, since we know the distributions of all random epochs $T(n)$, $n = 1, 2, \ldots$. The distribution function can be expressed as

$$F_{T(n)}(t) = 2\left(1 - \Phi\left(\frac{\alpha n}{\sqrt{t}}\right)\right), \qquad t > 0,$$

where $\Phi(x)$ is the standard Gaussian distribution function. Hence, from equation (21) we have, for any $t > 0$,

$$P(N(t) = n) = 2\left(\Phi\left(\alpha\frac{n+1}{\sqrt{t}}\right) - \Phi\left(\alpha\frac{n}{\sqrt{t}}\right)\right), \qquad n = 0, 1, \ldots,$$

which is the same as the distribution of the absolute value of a standard Gaussian random variable Z quantized by steps $\alpha/\sqrt{t}$, that is,

$$P(N(t) = n) = P\left(\frac{\alpha}{\sqrt{t}} n \leq |Z| < \frac{\alpha}{\sqrt{t}}(n+1)\right). \qquad \blacktriangle$$

STATIONARY RENEWAL PROCESSES. So far, we have restricted our attention to renewal point processes defined on a semi-infinite time domain $\mathcal{T} = (0, \infty)$. However, the very fact that such a renewal process is "switched on" at time $t = 0$ results in a transient behavior that typically begins to settle down to a steady state only as $t \to \infty$. It would then seem plausible that for studying the steady-state behavior of a renewal process it might be convenient to extend its time domain to $\mathcal{T} = (-\infty, \infty)$, thus pushing the era of its transient behavior into the infinitely remote past.

We have introduced a renewal process as a point process whose intensity process $\Lambda(t)$, $t \in \mathcal{T}$, depends on the past only through the time elapsed since the epoch of the last point-event occurring prior to the present time t and hence having the fundamental property that its waiting times $X(n)$ are independent random variables. In the case of the time domain $\mathcal{T} = (0, \infty)$, we have completed the definition by also requiring that the $X(n)$'s have common distribution function F_X, and it would seem that the same could be required also for a renewal process on $\mathcal{T} = (-\infty, \infty)$. But, as discussed in Section 7.1, to specify the probability law of a point process on the time domain $\mathcal{T} = (-\infty, \infty)$ the doubly infinite waiting times sequence

$$X(n), \; n = \ldots, -1, 0, 1, \ldots \tag{22}$$

must somehow be anchored to the time domain, and we have decided to do so by letting $X(0)$ be the waiting time between the two point-events whose epochs embrace the origin $t = 0$. Furthermore, it was also shown in Section 7.1 that we need to define more, for example, the first positive random epoch $T(1)$, to complete the specification of the point process. However, this particular choice of indexing the waiting time sequence (22) sets waiting time $X(0)$ apart from the rest, and thus it may be wise not to insist upon $X(0)$ sharing the same distribution with the rest of the $X(n)$'s in (22).

Thus, we will assume that the waiting time sequence $X(n)$, $n = \ldots, -1, 0, 1, \ldots$, consists of independent positive random variables, all except $X(0)$ having the same distribution with density $f_X(x)$. In addition (see Fig. 7.8.3) we have to define

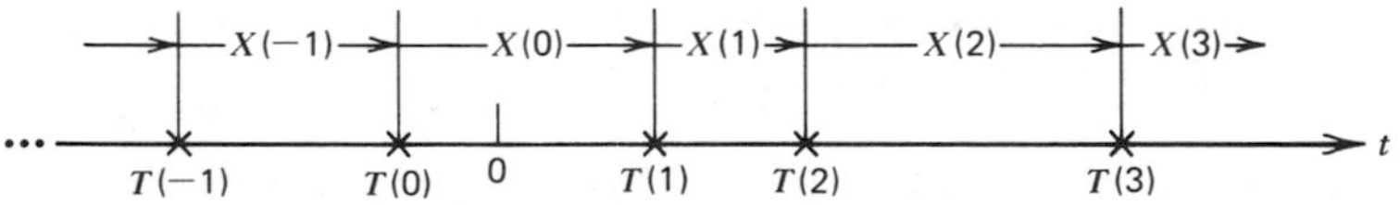

Figure 7.8.3

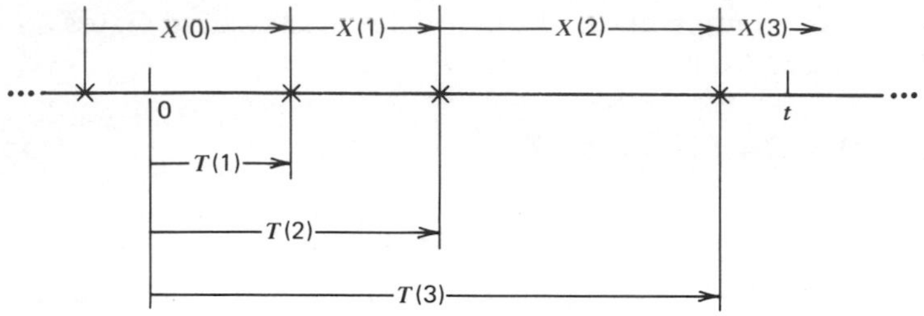

Figure 7.8.4

the distribution of the two epochs $T(0) \leq 0 \leq T(1)$, which we assume to be independent of the waiting times $X(n)$, $n = \pm 1, \pm 2, \ldots$. Their joint density $f_{T(0), T(1)}(t_0, t_1)$, together with the density $f_X(x)$, will thus provide a complete specification of the probability law of the renewal point process on $\mathcal{C} = (-\infty, \infty)$, including the waiting time $X(0)$, since clearly $X(0) = T(1) - T(0)$.

Now, let $N(t)$, $-\infty < t < \infty$, be the counting process associated with our renewal point process with the usual convention $N(0) = 0$. We wish again to determine an equation for the renewal function

$$\mu(t) = \mathrm{E}[N(t)], \qquad -\infty < t < \infty.$$

We do this by first considering the two portions $\mu(t)$, $t \geq 0$, and $\mu(t)$, $t \leq 0$, separately, and then joining them together in an appropriate fashion.

Let $t > 0$, and recall again that $N(t) \geq n$ if and only if $T(n) \leq t$, so that (see Fig. 7.8.4)

$$\begin{aligned} \mu(t) = \mathrm{E}[N(t)] &= \sum_{n=1}^{\infty} P(N(t) \geq n) = \sum_{n=1}^{\infty} P(T(n) \leq t) \\ &= P(T(1) \leq t) + P(T(1) + X(1) \leq t) + P(T(2) + X(3) \leq t) + \cdots \\ &= F_{T(1)}(t) + \sum_{n=1}^{\infty} P(T(n) + X(n) \leq t). \end{aligned} \tag{23}$$

However, for each $n = 1, 2, \ldots$, $T(n)$ and $X(n)$ are independent positive random variables with distribution functions $F_{T(n)}$ and $F_{X(n)} = F_X$, respectively, since the waiting times $X(n)$ are i.i.d. random variables with common distribution function F_X. It follows from the convolution formula for the sum of two independent random variables that for all $n = 1, 2, \ldots$

$$P(T(n) + X(n) \leq t) = \int_0^t F_{T(n)}(t - \tau) f_X(\tau)\, d\tau.$$

Thus, substituting into (23) and interchanging integration and summation (which can be justified on the basis of (19)), we get

$$\mu(t) = F_{T(1)}(t) + \int_0^t \left[\sum_{n=1}^{\infty} F_{T(n)}(t-\tau)\right] f_X(\tau)\, d\tau.$$

However, by (23), we also have

$$\sum_{n=1}^{\infty} F_{T(n)}(t-\tau) = \sum_{n=1}^{\infty} P(T(n) \le t-\tau) = \mu(t-\tau)$$

and thus we have derived the renewal equation

$$\mu(t) = F_{T(1)}(t) + \int_0^t \mu(t-\tau) f_X(\tau)\, d\tau, \qquad t \ge 0 \tag{24}$$

for the portion of the renewal function on $t \ge 0$. Notice that in the case of a renewal point process on $\mathscr{T} = (0, \infty)$, we have $T(1) = X(0)$. Then equation (24) would reduce to the renewal equation (9), so that as a special case we have here another derivation of the renewal equation (9).

Now for $t \le 0$, exactly the same reasoning can be applied, yielding the renewal equation for the remaining half of the renewal function

$$\mu(t) = F_{T(0)}(t) - 1 + \int_0^{-t} \mu(t+\tau) f_X(\tau)\, d\tau, \qquad t \le 0. \tag{25}$$

(A simple way to obtain this from (24) is to replace $T(1)$ by $-T(0)$, and $\mu(t)$ by $-\mu(-t)$.)

Since we wish the renewal point process to be stationary the renewal function must have the form

$$\mu(t) = \lambda t, \qquad -\infty < t < \infty,$$

where $\lambda > 0$ is a constant, the intensity of the process. If we substitute this into renewal equation (24), we obtain, upon integrating by parts,

$$\begin{aligned}
\lambda t &= F_{T(1)}(t) + \lambda \int_0^t (t-\tau) f_X(\tau)\, d\tau \\
&= F_{T(1)}(t) + \lambda(t-\tau) F_X(\tau)\Big|_{\tau=0}^{\tau=t} + \lambda \int_0^t F_X(\tau)\, d\tau \\
&= F_{T(1)}(t) + \lambda \int_0^t F_X(\tau)\, d\tau, \qquad t \ge 0,
\end{aligned}$$

since $F_X(0) = 0$. Hence,

$$F_{T(1)}(t) = \lambda t - \lambda \int_0^t F_X(\tau)\, d\tau = \lambda \int_0^t (1 - F_X(\tau))\, d\tau, \qquad t \ge 0,$$

and since any distribution function must tend to 1 as $t \to \infty$, we obtain from this condition

$$1 = \lambda \int_0^\infty (1 - F_X(\tau))\, d\tau.$$

However, the integral

$$\int_0^\infty (1 - F_X(\tau))\, d\tau = \int_0^\infty x f_X(\tau)\, d\tau = \mu_X$$

so that

$$\lambda = \frac{1}{\mu_X}$$

provided $\mu_X < \infty$. The distribution function and the density of the epoch $T(1)$ are then

$$F_{T(1)}(t) = \frac{1}{\mu_X} \int_0^t (1 - F_X(\tau))\, d\tau, \qquad t \geq 0$$

$$f_{T(1)}(t) = \frac{1}{\mu_X}(1 - F_X(t)) = \frac{1}{\mu_X} \int_t^\infty f_X(x)\, dx, \qquad t \geq 0. \tag{26}$$

If $\mu_X = \infty$, the renewal equation (24) cannot have a solution of the desired form $\mu(t) = \lambda t$.

For $t \geq 0$, we again substitute $\mu(t) = \lambda t$ into the renewal equation (25), and again using integration by parts, we end up with

$$1 - F_{T(0)}(t) = \lambda \int_0^{-t} (1 - F_X(\tau))\, d\tau, \qquad t \leq 0,$$

from which, again, $\lambda = 1/\mu_X$, and the density of $T(0)$ is

$$f_{T(0)}(t) = \frac{1}{\mu_X}(1 - F_X(-t)) = \frac{1}{\mu_X} \int_{-t}^\infty f_X(x)\, dx. \tag{27}$$

It follows that $-T(0)$ and $T(1)$ are identically distributed. However, they are not necessarily independent, and it remains for us to find their joint distribution. Let $f_{T(0),\,T(1)}(t_0, t_1)$ be their joint density, which because of $T(0) \leq 0 < T(1)$ can be nonzero only for $t_0 < 0 < t_1$. If the renewal point process is to be stationary, that is, invariant with respect to a time shift, this joint density can actually depend only on the difference $t_1 - t_0$. Thus, we must have

$$f_{T(0),\,T(1)}(t_0, t_1) = f(t_1 - t_0), \qquad t_0 < 0 < t_1,$$

where the function f on the right-hand side must be such that

$$\int_{-\infty}^{0} f(t_1 - t_0)\, dt_0 = f_{T(1)}(t_1), \qquad 0 < t_1, \tag{28}$$

and

$$\int_{0}^{\infty} f(t_1 - t_0)\, dt_1 = f_{T(0)}(t_0), \qquad t_0 < 0, \tag{29}$$

to yield the marginal densities obtained earlier. After making the substitution $x = t_1 - t_0$, the integrals on the left-hand side of (28) and (29) become

$$\int_{t_1}^{\infty} f(x)\, dx \qquad \text{and} \qquad \int_{-t_0}^{\infty} f(x)\, dx$$

respectively, and comparison with (26) and (27) reveals immediately that the function f must be equal to $1/\mu_X$ times the density f_X. Hence,

$$f_{T(0),\,T(1)}(t_0, t_1) = \frac{1}{\mu_X} f_X(t_1 - t_0), \qquad t_0 < 0 < t_1 \tag{30}$$

is the joint density of $T(0)$ and $T(1)$. The density of the waiting time $X(0) = T(1) - T(0)$ can now be evaluated from (30):

$$f_{X(0)}(x) = \int_{-\infty}^{\infty} f_{T(0),\,T(1)}(t - x, t)\, dt = \frac{1}{\mu_X} \int_{0}^{x} f_X(x)\, dt$$

that is,

$$f_{X(0)}(x) = \frac{1}{\mu_X} x f_X(x), \qquad x > 0. \tag{31}$$

We now summarize these findings as a theorem:

THEOREM 7.8.3. *A renewal point process defined on the time domain* $\mathscr{C} = (-\infty, \infty)$ *is a stationary point process if and only if the waiting time distribution* $F_X(x)$ *has finite expectation*

$$\mu_X = \int_0^{\infty} x f_X(x)\, dx$$

and the epochs $T(0) \le 0 < T(1)$ *have density*

$$f_{T(0),\,T(1)}(t_0, t_1) = \begin{cases} \dfrac{1}{\mu_X} f_X(t_1 - t_0) & \text{if} \quad t_0 < 0 < t_1, \\ 0 & \text{otherwise.} \end{cases}$$

The process then has intensity $\lambda = 1/\mu_X$, and the densities of the waiting time $X(0)$ *and the epochs* $T(0)$ *and* $T(1)$ *are given by* (31), (26), *and* (27), *respectively.*

Once a point process on $\mathcal{T} = (-\infty, \infty)$ is stationary, there is nothing special about the time $t = 0$, since the entire probability law of the process is invariant with respect to a shift of time. It follows that for any fixed time instant $t \in \mathcal{T}$, the spent, excess, and current waiting times at t, as defined at the end of Section 7.4, will have the same distribution as if $t = 0$.

We thus have the following corollary.

Corollary

For a stationary renewal point process on $\mathcal{T} = (-\infty, \infty)$ and any fixed $t \in \mathcal{T}$, the spent waiting time $Y_-(t)$, the excess waiting time $Y_+(t)$, and the current waiting time $Y(t)$ have the same distribution as the random variables $-T(0)$, $T(1)$, and $X(0)$, respectively.

Since a renewal process is a reasonable model of successive replacements of some component, these results are of considerable practical interest.

Example 7.8.9

Consider a stationary renewal point process on $\mathcal{T} = (-\infty, \infty)$ with linearly increasing hazard function $h(t) = \alpha t$, $\alpha > 0$. In Example 7.8.2, we found that the corresponding waiting time distribution is Rayleigh with density

$$f_X(x) = \alpha x e^{-\alpha x^2/2}, \qquad x > 0. \tag{32}$$

According to Theorem 7.8.3 the joint density of the random epochs $T(0)$ and $T(1)$ must be chosen as

$$f_{T(0),\,T(1)}(t_0, t_1) = \frac{1}{\mu_X}\alpha(t_1 - t_0)e^{-\alpha(t_1 - t_0)^2/2}, \qquad t_0 \le 0 < t_1,$$

where μ_X is the mean of the Rayleigh distribution (32),

$$\mu_X = \int_0^\infty \alpha x^2 e^{-\alpha x^2/2}\, dx = \sqrt{\frac{\pi}{2\alpha}}.$$

The resulting stationary renewal process then has intensity

$$\lambda = \frac{1}{\mu_X} = \sqrt{\frac{2\alpha}{\pi}}.$$

The corollary to Theorem 7.8.3 then implies that for any fixed $t \in \mathcal{T}$, the joint density of the spent time $Y_-(t)$ and the excess time $Y_+(t)$ is

$$f_{Y_-(t),\, Y_+(t)}(y_1, y_2) = \sqrt{\frac{2}{\pi}}\,\alpha^{3/2}(y_1 + y_2)e^{-\alpha(t_1 - t_2)^2/2},$$

$$y_1 > 0,\ y_2 > 0.$$

The spent and excess times are identically distributed with density

$$f_{Y_-(t)}(y) = f_{Y_+(t)}(y) = \sqrt{\frac{2\alpha}{\pi}}\,e^{-(\alpha/2)y^2}, \qquad y > 0.$$

Note that this is the density of the absolute value of a Gaussian random variable with zero mean and variance $1/\alpha$ and that the expectation $\mathrm{E}[Y_-(t)] = \mathrm{E}[Y_+(t)] = \sqrt{2/\pi\alpha} = (2/\pi)\mu_X$, is a constant fraction of μ_X regardless of the value of x.

The current waiting time $Y(t)$ has the Maxwell density

$$f_{Y(t)}(y) = \sqrt{\frac{2}{\pi}}\,\alpha^{3/2}y^2e^{-(\alpha/2)y^2}, \qquad y > 0,$$

and expectation

$$\mathrm{E}[Y(t)] = 2\sqrt{\frac{2}{\pi\alpha}} = 2\mu_X.$$

We have here another manifestation of the waiting time paradox discussed earlier in connection with a Poisson process. If we had a large number of independent replicas of our stationary renewal process and attempted to estimate the intensity $\lambda = 1/\mu_X$ by selecting some fixed time t and calculating the average current waiting time at this t, our observed average would be close to $2\mu_X$ so that we would have underestimated the true intensity λ by a factor of 2.

Finally, since the waiting time distribution (32) has finite variance, we see that

$$\sigma_X^2 = \frac{1}{\alpha}\left(2 - \frac{\pi}{2}\right).$$

Theorem 7.8.2 implies that the number of point-events with epochs in a time interval $(t, t + \tau]$ of sufficiently large length $\tau > 0$ will, for any $t \in \mathcal{T}$, be approximately Gaussian with mean

$$\frac{\tau}{\mu_X} = \tau\sqrt{\frac{2\alpha}{\pi}}.$$

and variance

$$\frac{\tau\sigma_X^2}{\mu_X^3} = \tau\sqrt{\frac{2\alpha}{\pi}}\left(\frac{4}{\pi} - 1\right). \qquad \blacktriangle$$

Remark: The results obtained here for a stationary renewal process give an indication of why the renewal theorem (Theorem 7.8.1) should be true. Suppose we have a renewal process on $\mathcal{T} = (0, \infty)$ and together with it a stationary renewal process with the same waiting time distribution. Then it is at least plausible that sooner or later there will be a time when both of these processes will produce point-events arbitrarily close to each other. Once that happens, the two processes will be probabilistically indistinguishable and hence the renewal process on $\mathcal{T} = (0, \infty)$ will have, for large t, intensity $\lambda(t)$ almost identical with the intensity $\lambda = 1/\mu_X$ of the stationary renewal process. Note that this would be true even if the first waiting time $X(0)$ of the renewal process had a distribution different from the common distribution of the rest of the waiting times, that is, if it were the so-called *delayed renewal process*.

Thus, we may say that as long as $\mu_X < \infty$, every renewal process tends to a steady state as $t \to \infty$.

Я

The theory of renewal processes is fairly extensive, and many further results and generalizations have been obtained. Unfortunately, we have to abandon this topic after barely scratching the surface, both because of lack of space and, more importantly, because we lack techniques for a more rigorous and detailed treatment.

MARKOVIAN POINT PROCESSES. Another important class of past-dependent point processes is obtained by allowing the intensity process to depend on the past evolution only through the *number* of point-events that have occurred up to the present time instant.

That is, if

$$N(t), \qquad t \geq 0$$

is the associated counting process, we now assume that the intensity process satisfies

$$\Lambda(t) = g(t, N(t)), \qquad t > 0, \tag{33}$$

where $g(t, n)$, $t > 0$, $n = 0, 1, \ldots$, is a nonnegative function, integrable with respect to t over any finite interval, and such that for all $n = 0, 1, \ldots$

$$\int_0^t g(\tau, n)\, d\tau \to \infty \qquad \text{as } t \to \infty.$$

Since the future evolution of a past-dependent point process depends on its past only through the present value of the intensity process, it is intuitively obvious that if the latter satisfies (33) the counting process $N(t)$, $t \geq 0$, will be a *Markov process*.

We will not give a formal proof of this statement but only remind the reader of the heuristic interpretation of the random intensity process:

$$P(dN(t) = 1 \mid N(t'), 0 < t' \leq t) \doteq \Lambda(t)\, dt,$$

and

$$P(dN(t) = 0 \mid N(t'), 0 < t' \leq t) \doteq 1 - \Lambda(t)\, dt.$$

Thus if $\Lambda(t) = g(t, N(t))$, then these relations become

$$P(dN(t) = 1 \mid N(t'), 0 < t' \le t) \doteq g(t, N(t))\, dt \tag{34}$$

and

$$P(dN(t) = 0 \mid N(t'), 0 < t' \le t) \doteq 1 - g(t, N(t))\, dt, \tag{35}$$

which suggests that the conditional distribution of $dN(t)$, given the past $N(t')$, $0 < t' \le t$, depends only on the present value $N(t)$, that is, the Markov property.

We can get further corroboration of this from the distribution of the waiting time sequence (3a,b). With the functions g in (2) depending only on the first two arguments t and n, we see from (3b) that, conditional upon the nth point-event occurring at epoch t_n, the distribution function of the subsequent waiting time $X(n)$ equals

$$1 - e^{-\int_0^x g(\tau + t_n,\, n)\, d\tau}, \qquad x > 0. \tag{36}$$

(For the first waiting time the distribution function is also given by (36) if we set $t_0 = 0$.) Since for a point process evolving without aftereffects, that is, an inhomogeneous Poisson process with intensity function $\lambda(t)$, $t > 0$, the corresponding distribution function is

$$1 - e^{-\int_0^x \lambda(\tau + t_0)\, d\tau}, \qquad x > 0.$$

We see that the only difference in this present case is the dependence on the number of point-events n.

Thus our past-dependent point process is almost like an inhomogeneous Poisson process except that upon each occurrence of a point-event its intensity may change, depending on the total number of point-events that have occurred up to that moment.

These processes provide a simple model for the growth of populations, for example, bacterial colonies, where it is assumed that the current birth rate depends only on the current population size (and possibly time) and that individuals do not die.

For this reason, a past-dependent point process with intensity process satisfying (33) is called an (inhomogeneous) *pure birth process.*

Example 7.8.10

A simple example of such a pure birth process is obtained by considering a point process on $\mathcal{T} = (0, \infty)$ that begins as a homogeneous Poisson process with intensity $\lambda > 0$ and, upon each occurrence of a point-event, produces its own probabilistic replica, that is, gives birth to a new homogeneous Poisson process with the same intensity λ. This is the same mechanism as in Example 7.7.2, except that this time all the intensities are constants and hence independent of the time of birth. Thus, if for some $t > 0$, we have $N(t) = n$, then during an interval $(t, t + dt]$ there are $n + 1$ independent homogeneous Poisson processes running together, each with intensity λ, and hence

$$\Lambda(t) = \lambda + \lambda N(t), \qquad t > 0,$$

depends only on $N(t)$. Clearly, in this case the term "birth process" is fully justified.

Sometimes, however, this term may not seem appropriate. Suppose we have a system that alternates between two possible states 0 and 1 (e.g., "off" and "on"). Each time the system is in state 0, it stays there for a random amount of time Y_0 exponentially distributed with parameter λ_0 before switching to state 1. Similarly, it stays in state 1 for a random time Y_1, again exponentially distributed but with a different parameter λ_1 before switching to state 0. Furthermore, these consecutive random times form a sequence of independent random variables. The instants at which the system is changing its state can thus be regarded as epochs of a point process. Assuming that the initial state of the system is 0, the associated waiting time sequence

$$X(n), \qquad n = 0, 1, \dots$$

is then a sequence of independent exponentially distributed random variables, the exponential distribution of $X(n)$ having parameter λ_0 or λ_1, depending on whether n is even or odd. The memoryless property of the exponential distribution then implies that for any $t > 0$, the conditional probability

$$P(dN(t) = 1 \mid N(t'), 0 < t' \le t) \doteq \begin{cases} \lambda_0\, dt & \text{if} \quad N(t) \text{ is even,} \\ \lambda_1\, dt & \text{if} \quad N(t) \text{ is odd.} \end{cases}$$

Thus, $\Lambda(t) = g(N(t))$ depends on $N(t)$ only, that is, (33) is satisfied, and our point process is, by definition, a pure birth process even if we may not like the term in this case. ▲

Granted that the counting process $N(t)$, $t \ge 0$, is Markov, it is only necessary to determine its transition probabilities to completely characterize its probability law, since by convention, $N(0) = 0$. Recall that $N(t)$, $t \ge 0$, is a counting process, and as such, all its sample paths must be step functions beginning at zero and increasing by unit jumps only. Thus, we only need to find the conditional probabilities

$$P(N(t_2) = n_2 \mid N(t_1) = n_1) \tag{37}$$

for any $0 \le t_1 \le t_2$ and nonnegative integers $0 \le n_1 \le n_2$.

Let $(t_2, t_2 + dt_2]$ be an infinitesimal interval so that we can assume that the probability of having more than one point-event occurring is negligible. Then we can write, using our customary differential notation,

$$\begin{aligned} &P(N(t_2 + dt_2) = n_2 \,|\, N(t_1) = n_1) \\ &= P(dN(t_2) = 0,\ N(t_2) = n_2 \,|\, N(t_1) = n_1) \\ &+ P(dN(t_2) = 1,\ N(t_2) = n_2 - 1 \,|\, N(t_1) = n_1), \end{aligned} \tag{38}$$

where the last term is zero if $n_2 = n_1$. Now

$$\begin{aligned} &P(dN(t_2) = 0,\ N(t_2) = n_2 \,|\, N(t_1) = n_1) \\ &\quad = P(dN(t_2) = 0 \,|\, N(t_1) = n_1,\ N(t_2) = n_2) P(N(t_2) = n_2 \,|\, N(t_1) = n_1). \end{aligned}$$

However, $N(t)$, $t \geq 0$, is a Markov process, and hence

$$\begin{aligned} &P(dN(t_2) = 0 \,|\, N(t_1) = n_1,\ N(t_2) = n_2) \\ &\quad = P(dN(t_2) = 0 \,|\, N(t_2) = n_2) \doteq 1 - g(t_2, n_2)\, dt_2, \end{aligned}$$

where we have used (35) for the last approximate equality. Applying the same procedure to the second term on the right-hand side of (38), but using (34) instead of (35), we get, after substitution back into (38), the equation

$$\begin{aligned} P(N(t_2 + dt_2) = n_2 \,|\, N(t_1) = n_1) &= [1 - g(t_2, n_2)\, dt_2] P(N(t_2) = n_2 \,|\, N(t_1) = n_1) \\ &+ g(t_2, n_2 - 1)\, dt_2\, P(N(t_2) = n_2 - 1 \,|\, N(t_1) = n_1). \end{aligned} \tag{39}$$

Let us denote the conditional probabilities (37) by

$$p(t_2, n_2 \,|\, t_1, n_1), \qquad 0 \leq t_1 \leq t_2; \qquad n_1 = 0, 1, \ldots; \qquad n_2 = 0, 1, \ldots,$$

and let us agree to regard this symbol as zero whenever $n_2 < n_1$. Removing the brackets on the right-hand side of (39), transferring the term with coefficient 1 to the left, and dividing by dt_2, we obtain the differential equation

$$\begin{aligned} \frac{dp(t_2, n_2 \,|\, t_1, n_1)}{dt_2} &= -[g(t_2, n_2) p(t_2, n_2 \,|\, t_1, n_1) \\ &+ g(t_2, n_2 - 1) p(t_2, n_2 - 1 \,|\, t_1, n_1)], \end{aligned} \tag{40}$$

defined for $t_2 \geq t_1 \geq 0$ (t_1 fixed), with initial conditions

$$p(t_2, n_2 \,|\, t_1, n_1)\Big|_{t_2 = t_1} = \begin{cases} 1 & \text{if} \quad n_1 = n_2, \\ 0 & \text{if} \quad n_1 \neq n_2. \end{cases}$$

This is the *Kolmogorov-Feller forward equation* for the pure birth process. In particular, for $t_1 = 0$ and $n_1 = 0$, we get the differential equation for the unconditional probability

$$p(t, n) = P(N(t) = n),$$

sometimes referred to as the *Fokker-Planck equation*:

$$\frac{dp(t, n)}{dt} = -g(t, n)p(t, n) + g(t, n-1)p(t, n-1), \qquad t \geq 0, \tag{41}$$

$n = 0, 1, \ldots$, where $p(t, -1) = 0$ and

$$p(0, n) = \begin{cases} 1 & \text{if} \quad n = 0 \\ 0 & \text{if} \quad n > 0. \end{cases}$$

It should be noted that both (40) and (41) are actually infinite systems of differential equations (one for each value of $n = 0, 1, \ldots$), or in other words, these are partial difference-differential equations for a function of two variables t and n, and obtaining a general solution is not a simple matter.

Nevertheless, it may be worthwhile to take a closer look at the special case of a *homogeneous pure birth process*, that is, one such that

$$g(t, n) = g(n), \qquad t > 0, \qquad n = 0, 1, \ldots,$$

depends on n only. The conditional probabilities

$$P(N(t_2) = n_2 \mid N(t_1) = n_1), \qquad 0 < t_1 \leq t_2$$

will, in this case, depend only on the time difference $t_2 - t_1$; in other words, the Markov counting process $N(t)$, $t \geq 0$, will have stationary increments. For any fixed $t_0 \geq 0$, call the conditional probability

$$P(N(t_0 + t) = n + m \mid N(t_0) = n) = p(t, m \mid n), \qquad t \geq 0; \qquad n = 0, 1, \ldots; \tag{42}$$

$m = 0, 1, \ldots$, and then the Kolmogorov-Feller differential equation (40) takes the form

$$\dot{p}(t, m \mid n) + g(n + m)p(t, m \mid n)$$
$$= \begin{cases} 0 & \text{if} \quad m = 0, \\ g(n + m - 1)p(t, m - 1 \mid n) & \text{if} \quad m = 1, 2, \ldots, \end{cases} \tag{43a}$$

$t \geq 0$, with initial condition

$$p(0, m \mid n) = \begin{cases} 1 & \text{if} \quad m = 0, \\ 0 & \text{if} \quad m = 1, 2, \ldots, \end{cases} \tag{43b}$$

with $\dot{p} = (d/dt)p$ denoting the time derivative. For $m = 0$, we then have immediately

$$p(t, 0 \mid n) = e^{-g(n)t}, \qquad t \geq 0, \qquad n = 0, 1, \ldots. \tag{44}$$

For $m > 0$ the solution of (43) can be expressed as

$$p(t, m \mid n) = g(n + m - 1)e^{-g(n+m)t} \int_0^t e^{g(n+m)\tau} p(\tau, m - 1 \mid n)\, d\tau; \tag{45}$$

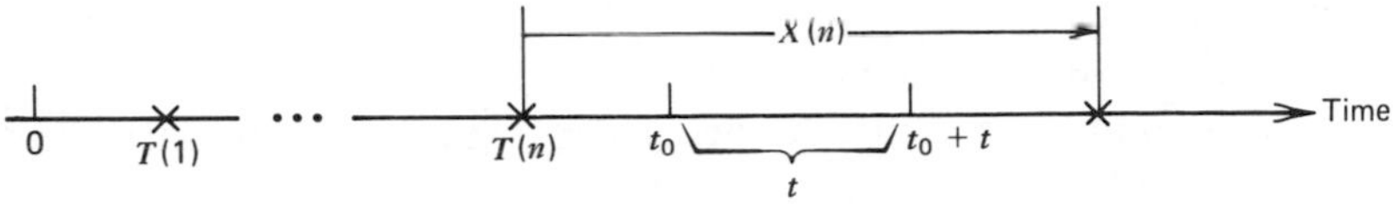

Figure 7.8.5

$t \geq 0$, $n = 0, 1, \ldots$, as is easily verified by substitution into (43a,b). This represents a recursive formula from which one can successively obtain $p(t, 1 \mid n)$, $p(t, 2 \mid n), \ldots$, starting with $p(t, 0 \mid n)$ given by (44). We will illustrate this shortly with a simple example, but first we would like to point out a particular consequence of equation (45). Recall that

$$p(t, 0 \mid n) = P(N(t_0 + t) = n \mid N(t_0) = n),$$

which, as a result of the Markov property, is the same as

$$P(N(t_0 + t) - N(t_0) = 0 \mid T(1), \ldots, T(n), N(t_0) = n)$$

and we conclude that $e^{-g(n)t}$ is the conditional probability of having no point-event with epoch in the interval $(t_0, t + t]$, given that exactly n point-events have occurred prior to time t_0. But (see Fig. 7.8.5) this is the same as saying that the nth waiting time $X(n)$ has the property that

$$P(X(n) > x + t \mid X(n) > x) = e^{-g(n)t}, \qquad t \geq 0$$

for any $x \geq 0$, since t_0 was an arbitrary time instant. However, since $X(n)$ is a positive random variable, the only continuous distribution with this property is the exponential distribution, and hence $X(n)$ must be exponentially distributed with parameter $\lambda_n = g(n)$. Furthermore, it can also be deduced from the Markov property that thc waiting times are independent, and hence we come to an important conclusion:†

For a homogeneous pure birth process the waiting time sequence $X(n)$, $n = 0, 1, \ldots$, *is a sequence of independent, exponentially distributed random variables with parameters* $\lambda_n = g(n)$.

This statement, in effect, completely characterizes the interval properties of a homogeneous pure birth process. But it can also be useful in other respects. With a pure birth point process, there is always the danger that if the function $g(n)$ increases with $n = 0, 1, \ldots$ sufficiently fast, the point-events may be occurring with increasing frequency, so that eventually an infinite number of them may occur within a finite time interval $(0, t]$. In other words, without some condition restric-

† This should be no surprise to the reader who has already studied Chapter Three.

ting the growth of the function $g(n)$, our assumption (ii) of Section 7.1 may be violated, and the process may explode.

To find such a condition, consider an increasing sequence of random epochs

$$0 < T(1) < T(2) < T(3) < \cdots$$

Clearly, an explosion will occur if and only if

$$\lim_{n\to\infty} T(n) < \infty,$$

which will certainly be the case if

$$\lim_{n\to\infty} \mathrm{E}[T(n)] < \infty.$$

However,

$$\mathrm{E}[T(n)] = \sum_{k=0}^{n-1} \mathrm{E}[X(n)] = \sum_{k=0}^{n-1} \frac{1}{g(k)}$$

since $X(k)$ is an exponential random variable with $\lambda_k = g(k)$. It follows that if the infinite series with positive terms

$$\sum_{k=0}^{\infty} \frac{1}{g(k)}$$

converges to a finite sum the homogeneous pure birth process is certain to explode. Conversely, it can be shown that if

$$\sum_{k=0}^{\infty} \frac{1}{g(k)} = +\infty$$

the homogeneous pure birth process will have zero probability of explosion. Thus, divergence of the infinite series is both necessary and sufficient to guarantee that the homogeneous pure birth process has only a finite number of point-events occurring in any finite interval, that is, that assumption (ii) is satisfied.

Example 7.8.11

Consider a homogeneous pure birth process with

$$g(n) = (n+1)\lambda, \qquad n = 0, 1, \ldots,$$

where $\lambda > 0$ is a constant. This is known as the *Yule-Furry process*, and in the first part of Example 7.8.10 we described one particular case in which such a process may be encountered.

Since the infinite series

$$\sum_{k=0}^{\infty} \frac{1}{g(k)} = \frac{1}{\lambda}\left(1 + \frac{1}{2} + \frac{1}{3} + \frac{1}{4} + \cdots\right)$$

diverges to $+\infty$, the process cannot explode. Let us now calculate the conditional probabilities $p(t, m \mid n)$ for the associated Markov counting process $N(t)$, $t \geq 0$, as defined by (42). With $g(n) = (n+1)\lambda$, we have from (44)

$$p(t, 0 \mid n) = e^{-(n+1)\lambda t}, \qquad t \geq 0, \qquad n = 0, 1, \ldots.$$

The recurrence formula (45) is, in this case,

$$p(t, m \mid n) = (n+m)\lambda e^{-(n+m+1)\lambda t} \int_0^t e^{(n+m+1)\lambda t} p(\tau, m-1 \mid n)\, d\tau.$$

Setting $m = 1$, we get

$$\begin{aligned} p(t, 1 \mid n) &= (n+1)\lambda e^{-(n+2)\lambda t} \int_0^t e^{(n+2)\lambda\tau} e^{-(n+1)\lambda\tau}\, d\tau \\ &= (n+1)\lambda e^{-(n+2)\lambda t} \int_0^t e^{\lambda\tau}\, d\tau \\ &= (n+1)e^{-(n+2)\lambda t}(e^{\lambda t} - 1). \end{aligned}$$

Setting $m = 2$ and using the result just obtained, we have

$$\begin{aligned} p(t, 2 \mid n) &= (n+2)\lambda e^{-(n+3)\lambda t} \int_0^t e^{(n+3)\lambda\tau}(n+1)e^{-(n+2)\lambda\tau}(e^{\lambda\tau} - 1)\, d\tau \\ &= (n+1)(n+2)\lambda e^{-(n+3)\lambda t} \int_0^t e^{\lambda\tau}(e^{\lambda\tau} - 1)\, d\tau \\ &= \frac{(n+1)(n+2)}{2} e^{-(n+3)\lambda t}(e^{\lambda t} - 1)^2. \end{aligned}$$

Sensing an emerging pattern, let us make an induction hypothesis:

$$p(t, m-1 \mid n) = \frac{(n+1)(n+2)\cdots(n+m-1)}{(m-1)!} e^{-(n+m)\lambda t}(e^{\lambda t} - 1)^{m-1},$$

and substitute into the recurrence formula. We obtain

$$p(t, m \mid n) = \frac{(n+1)(n+2)\cdots(n+m)}{m!} e^{-(n+m+1)\lambda t} \int_0^t m\lambda e^{\lambda\tau}(e^{\lambda\tau} - 1)^{m-1}\, d\tau,$$

where we have divided and multiplied by m and inserted λ behind the integral sign. The reason for doing this is that the integrand is the derivative

$$m\lambda e^{\lambda\tau}(e^{\lambda\tau} - 1)^{m-1} = \frac{d}{d\tau}(e^{\lambda\tau} - 1)^m,$$

and thus the integral equals $(e^{\lambda\tau} - 1)$ and the induction hypothesis is confirmed.

We have found the general formula

$$p(t, m \mid n) = \frac{(n+1)\cdots(n+m)}{m!} e^{-(n+m+1)\lambda t}(e^{\lambda t} - 1)^m,$$

which we now rewrite in a more elegant form:

$$p(t, m \mid n) = \binom{n+m}{m} e^{-(n+1)\lambda t}(1 - e^{-\lambda t})^m,$$

$t \geq 0, m = 0, 1, \ldots, n = 0, 1, \ldots$. Note that for t and $n > 0$ fixed, this is a modified negative binomial distribution with parameters $n + 1$ and $e^{-\lambda t}$.

With $n = 0$, we get the probability

$$p(t, m \mid 0) = P(N(t) = m) = e^{-\lambda t}(1 - e^{-\lambda t})^m, \qquad t \geq 0, \qquad m = 0, 1, \ldots,$$

so that $N(t)$ has a geometric distribution with parameter $e^{-\lambda t}$. It follows that the mean function of the Yule-Furry process is

$$\mu(t) = e^{\lambda t} - 1, \qquad t \geq 0.$$

Also from $P(T(m) \leq t) = P(N(t) \geq m) = (1 - e^{-\lambda t})^m$, $t \geq 0$, $m - 1, 2, \ldots$, we obtain the density of the mth random epoch

$$f_{T(m)}(t) = m\lambda e^{-\lambda t}(1 - e^{-\lambda t})^{m-1}, \qquad t \geq 0.$$

Interestingly, this is the same as the density of the maximum of m i.i.d. exponential random variables with parameter λ. ▲

Remark: A considerable generalization of the type of past-dependent point processes discussed above is obtained if we equip the point process with a mark sequence

$$U(n), \qquad n = 1, 2, \ldots.$$

As before, let

$$N(t), \qquad t \geq 0$$

be the counting process associated with the unmarked point process, and let

$$N_U(t), \qquad t \geq 0$$

be the associated mark accumulator process as defined in Section 7.5. Suppose now that we allow the intensity process $\Lambda(t)$, $t > 0$, (defined by means of the counting process $N(t)$ in the usual way) depend on the current value of the *mark accumulator process*

$$\Lambda(t) = g(t, N_U(t)), \qquad t \geq 0. \tag{46}$$

We also allow the distribution of the nth mark $U(n)$ to depend on the sum of all previous marks or, in other words, on the value of the mark accumulator process just prior to the occurrence of the nth point-event, the one carrying the mark $U(n)$.

It can then be shown that the *mark accumulator process*

$$N_U(t), \qquad t \geq 0, \tag{47}$$

is again a *Markov process*. This, after all, is quite plausible. The waiting times between consecutive point-events will still resemble those of an inhomogeneous Poisson process except that its intensity is changing after each point-event, depending on the current sum of all previous marks. The sequence of mark sums

$$S(n) = U(1) + \cdots + U(n); \qquad n = 1, 2, \ldots; \qquad S(0) = 0,$$

that is, the sequence of consecutive values of the mark accumulator process $N_U(t)$ between jumps is clearly a Markov sequence by the assumption made above.

In fact, barring irregularities like those excluded in Section 7.1 (explosions, extinctions, etc.) the mark accumulator process (47) is the most general continuous-time Markov jump process,† that is, a process whose sample paths are constant except for randomly occurring jumps of random magnitude.

A simple example of such a process is obtained if the mark sequence

$$U(n), \qquad n = 1, 2, \ldots$$

is a sequence of i.i.d. random variables taking only values $+1$ and -1 with probabilities α and $\beta = 1 - \alpha$, respectively. In this case the mark accumulator process

$$N_U(t), \qquad t \geq 0$$

has as its state space the set of all integers $n = \ldots, -1, 0, 1, \ldots$, and it only remains to specify the function

$$g(t, n)$$

in (46). The resulting process is called an (inhomogeneous) *birth and death process*, and it can be easily shown that its transition probabilities

$$p(t_2, n_2 | t_1, n_1) = P(N_U(t_2) = n_2 | N_U(t_1) = n_1),$$

satisfy the Kolmogorov-Feller forward equation:

$$\frac{dp(t_2, n_2 | t_1, n_1)}{dt} = -g(t_2, n_2)p(t_2, n_2 | t_1, n_1) + \alpha g(t_2, n_2 - 1)p(t_2, n_2 - 1 | t_1, n_1)$$
$$+ \beta g(t_2, n_2 + 1)\, p(t_2, n_2 + 1 | t_1, n_1), \qquad t_2 \geq t_1 \geq 0, \qquad (t_1 \text{ fixed}),$$

again with

$$p(t_2, n_2 | t_1, n_1)\Big|_{t_2 = t_1} = \begin{cases} 1 & \text{if} \quad n_1 = n_2 \\ 0 & \text{if} \quad n_1 \neq n_2. \end{cases}$$

† For the case of integer-valued marks, it is a continuous-time Markov chain studied in Chapter Three.

The derivation proceeds along the same lines as that for the pure birth process, except that this time we have to begin with three possible values for the increment $dN_U(t_2)$. We have either

$$dN_U(t_2) = 0 \quad \text{if} \quad dN(t_2) = 0,$$

that is, if there is no point-event with epoch in $(t_2, t_2 + dt]$, or

$$dN_U(t_2) = +1 \quad \text{or} \quad -1,$$

the mark value, if $dN(t_2) = 1$, that is, if there is a point-event with epoch in this interval. Consequently, we now have

$$P(dN_U(t) = 0 \mid N_U(t) = n) \doteq 1 - g(t, n)\, dt,$$

$$P(dN_U(t) = +1 \mid N_U(t) = n) \doteq \alpha g(t, n)\, dt,$$

$$P(dN_U(t) = -1 \mid N_U(t) = n) \doteq \beta g(t, n)\, dt,$$

and the rest is easy.

It should be clear from this example that a Kolmogorov-Feller forward equation (as well as the Fokker-Planck equation) can be constructed for any Markovian mark accumulator process (47). Of course, to actually solve such an equation is another matter. Я

DOUBLY STOCHASTIC POINT PROCESSES. Suppose we are given an ordinary continuous-time stochastic process

$$S(t), \qquad t > 0,$$

and a causal transformation $\mathcal{H}$ that maps the process $S(t)$, $t > 0$, into a new process

$$Y(t), \qquad t > 0. \tag{48}$$

Suppose further that the transformation $\mathcal{H}$ is such that all sample paths $y(t)$, $t > 0$, of the process (48) are nonnegative functions of t, integrable over any finite time interval, and such that

$$\int_0^t y(\tau)\, d\tau \to \infty \qquad \text{as } t \to \infty.$$

Then each sample path $y(t)$, $t \to 0$, qualifies as the intensity function for an inhomogeneous Poisson process, and thus for each of these sample paths, we can generate an inhomogeneous Poisson process with time domain $\mathcal{T} = (0, \infty)$ by setting its intensity function

$$\lambda(t) = y(t), \qquad t \geq 0.$$

This entire scheme is depicted in Figure 7.8.6. Clearly, this represents a basic scheme for intensity modulation of a Poisson point process that is typical in optical communication systems and in similar systems where the carrier is of an

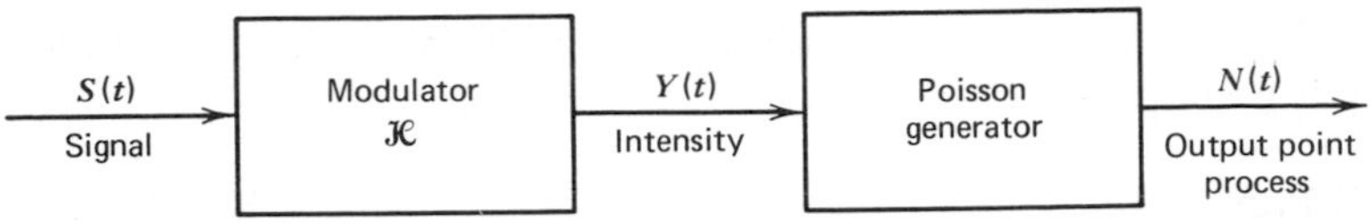

Figure 7.8.6

impulsive nature. The process $S(t)$, $t > 0$, is an information-carrying *signal*, and the transformation $\mathcal{H}$ is a modulator, which processes the signal to control the rate of emission of photons from the transmitter (Poisson generator), for example, a laser.

We will assume here that the probability law of the signal process $S(t)$, $t > 0$, above does not depend on the output point process, although if a feedback link between the output and the signal is present, this need not be the case. We will also assume for simplicity that the modulator $\mathcal{H}$ performs just a memoryless transformation, that is, that

$$Y(t) = h(S(t)), \qquad t > 0.$$

For each fixed sample path $s(t)$, $t > 0$, of the signal process the output point process will just be an inhomogeneous Poisson process, that is, the *conditional* probability law of the output, given $S(t) = s(t)$, $t > 0$, will be that of a Poisson process with intensity function $\lambda(t) = h(s(t))$. Thus, for instance, if $N(t)$, $t \geq 0$, is the associated counting process of the output point process, we have for any $0 < t_1 < t_2$, $n = 0, 1, \ldots,$

$$P(N(t_2) - N(t_1) = n \,|\, S(t') = s(t'),\ t' \geq 0) = \frac{[\int_{t_1}^{t_2} h(s(\tau))\, d\tau]^n}{n!} e^{-\int_{t_1}^{t_2} h(s(\tau))\, d\tau}, \qquad (49)$$

and *conditional* upon $S(t) = s(t)$, $t > 0$, the output process will evolve without aftereffects. However, the *unconditional probability law* of the output process will no longer evolve without aftereffects, since an information-carrying signal $S(t)$, $t > 0$, by its very nature, exhibits a considerable dependence between its values at different times, and this will make the output process a past-dependent point process.

Of course, one may say that all that must be done to turn a conditional probability law into an unconditional one is to remove the conditioning by taking its expectation. This is true, for instance, from (49), we can write immediately

$$P(N(t_2) - N(t_1) = n) = \mathrm{E}\left[\frac{[\int_{t_1}^{t_2} h(S(\tau))\, d\tau]^n}{n!} e^{-\int_{t_1}^{t_2} h(S(\tau))\, d\tau}\right], \qquad (50)$$

$0 < t_1 < t_2$, $n = 0, 1, \ldots,$ but the problem is to actually evaluate this expectation. In fact, except for extremely simple kinds of signal processes, it is virtually impossible to obtain an explicit form for this expression, and various advanced techniques must be employed to obtain at least some properties of the probability law of the output point process.

Example 7.8.12

Suppose that the signal $S(t)$ is a random sine wave of unit frequency

$$S(t) = A\cos(2\pi t + \Theta), \qquad -\infty < t < \infty,$$

where A is a Rayleigh random variable with density

$$f_A(a) = \frac{a}{\sigma^2} e^{-a^2/2\sigma^2}, \qquad a > 0,$$

while Θ is uniformly distributed over the interval $(0, 2\pi)$. The modulator is just a squaring device,

$$Y(t) = S^2(t),$$

and so by (50) the probability

$$P(N(t) = n) = \mathrm{E}\left[\frac{\left(\int_0^t S^2(\tau)\,d\tau\right)^n}{n!} e^{-\int_0^t S^2(\tau)\,d\tau}\right]$$

$$= \int_0^\infty \int_0^{2\pi} \frac{\left[\int_0^t a^2\cos^2(2\pi\tau + \theta)\,d\tau\right]^n}{n!} e^{-\int_0^t a^2\cos^2(2\pi\tau+\theta)\,d\tau} \frac{1}{2\pi}\frac{a^2}{\sigma^2} e^{-a^2/2\sigma^2}\,da\,d\theta,$$

$$n = 0, 1, \ldots,$$

$t > 0$. The integral is rather hard to evaluate, but if we are satisfied with finding the above probability only for times $t > 0$ that are integer multiples of $\frac{1}{2}$, $t = k/2$ for some $k = 1, 2, \ldots$, there is an easy way to do it. If we write the random sine wave in its quadrature decomposition

$$S(t) = Z_1 \cos 2\pi t - Z_2 \sin 2\pi t,$$

where now Z_1 and Z_2 are i.i.d. Gaussian random variables with zero mean and variance σ^2, we find that

$$S^2(t) = Z_1^2 \cos^2 2\pi t - Z_1 Z_2 \sin 4\pi t + Z_2^2 \sin^2 2\pi t.$$

With $t = k/2$ the integral

$$\int_0^t \sin 4\pi\tau\,d\tau = 0,$$

and

$$\int_0^t \cos^2 2\pi\tau\,d\tau = \int_0^t \sin^2 2\pi\tau\,d\tau = t,$$

so that

$$\int_0^t S^2(\tau)\, d\tau = t(Z_1^2 + Z_2^2).$$

The characteristic function

$$\psi_{N(t)}(\zeta) = \mathrm{E}[e^{\imath\zeta N(t)}], \qquad -\infty < \zeta < \infty,$$

can be evaluated by conditioning on Z_1 and Z_2. Since $N(t)$, given Z_1, Z_2, is Poisson with parameter $t(Z_1^2 + Z_2^2)$,

$$\begin{aligned} \mathrm{E}[e^{\imath\zeta N(t)} \mid Z_1, Z_2] &= e^{(e^{\imath\zeta}-1)t(Z_1^2+Z_2^2)} \\ &= e^{(e^{\imath\zeta}-1)tZ_1^2} e^{(e^{\imath\zeta}-1)tZ_2^2}. \end{aligned}$$

Since Z_1 and Z_2 are i.i.d. Gaussian random variables, we obtain, by taking the expectation of the above equation,

$$\begin{aligned} \psi_{N(t)}(\zeta) &= \mathrm{E}[e^{(e^{\imath\zeta}-1)tZ_1^2}]\mathrm{E}[e^{(e^{\imath\zeta}-1)tZ_2^2}] \\ &= \left(\int_{-\infty}^{\infty} e^{(e^{\imath\zeta}-1)tz^2} \frac{1}{\sqrt{2\pi\sigma^2}} e^{-z^2/2\sigma^2}\, dz\right)^2 \\ &= \left(\int_{-\infty}^{\infty} \frac{1}{\sqrt{2\pi\sigma^2}} e^{-z^2(1-2\sigma^2(e^{\imath\zeta}-1))/2\sigma^2}\, dz\right)^2 \\ &= (1 - (e^{\imath\zeta} - 1)t\sigma^2)^{-1}. \end{aligned}$$

Calling $p = (1 + t\sigma^2)^{-1}$ and $q = 1 - p$, the characteristic function can be written as

$$\psi_{N(t)}(\zeta) = \frac{p}{1 - qe^{\imath\zeta}}, \qquad -\infty < \zeta < \infty,$$

which is the characteristic function of a geometric distribution. Consequently,

$$P(N(t) = n) = pq^n = \frac{(t\sigma^2)^n}{(1 + t\sigma^2)^{n+1}}, \qquad n = 0, 1, \ldots,$$

which, however, has been derived only for $t = k/2$, a multiple of $\frac{1}{2}$, the half period of the random sine wave. ▲

There is another approach to the problem which, although not at all easier, provides an alternative to the above. Since the probability law of a past-dependent point process is (under some conditions) determined by the probability law of its intensity process

$$\Lambda(t), \qquad t > 0,$$

we may attempt first to get some information about the latter. Since in the present case, we have

$$\mathrm{E}[dN(t) \mid N(t'), S(t'), 0 < t' \le t] \doteq h(S(t))\, dt, \tag{51}$$

$h(s(t))$ being the intensity function of the Poisson process, given $S(t') = s(t')$, $0 < t' \le t$, and since the intensity process $\Lambda(t)$ is defined by the conditional expectation

$$\mathrm{E}[dN(t) \mid N(t'), 0 < t' \le t] \doteq \Lambda(t)\, dt,$$

removing the conditioning on $X(t')$, $0 \le t' \le t$ from (51) gives us the intensity process

$$\Lambda(t) = \mathrm{E}[h(S(t)) \mid N(t'), 0 < t' \le t], \qquad t \ge 0.$$

This does not look simpler than (50), and indeed it is not. However, it is the quantity that is of prime interest in most applications, because it is the best MS estimate of $h(S(t))$ based on the observation $N(t')$, $0 \le t' \le t$ (since it is the conditional expectation). This is exactly what we want to get in the communication problem—we want to recover the signal from the observation made on the modulated point process. (Presumably, the modulator does not destroy information, so once $y(t) = h(s(t))$ is estimated, $s(t)$ itself can be recovered.)

Example 7.8.13

A rather trivial example of a doubly stochastic point process is obtained if the signal process is just a random variable

$$S(t) = S \qquad \text{for all } -\infty < t < \infty.$$

Then the process $Y(t) = h(S(t)) = Y > 0$ is also a random variable, and we will assume that its density is $f_Y(y)$, where, of course, $f_Y(y) = 0$ for $y \le 0$. Thus, for any $t > 0$ and $n = 0, 1, \ldots$

$$P(N(t) = n) = \int_0^\infty e^{-yt} \frac{(yt)^n}{n!} f_Y(y)\, dy$$

since the process can be looked upon as a homogeneous Poisson process whose intensity $\lambda = y$ has been chosen at random according to the density $f_Y(y)$.

This time, however, we wish to find an expression for the intensity process

$$\Lambda(t) = \mathrm{E}[Y \mid N(t'), 0 < t' \le t].$$

Conditional upon $Y = y$ the process is Poisson, and hence

$$f_{T(1), \ldots, T(N(t)), N(t) \mid Y}(t_1, \ldots, t_n, n \mid y) = y^n e^{-yt},$$
$$0 < t_1 < \cdots < t_n \le t, \quad n = 0, 1, \ldots, \quad y > 0.$$

The joint density then is

$$f_{T(1), \ldots, T(N(t)), N(t), Y}(t_1, \ldots, t_n, n, y) = y^n e^{-yt} f_Y(y)$$

and thus the conditional density of Y, given the past $T(1), \ldots, T(N(t)), N(t)$, is

$$f_{Y|T(1), \ldots, T(N(t)), N(t)}(y \mid t_1, \ldots, t_n, n) = \frac{y^n e^{-yt} f_Y(y)}{\int_0^\infty y^n e^{-yt} f_Y(y)\, dy}.$$

Consequently, the conditional expectation

$$\Lambda(t) = \mathrm{E}[Y \mid N(t'), 0 < t' \leq t]$$
$$= \frac{\int_0^\infty y^{n+1} e^{-yt} f_Y(y)\, dy}{\int_0^\infty y^n e^{-yt} f_Y(y)\, dy}, \tag{52}$$

which depends on the past $N(t')$, $0 < t' \leq t$, only via the number $n = N(t)$ of the point-events occurring in $(0, t]$. It follows that the resulting doubly stochastic point process is, in this case, an inhomogeneous pure birth process with the function $g(t, n)$ given by the right-hand side of (52). In particular, if

$$f_Y(y) = e^{-y}, \qquad y > 0,$$

we get

$$g(t, n) = \frac{n+1}{t+1}, \qquad n = 0, 1, \ldots, \qquad t > 0$$

that is,

$$\Lambda(t) = \frac{N(t) + 1}{t + 1}.$$

▲

We will not discuss this problem any further because of its inherent difficulty. Our main reason for including this topic was to introduce the reader to this problem and to stress its importance in applications.

We conclude our discussion by reminding the reader that the three classes of past-dependent point processes discussed here by no means exhaust all possibilities. There are many other types, each with its own specific constructions and techniques and each quite important in a variety of applications. However, even a classification of these types would require a great deal more space than we can afford in this volume.

EXERCISES

Exercise 7.8

1. Calculate and sketch the hazard function $h(t)$, $t > 0$, for a renewal point process whose waiting times have the density:

(a) $f_X(x) = \frac{1}{\sqrt{2x}} e^{-\sqrt{2x}}, \qquad x \geq 0.$

(b) $f_X(x) = 4x^2e^{-2x}, \qquad x \geq 0.$

(c) $f_X(x) = e^{-(e^x - x - 1)}, \qquad x \geq 0.$

If one time unit has elapsed since the last point-event, what is the probability that no point-event will occur during the subsequent time unit?

2. Consider a renewal point process on $\mathcal{T} = (0, \infty)$ with the hazard function $h(\tau) = \tau(1 + \tau)^{-1}$, $\tau \in \mathcal{T}$. Calculate the conditional density of the intensity process $\Lambda(t)$ at $t > 0$ given that $N(t) = n$, $n = 0, 1, \ldots$. (*Hint:* Express $\Lambda(t)$ as a function of $t - T(n)$, and identify the density of $T(n)$.)

3. Write the renewal equation for the renewal point process with the hazard function $h(\tau) = \sqrt{\tau}$, $\tau > 0$. Determine the asymptotic behavior of the mean function $\mu(t)$ as $t \to \infty$.

4. Consider a renewal point process on $\mathcal{T} = (0, \infty)$ with waiting time density $f_X(x) = \alpha^2 xe^{-\alpha x}$, $x \geq 0$, where $\alpha > 0$ is a parameter. Use the fact that the random epochs $T(n)$ have Erlang distributions to obtain an explicit expression for the mean function $\mu(t)$, $t > 0$. Verify by substitution that your mean function satisfies the renewal equation (9). (*Hint:* Take the derivative of (19) and sum the series.)

5. Calculate the variance of the random variable Z of Example 7.8.5. Use the result to obtain an approximate expression for the mean and the variance of the number of particle registrations during $(0, t]$, where t is large.

6. Unlike the paralyzable counter described in Example 7.8.4, in some scintillation detectors an input impulse locks the counter only if it arrived while the counter was unlocked. In other words, pulses arriving while the counter is locked have no effect on the counter except that they are not registered. Assuming that the input pulses are arriving according to a homogeneous Poisson process and that at $t = 0$ the counter is locked, show that the time instances of pulse registrations are epochs of a renewal process. Find the hazard function $h(t)$, $t > 0$, for this process. Also calculate the expected number of registrations per unit of time if the process is assumed to have reached a steady state.

7. A stationary renewal process on $\mathcal{T} = (-\infty, \infty)$ has hazard function $h(\tau) = (1 + \beta\tau)^{-1}$, $\tau > 0$, where β is a positive constant. Find the density of the epoch $T(1)$ and of the waiting time $X(0)$.

8. It is desired to simulate a stationary renewal point process by recording the point-events as pulses on a long magnetic tape. In future use, the tape will always be started at some arbitrary place and will never run the entire length. If the waiting time density is to be

$$f_X(x) = \frac{4}{3}e^{-2x} + \frac{1}{3}e^{-x}, \qquad x > 0,$$

how should the density of the first epoch $T(1)$ be chosen? Does it matter if the tape is run backward from its end as the starting point?

9. Consider a particle counter that is just the opposite of the locking counters; it cannot register a single particle unless its arrival was preceded by another arrival within a time interval not greater than Δ. Figure 7.8.E.1 illustrates the input and output point

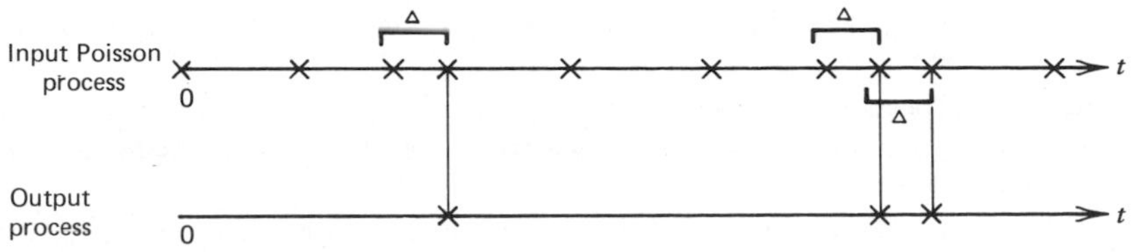

Figure 7.8.E.1

processes of such a counter. Assuming that the input is a homogeneous Poisson point process, show that the output is a renewal point process. Imitate the reasoning of Example 7.8.4 to derive a renewal-type equation for the waiting time density of the output process. (Assume $t = 0$ is an arrival epoch.)

10. Under the assumptions of Exercise 7.8.9, use the method of Example 7.8.5 to obtain the characteristic function of the waiting time distribution. Use the result to obtain the expected number of particle registrations per unit of time, assuming that the process has reached a steady state.

11. In the counter described in Exercise 7.8.9, each arriving particle may be thought of as delivering a small instantaneous charge to the counter sensor. The charge begins to decay immediately, and so another charge delivered within no more than Δ is needed to activate the counter circuit. Once the circuit is activated, it depletes the charge instantaneously, and so, unlike Exercise 7.8.9, *two* new arrivals within time interval Δ are needed to produce the next output point event. (See Fig. 7.8.E.2 for the effect of this modification.) Now repeat Exercises 7.8.9 and 7.8.10 for this type of counter.

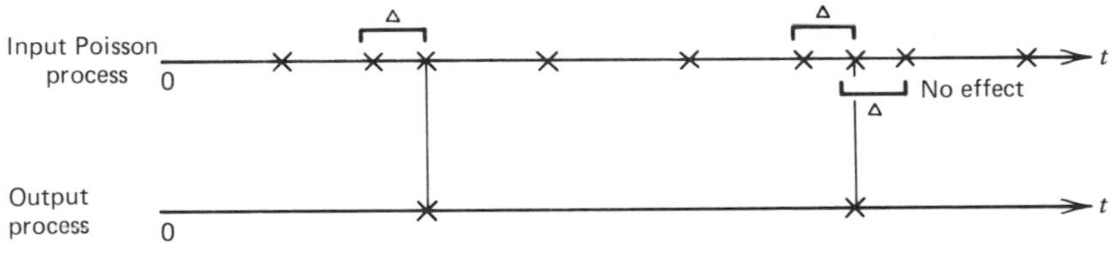

Figure 7.8.E.2

12. Consider a stationary renewal process on $\mathcal{T} = (-\infty, \infty)$ whose waiting time density is gamma,

$$f_X(x) = \frac{\lambda^\nu}{\Gamma(\nu)} x^{\nu-1} e^{-\lambda x}, \qquad x > 0,$$

with $\nu > 0$ and $\lambda > 0$ arbitrary parameters. Obtain an expression for the means and variances of the spent waiting time $Y_-(t)$, excess waiting time $Y_+(t)$, and current waiting time $Y(t)$ at $t \in \mathcal{T}$. Also calculate the covariance between $Y_-(t)$ and $Y_+(t)$.

13. Consider a Markovian point process on $\mathcal{T} = (0, \infty)$ at the end of Example 7.8.10, that is, such that

$$g(t, n) = \begin{cases} \lambda_0 & \text{if} \quad n \text{ is even (including } n = 0), \\ \lambda_1 & \text{if} \quad n \text{ is odd,} \end{cases}$$

where $0 < \lambda_0 < \lambda_1$. Use the Fokker-Planck equation to evaluate the probabilities $P(N(t) = n)$ for $n = 0, 1, 2$.

14. In the point process of Exercise 7.8.13, suppose every odd-numbered point-event is deleted. Show that the result is a renewal point process, and evaluate its waiting time density.
15. Consider a homogeneous pure birth process where $g(n) = \ln(2 + n)$, $n = 0, 1, \ldots$. Use the recurrence (45) to obtain the transition probabilities $p(t, 0 | 0)$, $p(t, 1 | 0)$, and $p(t, 2 | 0)$. Check whether or not the process explodes.
16. For the Yule-Furry process of Example 7.8.11, find the conditional expectation $E[N(t_2) | N(t_1) = n]$, $n = 0, 1, \ldots, 0 < t_1 < t_2$. Use the result to find the correlation function $R_N(t_1, t_2)$ of the counting process $N(t)$, $t > 0$.
17. Consider again the Yule-Furry process of Example 7.8.11, but assume now that each point-event carries a mark $U(n)$, $n = 0, 1, \ldots$. The marks are independent both mutually and of the Yule-Furry process and are all exponentially distributed with the same parameter α. Find the density of the mark accumulator process $N_U(t)$ at a fixed time $t > 0$.
18. Consider a doubly stochastic point process on $\mathscr{C} = (0, \infty)$, where the process $Y(t)$, $t > 0$, at the output of the modulator is defined by

$$Y(t) = 1 + M \cos \omega t, \qquad t > 0, \qquad \omega > 0,$$

with the modulation index M being a random variable uniformly distributed over $(0, 1)$. Use formula (49) to find the probability mass distribution $P(N(t) = n)$, $n = 0, 1, \ldots$, for $t > 0$.
19. Repeat Exercise 7.8.18 for the case when

$$Y(t) = \frac{2Zt}{1 + t^2}, \qquad t > 0,$$

with Z a random variable having the Erlang density

$$f_Z(z) = \frac{\alpha^3}{2} z^2 e^{-\alpha z}, \qquad z \geq 0, \qquad \alpha > 0.$$

20. Show that the doubly stochastic point process of Exercise 7.8.19 is, in fact, a Markovian point process with intensity process $\Lambda(t) = g(t, N(t))$, $t > 0$, and find the function $g(t, n)$.

APPENDIX ONE

THE DIRAC DELTA-FUNCTION

In the natural sciences and engineering it is often convenient, and sometimes even necessary, to consider finite masses, charges, impulses, and the like that are concentrated at a single point in space or time. A useful mathematical device to represent such singularities is the so-called Dirac delta function, denoted by the symbol δ. Formally, the delta function is defined by the equation

$$f(y) = \int_{-\infty}^{+\infty} \delta(x) f(y-x)\, dx = \int_{-\infty}^{+\infty} \delta(x-y) f(x)\, dx \tag{1}$$

which is assumed to be valid for any function f and any real number y, provided only that f is continuous in some open interval $(y-\varepsilon, y+\varepsilon)$ containing the point y.

Taking the function $f(y) = 1$ for all $-\infty < y < +\infty$, this implies that

$$\int_{-\infty}^{+\infty} \delta(x)\, dx = 1. \tag{2}$$

On the other hand, since the value of the integral $\int_{-\infty}^{+\infty} \delta(y-x) f(x)\, dx$ is required to depend only on the values of the function f in an arbitrarily small neighborhood of the point y, we ought to have

$$\delta(x) = 0 \qquad \text{for all } x \neq 0. \tag{3}$$

Clearly, no such function can exist in the realm of standard calculus, since no matter what value we would assign to $\delta(0)$, property (3) would force us to conclude that $\int_{-\infty}^{+\infty} \delta(x)\, dx = 0$ rather than 1 as required by (2). Nevertheless the intention behind the idea of the delta function becomes clear if we regard it as an idealized impulse of unit area

$$\delta(x) = \begin{cases} \dfrac{1}{\Delta x} & \text{if} \quad -\dfrac{\Delta x}{2} < x < \dfrac{\Delta x}{2}, \\ 0 & \text{otherwise} \end{cases} \tag{4}$$

where $\Delta x > 0$ is an "arbitrarily small" positive number. (See Fig. A.1.1.)

With such an interpretation the first integral (1) becomes approximately

$$\int_{-\infty}^{+\infty} \delta(x) f(y-x)\, dx \doteq \frac{1}{\Delta x} f(y-0)\, \Delta x = f(y),$$

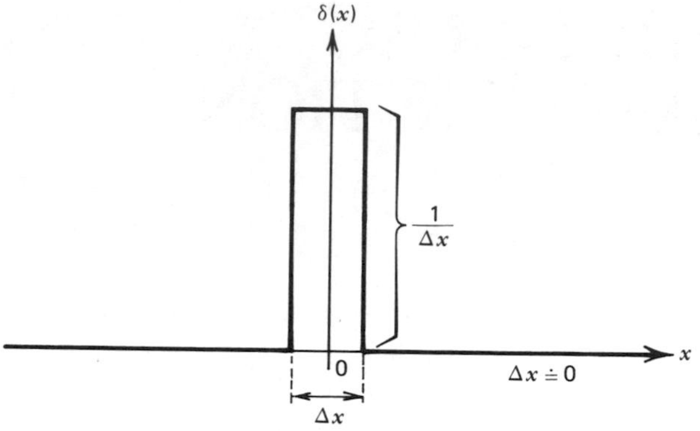

Figure A.1.1

the approximation being better the less $f(y)$ varies over the interval $(y - \Delta x/2, y + \Delta x/2)$. Although, as pointed out, we cannot let $\Delta x \to 0$ without running into the contradicting requirement (2), it is possible to define the delta function rigorously either by regarding it as a generalized function† or even perhaps by using the definition (4) with Δx a genuine infinitesimal as in the so-called nonstandard analysis. However, for our present purposes, we will be satisfied with the intuitive interpretation as described above. Note that we could use an impulse of any shape as the approximating impulse for the delta function as long as its area is 1 and more and more of it tends to concentrate closer and closer to zero. A double exponential impulse in Figure A.1.2, with $\varepsilon > 0$ but "almost zero," is an example.

The most important relation involving the delta function is equation (1),

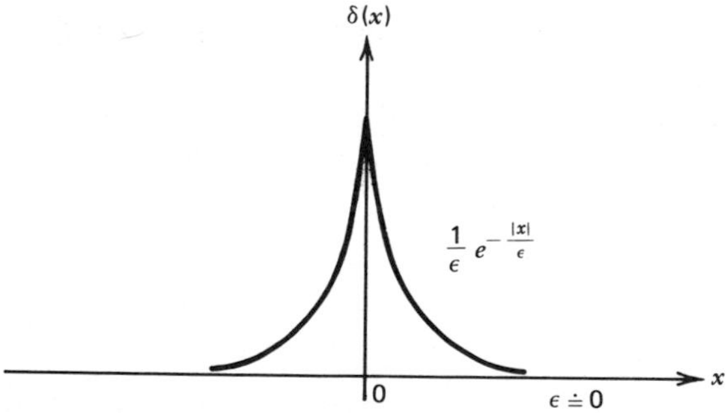

Figure A.1.2

† Defined as a linear functional on a certain abstract space of functions.

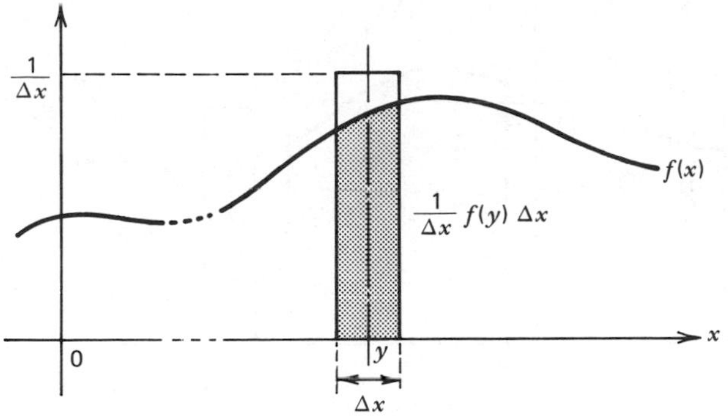

Figure A.1.3

sometimes referred to as a *sifting property*. This is because the term $\delta(x - y)$ in the integrand of

$$\int_{-\infty}^{+\infty} \delta(x - y) f(x)\, dx = f(y)$$

represents the idealized impulse located at the point y, which then sifts the function $f(x)$ by leaving only its value $f(y)$ at y (Fig. A.1.3). Several useful properties of the delta function can also be deduced from the idea of an idealized impulse, for instance, its symmetry,

$$\delta(-x) = \delta(x),$$

its transformation under a change of scale,

$$\delta(ax) = \frac{1}{|a|}\,\delta(x), \qquad a \neq 0$$

or, more generally, for a strictly monotone function $g(x)$ differentiable at zero and such that $g(0) = 0$

$$\delta(g(x)) = \delta(x)\left|\frac{dg(0)}{dx}\right|^{-1}.$$

The sifting property of the delta function is useful in representing sums or convergent infinite series in an integral form; for instance, we can write

$$\sum_{k=1}^{\infty} \frac{1}{k^2} = \int_{-\infty}^{+\infty} f(x)\, dx,$$

where

$$f(x) = \sum_{k=1}^{\infty} \delta(x - k)\frac{1}{x^2}.$$

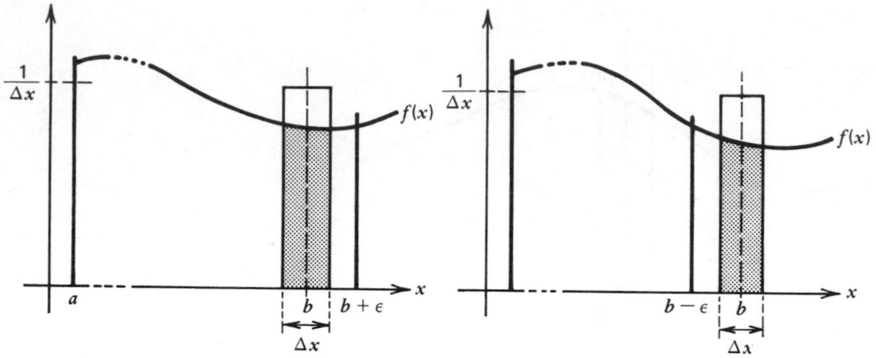

Figure A.1.4

If I is an interval with endpoints $a < b$, and if $f(x)$ is a continuous function, then again

$$f(y) = \int_a^b \delta(x - y) f(x)\, dx = \int_a^b \delta(x) f(y - x)\, dx$$

provided that the point y is in the interior of this interval, $a < y < b$. If, however, the point y coincides with one of the endpoints, for example, $y = b$, ambiguity could arise. We resolve this ambiguity by using the notation

$$\int_a^{b+} \delta(x - b) f(x)\, dx = \int_a^{b+} \delta(x) f(b - x)\, dx = f(b), \tag{5}$$

$$\int_a^{b-} \delta(x - b) f(x)\, dx = \int_a^{b-} \delta(x) f(b - x)\, dx = 0, \tag{6}$$

where $b+$ and $b-$, respectively, indicate that the upper limit of integration is actually $b + \varepsilon$ or $b - \varepsilon$, with $\varepsilon > 0$ arbitrarily small. A glance at Figure A.1.4 shows that equations (5) and (6) indeed agree with the intuitive picture of the delta function as an idealized impulse. Similarly, if $y = a$, we have analogously

$$\int_{a+}^{b} \delta(x - a) f(x)\, dx = \int_{a+}^{b} \delta(x) f(a - x)\, dx = 0,$$

and

$$\int_{a-}^{b} \delta(x - a) f(x)\, dx = \int_{a-}^{b} \delta(x) f(a - x)\, dx = f(a).$$

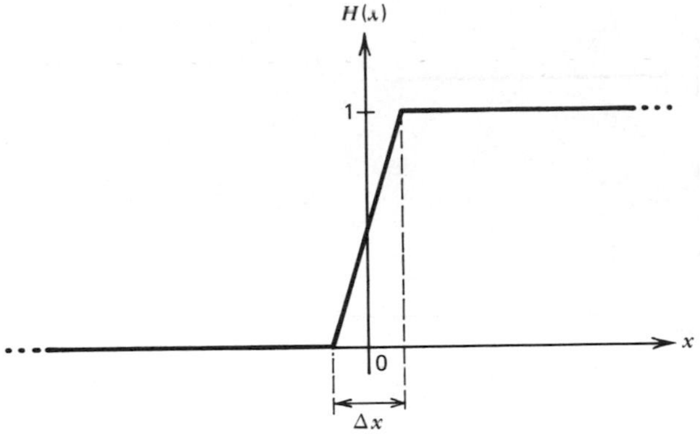

Figure A.1.5

As long as $\Delta x > 0$, the idealized impulse shown in Figure A.1.1 can be regarded as a derivative of a continuous function $H(x)$:

$$H(x) = \begin{cases} 1 & \text{if} \quad x > \dfrac{\Delta x}{2}, \\[2ex] \dfrac{x}{\Delta x} + \dfrac{1}{2} & \text{if} \quad |x| \le \dfrac{\Delta x}{2}, \\[2ex] 0 & \text{if} \quad x < -\dfrac{\Delta x}{2}, \end{cases}$$

which is shown in Figure A.1.5.

If Δx is now considered infinitesimal, the function $H(x)$ becomes a unit-step function discontinuous at the origin (the so-called Heaviside function, pictured in Fig. A.1.6), which we take to be continuous from the right at zero, that is,

$$H(x) = \begin{cases} 1 & \text{if} \quad x \ge 0, \\ 0 & \text{if} \quad x < 0. \end{cases}$$

This suggests a pair of symbolic differential relations

$$\frac{dH(x)}{dx} = \delta(x) \qquad \text{and} \qquad \int_{-\infty}^{x+} \delta(y)\, dy = H(x).$$

Since any function $F(x)$ with at most a discrete set of jump discontinuities can be written as a sum of a continuous function $G(x)$ and a linear combination of unit-step functions

$$F(x) = G(x) + \sum_k c_k H(x - y_k)$$

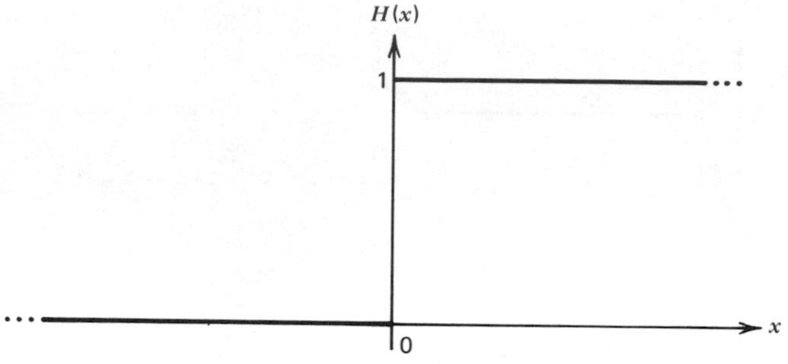

Figure A.1.6

we can thus define a derivative of such a function

$$\frac{dF(x)}{dx} = g(x) + \sum_k c_k \, \delta(x - y_k) = f(x)$$

where

$$g(x) = \frac{dG(x)}{dx}.$$

Assuming $F(x)$ to be continuous from the right at each of its jump points, we get a formally correct inverse relation

$$F(x) = \int_a^{x+} f(x)\, dx, \qquad x > a.$$

To summarize, if $F(x)$ has an isolated jump of height c at a point y, then its derivative there can be considered as the height of the jump times the Dirac delta function at y; in symbols,

$$\left.\frac{dF(x)}{dx}\right|_{x=y} = c\,\delta(x - y) \tag{7}$$

We can even introduce a derivative $d\delta(x)/dx$ of the Dirac delta function itself. Denoting this derivative by $\dot{\delta}(x)$, it can be shown that if $f(x)$ is a function continuously differentiable in a neighborhood of a point y, then

$$\int_{-\infty}^{+\infty} \dot{\delta}(x - y) f(x)\, dx = -\frac{df(y)}{dy} \tag{8}$$

that is, except for the minus sign, $\dot{\delta}(x - y)$ sifts the derivative of f at the point y. The derivative $\dot{\delta}(x)$ can be visualized as an idealized double impulse (Fig. A.1.7), which can be justified, for instance, by expressing the original impulse in Figure

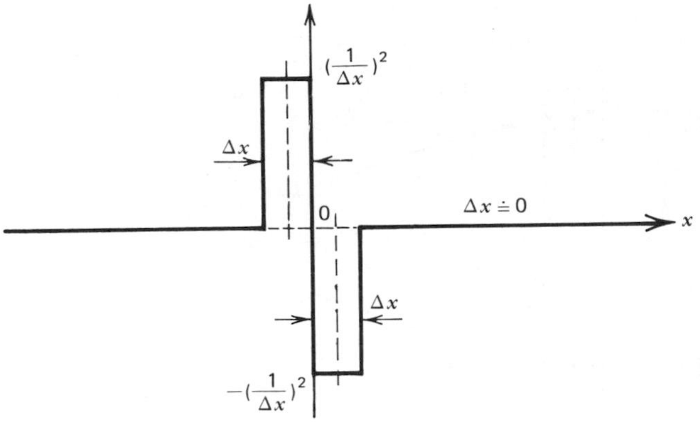

Figure A.1.7

A.1.1 as a difference

$$\frac{1}{\Delta x} H\left(x + \frac{\Delta x}{2}\right) - \frac{1}{\Delta x} H\left(x - \frac{\Delta x}{2}\right)$$

and applying formula (7). If we imagine such a double impulse located at the point y, we get, for small Δx, for the integral in (8) the difference

$$\left(\frac{1}{\Delta x}\right)^2 f\left(y - \frac{\Delta x}{2}\right)\Delta x - \left(\frac{1}{\Delta x}\right)^2 f\left(y + \frac{\Delta x}{2}\right)\Delta x = -\frac{f\left(y + \frac{\Delta x}{2}\right) - f\left(y - \frac{\Delta x}{2}\right)}{\Delta x}$$

(see Fig. A.1.8), which is for infinitesimal Δx the derivative $-df(y)/dy$. Similarly for the second derivative $d^2\,\delta(x)/\partial x^2 = \ddot{\delta}(x)$, we have

$$\int_{-\infty}^{+\infty} \ddot{\delta}(x - y) f(x)\, dx = \frac{d^2 f(y)}{dy^2}$$

and, more generally, the nth derivative of the Dirac delta function $\delta^{(n)}(x)$ would sift the nth derivative of f with a plus or a minus sign, depending on whether n is even or odd.

We can also formally define a Fourier transform of the Dirac delta function. From the definition of a Fourier transform $g = \mathcal{F}\{f\}$

$$g(v) = \int_{-\infty}^{+\infty} e^{-\iota\omega t} f(t)\, dt, \qquad \infty < v < +\infty,$$

$\omega = 2\pi v$, we get, by substituting $f(t) = \delta(t)$, by the sifting property of the delta function,

$$\int_{-\infty}^{+\infty} e^{-\iota\omega t}\, \delta(t)\, dt = e^{-\iota\omega 0} = 1 \qquad \text{for all } -\infty < v < +\infty.$$

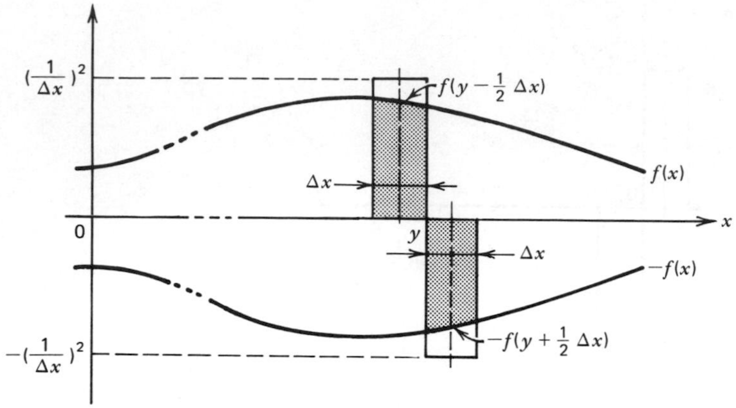

Figure A.1.8

Thus, the Dirac delta function forms a Fourier transform pair with the constant 1

$$1 = \mathcal{F}\{\delta\}, \qquad \delta = \mathcal{F}^{-1}\{1\},$$

so that we can also formally define the Dirac delta function as an integral

$$\delta(t) = \int_{-\infty}^{+\infty} e^{\imath\omega t}\, dv, \qquad \omega = 2\pi v.$$

Remark: In the defining property

$$\int_{-\infty}^{+\infty} \delta(x-y)f(x)\, dy = f(y) \tag{9}$$

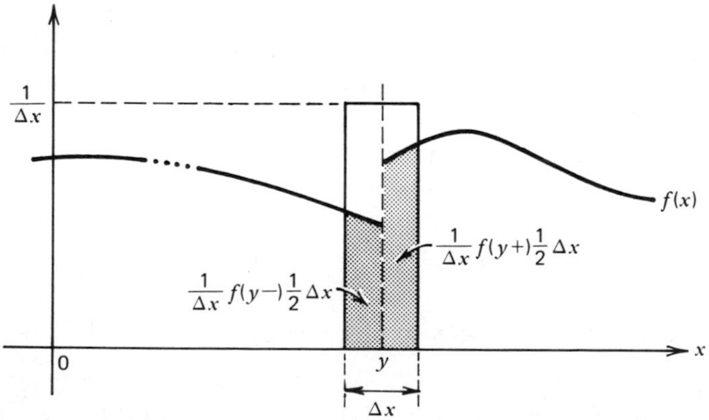

Figure A.1.9

for the delta function, we have assumed that the function f is continuous in a neighborhood of the point y. If the function f has a discontinuity there such that the one-sided limits

$$\lim_{x \to y-} f(x) = f(y-) \qquad \text{and} \qquad \lim_{x \to y+} f(x) = f(y+)$$

exist, then instead of (9) we get a more general equation

$$\int_{-\infty}^{+\infty} \delta(x - y) f(x)\, dx = \frac{1}{2}(f(y+) + f(y-))$$

as can be justified from Figure A.1.9. Я

APPENDIX TWO

TABLE OF COMMONLY USED PROBABILITY LAWS AND THE MULTIVARIATE AND BIVARIATE GAUSSIAN LAWS

Table A.1. Discrete Probability Laws

1. Name	Bernoulli	Symmetric Bernoulli	Binomial	Hypergeometric
2. Parameters	$0 < p < 1$; $q = 1 - p$		$n = 1, 2, 3, \ldots$ $0 < p < 1$ $q = 1 - p$	$N = 1, 2, 3, \ldots$ $R = 0, 1, \ldots, N$ $n = 0, 1, \ldots, N$
3. Probability mass distribution	$P(X = x) = \begin{cases} q, & x = 0 \\ p, & x = 1 \end{cases}$	$P(X = x) = \begin{cases} \frac{1}{2}, & x = -1 \\ \frac{1}{2}, & x = 1 \end{cases}$	$P(X = x) = \binom{n}{x} p^x q^{n-x}$ $x = 0, 1, \ldots, n$	$P(X = x) = \frac{\binom{R}{x}\binom{N-R}{n-x}}{\binom{N}{n}}$, $x = 0, 1, \ldots, n$
4. Mean	p	0	np	$n\frac{R}{N}$
5. Variance	pq	1	npq	$n\frac{R}{N}\left(1 - \frac{R}{N}\right)\frac{N-n}{N-1}$
6. Characteristic function	$q + pe^{\iota\zeta}$	$\cos \zeta$	$(q + pe^{\iota\zeta})^n$	Polynomial in $e^{\iota\zeta}$
7. Remark	Named in honor of Jacques Bernoulli	Linear function of Bernoulli	Number of successes in n trials. Sum of n independent Bernoulli variables	Sampling without replacement

Table A.1. Discrete Probability Laws (Continued)

1. Name	Negative binomial	Geometric	Modified negative binomial
2. Parameters	$r = 1, 2, 3, \ldots;$ $0 < p < 1;\ q = 1 - p$	$0 < p < 1$ $q = 1 - p$	$r = 1, 2, 3, \ldots;$ $0 < p < 1$
3. Probability mass distribution	$P(X = x) = \binom{x-1}{r-1} p^r q^{x-r}$ $x = r, r+1, \ldots$	$P(X = x) = pq^{x-1}$ $x = 1, 2, 3, \ldots$	$P(X = x) = \binom{x+r-1}{r-1} p^r q^x$ $x = 0, 1, 2, \ldots$
4. Mean	$\frac{r}{p}$	$\frac{1}{p}$	$\frac{rq}{p}$
5. Variance	$\frac{rq}{p^2}$	$\frac{q}{p^2}$	$\frac{rq}{p^2}$
6. Characteristic function	$\left(\frac{pe^{\iota\zeta}}{1 - qe^{\iota\zeta}}\right)^r$	$\frac{pe^{\iota\zeta}}{1 - qe^{\iota\zeta}}$	$\left(\frac{p}{1 - qe^{\iota\zeta}}\right)^r$
7. Remark	Number of trials to get r successes. Sum of r independent geometric random variables	Negative binomial with $r = 1$	Number of failures *before* the rth success

Table A.1. Discrete Probability Laws (Continued)

1. Name	Poisson	Multinomial
2. Parameters	$\lambda > 0$	$n = 1, 2, 3, \ldots; 0 < p_i < 1, i = 1, 2, \ldots, k$ $\sum p_i = 1$
3. Probability mass distribution	$P(X = x) = \frac{(\lambda)^x}{x!} e^{-\lambda}$ $x = 0, 1, 2, \ldots$	$P(X_1 = x_1, \ldots, X_k = x_k) = n! \prod_{i=1}^{n} \frac{p_i^{x_i}}{x_i!}$ $x_i = 0, 1, \ldots, n, i = 1, 2, \ldots, k; \sum x_i = n$
4. Mean	λ	$\mu_i = np_i, i = 1, 2, \ldots, k$
5. Variance	λ	$\sigma_{ii}^2 = np_i(1 - p_i), \sigma_{ij}^2 = -np_i p_j$
6. Characteristic function	$\exp(\lambda(e^{\imath\zeta} - 1))$	$\left(\sum_{i=1}^{k} p_i e^{\imath\zeta_i}\right)^n$
7. Remark	Number of random time instances in a unit interval. Limiting form of binomial with $\lambda = np$	Number of outcomes in each of k classes in n independent trials. With $k = n$, $p_i = 1/n$, this is the *Maxwell-Boltzman* law

Table A.2. Continuous Probability Laws

1. Name	Gaussian	Uniform	Erlang	Exponential
2. Parameters	$-\infty < \mu < \infty$, $\sigma > 0$	$-\infty < a < b < \infty$	$\lambda > 0$, $r = 1, 2, 3, \ldots$	$\lambda > 0$
3. Density function	$f(x) = \frac{1}{\sigma\sqrt{2\pi}} e^{-(x-\mu)^2/2\sigma^2}$ $-\infty < x < \infty$	$f(x) = \frac{1}{b-a}$ $a < x < b$	$f(x) = \frac{\lambda^r x^{r-1}}{(r-1)!} e^{-\lambda x}$ $x > 0$	$f(x) = \lambda e^{-\lambda x}$ $x > 0$
4. Mean	μ	$\frac{a+b}{2}$	$\frac{r}{\lambda}$	$\frac{1}{\lambda}$
5. Variance	σ^2	$\frac{(b-a)^2}{12}$	$\frac{r}{\lambda^2}$	$\frac{1}{\lambda^2}$
6. Characteristic function	$e^{\iota\zeta\mu - \sigma^2\zeta^2/2}$	$\frac{e^{\iota\zeta b} - e^{\iota\zeta a}}{\iota\zeta(b-a)}$	$\left(\frac{\lambda}{\lambda - \iota\zeta}\right)^r$	$\frac{\lambda}{\lambda - \iota\zeta}$
7. Remark	Limiting distribution for sums of independent random variables. Standard Gaussian has $\mu = 0$, $\sigma = 1$	Probability proportional to length of interval, "equally likely" case	Time of rth random time instance in Poisson process	Erlang with $r = 1$

Table A.2. Continuous Probability Laws (Continued)

1. Name	Gamma	Maxwell	Rayleigh	Inverse Gaussian
2. Parameters	$\lambda > 0,\ \nu > 0$	$\sigma > 0$	$\sigma > 0$	$a > 0$
3. Density function	$f(x) = \frac{\lambda^\nu x^{\nu-1}}{\Gamma(\nu)} e^{-\lambda x}$ $x > 0$	$f(x) = \sqrt{\frac{2}{\pi}} \frac{x^2}{\sigma^3} e^{-x^2/2\sigma^2}$ $x > 0$	$f(x) = \frac{x}{\sigma^2} e^{-x^2/2\sigma^2}$ $x > 0$	$f(x) = \frac{a}{\sqrt{2\pi}} x^{-3/2} e^{-a^2/2x}$ $x > 0$
4. Mean	$\frac{\nu}{\lambda}$	$\frac{2\sigma}{\sqrt{\pi}}$	$\sigma\sqrt{\frac{\pi}{2}}$	Does not exist
5. Variance	$\frac{\nu}{\lambda^2}$	$\sigma^2\left(\frac{3\pi - 8}{2\pi}\right)$	$\sigma^2\left(2 - \frac{\pi}{2}\right)$	Does not exist
6. Characteristic function	$\left(\frac{\lambda}{\lambda - \imath\zeta}\right)^\nu$	$\frac{\imath\sigma\zeta}{\sqrt{\pi}} + \left(1 + \frac{\imath\sigma\zeta}{2}\right)$ $\times \left(1 + \mathrm{erf}\left(\frac{\imath\sigma\zeta}{\sqrt{2}}\right)\right) e^{-\sigma^2\zeta^2/4}$	$1 + \imath\sigma\zeta\sqrt{\frac{\pi}{2}}$ $\times (1 + \mathrm{erf}(\imath\sigma\zeta)) e^{-\sigma^2\zeta^2/2}$	$\exp(-a\sqrt{2\zeta}\, e^{-\imath\pi/4})$
7. Remark	Generalization of Erlang to $r = \nu$, a positive real number	Velocity of a typical molecule of ideal gas. X^2 is gamma, $\lambda = 1/2\sigma^2$, $\nu = \frac{3}{2}$	Square root of the sum of squares of independent, Gaussian random variables, each with $\mu = 0$, variance σ^2. Named in honor of Lord Rayleigh	Time at which the standard Wiener process first reaches level a

Table A.2. Continuous Probability Laws (Continued)

1. Name	Bilateral exponential	Beta	Cauchy	Weibull
2. Parameters	$\lambda > 0$	$\alpha > 0, \beta > 0$	$\alpha > 0$	$\alpha > 0, \beta > 0$
3. Density function	$f(x) = \frac{\lambda}{2} e^{-\lambda\|x\|}$ $-\infty < x < \infty$	$f(x) = \frac{\Gamma(\alpha+\beta)}{\Gamma(\alpha)\Gamma(\beta)} x^{\alpha-1}(1-x)^{\beta-1}$ $0 < x < 1$	$f(x) = \frac{\alpha}{\pi(x^2+\alpha^2)}$ $-\infty < x < \infty$	$f(x) = \alpha\beta x^{\beta-1} e^{-\alpha x^{\beta}}$ $x > 0$
4. Mean	0	$\frac{\alpha}{\alpha+\beta}$	Does not exist	$\alpha^{-1/\beta}\Gamma\left(1+\frac{1}{\beta}\right)$
5. Variance	$\frac{2}{\lambda^2}$	$\frac{\alpha\beta}{(\alpha+\beta)^2(\alpha+\beta+1)}$	Does not exist	$\alpha^{-2/\beta}\left(\Gamma\left(1+\frac{2}{\beta}\right) - \Gamma^2\left(1+\frac{1}{\beta}\right)\right)$
6. Characteristic function	$\frac{\lambda^2}{\lambda^2+\zeta^2}$	$\frac{\Gamma(\alpha+\beta)}{\Gamma(\alpha)} \sum_{k=0}^{\infty} \frac{\Gamma(\alpha+k)}{\Gamma(\alpha+\beta+k)} \frac{(\iota\zeta)^k}{k!}$	$e^{-\alpha\|\zeta\|}$	Complicated expression
7. Remark	First law of Laplace	Order statistics of i.i.d. uniform (0, 1) random variables (generalized)	Distribution of the projection of a random angle onto a line	Frequently used in life testing and reliability

THE MULTIVARIATE GAUSSIAN PROBABILITY LAW

1. Let $\mathbf{Z}$ be an $n \times 1$ vector whose components are independent standard Gaussian ($\mu = 0$, $\sigma^2 = 1$) random variables, and let the $m \times 1$ vector $\mathbf{X}$ be defined by

$$\mathbf{X} = \mathbf{aZ} + \boldsymbol{\mu}_{\mathbf{X}}$$

where $\mathbf{A}$ is an $m \times n$ matrix of constants of rank m, and $\boldsymbol{\mu}_{\mathbf{X}}$ is a vector of constants. Then $\mathbf{X}$ is a multivariate Gaussian random vector with mean vector

$$E[\mathbf{X}] = \boldsymbol{\mu}_{\mathbf{X}}$$

and covariance matrix

$$\boldsymbol{\sigma}_{\mathbf{X}}^2 = \mathbf{aa}'.$$

2. The multivariate Gaussian density function is

$$f_{\mathbf{X}}(\mathbf{x}) = \frac{(\det \boldsymbol{\sigma}_{\mathbf{X}}^2)^{-1/2}}{(2\pi)^{m/2}} e^{-1/2(\mathbf{x}-\boldsymbol{\mu}_{\mathbf{X}})'(\boldsymbol{\sigma}_{\mathbf{X}}^2)^{-1}(\mathbf{x}-\boldsymbol{\mu}_{\mathbf{X}})}$$

and its characteristic function is

$$\psi_{\mathbf{X}}(\boldsymbol{\zeta}) = e^{\iota\boldsymbol{\zeta}'\boldsymbol{\mu}_{\mathbf{X}} - (1/2)\boldsymbol{\zeta}'\boldsymbol{\sigma}_{\mathbf{X}}^2\boldsymbol{\zeta}}$$

3. If $\mathbf{X}$ is multivariate Gaussian and

$$\mathbf{Y} = \mathbf{bX} + \mathbf{c}$$

where $\mathbf{b}$ is any matrix of constants and $\mathbf{c}$ is a vector of constants, $\mathbf{Y}$ is multivariate Gaussian with mean vector

$$\boldsymbol{\mu}_{\mathbf{Y}} = \mathbf{b}\boldsymbol{\mu}_{\mathbf{X}} + \mathbf{c}$$

and covariance matrix

$$\boldsymbol{\sigma}_{\mathbf{Y}}^2 = \mathbf{b}\boldsymbol{\sigma}_{\mathbf{X}}^2\mathbf{b}'.$$

4. If $\mathbf{X}$ is multivariate Gaussian, the components of $\mathbf{X}$ are independent (Gaussian) random variables if and only if $\boldsymbol{\sigma}_{\mathbf{X}}^2$ is diagonal.
5. If $\mathbf{X}$ is multivariate Gaussian, then the marginal probability law for any $k < m$ components of $\mathbf{X}$ is again multivariate Gaussian.
6. If $\mathbf{X}$ is multivariate Gaussian, then the conditional probability law for any subset of the components of $\mathbf{X}$, given values for any of the remaining components, is again multivariate Gaussian.

THE BIVARIATE GAUSSIAN PROBABILITY LAW

1. Two random variables, X and Y, have the bivariate Gaussian distribution if and only if their joint density function is

$$f(x, y) = \frac{1}{2\pi\sqrt{\sigma_X^2\sigma_Y^2 - \sigma_{XY}^2}} e^{-a/2}$$

where

$$a = \frac{\sigma_X^2\sigma_Y^2}{\sigma_X^2\sigma_Y^2 - \sigma_{XY}^2}\left[\frac{(x - \mu_X)^2}{\sigma_X^2} - 2\sigma_{XY}\frac{(x - \mu_X)}{\sigma_X^2}\frac{(y - \mu_Y)}{\sigma_Y^2} + \frac{(y - \mu_Y)^2}{\sigma_Y^2}\right].$$

The means of X and Y are μ_X, μ_Y, respectively, and their variances are σ_X^2, σ_Y^2. The covariance between X and Y is denoted by σ_{XY}, where $|\sigma_{XY}| < \sigma_X\sigma_Y$.

2. The bivariate Gaussian density function has a unique maximum value at $x = \mu_X$, $y = \mu_Y$; contours of constant density are concentric ellipses whose major axes are all at angle

$$\theta = \frac{1}{2}\arctan\left(\frac{2\sigma_{XY}}{\sigma_X^2 - \sigma_Y^2}\right)$$

to the x-axis.

3. X and Y are independent if and only if $\sigma_{XY} = 0$. Any linear combination

$$V = aX + bY$$

is Gaussian with

$$\mu_Y = a\mu_X + b\mu_Y$$

$$\sigma_Y^2 = a^2\sigma_X^2 + b^2\sigma_Y^2 + 2ab\sigma_{XY}.$$

4. The conditional density of Y, given $X = x$, is Gaussian with mean

$$\mu_Y + \frac{\sigma_{XY}}{\sigma_X^2}(x - \mu_X)$$

and variance

$$\sigma_Y^2 - \frac{\sigma_{XY}^2}{\sigma_X^2} \le \sigma_Y^2.$$

The conditional expectation

$$\mathrm{E}[Y|X = x] = \mu_Y + \frac{\sigma_{XY}}{\sigma_X^2}(x - \mu_X)$$

is linear in x and its graph is called the *regression* line. The angle between the regression line and the x-axis is arctan (σ_{XY}/σ_X^2). As a random variable $\mathrm{E}[Y|X]$ is Gaussian with mean μ_Y and variance $(\sigma_{XY}/\sigma_X)^2$.

APPENDIX THREE

SUMMARY OF CALCULUS FACTS

This appendix contains some formulas and theorems of calculus that are either used in the text or may be helpful in solving the exercises. The letter z always stands for a complex number, and the letters x, y, t denote real numbers.

COMPLEX NUMBERS

1. $z = x + \imath y$, where $x = \mathcal{Re}\{z\}$ and $y = \mathcal{Im}\{z\}$ are the real and imaginary parts, respectively, and $\imath = \sqrt{-1}$ is the imaginary unit.
2. Polar form: $z = |z| e^{\imath\theta}$, where $|z| = \sqrt{x^2 + y^2}$ is the modulus and θ the argument, $\tan \theta = y/x$, of z.
3. Complex conjugate:

$$z^* = x - \imath y = |z| e^{-\imath\theta},$$

$$|z|^2 = z \cdot z^*, \qquad x = \frac{1}{2}(z + z^*), \qquad y = \frac{1}{2}(z - z^*).$$

4. Euler formula:

$$e^{\imath\theta} = \cos\theta + \imath \sin\theta \quad \text{whence } |e^{\imath\theta}| = 1 \text{ for any } -\infty < \theta < \infty.$$

5. Triangle inequality:

$$|z_1 + z_2| \leq |z_1| + |z_2| \text{ for any two complex numbers } z_1 \text{ and } z_2.$$

HYPERBOLIC FUNCTIONS

1. Definitions:

$$\sinh x = \frac{1}{2}(e^x - e^{-x}), \qquad -\infty < x < \infty,$$

$$\cosh x = \frac{1}{2}(e^x + e^{-x}), \qquad -\infty < x < \infty,$$

$$\tanh x = \frac{\sinh x}{\cosh x} = \frac{1 - e^{-2x}}{1 + e^{-2x}}, \qquad -\infty < x < \infty,$$

$$\operatorname{arcsinh} x = \ln(x + \sqrt{1 + x^2}), \qquad -\infty < x < \infty,$$

$$\operatorname{arctanh} x = \frac{1}{2} \ln \frac{1 + x}{1 - x}, \qquad -1 < x < 1.$$

2. Derivatives:

$$\frac{d}{dx} \sinh x = \cosh x,$$

$$\frac{d}{dx} \cosh x = \sinh x,$$

$$\frac{d}{dx} \tanh x = \frac{1}{(\cosh x)^2},$$

$$\frac{d}{dx} \operatorname{arcsinh} x = \frac{1}{\sqrt{1 + x^2}},$$

$$\frac{d}{dx} \operatorname{arctanh} x = \frac{1}{1 - x^2}.$$

3. Basic identity:

$$(\cosh x)^2 - (\sinh x)^2 = 1$$

Other identities are obtained from the corresponding trigonometric identity by using the relations

$$\imath \sinh x = \sin \imath x, \qquad \cosh x = \cos \imath x.$$

4. McLaurin series:

$$\sinh x = \sum_{k=0}^{\infty} \frac{(x)^{2k+1}}{(2k+1)!},$$

$$\cosh x = \sum_{k=0}^{\infty} \frac{(x)^{2k}}{(2k)!}$$

GAMMA AND ERROR FUNCTIONS

1. Definitions:

$$\Gamma(t) = \int_0^\infty x^{t-1} e^{-x}\, dx, \qquad t > 0,$$

$$\operatorname{erf} z = \frac{2}{\sqrt{\pi}} \int_0^z e^{-u^2}\, du, \qquad |z| < \infty.$$

2. Specific values:

$$\Gamma(n+1) = n!,$$

$$\Gamma\left(n+\frac{1}{2}\right) = \frac{(2n)!}{n!}2^{-2n}\sqrt{\pi}, \qquad n = 0, 1, \ldots.$$

$$\operatorname{erf} x = 2\Phi(x\sqrt{2}) - 1, \qquad x > 0,$$

where $\Phi(x)$ is the standard Gaussian distribution function.

$$\imath \operatorname{erf} \imath x = \frac{2}{\sqrt{\pi}}\int_0^x e^{t^2}\,dt, \qquad x > 0.$$

3. Gamma recurrence:

$$t\Gamma(t) = \Gamma(t+1), \qquad t > 0.$$

4. Asymptotic behavior:

$$\lim_{t\to 0+} t\Gamma(t) = 1,$$

$$\lim_{x\to\infty} x\sqrt{\pi}\,e^{x^2}(1 - \operatorname{erf} x) = 1,$$

$$\lim_{x\to\infty} x\sqrt{\pi}\,e^{-x^2}\imath \operatorname{erf} \imath x = 1.$$

FINITE SUMS

1. $\sum_{k=1}^{n} k = \binom{n+1}{2}, \qquad n = 1, 2, \ldots.$

2. $\sum_{k=1}^{n} k^2 = \frac{2n+1}{3}\binom{n+1}{2}, \qquad n = 1, 2, \ldots.$

3. $\sum_{k=0}^{n} z^k = \frac{1 - z^{n+1}}{1 - z}, \qquad |z| \neq 1, \qquad n = 0, 1, \ldots$

4. $\sum_{k=0}^{n} kz^k = z\frac{1 - (n+1)z^n + nz^{n+1}}{(1-z)^2}, \qquad |z| \neq 1, \qquad n = 0, 1, \ldots.$

5. Binomial theorem:

$$\sum_{k=0}^{n} \binom{n}{k} z_1^k z_2^{n-k} = (z_1 + z_2)^n, \qquad n = 0, 1, \ldots$$

6. Multinomial theorem:

$$\sum \binom{m}{k_1, \ldots, k_m} z_1^{k_1} \cdots z_m^{k_m} = \left(\sum_{j=1}^{m} z_j \right)^n,$$

where $n = 0, 1, \ldots; m = 1, 2, \ldots; k_j = 0, 1, \ldots; j = 1, \ldots, m$; and the summation on the left-hand side is over all k_j's such that $k_1 + \cdots + k_m = n$.

7. Vandermonde convolution:

$$\sum_{k=0}^{n} \binom{m_1}{k} \binom{m_2}{n-k} = \binom{m_1 + m_2}{n},$$

$n = 0, 1, \ldots; m_1 = n, n+1, \ldots; m_2 = n, n+1, \ldots$; in particular,

$$\sum_{k=0}^{n} \binom{n}{k}^2 = \binom{2n}{n}.$$

POWER SERIES

1. Geometric series:

$$\sum_{k=0}^{\infty} z^k = \frac{1}{1-z}, \qquad |z| < 1.$$

2. Arithmetic-geometric series:

$$\sum_{k=0}^{\infty} k z^k = \frac{z}{(1-z)^2}, \qquad |z| < 1.$$

3. Binomial series:

$$\sum_{k=0}^{\infty} \binom{n+k}{k} z^k = \frac{1}{(1-z)^{n+1}}, \qquad |z| < 1, \qquad n = 0, 1, \ldots$$

4. Exponential and trigonometric series:

$$\sum_{k=0}^{\infty} \frac{z^k}{k!} = e^z, \qquad \sum_{k=0}^{\infty} (-1)^k \frac{x^{2k}}{(2k)!} = \cos x, \qquad \sum_{k=0}^{\infty} (-1)^k \frac{x^{2k+1}}{(2k+1)!} = \sin x.$$

5. Logarithmic series:

$$\sum_{k=1}^{\infty} \frac{x^k}{k} = \ln \frac{1}{1-x}, \qquad -1 \le x < 1.$$

6. Bessel series:

$$\sum_{k=0}^{\infty} \frac{1}{k!\,(n+k)!} \left(\frac{x}{2} \right)^{2k+n} = I_n(x), \qquad n = 0, 1, \ldots,$$

where $I_n(x)$ is the modified Bessel function of order n.

LIMITS AND INFINITE SERIES

1. Cauchy criterion:

   $\lim_{n\to\infty} z_n$ exists if and only if $|z_n - z_m| \to 0$ as both $n \to \infty$ and $m \to \infty$.

2. Kronecker lemma:

$$\text{If} \quad \sum_{k=1}^{\infty} \frac{|z_k|}{k} < \infty \quad \text{then} \quad \lim_{n\to\infty} \frac{1}{n} \sum_{k=1}^{n} z_k = 0.$$

3. Harmonic series:

$$\sum_{k=1}^{\infty} \frac{1}{k} = \infty, \qquad \sum_{k=1}^{\infty} \frac{(-1)^k}{k} = \ln \frac{1}{2}, \qquad \lim_{n\to\infty} \frac{1}{\ln n} \sum_{k=1}^{n} \frac{1}{k} = 1.$$

4. Euler's limit:

$$\text{If} \quad \lim_{n\to\infty} z_n = z \quad \text{then} \quad \lim_{n\to\infty} \left(1 + \frac{z_n}{n}\right)^n = e^z.$$

   In particular,

$$\lim_{n\to\infty} \left(1 + \frac{z}{n}\right)^n = e^z.$$

5. Stirling formula:

$$\lim_{n\to\infty} \frac{\sqrt{2\pi}\, n^{n+1/2} e^{-n}}{n!} = 1.$$

6. Square inverse series:

$$\sum_{k=1}^{\infty} \frac{1}{k^2} = \frac{\pi^2}{6},$$

$$\sum_{k=1}^{\infty} \frac{1}{(2k)^2} = \frac{\pi^2}{24},$$

$$\sum_{k=1}^{\infty} \frac{1}{(2k-1)^2} = \frac{\pi^2}{8}.$$

7. Trigonometric function:

$$\lim_{x\to 0} \frac{\sin \alpha x}{x} = \alpha, \qquad \lim_{x\to 0} \frac{1 - \cos \alpha x}{x} = 0,$$

   More generally,

$$\lim_{x\to 0} \frac{1 - e^{-\alpha x}}{x} = \alpha \quad \text{for complex } \alpha.$$

8. Exponential and logarithmic functions:

$$\lim_{x\to\infty} x^\alpha e^{-x} = 0, \qquad \lim_{x\to 0} \frac{1}{x^\alpha} e^{-1/x} = 0,$$

$$\lim_{x\to\infty} \frac{1}{x^\alpha} \ln x = 0, \qquad \lim_{x\to 0+} x^\alpha \ln x = 0,$$

where $\alpha > 0$.

9. Fourier series:

$$\sum_{k=1}^{\infty} \frac{\sin(2k-1)\pi x}{2k-1} = \frac{\pi}{4} \operatorname{sign} x, \qquad |x| < 1,$$

$$\sum_{k=1}^{\infty} \frac{\cos(2k-1)\pi x}{(2k-1)^2} = \frac{\pi^2}{4}\left(\frac{1}{2} - |x|\right), \qquad |x| < 1,$$

$$\sum_{k=1}^{\infty} \frac{(-1)^k}{k^2} \cos k\pi x = \frac{\pi^2}{4}\left(x^2 - \frac{1}{3}\right), \qquad |x| < 1.$$

DEFINITE INTEGRALS

1. Leibnitz' rule:

$$\frac{d}{dt}\int_0^t g(t, x)\, dx = g(t, t) + \int_0^t \frac{\partial g(t, x)}{\partial t}\, dx, \qquad t \geq 0.$$

2. Powers of sine and cosine:

$$\int_0^\pi (\cos x)^n\, dx = \sqrt{\pi}\, \frac{\Gamma\left(\dfrac{n+1}{2}\right)}{\Gamma\left(\dfrac{n}{2} + 1\right)}, \qquad n = 0, 1, \ldots,$$

$$\int_0^\infty \left(\frac{\sin x}{x}\right)^2 dx = \int_0^\infty \frac{\sin x}{x}\, dx = \frac{\pi}{2}$$

3. Decaying oscillations:

For $\alpha > 0$ and $-\infty < \omega < \infty$

$$\int_0^\infty e^{-\alpha x} \cos \omega x\, dx = \frac{\alpha}{\alpha^2 + \omega^2},$$

$$\int_0^\infty e^{-\alpha x} \cos \omega x\, dx = \frac{\omega}{\alpha^2 + \omega^2}.$$

More generally,

$$\int_0^\infty x^n e^{-\alpha x} \cos \omega x \, dx = \mathfrak{Re}\left\{\left(\frac{\alpha + \imath\omega}{\alpha^2 + \omega^2}\right)^{n+1}\right\},$$

$$\int_0^\infty x^n e^{-\alpha x} \sin \omega x \, dx = \mathfrak{Im}\left\{\left(\frac{\alpha + \imath\omega}{\alpha^2 + \omega^2}\right)^{n+1}\right\},$$

$n = 0, 1, \ldots .$

4. Laplace's integral:

$$\int_0^\infty e^{-\alpha x^2 + \beta x} \, dx = \frac{1}{2}\sqrt{\frac{\pi}{\alpha}}\, e^{-\beta^2/4\alpha},$$

and

$$\int_0^\infty e^{-\alpha x^2 + \imath\beta x} \, dx = \frac{1}{2}\sqrt{\frac{\pi}{\alpha}}\, e^{-\beta^2/4\alpha}\left(1 + \operatorname{erf}\frac{\imath\beta}{2\sqrt{\alpha}}\right),$$

where $\alpha > 0$, $-\infty < \beta < \infty$.

5. Gamma-type integrals:

$$\int_t^\infty x^n e^{-\alpha x} \, dx = \frac{n!}{\alpha^{n+1}} e^{-\alpha t} \sum_{k=0}^{n} \frac{(\alpha t)^k}{k!},$$

$t \geq 0$, where $\alpha > 0$ and $n = 0, 1, \ldots .$

$$\int_0^\infty x^{r-1} e^{-\alpha x^\kappa} \, dx = \frac{\alpha^{-r/\kappa}}{\kappa} \Gamma\left(\frac{r}{\kappa}\right)$$

where $\alpha > 0$, $r > 0$, $\kappa > 0$.

6. Beta integral:

$$\int_0^1 x^{\alpha-1}(1 - x)^{\beta-1} \, dx = \int_0^\infty \frac{x^{\alpha-1}}{(1 + x)^{\alpha+\beta}} \, dx = \frac{\Gamma(\alpha)\Gamma(\beta)}{\Gamma(\alpha + \beta)},$$

where $\alpha > 0$, $\beta > 0$.

7. Bessel integral:

$$\int_0^{2\pi} e^{x \cos t} \, dt = 2\pi I_0(x),$$

where $I_0(x)$ is the modified Bessel function of order zero.

TAYLOR EXPANSION

1. Taylor formula with remainder:

$$g(x + \Delta x) = g(x) + \left(\frac{d}{dx} g(x)\right)\frac{\Delta x}{1!} + \left(\frac{d^2}{dx^2} g(x)\right)\frac{(\Delta x)^2}{2!} + \cdots + \left(\frac{d^{n-1}}{dx^{n-1}} g(x)\right)\frac{(\Delta x)^{n-1}}{(n-1)!} + r_n(\Delta x),$$

where

$$r_n(\Delta x) = \left(\frac{d^n}{dx^n} g(x_0)\right)\frac{(\Delta x)^n}{n!} \qquad \text{and} \qquad x \le x_0 \le x + \Delta x.$$

2. Behavior of the remainder:

If the $(n-1)$st derivative $(d^{n-1}/dx^{n-1})g(x)$ is continuous at x, then

$$\frac{r_n(\Delta x)}{(\Delta x)^{n-1}} \to 0 \qquad \text{as } \Delta x \to 0.$$

3. Bound on the remainder for the complex exponential function:
If $g(z) = e^{\alpha z}$, $-\infty < \alpha < \infty$, then

$$|r_n(\Delta z)| \le \frac{|\alpha \, \Delta z|^n}{n!}.$$

In particular, $|e^{\alpha z} - 1| \le \alpha |z|$ for all z.

4. Taylor expansion of a function of two variables:

$$g(x + \Delta x, y + \Delta y) = g(x, y) + \sum_{n=1}^{\infty} \frac{d^n g(x, y)}{n!},$$

where $d^n g(x, y)$ is the nth total differential

$$d^n g(x, y) = \sum_{k=0}^{n} \binom{n}{k}\left(\frac{\partial^n}{\partial x^k \, \partial y^{n-k}} g(x, y)\right)(\Delta x)^k (\Delta y)^{n-k}.$$

In particular, the expansion up to the second differential is

$$g(x + \Delta x, y + \Delta y) = g(x, y) + \frac{\partial}{\partial x} g(x, y)\, \Delta x + \frac{\partial}{\partial y} g(x, y)\, \Delta y + \frac{1}{2}\frac{\partial^2}{\partial x^2} g(x, y)(\Delta x)^2 + \frac{\partial^2}{\partial x \, \partial y} g(x, y)\, \Delta x \, \Delta y + \frac{1}{2}\frac{\partial^2}{\partial y^2} g(x, y)(\Delta y)^2 + \cdots$$

SOME FUNCTIONAL AND DIFFERENTIAL EQUATIONS

1. Additive functional equation:
   The only function $g(x)$ continuous on $-\infty < x < \infty$ and such that $g(x + y) = g(x) + g(y)$ is the linear function

$$g(x) = \alpha x,$$

   where α is a constant.

2. Multiplicative functional equation:
   The only function $g(x)$ not identically zero, continuous on $-\infty < x < \infty$, and such that $g(x + y) = g(x)g(y)$ is the exponential function

$$g(x) = e^{\alpha x},$$

   where α is a constant.

3. Homogeneous first order linear differential equation:
   If $dx(t)/dt + \alpha(t)x(t) = 0$, $t \in \mathcal{T}$, where $\int \alpha(t)\,dt$ exists on $\mathcal{T}$, then $x(t) = Ce^{-\int \alpha(t)\,dt}$, $t \in \mathcal{T}$, is the general solution.

4. Inhomogeneous first order linear differential equation:
   The general solution of the equation

$$\frac{dx(t)}{dt} + \alpha(t)x(t) = g(t), \qquad t \in \mathcal{T},$$

   is the function

$$x(t) = \left(C + \int g(t)e^{\int \alpha(\tau)\,d\tau}\,dt\right)e^{-\int \alpha(t)\,dt},$$

   provided the indefinite integrals exist on $\mathcal{T}$.

5. Second order homogeneous linear differential equation with constant coefficients:
   The equation

$$\frac{d^2x(t)}{dt^2} + a_1\frac{dx(t)}{dt} + a_2\,x(t) = 0, \qquad t \in \mathcal{T},$$

   has the general solution

$$x(t) = \begin{cases} c_1 e^{\lambda_1 t} + c_2 e^{\lambda_2 t} & \text{if} \quad \lambda_1 \neq \lambda_2, \\ (c_1 + c_2 t)e^{\lambda_1 t} & \text{if} \quad \lambda_1 = \lambda_2, \end{cases}$$

   where λ_1, λ_2 are roots of the quadratic equation

$$\lambda^2 + a_1\lambda + a_2 = 0$$

that is,

$$\lambda_{1,2} = \frac{a_1}{2} \pm \sqrt{\left(\frac{a_1}{2}\right)^2 - a_2}\,.$$

FOURIER SERIES

1. Fourier series:
If $g(t)$ is a real-valued periodic function with period t_0 and such that

$$\int_0^{t_0} |g(t)|^2 \, dt < \infty$$

then its Fourier series is

$$S(t) = \sum_{n=-\infty}^{\infty} c_n e^{-\imath 2\pi n t}, \qquad -\infty < t < \infty,$$

where

$$c_n = c^*_{-n} = \frac{1}{t_0} \int_{-t_0/2}^{t_0/2} g(t) e^{\imath 2\pi n t/t_0} \, dt, \qquad n = 0, \pm 1, \ldots.$$

2. Convergence of Fourier series:

$$\int_t^{t+t_0} |g(\tau) - S(\tau)|^2 \, d\tau = 0 \qquad \text{for every } -\infty < t < \infty.$$

3. Parseval identity:

$$\frac{1}{t_0} \int_{-t_0/2}^{t_0/2} g(t) g(t + \tau) \, dt = \sum_{n=-\infty}^{\infty} |c_n|^2 e^{\imath 2\pi n \tau/t_0},$$

in particular with $\tau = 0$,

$$\frac{1}{t_0} \int_{-t_0/2}^{t_0/2} |g(t)|^2 \, dt = \sum_{n=-\infty}^{\infty} |c_n|^2.$$

FOURIER TRANSFORM

1. Definition:
The generally complex-valued functions g and G defined by

$$G(v) = \int_{-\infty}^{\infty} e^{-\imath \omega t} g(t) \, dt, \qquad \text{symbolically: } G = \mathcal{F}\{g\}$$

$$g(t) = \int_{-\infty}^{\infty} e^{\imath \omega t} G(v) \, dv, \qquad \text{symbolically: } g = \mathcal{F}^{-1}\{G\},$$

where $-\infty < t < \infty$, $-\infty < \nu < \infty$, $\omega = 2\pi\nu$, are called a Fourier transform pair.

2. Sufficient conditions for existence:
If

$$\int_{-\infty}^{\infty} |g(t)|\, dt < \infty \qquad \text{then} \qquad G = \mathcal{F}\{g\}$$

exists and is a bounded and continuous function. If

$$\int_{-\infty}^{\infty} |g(t)|^2\, dt < \infty \qquad \text{then} \qquad G = \mathcal{F}\{g\}$$

exists and

$$\int_{-\infty}^{\infty} |G(\nu)|^2\, d\nu < \infty, \qquad \text{so} \qquad g = \mathcal{F}^{-1}\{G\}$$

also exists.

3. Convolution theorem:
If $g = g_1 \circledast g_2$, that is, $g(t) = \int_{-\infty}^{\infty} g_1(\tau)g_2(t-\tau)\, d\tau$, then $G = \mathcal{F}\{g\}$ is the product $G(\nu) = G_1(\nu)G_2(\nu)$ of the Fourier transforms $G_1 = \mathcal{F}\{g_1\}$, $G_2 = \mathcal{F}\{g_2\}$.

4. Parseval identity:

$$\int_{-\infty}^{\infty} g(t)g^*(t+\tau)\, dt = \int_{-\infty}^{\infty} |G(\nu)|^2 e^{\iota\omega\tau}\, d\nu,$$

in particular, with $\tau = 0$

$$\int_{-\infty}^{\infty} |g(t)|^2\, dt = \int_{-\infty}^{\infty} |G(\nu)|^2\, d\nu.$$

5. Change of scale:
If $g_1(t) = g(\alpha t)$, with α a constant, then

$$G_1(\nu) = \frac{1}{\alpha} G\left(\frac{\nu}{\alpha}\right).$$

6. Time shift:
If $g_1(t) = g(t+\tau)$, with τ a constant, then

$$G_1(\nu) = e^{\iota\omega\tau} G(\nu), \qquad \omega = 2\pi\nu.$$

7. Modulation:
If $g_1(t) = g(t)e^{\iota\omega_0 t}$, with $\omega_0 = 2\pi\nu_0$ a constant, then

$$G_1(\nu) = G(\nu - \nu_0).$$

INDEX